A. V. Narlikar

High Temperature Superconductivity 2

Springer
Berlin
Heidelberg
New York
Hong Kong
London
Milan
Paris
Tokyo

A. V. Narlikar (Ed.)

High Temperature Superconductivity 2

With 347 Figures and 21 Tables

Springer

Prof. Dr. Anant V. Narlikar

Inter-University Consortium
for DAE Facilities
University Campus
Khandwa Road
Indore 452 017 (MP), India

ISBN 3-540-40639-5 Springer-Verlag Berlin Heidelberg New York

Cataloging-in-Publication Data applied for.
Bibliographic information published by Die Deutsche Bibliothek. Die Deutsche Bibliothek lists this publication in the Deutsche Nationalbibliografie; detailed bibliographic data is available in the Internet at <http://dnb.ddb.de>.

Springer-Verlag Berlin Heidelberg New York
a member of BertelsmannSpringer Science+Business Media GmbH

http://www.springer.de

Printed in Germany

Typesetting: Dataconversion by author
Cover-design: medio, Berlin
Printed on acid-free paper 62 / 3020 hu – 5 4 3 2 1 0

CONTRIBUTORS

BARTHELMESS, Henry-Jobes
Magnicon GbR, Wachtelweg 7
22869 Schenefeld, GERMANY

CAMPBELL, A.M.
University of Cambridge
IRC in Superconductivity
Madingley Road,
Cambridge CB3 0HE, UK

CHALOUPKA, Heinz J.
University of Wuppertal,
Dept. Electrical & Information Engineering,
D-42097 Wuppertal, GERMANY

DEW-HUGHES, David
Department of Engineering Science,
Oxford University
Parks Road,
Oxford OX1 3PJ, UK

DRUNG, Dietmar
Physikalisch-Technische Bundesanstalt,
Abbestrasse 2-12,
10587 Berlin, GERMANY

FARDMANESH, Mehdi
Electrical and Electronics Engineering Dept.
Bilkent University
06800 Ankara, TURKEY

FUNAKI, Kazuo
Research Institute of Superconductor Science and Systems,
Kyushu University
6-10-1 Hakozaki, Higashi-ku,

Fukuoka 812-8581, JAPAN

GUDOSHNIKOV, S.A.
Institute of Terrestrial Magnetism
Ionosphere and Radio Wave Propagation RAS
(IZMIRAN) 142190 Troitsk
Moscow Region, RUSSIA

HOTT, Roland
Forschungszentrum Karlsruhe,
Institut für Festkörperphysik,
P.O. Box 3640, D-76021 Karlsruhe, GERMANY

HULL, John R.
Argonne National Laboratory
ET-335, Argonne,
IL 60439, USA

IWAKUMA, Masataka
Research Institute of Superconductor Science and Systems,
Kyushu University
6-10-1 Hakozaki, Higashi-ku,
Fukuoka 812-8581, JAPAN

KAESSER, Tobias
Tesat-Spacecom GmbH & Co KG, Gerberstr. 49
D-71522 Backnang, GERMANY

KOBAYASHI, Hiroyuki
Department of Mechanical Engineering,
Gunma University
Kiryu, Gunma 376-8515, JAPAN

KREUTZBRUCK, Marc von
Institute of Applied Physics
Justus-Liebig-University Giessen
Heinrich-Buff-Ring 16,
35390 Giessen, GERMANY

LUDWIG, Frank
Technische Universität Braunschweig,
Institut für Elektrische Messtechnik und
Grundlagen der Elektrotechnik
Hans-Sommer-Str. 66,
38106 Braunschweig, GERMANY

MATVEETS, L.V.
Institute of Terrestrial Magnetism
Ionosphere and Radio Wave Propagation RAS
(IZMIRAN) 142190 Troitsk
Moscow Region, RUSSIA

McCULLOCH, Malcolm
Department of Engineering Science,
Oxford University
Parks Road,
Oxford OX1 3PJ, UK

MUTA, Itsuya
Department of Electrical Engineering,
Graduate School of Engineering, Kyoto University
Yoshida-Honmachi, Sakyo-ku,
Kyoto 606-8501, JAPAN

NAGAYA, Kosuke
Department of Mechanical Engineering,
Gunma University
Kiryu, Gunma 376-8515, JAPAN

NAKAMURA, Taketsune
Department of Electrical Engineering,
Graduate School of Engineering, Kyoto University
Yoshida-Honmachi, Sakyo-ku,
Kyoto 606-8501, JAPAN

OSTERGAARD, Jacob
DEFU,
Rosenorns Alle 9,
DK-1970 Frederiksberg C,
DENMARK

SCHILLING, Meinhard
Technische Universität Braunschweig,
Institut für Elektrische Messtechnik und
Grundlagen der Elektrotechnik
Hans-Sommer-Str. 66,
38106 Braunschweig, GERMANY

SCHURIG, Thomas
Physikalisch-Technische Bundesanstalt,
Abbestrasse 2-12,
10587 Berlin, GERMANY

SHINTOMI, Takakazu
High Energy Accelerator Research Organization
Oho, Tsukuba-shi, Ibaraki 305-0801, JAPAN

SUZUKI, Takaaki
Department of Mechanical Engineering,
Gunma University
Kiryu, Gunma 376-8515, JAPAN

TAKAHASHI, Noriaki
Department of Mechanical Engineering,
Gunma University
Kiryu, Gunma 376-8515, JAPAN

TIXADOR, Pascal
CNRS-CRTBT/LEG, B.P.166
38042 Grenoble Cedex 09, FRANCE

TONNESEN, Ole
Oersted-DTU, Section Electric Power Engineering
Building 325,
Technical University of Denmark,
DK-2800 Kgs. Lyngby,
DENMARK

WÖRDENWEBER, Roger
Institut für Schichten und Grenzflächen (ISG),
Forschungszentrum Jülich,
D-52425 Jülich, GERMANY

PREFACE

The discovery of high temperature superconductors (HTS) in 1986 brought in its wings a new set of puzzles, but also exciting promises and problems of a multidisciplinary nature. Popular imagination was enraptured with the novel possibilities and expectations from new materials that could potentially revolutionise the very basis of modern existence. The central challenge was how to exploit the new materials in the form of bulk samples, wires, tapes and thin films in varied electrical installations, appliances and electronic devices, all functioning in a superconducting way at liquid nitrogen temperature. The underlying problem was that these new materials were hard and brittle ceramics, and moreover many of their transport properties too were generally unfriendly for their conductor fabrication or making device structures. As a result of these new opportunities and challenges, a whole new inter-disciplinary agenda swung into action.

Today, even after over fifteen years of the discovery, the mechanism driving the superconductivity at high temperature, remains shrouded in mystery and controversy. But at the more practical level, recent years have seen a dramatic development in the fabrication of these materials and their applications. There is a need for an extended source of information on applied superconductivity focusing on HTS Materials and Engineering Applications. The information in research journals is generally not accessible in a consolidated form to bring out the actual overall status of the field. Here we have two books, the Book-1 *on* Materials *and the other,* the Book-2, *on* Engineering Applications, *both carrying the common title of* Advances in High Temperature Superconducting Materials and Engineering Applications. *The books take stock of the recent advancements in the state of art materials technology and engineering applications of HTS at the global level. The books carry extended original research papers and technical review articles written by physicists, chemists, materials scientists and engineers, all noted experts in the field from different countries. The interdisciplinary coverage of the volumes should interest and benefit graduate students, researchers and specialists from aforesaid academic faculties and also from the electrical and electronic industries. At the same time, the topical reviews, written by experienced researchers, should interest the undergraduate and the non-specialist reader. Wherever possible, the necessary attention has also been paid to actual future applications in a digestive way and tasks for future work have been elaborated. Every care has been taken to ensure that the prime focus of the two volumes remains firmly anchored to the theme of interrelation between materials fabrication and engineering applications of HTS and there is no digression to various other superconductivity areas not directly relevant to applications issues. A major part of literature presently available on applied superconductivity, mostly as conference proceedings and a few edited books, seems to significantly lack this focus. These two books together are perhaps the first attempt, to bring out a comprehensive status of simultaneous advancements in materials engineering and engineering applications of HTS in a rapidly changing scenario of applied superconductivity.*

Book-1 comprises 13 chapters focusing on major technical advancement in HTS materials processing for applications. At the start there is an interesting overview of HTS for applications. Advances in bulk materials, wires, tapes, conductors in varied forms and thin films of all the technically important HTS systems, e.g., YBCO, BSCCO, Hg- and Tl-based cuprates, MgB_2, *are presented in relation to their various applications. The relevant areas like microstructures, flux pinning, A.C. losses etc., having a direct bearing on applications, have received an adequate coverage in various chapters. A successful exploitation of HTS in superconducting power engineering installations often demands compatible high quality insulating materials which have to withstand very large magneto-mechanical stresses at cryogenic temperatures. There is an extended paper in this book, covering this important area, interfacing HTS with practical engineering applications.*

The book on applications, i.e., Book-2, *carries 16 chapters covering numerous application areas where HTS is making a frontal impact. As with the other book, the two initial chapters highlight and consolidate some prominent happenings in HTS applications. Advances in SQUIDS, SQUID applications, SQUID microscopy, microwave filters, thin film devices, bolometers, HTS magnets – both electromagnets as well as permanent magnets, magnetic energy storage, levitation applications, power cables, power transformers, a.c. machines, motors, fault current limiters etc., form the mainstay of this book.*

Contents of the two books together, I feel, make an unbelievable success story for applied superconductivity. It is astonishing to see such a spectacular progress occurring within a short span of 15 years. Formidable challenges originally perceived by HTS materials have been effectively met and the work has gone a long way in substantially fulfilling the initial promises and expectations for their exciting engineering applications. Although the choice of materials systems available for applications seems restricted, there does not seem any real need of a fresh discovery of a new HTS material, unless that happens to be the room temperature superconductor ! With remarkable progress achieved in understanding the complex phase relationships in existing cuprates, ingenious ways of their processing and dramatic improvements in their properties through dopings and additives, it has now been possible to produce these materials in wires, tapes, thin films or bulk form to suit the desired application requirements. Extraordinary gains of HTS technology in respect of several applications listed above are presented in chapters of Book-2. HTS microwave filters and HTS SQUIDS seem to emerge as niche market areas for HTS electronics. Similarly, the fault current limiters, power cables, a.c. machines, motors etc. have attended a competitive marketable status. It is important for the industries to take stock of these developments for their possible large scale commercialisation. Most electronic industries today, however, seem to look askance at cooling to cryogenic temperatures and want to confine cooling at best to the level of a fan unless the performance advantages are immense. The latter kind of situation is expected to hold for HTS technology where the liquid nitrogen temperatures can be realised with remarkable ease. A changeover to any new technology invariably carries higher initial costs that are generally outweighed by long term gains. This should happen also with the HTS technology. The present costs of HTS cables and conductors and also of HTS SQUID elements, for example, are high, but they should go down with time as the materials further improve and their demand rises.

I am thankful to more than 70 contributors from 12 countries for their most commendable efforts in preparing and timely sending their contributions. I am grateful to Roland Hott for making several interesting suggestions and sending various useful information that helped me to organize them. Thanks are further extended to IUC-DAEF, Indore for providing facilities and the required infrastructure for doing this work.

May, 2003 *Anant V. Narlikar*

CONTENTS

THE APPLICATION OF HIGH TEMPERATURE SUPERCONDUCTORS IN BRUSHLESS AC MACHINES

David DEW-HUGHES and Malcolm McCULLOCH

CURRENT STATUS OF HIGH-T_c SUPERCONDUCTING BULK ROTATING ELECTRIC MOTORS APPLICATION

Itsuya MUTA and Taketsune NAKAMURA

A MOTOR WITH HIGH-TEMPERATURE SUPERCONDUCTING LEVITATION AND ITS VIBRATION CONTROL

Kosuke NAGAYA, Takaaki SUZUKI, Noriaki TAKAHASHI and Hiroyuki KOBAYASHI

LEVITATION APPLICATIONS OF HIGH TEMPERATURE SUPERCONDUCTORS

John R. HULL

FIELDS AND FORCES FROM SUPERCONDUCTORS AND PERMANENT MAGNETS

A.M.CAMPBELL

TREND OF APPLICATIONS TO HIGH-Tc SUPERCONDUCTING POWER TRANSFORMERS IN JAPAN

Kazuo FUNAKI and Masataka IWAKUMA

APPLICATIONS OF HIGH - Tc SUPERCONDUCTORS TO SUPERCONDUCTING MAGNETIC ENERGY STORAGE (SMES)

Takakazu SHINTOMI

UNRIVALLED SENSITIVITY -

SQUIDS IN NONDESTRUCTIVE TESTING

Marc von KREUTZBRUCK

DC SQUID MAGNETOMETERS FROM $YBa_2Cu_3O_{7-\delta}$ FOR BIOMAGNETIC APPLICATIONS

Henry-Jobes BARTHELMESS, Frank LUDWIG, Meinhard SCHILLING, Dietmar DRUNG and Thomas SCHURIG

HTS APPLICATIONS: PROGRESS IN SQUID MICROSCOPY AND HIGH RESOLUTION NON-DESTRUCTIVE EVALUATION

S.A. GUDOSHNIKOV and L.V. MATVEETS

VORTEX MATTER AND SUPERCONDUCTING ELECTRONIC DEVICES

Roger WÖRDENWEBER

HTS MICROWAVE FILTERS: PROPERTIES, DESIGN AND SYSTEM APPLICATIONS

Heinz J. CHALOUPKA and Tobias KAESSER

RESPONSE ANALYSIS AND MODELING OF HIGH TEMPERATURE SUPERCONDUCTOR EDGE TRANSITION BOLOMETERS

Mehdi FARDMANESH

HIGH TEMPERATURE SUPERCONDUCTING CABLES

Ole TONNESEN and Jacob OSTERGAARD

HTS APPLICATIONS : PRESENT AND FUTURE PROSPECTS

Pascal TIXADOR
CNRS-CRTBT/LEG
B.P. 166
38 042 GRENOBLE Cedex 09, FRANCE

1. INTRODUCTION

When the high temperature superconductors (HTS) were discovered from 1986, a technical revolution in electrical engineering was announced with strong enthusiasm. The superconducting applications should develop rapidly and impose themselves in many fields. After 25 years, we can observe that no technical revolution occurred and that the LTS (Low Temperature Superconductors) still remains far and away the most used superconducting (SC) materials. Several explanations can be put forward. HTS reveal to be extremely complex and their elaboration under long length wires at a competitive cost has set and still set numerous technical difficult problems. The HTS are ceramic materials, a major obstacle to make wires. A new technology takes always time to develop and to come to market. Moreover, most of the time, HTS applications compete with conventional devices, which are already very efficient.

Nevertheless, the advantages and benefits brought by superconductors are so important that they will revolutionize the electrical engineering. Only time is required so that the SC technology comes to maturity, becomes economically competitive and penetrates the electrical market. HTS will play undoubtedly an important role in the future after a first development in niche markets.

Moreover superconductors bring not only an evolution but a revolution into the electrical engineering ranging from the manufacturers to the users via the utilities. To benefit fully from SC devices all the electric system should be rethought : voltage level, short-circuit and fault currents, ... The challenges to be taken up are numerous whereas only a few people would accept such a revolution. The introduction of superconductivity will be rather progressive.

To be complete, before a view of the main HTS applications, a paragraph about refrigeration systems for the HTS applications will give some data. So a few lines will summary informations about HTS materials for the applications.

2. REQUIRED ELEMENTS

2.1. Cryogenics

A refrigeration system is necessary to cool the HTS devices and to remove all the losses (thermal and electrical) dissipated at the cold temperature. In general the refrigeration system uses a working fluid, which follows a thermodynamic cycle (Stirling, Brayton, Claude, ...) depending on the refrigerator type. The working fluid is very often helium. Generally compressed at ambient temperature, it is expanded within an engine to produce the cooling effect. The refrigerator power is mainly absorbed by the compressor. The required power (P_a) at ambient temperature (T_a) to extract losses (P_c) at the low temperature T_c is expressed by a fraction of the ideal Carnot efficiency (equation (1)). The ratio P_a/P_c is called the specific consumption and its inverse the coefficient of performance.

$$P_a = P_c \frac{1}{\eta_{ref}} \frac{T_a - T_c}{T_c} \tag{1}$$

η_{ref} : refrigerator efficiency compared to the Carnot efficiency

J. R. Strowbrige [1] experimentally noted that the refrigerator efficiency (η_{ref}) mainly depends on the cold power, and little on the cold temperature (fig. 1). This efficiency, very low for small powers (2 % for 1 W), increases with the cold power to reach approximately 40 % as asymptotic value (fig. 1).

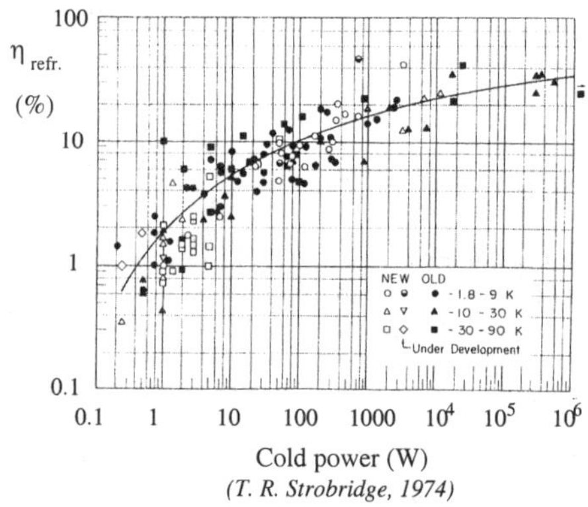

Fig. 1 : efficiency in function of the cold power [1].

The interest to operate at "high temperature" appears very clearly. The power requirement increases rapidly as the temperature goes down. A refrigerator consumes 2.9 kW to extract 100 W ($\eta_{ref} = 10$ %) at 77 K but 66 kW for the same power at 4.5 K.

Nevertheless the real issue of cryogenics is not its power requirement, but its cost since the losses can be kept to a very low level. Fortunately there is also a strong reduction of the costs when the temperature goes up. The reference [2] gives some costs for refrigeration systems according to their cold power (P) :

$$\mathrm{Cost}_{(k\$)}\ (4.5\ \mathrm{K}, P) = 1500\ P_{(kW)}^{0,6} \tag{2}$$

$$\mathrm{Cost}_{(k\$)}\ (65\ \mathrm{K}, P) = 15\ P_{(kW)}^{0,6} \tag{3}$$

The cost of 4 K cryogenics remains the real issue for the economic development of the SC power applications except some niches like the MRI (Magnetic Resonance Imaging) or NMR (Nuclear Magnetic Resonance) spectrometry. Many projects were great scientific successes, but the industrial developments did not follow because of 4 K cryogenics, too much expensive. The reliability is on the other hand not concerned, many installations showed a remarkable reliability. In addition the cryogenic systems are now completely automatic apparatuses ("plug and use") and do not require any cryogenic ability. The maintenance is rather low and easy.

At 65 K, the cost, two orders of magnitude lower than at 4.5 K, is not any more a real issue. In the same way, the liquid nitrogen is an industrial cryogenic fluid whereas liquid helium is of a use much more specific. The figure of 15 k$ per cold kW at 65 K (equation (3)) could be still divided by a factor approximately ten for mass production.

The typical loads are from some hundreds to thousand Watts at 65-80 K for transformers and at 20-35 K for rotating machines. Fault current limiters need lower cold powers (hundreds of watts at 65-80 K). Cables require typically 1-5 W/m at 65-80 K. The inquiries for dc SC magnets (SMES, magnetic separation, MRI) are much lower, from tenth Watts to tens of Watts at 20-30 K. The low range of power (under 500 W at 80 K, under 100 W at 25 K) are well adapted to cryocoolers [3] whereas turbine refrigerators suit better for higher powers, showing higher efficiencies. The turbine miniaturization is difficult without degrading their efficiencies.

The Gifford McMahon cryocoolers are now on the market since several years and are widely industrialised. Continuous improvements are still brought. The pulse tube cryocoolers are more recent and know large developments. Their main advantage is the absence of moving parts at low temperatures leading to high reliability, low maintenance and no induced vibration.

2.2. HTS materials

2.2.1. A.C. Losses

The absence of any dissipation in a superconductor is only valid when its electromagnetic environment is time constant. Submitted to a magnetic field or a transport current varying in time, any superconductor shows losses, known as A.C. losses. That is included in the Maxwell-Faraday law ($\nabla \times \boldsymbol{E} = -\, \partial \boldsymbol{B}/\partial t$). A time variable magnetic induction (external or created by the transport current) induces an electric field, which creates losses. These losses should be multiplied by the specific consumption to take into account the cryogenic cost and are rapidly very costly if they are not kept at an ultra low level.

The A.C. losses can be reduced by an appropriate structure of the SC wire. The suitable structure depends on the field conditions. However, in general, a wire for A.C. uses should be of reduced size with very fine filaments twisted with a short pitch whereas the matrix must be highly resistive or contain resistive barriers [4].

So superconductors suit better for D.C. devices since they show then absolutely no loss and do not require even a power supply as soon as the operating current is reached (short-circuit or persistent operation). The refrigerator provides only the thermal losses, which can be very low through a careful design of the thermal insulation. Used in A.C. applications, the superconductors show, even if it is slight, dissipation that must be removed by the refrigerator. Moreover the SC element design must allow the loss removal and limit the temperature rise (large exchange surfaces, ...). Table 1 illustrates these remarks by comparing the parameters of two SC power links one D.C. and the other A.C.

	Losses	Distance between refrigerators
D.C.	30 W/m	34 500 m
A.C.	5 700 W/m	1 200 m

Table 1 : 3000 MVA D.C. and A.C. superconducting power links.

2.2.2. HTS Wires

2.2.2.1. Bi PIT tapes

The first generation of HTC wires is the PIT (Powder In Tube) tapes with BiSrCaCuO as superconductor [5, 6]. A powder of precursors containing BiSrCaCuO is placed in a pure silver or a silver alloy tube. Extrusion, rolling, heat treatments and oxygenation form the SC phase with the suitable uniaxial texturc. Stainless steel layers on the tape improve the mechanical properties. The critical current is then only slightly reduced up to a tensile stress of 250 - 300 MPa in comparison to 75 MPa for non reinforced tapes.

This industrial process makes it possible to manufacture kilometer lengths with good performances. The overall critical current density (engineering critical current density J_e) is about 100 to 150 MA/m^2 at 77 K and 0 T for Bi-2223 tapes. These tapes have reached a pre-commercial stage, in particular with the commissioning of a production unit of 20 000 kilometers per year [5].

Unfortunately the transport current capacities of Bismuth compounds decrease dramatically under magnetic induction over 0.1 T at high temperature, 77 K in particular (fig. 2). Except for low magnetic field applications (power cables, transformers), the PIT tapes must be cooled below approximately 30 K.

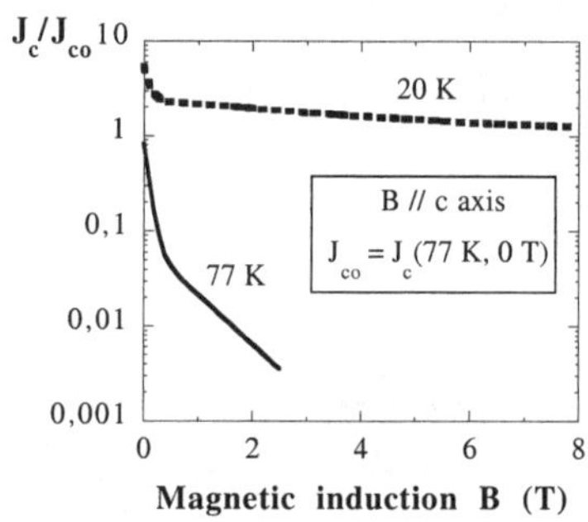

Fig. 2 : critical characteristics of PIT Bi-2223 tape at two temperatures ($J_{co} \approx$ 100 MA/m^2).

But the main issue of the PIT wires is their high cost for commercial applications. The measuring unit is the $/kA/m, that is to say the cost of one meter conductor carrying 1 kA. For a use in the electrical devices, the threshold value of 10 $/kA/m is often quoted. By way of comparison, for NbTi used with D.C. current, the cost is 0.8 to 2 $/kA/m. The silver is for example expensive but essential for PIT and the 10 $/kA/m target appears difficult to meet. The current figure is approximately 300 $/kA/m with an objective of 50 $/kA/m in five years. Silver suits because it does not react with the superconductor and allows the oxygen diffusion for the right oxygenation of the superconductor, which is a sensitive parameter.

Their use under A.C. high fields raises some difficulties because their A.C. losses are high. Developments are carried out to manage this issue (twisting, resistive barriers, ...).

2.2.2.2. Coated conductors

The YBaCuO compound (Y-123) shows higher critical current density than Bi, being less sensitive to magnetic field, at "high" temperature, 77 K. However, to transport high current densities, YBaCuO requires a biaxial texture, a huge obstacle for kilometer long wires [7]. The technique used to overcome that difficulty is to deposit biaxially YBaCuO (μm thickness) on a flexible metallic substrate (50 μm typically) through a buffer layer. A shunt layer over the Y-123 coating protects the superconductors from hot spots (local overheatings) and makes the current transfers easier. The wire with a biaxial Y-123 layer is the second generation HTS wire and is called a coated conductor.

In the most industrial process for coated conductors (RABiTS technique (Rolling-Assisted, Biaxially-Textured Substrates)), the biaxial texture of Y-123 is achieved by the highly aligned substrate. Deformation and heat treatment processing produce the desired texture of the substrate. The Y-123 layer can be deposited by numerous methods but the chemical techniques such as MOD (Metal Organic Deposition) suit because they are high volume and low cost manufacturing process.

They carry great hopes but their development and industrialization raise enormous difficulties. Currently the best performances over some meter length reach 100 to 200 A/cm (width of the tape) [5], [8]. These conductors, in addition to their high performances, should not be very expensive to produce and they have the potential to meet the 10 $/kA/m cost objective.

The coated conductors are certainly the industrial future of the HTS wires and they will replace the Bi-PIT when they will be available. Their cost is expected to be two to five times lower and they operate at higher temperature under field.

2.2.3. HTS Bulk material

For some applications, bulk superconductors suit and even are required. In addition, by cutting or machining, long length SC elements can be elaborated from a bulk piece : a solenoid in a tube, a meander in a plate. All the HTS materials can be synthesized in bulk form, in particular by the conventional sintering techniques but the critical current density performances are low (MA/m^2 at 77 K, 0 T). An alignment of the elementary BiSrCaCuO grains, by thermo-mechanical processes for example, led to much better performances in current densities (up to 100 MA/m^2 at 77 K, 0 T). The critical currents are still much higher in single domain YBaCuO samples [9]. A single domain is a single crystal but with parasitic phases. One of the most efficient techniques to synthesize YBaCuO single domains is the Top Seed Melt Texturing technique [10] proposed by Morita in 1991. Samples as large as 100 mm in diameter are elaborated [11, 12] but techniques makes possible to increase this size. Single domains can be soldered for example to form a large single domain.

3. LTS APPLICATIONS

If the LTS have not revolutionized the electrical engineering, they are essential in several fields [13]. The LTS market applications are mainly MRI or NMR spectroscopy. The high-energy physics and more recently thermonuclear fusion play a considerable driving role for the SC electrical engineering but are not commercial products. These applications for physics have been very important for the development of SC techniques through the large programmes carried out to take up the challenges proposed by the physicians. Thank to the high physics programmes, the NbTi conductors are now a mature and totally industrial material. The Nb_3Sn has progressed through the thermonuclear fusion programmes.

The main advantages of the superconductors are the total absence of Joule losses in D.C. conditions and the very high achievable current densities. The non-dissipative ampere turns are then considerable and it becomes possible to produce, with ultra low powers, high magnetic inductions in large volumes. A current induced in a carefully shorted-circuit SC coil (persistent mode) is practically constant on a human scale. The MRI magnet specifications give a reduction of the magnetic field lower than 10 % in a century.

The SC applications have been easily developed when they are not in real competition with existing technologies or when they appear in a new market. The first case is a high magnetic induction in a large volume, the second is coils for MRI. In these devices, the necessary cryogenics is not a real drawback since it is a costly but perfectly controlled advanced technique. The BEBC bubble chamber at CERN was cold during ten years between 1974 and 1984 without any stop and operated at rated current during 24 000 hours. A lot of advances have been made in cryogenics in twenty years, so cryogenics is no more a problem, except its cost.

The superconductors are used mainly to produce D.C. or very slowly variable magnetic inductions in time.

MRI was the first commercial market for superconductivity [14] and remains important (30 % of the NbTi conductor production). It still is a growing market. The field produced by the SC coil is used to polarize the core of some atoms, like the hydrogen proton. This one, excited by a radio frequency field, makes it possible to build three-dimensional images of high quality, in particular for soft tissues of the human body. MRI is non invasive and an invaluable help for the medical diagnosis. Magnetized volume is significant to make human body imagings and the ratio signal over noise increases with the induction amplitude. Due to these two characteristics, more than 80 % of full body MRI devices uses SC windings for the polarization magnetic induction, approximately 1 T. More than 10 000 SC MRI systems operate routinely throughout the world with a remarkable reliability. It is a market always in expansion.

The data analysis of the polarized atom cores leads to other information like nature, the chemical composition and the structure of the studied samples. These techniques are under fully development in the chemical and pharmaceutical industries and uses NMR spectrometers. The researchers into biomedical use the spectroscopy to analyze the three dimensional structure of molecules (search of new drugs) and proteins for the human genome studies. So NMR spectrometers are a growing market. If the required volume can be reduced (analyzes of small samples), the needed induction is considerable. The induction is proportional to the frequency, which determines the smoothness of analysis and its sensitivity. For 1 GHz, induction is 23.5 T. The commercial devices operate with frequencies up to 900 MHz. The developments are under progress for 1 000 MHz spectrometer [15]. The SC coils use NbTi in the areas where induction is moderate (lower than approximately 10 T) and Nb_3Sn conductors for the areas with strong field. The developments underway concern HTS inserts but used at very low temperature (lower than 5 K) for the areas with maximum field.

A shorted-circuit SC coil can store magnetic energy (fig. 3) and is one of the few means to store directly electricity. This device is the SMES (Superconducting Magnetic Energy Storage) [16]. Its efficiency is excellent, higher than 95 %, including cryogenic cost. If the volume density of stored energy is rather poor ($B^2/2\mu_0$ so 20 MJ/m^3 for a reasonable magnetic induction of 5 T), it is available very quickly and the power density can be considerable. The cyclability is almost infinite.

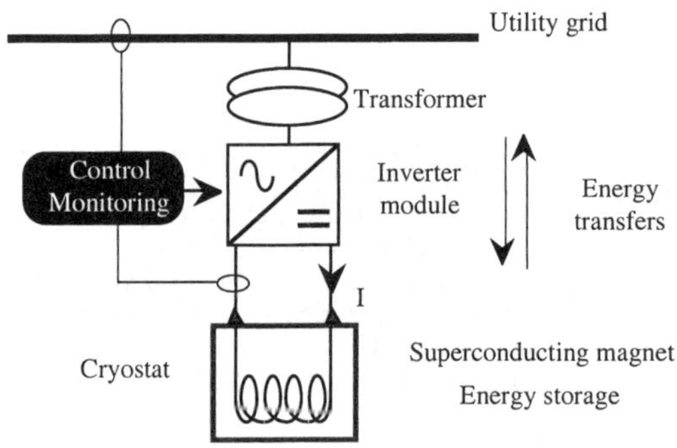

Fig. 3 : principle of a SMES connected to an electric network.

The SMES are mainly used for :

- energy storage for power quality uninterruptible power supplies for sensitive loads to perturbations, voltage sags for example [17],
- stabilization of an electrical networks by controlled energy transfers between the SMES and the network and to thus increase capacities of the existing systems, pushing back

the limits of stability and improving quality, the SMES is then a FACTS (Flexible Alternative Current Transmission System) [18],
- providing ultra-fast energy pulses (fundamental physics, electromagnetic weapons or launcher, magnetic forming...).

Several SMES are in operation and are the first commercial SC application in the electric system [17]. They all use NbTi wires. Stored energies are some MJ with powers of the MW order of magnitude.

SC magnets are used for mineral and chemical processing [19]. About 20 systems provide the magnetic separation of mineral, e.g. kaolin. For chemistry (polymerization, water cleaning, ...) the use of SC magnets is in an early exploration stage.

The magnetic field is used in high energy physics to guide, focus and deviate the particles. As soon as inductions or volumes become significant, superconductivity becomes impossible not to use. The circular particle accelerators require a magnetic induction to keep the particles on an accurate circular orbit. A high value of induction makes it possible to reduce the accelerator radius at a given particle energy. The next large particle collider, the LHC [20] (Large Hadron Collider) is under construction at the CERN in Geneva. The LHC will use a magnetic induction of 8.4 T created by 1232 NbTi dipoles cooled with He II at 1.9 K and distributed on the 26.7 km of the machine. Only the dipoles and quadripoles will use 1125 tons of NbTi conductors (about 200 000 km of strand). The LHC will use a new generation of detector magnets to deviate the particles produced at the collision times. These detectors (ATLAS [21] and CMS [22]) will use NbTi magnets producing significant inductions, up to 4 T in huge volumes, several hundred m^3. The magnetic energies stored by these magnets will reach GJ (2.6 GJ for the CMS solenoid and 1.2 GJ for the toroidal ATLAS system).

The most advanced concept for thermonuclear fusion is tokamak. Its principle is to magnetically confine the plasma, which temperature oversteps 100 million degrees. The ignition, i.e. the operation where the produced power compensates the losses, needs high magnetic inductions, approximately 10 T, in huge volumes (1000 m^3). There too, the superconductors are required for continuous operation. Tore Supra built at CEA-Cadarache in France opened the way of the SC tokamaks [23]. With a reasonable size (600 MJ stored), it operates routinely since 1986 and made it possible to capitalize a large number of knowledge. The large international project ITER (International Experimental Thermonuclear Reactor), started at the beginning of the nineties, aims to prove the feasibility of fusion for energy source [24]. It includes the test of toroidal model coils and the central solenoid. The level of required inductions (13 T) implies the use of Nb_3Sn coils with very high currents (40 to 60 kA). The developed conductor is a "cable in conduct" (fig. 4).

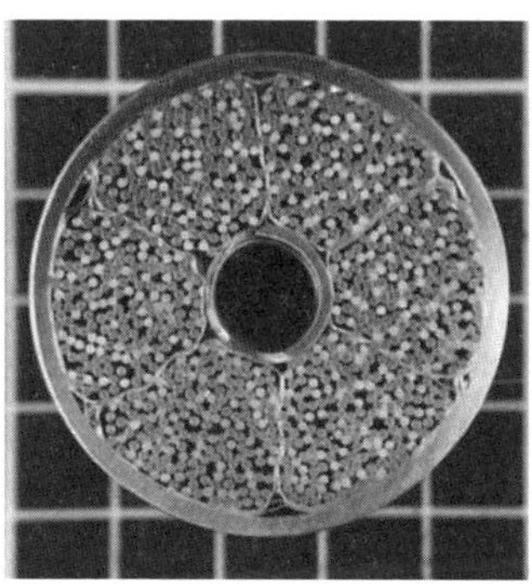

Fig. 4 : Nb_3Sn cable in conduct cross section developed for fusion [24].

The stellerator offers a structure of magnetic confinement alternate to tokamak. Germany is building a model machine, W7-X [25], which uses NbTi coils with a very specific shape to provide the right plasma magnetic confinement.

Magnetic induction is a invaluable tool for the physicists to study materials and physical phenomena. As soon as induction exceeds 2 Teslas, the SC solution quickly becomes more interesting in terms of operating costs. Table 2 compares two magnets from the GHMFL (Grenoble High Magnetic Field Magnet Laboratory). The SC coil is the NbTi external winding of the GHMFL hybrid magnet [26]. The gains in term of power supply are very important especially when the magnetic energies are compared. The supply powers can be compared since the cooling powers are similar. The water pumps for the resistive magnets cooling (100 m^3/h with a pressure drop of 23 bars) consume indeed as much as the refrigeration system of the SC magnet.

	Bitter magnet	NbTi magnet
Magnetic induction (T)	10	11
Field core (mm)	50	500
Magnetic energy (MJ)	0.105	23
Power supply (MW)	2.1 (7 000 A, 300 V)	0.005 (1000 A, 5 V)
Cooling (kW)	100 (pumps)	100 (liquefier)

Table 2 : Bitter and superconducting magnet comparisons.

These numerous applications have demonstrated their technical feasibility as well as their high performances. But their developments are slowed down by economic considerations. For LTS, cryogenics remains the main economic issue except for some niches described above. This issue is partially lowered with HTS, which requires a much "lighter" cryogenics and especially much less expensive but the cost and the electrical performances of the HTS conductors remain the critical issue.

4. HTS APPLICATIONS

4.1. Superconducting Fault Current Limiter

4.1.1. Introduction

In an electrical network, at the time of a fault such as a short-circuit, the current increases very quickly and can reach 20 to 30 times the rated current. The electrodynamic forces are then 400 to 900 times higher than in normal operation. The fault current can be only limited by the short-circuit impedance at the node of the fault. For very high voltage grids, the fault current is not reduced in amplitude but only interrupted by a switchgear. This interruption occurs at a natural zero crossing for an alternating current (A.C.). The minimum isolating time is then approximately 20 ms (one cycle) for an asymmetrical short-circuit (fig. 5). But in general the isolating time is several cycles and 50 ms is a typical value in very high voltage grids. All the components must withstand the extremely severe electromechanical forces during all this duration. The short-circuit current is the designing value for the devices and its leads to a large oversizing compared to the normal operation and thus to a significant overcost.

A efficient technique to secure the electrical supply of a consumer, is to multiply the supplying feeds. In addition this higher interconnection leads to an improved quality of the voltage wave. The voltage variations are reduced in amplitude, in particular at the time of sudden large overloads. The voltage drop is indeed proportional to the short-circuit impedance, which decreases with the number of parallel paths.

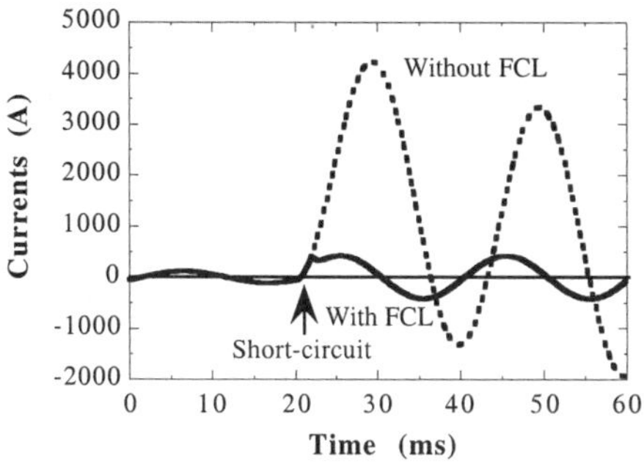

Fig. 5 : current evolutions at the event of a short-circuit.

But a lower short-circuit impedance leads to higher fault currents. So the improvement of energy quality is limited by the breaking capacities of the switchgears. So a device, which makes possible the increase of the short-circuit current without increasing the fault current is extremely attractive. The Fault Current Limiter (FCL) is such a device and only superconductors make possible to design FCL under very high voltages. Their non linear electric field versus current characteristic shows an ideal performance for current limitation. Under its critical current a superconductor shows a very low electric field but this increases by several orders of magnitude as soon as the critical current is overstepped. The length of the superconductor must be adapted so that the developed voltage equilibrates the grid voltage. A switchgear is always associated to the FCL to provide inter alia the fault isolation. As soon as the FCL is isolated, the superconductor cools down and recovers its SC state after a delay, the recovery time. The FCL can be then connected again to the grid. But if the loss of the SC state (quench) can be a very quick phenomena (μs), the recovery time, linked to thermal transfers, is much longer, from hundreds of milliseconds to minutes. It depends on the maximum raised temperature and on the cooling conditions. Long recovery times are a problem for the FCL in some network locations where the reclosing of the line must be quick for the continuity of service. Some electric schemes make possible to meet that problem.

Superconductors integrate the fault current detection, trigger themselves the limitation process and recover automatically the SC state. These natural favourable properties make the SC FCL very attractive and well studied throughout the world all the more that the electric power quality improvement is a real need today. In addition, the electricity market liberalization should reinforce the FCL interest. The additional feeds by independent power producers increases the short-circuit power and in some cases only a SC FCL could manage this short-circuit power.

The FCL brings not only technical advantages such as the improvements of the electricity supply quality and the voltage wave but also economic advantages by reducing for example the over design of the devices. They should withstand only the limiting current and no more the short-circuit current. But the FCL cost must be obviously not too high.

The first studies on SC FCL date from the sixties [27] and the first demonstrator appeared in the eighties and nineties when the NbTi A.C. composites were developed [28]. Alstom and EDF successfully realised and tested a 36 kV device [29]. The high Tc discovery stopped the NbTi FCL projects. All the studies are carried out now on to the SC oxides, which require a much simpler cryogenics and show lower recovery times.

4.1.2. Different kinds of superconducting fault current limiters

4.1.2.1 Resistive limiting device

The resistive limiter is the simplest device. The SC element with a low self inductance is inserted directly in series in the circuit to protect. The fault current is limited by the resistance developed by the SC element above a given value of current. A resistance, is often in parallel to the SC element in order to reduce over voltages [30] and to adapt the limitation current to the network requirements. Depending of the FCL location in the network, the limiting current must be high enough so that the amperemetric protections, based on the current amplitude, see the fault.

4.1.2.2 Inductive limiting device

In the inductive limiter, the superconductor through its quench only triggers the limitation, which is performed by an inductance so its name. The inductive FCL uses in general a series transformer inserted in the line to protect. Its primary winding is most of the time resistive. The SC element is the secondary winding or connected to it. The strong variation of the secondary resistance after the quench, makes that the transformer changes from a short-circuit operation with a low impedance to a nearly no-load operation with a much higher impedance, mainly an inductance.

Different solutions are possible for the secondary. It can consist in particular in a or several SC tubes especially Bi bulk cylinders (fig. 6) [31]. An advantage is, inter alia, the absence of current leads and the corresponding losses. ABB installed during one year such a limiting device of 1.2 MVA in a power plant in Switzerland for industrial tests [32]. This apparatus used Bi-2212 rings.

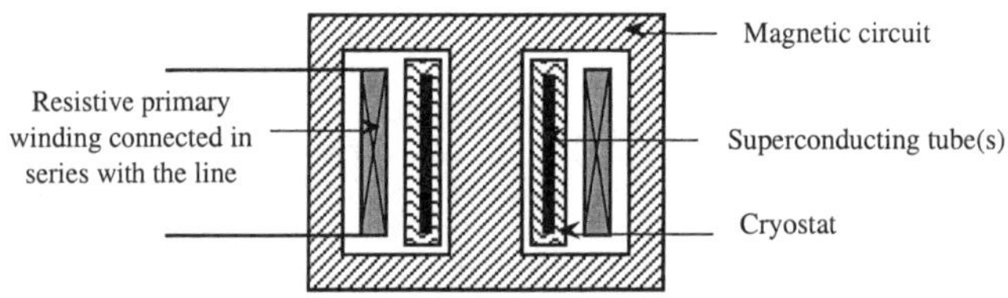

Fig 6 : transformer or "shielded core" type SC FCL (one phase).

By varying the transformer series turn ratio, the current in the SC element(s) can be easily adapted. However except in the special case with a variable magnetic coupling [30], the product voltage by the current remains constant in the transformer and the needed SC volume is nearly the same in a inductive limiter compared to a resistive one. The series transformer is a issue in terms of weight, volume and losses and it introduces a higher voltage drop in normal operation compared to the resistive limiter. These reasons make that the resistive limiter is the favourite solution for the SC FCL.

4.1.2.3 Limiting device not based on a transition

Some limiters use a SC element but are not based on its quench. The oldest is the saturable inductance. A magnetic circuit is easily saturated by a SC coil. The fault currents, by reducing the saturation, increase then the inductance, which limits the fault current.

Another limiter uses a SC coil inserted in a rectifier bridge through a D.C. voltage source [33, 34] (fig. 7). This source creates an adjustable current I_0, which separates itself symmetrically in the two bridge arms. As long as the line current is lower than $|I_0/2|$, the diodes are conducting and the voltage drop is reduced. But, as soon as the instantaneous line current exceeds $|I_0/2|$, one of the diodes is reverse biased and the current is forced to circulate through the inductance, which limits it. The use of thyristors makes it possible to sophisticate the system. General Electric in the United States carried out a device of 30 MVA with a Bi-2223 coil.

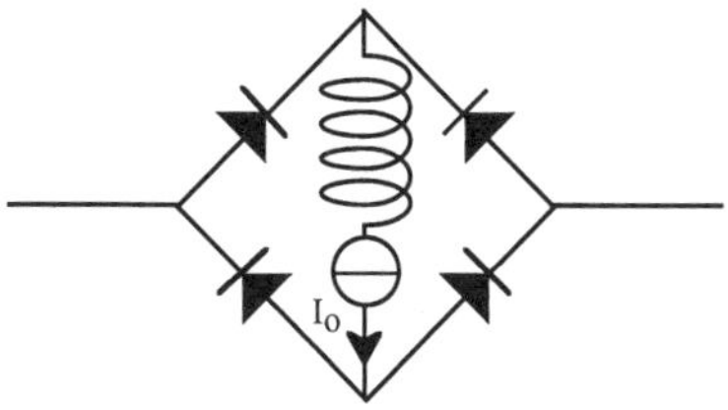

Fig. 7. Schematic of bridge concept FCL using a SC coil (single phase).

The SC FCL, whatever its type, must fulfil several extremely severe conditions related to the network.

4.1.3. FCL location in the network

The FCL can be localised mainly in two places of an electrical grid (fig. 8) [35, 36].

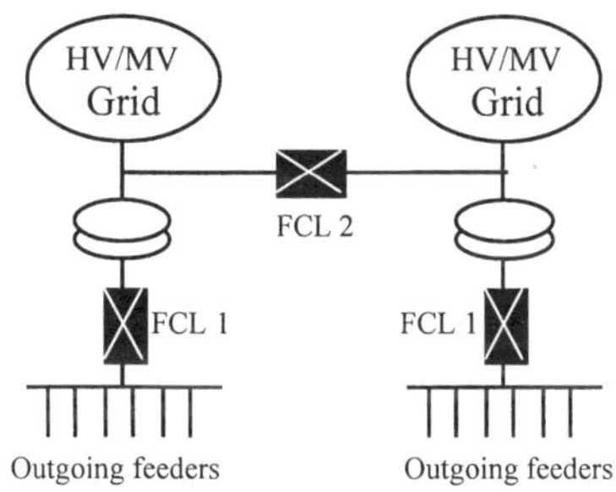

Fig. 8 : FCL possible locations in an electric network.

The first location (FCL 1) is on the medium voltage feeder. The fault currents can take extremely high values there and the FCL would reduce the thermal and mechanical stresses of the equipments. The second FCL location in bus couplers (FCL 2) appears particularly interesting for the grid [35]. It had already been identified for NbTi FCL [29]. The FCL would connect two parts of the network. These cannot be linked today since this connection leads to fault currents surpassing the breaking capacity of the existing switching devices. But this connection would advantageously secure the network and improve the voltage quality. The recovery time can be long since the FCL isolation leads to the present state with the two parts not inter-connected as in the case of the FCL failure. This last point should put the utility mind at ease since they always fear the reliability of a new system. The limitation current of the FCL in this FCL 2 location must be as low as possible, not higher than twice the rated current. This is the condition so that the fault current does not exceed 10 % the fault value of only one network. The fault current is the sum of the fault currents of each part of the network. Lastly, the current circulating in the FCL in normal operation is low, even zero when the two grid parts are balanced. The energy dissipated by the A.C. losses of the SC element is consequently low and this leads to a reduced cryogenic load.

The deregulation of the electricity market will reinforce doubtless the FCL interest. The deregulation induces the multiplication of the electricity producers on the network, therefore an automatic increase in the short circuit power, which leads to the problem of the fault current amplitudes. This leads sometimes to complex connection schemes to meet the fault currents. Only the FCL makes it possible a growth of the short-circuit power without an increase of the fault current values.

The development of SC devices such as power electric cables or transformers is also favourable to SC FCLs. These devices must be protected and the FCL is essential and rather natural. The limitation function can be integrated into the transformer itself [37, 38].

4.1.4. Operations

4.1.4.1. No fault operation

In normal operation, without fault, the limiter must be as invisible as possible for the network. Therefore its impedance must be very small to lower the voltage drop. This must be typically lower than one percent of the rated voltage. The superconductor must be assembled in a little inductive way for a resistive limiter. The losses must be also very low, especially the cold losses. The cold losses must be kept down to an extremely low level, not only to reduce the operation cost but also and above all to lower the investment cost. The cost of the cooling system increases indeed with the cold power (equations (2) and (3)) whereas the size and the weight growth too. The main contributions to the cold losses are the cryostat, the current leads and the SC A.C. losses. The A.C. losses are essentially self field losses since the field is low to reduce the limiter inductance. The Norris formula [39] gives a good order of magnitude of the self field losses per unit length. $G(i_m)$ depends of the wire geometry and is given only for an elliptic section.

$$P_{sf} = \frac{\mu_o f I_c^2}{\pi} G(i_m) \qquad i_m = \frac{I_{max}}{I_c} \tag{4}$$

$$G(i_m) = \left(1 - i_m\right) \log\left(1 - i_m\right) + i_m - \frac{1}{2} i_m^2 \qquad \text{(elliptic section)}$$

The self field reduction comes through the subdivision of the conductor (lower I_c).

The A.C. losses are the main contribution to the total losses except for very low currents (no load operation). The current lead losses are proportional to the rated current (40 W/kA for non cooled current lead from 77 K to 300 K [40]) and depend little on the current amplitude.

The rated current is the maximum current in steady state but any line must withstand higher currents for limited durations. These transient overcurrents are due for example to motor startings or transformer connections. The connection of a no load transformer can induce very high asymmetrical currents, but, which decrease rather quickly (lower than 0.5 s). The motor startings lead to longer overcurrents but with lower amplitudes. These transient overcurrents should not induce an irreversible limitation process, with the opening of the line. Furthermore for the motor startings, the voltage drop of the limiter must be low not to slow down the starting. The torque of an asynchronous machine is indeed proportional to the square of the voltage. On the other hand the limiter plays a beneficial part for the transformer connection since it limits the inrush current and accelerates its damping.

4.1.4.2. Fault operation

The limiting effect is based of the non linear E(J) superconductor characteristic (E : electric field, J : current density). A good empirical relation is given by a power law :

$$E = E_c \left(\frac{J}{J_c(T)}\right)^{n(T)} \tag{5}$$

This relation is valid under the critical temperature but the electric field is limited in amplitude, or the exponent decreases for very high electric fields. The Ohm's law replaces it above the critical temperature :

$$E = \rho_n(T)\,J \tag{6}$$

The thermics plays an essential part in the limitation process except for superconductors with very low critical current densities and electric fields. The general thermal equation is given by :

$$\nabla\cdot(\lambda\,\nabla T) + \boldsymbol{E} \bullet \boldsymbol{J} - p = c_p \frac{\partial T}{\partial t} \tag{7}$$

λ : thermal conductivity ; p : external exchange power ; c_p : specific heat per unit volume

For SC films all the terms are in general important since the exchanges with the substrate are very important. The substrate plays the part of an enthalpic tank. When its thermal conductivity is high (Al_2O_3) the substrate evacuates also the heat.

For bulk materials the thermal conditions can be supposed in first approximation adiabatic because, among other things, they are bad thermal conductors (λ_{YBCO} = 9 W/m/K = 0.015 λ_{cu} (77 K)). The thermal equation simplifies itself a lot :

$$\boldsymbol{E} \bullet \boldsymbol{J} = c_p \frac{\partial T}{\partial t} \tag{8}$$

The thermics gives the maximum performances in terms of electric field and current density. The designing constraint is the maximum temperature rise (ΔT_{max}) reached when the SC element is isolated by the opening of the switchgear, a delay Δt after the fault. The integration of equation (8) above the critical temperature assuming either a current or a voltage source gives the maximum values for the electric field (E_{max}) and the current density (J_{max}), being also the limiting current density. These expressions are obtained by neglecting the temperature dependencies of the specific heat and the normal state resistivity so their averaged values have been considered (ρ_n^{av}, c_p^{av}).

$$E_{max} \approx \sqrt{\frac{2\,\rho_n^{av}\,c_p^{av}\,\Delta T_{max}}{\Delta t}} \tag{9}$$

$$J_{max} \approx \sqrt{\frac{\dfrac{2\,c_p^{av}\,\Delta T_{max}}{\rho_n^{av}\,\Delta t}}{}} = \frac{E_{max}}{\rho_n^{av}} = J_{lim} \tag{10}$$

From these two parameters the required length (L) and cross section (S) can be deduced and then the minimum SC volume for a single phase :

$$\vartheta_{min} = S\,L \geq \frac{I_{lim}\sqrt{2}}{J_{max}}\,\frac{V\sqrt{2}}{E_{max}} \approx \frac{V\,I}{\dfrac{c_p^{av}\,\Delta T_{max}}{\Delta t}}\left(\frac{I_{lim}}{I}\right) \tag{11}$$

V is the phase to neutral rated voltage (line impedance neglected).

This volume is simply the volume needed to limit the temperature rise to an allowable value. For a maximum temperature rise of 100 K, a isolation delay of 20 ms and a limiting current five times the rated value, the SC volume yields to about 0.5 dm^3/MVA or 3 kg/MVA ($c_p^{av} \approx 1.8$ MJ/m^3/K). It does not depend on the normal state resistivity, which defines on the other hand the section and length (equations (9) and (10)). Figure 9 gives the maximum electric field and current density with respect to the normal state resistivity. The normal state resistivity is preferably high to reduce the length of the SC elements in particular for high voltage devices. A large length introduces some problems of assembling and self inductance.

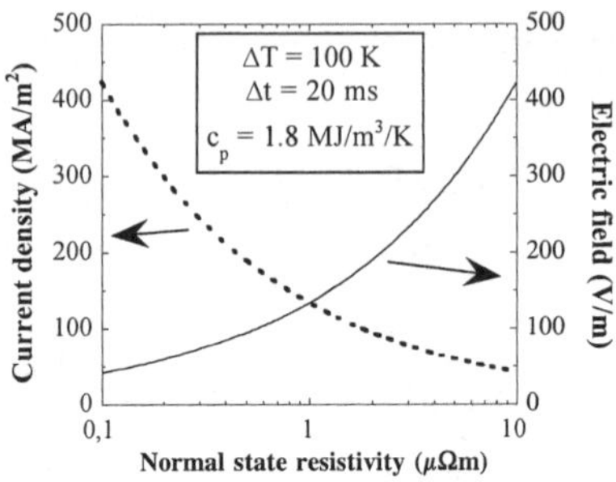

Fig. 9 : maximum current density and electric field versus normal state resistivity. (temperature rise of 100 K after an isolation delay of 20 ms, c_p = 1.8 MJ/m^3/K)

Equations (9) and (10) give the maximum values but the superconductor performances can be insufficient in particular in terms of critical current density. In no fault operation the rated current (I_n) density must be lower indeed than the critical value for low losses :

$$J_n = \frac{I_n}{S} = \frac{I_n}{I_{max}} J_{max} < J_c \tag{12}$$

If these values are easily reached in YBCO single domains, bulk BSCCO materials often show lower Jc values at liquid nitrogen temperature. BSCCO PIT tapes do not suit since their normal state resistivity is too low due to the silver matrix leading to an ultra low electric field under limitation (ten V/m).

For some locations in the network it can be interesting to use low critical current materials. In that case the superconductor heating is reduced and the ultimate case is the operation at constant temperature on the isothermal E(J) characteristic. The advantages are the absence of thermal stresses on the superconductor and the nearly instantaneous recovery of the SC state after a fault. On the other hand the volume of required material is very large. The related problem is not really the cost since they are low cost materials (sintered Bi for example) but the A.C. losses under normal operation.

4.1.5. Some realizations

4.1.5.1. Limiters with BiSrCaCuO

4.1.5.1.1. BiSrCaCuO sintered - ASTER Project

The samples containing Bi compounds (BiSrCaCuO) are rather simple to synthesize, especially under the sintered form. Within the context of the ASTER project (Schneider Electric company), an assembly of 60 sintered $Bi_2Sr_2Ca_2Cu_3O$ bars (fig. 10) was tested under a voltage of 1.1 kV [41]. Each bar has a diameter of 4 mm and a length of 120 mm. Its critical current at 77 K in zero field is approximately 35 A that is a critical current density of 2.8 MA/m^2. To increase the current capacity in no fault operation, two bars were put in parallel. Thirty sets of two bars in parallel were connected in series.

The large number of connections led to a total assembly resistance of 760 μΩ and to high losses under no fault conditions. They have to be carefully reduced. The A.C. losses in the bars themselves are low, in agreement with the formula of Norris. Figure 11 shows the time evolution of the current and the voltage for the high voltage test. The test bench of Schneider Electric consists in a capacitive discharge through an inductance and the SC assembly. The current reaches 1080 A instead of 5000 A without the SC elements. The limitation is thus effective, but occurs for a current definitely higher than the critical current, approximately 24 times.

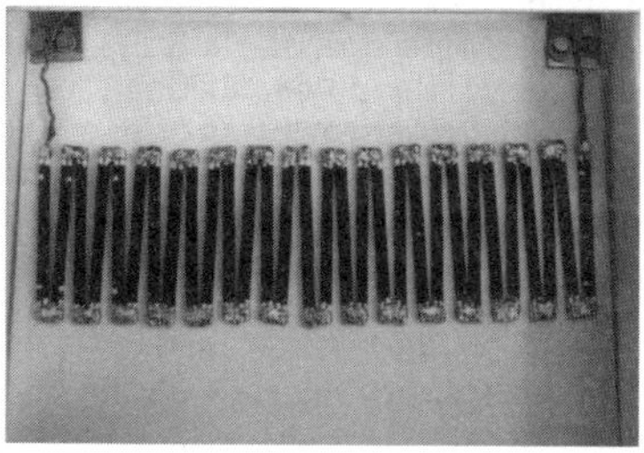

Fig. 10 : 60 sintered Bi bar assembling for the ASTER project. (each bar : l = 120 mm ; Ø = 4 mm)

The averaged maximum electric field (total voltage divided by the total assembly length) reached 300 V/m in this test. The normal state (fig. 12, $E = \rho_n J$; $T > T_c$) is reached only after 15 ms approximately. The voltage is never zero because of the high connection resistance.

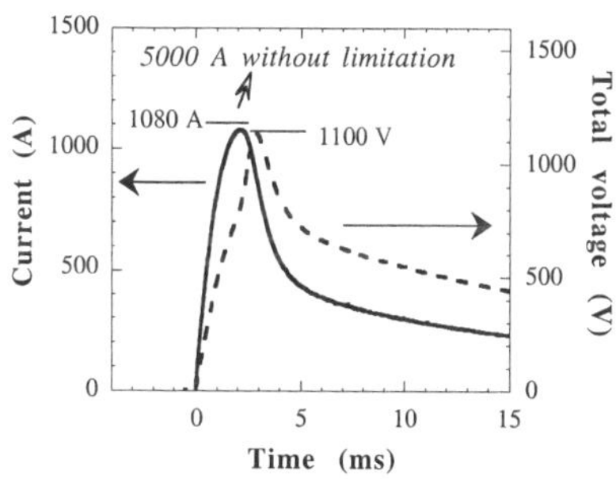

Fig. 11 : time evolutions of the current and voltage for a capacitor discharge under 1.1 kV.

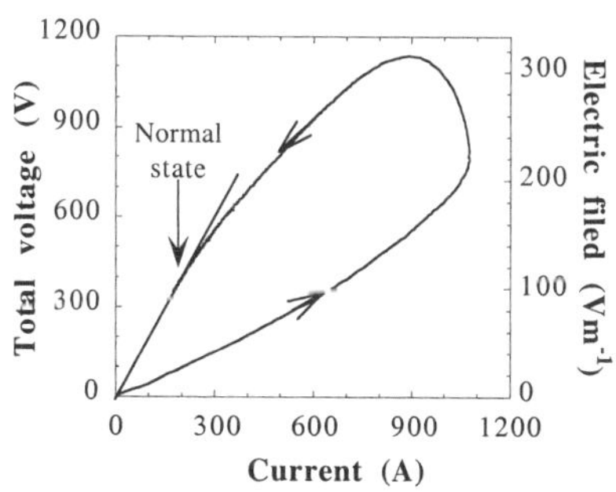

Fig. 12 : E(J) characteristic of the 1.1 kV test.

4.1.5.1.2. BiSrCaCuO sintered - ABB Project

ABB demonstrated the one year satisfying operation of a transformer type SC FCL [32]. The transformer remains an issue in terms of volume and weight so they developed a much more compact resistive FCL system based on Bi-2212 bulk meanders [42]. They successfully tested a single phase 6.4 MVA (800 A, 8 kV) demonstrator using 100 elementary modules. The meander shape module is performed in a composite sheet (0.3 x 0.4 m^2) made of a sandwich of the Bi-2212 bulk superconductor, of glass fibre mechanical reinforcement and a stainless steel bypass jointed to the HTS with conducting join. This bypass reduces the local temperature rises ("hot spots") and the requirements in term of HTS I_c uniformity. The Bi-2221 is elaborated using partial melt processing and shows J_c up to 50 MA/m^2 (77 K, self field). The A.C. losses are reduced, about 0.001 % of the power at the critical current and over currents are possible, even for long overloads. Under 8 kV the prospective short-circuit current of 28 kA_{max} was limited to a peak value of 10.6 kA or 9.4 $I_n\sqrt{2}$ (800 A). The limitation is more effective after due to the temperature rise of the HTS, 4.5 kA_{max} after 20 ms and 3.8 kA_{max} after 100 ms. The recovery time depends on the fault conditions but the order of magnitude is some seconds. The recovery is even possible under partial load and for a reduced limited electric field.

4.1.5.1.3. BiSrCaCuO - ACCEL/NEXANS CURL 10 Project

Melt cast $Bi_2Sr_2CaCu_2O$ can be processed under the shape of tubes. They can be then cut to bifilar coils to have a compact and little inductive SC element. The CURL 10 aims a three phase 10 MVA - 10 kV - 600 A FCL [43]. The operating temperature is 65 K to benefit from the enhancement of the critical current density when the temperature is lowered. The critical current density is then 40 MA/m^2. The tube typical dimensions for the CURL 10 project are an outer diameter of 50 mm, a height of 300 mm and a thickness of 4 mm. With a cross section of 24 mm^2 (rated current of 600 A and maximum current of about 850 A), the active length yields 5.4 m per tube.

To cope with the inhomogeneity (± 5 %) of the superconductor a metallic shunt is contacted continuously over the whole length. The contact resistance between BiSrCaCuO and the CuNi shunt is low (10 μVcm^2). A fibre glass tube is glued also into the SC tube for mechanical reinforcement. The designed electric field under limitation is 50 V/m and 30 tubes are needed for the voltage of 10 kV. So a special care was taken to the electrical contacts between the tubes to reduce the power dissipation. The achieved resistance between two element is lower than 1 $\mu\Omega$ at 65 K.

The quench develops rather slowly and, if the first current peak is already limited, the limitation really occurs for the following peaks. Under an electric field of 56 V/m, the assembling of three modules (tubes) in series limited the first current peak at 7.7 kA (17 kA prospective current) whereas the second peak current was only 3 kA.

4.1.5.2 Limiting devices with YBaCuO

4.1.5.2.1. YBaCuO bulk - BYFAULT Project

These studies were undertaken mainly within the context of the European project BYFAULT (Bulk YBCO and High Voltage Fault Current Limiter). The objective was the realization of one or two three-phase limiting device of 17 MVA (25 kV - 400 A). An intermediate stage included two single-phase demonstrators with a rating of 100 kVA (1 kV - 100 A), one resistive [44] and the other hybrid [45]. Only the resistive solution with YBaCuO meanders will be presented.

The ab plan resistivity of the YBaCuO pellets at 140 K is approximately 1.75 $\mu\Omega$m. Equation (10) gives the maximum current density for thermal considerations : 90 MA/m^2 (ΔT = 100 K, Δt = 20 ms). By considering a limiting current four times the critical value, the critical current density should not consequently exceed about 20 MA/m^2. Figure 13 shows that the operating temperature must be around 90.5 K, which was the selected operating temperature. Figure 13 shows also that the homogeneity is rather good in term of critical current density.

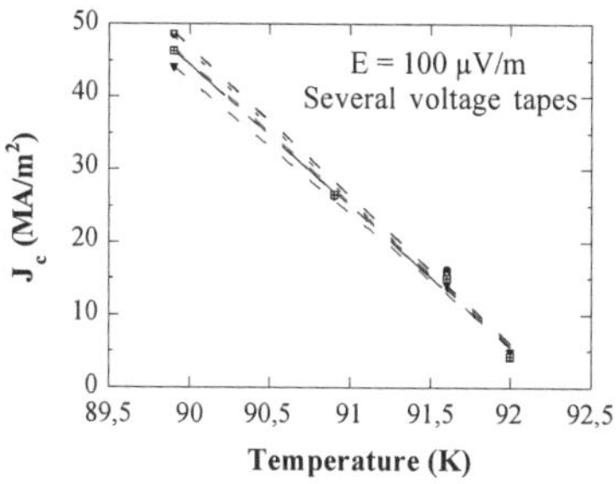

Fig. 13 : critical current densities of a single domain YBCO near the critical temperature.

The meanders (figure 14) are first cut under the form of discs from pellets 45 mm in diameter. The section of the straight parts is approximately 4 mm^2. The current is brought by silver wires fixed to YBaCuO by silver paste using an suitable thermal treatment. Due to the low contact resistance (1 $\mu\Omega$cm^2) and the large contact surface, the transition never starts at the current leads. Currents of 2000 A were injected at 77 K without problem at the contacts. These tests showed also that a meander mechanically withstands very high currents. For a maximum temperature rise of 100 K, the electric field yields to approximately 160 V/m. The active length (straight parts only) of the meanders can reach 300 mm, but the average value is 210 mm because the pellet parts with cracks were not used.

Under these conditions, about thirty meanders were necessary. We assembled in reality 43 meanders for the tests under 1 kV. The active length is 9 m, so an average electric field of 110 V/m. The assembly of the meanders was designed in a modular way. Four meanders are connected in series on a composite material plate and the plates are stacked vertically (fig. 14). In parallel with each meander, a stainless steel resistance is connected for voltage limitation among other things.

Fig. 14 : YBCO meander modules assembled for the 50 kVA demonstrator.

The operating temperature of 90.5 K is obtained by heating the nitrogen bath thanks to a resistance at the vessel bottom and by controlling the bath pressure. Several temperature sensors showed a good homogeneity in temperature, better than 0.1 K, on the height of the meander assembly. The tests under 1 kV were carried out at Schneider Electric. The test loop is a 1 kV voltage source with a maximum short-circuit current of 11 000 A. The short-circuit was preceded by a two periods steady state operation with a maximum current of 100 A. The voltage drop across the limiter is essentially resistive and lower than 0.2 % of the network voltage. Reactance is indeed lower than 3 mΩ, but the resistance reaches 18 mΩ, due to the connections between the meanders. They can be easily improved.

Figures 15 and 16 show the 1 kV test carried out at 90.5 K with the 43 meanders. The current was limited to 740 A instead of 6 500 A without quench. The transition was rather homogeneous along the meanders since the time evolutions of the resistances divided by their 300 K value are close for various groups of meanders and the total assembly (fig. 16). The resistances are calculated by dividing the voltage by the current so the errors around the current zero crossing.

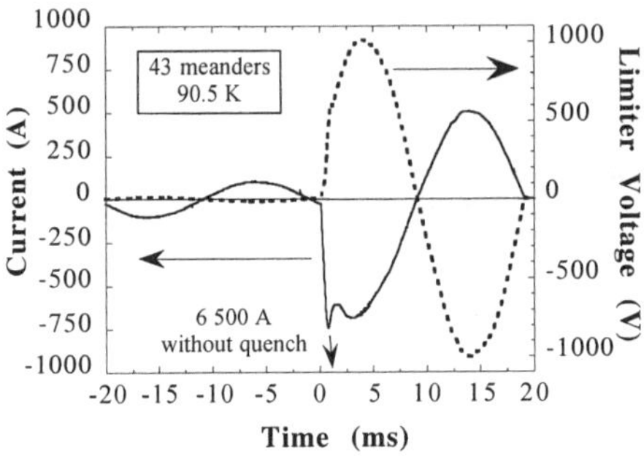

Fig. 15 : current and voltage time evolutions during a short-circuit test.

Some meanders (12) were destroyed during this test, but little preliminary selection was done before these tests. The large majority of the meanders successfully survives this test, validating the use of YBaCuO meanders to effectively limit the fault currents. The reasons for the destructions are certainly materials defects along the meanders such as cracks. The YBaCuO single domains must be improved in terms of homogeneity and defect reduction. Important works have been carried out for an enhanced pellet synthesis [46]. A lot of developments are still necessary to design an industrial device. A spiral cutting of the YBaCuO slices would be doubtless an improvement with an increase of the active length. This enhancement is a key issue for an industrial device in order to reduce the number of elements to be assembled. Single domain pellets with a diameter of 100 mm were successful elaborated [11, 12]. Such pellets with a height of 20 mm show a limitation power greater than 300 kVA (ΔT_{max} = 100 K ; Δt = 20 ms).

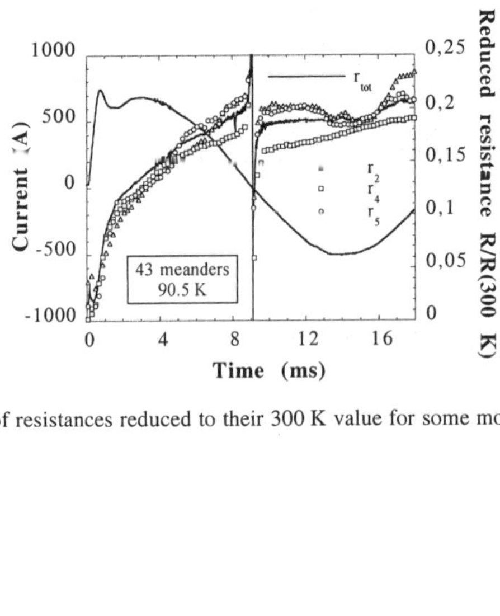

Fig. 16 : time evolutions of resistances reduced to their 300 K value for some modules after the short circuit.

4.1.5.2.2. YBaCuO film - SIEMENS Project [35]

SIEMENS chose YBaCuO thin film deposited on sapphire substrate for their SC FCL programme [35]. The elementary module is typically a 4" wafer (100 mm in diameter) with a YBaCuO coating of 250 nm in thickness. The YBaCuO is covered by a thin gold layer (about 100 nm) to avoid hot spots due to inhomogeneities and make the electric connections easier. The depositions use the physical evaporation techniques. A spiral shape path is realised for the current (fig. 17). With a path width of 7 mm the length is 0.8 m per wafer.

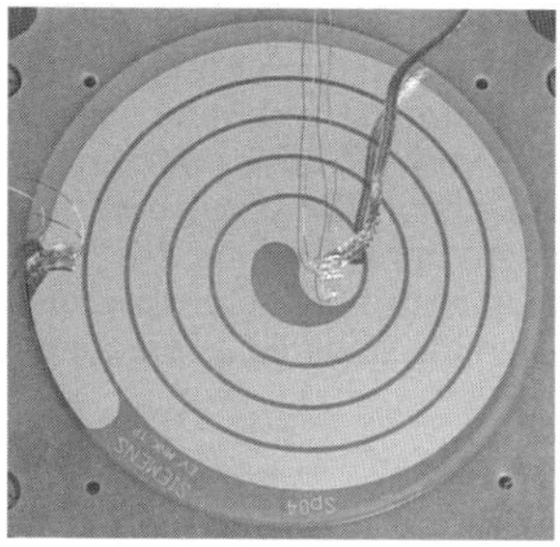

Fig. 17 : coated YBaCuO FCL element (spiral shape SC path on a 4" wafer). (courtesy of Siemens)

The critical current is around 40 A to 50 A, it means a critical current density of 20 to 30 GA/m^2. The electric field under limitation is high with YBaCuO film, about 800 V/m. This value corresponds to a reached temperature of 200 K after a fault of 50 ms in duration (opening of the switch gear). In these conditions a single wafer shows a short circuit power of 15 to 20 kVA (30 A_{rms} - 600 V_{rms}). The rated current is chosen to be approximately the critical value in amplitude ($I_n \sqrt{2} = I_c$). The current is limited very quickly reduced in amplitude. The peak limited current reaches 4 times the critical value and decreases very rapidly and the amplitude of the limited current is after about the critical current.

Siemens successfully tested FCL models up to 1.2 MVA (7.3 kV - 80 A three phase device, fig. 18). The 1.2 MVA model limited the prospective current of 5 kA_{max} under 300 A, which is approximately 3 times the critical current. The limiting performances are very effective and high. The recovery times are some seconds.

Fig. 18 : 1 MVA YBaCuO film FCL Siemens device. (courtesy of Siemens)

Nevertheless the very high costs of the YBaCuO coated modules remains an issue for the economic development of a FCL based on them. Another issue is also the large number of module to be assembled for high powers not to mention the problem of the series and parallel careful connections for a good equilibrium between the different modules.

4.1.6. Conclusion

The SC fault current limiter would bring a new and useful function in the networks, in particular in high or very high voltages. The deregulation of the electric utilities industry around the world and the request for a better quality of energy (voltage and supply continuity) will reinforce its interest. FCL makes also possible to increase the capacity of existing grids. The necessary minimum superconductor quantity is reasonable (25 kg/MVA - 4 litters/MVA). Several FCL model have proven the very effective operation of SC FCL, at least for voltages up to tens of kV. The current and voltage scaleups involve difficulties and require developments.

Nevertheless even if the scaleups are solved, the SC FCL has a high potential market only if its cost remains acceptable for the utilities. The superconductor cost plays a part but the cryogenics as well as the implementation (module assembling) are also important to consider.

4.2. Superconducting power cables

4.2.1. Introduction

Sometimes electric cables transmit considerable amounts of energies at the entrance of very large urban areas. The Joule effect in these cables wastes energy and the maximum possible temperatures limit the transmission capacities. The replacement of the resistive wires by superconductors is a promising and a logical solution [47]. Moreover, a cable is not a very complex system and the electromagnetic stresses in terms of magnetic fields are low on the conductor. However, if the concept is rather simple, there are some technical difficulties, in particular due to the cryogenics required for superconductivity. In addition, the cable is an essential part of an electrical network and it must be correctly interfaced to it. A SC cable must not disturb the grid operation and must withstand some severe fault conditions, short-circuits for example. At least it is necessary to carefully study some aspects such as exploitation, maintenance, repairing and reliability. Power transmission cables with low Tc superconductors successfully proved their technical feasibility [48]. However 4 K cryogenics prevents these applications from any further developments due to its much too high cost. The discovery of the high Tc superconductors, usable in liquid nitrogen, revived the studies since cryogenics becomes acceptable then even if it remains a critical issue for any commercial development.

In the state of the art, all the HTS cables use PIT Bi-2223 tapes. They are available in kilometer lengths and the critical current density at 77 K is rather high since they operate in a moderate field. The coated conductors will be attractive in term of wire cost reduction. Projects will use them for short length cable models.

4.2.2. Technical interest of superconducting cables

The achievable current densities in the superconductors (100-500 MA/m^2 to be compared with 1-2 MA/m^2 in resistive wires for power cables) increase the transmission capacity for a given cross section. The cryogenic vessels must be taken into account for the comparison so the real gain is between three and five in function of the configurations and the power levels. A SC solution is always more favourable when the power increases. Electricity transport is mainly achieved with A.C. currents at 50-60 Hz. The SC wires experience then A.C. losses. With an appropriate structure of the SC cable, the A.C. losses can be kept to a moderate level. Nevertheless they should be evacuated by the cooling fluid to limit the temperature rise since the thermal margins are low. It is so necessary to distribute regularly along the cable coolers to remove the heat. The A.C. losses lead to relatively short distances between the coolers (table 1) in comparison to those necessary for D.C. power cables, which experience only thermal losses except the joints between the superconductors. The A.C. losses in cables strongly depend on the transport current, they increase with the cube of the current. So they are a significant contribution to the total thermal load above all for currents close to the rated value. The other contributions to the thermal load are the cryostat losses, the dielectric losses, the losses related to the circulation of nitrogen, the joints, ... These different contributions depend little on the cable current, except the joints. The cost of the total losses brought back to room temperature using the coefficient of performance is not very high and is lower than those of a resistive cable except for very low transmitted powers. The SC cables are particularly interesting for the power cables with a very high mean electric load. Table 3 gives an example of the different losses. The concerned cable is based on the cold cryogenic design (see paragraph 4.2.5.), which uses a shield or a return conductor around each phase. A.C. losses are dissipated in the return wires.

This SC cable shows lower losses than the equivalent resistive cable for currents higher than 30 % of the rated value. The gain in terms of losses reaches approximately a factor three at the rated operation. If the cost of the thermal losses, cryogenics efficiency included, remains low, the investment cost of the refrigerator growths with the cold power and becomes rapidly an issue.

Load	A.C.. losses	Shield	Cryostat	Dielectric	Joints	Miscellaneous	**Total**
Full (I_n)	5.9 (34 %)	3.7 (22 %)	5.2 (30 %)	0.9 (5 %)	0.7 (4 %)	0.8 (5 %)	**17.2**
Half (I_n /2)	0.7 (9 %)	0.9 (11 %)	5.2 (63 %)	0.9 (11 %)	0.3 (4 %)	0.2 (2 %)	**8.2**

Table 3 : losses per unit length and per phase (W/m) for a 132 kV - 2000 A SC cable [49]

4.2.3. SC cables and network

A SC cable must withstand without damage some severe fault modes of the electrical network like short-circuits. The standards relating to these modes are drastic since a cable must carry high currents (approximately 20 times the rated value) during long durations (second) and carry immediately after the fault the rated load without interruption. To fulfil such a requirement a SC cable has in general a resistive core (Cu), which carry the over currents. It is much more economically interesting than increasing the SC volume. Other solutions can be considered such as a fault current limiter. The cable itself can play the limiting part but with an appropriate wire. This solution was studied for example for the connection between the generator and the set-up transformer [50].

4.2.4. Economical interest of SC cables

A solution should not be only interesting on the technical and scientist levels, it must be above all economically viable. The SC cable is penalized by cryogenics (cryostat and refrigeration) and by the SHTC wire cost. The accurate evaluation of the different costs is difficult since it is nearly a completely new technology without a lot of reliable references. The running losses over the lifetime of the cable lead however for large power cables, above GVA, and strongly loaded to a total cost in favour to the SC system, with the assumption of equivalent reliability.

Nevertheless the economical interest concerns above all the gains about civil engineering, in particular for retrofits. Retrofit is the replacement of a conventional cable by a SC cable. The transport capacity of some existing power cables in very dense urban area, particularly in the United States or Japan, must be increased to cope with the increasing energy demand. It is not possible to enlarge the capacity of resistive cables in the existing ducts except by increasing the voltage but that implies the change of all the equipments. New more powerful resistive cables would lead to huge civil engineering costs. A SC cable shows in that case the advantage to significantly increase the power (three to five times) using the existing ducts, then with no or little civil engineering. The distances between the different resistive cables needed for their cooling disappears with SC cables. The economic interest is then very clear. Even in the case of a new underground cable, the higher compacity of the SC solution leads to a significant gain in term of civil engineering and justifies the higher investment cost.

In some locations, a resistive power cable requires at its both ends two transformers to increase the voltage for a more economical transport. A SC cable could suppress these two transformers, reinforcing the savings in space and reducing the total cost of the connection. Superconductivity indeed allows a transport of large powers under medium voltages and not necessarily under high voltages.

4.2.5. Warm and cold dielectric designs

The SC cables use two different designs in function of the operating temperature of the electric isolation (fig. 19).

In the warm dielectric concept only the conductor is at low temperature whereas the insulation operates at ambient temperature over the cryostat. So the insulation is conventional and the existing proven technologies can be applied without any developments. The operation conditions of a SC cable insulation are less severe compared to a conventional cable. Its temperature can reach indeed 95 °C. The margin and operating life must then be improved. On the other hand the cryogenic takes up more space and the three phase cables are less compact compared to the cold dielectric.

In the cold dielectric concept, the wire and its insulation operate at the same cryogenic temperature. It is particularly interesting in case of a triple-core configuration (common cryostat) since the gains in compactness are then maximum compared to the warm dielectric solution. The operation of electric insulation at cryogenic temperature is less known compared to ambient temperature and even if no problem should occurs some developments are needed to check all the dielectric properties. The cold dielectric configuration also includes a return conductor which act as a shield in the equilibrate three phase operations. This return conductor increases the superconductor volume (about twice) and then the initial cable cost. But it decreases the electric losses, the magnetic signature and the cable inductance since the magnetic field is better confined. There are no or little eddy current losses in the metallic vessels. The electric insulation is between the phase and the return conductors. To meet the thermal contractions the insulation is laminar or tape dielectric impregnated by liquid nitrogen which removes also the dielectric losses whose contribution is nevertheless in general low.

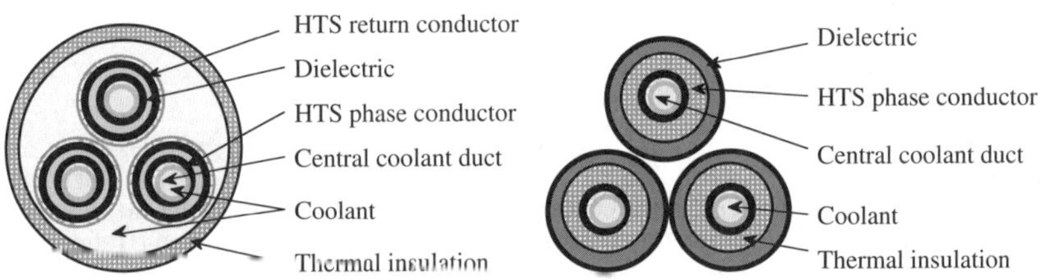

Fig. 19 : cold dielectric (a) and warm dielectric (b) configurations.

4.2.6. Cryogenics

The cryogenics of a SC cable must be sufficiently sophisticated to show very efficient thermal isolation but at a moderate cost. Another requirement is a sufficient mechanical resistance with respect to curvature, nitrogen pressure, differential contractions ... Reliability must be excellent. The objective in terms of thermal losses is approximately 1 W/m for a flexible cryostat. The thermal insulation is performed by using multi-layer super insulation in a high vacuum chamber. It represents the major space of the SC cable. Dielectric isolation takes some space too whereas the SC conductor itself represents only a little part of the whole. The cryostat consists in a central channel where the coolant, liquid nitrogen in general, flows. The HTS conductor is wound directly on it. The central channel can play the part of electric shunt for overcurrents for example.

The refrigerators are distributed all a few hundreds of meters or kilometers and pumps provide the circulation of liquid nitrogen. It is the most used cooling fluid. Rather inexpensive, the liquid nitrogen has a high specific heat and good electrical isolation properties. The operating temperature varies typically between 65 K and 80 K. Liquid nitrogen cannot be used under 65 K since it becomes solid under 63 K. The refrigeration systems distributed regularly along the SC cable length consists in several standard large capacity cryocoolers (Gifford McMahon, Pulse Tubes) to provide redundancy and variation of the cold power in function of the cable load. The required power is about 3-5 kW/km. Another option for the coolers is Brayton or Claude cycle refrigerators, which shows higher efficiency and compactness but uses less standard components. This is a key factor for cost reduction, one of the critical issues for the SC applications.

A key component is the terminations at both end of the HTS cable. They must manage all the electrical and fluid connections together with the electrical insulation difficulties. The current leads bring losses (Joule and conduction). A conduction cooled copper leads between 77 K and 300 K consumes about 40-45 W/kA [40].

The main difficulty with cryogenics remains primarily the maintenance of the refrigerators since a lot of installations showed their excellent reliability.

4.2.7. Some superconducting cable projects

The programs on SC cables were and still are numerous around the world [51, 52]. Table 4 summarizes some parameters of the major operating HTS SC cables. In the United state the Detroit 100 MVA cable has some cryogenic problems (vacuum vessel) and does not operate properly. Three SC warm dielectric cables should replace nine existing resistive cables.

	Southwire-IGC	NKT-NST	Tepco-SEI	Pirelli-EDF
Rating	27 MVA 3 phases 12 kV / 1.25 kA	27 MVA 3 phases 12 kV / 1.25 kA	114 MVA 3 phases 66 kV / 1 kA	1000 MVA 3 phases 225 kV / 2.6 kA
Length	30 m	30 m	100 m	50 m
Dielectric type	Cold dielectric	Warm dielectric	Cold dielectric	Cold dielectric
Test status	In operation since January 2000	Tested since May 2001	Tests started in June 2001	Partial tests in 2002

Table 4 : major HTS cable and their characteristics.

The Southwire cold dielectric cable is tested since the beginning of 2000 on their site of Carrollton in the United States. It has operated successfully more than 18 000 hours, supplying reliable power to the Southwire plants, with more than 5 000 hours at full load.

The NKT-NST warm dielectric cable (fig. 20) operates since May 2001 in a public electric network in a Copenhagen substation, delivering more than 150 GWh.

These two company established a joint venture called ULTERA in 2002 for SC cables. They have some HTS cables projects both in Europe and in the U.S., in particular with the American Electric Power Company in Ohio.

In Japan, the largest project is led by TEPCO (Tokyo Electric Power Company, utility) and SEI (Sumitomo Electric Industries, cable and wire). Figure 21 shows the 100 m 114 MVA SC three phase cable at CRIEPI (Central Research Institute of the Electric Power Industry) with a 180° bend. The external diameter is only 136 mm. The tests successfully started in June 2001.

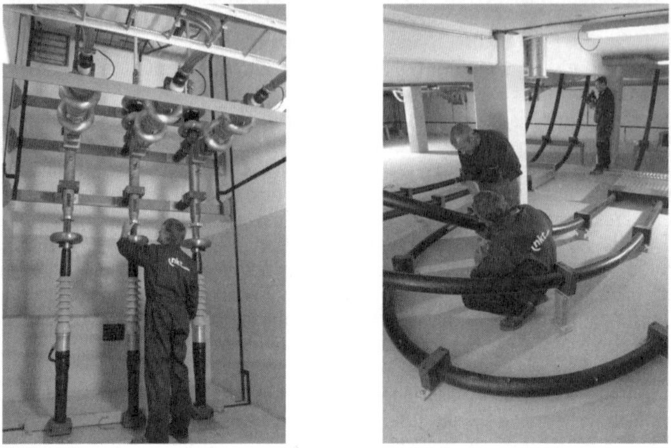

Fig. 20 : NKT-NST SC 3-phase cable with one of its termination.
(used by permission of Ultera)

Fig. 21 : TEPCO/Sumitomo SC cable and its cross-sectional view of 3-phase HTS cable.
(Courtesy of Tokyo Electric Power Company, Sumitomo Electric Industries, Ltd. and Central Research Institute of Electric Power Industry)

4.2.7. Conclusions

The SC cables know a real interest throughout the world following the emergence of the HTS wires operating in liquid nitrogen (Bi-2223 PIT tapes for instance, Y Coated Conductors in a near future). Using a rather simple cryogenics HTS cables are able to transmit more power than conventional cables and with a better efficiency especially when the load is high. The retrofit (upgrading existing underground cable) appears as one of the first niche applications of HTS cable.

Several programs demonstrates the technical feasibility at 100 metre scale and provide the necessary feedback for their development. The next step concern HTS cable with lengths of several hundred of meters, both in Europe and in U.S.

4.3. Superconducting Machines

4.3.1. Introduction

Thank to the very high achievable current densities and the absence of the Joule losses, superconductors in electrical machines reduce their weight, their losses and, to a lesser degree, improve their compactness. Constant innovations and improvements have been brought in manufacturing and operation of electric machines. The technological limits in terms of power density and efficiency are nearly reached using conventional means and only superconductivity could give a major technological leap. The first studies about SC machines started with the emergence of SC wires able to carry large currents in a magnetic field [53] and a lot of work were carried out with low T_c superconductors [54, 55].

The weight saving is very attractive for the numerous applications with high and critical weight constraints, such as moving systems, ship propulsion in particular. The enhanced efficiency brings savings in running costs and the gain over the machine life is important. So SC machines are attractive also from an economical point of view all the more that they are expected to be less expensive to manufacture than conventional machines of equivalent rating [56].

These advantages appear only above a critical (break-even) size or rating due to the required cryogenics system, i.e. where the refrigeration penalty becomes negligible (size, weight and costs). The critical size is reduced when the operating temperature increases. The complexity and cost of 4 K cryogenics have hindered the commercial development of LTS machines even if several demonstrators have successfully demonstrated their technical feasibility at rating up to 80 MW within the Japanese Super GM programme [57]. HTS are enabling to lower a lot the critical size and so to bring the HTS machines to market.

This part will concentrate on the A.C. synchronous machine, the most used structure. HTS in particular the YBaCuO single domain pellets can be used with high interest in special machines such as reluctance, hysteresis or trapped field motors [58]. These machines are however less developed and are presented elsewhere. HTS homopolar D.C. machines have been studied [59] as well for marine propulsion but the very high current collection problems remain a very difficult issue so these studies have not been carried on.

4.3.2. Electromagnetic structure and limitation of conventional machine

The maximum torque (Γ_{max}) per unit volume of an A.C. synchronous machine is proportional to the product of the excitation field component perpendicular to the current with the armature ampere turn loading (equation (13) for a cylindrical machine). The armature ampere turn loading is the total armature current divided by half of the mean armature circumference.

$$\Gamma_{max} = \sqrt{2}\; B^o\, K\, \pi\, r_o^2\, L \quad (13)$$

$$K = \frac{3\, N_s\, k_d\, I}{\pi\, r_o} \quad \textit{(three phase machine)}$$

$$\pi\, r_o^2\, L \approx \vartheta_{mach}$$

- B^o : excitation field component
- K : armature ampere turn loading
- r_o : mean armature radius
- L : active length
- N_s : total series turns per phase
- k_d : winding factor
- I : rated armature current
- ϑ_{mach} : approximate machine volume

The excitation field of an A.C. synchronous machine is created either by permanent magnets or by an D.C. electro-magnet. The mobile part is either the armature or the excitation, but due to current transfer serious problems at high ratings, the armature is preferably stationary and then the excitation rotating.

Permanent magnets are expensive materials, limit the possible magnetic induction and make the excitation field variations difficult but they are not dissipative. They are well suited particularly for small machines (kW), even if their use for higher ratings is spreading up.

High ampere turns are required to create a large excitation field by an electro-magnet. Using resistive wires these ampere turns are limited by the current density and the Joule losses. The allowable current density depends of the cooling conditions but is of the order of 5 to 10 MA/m^2. A magnetic circuit significantly lowers the needed ampere turns for a given magnetic induction. So practically all electrical machines have an iron core structure with slots where are embedded the windings. Moreover the magnetic circuit confines the flux within the machine. The stray field outside is negligible and prevent other equipments from magnetic disturbances. In a slotted magnetic structure, the torque is mainly applied to the magnetic circuit and very little to the conductors. These are submitted only to a reduced field since the main part concentrates itself in the teeth. The magnetic circuit is then very useful from a mechanical point of view all the more that the mechanical properties of copper are poor. Likewise the conductors, submitted only to stray fields, do not need to be finely divided in order to reduce the eddy current losses. But a magnetic circuit is heavy, limits the magnetic induction value due to the saturation phenomena and creates pulsating torques. The alternance of magnetic teeth and slots produces indeed local magnetic variations resulting in torque fluctuations.

The armature ampere turn loading is limited by Joule losses, by the current density in the conductors and by the space available due to the magnetic teeth in particular.

4.3.3. Superconducting A.C. synchronous machines

Superconductors suits particularly to produce magnetic fields constant in time and are then very appropriate for the field winding of A.C. synchronous machines. Due to their very high and lossless current densities (up to hundred times the allowable value in copper, that is hundreds MA/m^2) superconductors produce a strong magnetic field with a reduced conductor volume and without magnetic materials. The magnetic circuit subsists only in general under the form of an external magnetic shield to avoid large stray fields outside the machine. The machine structure becomes "air-cored" and no more "iron-core". An air cored structure is intrinsically not only light but also better for electromagnetic conversion since the stored magnetic energy is higher. The magnetic energy is :

$$W_{mag} = \frac{1}{2} \iiint_{space} \boldsymbol{B} \bullet \boldsymbol{H} \, dx \, dy \, dz \tag{14}$$

In an iron-cored structure the magnetic energy is stored only in the airgap because the magnetic field H is negligible in a non saturated magnetic material. In an air-cored structure, all the space contributes to the magnetic energy.

In an air-cored structure, the magnetic induction is no more limited by the saturation of the magnetic materials. The absence of magnetic teeth frees space for the armature conductors, increasing so the ampere turn loading in comparison with a conventional iron-gap armature, even with resistive conductors (see figures at 4.3.3.2).

In summary a SC field winding, even with a resistive armature, improves the torque of an A.C. synchronous machine through an enhancement not only of the excitation field but also of the armature ampere turn loading.

The winding inductance is much lower in an air-cored structure in comparison to an iron cored one. So the synchronous reactance of an air-cored A.C. synchronous machine is decreased in comparison to a conventional machine even if the ampere turn loading is increased. Fully SC synchronous machine are an exception since the ampere turn loading is very high and the field limited for low A.C. losses [60]. A lower synchronous reactance reduces the internal angle (angle between the total and the excitation fields) and improves the stability of the synchronous generators connected to power grids. Their capacity to provide reactive power is higher too.

The torque ripples are greatly reduced due to the absence of iron teeth so the machine can be very acoustically quiet. But the armature conductors are submitted to the total field so the torque is directly applied on the conductors, which must be hold by a suitable structure. The high field on the conductor leads also to large eddy current losses without an appropriate conductor structure. This is achieved by a finely divided Litz wire configuration.

4.3.3.1. Superconducting field winding

Even in an equilibrate synchronous operation, the field winding is submitted to time varying fields. The fundamental armature field rotates at the same speed than the rotor but all the space harmonics as well as the current harmonics create fields, which does not rotate at the synchronous speed and then create A.C. losses in the field winding. Nevertheless through a careful design of the armature winding it is possible to reduce the space harmonics. The air-cored structure makes this reduction easier in comparison to an iron core structure. So the A.C. losses in the field winding can be kept to a low level in synchronous equilibrate operation.

But as soon as the operation is not equilibrate (negative sequence current) or not strictly synchronous, A.C. losses appear in the field winding. These can be high and degrade the operation (high cryogenic load, increase of the wire temperature). In order to decrease the A.C. fields on the field winding an electromagnetic shield surrounds in general it. It consists in a high electric conducting cylinder or a squirrel cage where eddy currents are induced to protect the field winding from A.C. fields. Lower is the shield resistivity, better is the protection or lower is the cutoff frequency of the shield, which can be considered as a filter. This is why the electromagnetic shield operate at low temperature to benefit from the lower resistivities when the temperature decreases and so to show a very low cutoff frequency. Nevertheless the lower limit of this frequency is given by the time response of the field current. The electromagnetic shield reduces the external perturbations but shields also the variations of the field winding. The minimum cutoff frequency is fixed by the transient stability requirements of the grid.

When faults like short circuits occur very high currents are induced in the electromagnetic shield, which must withstand severe mechanical stresses. It must be designed in consequence, complicating so its realisation.

The rotor of a SC generator has often a second shield. The absence of losses provides a lot of advantages but does no more afford any damping of oscillations after a perturbation. Such an operation is not possible for a generator connected to a grid. The second shield provides this necessary damping. The damping is optimum when the frequency of the damper (inverse of its time constant) equals the oscillation frequency, which is about 1 to 2 Hz for a generator. This is why the electromagnetic shield protecting the field winding cannot be used as a damper : its cutoff frequency is too low. The damper shield works so at room temperature. This shield makes also possible the asynchronous operation of the motor in case of a total failure of the field winding [56]. The operation is of course degraded but possible. Other means are possible to damp the oscillations. A SMES connected to the armature can act as a very effective damper and improves also the transient stability of the generator [61, 62]. A SMES could even protect the generator from any external perturbation but this is limited by the energy capacity of the SMES.

Figure 22 depicts the main differences between a conventional and a SC machine. Figure 23 gives another cross section of a SC A.C. machine.

Cryogenics is now a well mastered technique but the rotation of a SC field winding complicates the cryostat and some difficulties occur. The high fault torques must be withstood by a rotating structure of high structural strength but which must not bring too much thermal losses and cope with high thermal differential contractions. So torque transfer must be designed with great care. Another issue is the cooling fluid (helium in general) transfer system with especially the seals, which must show long-life times. The SC field winding is generally cooled by a circulation of helium in a closed loop. The cooling can be direct or via cooling tubes in good contact with the coil. Helium is cooled by an external cryogenic machine, for example several cryocoolers (Gifford-McMahon) to achieve the cooling power but also for redundancy.

Liquid Neon (27 K) has been used also for the ASC 3.6 MW 1800 rpm machine [63]. The SC coils were cooled by conduction from a rotating Neon evaporator. Neon was condensed by external stationary cryocoolers (two Gifford McMahon) and worked in a closed loop.

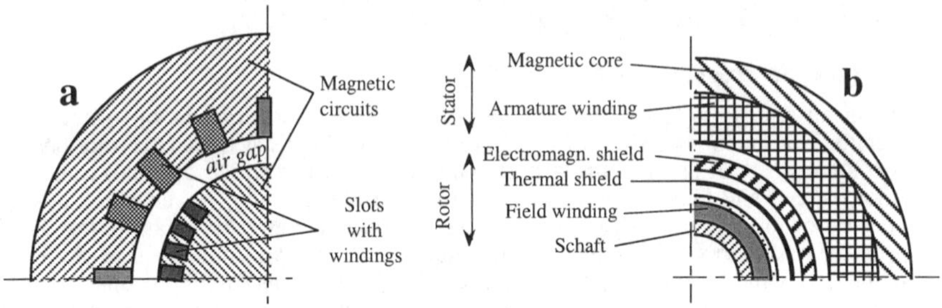

Fig. 22 : Schematic cross sections of A.C. generators.
a : classical ; b : superconducting field winding

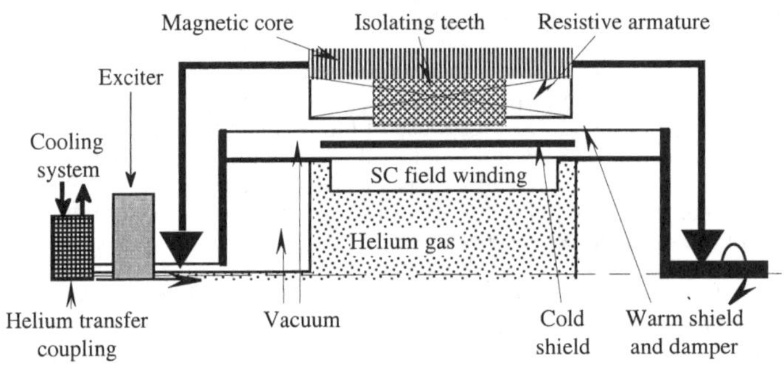

Fig. 23 : schematic cross section of an A.C. SC synchronous machine.

4.3.3.2. Armature

A SC armature would be very interesting especially in terms of ampere turn loading (K in equation (13)) due to the high current densities achievable in superconductors. The suppression of the magnetic teeth in a air-cored armature still increases the ampere turn loading in comparison to a conventional slotted armature. The following figures give some orders of magnitude :

- conventional machines : K = 100 - 250 kA/m,
- air-cored armature : K = 300 - 350 kA/m,
- superconducting armature : K = 600 - 800 kA/m.

An air-cored armature experiences large A.C. fields then A.C. losses in the superconductors. A.C. NbTi wire were developed in the eighties [28]. Their A.C. losses are very low thank to an appropriate structure despite the very high cryogenic penalty at 4 K. NbTi armatures were successfully developed [64] but the 4 K complex and expensive cryogenics prevented from commercial development. The A.C. losses in HTS are still much too large to design a HTS armature and resistive wires are then used. Nevertheless they must be Litz conductors to keep eddy-current losses low. Freshwater is generally used to extract the losses for high ratings. The coils must be hold by a high strength structure since the full torque is applied to them. The absence of magnetic grounded teeth simplifies on the other hand the electrical isolation and can extend the voltage to (very) high levels. The elimination of the set-up transformer was proposed [65]. ABB has also presented a conventional machine with high voltages [66].

The magnetic confinement for the environment is performed by an external magnetic shield. A conducting shield has been proposed to lower the weight. The shielding is performed then by the eddy currents induced in the conducting cylinder around the armature. It is efficient for a constant speed generator even it shows higher Joule losses in comparison to the iron losses in a magnetic shield. It does not suit for a variable speed motor since no there is no shielding at the starting.

4.3.3.3. Programmes under progress

Siemens has successfully developed a 400 kW - 1500 rpm - four pole machine [67] (fig. 24) with the first test in 2001. The field winding uses 48 PIT Bi-2223 conductor coil. The coils are stacked on a magnetic support (steel). The operating temperature range is between 25 and 30 K. He gas from a liquid helium tank was used at the beginning but Neon in a close loop condensed by a cryocooler has been used after. The cooling requirements are about 20-30 W at 25 K. The coils are conduction cooled.

Fig. 24 : 400 kW HTS Siemens machine on its testing bench (cryocooler on the left side) (Courtesy of Siemens)

The main program under progress is carried out by American Superconductor since 1995. In 2001 a 3.6 W at 1800 rpm motor demonstrated a satisfying operation [63] and in March 2003 a 5 MW - 230 rpm torque machine began its first tests. This realisation is a major step in the development of HTS machines. Its power and speed are fully representative of some ship applications (ferries and small cargo ships). Nevertheless the real demands are for higher power and a 36.5 MW 120 rpm motor will be the next milestone. This motor should weight 69 tons, about a third of the advanced resistive motors and the size should be reduced by a factor two.

4.3.4. Conclusion

The HTS motors are the most advanced applications of high T_c materials. The power of the last realisation (5 MW - 230 rpm) is already representative of some applications and demonstrates the technical feasibility. This ASC machine use components already designed for more powerful (25 MW) device so the scaleup should be simple. The next machine with a power of 36.5 MW at 120 rpm should fully demonstrate this innovative technology, bringing large savings in weigh, size, efficiency, noise reduction and costs. These super machines could be first used in naval applications with high interest. Improvements and progress in HTS wires and refrigeration would favourably impact the further developments and the commercialization.

5. GENERAL CONCLUSIONS

A lot of research and development activities concern today large scale application of HTS. They are motivated by the high potential prospects of using HTS in power devices : large achievable current densities, absence of D. C. losses, fast transition from a no or little dissipative state to a high resistive state. These remarkable properties lead to significant improvements in terms of compacity and weight reduction but also perform new interesting functions hitherto inaccessible such as FCL, energy storage (SMES) or passive magnetic bearings.

The cryogenic environment needed for the operation of HTS is no more an economical obstacle as it was for LTS but further improvements in cost, efficiency and reliability are necessary to meet the requirements for a real market development. The cooling systems are now nearly invisible to the user, which is essential.

The main issue remains the cost and performances of HTS conductors but some bolts are under solving in particular with the recent significant advances in coated conductor technology.

The main studied applications are the power cables, the electric machines and the FCLs. The two first applications have reached a significant size with ratings near those required for market applications. The first model devices are already under development. A large scaling step has to be made for the FCL and more time is needed for their commercial development.

The first HTS devices will appear in some niche applications (naval ship propulsion, cable retrofitting). They will be very important to prove the reliability of this new technology, which is a key issue for further extensions in the market.

The next years will be then very important for the future of HTS applications. The material challenges must be overcome with a wire production at a competitive cost. Efforts must be still carried out to achieve more effective cooling systems. The first HTS devices of representative size must prove their real advantages including cost, reliability, maintenance and availability over conventional technology.

Nomenclature

A.C. :	Alternating Current
c_p :	specific heat per unit volume (J/m^3/K)
D.C. :	Direct Current
E :	electric field (V/m)
FCL :	Fault Current Limiter
HTS :	High Temperature Superconductor
LTS :	Low Temperature Superconductor
λ :	thermal conductivity (W/m/K)
J :	current density (A/m^2)
J_c :	critical current density (A/m^2)
MRI :	Magnetic Resonance Imaging
NMR :	Nuclear magnetic Resonance
PIT :	Powder In Tube
ρ_n :	normal state resistivity (Ωm)
RABiTS :	Rolling-Assisted, Biaxially-Textured Substrates
SC :	superconducting
SMES :	Superconducting Magnetic Energy Storage
T :	temperature (K)

References

[1] T. R. Strowbridge, "Cryogenic refrigerators - An updated survey", NBS report NO. NBS TN-655, 1974.

[2] A. Ravex, "Cooling and Interfacing for Superconducting devices", presented at EUCAS'2001, Copenhagen, 2001.

[3] A. Ravex, "Recent developments in cryocoolers", Proceedings of ICEC'18, Mumbai, India, 2000, pp. 41-52.

[4] C. M. Friend, "AC losses of HTS tapes and wires", in Studies of High Temperature Superconductor, Vol. 32, A.V. Narlikar, Ed. New York : Nova Science Publischers, 2000.

[5] L.J. Masur, D. Buczek, E. Harley, T. Koenkandath, X. Li, J. Lynch, N. Nguyen, M. Rupich, U. Shoop, J. Scudier, E. Siegal, D. Verebelyi, W. Zhang, J. Kellers, "The status of commercial and developmental HTS wire", International Synmposium on Superconductivity (ISS), Yokohama, Japan, November 2002.

[6] P. Vase, R. Flükiger, M. Leghissa, B. Glowacki, "Current status of High-T_c wire", Superconductor Science and Technology, Vol. 13, 2000, pp. R71-R84.

[7] Y. Iijima, K. Matsumoto, "High temperature superconductor coated conductors : technical progress in Japan", Superconductor Science and Technology, Vol. 13, 2000, pp.68-81.

[8] A. Usoskin, J. Dzick, A. Issaev, J. Knoke, F. Garcia-Moreno, K. Sturm, H. C. Freyhardt, "Critical currents in long-length YBCO-coated conductors", Superconductors, Science & Technology, Vol. 14, 2001, pp. 676-679.

[9] M. Murakami, "Progress in applications of bulk high temperature superconductors", Superconductor Science and technology, Vol. 13, 2000, pp. 448-450.

[10] M. Morita, S. Takebayashi, M. Tanaka, K. Kimura, K. Miyamoto, K. Sarvano, "Quench and melt growth (QMG) process for large bulk superconductor fabrication", Advances in Superconductivity, Vol. 3, 1991, pp. 733-736

[11] T. Fujimoto, M. Mitsuru, N. Masahashi, T. Kaneko, "Fabrication of 100 mm-diametre Y-Ba-Cu-O bulk QMG superconductors with large levitation forces", Inst. Phys. Conf. Series N 167, Vol. 1, 1999, pp. 79-82.

[12] X. Chaud, D. Isfort, E. Beaugnon, R. Tournier, "Isothermal growth of large $YBa_2Cu_3O_{7-d}$ single domains up to 93 mm", Physica-C, Vol. 341-348, 2000, pp. 2413-2416.

[13] P. Komarek, Advances in large scale applications of superconductors », Superconductor Science and Technology, Vol. 13, 2000, pp. 17-22.

[14] G. Morrow, "Progress in MRI magnets", IEEE Transactions on applied superconductivity, Vol. 10, 2000, pp. 744-751.

[15] W.D. Markiewicz, J. Scwartz, H.J. Schneider Muntau, "The approach to 1 GHz plus high resolution NMR Progress in MRI magnets", IEEE Transactions on applied superconductivity, Vol. 10, 2000, pp. 724-727.

[16] C. A. Luongo, "Superconducting Storage Systems : an overview", IEEE Transactions on Magnetics, Vol. 32, 1996, pp. 2214-2223.

[17] M. Parizh, A. K. Kalafala, R. Wilcox, "Superconducting Magnetic Energy Storage for Substation Applications", IEEE Transactions on Applied Superconductivity, Vol. 7, 1997, pp. 849-852.

[18] http://www.amsuper.com/html/products/library/WPScs.pdf

[19] J. Boehm, "Magnetic separation - Superconductivity in industry", IEEE Transactions on applied superconductivity, Vol. 10, 2000, pp. 710-715.

[20] P. Lebrun, "Advanced superconducting technology for global science: the Large Hadron Collider at CERN", AIP Conference Proceedings, 2002, Vol. 613A, pp. 3-14.

[21] A. Dael, E. Acerbi, F. Alessandria, C. Berriaud, R. Berthier, F. Broggi, C. Mayri, Y. Pabot, J.M. Rey, M. Reytier, L. Rossi, M. Sorbi, H. Van-Hille, G. Volpini, Z. Sun, "Construction of the ATLAS B0 model coil", IEEE Transactions on Applied Superconductivity, 2001, Vol. 11, pp. 1597-1600.

[22] F. Kircher, P. Bredy, D. Campi, P. Fabbricatore, S. Farinon, H. Gerwig, A. Herve, I.L. Horvath, B. Levesy, R. Musenich, Y. Pabot, A. Payn, L. Veillet, "CMS coil design and assembly", IEEE Transactions on Applied Superconductivity, 2002, Vol. 12, pp. 395-398.

[23] B. Turk, "Six years of operating experience with tore supra, the largest Tokamak with superconducting coils", IEEE Transactions on Magnetics, Vol. 32, 1996, pp. 2264-2267.

[24] R. Eymard, "Cryogenics and the next step in magnetic fusion", Proceedings of ICEC'19, Grenoble, France, 2002.

[25] J. Sapper, W7-X team, "The superconducting magnet system for the Wendelstein7-X stellarator", Superconductor Science and Technology, Vol. 132, 2000, pp. 516-518.

[26] M. Ohl, P. Sala, H.J. Schneider-Muntau, J.C. Vallier, "Operating experience with the Grenoble Hybrid Magnet system", IEEE-Transactions-on-Applied-Superconductivity, Vol. 3, 1993, pp. 66-70.

[27] K. E. Gray, D.E. Fowler, "A superconducting fault current limiter", Journal Applied Physics, Vol. 49, 1978, pp. 2546-2550.

[28] A. Février, Y. Laumond, " Prospective uses of superconductors for 50/60 Hz applications ", Proceedings of ICEC 10, Berlin, 1986, pp. 139-152.

[29] T. Verhaege, C. Cottevieille, P. Estop, M. Quemener, J.P. Tavergnier, M. Bekhaled, C. Bencharab, P. Bonnet, Y. Laumond, V.D. Pham, C. Poumarede, P.G. Thérond, "Experiments with a high voltage (40 kV) superconducting fault current limiter", Cryogenics, 1996, Vol. 36, pp. 521-526.

[30] P. Tixador, "Superconducting current limiter - Some comparisons and influential parameters", IEEE Transactions on Applied Superconductivity, Vol. 4, 1994, pp. 190-198.

[31] J. Cave, D. Willen, R. Nadi, Y. Brissette, "Development of inductive fault current limiters up to 100 kVA class using bulk HTS materials", IEEE Transactions on Applied Superconductivity, Vol. 9, 1999, pp. 1335-1338.

[32] W. Paul, M. Lakner, J. Rhyner, P. Unternährer, T. Baumann, M. Chen, L. Widenhorn, A. Guérig, "Test of 1.2 MVA high Tc superconducting fault current limiter", Superconductor Science and Technology, Vol. 10, 1997, pp. 914-918.

[33] E.M. Leung, I. Rodriguez, G.W. Albert, B. Burley, M. Dew, P. Gurrola, D. Madura, G. Miyata, K. Muehleman, L. Nguyen, S. Pidcoe, S. Ahmed, G. Dishaw, C. Nieto, I. Kersenbaum, B. Gamble, C. Russo, H. Boenig, D. Peterson, L. Motowildo, P. Haldar, "High temperature superconducting fault current limiter development", IEEE Transactions on Applied Superconductivity, Vol. 7, 1997, pp. 985-988.

[34] T. Yazawa, E. Yoneda, J. Matsuzaki, M. Shimada, T. Kuriyama, S. Nomura, T. Ohkuma, Y. Sato, Y. Takahashi, "Design and test results of 6.6 kV high-Tc superconducting fault current limiter", IEEE Transactions on Applied Superconductivity, Vol.11, 2001, pp. 2511-2514.

[35] S. Fischer, H. Schmidt, R.R. Volkmar, "System requirements and test of superconducting fault current limiters", Conférence Internationale des Grands Réseaux Electriques (CIGRE), session 2000, paper 13-207.

[36] M. Noe, B. R. Oswald, "Technical and economical benefitsof superconducting fault current limiters in power systems", IEEE Transactions on applied superconductivity, Vol. 9, 1999, pp. 1347-1350.

[37] E. Serres, C. Levillain, F.X. Camescasse, J.F. Picard, M. Lakner, J. Rhyner, M. Chen, W. Paul, H. Zueger, J. Backes, D. Bonmann, S. Fleshler, A. Otto, R. Diehl, C. Christopherson, R. Mason, K. de Moranville, J. Daniel, "Current limitation in high temperature superconducting transformers and impact on the grid", Conférence Internationale des Grands Réseaux Electriques (CIGRE), session 2000, paper 12-205.

[38] N. Hayakawa, H. Kagawa, H. Okubo, N. Kashima, S. Nagaya, "A system study on superconducting fault current limiting transformer (SFCLT) with the functions of fault current suppression and system stability improvement", IEEE Transactions on Applied Superconductivity, Vol. 11, 2001, pp. 1936 1939.

[39] W.T. Norris, "Calculation of hysteresis losses in hard superconductor carrying ac : isolated conductors and edges of thin sheets", Journal Physics D : Applied Physics, Vol. 3, 1969, pp. 489-507.

[40] P.H. Herrmann,, C. Cottevieille, G. Duperray, A. Leriche, T. Verhaege, C. Albrecht, J. Bock, "Cryogenic load calculation of high T_C current lead", Cryogenics, Vol. 33, 1993, pp. 555-562.

[41] J. G. Noudem, P. Tixador, J.M. Barbut, O. Belmont, J. Sanchez, R. Tournier, "Current limitation at 1080 A under 1100 V with bulk Bi-2223", IEEE Trans. on Applied Superconductivity, Vol. 9, 1999, pp. 664-667.

[42] M. Chen, W. Paul, M. Lakner, L. Donzel, M. Hoidis, JP. Unternaehrer, R. Weder, M. Mendik, "6.4 MVA resistive Fault Current limiter based on Bi-2212 Superconductors", Physica C, Vol. 372-376, 2002, pp. 1657-1663.

[43] S. Elschner, F. Breuer, M. Noe, A. Wolf, J. Bock, "Qualification of MCP BSCCO 2212 bulk material for use in resistive current limiter, Physica C, Vol. 372-376, 2002, pp. 1668-1672.

[44] P. Tixador, L. Porcar, D. Buzon, E. Floch, D. Isfort, D. Bourgault, X. Chaud, R. Tournier, "Fault Current Limitation with Bulk YBCO", IEEE Transactions on Applied Superconductivity, Vol. 11, 2001, pp. 2034-2037.

[45] X. Granados, X Obrador, T. Puig, E. Mendoza, V. Gomis, S. Pinol, L. Garcia Tabares, J. Calero, "Hybrid superconducting fault current limiter based on bulk melt textured YBaCuO ceramic composites", IEEE Transactions on Applied Superconductivity, Vol. 9, 1999, pp. 1308-1311.

[46] R. Tournier, D. Bourgault, D. Buzon, X. Chaud,, E. Floch, D. Isfort, L. Porcar, P. Tixador, J. Bach, X. Obrador, T. Puig, X. Granados, E. Mendoza, J. Teva, E. Calleja, M. Segarra, I.G. Serradilla, H. Keschtkar, M. Sander, "YBCO single domain growth and shaping for fault current limiting applications", CIMTEC, 10th International Ceramic Congress, 3rnd Forum on new materials, Florence, italie, July 14-18 2002.

[47] J. Gerhold, "Power transmission", Handbook of applied superconductivity, IOP, 1998, Vol. 2, pp. 1627-1690.

[48] E.B. Forsyth, "The 60 Hz performance of superconducting power transmission cable rated for 333 MVA per phase", IEEE Transactions on Power Apparatus and System, Vol. 103, 1984, pp. 2023-2033.

[49] J. Oestergaar, "Superconducting power cables in Denmark - A case study", IEEE Transactions on Applied Superconductivity, Vol. 7, 1997, pp. 719-722.

[50] T. Verhaege, P.F. Herrmann, J. Bock, L. Cowey, G. Moulaert, H.C. Freyhardt, A. Usoskin, J. Paasi, M. Collet, "European Superconducting on a self-limiting superconducting power link", Superconductor Science and Technology, Vol. 13, 2000, pp. 488-492.

[51] M. Nassi, "HTS prototype for power transmission cables : recent results and future programmes", Superconductor Science and Technology, Vol. 13, 2000, pp. 460-463.

[52] K. Sato, S. Isolima, M. Watanabe, T. Masuda, M. Hirose, "The latest development of high Tc superconducting power cables", proceeding of ICEC'19, Grenoble, July 2002.

[53] H.H. Woodson, Z.J.J. Steckly, E. Hallas, "A study of alternators with superconducting field winding : I -Analysis", IEEE Transactions on Power Apparatus and Systems, Vol. 85, 1966, pp. 264-280.

[54] M. Rabinowitz, "Cryogenic power generation", Cryogenics, Vol. 17, 1977, pp. 319-330.

[55] D. Lambrecht, "Superconducting turbogenerators : status and trends", Cryogenics, Vol. 25, 1985, pp. 620-627.

[56] S. Kalsi, B. Gamble, N. Henderson, S. Karon, M. Mulholland, G. Snitchler, R. Howard, W. Sand, P. Winn, C. Gold, "Status of the navy HTS Superdrive motor for ship propulsion development", Thirds Naval Symposium on Electric Machines, Philadelphia, December 2000.

[57] T. Shimada, M. Shibuya, R. Takahashi, Y. Imai, H. Kusafuka, R. Shiobara, K. Yamaguchi, M. Takahashi, K. Suzuki, K. Miyaike, "Recent progress on 70 MW class superconducting generators", IEEE Transactions on Applied Superconductivity, Vol. 9, 1999, pp. 1189-1192.

[58] L.K. Kovalev, K.V. Ilushin, V.T. Penkin, K.L. Kovalev, S.M.A. Koneev, V.N. Poltavets, A.E. Larionoff, K.A. Modestov, S.A. Larionoff, W. Gawalek, T. Habisreuther, B. Oswald, K.J. Best, T. Strasser, "Hysteresis and reluctance electric machines with bulk HTS elements, recent results and future development", Superconductor Science and Technology, Vol. 13, 2000, pp. 498-502.

[59] H.O. Stevens, M.J. Superczinsky, T.J. Doyle, J.H. Harrison, H. Messinger, Superconducting machinery for naval ship propulsion, IEEE Transactions on Magnetics, Vol. 13, 1977, pp. 269-274.

[60] P. Tixador, "Fully superconducting machines", Handbook of Applied Superconductivity, Institute of Physics Publishing, 1998, pp. 1553-1577.

[61] J. L. Sabrié, "Feasibility of large AC superconducting equipment", Journal de Physique, Vol. 45, 1984, pp. 717-720.

[62] Y. Mitani, K. Tsuji, Power system stabilisation by superconducting magnetic energy storage connected to rotating exciter, IEEE Transactions on Applied Superconductivity, Vol. 3, 1993, pp. 219-222.

[63] D. Madura, M. Richardson, D. Bushko, G. Snitchler, P. Winn, S. Kalsi, B. Gamble, "Test results of a 5000 HP HTS motor", Applied Superconductivity conference, Houston, August 2002 (paper 2LB02).

[64] P. Tixador, F. Simon, H. Daffix, M. Deleglise, "150 kW Experimental Superconducting Permanent Magnet Motor", IEEE Trans. on Applied Superconductivity, Vol. 9, 1999, pp. 1205-1208.

[65] C. Flick, "New armature winding concepts for EHV and high CFCT applications of superconducting turbine generators", IEEE Transactions on Power Apparatus and System, Vol. 98, 1979, pp. 2190-2200.

[66] M. Leijon, F. Owman, T. Sorqvist, C. Parkegren, S. Lindahl, T. Karlsson, "Powerformer(R) : a giant step in power plant engineering", Proceedings of IEEE International Electric Machines and Drives Conference (IEMDC'99), 1999, pp. 830-832.

[67] W. Nick, G. Nerowski, H.W. Neumüller, M. Frank, P. van Hasselt, J. Frauenhofer, F. Steinmeyer, "380 kW synchronous machine with HTS rotor winding - development at Siemens and firts results", Physica C, Vol. 372-376, 2002, pp. 1506-1512.

APPLICATION FIELDS OF HIGH-TEMPERATURE SUPERCONDUCTORS

Roland HOTT

Forschungszentrum Karlsruhe, Institut für Festkörperphysik,
P. O. Box 3640, 76021 Karlsruhe, GERMANY

Email: roland.hott@ifp.fzk.de

1. Introduction

For classical superconductors it took about half a century from their discovery to arrive at technically applicable materials. Only about a third of this time has elapsed for High-Temperature Superconductors ("HTS") where a number of technically applicable materials species is now already available.

Epitaxial HTS *thin films* achieve excellent superconducting properties (critical temperature $T_c > 90$ K; critical current density J_c (77 K, 0 T) > 10^6 A/cm^2; microwave surface resistance R_s(77 K, f = 10 GHz) < 500 μΩ, $R_s(\mathrm{T}, f) \propto f^2$) that are well-suited for superconductive electronics. They are already in use in commercial and military microwave filter systems. HTS *Josephson junctions* are available which can be used for the construction of highly sensitive magnetic field sensors ("Superconducting QUantum Interference Devices", "SQUIDs") and are also tested for active electronic devices that may broaden the range of HTS thin-film applications. *Melt-textured* HTS *bulk material* shows superb magnetic pinning properties and may be used as high-field permanent magnets. In spite of the ceramic nature of the cuprate oxides, flexible HTS *wire* or *tape conductor* material is obtained either by embedding HTS as thin filaments in a silver matrix or by coating of metal carrier tapes.

Among these technical HTS materials, Ag-sheathed Bi-HTS conductors represent the only exception to the rule that strong biaxial texture is necessary to achieve technically meaningful currents. A main reason is here the good mechanical and electrical contact with the Ag matrix which allows high current flow under the relaxed condition that a little detour via Ag is possible in case the direct current transfer between the HTS grains is blocked. These arguments apply as well to "Ag-impregnated" Bi-2212 bulk which is commercially available in sizes of several 10 cm with homogeneous J_c(77 K, 0 T) of several kA/cm^2 [1]. Much higher critical current densities are achieved in polycrystalline HTS materials, but their use implies a higher risk for applications: In all these cases, substantial J_c reduction compared to the J_c (77 K, 0 T) ~ 1 MA/cm^2 encountered in well-textured material with a sufficient amount of pinning centers always indicates a high degree of structural and electrical inhomogeneity with potentially devastating consequences: Forced current flow through defective materials regions may lead to a local quenching of superconductivity and the creation of "hot spots" [2]. The deposited quench energy leads then again to a structural and chemical modification of the HTS material in the neighborhood of these "hot spots". This enlarges the quench zone and may finally end up in the destruction of the superconductor. This quench mechanism is different from the situation in classical superconductors where the quenching is caused by insufficient heat transfer due to the freezing of the phonon mechanism at LHe temperature. At the much higher operating temperature of HTS this is not such a critical issue since the heat distribution by phonon is here still very effective [3].

Cooling is the main concern for a future breakthrough of HTS applications. The market acceptance for HTS-based systems will depend critically on the availability of reliable and inexpensive cooling systems that are "invisibly" integrated in these systems.

2. Electronics

The low microwave losses of HTS thin films [4] enables the coupling of an unprecedentedly large number of resonators to *microwave filter* devices with much sharper frequency characteristics than conventional compact filters [5, 6].

In the US, the military interest in HTS filters in aircraft electronics, e.g. for better rejection of interference noise in aircraft radar systems, is still strong. In US mobile phone communication systems, HTS microwave filter subsystems are already a commercially available solution for problematic radio reception situations [7]: Improved noise data compared to conventional solutions result in a lower percentage of dropped calls; in rural areas rf coverage can be achieved by a smaller number of base stations; the use of HTS filters could allow to reduce the rf power of the handsets in urban areas. The miniaturization potential of HTS filters should be most appreciated in communication satellites [8], however, respective projects were not very fortunate [9, 10].

In *Josephson samplers*, the fast voltage reaction of a Josephson junction on a signal current that exceeds the critical current allows the determination of (repeated) signal forms with ps resolution. A first commercial system fabricated in Niobium technology disappeared again from the market soon after its introduction in 1987 due to problems with the complexity of the LHe cooling and high maintenance cost (as well as the introduction of semiconductor samplers with comparable bandwidth). An HTS Josephson sampler system under development at NEC has already demonstrated the resolution of a 5.9 Gbps digital waveform [11].

The periodicity of the electric characteristics of superconducting loops as a function of the introduced magnetic flux can be used for the construction of superconducting *AD converters.* Flux quantization can help to implement feedback loops with quantum accuracy [12]. First test of components [14,15,16] and complete devices [13] based on ~ 10 HTS Josephson contacts demonstrated technical feasibility.

Recent simulations of the noise-induced bit error rate in HTS *digital circuits* indicate good chances even for LN_2 temperature operability [17] for suitable circuit technology [18] since connecting Josephson junctions in a circuit results in improved energetic separation of the different switching states as compared to the isolated junction behavior [19].

An HTS programmable *voltage standard* based on an array of 136 YBCO bicrystal junctions operated at 64 K could be synchronized at 25 - 40 GHz showing stable voltage steps up to 9 mV with an accuracy better than 10^7 [20].

Among all the tested HTS *transistor* concepts [21] only Vortex-Flow Transistors (VFT) have demonstrated large gain-bandwidth potential [22]. Quasiparticle Injection devices [23] or Superconducting Field Effect Transistors ("SuFETs") [21,24,25,26,27] are still far from technical applications.

3. Sensors

Superconducting QUantum Interference Devices ("*SQUIDs*") are superconducting loops with integrated Josephson contacts which can be used as the most sensitive magnetic field sensors. The magnetic field resolution of HTS *SQUIDs* [28] operated at LN_2 temperature has arrived at values comparable to commercial LHe-cooled LTS SQUIDs, only one order of magnitude above the record values of LTS SQUIDs at 4 K [29]. However, a large commercial impact is only expected for HTS SQUID systems that are able to observe magnetic signals even in the presence of disturbing background fields without the burden of magnetic shielding [30,31,32,33,34]. Mobile non-shielded HTS SQUIDs were tested with success for biomedical [31,35], non-destructive evaluation [36,37] and geophysical applications [38].

In superconducting *bolometers*, the sharp superconducting transition as a function of temperature is used as very sensitive thermometer which allows to measure with high sensitivity the heating of a thermally connected absorber under electromagnetic irradiation. The high quantum efficiency of semiconductor detectors restricts yet the application interest to the infrared wavelength region $\lambda > 12$ μm where no adequate semiconductor material is available [39,40,41,42]. As LN_2 cooling is common practice for semiconductor IR radiation sensors, HTS bolometers could easily be inserted in such detector systems. A Gd-123 based bolometer has met the specifications for a sensor for IR observation of OH molecules in the atmosphere in a satellite project [39]. Hot-electron bolometers with small thermal relaxation time are tested as *mixers* for high-frequency signals in the range of several 100 GHz [43].

The interaction of an electromagnetic wave in a Josephson contact with the Josephson oscillations results in an analog way of doing a *Hilbert transformation of the radiation spectrum* where the Hilbert transform can be derived from the IV characteristics [44,45]. Experiments based on a YBCO bicrystal contact demonstrated that a spectral resolution of ~ 4 GHz can be achieved in a bandwidth from ~100 GHz to 3 THz [44]. The device has been tested with success at DESY for the determination of the bunch length of a pulsed electron beam by means of the emitted rf spectrum.

4. Magnets

The first commercial impact of HTS on superconducting magnet systems are *current leads* that substantially reduce the heat load to the cold magnet system. Whereas for normal metals the Wiedemann-Franz relation between electrical and thermal conduction results in a universal minimum value ~ 1 W/kA for charge transport to a 4 K heat load, HTS in the superconducting state offer a large current capability at the low thermal conductivity of oxide ceramics [46]. Bi-2212 bulk based leads demonstrated a reduction of the 4 K heat load by a factor ~ 5 compared to conventional leads in dc as well as in ac application [47,48]. Current leads based on Bi-2223 tapes with AgAu alloy sheath (with one order of magnitude lower thermal conductivity compared to Ag) achieve similar performance [49]. The intrinsic parallel shunt provided by the normal metal sheet and the reduced mechanical sensitivity are attractive features with respect to the emergency situation in case of a quench of the connected LTS coil. For big LTS research magnets the heat load due to the current leads amounts to 30 - 90 % of the total 4 K heat load. At present, the largest order for HTS current leads is expected from CERN where 64×13 kA, 310×6 kA and 750×600 A leads will be required for the Large Hadron Collider project [50]. Several types of HTS hybrid current leads have been tested [50]. At Fermilab 30 - 40 of the existing 50 pairs of 5 kA current leads are planned to be substituted by HTS hybrids. The gain of 30 % in heat load to the LHe bath will be used for lower operation temperature of the LTS magnets which will then allow stronger particle acceleration. HTS current leads for much higher currents in future fusion magnets (40 - 60 kA) are under development [51]. The reduction of the heat load by means of HTS current leads and the availability of efficient cryocoolers enabled "cryogen-free" LTS magnet systems which can be operated by means of conduction cooling without the requirement of LHe coolant [52].

The extraordinarily high critical magnetic fields of HTS (estimated: ~ 100 T) allow the flow of supercurrents in much higher magnetic background fields than in classical superconductors. Thus HTS magnets are able to generate an additional magnetic field even in the environment of a strong magnetic background field [53]. The critical issue is the mechanical stress in the HTS conductor as a consequence of the magnetic field. A Bi-2212 magnet successfully generated 5.42 T in a backup field of 18 T provided by a classical superconducting coil system [54]: The total magnetic field of the superconducting magnet system reached 23.42 T, which is the highest magnetic field ever achieved by a superconducting magnet. The goal of such activities are high field magnets for Nuclear Magnetic Resonance (NMR) experiments where the persistent mode operation of the superconducting coils is used to fulfill the extreme requirements of temporal stability. The respective time constants for the decay of the magnetic field in present HTS coils are still about an order of magnitude below these requirements.

The higher operation temperature of HTS magnets is of interest for many applications where classical superconducting coils can not be used due to strong heat generation or heat leakage to the cold part [55,56]. In a US project, DuPont assembled, installed and tested a 0.2 m bore 3 T HTS magnet for minerals and chemical separation[1] [57]. A full-scale pre-production unit with 0.5 m warm bore 3 T HTS magnet will be the next step in this project. In

[1] In a Japanese project, a mobile water purification system based on magnetic separation by strong melt-textured YBCO bulk cryomagnets is under construction [58].

another US project, Oxford Superconducting Technology plans to build a cost-effective, open-geometry MRI system based on a HTS 0.2 T magnet [57].

Superconducting magnetic bearings ("SMB") composed of melt-textured YBCO pellets and permanent magnets achieved a coefficient of friction (drag-to-lift ratio) of ~ 10^{-9} [59,60,61]. They are investigated for the levitation of rotors in flywheels [62] and motors [63]. Compared to active magnetic bearings based on permanent magnets and active rotor control, SMB can offer similar levitation forces [64], similar power consumption (cooling power / control power) and lower friction, but unfortunately only much lower stiffness [65]. In China, a first man-loading HTS Maglev test vehicle was tested successfully with up to five people and a total weight is 530 kg at a net levitation gap > 20 mm on a 15.5 m long guideway consisting of two parallel permanent magnetic tracks [66].

5. Power Applications

Superconductors can help to increase the efficiency of components for power applications by means of their lower losses as well as to reduce volume and weight by means of their potential for high power density [67]. However, reliability of these large scale devices is here of even higher importance than for other potential application fields. HTS demonstrators are still on their way of trying to convince power companies about their trustworthiness even over a long lifetime.

Power cables are in general about a factor of 10 more expensive than overhead lines. This restricts their application to urban areas. Experiments with LHe-cooled LTS cables in the 70's were technically successful [67], but their introduction into power grids turned out to be economically viable only for power transfer > 5 - 10 GVA. For LN_2-cooled HTS cables, this break-even point may be reduced to ~ 300 - 500 MVA, if the cables can be retrofitted to already existing cable ducts. This would allow for an increase of power transfer by a factor ~ 3 at a reduction of the voltage from ~ 400 kV to ~ 100 kV [68]. Therefore, such a retrofit is especially attractive for metropolitan areas with still growing power demand, such as Tokyo, where no more additional cables can be inserted into the cable ducts, and voltage has been increased right up to the highest justifiable level. Besides this civilian application, the US Navy is also interested in lightweight small-sized cables for battleships [69]. LN_2 cooling offers distinct advantages compared to conventional oil-cooling with respect to pollution in case of leakage and with respect to the risk of fire. The cable market amounts to 5 - 10 % of the total power equipment market. The low operational self-fields of ~ 0.1 T of power cables are within the acceptable limit of present commercial Bi-2223/Ag tapes at LN_2-cooled operation. In a Danish project, a 30 m 3-phase HTS cable with a voltage rating of 30 kV and a power rating of 104 MVA has been installed and tested under realistic conditions in a substation in the electric grid of Copenhagen [70]. In a Japanese cable project, a 100 m 3-core 66 kV/1 kA/114 MVA has been tested successfully for one year duration [71]. In a US cable project, Southwire installed a 30-m 3-phase cold dielectric cable system (12.5 kV, 1250 A_{rms}, 27 MVA) which is still in test operation since February 2000 [57]. In another US cable project, a 3-phase 120 m cable system has been installed in a power station [72] but severe problems with damage of the flexible cryostat of the warm-dielectric type cable have prevented until now operation of 2 of the 3 phases. Recently, two 300 m HTS cable projects have been approved in the US which are planned to be installed in electricity distribution systems in 2005 [73].

The *fault current limiter* ("FCL") is unanimously addressed by utilities as a very attractive component, integratable in existing power grids [74]. It prevents overloads from the grid components thus enabling longer lifetimes and avoiding investment cost due to the usually practiced overdimensioning. For FCLs rated at a nominal power of at least 10 - 20 MW as smallest practical size, an annual market volume of ~ 1 billion US-$ has been estimated. The current limitation is based on the quench of a superconductor due to a current exceeding its critical current, resulting in a very rapid tremendous increase of its electrical resistance. The material problem to be solved is the spatial homogeneity of the local critical current density on the length scale of the thermal diffusion length because otherwise the power is not dissipated homogeneously, but directed into hot spots, destroying the material. This homogeneity has been achieved up to now only for Bi-2212 bulk [75], Bi-2223 tapes [79] and YBCO tapes [80] and YBCO thin films [81]. A 1.2 MW-3 phase-FCL by ABB based on Bi-2212 bulk rings acting as secondary winding of a transformer-like device has been tested for one year in a Swiss power plant without major problems [75]. In a test series, a prospective fault current of 60 kA was limited to 700 A within the first 50 Hz-half-wave. The time for recovery to the full operational state after a quench amounted to a few seconds. Meanwhile, ABB presented a single phase 6.4 MVA device with a different FCL design based on 30×40 cm^2 sized Bi-2212 sheets structured into long length meanders [76]. In a German collaboration of industrial companies, research institutes and power utilities, a 15 MVA FCL demonstrator of the resistive type for the 10 kV voltage level based on Bi-2212 spirals is planned to be assembled and field-tested this year [77]. The limiter is designed for coupling two 125 MVA utility grids.

HTS *transformers* can be built at smaller volume and weight compared to their conventional counterparts [78]. With respect to the risk of fire and environmental aspects, LN_2 is a much more pleasant coolant than oil used today. However, the main advantage of SC transformers is their potential for an operation at a power level exceeding the rated power by up to 100 % even over a period of several hours: This exceptional operation mode merely requires more cooling power, but does not lead to increased wear as in the case of conventional transformers where even ~ 10 % overload causes thermal damage in the insulation. With regard to energy losses, a quarter of the losses of 5 - 10 % of the transported power in power distribution systems are due to transformers. The use of a winding based on a HTS conductor with fault current limiting functionality [79] is an additional option. ABB tested a 630 kW - 3 phase transformer based on Bi-2223/Ag tape conductor successfully at a Swiss utility for one year under regular operational conditions [82]. In a US project, a 5/10 MVA prototype based on Bi-2212 surface-coated conductor is under development [83]. Siemens developed a 1 MVA demonstrator transformer for railway applications which has been tested with success [84].

For large synchronous *motors*, 50 % volume and loss reduction compared to a conventional motor is expected [85, 86]. For the US Navy, this is a very attractive feature since their new generation of surface ships is based on electric drives due to lower operating and support cost [69]. The respective project for has meanwhile arrived at a 3.7 MW engine which was tested up to a peak power of 5.2 MW [87]. Furthermore, the rotor of a 5 MW engine has been completed which will now be assembled until mid of this year for the first HTS ship propulsion motor. The US Navy is also interested in a super-quiet homopolar engine with regard to improved stealth [69,88]. Siemens built a synchronous machine consisting of an HTS rotor and an air core stator that achieved a maximum continuous power of 450 kW and a short term maximum power of 590 kW at 1500 rpm [89]. Rotor cooling is provided by a GM "off the shelf" cryocooler, the typical requirement being about 30W @ 25K. The output power was limited by stator cooling. The machine was tested under different operational modes, including motor mode as well as generator mode. Siemens started now work on a 4 MW ship propulsion motor. Motors based on the strong magnetizability of melt-textured YBCO bulk material with an output power of 1-37 kW and current frequencies 50 and 400 Hz have demonstrated in LN_2 operation a 4–5 times better specific output power per weight than conventional electric machines [90]. Such motors could find an application in cryopumps for cryogenic liquids which could be used simultaneously as coolant for the HTS material.

HTS *generators* are of considerable military interest in the US: The US Navy is interested in lightweight small-sized HTS power systems (generators, cables, fault-current limiters, transformers, motors) for battleships [69]. The US Air Force develops a new generation of aircrafts where the hydraulics will be completely replaced by electronics. HTS generators would be extremely desired provided they are available within the next 3-5 years [91]. In a civilian US project, General Electric is developing a 100 MVA HTS power generator for commercial entry applications [57]. It is based on a previously constructed 15 MVA model.

Superconductivity offers two extremely different ways for *energy storage* [92]: In *Superconductive Magnetic Energy Storage* ("*SMES*") systems, a SC coil stores magnetic field energy. The energy can be transferred very rapidly via power electronics to or from the grid [93]. SMES systems based on classical superconductors are already in commercial use for the improvement of the power quality. HTS SMES systems are of interest with respect to size and volume reduction [94]. In *flywheels* based on superconductive magnetic bearings [95], electric power is transformed to kinetic energy via a motor-generator combination controlled by power electronics. Due to the mechanics involved, the energy can not be transferred as fast as by SMES. However, the density of storable energy can be much higher [92] and operational losses are expected to be significantly lower [96]. In a US project, a 10 kWh / 3 kW system has been assembled and installed [57]. In a Japanese project, a flywheel demonstrated a maximum energy storage of 1.4 kW h at 20 000 rpm [97]. A large-size bearing for a 25 kW h flywheel has been tested. In Germany, a 10 kWh / 1-2 MW system is under construction which is planned to be field-tested next year [98].

6. Cryocoolers

The most visible aspect of superconductivity for any new potential customer of superconductive devices is not superconductivity but the required cryocooling. As for the cost, as a rule of thumb about 10 % of the total system prize seems to be an accepted level for most of the applications. The goal of a US project for low price cryocoolers (~ 1000 US $ at annual sales of ~ 10 000 coolers) which should reliably provide a cooling power of some W at 50 - 150 K over a period of 3 years [99] has not been achieved yet[2]. The larger cooling requirements of HTS power applications can already be met by reliable commercially available cryocoolers, but, e.g. for transformers, only at a price which amounts to sales price of a conventional transformer. This economic mismatch may only be overcome by much higher sales figures

---

[2] An excellent topical survey on low-power cryocoolers can be found in [100].

and technical simplification of the cryocoolers. The technical boundary conditions for the selection of the cryocooler are minimal disturbance of the cooled device, e.g. by mechanical vibrations, and good cooling efficiency expressed by the ratio of cooling power at a certain operating temperature divided by the required electric power. In commercial US mobile phone microwave filter systems, cryocoolers with a cooling power of several W in the LN_2 temperature region have already demonstrated their utilizability in more than 20 million cumulative hours of operation in the field from which a Mean Time Before Failure ("MTBF") of ~ 800 000 hours has been estimated [101].

Different cooling principles are available for cooling to the LN_2 or even LHe temperature region. *Joule-Thomson* ("JT") coolers are based on the heat exchange during the expansion of a continuous gas flow. A commercial JT cooler with a cooling power of 3 W @ 77 K has been used successfully for HTS SQUID cooling [102].

Stirling, *Gifford-McMahon* ("GM") and *Pulse Tube* ("PT") coolers are all based on an oscillating gas flow and a regenerative heat exchange. The heat extraction depends on the oscillating frequency f of the gas flow and the phase angle between pressure p and volume V during an oscillation period [103,104]. In an analog electric network description these coolers can be regarded as RC phase shifters. In Stirling and GM coolers a piston is moving in the cold part in which the regenerator is usually already integrated. The PT cooler can be conceived as a GM or Stirling cooler where the gas flow is steered in such a way as if a "gas piston" was moved according to the particular cooler operation mode.

Stirling coolers are usually operated at frequencies f ~ 20 - 60 Hz and achieve minimal temperatures T_{min} = 30 - 55 K. The weight of coolers commercially available at 3000 to 20 000 US $ reaches from 300 g (150 mW @ 80 K, P_{el} = 3 W) to 35 kg (18 W @ 80 K, P_{el} = 375 W). The prize extremes are given by a 250 000 US $ cooler manufactured for the US HTS Space Experiment [9] (specified MTBF: 45 000 h) on one side and a cryocooler used in the commercial US mobile phone microwave filter systems with an estimated prize < 2000 US $ for annual sales numbers > 1000. For the US military aircraft HTS application plans, available Stirling coolers seem to satisfy already the specifications concerning efficiency, reliability and cost. A disadvantage of these coolers especially for SQUID applications is the small distance from the compressor to the cold head of at most 40 cm.

Coolers of the *Gifford-McMahon* type are sold worldwide in numbers of ~ 10 000 per year for cryopumps to the electronics industry [105] and for the assistance of the LHe cooling of LTS MRI magnets by means of 20 K heat shields. Compressor and cold head are usually separated units. The pressure oscillations are produced by a rotary valve in the cold head. Due to this construction principle GM coolers are on one side substantially heavier than Stirling coolers with the same cooling power and achieve only ~ 1/3 of their cooling efficiency. On the other side, they are more robust than Stirling coolers. This helped the GM coolers to establish themselves as standard coolers for temperatures < 30 K providing a cooling power of 0.5 - 3 W @ 4.2 K at an input power of 1 - 12 kW [106]. The suggested service intervals are 5 years for the compressor (with an annual oil exchange) and 1 - 3 years for the rotary valve and the seals. 4 K GM coolers are today available at a price of ~ 40 000 US $ which is estimated to be reducible to ~ 5000 US $ at drastically increased sales numbers.

Pulse Tube coolers have no moving parts in the cold region. Therefore no expensive high-precision seals are required and the cold head can be operated without any service inspection. 1-stage PT coolers achieve cooling powers of several W in the LN_2 temperature region (up to 166W @ 80 K at P_{el} = 3.42 kW [107]). With 2-stage PT coolers [108] 0.5 W @ 4.2 K at P_{el} = 6.3 kW and 0.17 W @ 4.2 K at P_{el} = 1.7 kW have been demonstrated. Already in 1993 a PT with 0.8 W @ 80 K at P_{el} = 33 W and a weight of 2 kg achieved the performance of a comparable Stirling cooler [109]. Meanwhile, Stirling-type PT (high operation frequency: $f \sim 50$ Hz) have arrived at an even better efficiency at 77 K than their conventional counterparts, GM-type PT (low operation frequency: $f \sim 5$ Hz) show at 4 K comparable efficiency [110]. Stirling-type PT with a cooling power of 12 W @ 80 K are commercially available where only the compressor requires servicing every 20 000 h. Miniature PT are in use for 10 year long-life space applications [109]. PT coolers were used with success for the cooling of HTS SQUID and rf devices [102, 111]. A completely non-metallic PT cooler has been tested with success for low magnetic noise SQUID cooling [111].

7. Conclusion

Recalling the situation in the late 1980 in the "prime time" of High-Temperature Superconductors immediately after their discovery, it is amazing to see which tremendous progress has been achieved in the meantime: HTS filters in mobile phone base stations, HTS SQUID operation in regular technical environment, HTS magnets generating a 5 T field in addition to a 18 T background, 100 m long HTS power cables are more than what could be expected on realistic grounds in such a short period of time. Steady improvement of the HTS materials basis will surely widen this spectrum of applications within near future.

Acknowledgement

I would like to thank T. Habisreuther, P. Komarek, M. Lakner, T. Scherer, C.W. Schneider, P. Seidel, M. Siegel, H. Töpfer and G. Thummes for critically reading the manuscript and for helpful comments.

References

[1] P. F. Herrmann, E. Beghin, J. Bock, C. Cottevieille, A. Leriche, T. Verhaege, Inst. Phys. Conf. Ser. No. **158** (1997) 825;
M. Noe, K. -P. Juengst, F. N. Werfel, S. Elschner, J. Bock, A. Wolf, F. Breuer, Physica C **372-376** (2002) 1626

[2] A. Usoskin, A. Issaev, H. C. Freyhardt, M. Leghissa, M. P. Oomen, H. -W. Neumueller, Physica C **378-381** (2002) 857

[3] T. Kiss, S. Noda, S. Nishimura, K. Ohya, D. Utsunomiya, Yu. A. Ilyn, H. Okamoto, Physica C **357-360** (2001) 1165;
Yu. A. Ilyn, V. S. Vysotsky, T. Kiss, M. Takeo, H. Okamoto, F. Irie, Cryogenics **41** (2001) 665

[4] M. A. Hein, Supercond. Sci. Technol. **10** (1997) 867

[5] R. R. Mansour, IEEE Trans. Microwave Theory Tech. **50** (2002) 750

[6] G. Tsuzuki, M. Suzuki, N. Sakakibara, IEEE Trans. Microwave Theory Tech. **48** (2000) 2519

[7] B. A. Willemsen, IEEE Trans. Appl. Supercond. **11** (2001) 60

[8] R. R. Mansour, T. Romano, S. Ye, S. Peik, T. Nast, D. Enlow, C. Wilker, J. Warner, IEEE Trans. Appl. Supercond. **11** (2001) 806

[9] M. Nisenoff, W. J. Meyers, IEEE Trans. Appl. Supercond. **11** (2001) 799

[10] T. Kässer, in “Supraleitung und Tieftemperaturtechnik“, 2000, p.65

[11] M. Hidaka, T. Satoh, N. Ando, M. Kimishima, M. Takayama, S. Tahara, Physica C **357-360** (2001) 1521;
M. Hidaka, M. Maruyama, S. Tahara, Physica C **378-381** (2002) 1435

[12] A. Yu. Kidiyarova-Shevchenko, D. E. Kirichenko, Z. Ivanov, F. Komissinsky, E. A. Stepancov, M. Khapaev and T. Claeson, Physica C **326-327** (1999) 83

[13] G. J. Gerritsma, M. A. Verhoeven, R. J. Wiegerink, H. Rogalla, IEEE Trans. Appl. Supercond. **7** (1997) 2987

[14] B. Ruck, Y. Chong, R. Dittmann, A. Engelhardt, E. Sodtke, M. Siegel, Supercond. Sci. Technol. **12** (1999) 701;
B. Ruck, Y. Chong, R. Dittmann, M. Siegel, Physica C **326-327** (1999) 170

[15] M. G. Forrester, B. D. Hunt, D. L. Miller, J. Talvacchio, R. M. Young, Supercond. Sci. Technol. **12** (1999) 698

[16] K. Saitoh, Y. Soutome, T. Fukazawa, K. Takagi, Supercond. Sci. Technol. **15** (2002) 280;
K. Saitoh, F. Furuta, Y. Soutome, T. Fukazawa, K. Takagi, Physica C **378-381** (2002) 1429

[17] T. Ortlepp, H. Toepfer, H. F. Uhlmann, Appl. Phys. Lett. **78** (2001) 1279

[18] D. Cassel, R. Dittmann, B. Kuhlmann, M. Siegel, T. Ortlepp, H. Toepfer, H. F. Uhlmann, Supercond. Sci. Technol. **15** (2002) 483

[19] H. Toepfer, T. Ortlepp, H. F. Uhlmann, D. Cassel, M. Siegel, Proceedings ISS 2002, to be published in Physica C

[20] A. M. Klushin, R. Behr, K.Numssen, M. Siegel, J. Niemeyer, Appl. Phys. Lett. **80** (2002) 1972

[21] J. Mannhart, Supercond. Sci. Technol. **9** (1996) 49

[22] J. Schuler, S. Weiss, T. Bauch, A. Marx, D. Koelle, R. Gross, Appl. Phys. Lett. **78** (2001) 1095

[23] C. W. Schneider, R. Moerman, D. Fuchs, R. Schneider, G. J. Gerritsma, H. Rogalla, IEEE Trans. Appl. Supercond. **9** (1999) 3648

[24] B. Mayer, J. Mannhart, H. Hilgenkamp, Appl. Phys. Lett . **68** (1996) 3031

[25] V. Talyansky, S. B. Ogale, I. Takeuchi, C. Doughty, T. Venkatesan, Phys. Rev. B **53** (1996) 14575

[26] R. Auer, R. Schneider, J. Appl. Phys. **81** (1997) 3237

[27] T. Nakamura, H. Inada, M. Iiyama, Physica C **300** (1998) 93

[28] D. Koelle, R. Kleiner, F. Ludwig, E. Dantsker, J. Clarke, Rev. Mod. Phys. **71** (1999) 631

[29] M. I. Faley, U. Poppe, K. Urban, V. Yu. Slobodchikov, Yu. V. Maslennikov, A. Gapelyuk, B. Sawitzki, A. Schirdewan, Appl. Phys. Lett. **81** (2002) 2406

[30] S. K. Lee, W. R. Myers, H. L. Grossman, H.-M. Cho, Y. R. Chemla, J. Clarke, Appl. Phys. Lett. **81** (2002) 3094;

H.-M. Cho, Y. T. Andresen, J. Clarke, M. S. DiIorio, K.-Y. Yang, S. Yoshizumi, Appl. Phys. Lett. **79** (2001) 2438

[31] T. Schurig, D. Drung, S. Bechstein, J. Beyer, F. Ludwig, Physica C **378-381** (2002) 1378

[32] K. Peiselt, S. Wunderlich, F. Schmidl, F. Schrey, C. Steigmeier, U. Hübner, P. Seidel, Physica C **378-381** (2002) 1385;

P. Seidel, F. Schrey, L. Dörrer, K. Peiselt, F. Schmidl, F. Schmidt, C. Steigmeier, Supercond. Sci. Technol. **15** (2002) 150

[33] V. Schultze, R. Ijsselsteijn, A. Chwala, N. Oukhanski, V. Zakosarenko, H. G. Meyer, Supercond. Sci. Technol. **15** (2002) 120

[34] A. Tsukamoto, T. Fukazawa, K. Takagi, K. Yokosawa, D. Suzuki, K. Tsukada, Appl. Phys. Lett. **79** (2001) 4405

[35] H. J. M. ter Brake, A. P. Rijpma, J. G. Stinstra, J. Borgmann, H. J. Holland, H. J. G. Krooshoop, M. J. Peters, J. Flokstra, H. W. P. Quartero, H. Rogalla, Physica C **368** (2002) 10

[36] H.-J. Krause, M. v. Kreutzbruck, Physica C **368** (2002) 70

[37] S. Chatraphorn, E. F. Fleet, F. C. Wellstood, J. Appl. Phys. **92** (2002) 4731

[38] C. P. Foley, D.L. Tilbrook, K. E. Leslie, R. A. Binks, G.B. Donaldson, J. Du, S. K. Lam, P. W. Schmidt, D. A. Clark, IEEE Trans. Appl. Supercond. **11** (2001) 1375

[39] M. J. M. E. de Nivelle, M. P. Bruijn, R. de Vries, J. J. Wijnbergen, P. A. J. de Korte, S. Sánchez, M. Elwenspoek, T. Heidenblut, B. Schwierzi, W. Michalke, E. Steinbeiss, J. Appl. Phys. **82** (1997) 4719

[40] L. Méchin, J.-C. Villégier, D. Bloyet, J. Appl. Phys. **81** (1997) 7039

[41] B. Lakew, J. C. Brasunas, A. Piqué, R. Fettig, B. Mott, S. Babu, G. M. Cushman, Physica C **329** (2000) 69

[42] L. R. Vale, R. H. Ono, D. G. McDonald, R. J. Phelan, Supercond. Sci. Technol. **12** (1999) 856

[43] O. Harnack, K. S. Il'in, M. Siegel, B. S. Karasik, W. R. McGrath, G. De Lange, Appl. Phys. Lett. **79** (2001) 1906;

K. S. Il'in, M. Siegel, J. Appl. Phys. **92** (2002) 361

[44] Y.Y. Divin, U. Poppe, O.Y. Volkov, V.V. Pavloskii, Appl. Phys. Lett. **76** (2000) 2826

[45] A. Kaestner, M. Volk, F. Ludwig, M. Schilling, J.Menzel, Appl. Phys. Lett. **77** (2000) 3057

[46] P. F. Herrmann, p.801 in "Handbook of Applied Superconductivity", Institute of Physics Publishing, 1998, Bristol and Philadelphia, ISBN 0 7503 0377 8

[47] P. F. Herrmann, E. Beghin, G. Bottini, C. Cottevielle, A. Leriche, T. Verhaege, J. Bock, Cryogenics **34** (1994) 543

[48] P. F. Herrmann, C. Cottevielle, A. Leriche, M. Quemener, J. Bock, IEEE Trans. Appl. Supercond. **7** (1997) 359

[49] T. Sasaoka, J. Sato, K. Fukushima, M. Okada, M. Endo, Y. Yaegashi, M. Nagano, Cryogenics **37** (1997) 409

[50] A. Ballarino, Physica C **372-376** (2002) 1413

[51] R. Heller, K. Takahata, G. Friesinger, T. Mito, M. Tasca, A. Nishimura, S. Satoh, S. Yamada, IEEE Trans. Appl. Supercond. **11** (2001) 2603

[52] J. Sakuraba, Y. Mikami, K. Watazawa, K. Watanabe, S. Awaji, Supercond. Sci. Technol. **13** (2000) 12;

K. Watanabe, S. Awaji, J. Sakuraba, K. Watazawa, T. Hasebe, K. Jikihara, Y. Yamada, M. Ishihara, Cryogenics **36** (1996) 1019

[53] S. W. Van Sciver, K. R. Marken, Physics Today, August 2002, p. 37

[54] M. Okada, K. Tanaka, T. Wakuda, K. Ohata, J. Sato, T. Kiyoshi, H. Kitaguchi, H. Wada, Physica C **335** (2000) 61

[55] T. Matsushita, Supercond. Sci. Technol. **13** (2000) 51

[56] K. Sato, T. Kato, K. Ohkura, S. Kobayashi, K. Fujino, K. Ohmatsu, K. Hayashi, Supercond. Sci. Technol. **13** (2000) 18

[57] R. Eaton, Viewgraphs "2002 DOE Wire Workshop", January 22, 2002, St. Petersburg, FL, http://www.eren.doe.gov/superconductivity/docs/05eaton_overview.ppt

[58] Superconductor Week, November 29, 2002

[59] H. J. Bornemann, T. Burghardt, W. Henning, A. Kaiser, IEEE Trans. Appl. Supercond. **7** (1997) 1805

[60] C. E. Rossman, J. I. Budnick, B. R. Weinberger, Appl. Phys. Lett. **70** (1997) 255

[61] J. R. Hull, Th. M. Mulcahy, J. F. Labataille, Appl. Phys. Lett. **70** (1997) 655

[62] S. Nagaya, N. Kashima, M. Minami, H. Kawashima, S. Unisuga, Y. Kakiuchi, H. Ishigaki, Physica C **357-360** (2001) 866

[63] B. Oswald, M. Krone, T. Straßer, K. -J. Best, M. Söll, W. Gawalek, H. -J. Gutt, L. Kovalev, L. Fisher, G. Fuchs, G. Krabbes, H. C. Freyhardt, Physica C **372-376** (2002) 1513

[64] H. Teshima, M. Sawamura, M. Morito, M. Tsuchimoto, Cryogenics **37** (1997) 505

[65] M. Komori, M. Murakami, IEEE Trans. Appl. Supercond. **11** (2001) 4065

[66] Jiasu Wang, Suyu Wang, Youwen Zeng, Haiyu Huang, Fang Luo, Zhipei Xu, Qixue Tang, Guobin Lin, Cuifang Zhang, Zhongyou Ren, Guomin Zhao, Degui Zhu, Shaohua Wang, He Jiang, Min Zhu, Changyan Deng, Pengfei Hu, Chaoyong Li, Fang Liu, Jisan Lian, Xiaorong Wang, Lianghui Wang, Xuming Shen, Xiaogang Dong, Physica C **378-381** (2002) 809

[67] P. M. Grant, IEEE Trans. Appl. Supercond. **7** (1997) 112

[68] M. M. Rahman, M. Nassi, IEEE Spectrum, July 1997, p.31

[69] M. Gouge, "2002 DOE Wire Workshop", January 22, 2002, St. Petersburg, FL, http://www.eren.doe.gov/superconductivity/docs/09dod_perspective_gouge.ppt

[70] D. Willén, F. Hansen, M. Däumling, C. N. Rasmussen, J. Østergaard, C. Træholt, E. Veje, O. Tønnesen, K.-H. Jensen, S. Krüger Olsen, C. Rasmussen, E. Hansen, O. Schuppach, T. Visler, S. Kvorning, J. Schuzster, J.Mortensen, J. Christiansen, S. D. Mikkelsen, Physica C **372-376** (2002) 1571

[71] M. Watanabe, T. Masuda, Y. Ashibe, M. Hirose, S. Isojima, S. Honjo, T. Uchiyama, M. Shimodate, Y. Takahashi, H. Suzuki, to be published in IEEE Trans. Appl. Supercond. **13** (2003);

T. Masuda, Y. Ashibe, M. Watanabe, C. Suzawa, K. Ohkura, M. Hirose, S. Isojima, S. Honjo, K. Matsuo, T. Mimura, Y. Takahashi, Physica C **372-376** (2002) 1580;

K. Ueda, Y. Ozawa, Y. Wachi, A. Kikuchi, T. Nakatsuka, K. Iwadate, Physica C **378-381** (2002) 1158

[72] S. Spreafico, M. Bechis, P. Caracino, G. Cavalleri, G. Coletta, P. Corsaro, P. Ladiè, M. Nassi, N. Kelley, Physica C **372-376** (2002) 1588;

P. Corsaro, M. Bechis, P. Caracino, W. Castiglioni, G. Cavalleri, G. Coletta, G. Colombo, P. Ladiè, A. Mansoldo, R. Mele, S. Montagnero, C. Moro, M. Nassi, S. Spreafico, N. Kelley, C. Wakefield, Physica C **378-381** (2002) 1168

[73] Superconductor Week, December 18, 2002

[74] E. Leung, IEEE Spectrum, July 1997, p.26

[75] W. Paul, M. Lakner, J. Rhyner, P. Unternährer, Th. Baumann, M. Chen, L. Widenhorn, A. Guérig, Supercond. Sci. Technol. **10** (1997) 914;
W. Paul, M. Chen, M. Lakner, J. Rhyner, D. Braun, W. Lanz, Physica C **354** (2001) 27

[76] M. Chen, W. Paul, M. Lakner, L. Donzel, M. Hoidis, P. Unternaehrer, R. Weder, M. Mendik, Physica C **372-376** (2002) 1657

[77] H. Salbert, R. Bussjaeger, D. Krischel, G. Heit, K.H. Weck, to be published in IEEE Trans. Appl. Supercond. **13** (2003)

[78] S. P. Mehta, N. Aversa, M. S. Walker, IEEE Spectrum, July 1997, p.43

[79] R. C. Diehl, S. Fleshler, R. Mason, C. Christopherson, K. DeMoranville, R. Harnois, A. Otto, G. N. Riley, Jr., E. Serres, J. O. Willis, H. J. Boenig, J. B. Schillig, IEEE Trans. Appl. Supercond. **11** (2001) 3265

[80] A. Usoskin, J. Dzick, A. Issaev, J. Knoke, F. García-Moreno, K. Sturm, H. C. Freyhardt, Supercond. Sci. Technol. **14** (2001) 676

[81] B. Gromoll, G. Ries, W. Schmidt, H.-P. Krämer, B. Seebacher, B. Utz, R. Nies, H.-W. Neumüller, E. Baltzer, S. Fischer, B. Heismann, IEEE Trans. Appl. Supercond. **9** (1999) 656

[82] H. Zueger, Cryogenics **38** (1998) 1169

[83] S. W. Schwenterly, S. P. Mehta, M. S. Walker, R. H. Jones, Physica C **382** (2002) 1

[84] M. Leghissa, B. Gromoll, J. Rieger, M. Oomen, H. -W. Neumüller, R. Schlosser, H. Schmidt, W. Knorr, M. Meinert, U. Henning, Physica C **372-376** (2002) 1688;
R. Schlosser, H. Schmidt, M. Leghissa, M. Meinert, to be published in IEEE Trans. Appl. Supercond. **13** (2003);
M. Meinert, M. Leghissa, R. Schlosser, H. Schmidt, to be published in IEEE Trans. Appl. Supercond. **13** (2003)

[85] G. Papst, B.B. Gamble, A.J. Rodenbush, R. Schöttler, Supercond. Sci. Technol. 10 (1997) 924

[86] R. D. Blaugher, IEEE Spectrum, July 1997, p.36

[87] A. Malozemoff, "2002 DOE Wire Workshop", January 22, 2002, St. Petersburg, FL, http://www.eren.doe.gov/superconductivity/docs/07malozemoff_industrial_perspective.ppt ;
D. Madura, M. Richardson, D. Bushko, G. Snitchler, P. Winn, S. Kalsi, B. Gamble, to be published in IEEE Trans. Appl. Supercond. **13** (2003)

[88] M. Superczynski, D. Waltman, IEEE Trans. Appl. Supercond. **7** (1997) 513

[89] M. Frank, W. Nick, H.-W. Neumueller, J. Frauenhofer, G. Nerowski, P. van Hasselt, to be published in IEEE Trans. Appl. Supercond. **13** (2003)

[90] L. K. Kovalev, K. V. Ilushin, V. T. Penkin, K. L. Kovalev, S. M.-A. Koneev, K. A. Modestov, S. A. Larionoff, W. Gawalek, B. Oswald, Physica C **354** (2001) 34

[91] J. Cloyd, "2002 DOE Wire Workshop", January 22, 2002, St. Petersburg, FL, http://www.eren.doe.gov/superconductivity/docs/10mil_perspective_cloyd.ppt

[92] G. Ries, H.-W. Neumüller, Physica C **357-360** (2001) 1306

[93] J. Colvin, T. Mann, S. Peck, P. Tissot, J. Zeigler, Superconductor Industry / Winter 1997 p.12

[94] K. Tasaki, Y. Sumiyoshi, M. Tezuka, H. Hayashi, K. Tsutsumi, K. Funaki, M. Iwakuma, Physica C **357-360** (2001) 1332;

H. Hayashi, K. Tsutsumi, K. Funaki, M. Iwakuma, K. Tasaki, Y. Sumiyoshi, M. Tezuka, Physica C **357-360** (2001) 1327;

A. Ishiyama, S. Akita, H. Kasahara, H. Sakaguchi, Physica C **357-360** (2001) 1311;

T. Nakano, T. Yamanaka, M. Minami, S. Akita, H. Kasahara, H. Sakaguchi, Physica C **357-360** (2001) 1315;

M. Minami, T. Nakano, S. Akita, H. Kasahara, H. Tada, M. Takahashi, Y. Nara, T. Yamanaka, H. Sakaguchi, Physica C **357-360** (2001) 1323

[95] J.R. Hull, Supercond. Sci. Technol. **13** (2000) R1

[96] N. Koshizuka, F. Ishikawa, H. Nasu, M. Murakami, K. Matsunaga, S. Saito, O. Saito, Y. Nakamura, H. Yamamoto, R. Takahata, T. Oka, H. Ikezawa, M. Tomita, Physica C **378-381** (2002) 11

[97] M. Minami, S. Nagaya, H. Kawashima, Fourth NES&C at Osaka University, 1999, p. 349;

S. Nagaya, N. Kashima, M. Minami, H. Kawashima, S. Unisuga, Y. Kakiuchi, H. Ishiga, Physica C **357-360** (2001) 866

[98] http://www.supraleiternet.de/datenbank/detail/id41.html#Leitprojekt

[99] Superconductor Week, July 28, 1997

[100] H. J. M. ter Brake, G. F. M. Wiegerinck, Cryogenics **42** (2002) 705

[101] Superconductor Week, October 14, 2002

[102] R. Hohmann, C. Lienerth, Y. Zhang, H. Bousack, G. Thummes, IEEE Trans. Appl. Supercond. **9** (1998)

[103] A. T. A. M. de Waele, P. P. Steijaert, J. Gijzen, Cryogenics **37** (1997) 313;

A. T. A. M. de Waele, P. P. Steijaert, J. J. Koning, Cryogenics **38** (1998) 329

[104] Y. L. Ju, C. Wang, Y. Zhou, Cryogenics **38** (1998) 169

[105] N. Tozai, Superconductor Industry / Winter 1996, p.50

[106] T. Sato, ISTEC Journal Vol. **9** Nr.3 (1996) 15

[107] S. Zhu, Y. Kakimi, Y. Matsubara, Cryogenics **37** (1997) 46

[108] C. Wang, G. Thummes, C. Heiden, Cryogenics **37** (1997) 159;

G. Thummes, C. Wang, C. Heiden, Cryogenics **38** (1998) 337

[109] E. Tward, C. K. Chan, C. Jaco, J. Godden, J. Chapsky, P.Clacy, Cryogenics **39** (1999) 717

[110] M. Gouge, "2002 DOE Wire Workshop", January 22, 2002, St. Petersburg, FL, http://www.eren.doe.gov/superconductivity/docs/11_cryogenics_gouge.ppt

[111] J. Gerster, L. Reißig, I. Rühlich, M. Thürk, P. Seidel, Inst. Phys. Conf. Ser. No. **158** (1997) 813

THE APPLICATION OF HIGH TEMPERATURE SUPERCONDUCTORS IN BRUSHLESS AC MACHINES

David DEW-HUGHES and Malcolm McCULLOCH
Department of Engineering Science, Oxford University
Parks Road, Oxford OX1 3PJ, UK

1. INTRODUCTION.

The practical application of high temperature superconductors in electrical machines has been impeded in part by the low critical current densities and sensitivity to magnetic field, which is typical of these materials. The situation is exacerbated by the difficulty of fabricating robust and flexible continuous conductor based on HTS; the manufacturing techniques required to overcome these difficulties result in a very expensive product. A final bar to their ready application is the strain-induced degradation, which limits the radius to which a conductor can be bent. The best commercial HTS tapes cannot be bent to radii less that about 100 mm. This inhibits their use in machine windings, except in very large machines, thus ruling out the construction of small experimental prototypes for developmental purposes. Current densities, which approach those typical of low temperature superconductors, i.e. in excess of 10^{10} Am^{-2}, are achieved only in epitaxial thin films of high temperature superconductors. Bulk materials have critical current densities that are several orders of magnitude below this value. A remarkable exception is the seeded, melt-textured, single domain samples of the rare-earth barium copper oxide (REBCO) 123 phase. These materials contain about 20% of a fine dispersion of the 211 phase, which contributes both to flux pinning, and to mechanical integrity. The best examples of these materials can carry critical current densities of 10^8 Am^{-2}, at 77K and in magnetic inductions of up to 1 Tesla. These materials can trap inductions of several Tesla at 77K, and may be regarded as rivals to permanent magnets. Many interesting and novel applications have been suggested for these materials. This article is concerned with their application in the rotors of brushless AC electrical machines.

All electric motors work on the same principle; that is to minimise the energy in the magnetic field. In singly excited machines, a magnetic field is produced by a single winding. Energy differences are generated by salient magnetic structures, which move relative to the winding. An example of such a machine is the synchronous reluctance motor. In doubly excited machines, a second magnetic field is attached to a structure that can move with respect to the first winding. Lowering the energy of the field is accomplished by aligning the two fields, resulting in mechanical work. Examples include the synchronous machine and the induction motor.

The materials which are commonly found in the construction of conventional electric machines can be classified either as good conductor, e.g. copper, hard magnetic material such as Alnico, SmCo, NeFeB, and soft magnetic material such as transformer steel. The former two are used as sources of magnetomotive force, while the latter is used to define the magnetic path and any saliencies. All materials have their limitations. The current density in copper is usually no greater than 10^7 Am^{-2}. The highest flux density generated by hard magnetic materials is about 1.2 Tesla. Soft magnetic materials saturate above about 1.5 Tesla. The power output of any machine is proportional to the intensity of the magnetic field in the air gap, and the armature loading, i.e. the actual current density in the armature or the current density equivalent to the magnetic induction on the armature. Superconductors, with their ability to both carry current densities several orders of magnitude greater than those in copper and to trap magnetic flux densities in excess of permanent magnets, are clearly an attractive choice for electrical machines [1].

The conversion of electromagnetic energy into mechanical energy requires a geometry in which the magnetic field, the electric current, and the direction of motion are all mutually perpendicular. Electrical machines have four broad classifications based on the geometry of these three quantities. In the tubular geometry the movement is rotational about a fixed axis, the magnetic field is radial and the current is axial. This is the most common geometry, and the majority of machines considered in this paper are constructed in this geometry. Unless specifically stated to the contrary, it may be assumed that tubular geometry prevails. Pancake geometry obtains when the magnetic field is axial, and the currents are radial, the motion as before being about a fixed axis. In a linear cylindrical machine, the direction of motion is along the axis, the field is radial and the currents are circumferential. The flat linear motor is essentially the cylindrical motor, with the cylinder cut parallel to the axis and laid out flat. The direction of motion is linear above the plane of the flat coils, the currents are in the same plane but perpendicular to the direction of motion, and the field is now normal to the plane of the coils. Examples of both pancake and flat linear geometries have been implemented with superconducting armatures, and these will receive attention in this paper.

Brushless machines are particularly suitable for exploiting superconductors, the absence of brushes being a distinct advantage in a cryogenic environment. Brushless AC machines consist of a system of stator coils, which produce a transverse magnetic induction **B** rotating with angular velocity ω_f, and a rotor with a magnetisation, either induced or pre-existing, of magnitude **M**. The rotor, generally cylindrical in outline, turns on an axis, parallel to the centreline of the driving coils and normal to the direction of the driving field, under the action of a torque $\mathbf{T} = \mathbf{B} \times \mathbf{M}$, with an angular velocity ω_r. The power of the machine is given by $\mathbf{P} = \omega_r \mathbf{T}$.

Non-superconducting *brushless* ac machines fall into four basic types divided between two categories. Induction and hysteresis motors are asynchronous machines, operating at rotational speeds less that that of the driving field, i.e. $\omega_r < \omega_f$. Permanent magnet and reluctance motors are synchronous machines which operate at the rotational speed of the driving field, i.e. $\omega_r = \omega_f$. There are superconducting equivalents of all of these machines; some superconducting machines show operational characteristics of more that one type of motor. Each type of machine will be examined in turn. The more complicated variants such as switched reluctance and written pole motors will not be considered.

The simple induction motor has a rotor of "squirrel cage" configuration, i.e. the rotor consists of two rings joined by bars, constructed from a conducting material such as copper or aluminium. Relative motion between the driving field and the rotor induces circulating currents in the rotor. The magnetic moment resulting from these currents interacts with the driving field to produce a torque. Slippage between the driving field and the rotor is essential to maintain the induced currents in the rotor. In the superconducting version of an induction machine the normal conductor is replaced by a superconductor.

In an hysteresis motor the rotor is made from a ferromagnetic material, and it is driven around its magnetisation loop by the driving field. Torque is produced by an interaction between the driving field and the magnetisation of the rotor. As with the induction motor, slippage between the driving field and the rotor is an essential part of the operation of an hysteresis motor. A Type II superconductor, like a ferromagnetic material, exhibits magnetic hysteresis in a changing magnetic field. Thus a solid superconducting rotor can act in exactly the same way as a ferromagnetic rotor. Relative motion of driving field and rotor will cause the superconductor to traverse its magnetisation curve. Again the interaction between magnetisation of the rotor and the driving field will produce a torque. In the case of a ferromagnetic rotor, the field drags the rotor after it. Because the superconductor tries to exclude magnetic flux, and has a negative susceptibility, the superconducting rotor is pushed ahead by the stator field.

If a superconducting version of either an induction or an hysteresis machine is made with a rotor of high current density and strong flux pinning, the induced currents or magnetic flux may become fixed in the rotor. On starting, the machine operates as either an hysteresis or an induction motor. Due to the equivalence of current and induction in a superconductor, it is not possible to differentiate between the two modes of operation. As the rotor speeds up and approaches synchronous speed the mode of operation changes as the flux becomes pinned and remains stationary with respect to the rotor. The machine then operates as a trapped-flux motor, and is in some sense the equivalent of a permanent magnet motor. The flux, which is now trapped in the rotor, locks on to the driving field; the rotor rotating in synchronism with the driving field. Provided that the external load on the machine is below a critical value, such that pinning forces in the superconductor are not exceeded, the machine will operate in synchronous mode. If the load is such as to cause depinning of the flux, then the mode of operation will revert to asynchronous.

Reluctance motors have a non-cylindrical ferromagnetically very soft rotor of high magnetic permeability. The rotor offers paths for the driving field flux, which are magnetically easy or magnetically difficult, depending on the attitude of the rotor to the field. In the superconducting versions of these machines the superconductor does not replace the ferromagnetic material, but reinforces it. The difference in permeability of the *direct* (easy) and *quadrature* (difficult) paths is enhanced by the use of composite ferromagnetic/superconducting rotors. In this type of machine it is the flux shielding properties of the superconductor that are exploited. The rotor tends to align itself to allow the stator field to traverse the *direct* path. Reluctance machines do not require relative motion between field and rotor for their operation and the rotor moves in synchronism with the stator field.

As has been mentioned above, synchronous behaviour can arise in an hysteresis machine if the stator field results in trapped flux in the rotor. As the trapped flux originates from the rotor field, clearly the trapped flux density cannot be greater than the air-gap induction. The magnitude of the airgap induction is usually a few tenths of a Tesla, and rarely exceeds one Tesla. Advantage can be taken of the ability of superconductors to trap large amounts of flux by pre-trapping flux in the rotor. This can be achieved by pulsing a current, which is much greater than the normal operating current, for a short period of time through the stator coils, or by having a separate exciting coil, again which can be pulsed with a large current for a brief period. Alternatively the rotor can be cooled from above T_c to the operating temperature in a steady magnetic field provided by a separate solenoid. The trapped-flux machine has the potential for a much higher power output than the other machines outlined above, but pulsing will require an auxiliary pulsed power source, adding to the complexity of the device.

The particular advantage of machines with superconducting rotors would appear to be a five to ten times increase in power density in the rotor. Such machines are therefore likely to find applications in which the power to size or mass ratio is of prime importance. These applications are to be found in aircraft and space vehicles. Superconducting machines may also find application in already existing cryogenic environments, such as the gas separation industry. Here tonnage quantities of cryogenic fluids are daily pumped around commercial plants. Pumps driven by superconducting motors, integrated in the construction of the pump, have been proposed for this service

2. HYSTERESIS MACHINES.

Several groups have developed superconducting hysteresis machines, in which the rotating field is provided by conventional copper three-phase stator coils, the rotor being a cylinder of superconductor. The pioneering work was carried out by Kovalev and his colleagues in the Russian Centre of Cryogenic Superconducting Electrical Machines of the Moscow Aviation Institute (MAI) [2]. This work was continued in a profitable collaboration with the Institut für Physikalische Hochtechnologie, Jena (IPHT) [3].

Early versions of the machines used rotors consisting of polycrystalline YBCO cylinders mounted on steel axles. Later versions had stacks of top-seeded melt-textured single domain YBCO pucks, and rotors built up from segments machined from such pucks. A steady progression in output power was achieved, from 5 Watts to 4 kW. Some examples of early machines are shown in figure 1. The machines performed as expected, and demonstrated the technical feasibility of superconducting motors. However, because the stator winding produces the field and the superconductor is a high reluctance path, a large current is needed to provide the magnetic field. This gives rise to the main disadvantage of the hysteresis type machine, its low power factor.

Figure 1. Examples of early hysteresis machines constructed by MAI. Courtesy L K Kovalev.

The power output and torque developed by a hysteresis machine can be estimated in a very simple way. The power into the machine, P_{in}, which is the time-averaged product of stator current and voltage, is equal to the power output, P_{out}, minus the hysteresis loss per second in the rotor, neglecting any other losses. The power out, P_{out}, is the torque, T, times the angular velocity of the rotor, ω_r. The hysteresis loss per second is the loss per cycle of the rotor, ΔW_c, multiplied by the number of cycles per second:

$$Loss = \frac{1}{2}\Delta W_C \frac{\omega_f - \omega_r}{2\pi} n$$

where ω_f and ω_r are the angular frequencies of field and rotor respectively. The difference in the two frequencies is the frequency with which the rotor experiences the changing field. n is the number of poles.

$$P_{out} = P_{in} - \frac{1}{4\pi}\Delta W_C(\omega_f - \omega_r) = T\omega_r$$

and

$$T = \frac{P_{in}}{\omega_f(1-s)} - \frac{1}{4\pi}\Delta W_C\left(\frac{s}{1-s}\right)$$

where s, the slippage, is equal to $(\omega_f - \omega_r)/\omega_f$.

Clearly the power output and torque will be maximised when the slippage is zero, i.e. when the machine is operating at synchronous speed. In the superconductor, the conversion of electromagnetic energy to mechanical energy is achieved via the electromagnetic, or Lorentz, force on the flux lines being resisted by the mechanical pinning force. The maximum torque that can be achieved occurs when the pinning forces are just about to be overcome, and the superconductor is in the critical state. In this case the torque is susceptible to calculation and is the product of local magnetisation and local magnetic induction, times the distance from the axis, integrated over the volume of the rotor:

$$T = \int_V r \times (J \times B) dV$$

The transverse field induces currents in the cylindrical rotor, which travel up on side of the rotor, across the top and down the other side, finally completing the circuit across the bottom of the rotor. The distribution of these currents, and their associated inductions, depends upon the strength of the flux pinning. It should be noted that the pinning should not, as might be at first expected, be as strong as possible. If the flux pinning is too strong, no flux will enter the rotor, and no torque will be generated. Conversely, if the flux pinning is too weak, the flux will just flow in the rotor, and again no torque will be generated.

The distribution of current and magnetic flux in the rotor, and the operation of these machines, has been explored by finite element modelling [4-6]. The rotor is modelled as a cylinder of infinite length, so as not to have to deal with the end distribution of current. This is not too bad an assumption for rotors fabricated from single-domain REBCO. The c-axis of the superconductor is aligned with the axis of the rotor; the end faces of the rotor comprising a-b planes. The critical current density in the a-b planes is some three times greater, and hence the flux penetration much less, than in the c-axis. The currents flowing across the end faces do not contribute to the torque. It is the current flowing parallel to the c-axis that is modelled and that provides the energising force. The superconductor is presumed to follow the simple Bean model of the critical state; with the current density everywhere is either equal to the critical current J_c, or zero.

The result of this modelling is that the torque is given by [5]:

$$T = \frac{4}{3} l R^3 J_C B_0 I(k)$$

where l is the active length of the rotor, R its radius, and I(k) is a dimensionless factor, depending on the parameter k which describes the degree of flux penetration into the superconductor.

$$k = \frac{B_0}{\mu_0 J_C R}$$

The variation of I(k) with k is shown in figure 2. Complete penetration occurs when k = 0.44. Maximum torque is generated when k = 0.56, see figure 3.

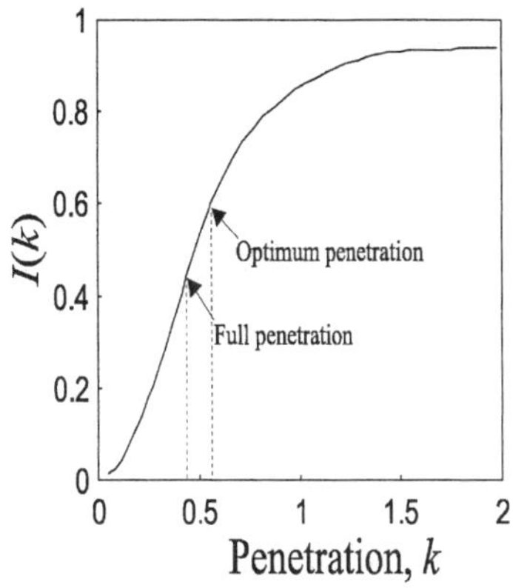

Figure 2. I(k) versus k.

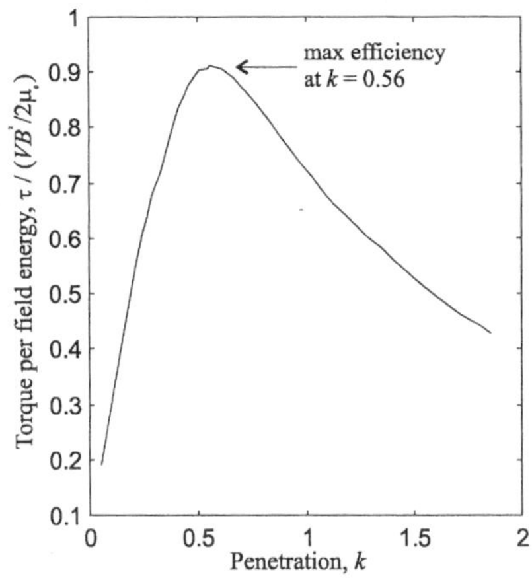

Figure 3. Torque versus k.

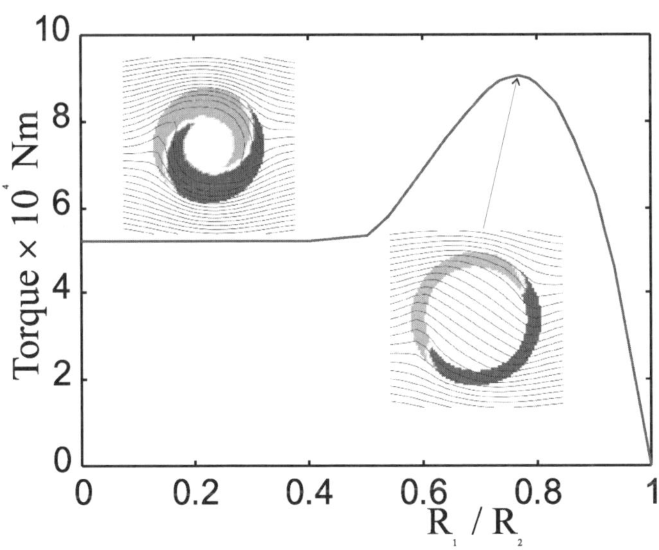

Figure 4. Torque versus ratio of inner to outer radii in a hollow cylindrical rotor. The insets show the distribution of current in solid and optimal hollow rotors. The dark and light areas represent currents flowing into and out of the pictures.

The pattern of current and flux distribution in a solid superconducting rotor, as the rotor is accelerated from the stationary position as the stator field is switched on, is shown in the left-hand inset in figure 4. The flux is seen to penetrate initially along a spiral path toward the axis of the rotor. Current flowing in the inner regions of the rotor produces torque, which acts in the opposite direction to the driving torque. The central region of the cylindrical rotor is therefore a liability; a hollow rotor gives a better performance. Modelling predicts that optimum performance is obtained with a hollow rotor when the inner radius is 0.75 of the outer radius, fig 4 [5].

Single domain pucks of top-seeded melt-textured REBCO rarely exceed about 50mm in diameter; the growth of larger artefacts is difficult. Larger rotors can be constructed from several segments joined by welding or brazing, but such joints tend to carry a lesser critical current density than the matrix material. However, the modelling has further shown that superconducting joints are not needed in a segmental rotor. The joints allow a greater penetration of the field, and hence a better performance [7]. Optimum current and field distributions in 3-segment and 6-segment rotors are shown in figure 5.

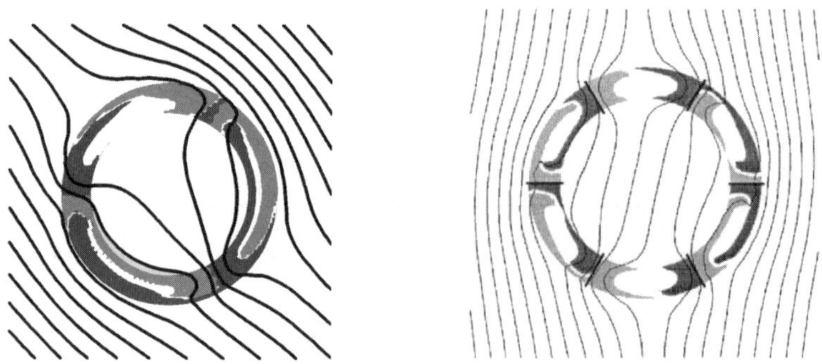

Figure 5. Current and field distribution for 3 and 6 segmental rotors.

If the superconductor has a high penetration, i.e. a low J_c, then a single hollow rotor is optimal. However the higher the J_c in the superconductor, the more advantageous it becomes to split the rotor in to a greater number of segments. In this way, large machines with high power output can be constructed with rotors built up from a large number of segments of high quality, high J_c superconductor.

Hysteresis machines have also been built in the pancake configuration. The rotor, a flat disc, containing either discrete pieces of superconductor or an annulus built-up from segments of superconductor, is suspended between pancake-shaped coils [8,9]. If the machine axis is vertical, then the levitating effect of the coils on the superconductor can be used to counteract end-thrust. This means that end-bearings need only support the rotor while stationary. Another advantage of the pancake configuration is that it is relatively easy to increase the power of the machine by stacking several rotor discs on the same axle. However the pancake motor is overall more bulky than a tube-geometry machine of the same power, see the middle machine in figure 1, and the coils are more complex.

The hysteresis principle can also be used for a flat linear motor. As with the pancake machine, levitation occurs in, and assists, the operation of such machines. A linear superconducting actuator has been designed and constructed by Tsuda et al. [10]. Levitation is achieved by permanent magnets in the track and YBCO bulks in the carriage. Longitudinal motion is derived from three-phase copper windings in the track, and an HTS secondary in the carriage. Work done in collaboration between the Department of Engineering Science, Oxford, and the University of the Witwatersrand, has also developed a proof of concept prototype magnetic levitation system [11]. This linear motor consists of a long stator, made from iron and copper windings and a short rotor, made from six YBCO pucks. The stator windings have both edges tilted at an angle of 45° to improve lateral stability of the rotor. This stability is further enhanced by the flux trapping nature of the superconductor. The prototype was developed on a three meter track, and was driven by a variable frequency inverter. It showed exceptional stability, with a clearance of 7mm. The system is shown in operation in figure 6.

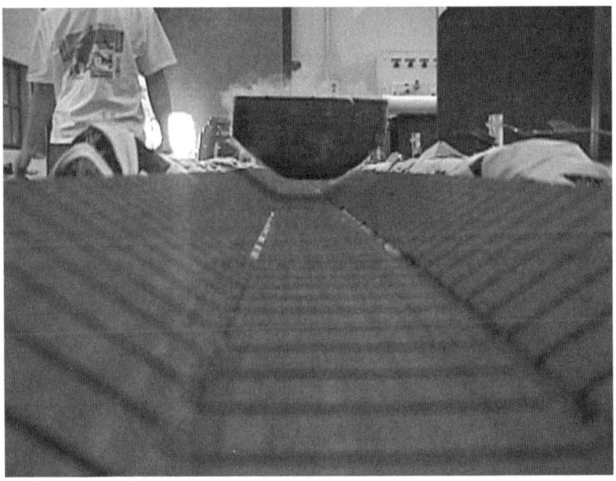

Figure 6. The linear motor in operation, clearly demonstrating levitation.

A finite element program, developed at the University of Oxford, was used to optimise the configuration of the pucks relative to the pole pitch at slot pitch of the stator winding [6]. Firstly the total detent forces were minimised and smoothed. Secondly the levitation forces on the rotor where maximised. Lastly the thrust was maximised. The results of this modelling are shown in figure 7.

All of the machines described in this and subsequent sections, unless otherwise noted, have been designed to operate at standard mains frequency (50-60 Hz) and by cooling with liquid nitrogen. Large commercial machines are expected to be combined with dedicated cryocoolers.

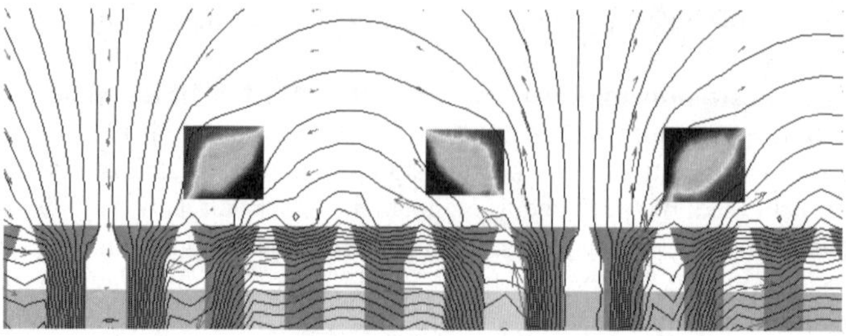

Figure 7. Current and field distribution in the stator windings and the HTS bulks, in the linear levitated motor.

3. RELUCTANCE MACHINES.

In a two-pole reluctance machine, low magnetic reluctance along the direct axis and high magnetic reluctance along the quadrature axis are achieved by a soft ferromagnetic, usually iron, rotor having two salient poles. Torque = $I^2(L_d - L_q)$, where I is the stator current and L_d and L_q are the permeabilities (inverse of the reluctance) in the direct and quadrature directions respectively. The difference in permeabilities is increased if two blocks of superconductor, which are not electrically connected, are attached to either side of the rotor. Induced currents in the superconductor prevent the flux from traversing the quadrature axis and encourage flux to traverse the direct axis. Two configurations of composite iron/HTS rotors for superconducting reluctance machines are shown in figure 8.

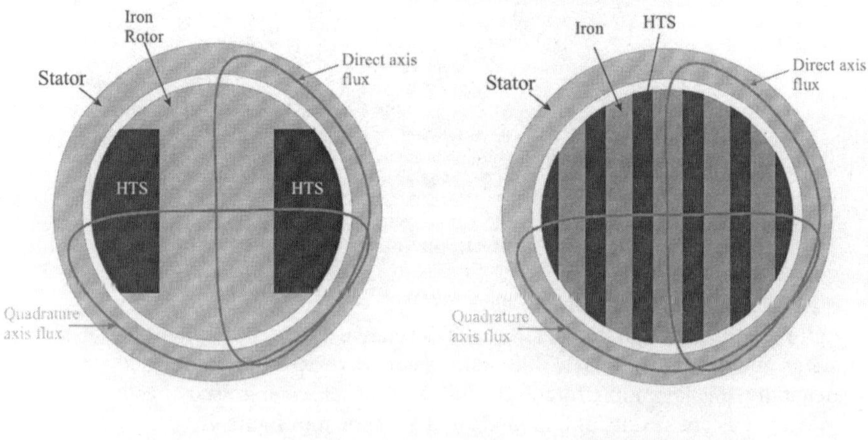

(a) 'Mushroom' design.

(b) 'Zebra' design.

Figure 8. Mushroom and Zebra schemes for HTS reluctance machines.

Modelling of this configuration shows that the superconductor improves machine performance both by the enhancement of the difference in permeabilities and by the superconductor providing some hysteretic effort. The degree of enhancement depends upon a parameter, S, the shielding parameter [12]. This is just the reciprocal of the penetration, k, given in the previous section:

$$S = \frac{\mu_0 J_C R}{B}$$

Relative output torque rises with increasing S to a maximum improvement of 55%. The dependence of torque upon S shows a sharp knee when S ~10; at this value the improvement is about 50% and there is little profit in requiring any higher value. As was the case for the hysteresis machine, there is no advantage in the superconductor having very high values of critical current. Again the pioneer in the development of this type of machine is the group of Kovalev at MAI [13,14].

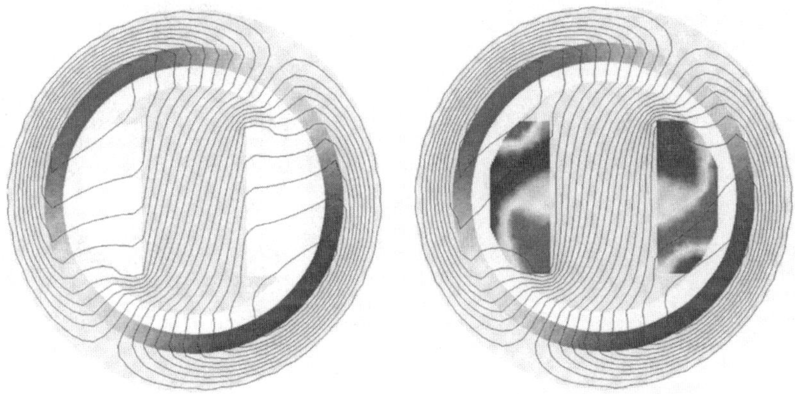

(a) shielding, $s = 0$ (b) shielding, $s = 10$

Figure 9. Field and current distribution in a mushroom type rotor without and with blocks of superconductor.

The modelled distribution of field and current in a mushroom type rotor is shown in figure 9 [12]. The hysteretic contribution due to currents in the HTS blocks is obvious from the right hand figure. A disadvantage of the conventional reluctance machine is the low starting torque, and to overcome this is usually built with a higher power output than is needed for steady operation. The low starting torque arises because full torque is generated when the stator field is exactly aligned with the direct axis of the rotor, and this occurs continuously only when the rotor is in synchronism with the stator field. The hysteretic contribution from the presence of superconductor in the rotor gives valuable assistance in starting the superconducting reluctance machine. This is perhaps more important than the enhancement of the reluctance effect, in reducing the size of machine for a given power output.

The MAI group is also responsible for an improved type of reluctance rotor, the Zebra rotor [13-15]. In this, the rotor is made up from alternate lamellae of soft iron and superconductor, figure 8b. The flux is directed, by the superconductor acting as a shield, along the direct axis. The modelled distribution of current and field for this type of rotor, without and with the superconducting plates, is shown in figure 10. As for the mushroom configuration, output torque rises with increased shielding, S. The maximum enhancement, however, is slightly greater than that for the mushroom rotor, at 60%, for the particular number and thickness of superconducting plates shown, and is reached when S exceeds a value of 30. This value can be improved by correct choice of plate number and thickness.

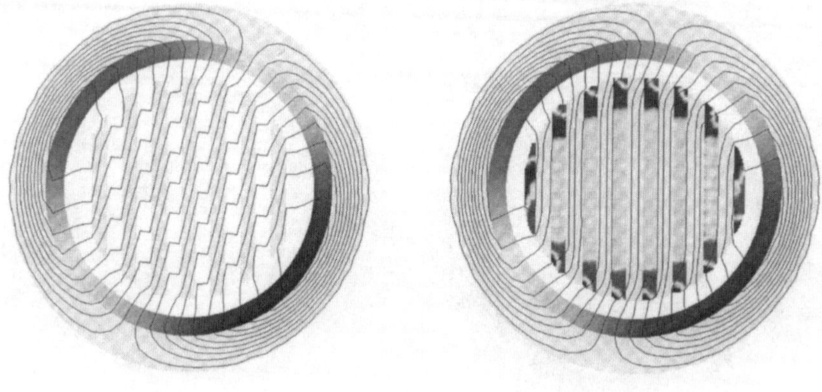

(a) shielding, $s = 0$ (b) shielding, $s = 20$

Figure 10. Current and field distribution for the zebra rotor. The load angle is at 45° to the horizontal, corresponding to a position of maximum torque.

It is the flux-shielding ability of the superconductor that is brought into play in this type of machine, so the minimum thickness of superconductor must be greater than the penetration depth. For this configuration of rotor, strong flux pinning is desirable. In a collaboration with several German institutions, including the motor manufacturer Oswald Elektromotoren GmbH, a series of zebra machines have been developed with output powers up to 200kW [16-19]. The superconductor in these machines is YBCO in plate form, cut from single-domain top-seeded melt-grown material. Techniques for joining plates to produce components of the required size have been developed, and include welding and brazing. It is essential, in order to preserve the integrity of the shielding, that the critical current density carried by these joints is not much degraded from the bulk value. Figure 11 shows a 25kW zebra motor under test at MAI.

Figure 11. 25kW reluctance machine on horizontal test-bed at Moscow Aviation Institute.

Figure 12. Combined 4-pole motor and pump operating in bath of liquid nitrogen. The stream of nitrogen can be seen issuing from the pump.

The MAI group has further developed the reluctance machine, by utilising BiSCCO/Ag composite foils. These foils consist of 5-20 layers of Bi-2223 HTS phase, sandwiched between, and clad with, silver sheets. The plates are produced by the All-Russian Scientific Research Institute of Inorganic Materials (VNIINM), by a technique based on that for the production of Bi-2223/Ag multi-filament tape conductor. This material was developed specifically for use in superconducting reluctance machines. Zebra-type machines with either a large number of thin composite foils, or a lesser number of thicker foils, were constructed and tested. The thicker foil variants were found to give the better performance, with power output of 4kW and an efficiency of 0.9. The hysteretic contribution would be larger in the thicker superconducting foils, and would account for this result. The silver will also contribute to the output via an inductive effect. This composite material offers a way forward for the construction of large machines, while avoiding the problems associated with the joining of blocks of HTS.

A 4-pole reluctance machine has also been constructed, using the composite BiSCCO/Ag plates [20]. This motor is integrated with a liquid hydrogen pump, and is designed for airborne operation, the high power to mass ratio making these machines particularly suitable for this duty. The standard aircraft power frequency is 400 Hz; the use of four poles halves the rotational speed, a consideration necessitated by the prevention of cavitation in the pumped fluid. This combined motor and pump is shown in operation, immersed in a bath of liquid nitrogen, in figure 12.

Oswald Motoren have presented a conceptual design for a cylindrical linear drive based on the reluctance principle [19]. This design does not yet seem to have been translated into an actual device.

4. PRE-TRAPPED FLUX MACHINES.

If, in the mushroom-type reluctance machine described in the previous section, the two superconducting blocks are electrically connected by more superconductor, top and bottom, a complete superconducting circuit will surround the iron core of the rotor. Currents introduced into this circuit will support a magnetic flux, normal to the plane of the circuit, in the iron, as shown in figure 13. Persistent currents will result in a permanent magnetic moment on the rotor. The motor will then act somewhat similar to a permanent magnet machine. Torques some 10 times greater than that of the equivalent reluctance machine can be generated. The machine will have hysteretic and reluctance components in its performance, in addition to that from the trapped flux, and its performance is therefore somewhat different from that of a conventional permanent magnet machine. Another difference is the tendency, due to flux creep, for the trapped flux to decrease. Thus some means for restoring the trapped flux during machine operation must be included in the design.

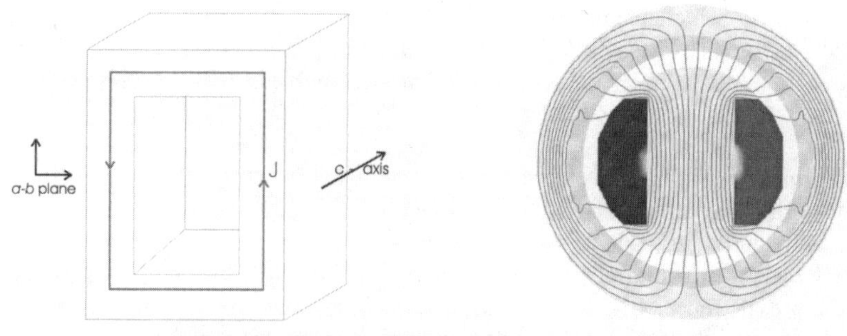

Figure 13. (Left) Window-frame configured from a single-domain block of YBCO, with currents flowing in the a-b plane, suitable for trapping flux parallel to the c-axis. (Right) Modelled distribution of maximum trapped field, S = 13.5.

As mentioned above, the persistent currents may be induced either by cooling the rotor in a large magnetic field, or by subjecting the rotor to pulsed fields once it is cooled. The first method requires a separate, DC coil to provide the magnetising field. In the second method, provided the pulses are of sufficiently short duration as to cause no damage, the pulses can be applied by one of the stator coils. In this case an external pulsed power source is also required. It is also important that the circuit is fabricated with joints that do not degrade the critical current. It would appear that there has been only one attempt to date to actually construct a trapped-flux machine. Again the pioneering work was carried out by the MAI group [20]. The superconducting circuit was made by cutting out rectangular "window frame" pieces from the BiSCCO/Ag composite sheets. Several of these were stacked together, and cooled to below their critical temperature in a DC magnetic field. The machine was run as a generator; in open circuit mode it produced an emf of ~230V. It was estimated that the level of trapped induction was 0.8 Tesla. It should be noted that, if flux is trapped in any of the previously described machines, they can be run as generators [21].

5. APPLICATIONS.

The use of superconductors in electrical machines is advantageous because the very high current densities lead to high power densities, a reduction in size for the same power or an increase in power for the same size. Physical size and weight are of prime importance in both aerospace and land transportation. Reduction in weight leads to an increase in payload, with implications for the economy of operations. A disadvantage is the requirement for operation at cryogenic temperatures. This disadvantage could be minimised in applications in environments that are inherently cold. One such environment is found in the gas separation industry, where an obvious application is in driving pumps for cryogenic fluids, such as liquid hydrogen, liquid oxygen and liquid nitrogen. Two configurations of simple liquid nitrogen pumps driven by superconducting rotors have been demonstrated at Oxford [22].

One particular application, combining both the needs for a high power to weight ratio and the existence of a cryogenic environment, would be that of driving pumps for liquid oxygen in kerosene/oxygen fuelled rockets. This is believed to have provided the original impetus for the programme at the Moscow Aviation Institute. A further application is that of pumping liquid hydrogen. Atmospheric pollution is currently acting as a faster driver than the depletion of resources for the implementation of alternative energy sources. Hydrogen is emerging as the favoured fuel, as its sole product of combustion is pure water. The preferred means of hydrogen storage is as a liquid. This will require the development of facilities for the production and handling of liquid hydrogen. Pumps driven by superconducting motors are likely to be an essential part of this infrastructure. Hydrogen powered road vehicles will require the provision of metered pumps on fuel station forecourts, an enormous potential market. In rail transport, the size of driving motor is limited by the space between the wheels; more power requires more powered axles. The greater power/size ratio of a superconducting motor can result in fewer powered axles. Provided that the weight and cost of on-board cryogenics is less than those of the deleted motors, there is a net advantage to be gained in the use of superconducting motors. The Private Stock Company "TUPOLEV", with the support and encouragement of the German and Russian Governments, has for some years explored the use of liquid hydrogen as an aircraft fuel [23]. Given the presence of liquid hydrogen, the use of superconducting motors to pump the fuel, superconducting actuators for control surfaces and landing gear, and superconducting alternators for on-board power, are distinct possibilities. These ideas are undergoing development in a consortium of the Moscow Aviation Institute (motor design and testing), VNIINM (materials development), PSC Tupolev (aircraft requirements) and the Russian Research

Institute of Electromechanics (manufacture and cryogenics), supported by NATO under its Science for Peace programme.

Linear machines are ideal for levitated transport installations, combining, as they do, both levitation and translation. These properties have also lead to their proposed application for materials handling, translation and positioning in industrial processes.

6. CONCLUSIONS.

The construction of hysteresis, reluctance and trapped-flux brushless AC machines with superconducting elements in the rotor has been shown to be technically feasible. These machines offer significant improvements in power to mass ratio over conventional machines. They can be implemented in a variety of configurations, and open up an exciting new area of advanced technology. The performance requirements of the rotor materials in these applications can be met by existing high temperature superconducting materials. The next stage in their development, as more advanced and reliable wire and tape conductors based on high temperature superconductors become available, will be to fabricate the field windings from high temperature superconducting wire or tape.

The development of techniques for modelling field and current has provided a powerful tool to assist in the design of machines incorporating superconductors. It is also possible, with modelling, to optimise the design for maximum performance.

7. ACKNOWLEDGEMENTS.

The authors' knowledge of the work at MAI was facilitated, firstly by a Royal Society Joint Projects grant, and subsequently by the NATO Science for Peace project SfP974241, coordinated by one of us (DDH). Some of the work described above was carried out within the framework of the above project.

8. REFERENCES.

[1] McCulloch,M.D., Barnes,G.J. and Dew-Hughes,D; *ICEM-2000 Proc. Int. Conf. on* Electrical *Machines* Helsinki Univ. Technol, Espoo, Finland (2000) Vol 2 p 812.
[2] Kovalev,L.K., Ilyushin,K.V., Penkin,V.T., and Kovalev,K.L.; *Electrical Technology* No.2, 145, (1994).
[3]. Kovalev,L.K. *et al.; Superconductivity, research and development* **9-10**, 69 (1998); Kovalev,L.K. *et al., Materials Science and Engineering* **B53**, 216 (1998); Kovalev,L.K. *et al.; IEEE Trans Appl. Superconductivty* **9** 1261 (1999).
[4]. Barnes,G.J., McCulloch,M.D. and Dew-Hughes,D.; *Superconductor Science and Technology* **12** 518 (1999).
[5]. Barnes,G.J., Dew-Hughes,D. and McCulloch,M.D.; *Superconductor Science and Technology* **13** 229 (2000).
[6]. Barnes,G.J., McCulloch,M.D. and Dew-Hughes,D.; *Superconductor Science and Technology* **13** 875 (2000).
[7]. Barnes,G.J., McCulloch,M.D. and Dew-Hughes,D.; *Physica C* **331** 133 (2000).

[8]. Márquez,I., *et al.; IEEE Trans Appl. Superconductivty* **9** 1249 (1999).
[9]. Muta,I., Jung,H., Nakamura,T. and hoshino,T.; *Physica C* **372-376** 1531 (2002).
[10]. Tsuda,M., Koike,T, Muramatsu,R and Haseyama,S; *Advances in Superconductivity XII,* T. Yamashita and K. Tanabe (eds) Springer Verlag, Tokyo (2000) pp 818 - 820.
[11]. Cruise,R.J., Vandenbroucke,K., Landy,C.F., Barnes,G.J. and McCulloch,M.D.; *Physica C* **341-348** 2627 (2000).
[12]. Barnes,G.J., McCulloch,M.D. and Dew-Hughes,D.; *Inst Physics Conf. Series No. 167,*IOP Publishing, Bristol (2000) p 1075.
[13]. Kovalev,L.K., *et al., ICEC17,* Dew-Hughes,D., Scurlock,R.G. and Watson,J.H.P. (Eds) IOP Publishing, Bristol (1998) p 527.
[14]. Kovalev,L.K. *et al., IEEE Trans. Applied Superconductivity* **9** 1261 (1999).
[15]. Kovalev,L.K., *et al., Advances in Cryogenic Engineering* **45B** 1659 (2000).
[16]. Oswald,B. *et al., ICEC17,* Dew-Hughes,D., Scurlock,R.G. and Watson,J.H.P. (Eds) IOP Publishing, Bristol (1998) p 547.
[17]. Oswald,B. *et al., IEEE Trans. Applied Superconductivity* **9** 1201 (1999).
[18]. Oswald, B., *et al., Advances in Cryogenic Engineering* **45B** 1653 (2000).
[19] Oswald,B. et al.; *Physica C* **372-376** 1513 (2002).
[20]. Kovalev,L.K., *et al., Physica* **C372-376** 1524 (2002).
[21]. Kovalev,L.K., *et al., ICEC17,* Dew-Hughes,D., Scurlock,R.G. and Watson,J.H.P. (Eds) IOP Publishing, Bristol (1998) p 379.
[22]. Dew-Hughes,D., *et al., Advances in Cryogenic Engineering* **45B** 1477 (2000).
[23]. Pohl,H.W. and Malychev,V.M., *Hydrogen Energy Progress 10,* Block,D.L. and Veziroglu,T.N. (Eds.) International Association for Hydrogen Energy, (1994) p 1969. Details of the cryogenic aircraft can be found in Janes' All the World's Aircraft, 2001-02, pages 459 and 462.

CURRENT STATUS OF HIGH-T_c SUPERCONDUCTING BULK ROTATING ELECTRIC MOTORS APPLICATION

Itsuya MUTA and Taketsune NAKAMURA
Department of Electrical Engineering, Graduate School of Engineering, Kyoto University
Yoshida-Honmachi, Sakyo-ku, Kyoto 606-8501, JAPAN

1. INTRODUCTION

One of the most tremendous differences between oxide high T_c superconductor (HTS) and practical low T_c superconductor (LTS) from a material aspect is the availability of HTS "bulk" shape due to its higher temperature operation compared to LTS. Since the critical temperature of most of the HTS is higher than the boiling point of atmospheric liquid nitrogen (77.3 K), the possible operating temperatures of the HTS power systems are much higher than those of the LTS counterparts, e.g., boiling point of liquid helium (4.2 K). Then, the thermal stability of such HTS systems are remarkably large, and then the quench phenomenon, which is the sudden transition from superconducting to normal conducting state due to the thermal instability, is not serious problems. Therefore, these bulk materials stimulate their power applications with the aim of high efficiency, compact size and low power dissipation.

HTS bulk rotating electric motor is surely one of such exciting candidates, and various types of such motors have extensively and intensively been investigated in order to increase the power density as well as compact systems compared to the conventional motors [1-16]. Hereafter, rotating electric motor is just stated as motor. Up to now, HTS bulk has only been considered to solid rotor, and the stator has been the conventional one that consists of copper windings with laminated iron core. Recently, small and middle-scale motors have been developed with the co-operation between Russia and Germany [17-24]. The HTS bulk motors have also been considered for the special purposes at cryogenic temperature environment such as circulation of liquid fuel [25], automobile [26] and aerospace industry [9].

This chapter reviews the recent trends of the HTS rotating bulk motor with various kinds of torque generation mechanisms. Up to now, large capacity motors, e.g., several MW motor, has not yet been available mainly because of the technical difficulty for the fabrication of large grain size HTS bulk rotor. Then fractional force, small capacity and middle capacity motors are subjected to this chapter. Although the HTS motors have also been studied with the use of HTS tape, it is not covered in this chapter. Other types of machines such as linear-type bulk motor and flywheel system are also not included.

2. FUNDAMENTAL ASPECTS

2.1 Mechanisms of Torque Generation

There are mainly three types of HTS bulk motors: i.e., hysteresis motor, reluctance motor and trapped-flux motor. The performances of these motors are strongly dependent on their solid rotor's flux pinning property. As is well-known, different kinds of HTS bulk materials have different pinning forces, i.e., typical pinning force of rare earth system ((RE)BaCuO-123: e.g., Y-123, Sm-123, Gd-123) is stronger than that of BiSrCaCuO system (e.g., Bi-2223, Bi-2212). These properties are extremely varied as a function of operational conditions such as temperature, magnetic field and field direction. Therefore, dependency of temperature as well as magneto motive force (ampere turns) upon the motor characteristics is prominent due to the flux trapping and/or shielding capability of the HTS bulk rotor. The differences of HTS bulk's property are also induced depending on the fabrication methods: i.e., melt-textured or sintered. For instance, the Y-123 bulk materials fabricated with melt-textured method have strong flux pinning property compared to those with sintered method. Melt-textured Bi-2223 bulk with large size has not been available yet.

Historically, small-scale hysteresis motor has firstly been considered for applicability study [27]. The basic mechanism of the torque generation is phase difference between the rotating magnetic flux generated by the stator and that in the HTS bulk rotor. The shielding current circulating in the HTS bulk rotor fixes the trapped flux at steady state operation, and induces the phase difference due to the nonlinear relation between the trapped flux and pinning center, i.e., hysteresis. Therefore, this type of motor is called hysteresis motor. Because of simpler structure and easier construction, this motor has widely been investigated especially for fundamental study.

Flux shielding capability of HTS bulk is utilized for the reluctance type motor. The torque is generated by the difference in reluctance between the direct and quadrature axes. That is, a rotating field along the quadrature axis is screened by HTS shield bulk, and is concentrated along the direct axis, resulting in a large torque generation. Iron is commonly utilized in order to allow the magnetic path along the direct axis, and some types of configuration for the composite structure of iron and HTS bulk have been proposed. In the recent studies, large-scale motors have been developed and tested successfully for this type of motor [17-24].

In a trapped flux motor, the HTS bulk is used as a pseudo- permanent magnet. Then, this type of motor is regarded as a superconducting permanent magnet motor. The magnetic flux density trapped in the bulk rotor, on the other hands, has to be larger than the saturation flux density of the conventional permanent magnet in order to compete with the conventional motor. Therefore, the magnetizing problem also exists for the realization of high-trapped flux magnet. Typical method for the magnetization of HTS bulk is that the bulk is cooled below its critical temperature in a high field, i.e., Field Cooling (FC). The magnetizing coil, however, must become larger in order to produce high field at a steady state condition. Then, this method is non realistic for the large scale motor systems. Pulsed field magnetization method may be one of the most efficient solutions in order to obtain the large trapped flux with compact facilities [16, 28]. As a new type of trapped flux motor, Halbach array for the rotor has also been proposed and discussed [29].

More detailed explanations about the aforementioned types of HTS bulk motors are also reviewed in [30].

2.2 Effects of Harmonics in Air-gap Flux Distribution

Except for a reluctance motor, the magnetic flux has to be adequately penetrated into the bulk without flux motion such as flux flow and/or flux creep, for the realization of large torque generation. On the other hands, this adequate flux trapping is also crucial from the point of view of the system performances as follows.

It is well known that the ideal shape of the magnetic flux density distribution of the air-gap is sinusoidal for the minimization of magnetic losses in a conventional hysteresis motor, in which a ferromagnetic material is utilized for the solid rotor [9]. Then, if such a distribution is distorted from the sinusoidal waveform, the spatial harmonics of the distribution influence the characteristics of the motor. On the other hands, the same detrimental effect is easily deduced in case of HTS bulk motor, too. The harmonic components will also be affected by the electromagnetic property of HTS bulk that depends on the magnetic processes.

In this sub-section, the effect of the spatial harmonics of the air-gap flux distribution on the HTS bulk motor characteristics is explained. The influences of shielding property of HTS bulk motor on these harmonics are also to be explained [31-33].

Two kinds of three-phase armature windings are considered for the investigation. Fig. 1 (a) is a photograph of salient and concentrated stator (two-poles) that consists of eight coils. In this stator, each coil consists of rectangular-cross sectional copper wires with the turn number of 200 wound around the laminated iron core. Another stator is a

(a) Concentrated windings (two-pole)

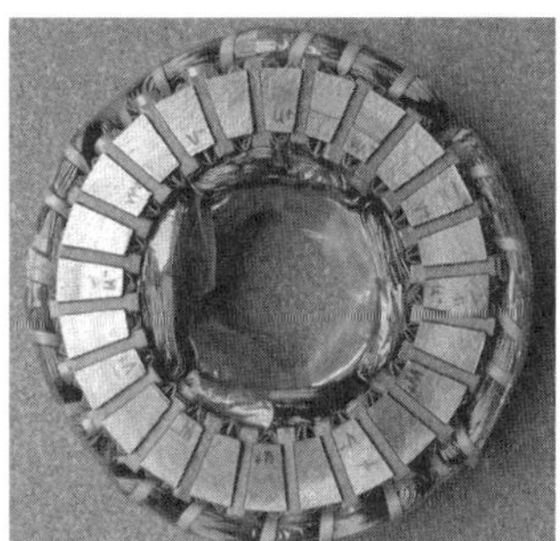

(b) Distributed windings (four-pole)

Fig. 1 Photographs of three-phase stators for an axial-type Bi-2223 bulk motor [31-33]

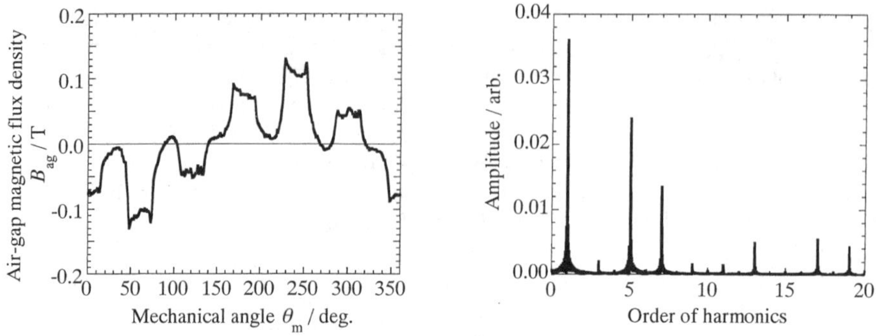

(a) Flux density distribution (b) FFT result

Fig. 2 Spatial distribution of the air-gap flux density at 77.3 K for Fig. 1 (a) [31-33]

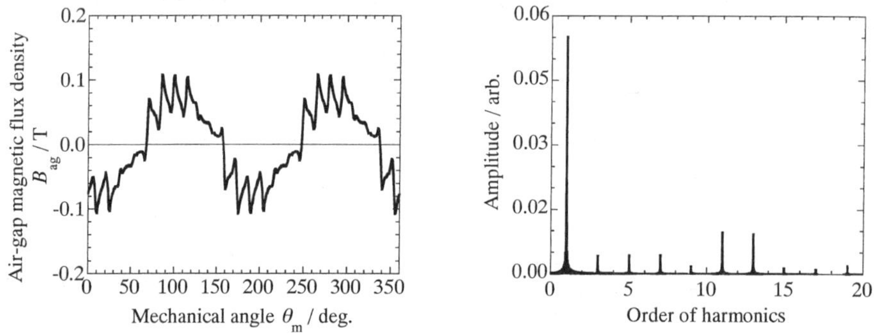

(a) Flux density distribution (b) FFT result

Fig. 3 Spatial distribution of the air-gap flux density at 77.3 K for Fig. 1 (b) [31-33]

distributed winding (four-poles) as shown in Fig. 1 (b). A rotational speed of the traveling magnetic field, N [rpm], can be calculated as follows:

$$N = \frac{120f}{P} \tag{1}$$

where, f [Hz] and P denote the line frequency and pole numbers, respectively. Then, N = 3600 and 1800 rpm for two-pole and four-pole motors at 60 Hz frequency, respectively.

In order to examine the effect of the HTS bulk rotor on the spatial harmonics of the air-gap magnetic flux distribution, Bi-2223 sintered disk, 120 mm in diameter and 7 mm in thickness, is sandwiched in between each pair of the stators with the air-gap length of 1.5 mm. Transverse type hall sensor is mounted on the Bi-2223 rotor, and the magnetic flux density is measured in a rotating magnetic field with the speed mentioned above. After the finish of one measurement, the Bi-2223 rotor is rotated at 1° and the same measurements are repeated.

Fig. 2 and 3 show the experimental results measured at 77.3 K: (a) Flux density distributions at phase angle of 0°, (b) Fast-Fourier-Transform (FFT) results. Bi-2223 bulk rotor is surely in the superconducting state at this temperature. As can be easily understood, the flux density distribution for the concentrated armature windings is largely

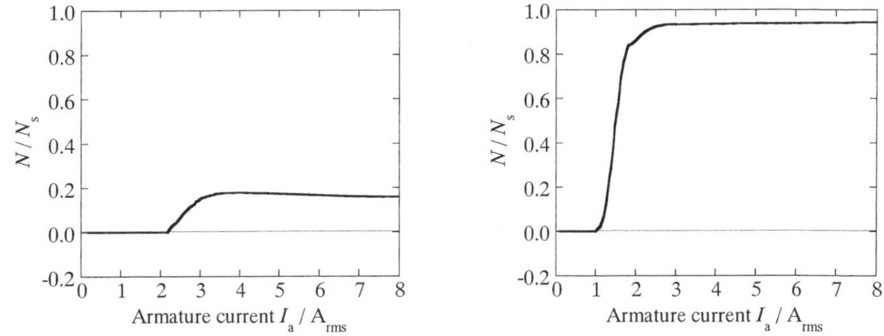

(a) Concentrated winding (b) Distributed winding

Fig. 4 No-load test results of axial-type Bi-2223 bulk motor at 77.3 K

distorted (Fig. 2 (a)), and large amounts of 5th and 7th spatial harmonics are included (Fig. 2 (b)). On the contrary, distributed stator has more like sinusoidal waveform in the air-gap (Fig. 3 (a), (b)). Of course, it can be deduced that these harmonic components influence the motor characteristics. As one of the examples, no-load test results obtained with these stators are shown in Fig. 4. In this figure, the rotational speeds of the motor are normalized by the synchronous speeds, N_s. As can be seen, the speed for the concentrated winding is extremely suppressed to the low value compared to the case of distributed winding. This means that the large slippage occurs for the concentrated armature windings, and large energy is dissipated. In case of distributed stator, however, rotating speed reaches nearly synchronous speed. As can be seen in Fig. 2 (b), large amount of 7th component is included in the rotating magnetic field for the concentrated stator, and the rotor is synchronized to this component. That is, the rotational speed is around $N_s/7$, and this effect is called crawling. Therefore, we have to reduce the harmonic components in order to avoid this detrimental effect.

Fig. 5 shows the flux density distribution for the distributed stator at (a) room temperature and (b) $T = 77.3$ K, respectively. The curve for $I_a = 2\ A_{rms}$ in Fig. 5 (b) is the

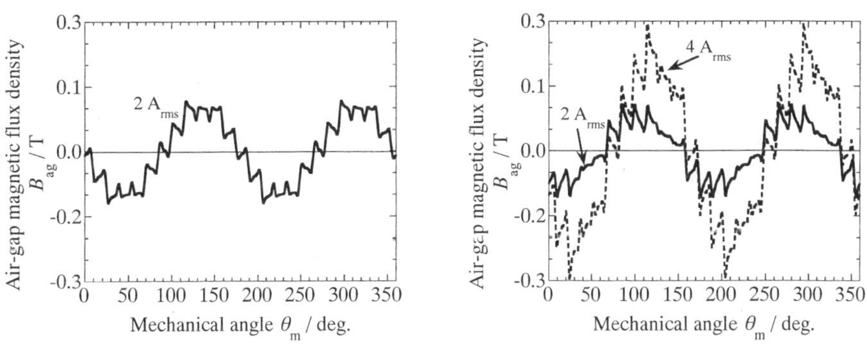

(a) Room temperature (b) T = 77.3 K

Fig. 5 Distributions of magnetic flux density in the air-gap for distributed stator (Fig. 1 (b)). Armature current values are indicated in the figures [31-33].

same with Fig. 3 (a). As compared between (a) and (b), the flux distribution is distorted due to the shielding property of Bi-2223 bulk rotor. In Fig. 5 (c), on the other hands, the distribution is recovered to be the sinusoidal waveform due to the flux invasion into the bulk. That is to say, the magnetic flux has to be adequately penetrated even from the point of view the air-gap flux distribution.

3. FRACTIONAL FORCE POWER MOTORS

Hysteresis and trapped-flux bulk motors are mainly explained in this section. Reluctance type motors are to be explained as small and middle class motors in section 4. Y-123 bulk materials are popularly been considered for fractional force power motors, because these materials are widely been fabricated and have strong flux pinning property at liquid nitrogen temperature, i.e., 77.3 K. There are mainly two types of HTS bulk motors from the structural point of view: axial-type motor and radial type motor. In an axial-type motor, a HTS disk is usually utilized and the rotating field is applied axially. In a radial-type motor, on the other hands, the field is applied to a HTS cylinder in a radial direction. It is well known that HTS has large anisotropy of the electromagnetic property with respect to the field direction. Then, the interaction between the rotating field and the flux trapped in the HTS bulk may be different between an axial-type and a radial-type motor when we consider the mechanisms of the torque generation because of differences of pinning property and shielding current path by this isotropic electromagnetic characteristics.

In a theoretical study of HTS bulk motors, finite element analysis (FEA) technique is widely utilized. The modeling of constitutive relation between electric field, E, vs. current density, J, is the most important issue for the expression of superconducting property. For a HTS bulk that has strong flux pinning force, critical state model will be better approximation for such relation. As a simple model, the following Bean model [34] is commonly used:

$$\boldsymbol{J} = \pm \boldsymbol{J}_c \tag{2}$$

where, J_c denotes the critical current density. J_c is usually defined by the electric field criterion, E_c, from the E-J curves, and typical value of E_c is from 10^{-5} to 10^{-4} $\mathrm{Vm^{-1}}$. As can be seen in eq. (2), only J is restricted to be three values, i.e., $+J_c$, $-J_c$ and 0. When magnetic field dependency of J_c is crucial, Kim model [35] is considered as follows:

$$J_c(|\vec{B}|) = \frac{J_{c0}}{1 + |\vec{B}|/B_0} \tag{3}$$

where, J_{c0} is a critical current density without magnetic field dependency. B and B_0 are the vector of magnetic flux density and the parameter describing the field effect, respectively. Some FEA analyses have been performed based on the critical state model [30, 36-38]. From these results, the segmented bulk rotor will be the better choice in order to penetrate adequately the flux into the bulk. The fabrication of the large bulk is technically very difficult so far. Then, this segmented rotor has a potential for the scale-up of the motor in a future. The magnetic process of Y-123 has also been examined experimentally [39]. By assuming the Bean model, the magnetic power density per cycle is expressed as follows [14]:

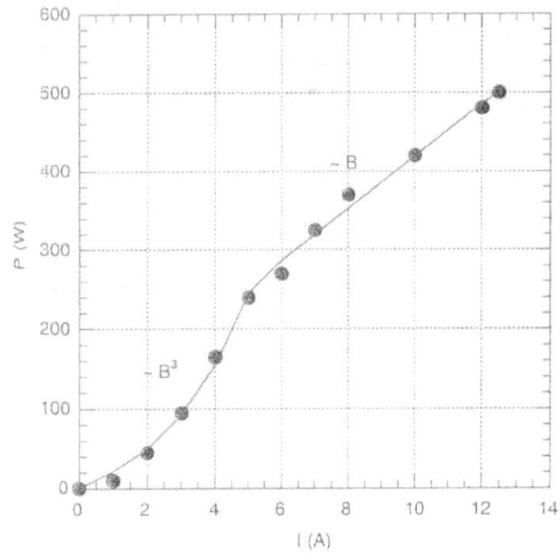

Fig. 6 Experimental result of output power, P, vs. armature current, I, in YBCO hysteresis motor operated at 77.3 K [39].

$$\boldsymbol{P} = \frac{64}{3\pi}\mu_0 \frac{\boldsymbol{H}_{\mathrm{m}}^{\ 3}}{\boldsymbol{H}_{\mathrm{p}}} \qquad ; \beta < 1, \tag{4}$$

$$\boldsymbol{P} = \frac{64}{3\pi}\mu_0 \boldsymbol{H}_{\mathrm{p}}(3\boldsymbol{H}_{\mathrm{m}} - \boldsymbol{H}_{\mathrm{p}}) \qquad ; \beta > 1, \tag{5}$$

where, μ_0 denotes permeability in vacuum. H_{m} and H_{p} are amplitude of applied magnetic field intensity and full penetration field, respectively. Parameter β is defined as $H_{\mathrm{m}} / H_{\mathrm{p}}$ that describes the rate of magnetic flux invasion in the bulk. That is, the magnetic flux is fully penetrated in the bulk when $\beta > 1$. As can be seen in eqs. (4) and (5), the magnetic amplitude dependence of the out power is changed from cubic to linear function at $\beta = 1$. Fig. 6 shows the experimental example of output power vs. armature current characteristics, and the curve follows these equations.

Researches with the use of other HTS bulk materials except for Y-123 are not so familiar, and fundamental studies with small-scale model have only been carried out at this moment. Authors' group has actively investigated the axial-type Bi-2223 as well as Sm-123 bulk motor as fundamental studies [31-33, 40-48]. The basic motivation of their study is that there will exist the optimum bulk materials as a solid rotor depending upon the operational conditions such as temperature. That is, the magnetic flux must be adequately penetrated or trapped without flux motion, i.e., flux flow and/or flux creep, for the torque generation with less dissipation. This flux penetration is also necessary from the point of view of the reduction of the spatial harmonics of air-gap magnetic flux distribution as discussed in 2.2. Then, large magneto-motive force is necessary in order to magnetize the HTS bulk when the flux pinning is stronger. This means that decreasing the operation temperature will enlarge the armature winding (and system size) due to the strong flux pinning force. Consequently, the optimal HTS bulk materials, i.e., flux pinning property, will appear from the point of view of the system design. As is well known, the main difference between Sm-123 and Bi-2223 is their flux pinning strength as

Fig. 7 Photograph of axial-type fractional force power HTS bulk motor. Bi-2223 or Sm-123 disk is used for the solid rotor. HTS rotor is sandwiched in between three-phase and four-pole distributed armature windings [31-33, 40-48].

Table I Specifications of HTS disk used for the axial-type bulk motor

	Bi-2223	Sm-123
Process	Sintered	Melt-textured
Diameter D / mm	120	120
Bore diameter d_b / mm	26.0	26.0
Thickness t_h / mm	7.0	8.0
Critical temperature T_c / K	106	92
Critical current density J_c / Acm^{-2} @77.3 K & self-field	3×10^3	Un-known

well as its magnetic field dependency especially at liquid nitrogen cooling. Therefore, the dependency of the pinning property on the motor characteristics is to be examined systematically.

Here, we concentrate on the HTS motor cooled with liquid nitrogen for an applicability study as well as academic interest. Fig. 7 shows the photograph of the axial-type model motor [31-33, 40-48]. The diameter of Bi-2223 as well as Sm-123 bulk rotor is 120 mm. The hole with a bore diameter of 26.0 mm is made in order to fixate the stainless shaft. The thickness of these rotors is slightly different: i.e., 7.0 mm for Bi-2223 and 8.0 mm for Sm-123. Other specifications of these bulk materials are listed in Table I. The crystallographic c-axis of these disks is perpendicular to their wide surfaces. Then, the magnetic field is applied to this axis, and the shielding current will be induced in the wide plane of the bulk when the flux is adequately penetrated.

Each solid rotor is sandwiched in between two armature windings those are connected in parallel, and the rotating magnetic field is applied axially. Three-phase and four-pole armature windings are utilized that has 24 laminated teeth (Fig. 1 (b)), and the corresponding rotating speed of the magnetic field at 60 Hz frequency is 1800 rpm. Bi-2223 motor is tested for different temperatures ranging from 64 K to 77.3 K. Sm-123 motor, however, is tested only at 77.3 K due to the difficulty of magnetization. In case of changing the operation temperature, liquid nitrogen in the metal cryostat is evacuated by a rotary vacuum pump in order to control the saturation pressure. All the measurements

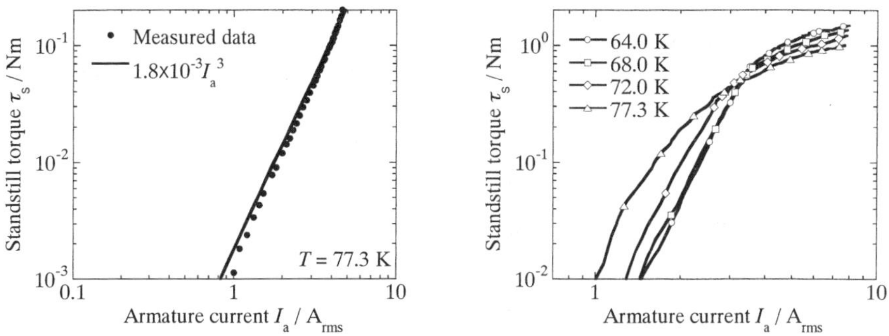

(a) Sm-123 bulk motor [47] (b) Bi-2223 bulk motor [46]

Fig. 8 Locked-rotor test results for axial-type bulk motor

are performed at steady state by finely changing the variables such as line voltage and load.

Fig. 8 (a) shows the locked rotor test results of Sm-123 bulk motor with double-logarithmic plots [47]. The standstill torque, τ_s, is measured as a function of I_a at fixated condition of the shaft. It can be seen that the measured points show the cubic dependence with respect to I_a as compared to the fitted solid line. The flux pinning force of Sm-123 is strong like Y-123, then this result agrees with eq. (4). In case of Bi-2223 bulk motor, however, this discussion is invalid due to the weak pinning property. Fig. 8 (b) shows the temperature dependence of τ_s-I_a curves [46]. As can be seen, the characteristics show complicated dependences depending on the temperature and eq. (4) does not meet any more. It has to be noted in this figure that the temperature dependency of the torque curves change at the point that the curvature changes. That is, the lower the temperature is, the higher the torque is at $I_a < 2$ A_{rms}. Opposite dependency, however, is observed at $I_a > 2$ A_{rms}. These behaviors can easily be explained by the temperature dependent pinning

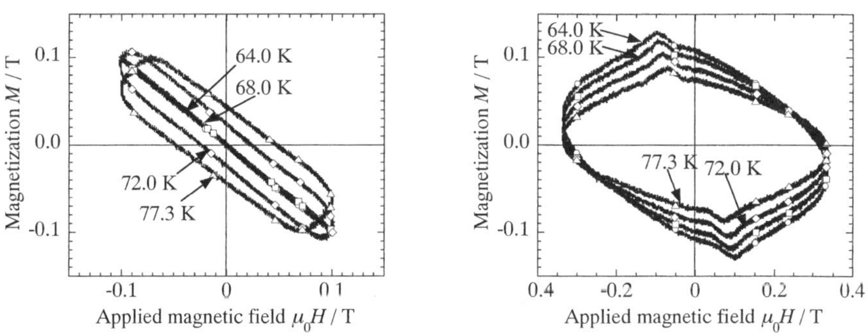

(a) $\mu_0 H = 0.13$ T ($I_a = 1.5$ A_{rms}) (b) $\mu_0 H = 0.35$ T ($I_a = 5.0$ A_{rms})

Fig. 9 Temperature dependence of hysteresis loops in a rotation magnetic field at 1800 rpm for Bi-2223 bulk rotor [45].

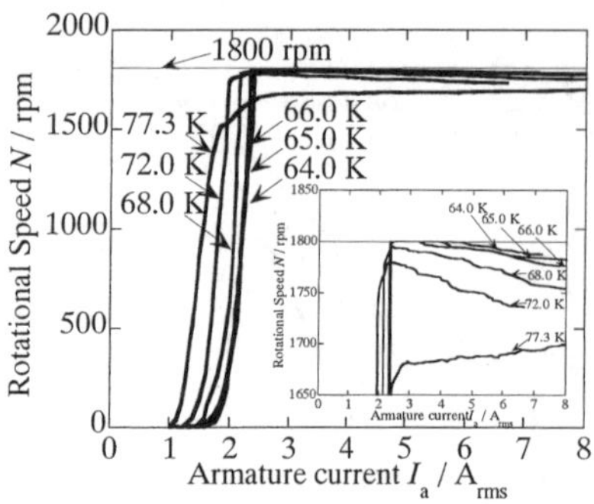

Fig. 10 No-load test results for Bi-2223 bulk motor. Inset: enlarged figure [46].

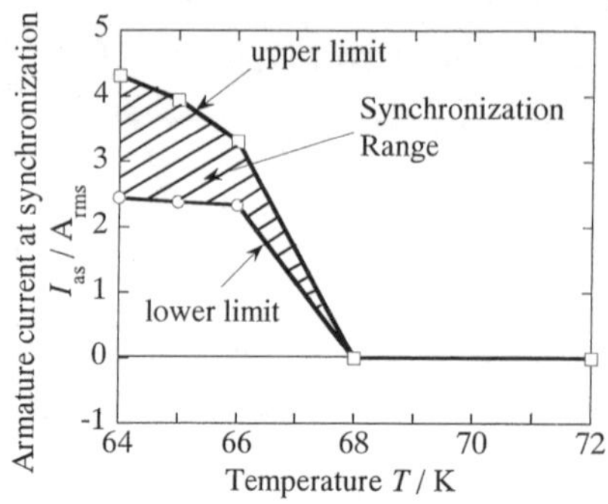

Fig. 11 Synchronized rotation range obtained from Fig. 10 [46].

property. Namely, when the flux is penetrated only partially, the pinning center impede the flux invasion from outside by decreasing the temperature. On the contrary, when the flux is fully penetrated, i.e., $I_a > 2$ A_{rms}, the penetrated flux is trapped strongly. These characteristics can be explained by the temperature dependent hysteresis curves. Fig. 9 shows temperature dependent hysteresis loops in a rotating field at 1800 rpm in the same Bi-2223 bulk rotor used for the above [45]. As can be seen, the areas of the hysteresis loops are opposite temperature dependencies (a) below and (b) above $\mu_0 H = 0.13$ T (I_a = 2.0 Arms). It is known that the area of the hysteresis loop translates the torque. Then, the tendency of the hysteresis loop characteristics is consistent with the locked rotor test results.

Next, No-load test results of Bi-2223 bulk motor are shown for different temperatures in Fig. 10 [46]. It should to be noted that the rotational speed of the motor becomes higher value by increasing the temperature when the current is smaller than around 2 A_{rms}, and opposite characteristics are observed above this current. The reason for this characteristic can be explained by the temperature dependent pinning property as is already explained for the locked rotor test results. As can be clearly seen, the motor does not reach synchronous rotating speed (1800 rpm) at 77.3 K. When the temperature is decreased, on the other hands, the saturation speed, at which the magnetic flux is fully penetrated in the bulk rotor, increases due to the stronger pinning force. And finally, it reaches 1800 rpm below the temperature of 68 K. Furthermore, the interesting point in this figure is that the small slippage also occurs when the armature currents are over than certain values (see the inset in Fig. 10). This will be due to the occurrence of the flux flow in Bi-2223 rotor. That is to say, there is synchronized rotation range of the operation temperature as well as armature current for Bi-2223 bulk motor that is cooled by liquid nitrogen. Fig. 11 shows the synchronization range of armature current as a function of temperature [46]. It can be seen that the synchronization range reduces by increasing the operation temperature, and vanishes above 68 K.

The no-load test results regarding Sm-123 and Bi-2223 bulk motor are also carried out for comparison. In order to investigate the hysteresis of the motor characteristics, we perform a test firstly by increasing the line voltage and then decreasing it. Although Sm-123 bulk motor is tested at 77.3 K, Bi-2223 bulk motor is done at 64.0 K because of the lowest temperature of liquid nitrogen cooling. Fig. 12 shows the results for (a) Sm-123

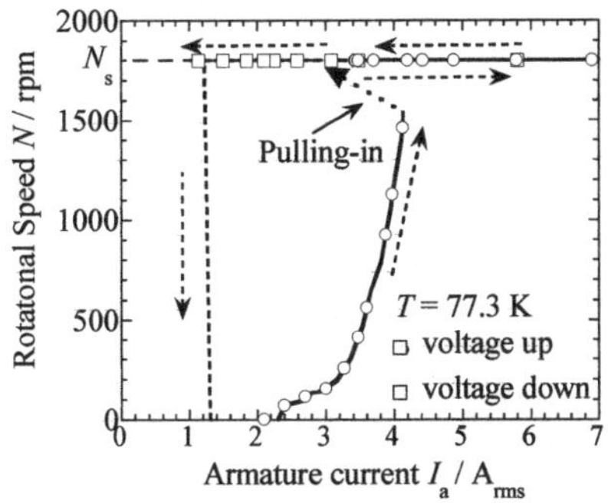

(a) Sm-123 bulk motor at 77.3 K [47]

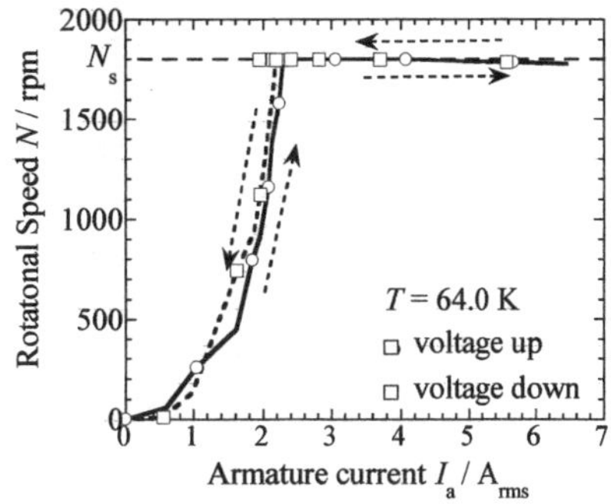

(b) Bi-2223 bulk motor at 64.0 K [48]

Fig. 12 Hysteresis characteristics of no-load test results for axial-type motor.

bulk motor [47] and (b) Bi-2223 bulk motor [48], respectively. As can be shown in Fig. 12 (a), the rotational speed of Sm-123 bulk motor is pulled-in to its synchronous speed when the speed is 1540 rpm. At this time, and the corresponding armature current reduces because the counter-electromotive force due to the eddy current with slippage vanishes. After the synchronization, the rotational speed remains to be the synchronous speed even at the armature current of 7 A_{rms}. However, the small slippage occurs after the synchronization in case of Bi-2223 bulk motor as is already explained (see the inset of Fig. 10). On the other hands, when the line voltage (armature current) is decreased, there exists extinguished difference between the Sm-123 bulk motor and Bi-2223 one. That is, synchronous speed is maintained even at smaller armature current in Sm-123 bulk motor, and then suddenly stops. However, in case of Bi-2223 bulk motor, the no-load characteristic follows almost the same curve with that obtained by increasing process of the line voltage. That is to say in these results, Sm-123 bulk motor behaves like trapped-flux motor after the rotational speed reaches the synchronous one. On the other hands, Bi-2223 bulk motor behaves as a hysteresis one, at least, in liquid nitrogen cooling conditions.

It should to be noted that aforementioned results are very informative when we discuss the mechanisms of the torque generation by taking into account the nonlinear electromagnetic characteristics of HTS bulk. Further studies are necessary for more detailed discussions.

4. SMALL AND MIDDLE CAPACITY MOTORS

The small- and middle-capacity motors have commonly been fabricated by means of Y-123 bulk. Russia and Germany group have developed the several kW class reluctance motors with various structures of the rotors as shown in Fig. 13 [24]. As can be seen in the figure, four kinds of combination of Y-123 blocks and steels are considered. In their study, it has been concluded that the specific output power of the HTS bulk motor is four to seven times better than that of the conventional motors [24]. These results will be encouraging the activity of the HTS bulk motor development. They have also developed 20 kW output power HTS reluctance motors and tested successfully at 3000 rpm and 77 K [18]. They are now going on to the middle capacity motor developments with 150 kW

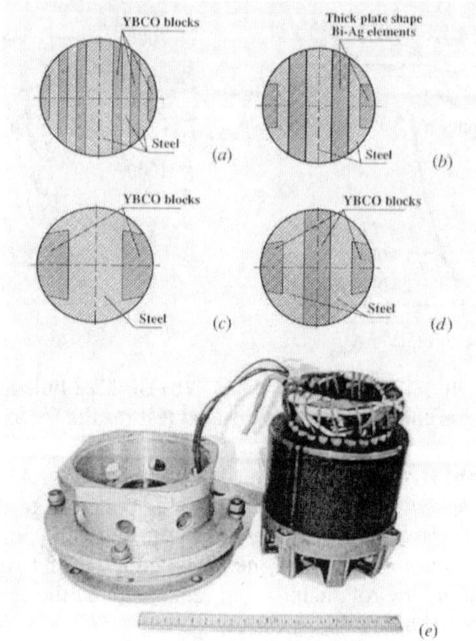

Fig. 13 Y-123 reluctance motor [24]. Various types of the rotor structure are shown from (a) to (d). Fabricated system of the motor is shown in (e).

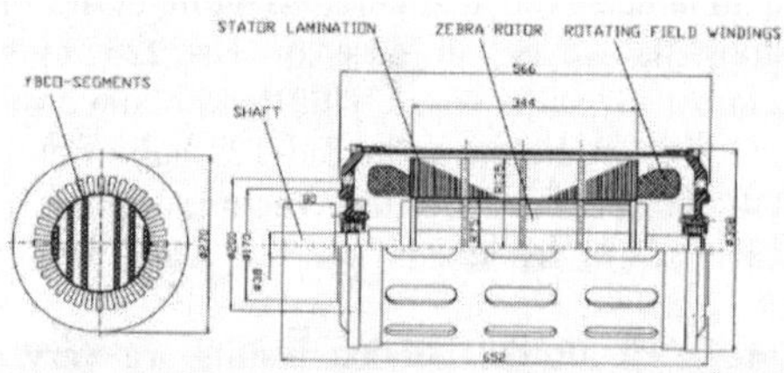

Fig. 14 Y-123 bulk reluctance motor designed for 150 kW mechanical output power at 3000 rpm [22].

mechanical output power at 3000 rpm [22]. Fig. 14 shows the schematic view of the designed motor, and will be cooled by liquid nitrogen.

Central Japan Railway Company in JAPAN has developed the motor with the use of conduction cooled trapped-flux bulk magnet [16]. Fig. 15 shows the schematic view of the fabricated motor, in which eight numbers of YBCO-123 bulk, 60 mm in diameter and 15 mm in thickness, are fixed on the stator with conducting copper bands. Unique points of this motor are that YBCO-123 bulks are cooled directly by the Gifford-Macmafon (GM) cryocooler and the magnetic flux can be trapped by so-called *in-situ* method after zero field cooling. That is, magnetizing coils are installed in the cryostat and each bulk is magnetized separately without breaking the vacuum space. Operation temperature is 30 K, and then large flux can be trapped in the bulk due to the strong flux pinning. Because the bulks are cooled by GM cryocooler, the armature rotating type is introduced. Table I

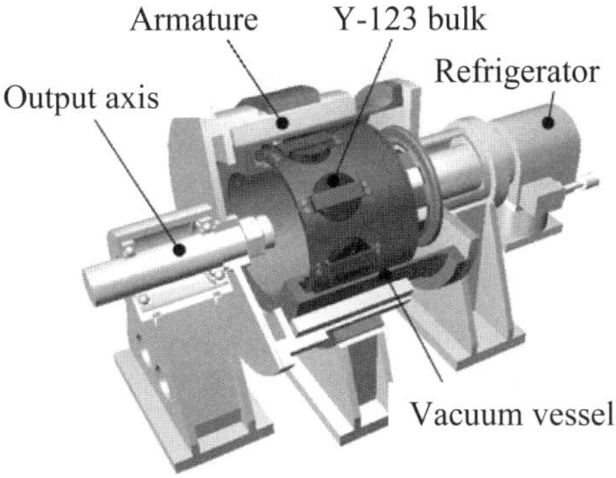

Fig. 15 Schematic view of YBCO-123 trapped-flux magnet motor [16]

Table II Specific parameters of the motor (Fig. 15) [16]

Parameters	Values
Rotational speed / rpm	600
Torque / Nm	25
Output power / kW	1.5
Number of poles	8
3-phase line voltage / V	200
3-phase armature current / A	22
Armature length / mm	210
Inner diameter of armature / mm	250

shows the specifications of the fabricated motor. Up to now, the output power of 1.5 kW has been successfully tested.

5. SUMMARY

In this chapter, present status of the R&D activities of HTS bulk motors was reviewed. Fundamental and applicability studies have widely been performed with the use of fractional force power motors. Fundamental aspects of HTS bulk motors were explained, and some interesting characteristics were reported with axial-type motor models with different kinds of HTS rotors. Numerical analyses for the electromagnetic property in a rotating magnetic field have also been carried out in order to clarify the torque generation mechanisms in the HTS bulk motors.

Based on the fundamental study, small and/or middle capacity reluctance motors have been developed and successfully tested. Small-scale trapped-flux motor has also been actively developed. In order to improve further the motor characteristics, HTS bulk's property has to be improved. That is, the shielding property is crucial for reluctance motor, and flux trapping property (pinning property) for the trapped-flux motor and hysteresis motor. Recently, high field bulk magnet that can trap over 17 T at 29 K has been developed [49], and this magnet will accelerate the development of high performance bulk motor.

REFERENCES

[1] Y. Itoh et al., Physica C 235-240 [1994] 3445
[2] H. Ohsaki et al., Adv. in Supercond. IX (Springer Verlag Tokyo) 2 [1996] 1373
[3] P. Tixador et al., IEEE Trans Appl. Supercond. 7 [1997] 896
[4] M.D. McCulloch et al., Inst. Phys. Conf. 158 [1997] 1519
[5] M.D. McCulloch et al., Mater. Sci. Eng. B53 [1998] 211
[6] M.D. McCulloch et al., Supercond. Sci. Technol. 15 [2002] 826
[7] I. Márquez et al., IEEE Trans. Appl. Supercond. 9 2 [1999] 1249
[8] W.-R Canders et al., Proc. ICEM'2000 [2000] 649

[9] A.Leão Rodrigues et al., Proc. ICEM'2000 [2000] 802
[10] M.D. McCulloch et al., Proc. ICEM'2000 [2000] 812
[11] H. Ohsaki et al., J. Mater. Process. Technol. 108 [2001] 148
[12] Y. Tsuboi et al., Physica C 357-360 [2001] 889
[13] X. Granados et al., Physica C 372-376 [2002] 1520
[14] Y. Tsuboi et al., Cryogenic Engineering (in Japanese) 37 11 [2002] 704
[15] Y. Tsuboi et al., IEEE Trans. Appl. Supercond., in press
[16] M. Hirakawa et al., Physica C, in press
[17] L.K. Kovalev et al., Mater. Sci. Eng. B53 [1998] 216
[18] B. Oswald et al., IEEE Trans. Appl. Supercond. 9 2 [1999] 1201
[19] L.K. Kovalev et al., Supercond. Sci. Technol. 13 [2000] 498
[20] L. K. Kovalev et al.: Physica C 354 (2001) 34
[21] L.K. Kovalev et al., Physica C 357-360 [2001] 860
[22] B. Oswald et al., Physica C 372-276 [2002] 1513
[23] L.K. Kovalev et al., Physica C 372-376 [2002] 1524
[24] L.K. Kovalev et al., Supercond. Sci. Technol. 15 [2002] 817
[25] D. Dew-Hughes et al., Adv. in Cryog. Eng. [1999]
[26] SUPERCONDUCTIVITY COMMUNICATIONS, 4 1 [1996] (URL: http://www.chem.t.u-tokyo.ac.jp/appchem/labs/kitazawa /SUPERCOM/20/s20a03.html (in Japanese))
[27] L.K. Kovalev et al., Electr. Technol. 2 [1994] 145
[28] U. Mizutani et al., Cryogenics 34 11 [1999] 654
[29] J.R. Hull et al., IEEE Trans. Appl. Supercond. 9 2 [1999] 1229
[30] G. Barnes et al., Supercond. Sci. Technol. 13 [2000] 875
[31] T. Nakamura, I. Muta et al.: Cryogenic Engineering (in Japanese) 37 11 [2002] 726
[32] T. Nakamura, et al.: IEEE Trans. Appl. Supercond., in press
[33] I. Muta, T. Nakamura et al.: Proc. EEMODS'02, Treviso, Italy [2003] in press
[34] C.P. Bean, Rev. Mod. Phys. 36 [1964] 31
[35] Y.B. Kim et al., Phys. Rev. 129 [1963] 528
[36] G. Barnes et al., Supercond. Sci. Technol. 12 [1999] 518
[37] G. Barnes et al., Physica C 331 [2000] 133
[38] G.J Barnes et al., Supercond. Sci. Technol., 13 [2000] 229
[39] T. Habisreuther et al., IEEE Trans. Appl. Supercond. 7 2 [1997] 900
[40] I. Muta, et al.: IEEE Trans. on Appl. Supercond. 11 [2001] 1964
[41] I. Muta, T. Nakamura et al.: Physica C 354 [2001] 100
[42] I. Muta, et al.: Physica C, 372-375 [2002] 1535
[43] I. Muta, et al.: Proc. ICEM 2002, Brugge, Belgium [2002] 196
[44] H.J. Jung, T. Nakamura et al.: IEEE Trans. Appl. Supercond., in press
[45] T. Nakamura, et al.: Physica C, accepted
[46] T. Nakamura, et al.: Supecond. Sci. Technol., to be submitted
[47] T. Nakamura, et al.: Physica C, to be submitted
[48] T. Nakamura, et al.: Physica C, to be submitted
[49] M. Tomita et al., Nature 421 [2003] 517

A MOTOR WITH HIGH-TEMPERATURE SUPERCONDUCTING LEVITATION AND ITS VIBRATION CONTROL

Kosuke NAGAYA, Takaaki SUZUKI, Noriaki TAKAHASHI and Hiroyuki KOBAYASHI
Department of Mechanical Engineering, Gunma University
Kiryu, Gunma 376-8515, JAPAN

1 IINTRODUCTION

A high temperature superconductor can levitate a magnet without control [1]-[5]. When it is used to a bearing, an extremely high-speed rotation is possible which cannot be realized in mechanical bearings because of existence of oil whirl. In addition, there is no friction in the superconducting bearing and so, it is used in an energy storage flywheel system [6]. In the system, however, large vibrations are generated due to the centrifugal force of the flywheel, so that a five-axes active magnetic bearing is used for controlling vibrations[6]. The active bearing is complicated. Moreover, it expends large energies.

This paper provides a new levitation system consisting of a superconductor and a pair of magnets. In the system, the super conductor supports both axial and radial forces at the lower end of the shaft, and a pair of permanent magnets supports the upper end. An attractive force acts in the downward direction along the shaft axis. The attractive force stabilizes the system in the horizontal direction, but it makes the levitation unstable in the axial direction. In this system, the axial pinning force of the high temperature superconductor cancels the unstable force in the axial direction. Hence, stable levitation is possible in both axial and radial directions without control. This is an advantage of the system as compared to previous systems with active magnetic bearings.

In order to have high-speed rotation, this article also provides a new driving system for the motor. The driving system consists of a 8-poles circular disc-type permanent magnet, and six driving coils. By using the magnet and coils, both attractive forces and repulsive forces between the magnet and the coils are used as a rotating torque. In addition, the repulsive force cancels a large part of the attractive force in the axial direction, and so exciting forces due to the electromagnetic driving system are small in comparison with the previous system [7].

Since the damping is small it is difficult to pass through critical speed. In order to pass through the critical speed, this paper provides a method for passing through the critical speeds. The control system consists of four coils and the permanent magnet. The levitation magnet, which lies at the upper end, is also used as an actuator for controlling vibrations. The control force is generated between coils and the levitation magnet.

2. MECHANISM OF THE MOTOR

2.1 Method of levitation

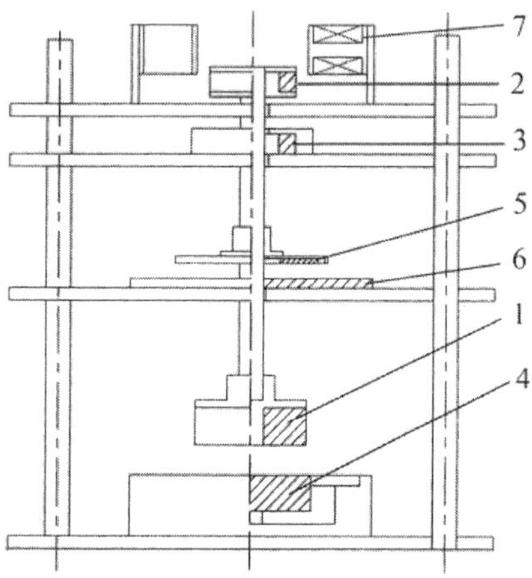

Fig.1 Geometry of the motor with superconducting levitation

Since the rigidity of the levitation system in our previous paper[7] is small, vibrations in the axial direction are large when the shaft rotates with high speed. In order to attenuate the disadvantage, the present paper discusses a new type levitation system as shown in Fig.1. In the system, the construction of the lower end of the shaft is the same as that in the previous levitation system [7]. A ring shaped permanent magnet 3 is fixed to the frame, and the shaft passes through the hole of the magnet. A ring shaped permanent magnet 2 is connected to the upper end of the shaft, and its N-pole faces S-pole of the fixed magnet. Hence, an attractive force applies in the downward direction along the shaft axis. Since the attractive force in the downward direction compresses the air gap between the permanent magnet 1 and the superconductor, the rigidity between the magnet 1 and the superconductor increases. Then, vibrations of the shaft will be suppressed.

The electromagnetic force due to the high temperature superconductor is the restoring force. The force varies with the initial air gap between the superconductor and the magnet, vibration amplitude and vibration frequency. When vibration amplitudes are small, the force is assumed to be linear with respect to the amplitude, but nonlinear with respect to the vibration frequency. According to our previous paper[8], the restoring force increases with the frequency, but saturates at large frequencies. Hence, the forces are assumed to be

$$k_{z1} = \alpha_z \{1 + \beta_z (1 - e^{-\gamma_z \omega})\} \tag{1a}$$
$$k_{r1} = \alpha_r \{1 + \beta_r (1 - e^{-\gamma_r \omega})\} \tag{1b}$$
$$k_{\theta 1} = \alpha_\theta \{1 + \beta_\theta (1 - e^{-\gamma_\theta \omega})\} \tag{1c}$$

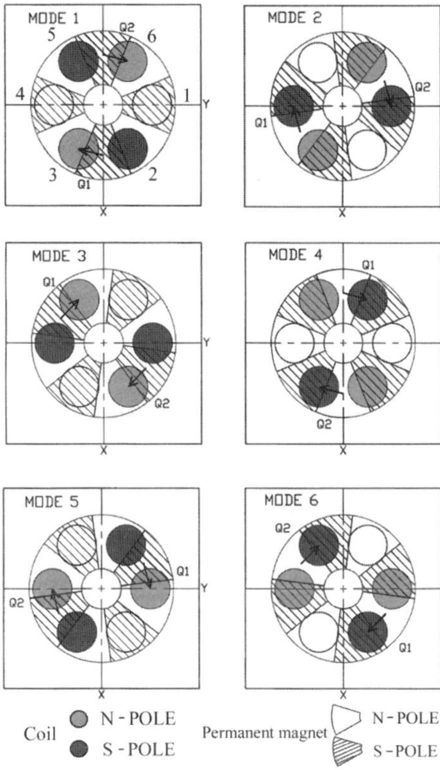

Fig.2 Principle of the driving system.

where k_{z1} is the spring constant in the axial direction, k_{r1} the spring constant in the direction perpendicular to the z-axis, k_θ the spring constant in the rotational motion, and α_z through γ_θ the constants which are obtained from the relation between the force and the displacement.

The attractive force acts between the permanent magnets, which lie at the top of the shaft. The force is the restoring force. The analysis for obtaining the force is given in the Appendix. The forces are approximately written in the following forms:

$$F_x = -\frac{e_1 x_2^2 + e_2 x_2}{(d - z)^2} \tag{2a}$$

$$F_y = -\frac{e_1 y_2^2 + e_2 y_2}{(d - z)^2} \tag{2b}$$

$$F_z = -\left(\frac{g_1}{(d+z)^2} - \frac{g_1}{d^2} \right) \tag{2c}$$

where d is the initial air gap between the permanent magnets, x, y the displacements in x- and y-direction, z the axial displacement. Using numerical fitting between Eqs.(2) and the results obtained by the magnetic field analysis gives the coefficients $g_1 \sim e_2$.

2.2 Motor driving system

Table 1 Magnetization pattern of the driving coil

MODE	1	2	3	4	5	6
COIL1		S	S		N	N
COIL2	S		N	N		S
COIL3	N	N		S	S	
COIL4		S	S		N	N
COIL5	S		N	N		S
COIL6	N	N		S	S	

Since the rigidity of the superconducting levitation system is small, vibrations will be generated, so that it is difficult to have high-speed rotation without control. To suppress vibrations, the five-axes active magnetic bearing was used in the previous flywheel system. The active magnetic bearing is complicated and spends much energy. In the present article, provides a new driving system, in which the vibration force due to the driver is small.

Figure 3 shows the proposed driving system. By using this system, unbalance forces, which generate vibrations, are attenuated as mentioned blow, so that stable rotation is possible without control. Moreover, the torque is large, and the resolution is small (15° per pulse) in comparison with our previous reluctance motor[7].

The present motor consists of an eight- poles circular permanent magnet and six driving coils as shown in Fig.3. The magnetization patterns for the coils are shown in Table1, in which S and N show the magnetic poles generated at the surfaces of the coils. The location of each permanent magnet and coil is shown in Fig.2. One cycle finish after six mode patterns shown in Table 1 (Mode1 through Mode 6). Consider Mode1 for example. When the negative current flows in coil 2 (S-pole) and positive current in coil 3 (N-pole), an attractive force F_a acts between the coil 3 and the S-pole of the permanent magnet, and repulsive force F_r acts between coil 2 and the N-pole. Hence, the force $F=F_a+F_r$ acts in the tangential direction of the circular magnet, which rotates the rotor in the clockwise direction. The same torque acts in the tangential direction due to the coil 5 and 6. The rotation angle due to one rectangular driving pulse voltage is 15°. When the pulse voltages are supplied to the coils, in which magnetization patterns are changed from

Mode 1 to Mode6 as shown in Table1, the rotor rotates in the clockwise direction. Repeating the operation maintains the rotor rotations.

The attractive force increases and repulsive force decreases in the rotation of the rotor in one pulse. This implies that the variation of the total force in the axial direction is small because the force is the subtraction of the attractive force and the repulsive force. The axial force is the exciting force, which induces vibrations. Since, the force is small compared to that of the reluctance type motor [7], vibrations will be small in the present driving system. This is one of the advantages of this motor. The torque also varies in one pulse, but the rotation angle is small (15°) in one pulse, and so the rotation speed can be assumed to be constant when the period of the driving voltage is the same. The rotor rotates until 90° after six modes. This implies that the angular velocity of electromagnetic force is 4ω, where ω is the angular velocity of the shaft. Then, the axial force due to the electromagnetic force $Q_z(I,t)$ can be written by

$$Q_z(I,t) = I(\frac{12\omega(a-b)}{\pi}t + b) \tag{3}$$

where I is the magnitude of electric current of the coil, t the time, a,b the constants. The time of one pulse is $t_0 = \pi/(12\,\omega)$, and so the force is

$$Q_z(I,t) = I\left\{\frac{12\omega(a-b)}{\pi}\left(t - \frac{n\pi}{12\omega}\right) + b\right\}$$

$$\text{for } \frac{n\pi}{12\omega} < t \le \frac{(n+1)\pi}{12\omega} \tag{4}$$

Arrangements of permanent magnets and the coils are not the same for all modes, the forces are not the same, but the difference will be small. Then, equation (4) can be assumed to all modes.

The electromagnetic force rotates with the velocity 4ω, and so the angle of rotation of the electromagnetic force Q_{za} is $\theta = 4\omega t$. Hence, the moments around x- and y-axis are

$$T_{yca} = Q_{za} R\cos 4\omega t, \quad T_{xca} = Q_{za} R\sin 4\omega t$$

where R is the length from the center of the rotor and the center of the coil. When one force act on point a and the other acts on point b, but the phase difference between them is 180° . Each torque is

$$\begin{aligned} T_{yc} &= \{Q_{za}(I,t) - Q_{zb}(I,t)\}R\cos 4\omega t \\ T_{xc} &= \{Q_{za}(I,t) - Q_{zb}(I,t)\}R\sin 4\omega t \end{aligned} \tag{5}$$

In this motor, two same forces with phase difference 180° act on the same circle of the rotor simultaneously when the currents flows sequentially as shown in Table 1. This implies that unbalance moments are cancelled. The unbalance moments are undesirable because they induce pitching vibrations. There is no unbalance moment in this motor as just mentioned. This is also an advantage of this motor.

3 METHOD FOR PASSING THROUGH THE CRITICAL SPEEDS

Stable levitation can be obtained by using the system proposed without control. At the resonance, however, it is difficult to pass through the critical speeds due to a rack of damping. In order to pass through the critical speeds, four coils lie near the upper levitation magnet as shown in Fig.3 . The coil has no iron core, and so there are no electric forces between the magnet and the coil when the control is not performed.

In the control, inertia force cancellation combining the PD control [10] is used for suppressing vibrations at the resonance frequencies. In this system, since the vibration at the first resonance frequency (10Hz) is significant, so that the system is approximated by a single degree of freedom system for obtaining the control force of the inertia force cancellation. This means that the inertia force cancellation attenuates the principal vibration and the PD control the remained vibrations. The equivalent mass is $m_e = k/p_1^{\,2}$, where $p_1(rad/s)$ is the first natural frequency, and k is the spring constant in the direction perpendicular to the shaft axis. The equation of motion under the assumption of the single degree of freedom is

$$m_e \frac{d^2u}{dt^2} + C_e \frac{du}{dt} + ku = m_e r\omega^2 \sin\omega t + Q\sin\omega t \tag{6}$$

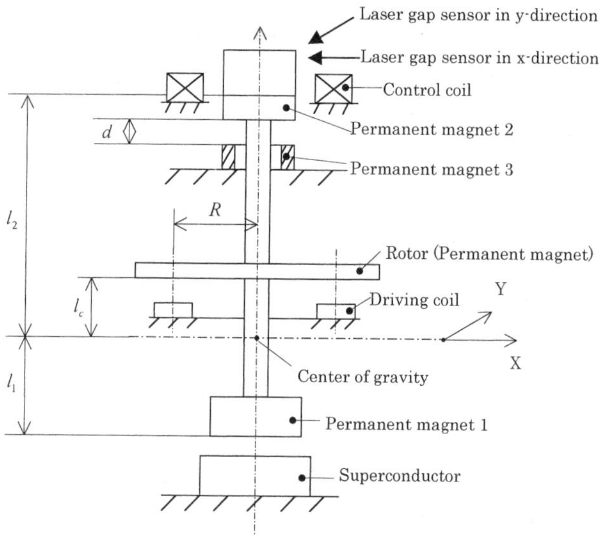

Fig. 3 Analytical model.

where u is the displacement at the top of the shaft, C_e the damping coefficient, k the spring constant, r the eccentricity, Q the inertia cancellation force, f_1 the displacement feedback coefficient, and f_2 the velocity feedback coefficient. The inertia force is the centrifugal force; $Q = -m_e r\omega^2$, and so the control force, which is in proportion to ω^2, cancels the inertia force. The displacement is

$$u = \frac{q}{p^2}\alpha \sin(\omega t - \beta) \tag{7}$$

where

$$\alpha = \frac{1}{\sqrt{(1-\omega^2/p^2)^2 + 4\zeta^2\omega^2/p^2}}$$

$$\tan\beta = \frac{2\zeta\,\omega/p}{1-\omega^2/p^2}$$

$$p^2 = \frac{k+f_1}{m_e}, q = r\omega^2、\quad \varsigma = \frac{C_e+f_2}{2\sqrt{km_e}} \tag{8}$$

The damping is removed because it varies in the rotation. Then,

$$u = \frac{q\alpha}{p^2} sin\omega\, t \tag{9}$$

From the above equation, eccentricity is obtained as

$$r = \frac{|u|}{\alpha}\left(\frac{p}{\omega}\right)^2 \tag{10}$$

where u is the displacement obtained by a sensor. The inertia cancellation force is

$$Q = -m_e \frac{p^2}{\alpha} u \tag{11}$$

Then, the control voltage is

$$V = \frac{-R}{Ka}\left(m_e \frac{p^2}{\alpha} u + f_1 u + f_2\,\dot{u}\right) \tag{12}$$

where K is the amplification coefficient of the power amplifier, R the resistance of the coil, a the proportional constant between the current to control force, and where $\alpha = 1/(1-\omega^2/p^2)$.

In order to have the optimal feedback coefficients for the PD control, the following cost function is considered:

$$J = \int_0^{\Omega} \{w_1(|x|+|y|) + w_2(|\dot{x}|+|\dot{y}|)\} d\omega \tag{13}$$

where J is the cost function, w_1 and w_2 the weight. The optimal coefficients f_1 and f_2 are obtained by minimizing the cost function [Eq.(13)].

4 RESPONSE ANALYSIS

4.1 Analytical model

Figure 3 depicts the analytical model. The origin of the coordinate x,y and z lies on the center of gravity. The control forces in the x- and y-axis are

$$Q_x = -m_e \frac{p^2}{\alpha} x_2 - f_1 x_2 - f_2 \dot{x}_2 \tag{14a}$$

$$Q_y = -m_e \frac{p^2}{\alpha} y_2 - f_1 y_2 - f_2 \dot{y}_2 \tag{14b}$$

The equations of motions are

$$m\frac{d^2x}{dt^2} + c_r \frac{dx}{dt} + k_{r1} x = f_x \tag{15a}$$

$$m\frac{d^2y}{dt^2} + c_r \frac{dy}{dt} + k_{r1} y = f_y \tag{15b}$$

$$m\frac{d^2z}{dt^2}+c_z\frac{dz}{dt}+k_{z1}z=f_z \tag{15c}$$

$$J\frac{d^2\theta_y}{dt^2}-J_p\omega\frac{d\theta_x}{dt}+c_\theta\frac{d\theta_y}{dt}+k_\theta\theta_y=T_y \tag{15d}$$

$$J\frac{d^2\theta_x}{dt^2}+J_p\omega\frac{d\theta_y}{dt}+c_\theta\frac{d\theta_x}{dt}+k_\theta\theta_x=T_x \tag{15e}$$

where

$$\begin{aligned} f_x &= F_x+Q_x+f_{xe} \\ &= -\frac{e_1x_2{}^2+e_2x_2}{(d-z)^2}-\left(m_e\frac{p^2}{\alpha}u+f_1x_2+f_2\dot{x}_2\right)+Me\omega^2\cos(\omega t+\phi) \end{aligned} \tag{16a}$$

$$\begin{aligned} f_y &= F_y+Q_y+f_{ye} \\ &= -\frac{e_1y_2{}^2+e_2y_2}{(d-z)^2}-\left(m_e\frac{p^2}{\alpha}u+f_1y_2+f_2\dot{y}_2\right)+Me\omega^2\sin(\omega t+\phi) \end{aligned} \tag{16b}$$

$$\begin{aligned} f_z &= F_z+2Q_z(I,t)= \\ &-\left(\frac{g_1}{(d+z)^2}-\frac{g_1}{d^2}\right)+2I\left\{\frac{12\omega(a-b)}{\pi}\left(t-\frac{n\pi}{12\omega}\right)+b\right\} \end{aligned} \tag{16c}$$

$$\begin{aligned} T_y &= F_xl_2+Q_xl_2-k_{r1}l_1x_1+f_{xe}l_c \\ &= -\left\{\frac{e_1x_2{}^2+e_2x_2}{(d-z)^2}\right\}l_2-\left(m_e\frac{p^2}{\alpha}u+f_1x_2+f_2\dot{x}_2\right)l_2-k_{r1}l_1x_1+Me\omega^2l_c\cos(\omega t+\phi) \end{aligned} \tag{16d}$$

$$\begin{aligned} T_x &= F_yl_2+Q_yl_2-k_{r1}l_1y_1+f_{ye}l_c \\ &= -\left\{\frac{e_1y_2{}^2+e_2y_2}{(d-z)^2}\right\}l_2-\left(m_e\frac{p^2}{\alpha}u+f_1y_2+f_2\dot{y}_2\right)l_2 \\ &\quad -k_{r1}l_1y_1+Me\omega^2l_c\sin(\omega t+\phi) \end{aligned} \tag{16e}$$

and where $\{n\pi/(12\omega)\}<t\le\{(n+1)\pi/(12\omega)\}, n=1,2,3,\cdots\infty$. There are the following relations:

$$\begin{aligned} x_1 &= x-l_1\theta_y, \quad x_2=x+l_2\theta_y \\ y_1 &= y-l_1\theta_x, \quad y_2=y+l_2\theta_x \\ \theta_x &= \frac{y_2-y_1}{l_1+l_2}, \quad \theta_y=\frac{x_2-x_1}{l_1+l_2} \end{aligned} \tag{17}$$

Solving above simultaneous differential equations give the x_1, y_1, x_2, y_2, x, and y.

5 EXPERIMENT

The experimental tests have been carried out for the system shown in Fig.1. The spring constant at the top of the shaft is obtained as $k=411N/m$. Figure 4 depicts the experimental results. Displacements at the resonance frequency are suppressed by using the present control. Figure 5 show the theoretical results. Experimental results are in good agreement with the theoretical.

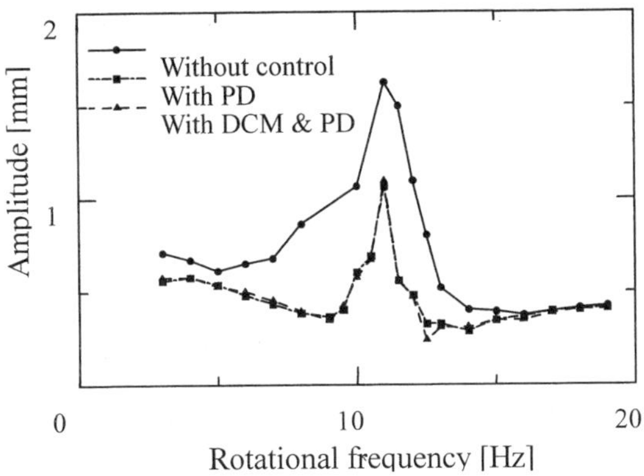

Fig.4 Frequency response of the displacements at the top of the shaft.

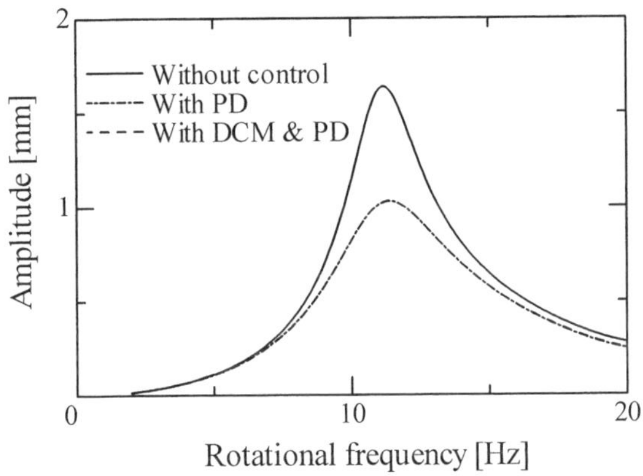

Fig. 5 Theoretical frequency response of the displacement at the top of the shaft

6 CONCLUSION

A new motor consisting of eight poles permanent magnet and six coils are presented. In order to have stable rotation without control, a new levitation system is presented, in which the super conductor lies at the base which supports gravity force of the shaft and the rotor. A pair of permanent magnet lies at the upper end of the shaft that supports the force in the direction perpendicular to the axis of the shaft. The pinning force of the high temperature superconductor cancels the unstable force in the axial direction due to the permanent magnets. Hence stable levitation is obtained without control. The driving method is presented with consideration of cancellation of vibration forces.

A control system and the control method for passing through the critical speed are presented. Analytical results are obtained. To validate the system and the analysis, experimental tests are also carried out. It is ascertained that the present system has enough rigidity, and attenuates vibrations as compared to the previous system. Hence, high-speed rotation is possible for the motor.

REFERENCES

[1] J.R. Hull, T.M.Mulcahy,K.L.Uherka and R.G.Abboud, Low rotational drag in high-temperature superconducting bearings, *IEEE Trans. Appli. Supercond.* Vol.5, No.2 (1995), pp.626-629.
[2] T. Hikihara, H.Adachi, S.Ohashi, and Y. Ueda, Levitation drift of flywheel and HTSC bearing system caused by mechanical resonance, *Physica*, C, 291, No.1-2 (1997), pp.34-40.
[3] T. Sugiura, M.Tashiro, Y.Uematsu and Y. Yoshizawa, Stability of a high Tc superconducting levitation system, *IEEE Trans. Appl. Supercond.* Vol.7, No.2 (1996), pp.386-389.
[4] M. Tsuchimoto and T. Honnma, Numerical evaluation of levitation force of HTSC flywheel, *IEEE Trans. Appl. Supercond.* Vol.4, No.4 (1994), pp.211-215.
[5] M.Komori, T.Matsushita, S.Fukata and A.Tsuruta, Dynamic characteristics of a high Tc superconducting bearing with a set of alternative polarity magnets, *J. Appled physics*, Vol.34, No.3A (1995), pp.L294-295.
[6] H.Higasa, F.Ishikawa, M.Shibayama and T.Ono, "A feasibility study of an 8-MWh flywheel type power storage system using oxide superconductor", *Electrical Engineering in Japan,* Vol.114, No.7, (1994), pp.20-31.
[7] K.Nagaya, Y.Kosugi, T.Suzuki, and I.Murakami, "Pulse motor with high-temperature superconducting levitation", *IEEE Trans. Appl. Superconduct.*, vol.9, no.4, pp.4688-4694, 1999.
[8] K. Nagaya and S. Shuto, Approximate boundary conditions in a circular conductor and their application to nonlinear vibration analysis of high-Tc superconducting levitation system, *IEEE Trans. Magn.*, Vol.32, No.3 (1996), pp.1888-1896.
[9] K. Nagaya, N. Hayashi, and K. Ohzeki, Control of resonance for rotating shaft levitated by the high temperature super conductor by using a rotating magnetic damper., *J. Material Processing Technology*, (2000), in press.
[10] K. Nagaya, "Method of control of flexible beams subjected to forced vibrations by use of inertia force cancellation", *Journal of Sound and Vibration*, Vol.184, No.2 (1995), pp.185-194.

LEVITATION APPLICATIONS OF HIGH-TEMPERATURE SUPERCONDUCTORS

John R. Hull
Argonne National Laboratory, ET-335, Argonne, IL 60439 USA

1. INTRODUCTION

Figure 1 shows a permanent magnet (PM) stably levitated over a high-temperature superconductor (HTS). Images such as this have become a well known symbol for HTS technology. Its popularity perhaps derives from the ease with which the demonstration can be performed – requiring a simple Styrofoam cup that holds liquid nitrogen, a common PM, and an HTS that can be fabricated in most ceramic kilns. At the same time, the image elicits a sense of awe and mystery, invoked by an object stably suspended by some invisible mechanism.

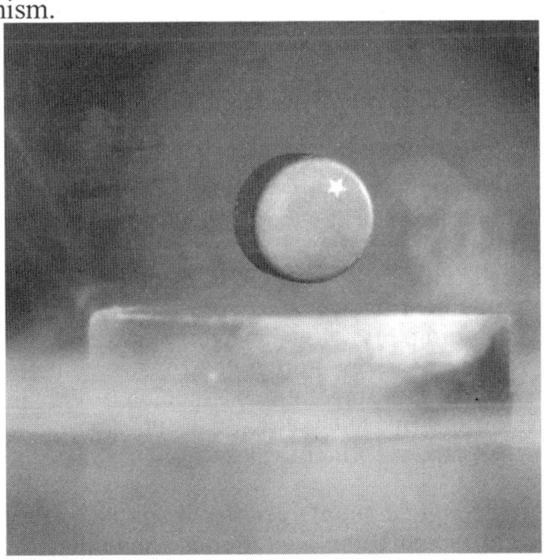

Figure 1. PM (with star) stably levitated over an HTS.

The levitation of a PM over an HTS also provides a tactile and visual indication of many of the basic phenomena associated with HTSs. The levitation is stable. If one pushes the PM up, down, or sideways, or tries to tilt it, a restoring force returns the PM to its initial position. The forces can also be highly hysteretic. If one pushes hard enough, the equilibrium position of the PM can be changed to almost any orientation, or the center of mass of the PM can be moved to a new equilibrium position. If the PM is a cylinder with a relatively symmetric magnetic field, it readily rotates about its axis of symmetry, as indicated by the blurred arc to the left of the star in Fig. 1. Such behavior suggests that the HTS could be used in the construction of a superconducting bearing.

Even before the discovery of the first HTS [1], the ability of superconductors to provide levitation was well established. The first stable levitation that involved a superconductor was reported in 1945 [2, 3], with the levitation of a small PM over a concave lead disk. In this experiment, the superconductor was the stator and the PM was the levitated part or rotor. By 1953 levitation of a small superconducting lead sphere in a coil structure and attempts to measure rotational drag on the superconducting rotor were reported [4]. This first bearing was the prototype for later use of levitated superconducting niobium spheres in sensitive gyroscopes [5]. The first experimental motor that employed superconducting bearings was demonstrated in 1958 [5-7]. For these bearings, superconducting stator coils produced a magnetic field that interacted with a superconducting foil on the rotor. Several years later, a superconducting ring carrying a persistent current had been levitated in a magnetic field, with active feedback coils supplying stability [8]. At the same time, superconductors were being proposed for levitation of large vehicles for transportation applications [9] and high-speed rocket sleds [10]. These early studies all involved the use of low-temperature superconductors (LTSs).

The discovery of HTSs was soon followed by a tabletop demonstration of levitation with the new material [11], and shortly thereafter experiments on levitation were reported by many investigators and later reviewed [12-15]. It was soon found that the PM could be suspended below the HTS [16], or even to the side of it [17]. Although the stable suspension of a superconductor below a PM should have been predictable from magnetization curves as early as 1936 [18], it was only after the experimental demonstration with an HTS [16] that the suspension phenomenon was observed for LTS superconductors at 4 K [19].

The early investigation of HTS levitational phenomena was accompanied by early interest in their use as bearings, and small stable rotors with rotational speeds >100,000 rpm were soon demonstrated [20-22]. This interest arose from the advantages of noncontacting surfaces without an active feedback system, ability to operate in a vacuum, and potential for extremely low rotational drag. In addition, the inconvenience of refrigerating the HTS was deemed more tolerable than that for LTS. Other advantages of HTSs were their relatively simple fabrication and use in bulk form and a much higher magnetic field before onset of flux-jump instabilities.

The development of levitational applications has progressed significantly since the discovery of HTSs and is reviewed in this article. I start with an introduction to basic levitational physics to motivate an appreciation of pertinent superconducting properties. Then, after a brief mention of HTS materials used in levitation, I discuss levitational methods, additional physics and phenomena associated with levitation, modeling techniques, and the phenomena associated with superconducting bearings. These sections are followed by discussions of individual applications, including levitated cryogenic structures, maglev transport, other linear transport systems, flywheel energy storage, and instrumentation. The discussion focuses mostly on those phenomena and applications associated with levitation involving HTSs. However, some discussion is afforded levitation usually associated with normal conductors, since an HTS conductor can, in principle, always be substituted for a normal conductor.

A number of excellent reviews provide a good supplement to this article. Brandt [12, 13] provided an early physical description of some basic levitational phenomena. Moon's monograph [23] gives background on relevant magnetomechanics, and a comprehensive discussion of magnetic transportation and early HTS levitational phenomenology. Hull has summarized developments in superconducting bearings [24] and superconducting levitation [25].

2. BASIC LEVITATIONAL PHYSICS

In a magnetic-levitational process, a superconductor can be used in several ways:

(1) A current will generate a magnetomotive force (i.e., a magnetic field);
(2) A current-carrying element that experiences a magnetic field will experience a force in a direction perpendicular to the current and the magnetic field; and
(3) A superconductor can respond to changes in the magnetic field in such a way as to provide stability.

The three ways are generally interdependent; however, it is useful to separate them conceptually. In the first instance, a superconducting coil carrying current generates a magnetic field. This type of magnet is similar to an electromagnet with the advantage that for a constant current, the coil has no resistance. The coil is also analogous with the Amperian currents that can be used to characterize the magnetization of a PM. The second is a result of the well known Lorentz force law. The third use of a superconductor – levitational stability – takes advantage of several special properties of a superconducting material, for which it is convenient to divide the discussion into the Meissner effect, the zero-resistance effect, and flux-pinning.

A. Diamagnetic Response

One of the fundamental properties of superconductors is their tendency to exclude magnetic flux from their interiors. This exclusion of magnetic flux (the Meissner effect) makes them behave like strong diamagnets. Accordingly, a superconductor with a PM positioned close above it, as shown in Fig. 2a, develops a shielding current, which excludes flux in such a way that the actual magnet "sees" its mirror image.

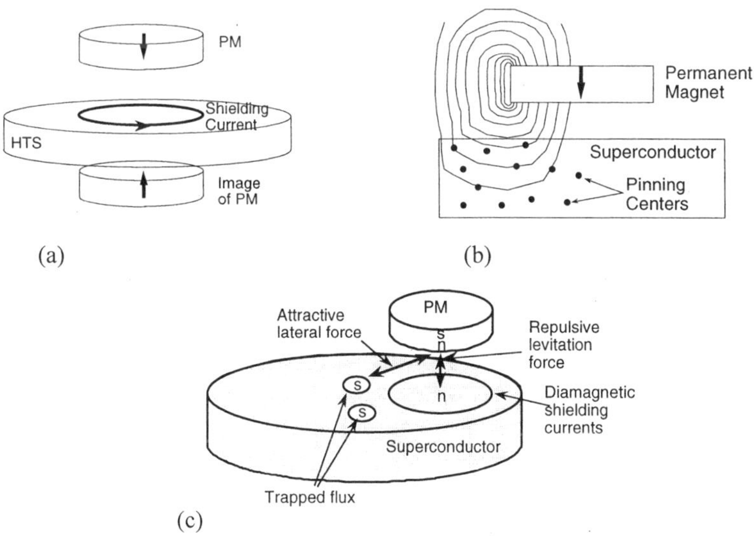

Figure 2. Schematic diagrams of levitation basics: (a) diamagnetic response, (b) flux pinning, and (c) flux trapping.

More specifically, if the magnetization of a PM is in the vertical direction, with its north pole down, the image will also be vertically magnetized, but with its north pole up, exerting a repulsive force on the real magnet. The closer the PM moves to the superconductor, the stronger the repulsive force; the farther away the magnet moves, the weaker the force. This setup can yield a levitation that is stable in the vertical direction. If the superconductor has infinite extent in the horizontal direction, then the setup is neutrally stable in the horizontal plane. If the superconductor has a flat surface and a finite horizontal extent, the setup is unstable in the horizontal direction. Horizontal stability may be obtained if the superconductor is given a concave shape, so that vertical superconducting walls are formed around the PM, as first demonstrated by Arkadiev [2, 3].

In the above discussion, I have assumed that the superconductor is Type-I. In this case, the initial position of the PM while the superconductor cools below its transition temperature makes no difference. The final equilibrium position is always the same, whether the PM is initially far away and brought to a position immediately above the superconductor, or whether the PM is initially resting on the superconductor surface. According to the Meissner effect, flux will always be excluded from the interior of the superconductor. The shielding currents in a Type-I superconductor are completely loss free, so that, in the absence of viscous drag from surrounding gas or eddy currents or hysteresis in the PM, oscillations of a PM above a Type-I superconductor will be undamped.

The same stable levitational state can be achieved from the zero-resistance property of a superconductor if the PM is initially far away and then slowly brought toward the superconductor surface after it has cooled. In this case, the shielding current arises due to Faraday's law to oppose any change of flux inside the superconductor. Because there is no resistance to the superconducting current, it persists indefinitely.

B. Flux Pinning

From a technological viewpoint, the most useful superconductors for levitational purposes are usually Type-II superconductors, in which, above a first critical field, H_{c1}, it is energetically favorable for magnetic flux to enter the interior of the superconductor in discrete localized regions that become normal (i.e., not superconducting) with each region of flux surrounded by a small vortex of superconducting shielding current. In Type-II superconductors, the stability of the levitational phenomena resulting from the diamagnetic response is greatly enhanced by the additional phenomena resulting from flux pinning, shown in Fig. 2b. A flux pinning center is a nonsuperconducting region, such as an inclusion, crack, or other crystalline defect. Because the superconducting region surrounding the nonsuperconducting center is strongly inclined to exclude magnetic flux, a magnetic flux line through the center often becomes trapped there. When a sufficient number of flux lines is trapped in the superconductor, the PM remains stably levitated in position, even over a flat surface.

The flux lines between the PM and the superconductor act in an imperfect analogy to mechanical springs with attachments on the PM and in the superconductor. If the magnet is moved vertically or horizontally, the "springs" will tend to pull the magnet back to its equilibrium position. If the flux pinning is sufficiently strong, the magnet may be stably suspended below the superconductor, or even along its side.

One of the distinctions between LTSs and HTSs is the behavior of the irreversibility field. In addition to the two temperature-dependent phase-transition fields, H_{c1} and H_{c2}, all superconductors have a magnetic irreversibility field, H_{irr}, that lies between H_{c1} and H_{c2}. H_{irr} is the field at which the magnetization M versus applied field H curve is no longer double-valued [26]. For the LTSs NbTi and Nb_3Sn, H_{irr} is extremely close to H_{c2}, and there is no important distinction between them. At higher temperatures, thermal activation is much greater, which leads to easier flux motion near H_{c2} for HTSs. H_{irr} marks a phase transition between the region where magnetic flux is solidly pinned in the superconductor and the region where flux may move. Sometimes the curve is said to denote the boundary between the region where flux is frozen and the region where flux is melted. For stable levitation, the flux must be frozen in the superconductor. Otherwise, the PM would slowly lose levitation height.

If the magnet moves far enough laterally that the flux lines move from their original pinning centers to new ones, then an additional stabilizing force, involving trapped flux, begins to act. In the superconductor, trapped flux appears as regions of induced magnetization of the same pole orientation as the levitated magnet. The trapped flux results from movement of flux lines from their original pinning centers so that the local flux decreases. As shown in Fig. 2c, this flux movement results in an attractive interaction with the PM that provides a lateral restoring force. The lateral motion of the PM also produces a slight increase in the vertical force because of the flux trapped in the original location.

For a Type-II superconductor, the equilibrium position of a levitated PM depends on its position during the time that the superconductor transitions below the critical temperature. One may distinguish two cases: zero-field cooling (ZFC), where the superconductor is cooled with the PM far away, and field cooling (FC), where the superconductor is cooled with the PM in close proximity. In the ZFC case, as the PM is brought close to the superconductor, the flux tends to be excluded or trapped near the superconductor's surface, and the repulsive force between the PM and the superconductor is strong. In the FC case, a substantial amount of flux is trapped in the superconductor and very little force is exerted on the PM when it is at the FC position.

3. HTS MATERIALS USED IN LEVITATION

A detailed discussion of HTS materials and their fabrication is beyond the scope of this article, and I will be concerned with HTS properties only insofar as they affect levitational phenomena. In principle, almost any HTS material and form could be used in a levitational process. HTS wires and thin films have been investigated somewhat, but most efforts have involved the use of bulk HTSs. Unlike superconducting wire applications, in which the supercurrent must pass from grain to grain along quite a distance, the supercurrent for levitation applications needs to circulate only within individual grains.

A. Bulk

The use of bulk HTSs for levitational purposes was invigorated when melt-texturing and other techniques allowed fabrication of single-domain components that were larger than 1 cm in size. Magnetization of the superconductor is proportional to the product of the critical current density (J_c) and the grain diameter [27]. Large grain diameter is important when sufficiently large magnetization must be achieved for useful levitational forces. Many excellent reviews [28-31] and original papers [32, 33] cover details about the fabrication issues of melt-textured HTSs.

The current material of choice for superconducting levitation is Y-Ba-Cu-O (YBCO) and the rare-earth Ba-Cu-O analogs (RE-Ba-Cu-O, where RE denotes the rare-earth elements Nd, Sm, Eu, Gd, Dy, Ho, Er, Tm, Yb, Lu, and La), because they exhibit a high H_{irr} at liquid nitrogen temperatures, and many have the ability to grow large grains. When compared with some of the members of the Bi, Tl, or Hg HTS families, YBCO exhibits a relatively low critical temperature of 92 K, but its irreversibility curve is one of the highest at 77 K and lower temperatures. When bulk YBCO materials are made by a melt-texturing process, grains can be grown to diameters of several centimeters [27]. In the present state of the art, the upper limit of the grain diameter produced by melt texturing is ≈10 cm. The ability to produce good-quality YBCO thin films is also limited to approximately this size. If techniques to grow large grains for other HTS materials with high H_{irr} curves could be developed, these materials would also be of interest for levitation applications.

Although state-of-the-art melt-textured YBCO is satisfactory for many levitation applications, researchers are continually striving to obtain higher J_c values, partially to more closely approach the theoretical maximum of levitational force, but primarily to improve the economics of using HTS materials in bearings and other applications. The J_c values can be substantially increased by techniques that involve radiation [34], and materials made by these methods are expected to be of interest in levitation designs that require greater magnetic fields and in trapped-flux applications.

The cost of the magnetomotive force for a bulk HTS is competitive with other superconductors. Consider a pellet that is 26 mm in diameter by 13 mm high and has a current density of 40 kA/cm^2. Such pellets can be made in large quantities for less than \$30 a pellet. The cost of the supercurrent is then about \$0.01/Am, an often-quoted cost target for wire HTS.

B. Wire

The use of superconducting wire to wind coils for the production of magnetic fields is a well established practice. Coils made from the LTSs NbTi and Nb3Sn are found in many laboratories and commercial practice. Efforts to develop HTS wire continue apace. Wires produced from the bismuth family of superconductors have been made into many prototypes and laboratory solenoids, some of which have been used in levitational and HTS trapped-field experiments. These HTSs must operate at temperatures lower than about 20-30 K to generate significant magnetic fields. Efforts to develop long lengths of coated conductors of Y-Ba-Cu-O for operation at 77 K are ongoing. The material MgB_2 is also a potential candidate for wire applications in the 20-30 K temperature range.

C. Film

A number of studies have been performed on the levitational properties of thin-film HTSs [35-39]. Although YBCO thin films often have J_c values of >1 MA/cm^2 at 77 K, the thickness of these films is only ≈1 μm, and they do not provide much levitation force. The magnetization quickly saturates, and very large hysteresis loops are possible if this occurs. However, when unsaturated, the stiffness of such films is often similar to that produced by bulk material of greater thickness [36, 38].

4. MAGNETIC LEVITATIONAL METHODS

Levitation using HTSs forms a subset of magnetic levitation, and general references to magnetic levitation techniques are pertinent [23, 40-49]. The levitation of a PM above or below an HTS has been discussed in Sec. 2. The inverse, whereby a cold HTS is levitated above or below a PM, is governed by the same physics.

A. Magnetic Forces

The basic forces involved in magnetic levitation are derived from the basic laws of electricity and magnetism. The first basic force is that between magnetic poles. Although isolated magnetic poles are not known to exist in nature, they constitute a convenient model and are discussed in most elementary science texts. For a pair of magnetic poles, opposite poles attract, like poles repel, and the force between point poles is proportional to the inverse square of the distance between them.

A force also occurs on a current-carrying wire in a magnetic field. The force is proportional to the current, the wire length, the strength of the magnetic field, and the sine of the angle between the current direction and the magnetic field. The direction of the force is perpendicular to both the direction of the current and the direction of the magnetic field.

A magnetic dipole may experience torque in a magnetic field, with the torque proportional to the dipole moment, the strength of the field, and the sine of the angle between the moment and the field. The torque always acts to make the direction of the magnetic moment the same as that of the field. A magnetic dipole can experience a net force in a gradient of the magnetic field. The force is proportional to the dipole moment, the field gradient, and the cosine of the angle between the directions of the moment and the field gradient.

Another basic force occurs when a conductor moves in a magnetic field. This force is derived from Faraday's law and Lenz's law, which indicate that a changing magnetic flux produces a voltage in a circuit in such a way that it opposes the change in flux. This effect is responsible for jumping rings and eddy currents [46].

The response of a material to magnetic fields also produces a magnetic levitation force. A diamagnetic material, such as graphite, aluminum, water, or a superconductor, will be repulsed by a magnetic pole, whereas a paramagnetic material, such as oxygen, will be attracted to a magnetic pole. Ferromagnetic materials, such as iron, are strongly attracted to magnetic poles.

A convenient parameter in the discussion of magnetic forces and their stability is the magnetomechanical stiffness. This stiffness is the negative of the spatial gradient of the magnetic force. It is directly analogous to a spring constant in a mechanical system. In general, a system is stable in a given direction if the stiffness in that direction is positive. Conversely, the system is generally unstable if the stiffness is negative.

B. Levitation of Permanent Magnets

Magnetic levitation of one PM by another is not stable according to Earnshaw's theorem, which is discussed in Sec. 5.A. Nevertheless, this arrangement is still useful if stability is provided by other means. A simple rendition of such a levitation, originally proposed by Evershed [50], is shown in Fig. 3a. Here the levitated object (the rotor) consists of a PM below which hangs a rigid rod with a point on the bottom. Each PM is magnetized, as shown by the dark arrow in Fig. 3a, with its north pole down. The PM of the rotor experiences an attractive force toward the stationary PM immediately above it. This magnetic system is statically stable in the radial direction but unstable in the vertical direction. The gap between the two magnets is adjusted in such a manner that the attractive force between the pair of magnets is just less than 100% of the rotor weight, so that the rotor would tend to fall. The remainder of the weight is supported by the small mechanical bearing at the bottom, which supplies sufficient stiffness for vertical stability and some additional radial stability.

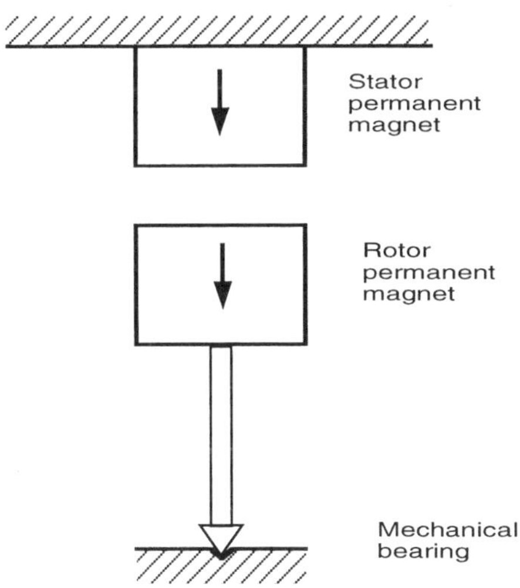

Figure 3. Evershed bearing design: (a) simple schematic of vertical orientation.

The Evershed design, either in the simple form shown in Fig. 3a, or in some modification, is used in applications ranging from simple toys and watt-hour meters to high-speed centrifuges. This basic design may be combined with other magnetic levitation techniques and is often referred to as "magnetic biasing." An example of an Evershed technique used in a simple toy is shown in Fig. 3b. PMs in the dumbbell-shaped rotor and in the black plastic platform create a configuration that is stable in the radial direction but unstable along the axis of rotation. The small glass sheet attached to the platform and metal point on the rotor provide a mechanical bearing that stabilizes the system along the rotational axis.

Figure 3. Evershed bearing design: (b) use of Evershed principle in simple toy in horizontal orientation;

A modification of the basic Evershed design is shown in Fig. 3c. Here, the PMs are arranged in a push-pull configuration to achieve a near-zero stiffness. The set of three PMs would replace the two upper PMs in Fig. 3a. It is convenient to consider the upper and lower PMs as stators that are mechanically connected and to consider the middle PM as the rotor. The attractive force between the upper PM and the middle PM is stable radially and unstable vertically. The repulsive force between the bottom PM and the middle PM is stable vertically and unstable radially. When combined, there is a net upward force on the middle PM, and the stiffness is zero when the middle PM is in the middle of the gap between the two stators.

Typically, the operating point of the rotor is chosen to be above the middle of the gap, such that the push-pull configuration is radially stable but vertically unstable. With radial stability provided at the top of the rotor by the push-pull set of PMs and at the bottom of the rotor by the mechanical

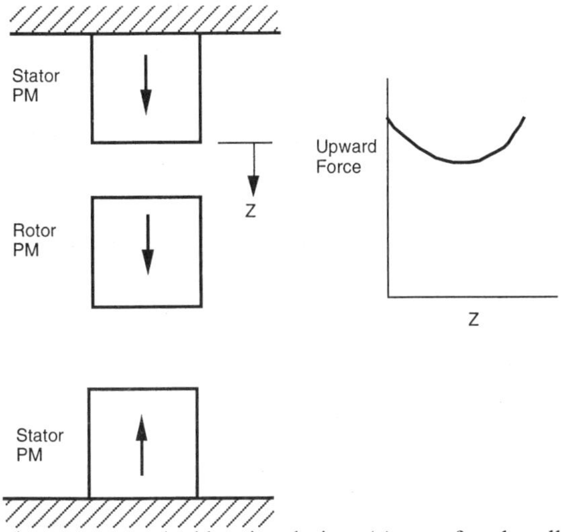

Figure 3. Evershed bearing design: (c) use of push-pull configuration to create near-zero stiffness levitation.

Truly contact-free magnetic levitation can be attained by replacing the mechanical bearing in Fig. 3a with an active magnetic bearing or with an HTS bearing [51-53]. The latter is shown in Fig. 3d. In practical HTS levitational systems, the relatively low levitational pressure available in the interaction between the PM and the HTS is often augmented by various hybrid schemes in which interactions between pairs of PMs provide the bulk of the levitational force.

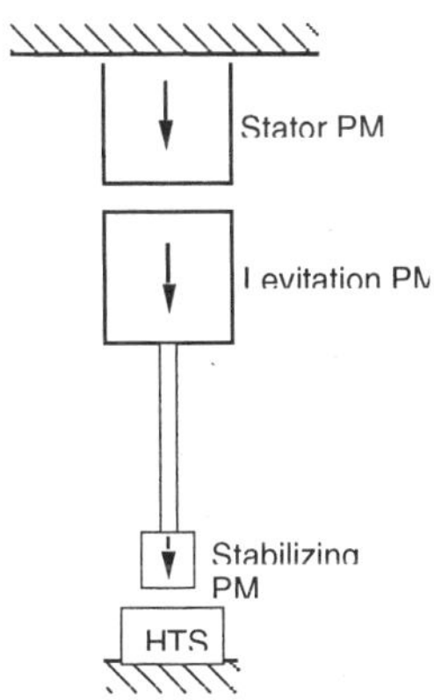

Figure 3. Evershed bearing design: (d) use of HTS bearing to stabilize Evershed configuration. Most of the weight is provided by attraction between permanent magnets, and the remaining weight and vertical stability are provided by a small mechanical bearing.

C. Electromagnetic Levitation

Contact-free magnetic levitation can also be attained by replacing the mechanical bearing in Fig. 3a with a small electromagnet and a feedback system, as shown in Fig. 4. As before, most of the levitation force is provided by the attractive force between the two permanent magnets. The coil in Fig. 4 is wired to magnetize the iron in the same sense that the permanent magnets are magnetized. Then, as in Fig. 3a, when the coil is energized, the system is stable in the radial direction but unstable in the vertical direction. The feedback system consists of two sensors that detect the top and bottom edge of the levitated permanent magnet. For example, the sensors may be photocells that detect light from a light source to the left of the magnet. When light to the bottom sensor is obscured by the falling magnet, the current to the coil is turned on, and the attractive force increases. When light to the top sensor is obscured by the rising magnet, current to the coil is turned off, and the attractive force decreases. In such a manner, the magnet may be stably levitated. This basic design forms the basis of most active magnetic-bearing concepts. The actual control algorithms and sensors may be more complicated than the simple system illustrated in Fig. 4. Several researchers have investigated the possibility of combining HTS bearings with active magnetic bearings [54-57] and with eddy current dampers [58].

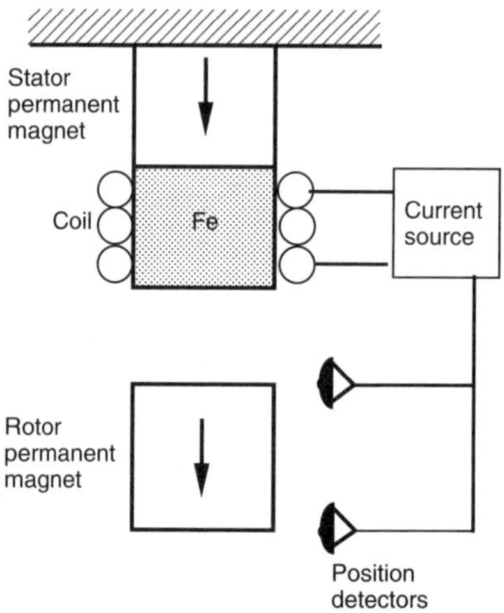

Figure 4. Attractive levitation of a permanent magnet with feedback control of an electromagnet.

Stationary electromagnets may also be used to levitate HTSs [59-62]. The electromagnets may be wound from normal conductors or superconductors. The levitation can be made stable without the use of an active feedback mechanism. Typically, as the electromagnet is energized, the diamagnetic nature of the HTS causes it to levitate above the magnet. If the HTS has a preexisting trapped field, then the HTS can be suspended below the electromagnet. A major advantage of this approach is a higher levitational pressure, in that ferromagnetic materials may achieve magnetizations up to ≈2.5 T, and, with a NbTi superconducting coil, the magnetic field at the surface of the HTS can easily be 5 T. A further advantage is that azimuthal homogeneity of the magnetic field produced by a coil can be much higher than that attainable in a PM; therefore, the potential for extremely low-loss bearings exists. Taking advantage of the homogeneity requires holding the energizing current of the coil constant in time to prevent changes in levitation height. The main disadvantage of this approach, when compared with levitating a PM, is that much lower temperatures are required, and the levitated HTS must be cooled over long periods of time.

D. Alternating-Current Levitation

The ability of electromagnetic forces to impart significant levitation forces through Faraday's law is well known from the jumping-ring experiments [46] that are performed in many introductory physics classes. Consider the system shown in Fig. 5, in which a coil is stationed above a conducting plate. If the coil is energized with a pulse so that current flows as shown, by Faraday's law, eddy currents will be induced in the plate in such a way that magnetic flux is expelled from the plate. The effect of these eddy currents is that the coil will "see" its mirror image, as shown in Fig. 5, and the interaction between the coil and its image will produce a repulsive levitation force. The system is stable in the vertical direction and neutrally stable in the horizontal direction. If the plate has a slight concave shape, the system will also be stable in the horizontal direction.

Because of the finite electrical conductivity of the plate, the eddy currents in the plate that arise from a current pulse in the coil will exponentially decay in time, as will the levitation force, after the pulse has ended. In the jumping-ring experiment, the conducting plate consists of a ring that sits on top of a coil. When the coil is pulsed, the mutual repulsive force causes the ring to accelerate upwards. To provide a more continuous levitation force, another current pulse is needed in the coil before the initial eddy currents have completely died away. This pulse is most conveniently accomplished by supplying the coil with an alternating current.

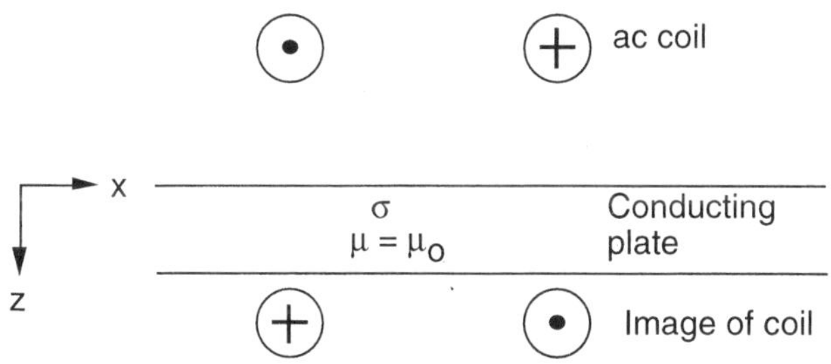

Figure 5. Repulsive levitation of an ac coil above a conducting sheet using method of images. σ is the electrical conductivity, μ is the magnetic permeability, and μ_0 is the magnetic permeability of vacuum.

E. Electrodynamic Levitation

In electrodynamic levitation, a moving magnet (PM, electromagnet, or superconducting magnet) interacts with a conducting sheet or a set of coils to produce a levitation force. A drag force, typically much higher than that associated with ferromagnetic suspensions, is associated with the eddy currents. However, above some speed, the drag force decreases as the reciprocal of the square root of the velocity. The system is passively stable, i.e., no feedback is required. The disadvantage is that there is a minimum speed below which the levitation force is not sufficient, so some mechanical support is needed on startup.

This phenomenon can be understood by applying the principle of images, as shown in Fig. 6. In the case of a plane conducting sheet, the imaginary system on the negative side of the sheet is not the simple image, positive or negative, of the real magnet on the positive side, but consists of a moving train of images [46, 63-66]. According to this model, when a magnet passes a point on the conducting plane, it induces first a "positive" image, then a "negative" image. These images propagate downward at a velocity R, which is proportional to the specific resistivity (and to the reciprocal thickness if the sheet is thin when compared with the skin depth). R is also the electrical resistance of a square portion of the conducting sheet; its value $R = \rho/2\pi h$ (ρ = resistivity, h = thickness) is independent of the size of the square. In electromagnetic units, R has the dimensions of velocity.

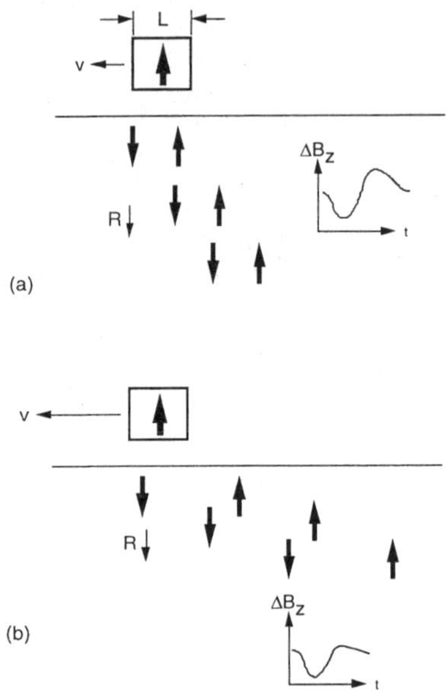

Figure 6. Maxwell's eddy current model applied to a magnet moving over a conducting plane: (a) low velocity and (b) high velocity. The vertical component of the induced images and the magnetic field due to the eddy currents are shown in each case.

Two examples that apply Maxwell's model are shown in Fig. 6. In the first example, the velocity v of the magnet is < R. The positive image has moved down a distance Rdt = RL/v when the negative image appears at the same location. Then, as the two images move away head-to-tail, the induced field falls toward zero. Because of the eddy currents, the vertical component of magnetic field ΔB_z is in one direction near the leading edge of the moving magnet and the opposite direction at the trailing edge, which leads to the waveform shown at the right of Fig. 6a.

In the second example (Fig. 6b), the velocity is considerably greater than R. The positive image has moved only a small distance RL/v away when the negative image appears, and the two images nearly cancel each other thereafter. Now, the magnetic field due to the eddy currents ΔB_z is predominantly in one direction, as shown in Fig. 6b.

If the conducting sheet is concave, the system can be made stable in the horizontal direction perpendicular to the direction of the velocity.

One may improve on the basic system of a magnet moving over a conducting plate by means of the null-flux-geometry system [9] shown in Fig. 7. When the moving magnet or coil is in the symmetry plane, no net flux threads the track loop so that lift and drag forces approach zero. Lift forces increase linearly, and drag forces rise as the square of the (small) displacement Δh from the symmetry plane. The velocity dependence of null-flux systems is the same as that of the eddy current systems.

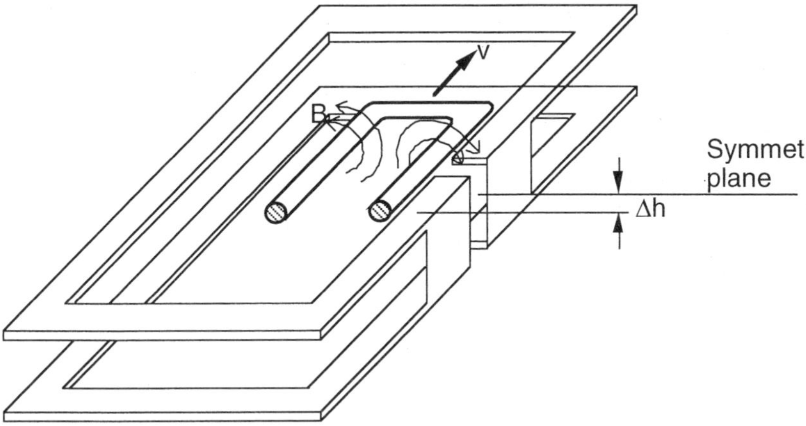

Figure 7. Null-flux geometry, showing full stationary null-flux coil and one-half of levitated coil moving with velocity v and displacement from the symmetry plane by Δh.

F. Levitation by Tuned Resonators

Several magnetic bearing concepts involve passive techniques. These concepts have the advantage of simplicity and the lack of a control system. One system achieves a stable stiffness characteristic by using an LC circuit excited slightly off resonance [41]. The LC circuit is formed with the inductance of the electromagnetic bearing coil and a capacitor. The mechanical displacement of the rotor changes the inductance of the electromagnet. The LC circuit is operated near resonance and tuned in such a way that it approaches resonance as the rotor moves away from the electromagnet. This increases the current from the ac voltage source and thus pulls the rotor back to its nominal position. The low forces and stiffnesses of this system, coupled with the necessity for continuous ac energization of the coils, are disadvantages. The system is also subject to a low-frequency negative-damping instability, and auxiliary damping is usually required for stability.

G. Diamagnetic Levitation

The possibility to stably levitate diamagnetic materials was predicted as early as 1882 [67]. In 1939, Braunbek [68], and later Boerdijk [69], experimentally demonstrated the stable levitation of small pieces of bismuth and graphite (these are slightly diamagnetic materials with relative magnetic permeability $\mu_r < \approx 1$) in strong magnetic fields. Interest in levitation of diamagnetic materials has increased with the availability of superconducting magnets that possess high fields extending over large volumes [70, 71]. A 16-T steady magnetic field has been used to levitate a living frog [72]. The magnetic field induces a magnetic moment m in the diamagnetic material that is in the opposite direction to the applied field B, where the magnitude of the moment is given by

$$m = |\chi| V B / \mu_o \qquad (1)$$

In this equation, χ is the susceptibility of the material (water, which constitutes the majority of the mass of most living creatures, is weakly diamagnetic with $\chi \sim -1 \times 10^{-5}$), V is the material volume, and μ_o is the permeability of free space. The magnetic moment of the diamagnetic material interacts with the gradient of the applied magnetic field to produce a levitation force. For a magnetic field that is predominantly in the vertical direction, the vertical component of the levitational force against gravity is given by

$$F_z = m\, dB/dz \qquad (2)$$

For the levitation to be stable, the levitated object must be placed where the energy is a minimum. This is only possible with diamagnetic objects. The levitation is also stable because the magnetic moment of the diamagnetic object rotates with the applied magnetic field.

Figure 8 shows how diamagnetic levitation can be combined with levitation by a PM pair to form a stable system. In Fig. 8, the levitated PM is attracted upward to a stator ferrite PM. Two disks of graphite, one immediately above and the other below the PM, provide the stabilization. The PM/PM interaction is stable radially but unstable vertically. The PM/diamagnet combination is neutrally stable radially and stable vertically.

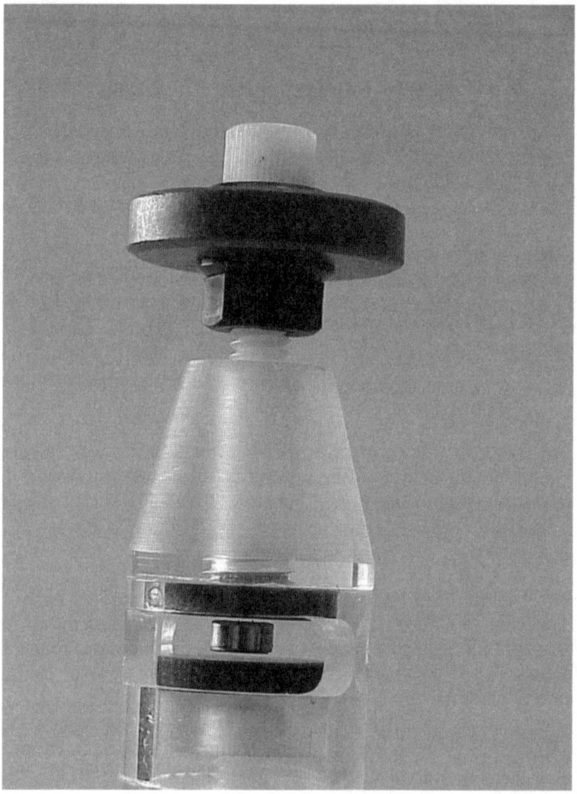

Figure 8. Example of use of diamagnetic levitation to stabilize a PM. The device was manufactured by M. D. Simon.

H. Mixed-μ Levitation

An alternative design to the conventional PM/HTS system is the mixed-μ levitation [73, 74], shown in Fig. 9. This type of levitation has analogies to the diamagnetic levitation shown in Fig. 8. In this levitation, a soft ferromagnetic cylinder (Fe) ($\mu_r > 1$, where μ is the relative magnetic permeability) is levitated in attractive levitation between two PMs and stabilized by two HTSs ($\mu_r < 1$), one HTS placed between the rotor and each of the PMs. The PM pair induces a magnetization in the Fe, which experiences a levitation force that depends on the gradient of the applied magnetic field from the PMs. For stabilization, the magnetization induced in the Fe produces diamagnetic images and trapped flux in the HTSs. In Fig. 9, the HTSs are housed in nonmetallic cryochambers, the lower PM is mounted on a movable stage, and the Fe cylinder is surrounded by a ceramic disk-type rotor. The PMs shown in Fig. 9 are approximately identical. In this case, to obtain a net upward force, the distance between the upper PM and the rotor must be less than the distance between the lower PM and the rotor.

One advantage of this design is that the rotor can be made of a material with high tensile strength, such as steel, and its maximum rim velocity should be significantly greater than that for a rotating PM or HTS. A second advantage is that losses in the HTSs are not affected by inhomogeneities in the PMs. The primary contribution to a changing magnetic field near the HTSs is due to radial oscillation of the rotor. To date, the mixed-μ bearing has displayed the lowest rotational loss of any HTS bearing tested in the author's laboratory. Disadvantages of the mixed-μ design include a stability that is weaker than the stability of the PM/HTS configuration and difficulty in scaling the design to larger sizes.

The general concept of mixed-μ levitation, in which a ferromagnetic and diamagnetic structure are combined, was also considered with low-temperature superconductors [75, 76]. With HTSs, many possible geometries are possible besides that depicted in Fig. 9 [77, 78].

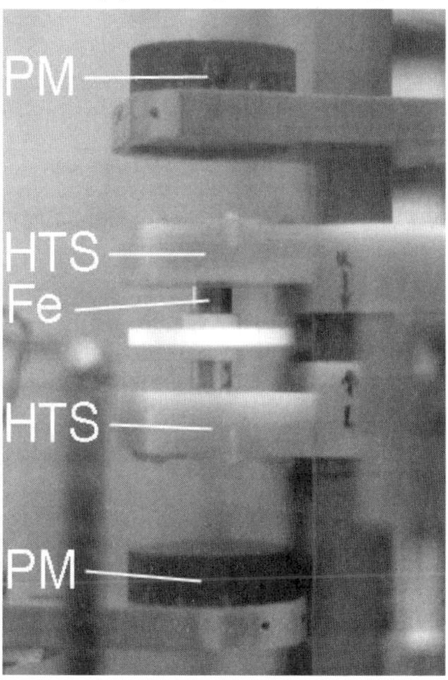

Figure 9. Mixed-mu HTS bearing.

I. Levitation with Flux Compression

Another levitational method uses the strongly diamagnetic properties of HTSs to provide surfaces between which magnetic flux can be compressed [79]. A schematic of the basic concept is shown in Fig. 10. The stator consists of a bulk HTS cylinder, as in conventional PM/HTS levitation, but the rotor consists of a cylindrical PM surrounded by an annular HTS component. In Fig. 10, R_m is the radius of the PM, R_s is the radius of the rotor HTS, L is the height of the rotor, and g is the gap between the rotor and the stator HTS. Unlike conventional PM/HTS levitation, in which the PM rotor can be at any temperature below the Curie point of the PM material, here the rotor must be cold enough that the HTS part of the rotor superconducts. Referring to Fig. 10, if both HTSs act as perfect diamagnets, then the flux that emanates from the bottom pole of the PM must pass between the top surface of the stator HTS and the bottom surface of the rotor HTS. This flux eventually makes its way to the top pole of the PM. Assuming that flux is excluded from the interior of both HTSs, then as the gap between the rotor and stator decreases, the magnetic flux in the gap compresses. The compressed magnetic flux results in a higher magnetic field in the gap and a higher levitation pressure.

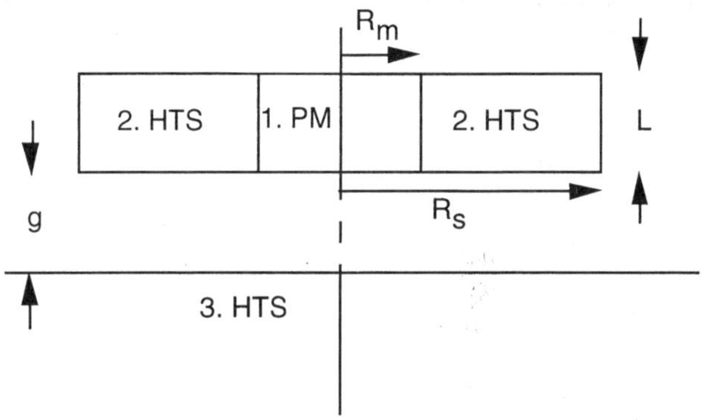

Figure 10. Schematic of cage-cooled superconducting bearing that uses flux-compression between the two HTS surfaces for levitation.

The maximum possible levitation pressure obtained with the geometry of Fig. 10 can potentially exceed that between a PM and its magnetic image [79]. Physically, this can be understood by analogy with the multiple images that occur when two mirror surfaces are placed parallel to each other. In our case, the PM creates a magnetic image in the stator HTS that is opposite in polarity to the real PM polarity and creates a repulsive force. This image creates its image in the rotor HTS, which has the same polarity as the PM and adds to the repulsive force. This rotor image, in turn, creates its image in the stator, etc. All of the images in the rotor have the same polarity as the PM, and all of the images in the stator have the opposite polarity as the PM. If the gap is relatively large, then the additional images occur at successively greater effective distances and do not contribute significantly to the levitation pressure. However, if the gap is small, the multiple images are relatively close to each other, and the levitation pressure can become quite high in the ideal case.

A difficulty in the fabrication of such a levitation system is the initial creation of the rotor. This difficulty was overcome in some early experiments by using a procedure called "cage cooling," in which the rotor is surrounded by a soft ferromagnetic shell while the rotor HTS is cooled [79]. The shell is removed prior to assembly of the bearing. So far, only a few experiments have been performed with this type of levitation.

J. Dynamically Stable Levitation

Several levitational techniques require some type of dynamical process in order to generate a levitational force or to maintain stability. One dynamical method was mentioned briefly in Sec. 4.E for the case of eddy currents used to induce a levitational force. Eddy current bearings have been investigated for a number of years [80, 81], and optimization of their design continues [82-84].

The Levitron® is a toy, manufactured by Fascinations in Seattle, that consists of a spinning permanent magnet in the form of a top. The spinning magnet is repelled by a magnet of opposite polarity in a stationary base beneath the top. The levitation height is such that the top is statically stable in the vertical direction and unstable in the radial direction. As designed, the top is also statically unstable to rotation of the magnetic moment. The operating principle of the Levitron is similar to that of other passive magnetic bearing concepts in that its stability is dynamical [85-87]. It is distinguished from other passive magnetic bearing concepts in that its stability is the result of the tendency of its precession axis to align with the local field direction, that is, when the top moves radially away from equilibrium, the magnetic moment of the top will rotate so as to point in the direction of the magnetic field from the stationary magnet. This rotation of the moment adds a term to the potential energy of the system that is not present if the magnetic moment is decoupled from the radial excursion. There is a rotational frequency range for the dynamic stability. The top must be spinning fast enough so the gyroscopic action prevents it from flipping over. However, if the top is spinning too fast, the gyroscopic action will maintain the direction of the moment vertical, and it cannot reorient fast enough to point in the direction of the magnetic field. If the top is able to precess several times around the field line during the time scale of a radial excursion, the system is said to be "adiabatically stable." The same concept is used to trap cold neutrons and produce Bose-Einstein condensation [87]. Under presently investigated conditions, the frequency range in which the top is stable is rather small, and the stability is rather fragile. It is not clear whether this concept would be useful in practical magnetic bearings.

Another magnetic levitation technique that is dynamically stable is analogous to the strong-focusing technique used in particle accelerators [88]. In this system the static stability of the magnetic system alternates as the levitated part rotates. During one portion of the rotation, the system is radially stable and vertically unstable. During the next portion of the rotation, the system is vertically stable and radially unstable. For some frequency ranges, such a system can be dynamically stable.

K. Current-Carrying Conductor in DC Field

Applications exist for which the levitation of a wire [89] or a coil [8,90] in an external magnetic field is desirable. In general, a wire in an external field is not stable. Also, similar to a PM, a constant-current coil is not stable in a magnetic field. These systems will need some sort of external constraint or active feedback mechanism to remain stable. However, a persistent-current superconducting coil can be made stable under some conditions. Consider a persistent-current HTS ring suspended in attractive levitation under a constant-current coil. The system is stable in the radial and tilt directions. The flux enclosed by the suspended coil wants to remain constant, so when the coil drops away from the stationary coil, the current in the suspended coil will increase and act to provide stability in the vertical direction. When the suspended coil moves toward the stationary coil, the current in the suspended coil will decrease and again act to provide vertical stability. In a real system, hysteretic losses in the suspended coil will occur as the current changes, and these losses will act to degrade the current over time.

L. Permanent Magnet Configurations

While a single PM is readily attainable and can provide a workable levitation system, optimization of the PM configuration usually results in a more complicated structure. One or more PM rings are often used, with both radial and axial magnetizations arranged in different combinations [91]. Ferromagnetic pole pieces and magnetic shunts are often employed to increase the levitational force of stiffness. Examples are shown in Fig. 11. The PM configurations shown in Fig. 11 can be used with other PMs or with HTSs. When PMs are used in the rotor part of a high-speed bearing, structural considerations can also enter the design process.

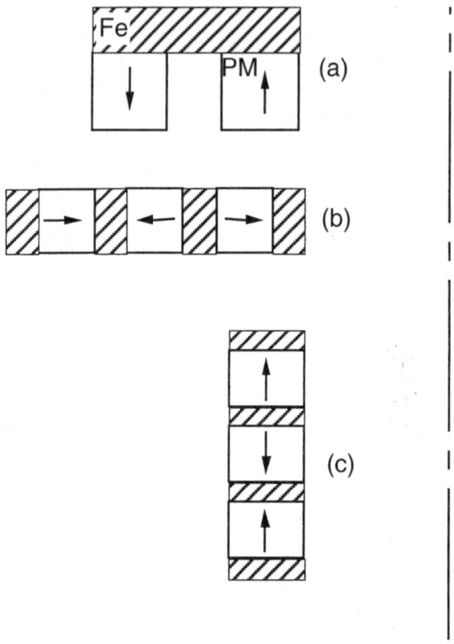

Figure 11. PM configurations involving ferromagnetic structures. See text for further description.

The magnetic field adjacent to a pole of a PM is largest when the aspect ratio of the PM is high, i.e., when the length is much greater than the width. To maximize the levitational pressure in a PM/HTS system, one may increase the magnetic field at the bottom of the PM by making the PM cylinder taller. This has the disadvantage of increasing the levitated mass and cost of the PM. A more acceptable design uses nested rings of PMs with alternating vertical polarity and includes a shunt of soft-ferromagnetic material on top of the rings [92-94]. In Fig. 11(a) the PM consists of a double ring with an iron shunt on top that connects the two rings. The purpose of this iron shunt is to increase the magnetic field produced below the PMs, effectively yielding the same magnetic field as if the PMs were infinitely high. The axial levitational force produced by such a configuration is maximal when a radial gap separates the two rings. Such a PM configuration was used in an HTS bearing for a high-speed flywheel [93, 95].

Radially magnetized nested rings of alternating polarity separated by soft ferromagnets, such as illustrated in Fig. 11(b), are also possible. Several studies have addressed the optimization of these designs [96-99], and HTS bearings of this type have also been used in a flywheel [100, 101]. Such designs are thought to have high stiffness. Typically, the rings are composed of segments, with each segment magnetized in a single horizontal direction. Thus, such designs often have more circumferential inhomogeneity than those that use axially magnetized segments.

A similar design that uses axially magnetized segments is shown in Fig. 11(c). This configuration is also used in HTS bearings [102]. In this case, the HTS is typically located radially outward from the PM. Whereas in Figs. 11(a) and (b) the axial levitational force is created by compressing flux between the PM and the HTS that is located below the PM, the configuration shown in Fig. 11(c) achieves the axial force by shearing the magnetic flux lines between the PM and the HTS. An advantage of this configuration is that large surface areas can be attained without increasing the outer radius of the PM, thereby keeping the centrifugal forces on the PM to a minimum. Such designs are often considered when large radial stiffness is deemed necessary.

M. Hybrid HTS Concepts

Several other hybrid HTS designs, shown in Fig. 12, deserve mention. Figure 12(a) shows a PM pair in repulsive levitation with an HTS located between them acting to stabilize the system [51, 103]. In this case, the forces between the PMs are stable vertically and unstable radially. The HTS adds additional vertical stability, and being closer to the rotor PM, its positive stiffness in the radial direction is larger than the negative stiffness in the radial direction between the PM pair. The configuration with an HTS between a PM pair can also be used with the PMs in attractive levitation [51, 103]. Various combinations of attractive-force and repulsive-force hybrids have been devised [51, 103, 104].

One may also establish a levitation system where the HTS has an initial trapped field, as shown in Fig. 12(b). Compared to the conventional PM and HTS configuration, one may increase the levitational force considerably by this method; however, the stability of the system may suffer [105, 106]. To create trapped flux, the HTS is usually field cooled in the presence of an external electromagnet or superconducting magnet. Alternatively, large pulsed fields can be applied after the HTS is cooled, although this method is the less effective of the two at trapping flux. The reported trapped fields are usually the maximum value measured at the surface of the HTS after the magnetizing fields have been removed. The average magnetization of the cylindrical HTSs is significantly less than these maximum fields. Trapped fields that are significantly >2.0 T have been reported at 77 K, and fields of up to 16 T have been reported at lower temperatures [106-109].

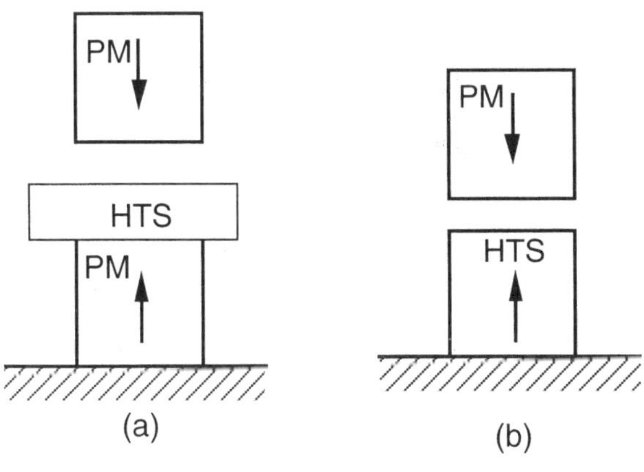

Figure 12. Examples of hybrid HTS levitation methods. See text for further description.

5. LEVITATIONAL PHYSICS – PART 2

A. Earnshaw's Theorem

To appreciate the stability of the PM/HTS system, one need only compare it with most other passive magnetic systems, which are statically unstable. Most magnetic systems are governed by Earnshaw's theorem, which states that there is no stable, *static* three-dimensional arrangement of a collection of poles (electric, magnetic, or gravitational) whose magnitudes do not change and where particles interact via a $1/r^2$ force law [110]. Particles that experience inverse-square-law forces obey Laplace's equation, and Earnshaw's theorem is developed from the property of the associated curl- and divergence-free fields that precludes the existence of local, detached, scalar-potential maxima or minima, and allows only saddle-type equilibria. In terms of magnetic stiffness K, the theorem for a collection of poles can be written as

$$K_x + K_y + K_z = 0 \, . \tag{3}$$

For axisymmetrical systems, the theorem may be rewritten as $2K_x = - K_z$, and the system will be unstable in the direction that exhibits the negative stiffness. Braunbek extended Earnshaw's result to show that electric or magnetic suspension is not possible when all materials have $\varepsilon_r > 1$ or $\mu_r > 1$, but that it is possible when materials with $\varepsilon_r < 1$ or $\mu_r < 1$ are introduced (where ε_r is the relative electrical permittivity, and μ_r is the relative magnetic permeability) [111]. These results collectively are often referred to as "Earnshaw's theorem." This theorem is grasped intuitively by most people when they release a permanent magnet next to the ferromagnetic door of their refrigerator. The magnet either moves to stick to the door, or it falls on the floor; it does not hover in space near the point where it was released.

Although superconductors probably do not strictly satisfy most formal definitions of a diamagnetic material ($\mu_r < 1$), the HTS often strongly mimics the behavior of a pure diamagnetic material ($\mu_r << 1$), and it is useful to invoke this picture when we strive to understand many of the levitation phenomena. Diamagnetic materials, such as superconductors, are not governed by Earnshaw's theorem, and this condition enables the possibility of creating stable levitation systems. The possible generalization or breakdown of Earnshaw's theorem for nonlinear media has received some attention [112]. We note that Earnshaw's theorem applies only to conditions of static stability, not to dynamic systems.

B. AC Levitation

The general phenomenon of levitation involving alternating currents or fields can be quantitatively understood by considering an alternating magnetic field, $\mathbf{B} = B\mathbf{i}$ (where $\mathbf{i}$ is the unit vector in the x direction, and $B = B_o \exp(j\omega t)$ incident on the surface of a half space $z > 0$, where j is the square root of -1), with electrical conductivity σ and magnetic permeability μ, ω is the radial frequency, and t is time. One then must solve the magnetic diffusion equation

$$\partial^2 B/\partial z^2 = \mu\sigma \, \partial B/\partial t \tag{4}$$

The solution to Eq. (4) in the conducting half space is

$$B = B_o \exp(-z/\delta) \exp[j(\omega t - z/\delta)] \tag{5}$$

where the skin depth δ is given by

$$\delta^2 = 2/\mu\sigma\omega$$

The current density **J** in the half space, given by Maxwell's equation $\mathbf{J} = \nabla \times \mathbf{H}$, is in the -y direction; its magnitude is given by

$$J = (B_o/\mu\delta)\,(1 + j)\,\exp(-z/\delta)\,\exp[j(\omega t - z/\delta)] \qquad (6)$$

Thus, we see that current density has the same exponential decay as the magnetic field but is phase-shifted by 45°. Here, the force per unit volume, given by $\mathbf{F} = \mathrm{Re}\{\mathbf{J}\} \times \mathrm{Re}\{\mathbf{B}\}$, is in the z direction, and its magnitude is

$$F(z) = (B_o^2/\mu\delta)\,\exp(-2z/\delta)\,[1/2\ - 2^{-1/2}\sin(2\omega t\ - 2z/\delta\ - \pi/4)] \qquad (7)$$

The force consists of a time-independent part plus a sinusoidal part that is twice the applied frequency. The mean force on the plate is downward, as expected, with a corresponding force upward on an ac coil above the plate. The pressure P at the surface of the plate is given by

$$P = \int_0^{\infty} F(z)\,dz \qquad (8)$$

$$P = (B_o^2/4\mu)\,[1 + \cos(2\omega t)] \qquad (9)$$

The average levitation pressure is independent of frequency and proportional to the square of the applied magnetic field. To achieve a relatively constant levitation height, the period of the applied field $\tau = 2\pi/\omega$ should be much smaller than the characteristic time of the mechanical motion. However, the frequency cannot be made arbitrarily high, because this type of levitation is associated with joule heating; the heating rate Q per unit volume is given by

$$Q = J^2/\sigma \qquad (10)$$

From Eqs. (6) and (10), we surmise that the maximum heating rate occurs at the surface and is proportional to the frequency, with the total heating rate proportional to the square root of the frequency.

C. Magnitude of Levitation Pressure

The levitation pressure of a magnetic system is considerably smaller than that of most mechanical systems. Here, we consider two systems: two magnetized objects, such as a pair of permanent magnets or a permanent magnet and a ferromagnet; and a coil traveling at some velocity v over a conducting sheet.

The maximum magnetic pressure P between two objects of magnetization M_1 and M_2 occurs at zero gap between the two objects and is given by

$$P = M_1\, M_2\, /\, 2\, \mu_o \qquad (11)$$

where $\mu_o = 4\pi \times 10^{-7}$ N/A^2. As convenient reference, for $\mu_o M_1 = \mu_o M_2 = 1.0$ T, the pressure is 400 kPa. In a sintered NdFeB permanent magnet, $\mu_o M$ is typically between 1 and 1.5 T; ferromagnetic materials may achieve magnetizations up to about 2.5 T.

For a set of dc coils of alternating polarity of spatial period L moving at a height h over a conducting sheet, the maximum levitation pressure is given by the pressure of the image force and is

$$P_I = (1/2\,\pi\,\mu_o)B_o^2\,\exp(-4\,\pi\,h/L) \qquad (12)$$

where B_o is the rms (spatially averaged) value of the magnetic induction in the plane of the magnets. With a NbTi superconducting coil, B_o can easily be 5 T.

D. Electrodynamic Levitation

Reitz [113] solved Maxwell's equation for several types of moving magnets with the geometry shown in Fig 6. In each case, he obtained a "wake of images," similar to those shown in Fig. 6, moving into the plate with a velocity $w = 2\rho/\mu_o h$, which is a surface resistance expressed in rationalized mks units [46].

The force on a magnet moving over a nonmagnetic conducting plane can be conveniently resolved into two components: a lift force (F_L) perpendicular to the plane and a drag force (F_D) opposite to the direction of motion. At low velocity, the drag force is proportional to velocity v and considerably greater than the lift force, which is proportional to v^2. As the velocity increases, however, the drag force reaches a maximum (referred to as the "drag peak") and then decreases as $v^{-1/2}$. On the other hand, the lift force, which increases with v^2 at low velocity, overtakes the drag force as velocity increases and approaches an asymptotic value at high velocity, as shown in Fig. 13. The lift-to-drag ratio, which is of considerable practical importance, is given by $F_L/F_D = v/w$.

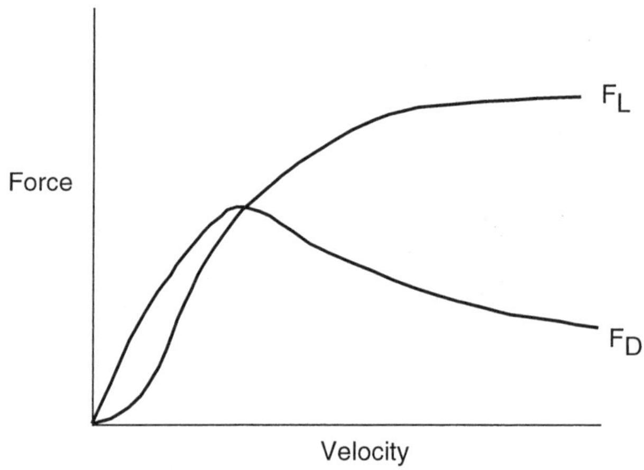

Figure 13. Velocity dependence of lift force F_L and drag force F_D.

Qualitatively, these forces can be understood by considering the diffusion of magnetic flux into the conductor. When a magnet moves over a conductor, the flux tries to diffuse into the conductor. If the magnet is moving rapidly enough, the flux will not penetrate very far into the conductor, and the flux compression between the magnet and the conductor causes a lift force. The flux that does penetrate the conductor is dragged along by the moving magnet, and the force required to drag this flux along is equal to the drag force. At high speeds, less of the magnetic flux has time to penetrate the conductor. At high speed, the lift force that is a result of flux compression approaches an asymptotic limit, and the drag force approaches zero.

The lift force on a vertical dipole of moment m moving at velocity v at a height z_o above a conducting plane is equal to [113]

$$F_L = 3\mu_o m^2/32\pi z_o^4\,[1 - w(v^2 + w^2)^{-1/2}] \qquad (13)$$

At high velocity, the lift force approaches the ideal lift from a single image: $3\mu_o m^2/32\pi z_o^4$; at low velocity, the factor in the brackets is approximately equal to $v^2/2w^2$, so the lift force increases as v^2.

The drag force, as already pointed out, is w/v times the lift force, so the drag force is proportional to v at low velocity. According to the thin-plate model that we have been discussing thus far, the drag force should fall off with 1/v as the lift force reaches its high-speed limit. However, at high velocity, penetration of the eddy currents and magnetic fields is limited to the skin depth, which is proportional to $v^{-1/2}$. As a first approximation, one might replace plate thickness by skin depth at high speed. The transition from thin-plate to skin-depth behavior should occur at about 30 m/s in a 1-cm-thick aluminum plate, for example.

E. Calculation of PM Forces

The calculation of forces between PM pairs is useful in determining the forces and stiffnesses of magnetic configurations. In general, this type of calculation may be carried out by finite-element computer codes, several of which are commercially available. Some geometries are amenable to analytical calculations. One such geometry consists of two cuboidal magnets with magnetizations in the same direction along one of the axes [114].

Consider two parallelpipedic magnets with parallel sides and constant magnetization along the positive z axis. Let the first magnet, of magnetization J, have its center at the origin and have sides of length 2a, 2b, and 2c along the x, y, and z axes, respectively. Let the second magnet, of magnetization J', have its center at coordinate (α, β, γ) and have sides of length 2A, 2B, and 2C along the x, y, and z axes, respectively. The force expressions are given by a sum of 256 terms that derive from taking every possible combination of 6 indices (i, j, m, n, p, q), with each index taking the value of 0 or 1. The force expressions are given in terms of the four auxiliary functions u, v, w, and r, where

$$u_{ij} = \alpha + (-1)^j A - (-1)^i a \tag{14}$$

$$v_{mn} = \beta + (-1)^n B - (-1)^m b \tag{15}$$

$$w_{pq} = \gamma + (-1)^q C - (-1)^p c \tag{16}$$

$$r_{ijmnpq} = [u_{ij}^2 + v_{mn}^2 + w_{pq}^2]^{1/2} \tag{17}$$

The three component forces are

$$F_x = [JJ'/(4\pi\mu_o)]\ [1/2\ (v^2-w^2)\ln(r-u) + uv\ln(r-v) + vw\tan^{-1}(uv/rw) + 1/2\ ru] \tag{18}$$

$$F_y = [JJ'/(4\pi\mu_o)]\ [1/2\ (u^2-w^2)\ln(r-v) + uv\ln(r-u) + uw\tan^{-1}(uv/rw) + 1/2\ rv] \tag{19}$$

$$F_z = [JJ'/(4\pi\mu_o)]\ [-uw\ln(r-u) - vw\ln(r-v) + uv\tan^{-1}(uv/rw) - rw] \tag{20}$$

To calculate the forces when circular geometries are used, expressions such as Eqs. (18)-(20) can be evaluated numerically, basic forces between magnetic poles can be evaluated numerically [115], or special functions must be used [116-118].

6. PM/HTS LEVITATIONAL PHENOMENA

To understand and predict the behavior of an HTS levitational system, one must know the forces that are generated between the superconductor material and a PM or a current distribution. Levitation, suspension, restoring, damping, and drag forces must be determined, and their effect on the dynamic performance of the bearing must be elucidated. Several reviews that describe these forces from a basic physics viewpoint and explore their use in levitation applications have been published [13, 23-26].

As discussed above, a very typical geometry for HTS levitation consists of a cylindrical PM, with its axis of symmetry and magnetization oriented vertically, that is levitated above an HTS with a flat horizontal upper surface. The diameter of the PM is usually less than that of the HTS. In this section, the forces, stiffnesses, and damping are examined in terms of this geometry, but the extension of the results to other geometries is usually straightforward.

A. Vertical Forces

The levitation capability of the PM/HTS system is determined by the vertical force F_z, which is hysteretic for movements in the vertical direction z, as illustrated in Fig. 14. In this example, a cylindrical, upright, vertically magnetized PM was kept with its bottom surface at a height of 10.0 mm above the top surface of a cylindrical YBCO bulk HTS. For this PM, a separation of 10 mm provides very little magnetic field near the HTS, a situation that is called "zero-field cooling" (ZFC). After the HTS was cooled in liquid nitrogen, the magnet was slowly brought down to a position 1 mm above the HTS, while F_z was measured at various points along this first descent. The PM was then moved away from the HTS, back to its original ZFC position. Then it followed a second descent, during which a minor reversal of 0.4 mm was effected at a distance of 2.0 mm above the HTS. A second ascent was identical with the first, and a third descent was identical with the second, etc.

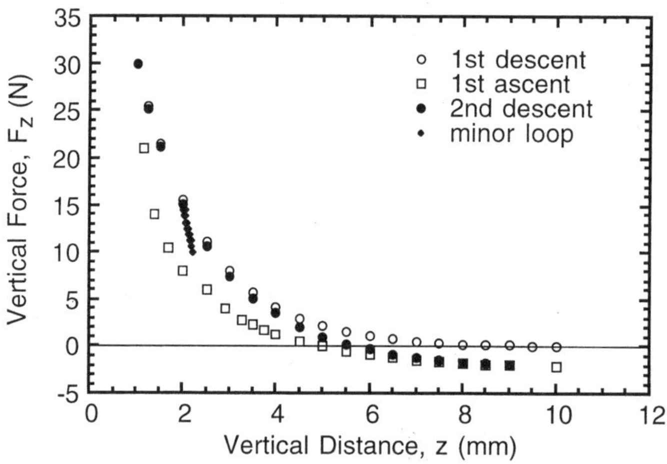

Figure 14. Vertical force vs. vertical distance between PM and HTS under ZFC.

As seen in Fig. 14, the force during the first descent is always larger than the force during the second. Upon reversal from 1 mm above the HTS during the first ascent, the force drops very quickly and even becomes negative, indicating an attractive force. A combination of the first ascent and second descent forms a major hysteretic loop, where the area under the curve is equivalent to the hysteretic energy loss. The form of the hysteresis loop leads to the possibility of levitation over a range of heights. For example, if the PM weighed 10 N, equilibrium could exist anywhere in the range of 1.8 to 2.6 mm for the example in Fig. 14. The

PM in this example weighed ≈0.02 N, so it is also clear that this type of bearing is capable of supporting a mass many times that of the PM.

Several other aspects of levitation phenomena can be deduced from the curves shown in Fig. 14. The force is stable in the vertical direction for the case with the PM above the HTS in repulsive levitation, i.e., when z is positive above the HTS, the vertical stiffness $K_z = -dF_z/dz$ is positive everywhere. The slope of the minor loop is significantly higher than that determined from the major loop. The stiffness represented by the slope of the minor loop is the dynamic stiffness of the system and would be used to calculate vibration frequencies [119]. The stiffness is positively correlated with the force on initial descent [120], and increases during descent as the separation distance decreases. The width of the minor loop is much smaller than that of the major loop. Thus, the PM/HTS acts as a nonlinear damper, and the damping coefficient increases with vibration amplitude.

For the first descent in ZFC at heights less than several millimeters above the HTS surface, the force is exponential with distance [120]. This relationship is probably due to the vertical and horizontal extension of the PM. In practice, it is difficult to measure the force immediately above the HTS, partly because surfaces are not perfectly flat, but mainly because the HTS surface is usually covered with liquid nitrogen. The exponential behavior allows extrapolation of the force to the surface, and various HTSs can be compared in this way if a reference PM of known strength is used. ZFC produces the largest repulsive force but may be impractical, because it requires the cooling of the superconductor before the bearing is assembled.

Hysteresis of vertical force F_z differs significantly when the HTS experiences field cooling (FC), i.e., when the HTS is cooled below its critical temperature while a substantial magnetic field is present, that is, when the PM is close to the superconductor. FC produces less repulsive levitational force than ZFC but can be used to make an attractive-force bearing and generally is more stable in the horizontal direction. The vertical forces experienced after FC are illustrated in Fig. 15, where the FC occurred at z = 1 mm. The force immediately after FC is almost zero. In principle, there will always be a small amount of repulsive force because of the Meissner effect; however, in practice, this force can usually be neglected because the Meissner fraction for melt-textured YBCO is almost zero [27]. As shown in Fig. 15, as the PM moves away from the HTS, F_z is negative, indicating that levitation via suspension, with the PM below the HTS, is possible. In this example, the suspension is only stable for z < 4.5 mm, where K_z is positive.

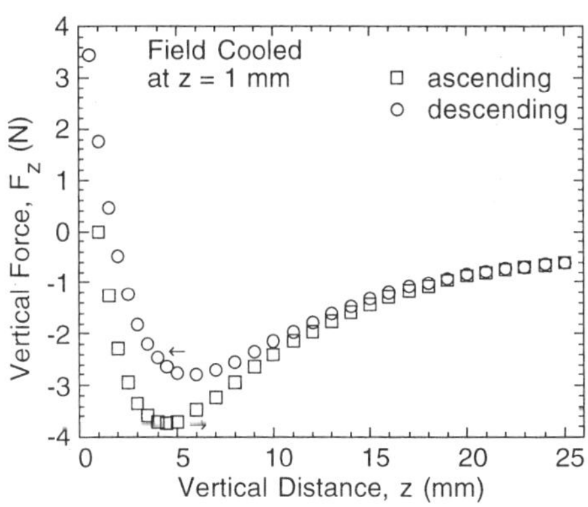

Figure 15. Vertical force vs. vertical distance between PM and HTS under FC at z = 1 mm.

The vertical stiffness has been correlated with the vertical force [120], and details of the vertical stiffness have been studied [35, 120, 121]. In ZFC, vertical stiffness is approximately independent of descent versus ascent, but slight differences may be observed and are attributed to differing amounts of trapped flux in the HTS during a particular vertical traverse [120]. In FC, the dynamic vertical stiffness is less than but greater than half that for ZFC at the same separating distance (i.e., $K_{Z,zfc}/2 < K_{Z,fc} < K_{Z,zfc}$) [120].

The levitational force is approximately independent of the speed at which the PM approaches or recedes from the HTS [23]; however, the measurements can experience force creep. Flux creep is an inherent phenomena in HTSs [12, 23]. A macroscopic manifestation in levitation is force creep, in which the repulsive levitational force decreases with time [122]. For the most part, the force creep is logarithmic over time and can be well correlated with thermally activated flux creep [123]. In this case, the creep is greater in higher magnetic fields and field gradients (higher levitational forces) but becomes smaller as flux pinning is improved [124]. With modern HTSs, the effect of approach velocity on the levitation force in a PM/HTS system is only noticeable when the final resting location of the PM is very close to the HTS [125]. In some experiments with thin films of HTSs, nonlinear force creep that exhibits force jumps has been observed [126].

B. Effects of Geometry

The geometry and other characteristics of the HTS dramatically influence the levitation force. The levitational force that the superconductor provides is proportional to its average magnetization, which is proportional to the product of its grain diameter times its critical current density J_c. Thus, one expects the levitational force to increase as temperature decreases and J_c increases. This relationship is observed experimentally, and the magnetization is well correlated with the levitational force at all temperatures [22]. It is generally believed that the magnetization of the HTS cannot be greater than that necessary to produce a pure diamagnetic image of the PM in ZFC. (A possible exception may be geometries that utilize flux compression between two HTSs [80].) Thus, beyond a certain J_c or grain size, all HTSs will yield similar bearing pressures, which are determined by the properties of the PM, as discussed in Sec. 5.C. and indicated by Eq. (11).

The J_c of typical melt-textured YBCO samples at 77 K is ≈40 kA/cm^2, which, together with a diameter of several cm, allows the levitational pressures between the YBCO and an NdFeB PM to reach ≈280 kPa (≈40 psi). Such pressures have been measured in the author's laboratory with very good YBCO samples. However, this pressure occurs at zero separation distance. In a practical system with a finite separation, the pressure will be at least a factor of 2 to 3 times lower.

The thickness of the HTS is also an important parameter in determining the levitational force. Because the current density is finite, for thin samples the levitational force increases linearly with thickness; however, beyond a certain thickness, the force is mostly independent of thickness [127-129]. This effect is partially a corollary to the discussion above in that the magnetization in the HTS is limited by the properties of the PM. It is also a result of the inverse-square law of forces between electrical or magnetic poles, in that currents flowing far from a PM will produce a negligibly small force on them.

HTSs are highly anisotropic, with superconducting currents capable of flowing at higher density in the ab plane of the crystal. With the ability to grow large single-domain HTSs, the angle dependence of crystal orientation on levitational force has been explored [130-133]. For melt-textured YBCO, this anisotropy results in a levitational force that is ≈3 times greater if the c axis is parallel to a uniform applied field, than if the c axis is perpendicular to the applied field. The force varies monotonically as a function of orientation angle between these two extremes. In actual HTS bearings, the applied field is often some complex combination of parallel and perpendicular orientations that varies over the surface of the HTS.

Several studies have been performed on the levitational properties of thin-film HTSs [35-39]. Although YBCO thin films often have J_c values of >1 MA/cm^2 at 77 K, the thickness of these films is only ≈1 μm, and they do not provide much levitation force. The magnetization quickly saturates, and very large hysteresis loops are possible when this occurs. However, when unsaturated, the stiffness of such films is often similar to that produced by bulk material of greater thickness [36, 38].

The geometry and condition of the PM also affect the levitation force. The effect of the size and shape of the PM, relative to that of the HTS, has been studied mainly by theoretical models that have been validated by some experimental measurements [39, 134-139]. The magnetization of the PM is temperature-dependent, with magnetization increasing with decreasing temperature in most types of PMs, but with FeBNd, the magnetization reaches a peak at ≈140 K and then decreases as the axis of easy magnetization changes. In measurements such as those depicted in Figs. 14 and 15 (where the PM moves in the air) a PM exposed to air will become colder as the distance to the HTS decreases. Unless the temperature of the PM is controlled, the temperature dependence of the PM can affect force measurements and has been known to affect levitation height in HTS bearings [52]. One way to avoid the complication of PM temperature dependence when forces are measured requires that both the PM and the HTS be kept in liquid nitrogen in ZFC. Because of the temperature dependence of the PM, the procedure would be more problematic under FC, depending on the thermal diffusivities of the PM and HTS. It may be necessary to precool the PM, bring it into the FC position, and then add liquid nitrogen to the system.

Let us assume that the HTS is a single grain of large diameter. Because the magnetization of the HTS is proportional to the diameter of the shielding currents, increasing the diameter of the PM should increase the levitational pressure within some range of the diameter. This effect is expected to saturate once the magnetization of the HTS is approximately equivalent to the diamagnetic image of the PM. The homogeneity of the PM magnetization also affects the levitational force [118] but is mostly relevant in understanding small differences between theoretical models and experimental measurements.

At present, no standardized method exists for measuring or reporting levitational forces. The measured levitation force has been related to the levitational force produced by a pure diamagnetic image, i.e., the force between the PM and another exactly like it in repulsive levitation. The levitation force is then reported as a percentage of the theoretical maximum [140]. To use this method, however, one must be careful to appropriately estimate the pure diamagnetic force, and knowledge of the magnetic field only at the surface of the PM is usually insufficient for this purpose [118, 141].

C. Horizontal Forces

If a PM is moved parallel to the surface of an HTS, the magnetic flux lines are sheared, resulting in a lateral magnetic drag force [23, 142]. This force consists of reversible and hysteretic components. Typical horizontal forces that might be encountered in superconducting bearing applications are shown in Fig. 16 for a common measurement height of 5 mm and a horizontal traverse from x = 0 to 12 mm, followed by a horizontal traverse to x = -12 mm, followed by a horizontal traverse to x = 0 mm. The data indicate that the behavior in ZFC differs distinctly from that in FC. For small displacements after FC, the PM is stable, i.e., the horizontal stiffness $K_x = -\partial F_x/\partial x > 0$. In ZFC, the PM is first moved vertically from 30 mm to 5 mm, and then lateral traverses are executed. In ZFC, as shown in Fig. 16, the PM is unstable to horizontal motion, i.e., the horizontal stiffness is negative. In experimental demonstrations in ZFC, the PM may be pushed down almost to the surface of the HTS and then allowed to freely rise to its equilibrium levitation height. Usually, enough flux becomes trapped in the HTS that the levitation is stable, although the horizontal stability is much weaker than in FC. If the HTS exhibits less flux pinning or has smaller grains than melt-textured YBCO, then enough flux may penetrate the system that the horizontal stiffness becomes stabilizing even in ZFC. Of course, the levitational force will also be lower. For the system shown in Fig. 16, edge effects of the HTS produce the force maxima. In FC, if a smaller-diameter PM is used, the force maximum becomes a plateau, and the result is a constant drag force for continued horizontal motion in the same direction [142].

One may show both experimentally and analytically that in FC, the horizontal stiffness is exactly half the field-cooled vertical stiffness (i.e., $2K_{x,fc} = K_{z,fc}$) [118]. This forms an interesting analog to the stiffnesses as determined by Earnshaw's theorem for magnetic systems (see Sec. 5.A), in which $2K_x = -K_z$. In the PM/HTS system in FC, all of the stiffnesses are positive.

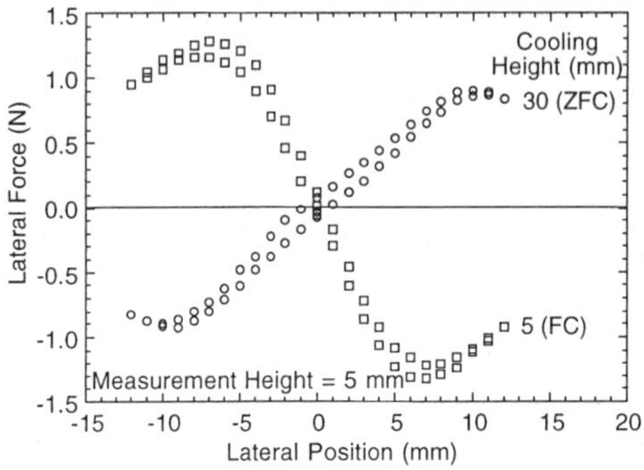

Figure 16. Lateral force vs. lateral displacement at measuring height of 5 mm in ZFC and FC. Data from [118].

D. Cross Stiffness

The PM/HTS system exhibits an anisotropic form of cross stiffness, which has only recently been investigated [118, 143, 144]. If the PM moves in the vertical direction, horizontal force is negligible. However, if the PM moves in the horizontal direction, the magnitude of the vertical force can change, as shown in Fig. 17, where the lateral traverse is the same as in Fig. 16. As in the case with horizontal forces, there is a difference in the sign of the cross-stiffness term, depending on whether the system was in ZFC or FC.

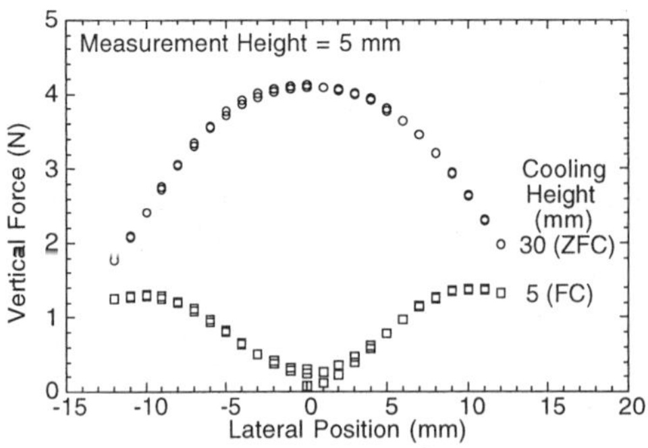

Figure 17. Vertical force versus lateral displacement at measuring height of 5 mm in ZFC and FC. Data from [118].

E. Nonlinearity of Forces

For HTS levitation applications, where the expected vertical or horizontal vibration amplitudes are typically <1 mm, the forces in the PM/HTS interaction can be divided into an elastic region that occurs at small amplitudes, and a hysteretic region that dominates for amplitudes that are larger than some threshold. In the elastic regime, the flux lines remain in their pinning centers, bend as the PM moves, and produce a restoring shear force, according to the Maxwell electromagnetic stress tensor. The hysteretic regime begins when the PM moves sufficiently far for some flux lines to move from their original pinning centers to new ones.

In the hysteretic regime, the force is nonlinear and has been compared with dry friction in mechanical problems [15, 145]. For HTS with an incomplete penetration of the magnetic field, the energy dissipation E_h in a complete hysteresis cycle is [27]

$$E_h = A\,(\Delta B)^3/J_c \tag{21}$$

where A is a geometric factor of order unity, and ΔB is the peak-to-peak variation of the magnetic field. Because ΔB is approximately proportional to the displacement of the PM, the energy loss is approximately proportional to the cube of the amplitude, and the friction force is amplitude-dependent and nonlinear.

The manifestation of the elastic/hysteretic behavior occurs in many parameters that affect superconducting bearings. The vertical and horizontal stiffnesses are dependent on the amplitude of vibration [146, 147]. For small amplitudes, the stiffness is constant, but, after some critical amplitude that depends on levitation height, the stiffness begins to decrease as the amplitude increases. The higher the current density of the superconductor, the higher the critical amplitude value for the onset of stiffness decrease. A transition between elastic and hysteretic behavior has also been observed in the width of minor loops in the force measurements [148]. The decay of oscillations from PMs, either freely levitated [149, 150] or attached to vibrating beams [151], is another method used to study this transition. The hysteresis loss is mostly frequency-independent. However, detailed studies of losses in PM/HTS systems and superconducting bearings show the presence of small velocity-dependent effects that appear to be intrinsic to the superconductors [152-154].

F. Vibrating Systems and Damping

An HTS levitation system will undergo vibrations at various frequencies during its operation. The study of stiffness and damping in vibrations of PM/HTS systems has been mentioned above. Because the loss mechanism responsible for damping is hysteretic with constant force, the damping closely resembles that of dry friction. Thus, an oscillation resulting from an impulse will damp out with an amplitude envelope that decays linearly, as opposed to the exponential decay exhibited by damping forces proportional to velocity.

The damping characteristics of a YBCO near 77 K are highly temperature-dependent [155], and this feature might be used as a switchable damper, which seems particularly attractive for HTS films [156]. The vibrations of a PM near an HTS have been shown to produce period doubling and chaos [157]. Another route to chaotic vibration is bifurcation of quasi-periodic oscillations [158], and in this case multifractality appears in the critical state of chaotic motion [159].

7. MODELING OF HTS LEVITATION

One may roughly divide modeling of the levitation process into efforts that assume the pure Meissner effect is applicable and those that focus on flux penetration. Modeling of the Meissner effect can be further divided into studies that assign a zero magnetic permeability to the superconductor and those that use the London equations to properly account for the effects of penetration. One expects the results of a London model and a Meissner model to agree when the HTS, PM and gap dimensions are large compared to the penetration length. Most efforts to model flux penetration use some form of the critical state model, although models based on dynamic equations have also been investigated. Image models can be used in both Meissner and flux penetration studies.

A. London Models

Within a superconductor, the London equations describe the magnetic field **B** by the equation

$$\mathrm{Del}^2\mathbf{B} - (1/\lambda^2)\mathbf{B} = 0 \tag{22}$$

where λ is the penetration depth. This equation is readily solved as a boundary-value problem for many geometric configurations. The levitation force above a semi-infinite superconductor has been calculated for cylindrical [160] and other PM geometries [161]. The force has also been calculated for a magnetic dipole in either vertical or horizontal orientation over a superconducting plane [162] or sheet of finite thickness [163-165]. The interaction energy between a magnetic dipole at an arbitrary angle either outside or within a superconducting plane has also been calculated [166]. Solutions for a magnetic dipole outside of a superconducting sphere have been given for a dipole with vertical [167] and horizontal [168] orientation.

One of the major motivations for using the London model is that, for some geometries, there is a simple relationship between the levitation force and the penetration depth λ [169]. The penetration depth varies as a function of temperature, and the nature of this temperature dependence is thought to distinguish different theoretical models of superconductivity. This observation has led to a number of studies on how to properly determine the penetration depth from different levitation geometries in magnetic force microscopy [170-173].

B. Meissner Models

In many cases in which hysteretic behavior is not important, the HTS can be treated as a pure diamagnet. Although flux pinning is most prominently associated with FC, strong flux pinning near the HTS surface can greatly extend the range over which the HTS behaves almost like a pure diamagnetic material in ZFC. These pure diamagnetic models do not allow for the calculation of dynamic stiffness or any losses.

The advantage of the Meissner approximation is that analytical solutions are available for a number of geometries [174, 175]. The model can be readily adapted to finite-element calculations by setting the relative permeability of the HTS to a very small value, and the effects of flux penetration can be finessed in a crude approximation by allowing the relative permeability to take on some finite value that is less than unity [134, 176, 177]. The effects of the finite size of the PM and the HTS can be accommodated in the solutions [132, 178].

C. Image Models

In the Meissner model, as in Fig. 2a, a diamagnetic image is formed in the superconductor, with each magnetic dipole of the PM represented by its mirror image dipole, which is oppositely magnetized and located the same distance below the superconductor surface as the magnet is above the surface. As the PM moves closer or further from the superconductor surface, the distance of the image changes accordingly. The calculation of levitational force often reduces to that between a PM and its mirror image. An advantage of this approximation is that analytical solutions are available for a number of geometries [174, 175].

A model that combines a "frozen" mirror image [179] with the diamagnetic mirror image may be used to describe many of the dynamic stiffness properties in FC [180]. In this model, shown schematically in Fig. 18, two images appear in the superconductor: a diamagnetic image of the PM (2) that follows the movement of the PM (1), and a frozen image (3) that is equal in magnitude and location to the diamagnetic image at the time of FC but oppositely polarized and remains in its initial location even if the PM moves. The frozen-image model can provide a good intuitive conceptualization of the relationship between lateral and vertical stability. In particular, it explains why, even if infinite pinning strength were present in the superconductor, there would be a finite lateral stiffness. Comparison of the predictions of this model with experimental results shows that, for PM movements of less than several millimeters over a melt-textured HTS, the forces and stiffnesses agree well enough to be useful in most HTS bearing calculations [118]. If the HTS can be treated as having infinite horizontal extent, then the frozen-image model also provides a good quantitative description of the stiffnesses; however, in real systems with finite HTSs and PMs with spatial extent, the three-dimensional geometry and edge effects will complicate the calculation, especially for lateral stiffness in ZFC [118]. As in the pure-diamagnetic model, the frozen-image model does not predict effects associated with hysteresis.

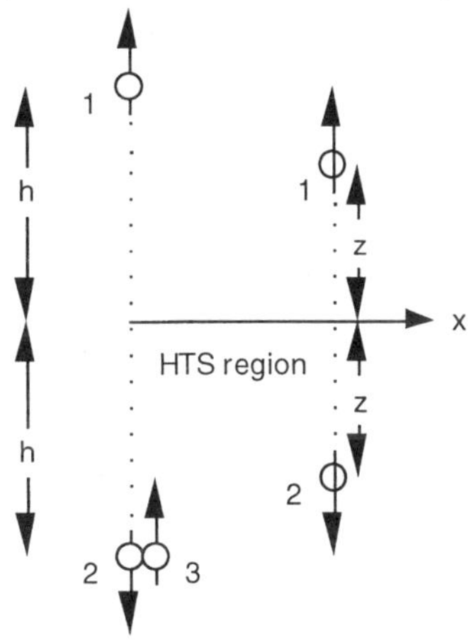

Figure 18. Schematic of frozen-image model, with semi-infinite HTS located below the x axis. The real PM is represented by dipole 1, the diamagnetic image by dipole 2, and the frozen image by dipole 3. The PM is field cooled at x=0 and height h above the HTS, represented by the three dipoles on the left. After cooling, the PM moves down to a distance z above the HTS and to the right to position x, with corresponding motion of the diamagnetic image.

D. Critical-State Model

The critical-state model [181, 182] is often invoked to explain many levitational phenomena [183-196]. While calculationally cumbersome for numerous levitational geometries, it is rooted in a physics that contains the essential features of flux pinning in Type-II superconductors and has been used successfully for many years to analyze ac losses and flux jumping in low-temperature superconductors and HTSs. In a critical-state model, one often assumes that at any location in the superconductor, the current is either flowing with magnitude J_c or is zero. J_c may be assumed constant or assigned a dependence on magnetic field. The critical-state model has had good success in characterizing many of the hysteresis behaviors of HTSs in levitational situations.

In HTSs that are composed of small grains, some models assume that the grains may be treated as cylinders, and that the magnetic field propagates from the cylinder wall to the interior [184, 188]. Other models calculate the performance under the assumption that the grains are completely penetrated [190]. If the more physically correct model of a J_c that depends on magnetic field, one must take into account the actual internal magnetic field of the HTS, which includes effects of self-demagnetization. How the supercurrents distribute themselves in the HTS is usually somewhat arbitrary in the models. In ZFC for large-grain HTSs, models usually assume that the shielding currents begin at the surface [195], although recent efforts have investigated the use of thermal annealing algorithms to calculate the J_c distribution *ab initio* [194]. In the axisymmetric cases investigated with most models, it is usually assumed that the ab plane of the HTS is horizontal, and that the current flows exclusively in this plane.

E. Preisach Model

Mayergoyz and Keim have suggested that the Preisach hysteresis model has a very close analogy with the hysteresis of flux motion in an HTS [197]. This is fortuitous in that an enormous effort has been devoted to refining the Preisach model for use in magnetics. The vector version of the Preisach model seems well adapted for use in modeling the anisotropy of critical current density in the HTS [198].

Similar to the critical state model, the Preisach model does not take into account the details of magnetic flux motion. In HTSs, the local electric field is related to the local current density by a power law with exponent n. While the Preisach model is capable of handling many types of nonlinear diffusion [199], the power law exponent must be relatively large for this model to be accurate [200].

F. T-J Model

One may also base models directly on the relationship between the electric field and the current density, e.g., as is the case with the T Method [185, 201-203]. In this model, one solves for the time dependence of the Maxwell stress tensor (T) in terms of the electric field due to the currents and the external magnetic field. The new current density is then derived from the gradient of stress tensor components.

8. HTS BEARINGS

The ease with which a cylindrical PM rotates while stably levitated over an HTS, as shown in Fig. 1, leads one to consider this configuration for use in loss-loss bearings. In the most common configuration for a basic HTS bearing design, the PM is levitated and free to rotate while the HTS is held stationary. This is configuration is employed mainly because of the convenience of cooling the HTS in boiling liquid nitrogen or by a good thermally conductive contact with a cold source. If the bearing is to operate in vacuum, the only heat transfer mechanism for the levitated component is radiation, which is governed by Stefan's law. Because radiative power from a surface is proportional to T^4, radiation losses are very low at cryogenic temperatures. Because the Curie temperature of commercially available PMs is significantly higher than the critical temperature of HTSs, the configuration in which the PM is levitated exhibits the greatest thermal stability. In situations where the bearing application allows gaseous cooling, or even circulation of a liquid cryogen, it may be preferable to hold the PM stationary and allow the HTS to levitate and rotate.

A. General Bearing Types

The two basic types of rotating bearing are journal, shown in Fig. 19(a), and thrust, shown in Fig. 19(b). In a thrust bearing, the major load force is in the direction of the rotational axis; in a journal bearing it is perpendicular to the rotational axis. A typical PM/HTS journal bearing configuration could consist of a rotating shaft with PM disks mounted as endcaps, and the shaft assembly supported by HTS bushings that completely or partially surround the PMs at both ends of the shaft assembly. In this article, the discussion focuses almost entirely on the thrust type, either of the repulsive or attractive configuration. Researchers have also demonstrated PM/HTS journal bearings [20, 23, 204], and much of the discussion here is directly applicable to this type of bearing. One may also classify a magnetic bearing as either a levitation bearing, which is expected to provide a mostly constant force, or a stiffness bearing, which provides a stabilizing force against disturbances and has an average net force of zero. Most practical magnetic bearings are a combination of the levitation and stiffness types.

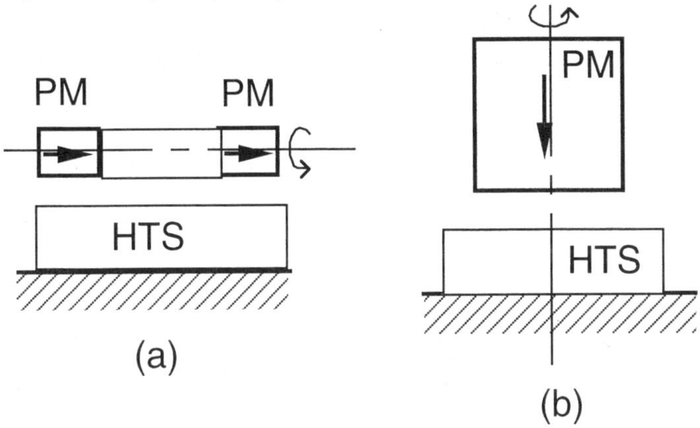

Figure 19. General HTS bearing types. See text for further description.

In a magnetic levitation bearing, the thrust bearing has a significant advantage over the journal bearing. To provide a levitation force, the magnetic pressure must be greater at the bottom of the levitated component than at the top. The magnetic pressure P is determined by the Maxwell stress tensor as

$$P = (B_t^2 - B_n^2) / (2\mu_o) \qquad (23)$$

where $\mu_o = 4\pi \times 10^{-7}$ N/A^2, and B is the magnetic field along the plane on which the pressure is to be determined, subscript t denotes the component that is tangential to the plane, and subscript n denotes the component that is normal to the plane. Therefore, a system with a levitation force must experience a different magnetic field at the top than at the bottom. In a journal bearing, the rotating component experiences the difference in magnetic field on every part of the perimeter during each rotation; this produces eddy currents and hysteretic losses. The use of nonconducting ferrite PMs would eliminate the problem of eddy currents in journal bearings, but the available magnetization of these materials is significantly lower than that of FeBNd or SmCo, and the levitational pressures obtainable are thus much lower. In a levitational thrust bearing, the magnetic field remains the same on the top and bottom horizontal surfaces during the rotation, unless the PM or HTS has some azimuthal inhomogeneity. In a pure stiffness bearing, there is usually no advantage of a thrust over a journal type.

B. HTS Bearing Phenomena

The parameters needed to characterize a bearing for use in rotating a mass are the levitational force, dynamic stiffness, damping, and rotational loss. The 280-kPa levitational pressure of an HTS bearing is lower than that achievable with a conventional electromagnetic bearing (≈1 MPa) and significantly lower than that typically achieved with mechanical roller bearings (>>10 MPa). However, even in the present early period of technological development of superconducting bearings, several laboratories have stably levitated and rotated masses >200 kg. In principle, the amount of mass that can be levitated is limited only by the number and size of PMs and HTSs that are available.

One of the advantages of noncontact in magnetic bearings is elimination of overheating as the determining factor for the maximum speed of the bearing. Instead, the speed of the bearings tends to be limited by the mechanical strength of the rotating bearing components. PMs and HTSs are brittle and exhibit low mechanical strength, both of which lead to structural failure under sufficiently large centrifugal forces. These materials can be radially banded by materials with greater tensile strength; however, designs that permit high velocities at the bearing perimeter demand a careful balance between mechanical strength and the requirements of the magnetics [93, 205].

C. Rotational Loss

The ease with which a PM disk spins when levitated over an HTS, and the absence of contact between the surfaces, produce the illusion that the rotation is without loss. In reality, small magnetic losses gradually slow the rotation. These losses are primarily the result of azimuthal inhomogeneities in the magnetization of the PM disk, which produce hysteretic loss in the superconductor. Typically, for PMs with the best homogeneity, the amplitude of the azimuthally varying component of the magnetic field at a fixed radius above the rotating surface is ≈1% of the average field at that radius. Although small, this inhomogeneity is sufficient to cause a detectable decay in rotational rate when the magnet spins in a vacuum.

A figure of note for the rotational decay of a bearing is the coefficient of friction (COF), defined as the rotational drag force divided by the levitational force (weight of the levitated rotor) [206],

$$\mathrm{COF} = -(2\pi\, R_\gamma^2\, df/dt) / (g\, R_D) \qquad (24)$$

where R_γ is the radius of gyration of the rotor, f is the rotational frequency, t is time, g is the acceleration of gravity, and R_D is the mean radius at which the drag force acts. In Eq. (24), all terms are easily measured or calculated except R_D, the value of which is somewhat ambiguous because, *a priori*, we do not know what the bearing loss mechanism is, nor how it is distributed over the radius of the bearing. To be definitive, the author always takes R_D to be the outer radius of the PM part of the rotor.

The COF is useful when comparing the performance of different bearing designs or designs that use various types of HTSs because it helps to normalize the effects of bearing size and differences in moment of inertia (R_γ term) among experimental devices. The COF for a mechanical roller bearing is of the order of 10^{-3}. The COF for an active magnetic bearing is $\approx 10^{-4}$ after factoring in the parasitic losses for the feedback circuits and power for the electromagnets. Measured COFs for simple HTS bearings are as low as 10^{-7}. The parasitic losses of a superconducting bearing are due to the power that is required to keep the superconductor cold. For refrigerators that operate at ≈30% of Carnot efficiency (the theoretical maximum) ≈14 W of electricity are required to remove 1 W of heat at liquid nitrogen temperatures. Thus, the equivalent COF for an HTS bearing with parasitics factored in is $\approx 2 \times 10^{-6}$, about two orders of magnitude lower than the best alternative bearing. To accurately measure the low COF values that can be attained with HTS bearings, chamber pressures must be less than ≈1 mPa to reduce air friction [207]; however, for most practical applications, the vacuum requirement is not so stringent.

Figure 20 shows the experimental apparatus used to measure the performance of HTS bearings in the author's laboratory. A simplified schematic is shown in Fig. 20(a), and a photo of one of the versions of the device is shown in Fig. 20(b). The cryogenic chamber top and cylindrical side wall consist of G-10 fiber/epoxy composite and are covered by several layers of aluminized mylar to reduce radiation heat transfer. Liquid nitrogen is gravity fed to the chamber from a reservoir about one meter above the cryochamber (not shown in the photo). This arrangement helps keep the chamber pressure constant, thereby keeping the boiling point of nitrogen and the HTS temperature nearly constant. The rotational rate is measured with a tachometer triggered by the brightness of a light beam that is reflected from a retroreflective tape on the side of the rotor. The accuracy of the measurement is slightly affected by vibration of the rotor in the horizontal direction perpendicular to the path of the beam. Before the gas employed for rotor acceleration is fed into the chamber, it passes through a heat exchanger resting in liquid nitrogen. Using cold gas for acceleration helps to prevent inadvertent warming of the cryochamber housing the HTSs.

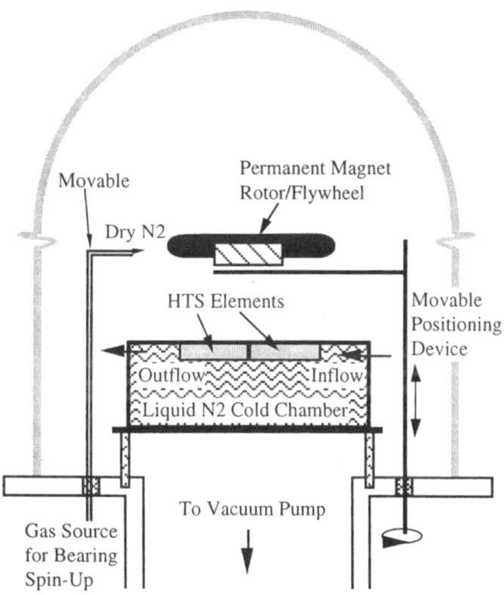

(a)

(b)

Figure 20 Experimental apparatus used to measure the performance of HTS bearings: (a) schematic and (b) photo of central part of apparatus.

Figure 21shows a data set from an experiment to determine the COF for an HTS bearing. After a coast period at very low speed, the rotor was accelerated to about 160 Hz with the gas jet. After the gas jet was turned off, it took almost a minute for the chamber pressure to reach a low enough value to measure the magnetic drag. The test then consisted of a series of ten periods of coast and deceleration. As seen in the expansion of one of the coast periods in Fig. 21(b), one measures the slope of the rotational rate versus time curve to determine df/dt. Because of the limited resolution of the tachometer, for low COFs, a sufficient coast time is needed to achieve good statistics on the value of the slope. The measured COF for the eleven coasting periods is shown as a function of the rotational rate in Fig. 21(c).

(b)

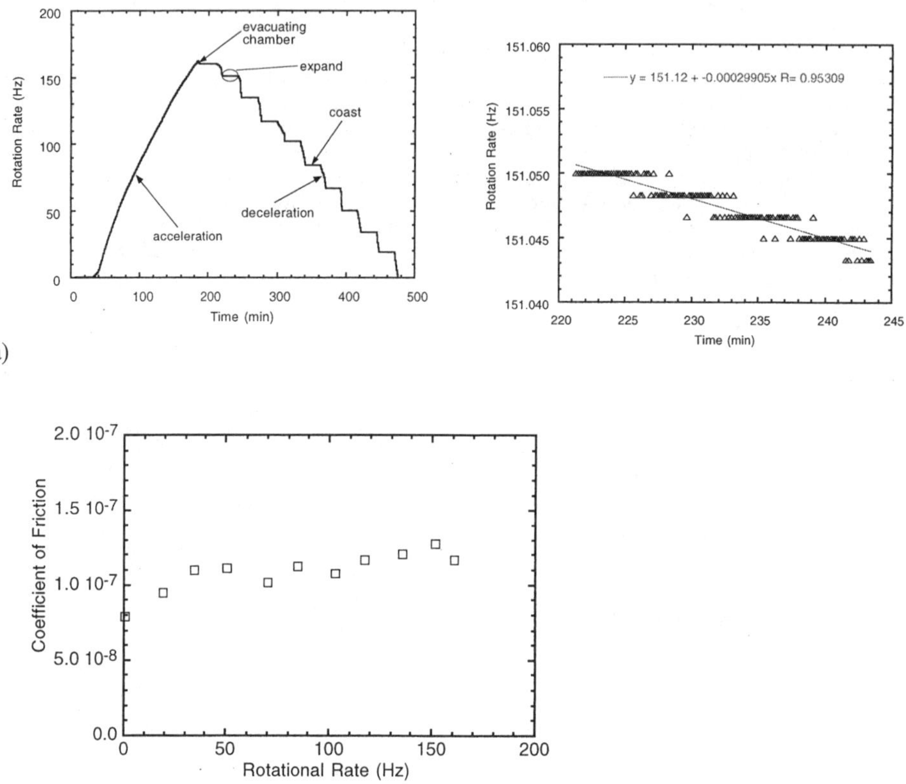

(a)

(c)

Figure 21. Example of coefficient of friction (COF) determination: (a) rotational rate versus time over complete experiment, (b) expansion of data segment from (a), and (c) COF versus rotational rate for coast periods.

Figure 22 shows the historical reduction in attainable COF for various types of HTS bearings at the author's laboratory. Similar progress has occurred at other laboratories in the world. The 1-36 g category represents a single PM disk levitated over a single HTS, and the progress in reducing the COF here is mainly due to continual increases of J_c and grain diameters and to better alignment of the c axis in melt-textured YBCO, as was demonstrated in one of the first studies of rotational decay in HTS bearings [208].

The 300-400 g category represents a single-piece PM ring levitated over an array of HTSs. Progress in reducing COF here can be attributed to optimization of the array geometry and better understanding of the cryochamber design [209], new methods for improving the homogeneity of the magnetic field of the single-piece PM ring [210], and effective use of an

Evershed-type design [52]. Starting at a speed readily attainable without mechanically reinforcing the PM ring, this type of bearing should be able to continuously rotate for more than 10 years before coming to rest.

The 5-15 kg category represents a double-ring composed of PM segments and levitated over an array of HTSs. Progress in reducing the COF in this category is mostly due to better construction techniques for the PM rings [205, 210, 211], as well as continued improvements in the HTS properties.

The 100-200 kg category represents a combination of PM/HTS bearings in an Evershed-type design [212]. The mixed-μ bearing was discussed in a previous section. The progress exhibited in Fig. 22 demonstrates that low rotational losses can be achieved in a wide variety of bearing types and that size does not seem to be a limit on attaining low COFs.

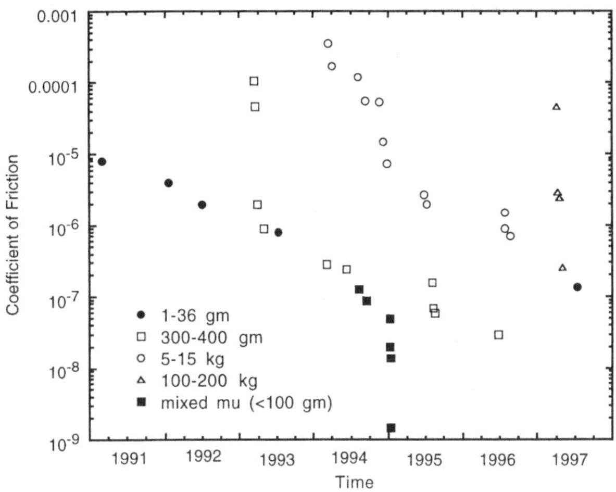

Figure 22. History of lowest COF attained when using HTS bearings for low-speed rotors at Argonne National Laboratory.

D. Low-Speed Rotational Loss

A number of experiments have shown that if one eliminates electrically conducting material from the HTS bearing system, then the COF is mostly independent of the rotational speed. The remaining loss seems most likely due to hysteresis. The generally accepted hysteretic loss mechanism for HTS bearings is based on the critical-state model. Hysteretic loss is produced whenever there is a cyclic change in applied magnetic field, and the energy loss per cycle is [213]

$$E_h = K\mu_0(\Delta H)^3/J_c \tag{25}$$

where E_h is the hysteretic energy loss per unit area per cycle, K is a geometric coefficient of order unity, μ_0 (= $4\pi \times 10^{-7}$ N/A^2) is the magnetic permeability of vacuum, ΔH (= $\Delta B/\mu_0$) is the peak-to-peak amplitude of the varying magnetic field, and J_c is the critical current density in the HTS. When a levitated PM spins over an HTS, rotational loss occurs because of azimuthal inhomogeneities of the magnetic field of the PM, which causes a ΔB at a surface location on the HTS over a complete rotation of the PM. The amplitude of the ΔB increases if there is any appreciable whirl amplitude of the PM, in which case the radial gradient of the magnetic field also contributes to the ΔB. One may conceptualize the magnetics of the bearing system as the PM providing a constant magnetic field B that interacts with the HTS to provide levitation force and a circumferentially varying magnetic field ΔB that produces the rotational loss. In general, the ΔB associated with inhomogeneity is much less than the constant magnetic fields B that provide the levitation force. The inhomogeneity ΔB is determined by the manufacturing process of the PM. This inhomogeneity can be improved to some degree by the strategic addition of soft ferromagnetic shims [207, 214].

The diameters of high-performance melt-textured HTSs are currently limited to a maximum of ≈100 mm. A large bearing system will thus need an array of superconductors [211]. In the case of a single magnet levitated over an array of superconductors, an additional loss arises from magnetization of the individual superconductors upon levitation. The magnetization of the array leads to a time periodic magnetic field that is seen by the rotating PM, and to eddy current losses that depend on the electrical conductivity of the PM [52, 209]. These losses can be minimized by appropriate use of one of the augmentation designs [52].

E. Bearing Dynamics I

The low stiffness of the HTS bearing has profound consequences in rotor dynamic behavior that is not exhibited with other types of bearings. Rigid-rotor bearing resonances, caused by the magnetomechanical interactions in the PM/HTS system, occur at very low frequencies on the order of several hertz. Thus, the operating speed regime of disk- or ring-type rotors is in the supercritical range, well above the bearing resonant frequencies. The frequencies are determined by the dynamic stiffnesses of the system and will generally involve some combination of FC and ZFC effects. Following the exact trend as the stiffness, the resonant frequencies decrease with increased separation between the PM and HTS. For a given bearing design, however, the frequency is mostly independent of the bearing load, with an increase in load producing a decrease in separation that increases the stiffness to mostly compensate for the increase in mass [215]. Consequently, there is usually a large operating speed range between the bearing resonances and the frequencies at which bending modes start to appear in the rotor structure. Further, the low-stiffness bearing allows the rotor to self-center at low speeds, with relatively low out-of-balance forces at these resonances, so rotor balancing is less critical than with other bearing designs.

The HTS bearings often contain large gaps between rotating and stationary surfaces and are capable of tolerating radial oscillations of several millimeters in amplitude. The hysteretic nature of a superconducting bearing makes damping of translational motion amplitude-dependent. For low-amplitude vibrations, damping is small, but it quickly increases as vibrational amplitude increases. The hysteretic nature also leads to a greater uncertainty of the equilibrium position of the rotor than is typical in most rotating machinery, and this uncertainty also requires larger running clearances. Before the HTSs are cooled, the bearing is statically unstable, as predicted by Earnshaw's theorem. Thus, some mechanical constraint is necessary before and while the HTSs are cooling, and this constraint must be removable after cooling for a contact-free levitation.

Two synchronous resonances occur in a magnetically suspended rotor with a polar moment of inertia greater than the transverse moment (i.e., a disk geometry): one is vertical and one is radial. Based on the ratio of HTS levitational and horizontal dynamical stiffnesses, one expects the frequency of the vertical mode to be higher than that of the radial mode. At a given rotational frequency, the COF is positively correlated with the amplitude of vibration [75]. In practice, the radial mode has the largest effect on the COF in most HTS bearing systems. Usually, a vibration amplitude at the natural frequency of the radial mode is present at all rotational speeds; however, the amplitude greatly increases when the rotational frequency is close to that of the natural radial frequency [74].

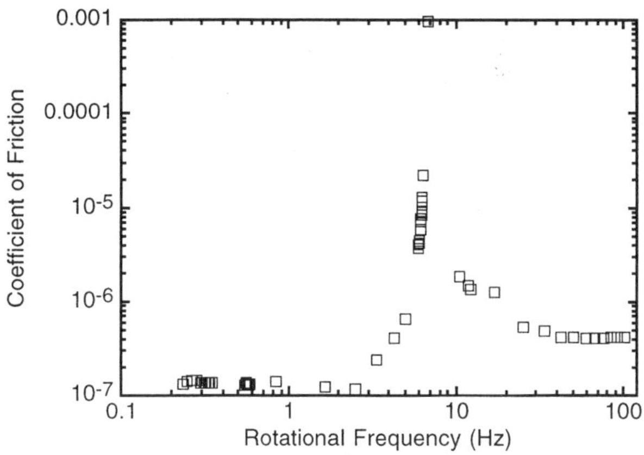

Figure 23. COF vs. rotational frequency for PM disk (25.4-mm diam., 9.6-mm high) levitated 10 mm above YBCO cylinder.

Figure 23 shows the COF as a function of rotational frequency for a disk PM that is levitated 10 mm above a single YBCO cylinder and spinning in a vacuum chamber. The behavior shown in Fig. 23 is typical for most HTS bearing designs in which rotors have disc-type moments of inertia and eddy currents have been nearly eliminated. One may divide the behavior into three regions: below the resonance, at the radial resonance, and above the resonance in supercritical operation. Below the resonance ($f < 3$ Hz), the levitated structure rotates about the center of magnetism of the bearing, and the losses are caused by the inhomogeneity of the PM's field. The resonance region (3-20 Hz) is relatively broad and probably includes some coupling to the vertical vibration mode. The high COF value is mainly caused by the large-amplitude radial oscillations in this frequency range. In some systems, especially those with well balanced rotors, the resonance is very narrow, and the vertical mode, if present at all, contributes very little [207].

F. High-Speed Rotational Loss

Above the resonance (>20 Hz), the rotational losses are affected by an additional factor, the rotation of the magnet about its center of mass rather than its center of geometry or center of magnetism. This transition of the rotation center in passing through a resonance is well understood in Jeffcott-rotor theory [216]. The rotation about the center of mass produces a whirling motion of the center of magnetism over the HTS, and the additional hysteresis loss is proportional to the radial gradient of the magnetic field of the PM and the amount of offset of the center of mass from the center of magnetism. This factor has been modeled by assuming that the bearing is a linear, elastic, isotropic spring with structural damping [217]. As shown in Fig. 23, above and below the resonance, the losses are mostly independent of velocity. However, detailed studies of losses in HTS bearings show the presence of small velocity-dependent effects that appear to be intrinsic to the HTSs [152-154]. Otherwise, if the rotor is in perfect balance, the COF above and below the resonance will be identical.

A more generic rigid rotor, such as a long PM cylinder levitated over an HTS, has six possible modes of vibration. These modes may be excited by external vibrations, e.g., the motion of the earth vibrating the HTS, or by synchronous excitation as the rotational frequency of the PM coincides with a vibrational mode frequency. In an axisymmetric system, the six modes usually degenerate into four frequencies: a very low (usually sub-Hz) rotational frequency about the axis of symmetry that can usually be ignored, a vertical oscillation of the mass center, a radial oscillation of the mass center (two degenerate modes), and a tilt of the axis of symmetry about the vertical axis (two degenerate modes). All of the modes are typically detectable when the PM is levitated and not rotating. All of the modes are somewhat dependent on rotational frequency, especially the tilt mode. If the transverse moment of inertia of the rotor is greater than its polar moment, there is an additional synchronous resonance, which exhibits the form of a conical vibration, i.e., the top of the rotor moves to one side while the bottom of the rotor moves to the opposite side. The frequency of the conical mode depends on the rotational frequency, and this type of rotor never truly enters the supercritical range, but rather the ratio of rotational frequency to conical-mode vibrational frequency approaches an asymptotic limit that depends on the ratio of the moments of inertia.

G. Bearing Dynamics II

A phenomenon related to bearing dynamics is the possibility for self-oscillation that may lead to rotation of a PM disk initially levitated without motion over an HTS [23, 218, 219]. The cause of the phenomenon appears to be a temperature gradient across the PM, in which the temperature dependence of the magnetization causes the center of levitation force to reside below the center of gravity, a situation that is mechanically unstable. If the oscillation frequency and the thermal diffusion times in the PM are appropriately matched, small oscillations can grow into larger amplitudes and eventually become a self-sustaining rotation.

A different phenomenon that also affects an HTS bearing is the need for a larger breakaway torque at startup than is required to keep the bearing rotating at constant speed [220]. This phenomenon, which is believed to be due to the azimuthal inhomogeneity of the PM, is analogous to the difference between static and dynamic friction in mechanical systems. A corollary to this phenomenon is that the drag torque is not constant as a function of rotation angle [221], and this behavior has been observed experimentally.

The PM/HTS system provides the opportunity to observe phenomena in nonlinear dynamics, which usually is overshadowed in systems with high damping, such as mechanical bearings. The dynamics of HTS bearings under the influence of the inherent nonlinear forces are still actively studied [143, 222, 223]. The cross-stiffness term in which the vertical force varies with radial displacement has been shown to cause a bifurcation of the solution for steady-state motion near the resonance [144], a prediction that has been verified by COF measurements [118]. A ramification of this result is that the torque needed to continue acceleration of the rotor requires a substantial step increase at a frequency slightly below the resonance peak. The precession of the levitated PM and the decay of tilt angle after disturbances have also been studied [224].

Another dynamics topic of critical relevance to the operation of HTS bearings is the change in levitation height caused by a varying magnetic field at the surface of the HTS, either from radial vibrations or pure rotation of the PM. An early work reported that the application of an ac field could produce a unique levitation height of an HTS levitated above a ring-shaped PM, independent of the initial height anywhere in the region allowed by the hysteresis loop [225]. An applied time-varying field was also observed to be capable of changing the orientation of a levitated HTS [226].

A drift in levitation height under either a variable magnetic field or a vibration has been studied by several researchers [227-234]. Such a drift is in addition to any possible change in levitation height due to force creep. The drift in levitation height can be made so severe that levitation is totally lost [232-234], but these extreme cases can usually be avoided by proper bearing design. In the many experiments performed in the author's laboratory, some rotors have experienced a drop in levitation height of several hundred micrometers immediately after passing through the resonance, but after operation at higher speeds this gap is usually recovered.

9. STATIC LEVITATED STRUCTURES

One of the most popular applications of HTS levitation is in the demonstration of the ability to levitate large masses, especially people [235]. Sumo wrestlers of total levitated mass of 200 kg have been levitated. At the time of this writing, the largest levitated masses with a PM/HTS interaction are 400 kg [236] and 530 kg [237].

The use of magnetic levitation to avoid mechanical contact has special application to cryogenic structures. Most cryogenic tanks and transmission lines already have vacuum jackets, so the heat leak into the cryogen is through radiation and conduction paths. If a cryogenic storage tank or transmission line is magnetically levitated, then the conduction path disappears, and it is possible to reduce the heat leak significantly. Because a cryogenic environment already exists in this application, the use of an HTS to provide the levitation seems very reasonable. One of the earliest of these studies was the use of HTS levitation in hydrogen storage tanks for automobiles [238, 239].

The hysteretic nature of HTSs has suggested their use in docking vehicles in space [240]. As one vehicle approaches another it would experience a decelerating repulsive force. After the relative velocities have disappeared at a small separation distance, the vehicles would experience an attractive force if their distances tended to separate.

One of the earliest investigations into superconducting levitation was a persistent current ring for magnetic confinement of plasmas [8]. This concept has received renewed interest for the investigation of steady-state, high-beta plasmas with near-classical magnetic confinement [241-243]. In this experiment, a BSCCO HTS double-pancake coil provides the magnetic field to levitate a 620-kg Nb_3Sn superconducting persistent-current coil. A feedback system senses the position of the levitated coil, and the current in the BSSCO coil is adjusted to keep the levitation stable.

The use of superconducting levitation has been proposed for construction of an elevator that reaches low-earth orbital altitude [244, 245]. Currently available NbTi superconductor and carbon-fiber structural materials enable the possibility of constructing a magnetically levitated space elevator from the earth's surface up to an altitude of $\approx$200 km. The magnetic part of the elevator consists of a long loop of current-carrying NbTi, composed of one length that is attached to the earth's surface in an east-west direction and a levitated-arch portion. The critical current density of NbTi is sufficiently high that these conductors will stably levitate in the earth's magnetic field. The magnetic self-field from the loop increases the levitational force and, for some geometries, assists levitational stability. The 200-km maximum height of the levitated arch is limited by the allowable stresses of the structural material. The loop is cryogenically cooled with helium, and the system utilizes intermediate pumping and cooling stations along both the ground and the levitated portion of the loop, similar to other large terrestrial cryogenic systems. Mechanically suspended from the basic loop is an elevator structure, upon which mass can be moved between the earth's surface and the top of the loop by a linear electric motor or other mechanical or electrical means. At the top of the loop, vehicles may be accelerated to orbital velocity or higher by rocket motors, electromagnetic propulsion, or hybrid methods.

10. MAGLEV TRANSPORT

A. Maglev Development History

Maglev transport involves the levitation of vehicles by one of the magnetic levitation principles discussed earlier. There is then no mechanical contact between the vehicle to be transported and the guideway that directs its travel. This application has been considered primarily for high-speed transport of people via trains, where damage to the tracks by a high-speed wheel-on-steel-rail system can be significant. The history of maglev transport has been detailed elsewhere [246-249].

As early as 1907 Robert Goddard, better known as the father of modern rocketry, but then a student at Worcester Polytechnic Institute, published a story in which many of the key features of a maglev transportation system were described. In 1912, a French engineer Emile Bachelet proposed a magnetically levitated vehicle for delivering mail. His vehicle was levitated by copper-wound electromagnets moving over a pair of aluminum strips. Because of the large power consumption, however, Bachelet's proposal was not taken very seriously, and the idea lay more or less dormant for half a century.

In 1963 J. R. Powell, a physicist at Brookhaven National Laboratory, suggested using superconducting magnets to levitate a train over a superconducting guideway. Powell and G. R. Danby proposed in 1967 a system that used a less expensive conducting guideway at room temperature. Later, they conceived the novel idea of a "null-flux" suspension system that would minimize the drag force and thus require much less propulsion power [9].

During the late 1960s, groups at the Stanford Research Institute and at Atomic International studied the feasibility of a Mach-10 rocket sled that employed magnetic levitation [10]. In 1972 a group at Stanford Research Institute constructed and demonstrated a vehicle that was levitated with superconducting magnets over a continuous 160-m-long aluminum guideway. The vehicle, weighing several hundred kilograms, demonstrated both passive and active electromagnetic damping of the vehicle's motion [250].

At about the same time, a team from MIT, Raytheon, and United Engineers designed the magneplane system, in which lightweight cylindrical vehicles are propelled in a curved aluminum trough by a synchronously traveling magnetic field [251]. One advantage of the curved trough is that the vehicle is free to assume the correct bank angle when negotiating curves, but the guideway itself is banked at only approximately the desired angle. The magneplane concept was tested with a 1/25-scale model system that used both permanent magnets and superconducting coils for levitation above a 116-m-long synchronized guideway.

Research groups at the Ford Motor Company Scientific Laboratories [113, 252], the University of Toronto, and McGill University [253, 254] carefully studied magnetic levitation and electromagnetic propulsion; although they did not construct test vehicles, these groups contributed immensely to our understanding of the basic physics and engineering principles involved. The initial research effort in the United States ended about 1975. Interest revived again around 1989, when four major conceptual designs were funded for several years by the federal government [255], after which government funding again disappeared.

Maglev systems have been studied in several other countries, most notably Japan, Germany, and the United Kingdom. The only maglev train ever in commercial service was a 600-m-long route in Birmingham, England, that connected the airport to a conventional rail line. This service operated from 1984 to 1995. Its service was terminated mainly because of lack of parts in a one-of-a-kind technology.

Electromagnetic suspension (EMS) depends on the attractive forces between electromagnets and a ferromagnetic (steel) guideway. One such system is under intensive investigation in Germany. The magnet-to-guideway spacing must be small (only a few centimeters at most), and this requires that the straightness tolerances of the guideway must be relatively small. On the other hand, it is possible to maintain magnetic suspension even when the vehicles are standing still, which is not true for electrodynamic suspension (EDS). In the EMS system, a separate set of electromagnets provides horizontal guidance force, and the levitation magnets, acted on by a moving magnetic field from the guideway, provide the propulsion force. The German Transrapid TR-07 vehicle is designed to carry 200 passengers at a maximum speed of 500 km/h. The levitation height is 8 mm, and power consumption is estimated to be 43 MW at 400 km/h. A full-scale prototype of the Transrapid has run for many years on a 30-km test track in Emsland, Germany.

In principle, part of the magnetomotive force to provide the attractive levitation force in EMS systems could be provided by a dc superconducting coil. The time-varying stabilizing fields would then be provided by normal coils.

EDS depends on repulsive forces between moving magnets and the eddy currents they induce in a conducting aluminum guideway or in conducting loops, as shown in Fig. 7. This system is being intensively investigated in Japan. The repulsive levitation force is inherently stable with distance, and comparatively large levitation heights (20 to 30 cm) are attainable by using superconducting magnets. Various guideway configurations, such as a flat horizontal conductor, a split L-shape conductor, and an array of short-circuit coils on the sidewalls, have been investigated. Each has its advantages and disadvantages. The proposed Japanese high-speed maglev system involves the use of interconnected figure-8 ("null-flux") coils on the sidewalls. The null-flux arrangement tends to reduce the magnetic drag force and thus the propulsion power needed. A prototype of this system has been operated for many years on a 7-km-long test track in Miyazaki, Japan. The Japanese currently plan to construct a commercial version that will operate between Tokyo and Osaka, with the first part of this route, a 42-km-long test track, in the Yamanashi prefecture. A three-car test train achieved the design speed of 550 km/h on this test track in December 1997 [256].

B. HTS and Maglev

HTS could be used in maglev applications in several ways. First, HTS wire could replace the NbTi wire currently used in the superconducting racetrack coils carried aboard the vehicle [257]. This would allow cryogenic refrigeration at higher temperatures than the present values near 4 K. Second, trapped-field HTSs could be used to replace the coils [258, 259]. In this case, the trapped-field HTSs would act as very powerful PMs that allow much higher levitation heights than can be achieved with conventional PMs on maglev vehicles. Third, the diamagnetic HTSs could be placed over a PM guideway [237, 260, 261]. Fourth, HTS wire or an HTS trapped-field magnet could provide dc magnetization in an EMS system [262].

11. OTHER LINEAR APPLICATIONS

The stable levitational force in HTS suggests application in magnetically levitated conveyor systems in clean rooms, where high- purity requirements mandate no mechanical contact [263-265]. Linear HTS bearings have also been proposed for microactuators [266]. Bulk HTS has also been suggested for use in electromagnetic launchers [267].

12. FLYWHEEL ENERGY STORAGE

The availability of HTS bearings that are so nearly friction-free naturally leads to their consideration for flywheel energy storage, and several major projects have investigated this application [56, 211, 268-275]. Flywheels with conventional bearings typically experience high-speed idling (that is, no power input or output) losses on the order of ≈1% per hour. With HTS bearings, losses as little as 0.1% per hour are believed to be achievable, even when the flywheel powers the refrigerator that cools the HTSs. When coupled with efficient motors/generators and power electronics (capable of losses as low as 4% on input and output), the potential exists for constructing flywheels with a 90% diurnal storage efficiency. Probably only one other technology is capable of achieving such high diurnal storage efficiencies: superconducting coils hundreds of meters in diameter.

Electric utilities have a great need for efficient diurnal energy storage, such as flywheels, because their inexpensive baseload capacity is typically underutilized at night and they must use expensive generating sources to meet their peak loads during the day. A distributed network of diurnal-storage devices could also make use of underutilized capacity in transmission lines at night and add robustness to the electric grid. These factors are expected to become more important in the forthcoming deregulation of the electric utility industry. Efficient energy storage would also be beneficial to renewable energy technologies, such as photovoltaics and wind turbines.

With modern graphite fiber/epoxy materials, the inertial section of a flywheel rotates with rim speeds well in excess of 1000 m/s and achieves energy densities greater than those of advanced batteries. The kinetic energy in a (large) Frisbee-sized flywheel with this rim speed is ≈1 kWh, and a person-sized flywheel could store 20-40 kWh. Experimental versions of flywheels that employ HTS bearings have already stored >2 kWh.

13. INSTRUMENTATION

The low magnetic drag of an HTS bearing has suggested its use as a sensitive detector of gas pressure [220]. The possibility of using low-loss superconducting bearings in sensitive gyroscopes has long been recognized [5]. Recently, HTS bearings have been proposed for sensitive gravimeters [276]. In this application, subradian oscillations or small angular rotations are observed. If the motion is small enough, the HTS should remain in an elastic regime, with all of the flux lines remaining in their pinning centers, and the dissipation should be extremely small. Such bearings should provide a much higher Q than is possible with a torsional fiber.

Recently, HTS bearings have been proposed for use in a sensitive detector to study the polarization of the cosmic background radiation (CMB) [277]. Information about the spatial anisotropy of the CMB has allowed cosmologists to distinguish between different evolutionary models of the universe, and information about the polarization could lead to further refinements. The proposed device would use a rotating half plate that modulates the CMB from a given patch of the sky passing into a polarization analyzer. The challenge in adapting the technique to CMB polarimetry is coupling the rotation mechanism to bolometric detectors, which have high sensitivity to microphonic noise. Standard mechanical rotation mechanisms, such as gears and bearings, have stick-slip friction which induces vibrations. The vibrations deposit mechanical energy in the detectors and cause wires to vibrate, thereby inducing excess noise. The low rotational drag and vibration isolation characteristics of the HTS bearing make it suitable for providing a low-noise environment. Cryogenic cooling is already available in the CMB experiment to detect the long wavelength radiation.

14. MISCELLANEOUS APPLICATIONS

Because superconducting levitation is versatile over a wide range of stiffness and damping, it has been suggested for numerous applications [23]. Superconducting bearings, like magnetic bearings, do not require a lubricant; this could be a major advantage in harsh chemical or thermal environments. In addition to the early applications considered for low-temperature superconductors mentioned in the introduction, such as electric motors and gyroscopes, our increased understanding of the hysteretic behavior of HTSs has suggested new possibilities.

HTS bearings are particularly attractive for cryogenic turbopumps [278]. Liquid hydrogen, which is typically pumped at a temperature of ≈30 K, seems the most likely candidate, but liquid oxygen is also a possibility, if materials with critical temperatures higher than that of YBCO can be used in HTS bearings. The tribological properties of mechanical bearings at cryogenic temperatures are poor, and experience with rocket engines indicates that the bearings of turbopumps are one of their most life-limiting components. In this application, the cryogen is already available and does not impose an extra cost; however, the low levitation pressures of present HTS bearings would require a significant redesign of any existing pump. Other applications for HTS bearings in which a cryogenic environment already exists include cryogenic turbine flow meters [279] and cryocoolers for space [23] or terrestrial use. Yet another application for which a cryogenic environment naturally exists is the use of HTS bearings in a lunar telescope [280]. The drive mechanism of such a telescope must be capable or exceedingly fine steps and repeatability, and it must survive and operate in the cold temperatures of the lunar night. Prototype HTS bearings for such a telescope have been constructed and tested [221]. Applications that create a high-value-added product per hour, such as high-speed machine tools or textile spindles, could also justify the added expense of providing a cryogenic environment for HTS bearings [23].

Hysteresis is also important in the potential use of HTS as torque couplers and vibration dampers [281, 282]. Apparently, the first successful vacuum tunneling experiment, which eventual led to the development of the scanning tunneling microscope, was made possible by suppression of vibrations with the use of a superconducting bearing [283].

CONCLUSIONS

The field of levitation applications of HTSs is large and expanding at a rapid rate. Only a small sample of the many works in this area were discussed here. There are a great many other good works that limited space and time did not permit me to discuss.

ACKNOWLEDGMENTS

This work was partially supported by the U.S. Department of Energy, Energy Efficiency and Renewable Energy, as part of a program to develop electric power technology, under Contract W-31-109-Eng-38.

REFERENCES

[1] M.K. Wu, J.R. Ashburn, C.J. Torng, P.H. Hor, R.L. Meng, L. Gao, Z.J. Huang, and C.W. Chu, *Phys. Rev. Lett.* **58** [1987] 908.
[2] V. Arkadiev, *J. Phys. USSR* **9** [1945] 148.
[3] V. Arkadiev, *Nature* **160** [1947] 330.
[4] I. Simon, *J. Appl. Phys.* **24** [1953]19.
[5] T.A. Buchhold, *Sci. Am.* **202 (3)** [1960] 74.
[6] K.F. Schoch, *Adv. Cryog. Engng.* **6** [1960] 65.
[7] T.A. Buchhold, *Cryogenics* **5** [1965] 216.
[8] J. File, G.D. Martin, R.G. Mills, and J.L. Upham, *J. Appl. Phys.* **39** [1968] 2623.
[9] J.R. Powell and G.T. Danby, *Cryogenics* **11** [1971] 192.
[10] C.A. Gudajahn, S.L Wipf, H.J. Fink, R.W. Boom, K.E. MacKenzie, D. Williams, and T. Downey, *J. Appl. Phys.* **40** [1969] 2133.
[11] E.A. Early, C.L. Seaman, K.N. Yang and M.B. Maple, *Am. J. Phys.* **56** [1988] 617.
[12] E.H. Brandt, *Science* **243** [1989] 349.
[13] E.H. Brandt, *Am. J. Phys.* **58** [1990] 43.
[14] F.C. Moon and J.R. Hull, *Proc. Intersoc. Energy Conv. Engng. Conf.* **3** [1990] 425.
[15] J.R. Hull, T.M. Mulcahy, L. Lynds, B.R. Weinberger, F.C. Moon and P.-Z. Chang, *Proc. Intersoc. Energy Conv. Engng. Conf.* **3** [1990] 432.
[16] P.N. Peters, R.C. Sisk, E.W. Urban, C.Y. Huang, and M.K. Wu, *Appl. Phys. Lett.* **52** [1988] 2066.
[17] W.G. Harter, A.M. Hermann, and Z.Z. Sheng, *Appl. Phys. Lett.* **53** [1988] 1119.
[18] D. Schoenberg, *Proc. Roy. Soc.* **A155** [1936] 712.
[19] R.J. Adler and W.W. Anderson, *Appl. Phys. Lett.* **53** [1988] 2346.
[20] F.C. Moon and P.-Z. Chang, *Appl. Phys. Lett.* **56** [1990] 397.
[21] D.E. Weeks, *Rev. Sci. Instrum.* **61** [1990] 195.
[22] B.R. Weinberger, L. Lynds, and J.R. Hull, *Supercond. Sci. Technol.*, **3** [1990] 381.
[23] F.C. Moon, *Superconducting Levitation* (Wiley & Sons, New York) [1994].
[24] J.R. Hull, *Supercond. Sci. Technol.*, **13** [2000] R1.
[25] J.R. Hull, in *Engineering Superconductivity*, ed. P. J. Lee, (Wiley & Sons, New York) [2001] 563.
[26] T.P. Sheahen, *Introduction to High-Temperature Superconductivity,* (Plenum, New York) [1994] Chap. 14.
[27] M. Murakami, *Melt Processed High-Temperature Superconductors* (World Scientific, Singapore) [1992].

[28] M. Murakami, N. Sakai, T. Higuchi, and S.I. Yoo, *Supercond. Sci. Technol.* **9** [1996] 1015.
[29] X. Yao and Y. Shiohara, *Supercond. Sci. Technol.*, **10** [1997] 249.
[30] M. Muraldihar, M.R. Koblishka, T. Saitoh and M. Murakami, *Supercond. Sci. Technol.* **11** [1998] 1349.
[31] C.-J. Kim, and G.-W. Hong G-W, *Supercond. Sci. Technol.* **12** [1999] R27.
[32] M. Murakami, (ed) *Appl. Supercond.* **6 (2-5)** [1998].
[33] D.A. Cardwell, M. Murakami, and K. Salama, (eds) *Mater. Sci. Engng.* **53B (1,2)** [1998].
[34] R. Weinstein, R. Sawh, Y. Ren and D. Parks, 1998 *Mater. Sci. Engng.* **B53** [1998] 38.
[35] B. Lehndorff, H.-G. Kurschner and B. Lucke, *Appl. Phys. Lett.* **67** [1995] 1932.
[36] B.R. Weinberger, *Appl. Supercond.* **2** [1994] 511.
[37] P. Schonhuber and F.C. Moon , *Appl. Supercond.* **2** [1994] 523.
[38] B. Lucke, H.-G. Kurschner, B. Lehndorff, M. Lenkens and H. Piel, *Physica C* **259** [1996] 151.
[39] T.H. Johansen, A.B. Riise, H. Bratsberg and Y.Q. Shen, *J. Supercond.* **11** [1998] 519.
[40] A.H. Boerdijk, *Philips Res. Rep.*, **11** [1956] 45.
[41] R.H. Frazier, P.J. Gilinson, Jr., and G.A. Oberbeck, *Magnetic and Electric Suspensions*, (MIT Press, Cambridge, Massachusetts) [1974].
[42] E.R. Laithwaite (ed.), *Transport without Wheels*, (Elek Science, London) [1977].
[43] R.G. Rhodes and B.E. Mulhall, *Magnetic Levitation for Rail Transport*, (Oxford, Clarendon Press) [1981].
[44] B.V. Jayawant, *Rep. Prog. Phys.*, **144** [1981] 411. Also, *Proc. R. Soc. London A,* **416** [1988] 245.
[45] F.C. Moon, *Magneto-Solid Mechanics* (Wiley & Sons, New York) [1984].
[46] T.D. Rossing and J.R. Hull, *Phys. Teacher*, **29** [1991] 552.
[47] H. Bleuler, *JSME Int. J. Ser. III.*, **35** [1992] 335.
[48] G. Schweitzer, H. Bleuler, and A. Traxler, *Active Magnetic Bearings*, (Hochschulverlag AG an der ETH Zurich) [1994].
[49] J.R. Hull, In *Encyclopedia of Electrical and Electronics Engineering*, (ed. J. G. Webster, John Wiley and Sons, New York) **11** [1999] 740.
[50] S. Evershed, *J. Inst. Elect. Eng.*, **29** [1900] 743.
[51] C.K. McMichael, K.B. Ma, M.A. Lamb, M.W. Lin, L. Chow, R.L. Meng, P.H. Hor, and W.K. Chu, *Appl. Phys. Lett.* **60** [1992] 1893.
[52] J.R. Hull, T.M. Mulcahy, and J.F. Labataille, *Appl. Phys. Lett.* **70** [1997] 655.
[53] T. Coombs, A.M. Campbell, R. Storey, and R. Weller *IEEE Trans. Appl.Supercond.*, **9** [1999] 968.
[54] M. Komori, M. Kumamoto, and H. Kobayashi, *IEEE Trans. Appl. Supercond.* **8** [1998] 79.
[55] K. Nagaya K *IEEE Trans. Magn.* **32** [1996] 445.
[56] H. Kameno, Y. Miyagawa, R. Takahata, and H. Ueyama, *IEEE Trans. Appl. Supercond.* **9** [1999] 972.
[57] T. A. Coombs, A.M. Campbell, and D.A. Cardwell, *IEEE Trans. Appl. Supercond.* **5** [1995] 630.
[58] H. Teshima, *Jpn. J. Appl. Phys.* **36** [1997] 68.
[59] T. Takamori, J.J. Bol, and D.B. Dove, *Appl. Phys. Lett.* **55** [1989] 1454.
[60] Y. Iwasa and H. Lee, *Cryogenics,* **37** [1997] 807.
[61] H. Lee, M. Tsuda and Y. Iwasa, *Cryogenics,* **38** [1998] 419.
[62] M. Tsuda, H. Lee, and Y. Iwasa, *Cryogenics,* **38** [1998] 743.
[63] J.C. Maxwell, *A Treatise on Electricity and Magnetism*, Vol. 2, Oxford, Clarendon Press, [1891]. Reprinted by Dover, New York [1954].
[64] J.C. Maxwell, *Proc. R. Soc. Lonon A,* **20** [1872] 160.
[65] W.M. Saslow, *Am. J. Phys.,* **59** [1991] 16.
[66] W.M. Saslow, *Am. J. Phys.,* **60** [1992] 693.

[67] W. Thompson, *Mathematical and Physical Papers,* Vol. 1 Cambridge at the University Press [1882] 88.
[68] W. Braunbek, *Zeit. Physik*, **112**, [1939] 764.
[69] H. Boerdijk, *Philips Res. Rep.,* **11**, [1956] 45.
[70] E. Beaugnon and R. Tournier, *J. Phys. III,* **1** [1991] 1423.
[71] M.D. Simon, L.O. Heflinger, and A.K. Geim, *Am. J. Phys.,* **69** [2001] 702.
[72] M.V. Berry and A.K. Geim, *Eur. J. Phys.,* **18** [1997] 307.
[73] J.R. Hull, J.L. Passmore, T.M. Mulcahy, and T.D. Rossing, *J. Appl. Phys.* **76** [1994] 577.
[74] J.R. Hull, E.F. Hilton, T.M. Mulcahy, Z.J. Yang, A. Lockwood, and M. Strasik, *J. Appl. Phys.* **78** [1995] 6833.
[75] G.J. Homer, T.C. Randle, C.R. Walter, M.N. Wilson, and M.K. Bevir, *J. Phys. D,* **10** [1977] 879.
[76] H. Joyce, J.T. Williams, R.J.A. Paul, G. Gallagher-Daggitt, and J. Brown, *IEEE Trans. Magn.* **17** [1981] 2154.
[77] H. Ohsaki, M. Takabatake, and E. Masada, *IEEE Trans. Appl. Supercond.* **7** [1997] 908.
[78] Y. Fukasawa and H. Ohsaki, *IEEE Trans. Appl. Supercond.* **9** [1999] 980.
[79] J.R. Hull, and M. Komori, *Supercond. Sci. Technol.*, **15** [2002] 763.
[80] K.A. Connor, and J.A. Tichy, *ASME J. Tribology*, **110** [1988] 320.
[81] J.A. Tichy, and K.A. Connor, *ASME J. Tribology*, **111** [1989] 209.
[82] A.V. Filatov, and E.H. Maslen, *IEEE Trans. Magn.* **37** [2001] 3913.
[83] A.V. Filatov, E.H. Maslen, and G.T. Gillies, *J. Appl. Phys.* **91** [2001] 2355.
[84] A.V. Filatov, E.H. Maslen, and G.T. Gillies, *J. Appl. Phys.* **92** [2002] 3345.
[85] M.V. Berry, *Proc. R. Soc. London A,* **452** [1996] 1207.
[86] T.B. Jones, M. Washizu, and R. Gans, *J. Appl. Phys.* **82** [1997] 883.
[87] M.D. Simon, L.O. Heflinger, and S.L. Ridgway, *Am. J. Phys.,* **65** [1997] 286.
[88] J.R. Hull, *J. Appl. Phys.* **58** [1985] 3594.
[89] J.R. Hull, T.M. Mulcahy, and R.C. Niemann. *Adv. Cryog. Engng*. **47**, [2002] 1711.
[90] A.V. Filatov, O.L. Poluschenko, and S.-C. Shin *Cryogenics,* **38** [1998] 595.
[91] J.P. Yonnet, *IEEE Trans. Magn.* **17** [1981] 1169.
[92] S. Zannella, F. Frangi, and E. Varesi, 1994 *Nuov. Cim.* **16D** [1994] 2061.
[93] T.M. Mulcahy, J.R. Hull, K.L. Uherka, R.G. Abboud, J.H. Wise, and D.W. Carnegie, *IEEE Trans. Magn.* **32** [1996] 2609.
[94] K. Nagashima, T. Otani, and M. Murakami, Physica C 328 [1999] 137.
[95] T.M. Mulcahy, J.R. Hull, K.L. Uherka, R.G. Abboud, and J. Juna, *IEEE Trans. Appl. Supercond.* **11** [2001] 1729.
[96] M.C. Marion-Pera and J.P. Yonnet, *IEEE Trans. Magn.* **30** [1994] 4743.
[97] M. Tsuchimoto, T. Kojima, H. Waki, and T. Honma, *Appl. Supercond.* **2** [1994] 549.
[98] M. Komori, T. Matsushita, and M. Takeo, 1993 *Cryogenics* **33** [1993] 1058.
[99] P. Tixador, P. Hiebel, Y. Brunet, X. Chaud, and P. Gautier-Picard, *IEEE Trans. Magn.* **32** [1996] 2578.
[100] H. Higasa, F. Ishikawa, N. Kawauchi, S. Yokoyama, S. Nakamura, H. Imaizumi, and N. Ito, *Internat. Symp. Superconduc.* **6**, [1994] 1249.
[101] A.C. Day, M. Strasik, K.E. McCrary, P.E. Johnson, J.W. Gabrys, J.R. Schindler, R.A. Hawkins, D.L. Carlson, M.D. Higgins, and J.R. Hull, *Supercond. Sci. Technol.* **15**, [2002] 838.
[102] H. Kameno, Y. Miyagawa, R. Takahata, and H. Ueyama, *IEEE Trans. Appl. Supercond.* **9** [1999] 972.
[103] Z. Xia, Q.Y. Chen, K.B. Ma, C.K. McMichael, M. Lamb, R.S. Cooley, P.C. Fowler and W.K. Chu, *IEEE Trans. Appl. Supercond.* **5** [1995] 622.
[104] W.K. Chu, K.B. Ma, C.K. McMichael, and M.A. Lamb, *Appl. Supercond.* **1** [1993] 1259.
[105] Y. Postrekhin, K.B. Ma, J.H. Yu, M. Lamb, and W.K. Chu, *IEEE Trans. Appl. Supercond.* **9** [1999] 2235.

[106] W. Hennig, D. Parks, R. Weinstein, and R.-P. Sawh, *Appl. Phys. Lett.* **72** [1998] 3059.
[107] R. Weinstein, R. Sawh, Y. Ren, and D. Parks, *Mater. Sci. Engng. B* **53** [1998] 38.
[108] M. Morita, K. Nagashima, S. Takebayashi, M. Murakami, and M. Sawamura, *Mater. Sci. Engng. B* **53** [1998] 159.
[109] S. Gruss, G. Fuchs, G. Krabbes, P. Schatzle, J. Fink, K.H. Muller, and L. Schultz, *IEEE Trans. Magn.* **34** [1998] 2099.
[110] S. Earnshaw, *Trans. Cambridge Philos. Soc.* **7** [1842] 98.
[111] W. Braunbek, *Zeit. fur Physik,* **112** [1939] 753.
[112] M. Milgrom, *Phys. Lett. A,* **243** [1998] 33.
[113] J.R. Reitz, *J. Appl. Phys.,* **41** [1970] 2067.
[114] G. Akoun, and J.-P. Yonnet, *IEEE Trans. Magn.* **20** [1984] 1962.
[115] E.P. Furlani, *IEEE Trans. Magn.* **29** [1993] 4165.
[116] K.-B. Kim, E. Levi, Z. Zabar, and L. Birenbaum, *IEEE Trans. Magn.* **32** [1996] 478.
[117] Z.J. Yang, and J.R. Hull, *Int. J. Electromag. Mech.* **10** [1999] 351.
[118] J.R. Hull and A. Cansiz, *J. Appl. Phys.* **86** [1999] 6396.
[119] F.C. Moon, K.-C. Weng, and P.-Z. Chang, *J. Appl. Phys.* **66** [1989] 5643.
[120] P.-Z. Chang, F.C. Moon, J.R. Hull, and T. M. Mulcahy, *J. Appl. Phys.* **67** [1990] 4358.
[121] A.B. Riise, T.H. Johansen, and H. Bratsberg, *Physica C* **234** [1994] 108.
[122] F.C. Moon, P.-Z. Chang, H. Hojaji, A. Barkatt, and A.N. Thorpe, *Jpn. J. Appl. Phys.* **29** [1990] 1257.
[123] A.B. Riise, T.H. Johansen, H. Bratsberg, and Z.J. Yang, 1992 *J. Appl. Phys.* **60** [1992] 2294.
[124] M. Murakami, S. Gotoh, N. Koshizuka, S. Tanaka, T. Matshushita, S. Kambe, and K. Kitazawa, *Cryogenics* **30** [1990] 390.
[125] E. Postrekhin, K.B. Ma, and W.K. Chu, *IEEE Trans. Appl. Supercond.* **11**, [2001] 1984.
[126] P. Schonhuber and F.C. Moon, *Appl. Supercond.* **2** [1994] 523.
[127] J. Wang, M.M. Yanoviak, and R. Raj, *J. Am. Ceram. Soc.* **72** [1989] 846.
[128] T.H. Johansen, H. Bratsberg, and Z.J. Yang, *Proc. 2nd World Congress on Superconductors*, Houston [1990].
[129] J. Unsworth, J. Du, B.J. Crosby, and J.C. Macfarlane, 1993 *IEEE. Trans. Magn.* **29** [1993] 108.
[130] D. Shi, D. Qu, A. Sagar, and K. Lahiri, *Appl. Phys. Lett.* **70** [1997] 3606.
[131] B.A. Tent, D. Qu, and D. Shi, *Physica C* **309** [1998] 89.
[132] B.A. Tent, D. Qu, D. Shi, W.J. Bresser, P. Boolchand, and Z.-X. Cai, *Phys Rev. B* **58** [1998] 11761.
[133] W.M. Yang, L. Zhou, Y. Feng, P.X. Zhang, S. Chen, M.Z. Wu, C.P. Zhang, J.R. Wang, Z.H. Du, F.Y. Wang, Z.M. Yu, X.Z. Wu, W. Gawalek, and P. Gornert, *Physica C* **307** [1998] 271.
[134] Y.S. Cha, J.R. Hull, T.M. Mulcahy, and T.D. Rossing, *J. Appl. Phys.* **70** [1991] 6504.
[135] M. Tsuchimoto, T. Kojima, H. Waki, and T. Honma, *Appl. Supercond.* **2** [1994] 549.
[136] Z.J. Yang and J.R. Hull, *J. Appl. Phys.* **79** [1996] 3318.
[137] H. Teshima, M. Morita, and M. Hashimoto, *Physica C* **269** [1996] 15.
[138] S. Sagar, K. Lahiri, D. Shi, and Z.J. Yangf, *IEEE Trans. Appl. Supercond.* **7** [1997] 1929.
[139] A. Badia and H.C. Freyhardt, *J Appl. Phys*. **83** [1998] 2681.
[140] D. Litzkendorf, T. Habisreuther, M. Wu, T. Strasser, M. Zeisberger, W. Gawalek, M. Helbig, and P. Gornet, *Mater. Sci. Engng*. **B53** [1998] 75.
[141] J.R. Hull, *Adv. Supercond.* **10** [1997] 1401.
[142] F.C. Moon, *Appl. Electromagn. Mater*. **1** [1990] 29.
[143] T. Sugiura and H. Fujimori, *IEEE Trans. Magn.* **32** [1996] 1066.
[144] T. Sugiura, K. Matsunaga, Y. Uematsu, T. Aoyagi, and M. Yoshizawa, *Adv. Supercond.* **10** [1997] 1349.
[145] M. Murakami (ed), *Appl. Supercond.* **6 (2-5)** [1998].
[146] S.A. Basinger, J.R. Hull, and T.M. Mulcahy, 1990 *Appl. Phys. Lett.* **57** [1990] 2942.

[147] J.R. Hull, T.M. Mulcahy, K. Salama, V. Selvamanickam, B.R. Weinberger, and L. Lynds, *J. Appl. Phys.* **72** [1992]2089.
[148] T.H. Johansen, H. Bratsberg, Z.J. Yang, S.J. Guo, and B. Loberg, *J. Appl. Phys.* **70** [1991] 7496.
[149] V.V. Nemoshkalenko, E.H. Brandt, A.A. Kordyuk, and B.G. Nikitin, *Physica C* **170** [1990] 481.
[150] A.A. Kordyuk and V.V. Nomeshkalenko, *J. Supercond.* **9** [1996] 77.
[151] A.N. Terentiev, H.J. Lee, C.-J. Kim, and G.W. Hong, *Physica C* **290** [1997] 291.
[152] Z.J. Yang and J.R. Hull, *IEEE Trans. Appl. Supercond.* **7** [1997] 318.
[153] A.A. Kordyuk and V.V. Nemosckalenko, *IEEE Trans. Appl. Supercond.* **7** [1997] 928.
[154] C.E. Rossman, J.I. Budnick, and B.R. Weinberger, *Appl. Phys. Lett.* **70** [1997] 255.
[155] A.N. Terentiev and J.P. Harrison, *Jpn. J. Appl. Phys.* **34** [1995] L294.
[156] A.N. Terentiev, S.H. Moffat, R.A. Hughes, J.S. Preston, J. van Lierop, and J.P. Harrison, *Cryogenics* **37** [1997] 113.
[157] F.C. Moon, *Phys. Lett. A* **132** [1988] 249.
[158] T. Hikihara and F.C. Moon, *Phys. Lett. A* **191** [1994] 279.
[159] T. Hikihara, T. Fujinami, and F.C. Moon, *Phys. Lett. A* **231** [1997] 217.
[160] Z.J. Yang, T.H. Johansen, H. Bratsberg, A. Bhatnagar, and A.T. Skjeltorp, *Physica C* **197** [1992] 136.
[161] Z.J. Yang, *J. Supercond.* **5** [1992] 259.
[162] Z.J. Yang, *Japn. J. Appl. Phys*. **31** [1992] L477.
[163] Z.J. Yang, *Japn. J. Appl. Phys*. **31** [1992] L938.
[164] M.W. Coffey, *Phys. Rev. B* **52** [1995] R9851.
[165] J.C. Wei, J.L. Chen, L. Horng, and T.J. Yang, *Phys. Rev. B* **54** [1996] 15429.
[166] Z.J. Yang, *Physica C* **234** [1992] 263.
[167] Z.J. Yang, *Solid State Comm.* **107** [1998] 745.
[168] M.W. Coffey, *Phys. Rev. B* **65** [2002] 214524.
[169] J.H. Xu, J.H. Miller, Jr., and C.S. Ting, *Phys. Rev. B* **51** [1995] 424.
[170] M.W. Coffey, *Phys. Rev. Lett.* **83** [1999] 1648.
[171] A. Badia, *Phys. Rev. B* **60** [1999] 10436.
[172] M.W. Coffey, *J. Appl. Phys.* **87** [2000] 392.
[173] Q. Lu, K. Mochizuki, J.T. Markert, and A. de Lozanne, *Physica C* **371** [2002] 146.
[174] Z.J. Yang, *Appl. Supercond.* **2** [1994] 559.
[175] Z.J. Yang and J.R. Hull, *J. Appl. Phys.* **79** [1996] 3318.
[176] M. Marinescu, N. Marinescu, J. Tenbrink, and H. Krauth, *IEEE Trans. Magn.* **25**, [1989] 3233.
[177] D. Camacho, J. Mora, J. Foncuberta, and X. Obradors, *J. Appl. Phys.* **82** [1997] 1461.
[178] A. Badia and H.C. Freyhardt, *J. Appl. Phys.* **83** [1998] 2681.
[179] M. Braun, P. Buszka, T. Motylewski, W. Przydrozny, and C. Sliwa, *Physica C* **171** [1990] 537.
[180] A.A. Kordyuk, *J. Appl. Phys.* **83** [1998] 610.
[181] C.P. Bean, *Phys. Rev. Lett.* **8** [1962] 250.
[182] C.P. Bean, *Rev. Mod. Phys*. **36** [1964] 31.
[183] L.C. Davis, E.M. Logothetis, and R.E. Soltis, 1988 *J. Appl. Phys*. **64** [1988] 4212.
[184] L.C. Davis, *J. Appl. Phys*. **67** [1990] 2631.
[185] T. Sugiura, H. Hashizume, and K. Miya, *Int. J. Appl. Electromag, Mater*. **2** [1991] 183.
[186] M. Tsuchimoto, T. Kojima, H. Takeuchi, and T. Honma, *IEEE Trans. Magn.* **29** [1993] 3577.
[187] T.H. Johansen, H. Bratsberg, A.B. Riise, H. Mestl, and A.T. Skjeltorp, *Appl. Supercond.* **2** [1994] 535.
[188] A.B. Riise, T.H. Johansen, and H. Bratsberg, *Physica C* **234** [1994] 108.
[189] P.O. Hetland, T.H. Johansen, and H. Bratsberg, *Appl. Supercond.* **3** [1995] 585.
[190] H. Kuze and A. Onae, *J. Appl. Phys*. **77** [1995] 770.
[191] A. Sanchez and C. Navau, *Physica C* **268** [1996] 46.

[192] P.O. Hetland, T.H. Johansen, and H. Bratsberg, *Cryogenics* **36** [1996] 41.
[193] A.O. Hauser, *IEEE Trans. Magn.* **33** [1997] 1572.
[194] L.R. Turner and M.W. Foster, *IEEE Trans. Magn.* **34** [1998] 3024.
[195] C. Navau and A. Sanchez, *Phys. Rev. B* **58** [1998] 963.
[196] I.D. Mayergoyz and T.A. Keim, *J. Appl. Phys.* **67** [1990] 5466.
[197] I.D. Mayergoyz, A.A. Adly, M.W. Huang, and C. Krafft, *IEEE Trans. Magn.* **36** [2000] 3505.
[198] I. Mayergoyz and C. Serpico, *Physica B* **275** [2000] 17.
[199] M. Sjostrom, *Physica B* **306** [2001] 256.
[200] T. Sugiura, H. Hashizume, and K. Miya, *Int, J. Appl. Electromagn. Mater.* **2** [1991] 183.
[201] M. Uesaka, Y. Yoshida, N. Takeda, and K. Miya, *Int, J. Appl. Electromagn. Mater.* **4** [1993] 13.
[202] Y. Yoshida, M. Uesaka, and K. Miya, *IEEE Trans. Magn.* **30** [1994] 3503.
[203] K. Demachi, A. Miuri, T. Uchimoto, K. Miya, H. Higasa, R. Takahata, and H. Kameno, *Physica C* **357-360** [2001] 882.
[204] C.K. McMichael, K.B. Ma, M.W. Lin, M.A. Lamb, R.L. Meng, Y.Y. Xue, P.H. Hor, and W.K. Chu, *Appl. Phys. Lett.* **59** [1991] 2442.
[205] T.M. Mulcahy, J.R. Hull, K.L. Uherka, R.C. Niemann, R.G. Abboud, J. Juna, and J.A. Lockwood, *IEEE Trans. Appl. Supercond.* **9** [1999] 297.
[206] J.R. Hull, J.L. Passmore, T.M. Mulcahy, and T.D. Rossing, *J. Appl. Phys.* **76** [1994] 577.
[207] J.R. Hull, J.F. Labataille, T.M. Mulcahy, and J.A. Lockwood, *Appl. Supercond.* **4** [1996] 1.
[208] B.R. Weinberger, L. Lynds, J.R. Hull, and U. Balachandran, *Appl. Phys. Lett.* **59** [1991] 1132.
[209] J.R. Hull, T.M. Mulcahy, K.L. Uherka, and R.G. Abboud, *IEEE Trans. Appl. Supercond.* **5** [1995] 626.
[210] T.M. Mulcahy, J.R. Hull, K.L. Uherka, R.G. Abboud, J.H. Wise, D.W. Carnegie, C.W. Gabrys, and C.E. Bakis, *IEEE Trans. Magn.* **32** [1996] 2609.
[211] J.R. Hull, T.M. Mulcahy, K.L. Uherka, R.A. Erck, and R.G. Abboud, *Appl. Supercond.* **2** [1994] 449.
[212] T.M. Mulcahy, J.R. Hull, K.L. Uherka, R.G. Abboud, and J. Juna, *IEEE Trans. Appl. Supercond.* **11**, [2001] 1729.
[213] J.R. Clem, *Adv. Supercond. X* **1** [1997] 23.
[214] H.W. Lorenzen, R. Zickermann, and D. Schafer, *IEEE Trans. Appl. Supercond.* **11**, [2001] 1793.
[215] H. Teshima, M. Tanaka, K. Miyamoto, K. Nohguchi, and K. Ninata, *Physica C* **256** [1996] 142.
[216] F.E. Ehrich, *Handbook of Rotor Dynamics* (McGraw-Hill, New York) [1992].
[217] A. Tonoli and H.J. Bornemann, *J. Sound Vibr.* **212** [1998] 649.
[218] G. Martini, A. Rivetti, and F. Pavese, *Adv. Cryog. Engng.* **35** [1990] 639.
[219] K.-B. Ma, J.-R. Liu, C. McMichael, R. Bruce, D. Mims, and W.-K. Chu, 1991 *J. Appl. Phys.* **70** [1991] 3961.
[220] A.D. Chew, A. Chambers, and A.P. Troup, *Appl. Supercond.* **3** [1995] 327.
[221] E. Lee, K.B. Ma, T.L. Wilson, and W.K. Chu, *IEEE Trans. Appl. Supercond.* **9** [1999] 911.
[222] T. Sugiura, M. Tashiro, Y. Uematsu, and M. Yoshizawa, *IEEE Trans. Appl. Supercond.* **7** [1997] 386.
[223] C.E. Rossman and J.I. Budnick, *Physica C* **295** [1998] 304.
[224] A.N. Terentiev, *Physica C* **166** [1990] 71.
[225] A.N. Ternetiev and A.A. Kuznetsov, 1990 *Physica C* **169** [1990] 112.
[226] A.N. Terentiev and A.A. Kuznetsov, *Physica C* **195** [1992] 41.
[227] N.N. Krasnyuk and V.P. Mitrofanov, 1992 *Supercond.* **5** [1992] 1638.

[228] E.V. Postrekhin, S.N. Koscheeva, and L.W. Zhou, *Physica C* **248** [1995] 311.
[229] T. Hikihara and F.C. Moon, *Physica C* **250** [1995] 121.
[230] T. Hikihara, H. Adachi, S. Ohashi, Y. Hirane, and Y. Ueda, *Physica C* **291** [1997] 34.
[231] T.A. Coombs and A.M. Campbell, *Physica C* **256** [1996] 298.
[232] B.M. Smolyak, M.V. Babanov, G.V. Ermakov, I.B. Smolyak, and S.V. Naumov, *Physica C* **302** [1998] 23.
[233] B.M. Smolyak, M.V. Babanov, G.V. Ermakov, E.V. Postrekhin, L.W. Shou, I.B. Smolyak, and S.V. Naumov, *Mater. Sci. Engng.* **B53** [1998] 177.
[234] M. Murakami, *Appl. Supercond.* **1**, [1993] 1157.
[235] F. Werfel et al., ASC 2002.
[236] S. Gauss, J.H. Albering, J. Bock, M. Kesten, H. Fieseler, W.R. Canders, H. May, H.C. Freyhardt, and M. Ullrich, *IEEE Trans. Appl. Supercond.* **9** [1999] 1004.
[237] R. Weinstein, D. Parks, R.-P. Sawh, V. Obot, J. Liu, and G.D. Arndt, *3rd Int. Symp. Magn. Susp. Tech.nol.* Tallahassee [1995].
[238] J.R. Hull and T.M. Mulcahy, *3rd International Symposium on Linear Drives for Industrial Applications*, pp. 42-47, Nagano, Japan (October 2001).
[239] J.R. Hull, T.M. Mulcahy, and R. C. Niemann, *Adv. Cryog. Engng.* **47**, [2002] 1711.
[240] T.D. Rossing and J. R. Hull, *Phys. Teacher*, **29** [1991] 552.
[241] B.V. Jayawant, *Rep. Prog. Phys.*, **144** [1981] 411. Also, *Proc. R. Soc. London A,* **416** [1988] 245-320, 1988.
[242] R.G. Rhodes and B.E. Mulhall, *Magnetic Levitation for Rail Transport*, Oxford, Clarendon Press [1981].
[243] E. R. Laithwaite (ed.), *Transport without Wheels*, London, Elek Science [1977].
[244] H.T. Coffey, J.C. Solinsky, J.D. Colton, and J. R. Woodbury, *IEEE Trans. Magn.,* **10** [1974] 451.
[245] R.H. Borcherts, L.C. Davis, J.R. Reitz, and D.F. Wilkie, *Proc. IEEE,* **61** [1973] 569.
[246] D.L. Atherton and A.R. Eastham, *J. Appl. Phys.,* **45** [1974] 1398.
[247] D.L. Atherton and A.R. Eastham, *IEEE Trans. Magn.,* **11** [1975] 627.
[248] H.T. Coffey, *IEEE Trans. Appl. Supercond.,* **3** [1993] 863.
[249] H. Nakao, T. yamashita, Y. Sanada, M. Yamaji, S. Nakagaki, T. Shudo, M. takahashi, Am Miura, M. Terai, M. Igarashi, T. Kurihara, K. Tomioka, and M. Yamaguchi, *IEEE Trans. Appl. Supercond.,* **9** [1999]1000.
[250] S. Yokoyama, K. Shimohata, T. Inaguchi, T. Kim, S. Nakamura, S. Miyashita, and F. Uchikawa, *IEEE Trans. Appl. Supercond.,* **5** [1995] 610.
[251] S.I. Yoo, T. Higuchi, N.Sakai, H Fujimoto, and M. Murakami, *Mater. Sci. Engng.* **53B** [1998] 203.
[252] H. Kamijo, T. Higuchi, H. Fujimoto, H. Ichikawa, and T. Ishigohka, *IEEE Trans. Appl. Supercond.* **9**, [1999] 976.
[253] J. Wang, S. Wang, Z. Ren, X. Dong, G. Lin, J. Lian, C. Zhang, H. Huang, C. Deng, and D. Zhu, *IEEE Trans. Appl. Supercond.* **9**, [1999] 904.
[254] J.S. Wang, S.Y. Wang, Z.Y. Ren, M. Zhu, H. Jiang, and Q.X. Tang, *IEEE Trans. Appl. Supercond.* **11**, [2001] 1801.
[255] A. Senba, H. Kitahara, H. Ohaski, and E. Masada, *IEEE Trans. Magn.* **32**, [1996] 5049.
[256] H. Ogiwara, T. Azukizawa, and M. Morishita, *Appl. Supercond.* **1** [1993] 1185.
[257] H. Minami and J. Yuyama, *Jpn. J. Appl. Phys.* **34** [1995] 346.
[258] Y. Sangawa, H. Ueda, M. Tsuda, A. Ishiyama, Sl. Kohayashi, and S. Haseyama, *IEEE Trans. Appl. Supercond.* **11**, [2001] 1797.
[259] Y.-K. Kim, M. Katsurai, and H. Fujita, *Sensors and Actuators* **20** [1989] 33.
[260] J.R. Hull, *Spectrum* **34 (7)** [1997] 20.
[261] H.J. Bornemann, T. Ritter, C. Urban, O. Zaitsev, K. Weber, and H. Rietschel, *Appl. Supercond.* **2** [1994] 439.
[262] Q.Y. Chen, Z. Xia, K.B. Ma, C.K. McMichael, M. Lamb, R.S. Cooley, P.C. Fowler, and W.K. Chu, *Appl. Supercond.* **2** [1994] 457.

[263] M. Minami, S. Nagaya, H. Kawashima, T. Sato, and T. Kurimura, *Adv. Supercond.* **10** [1997] 1305.
[264] Y. Miyagawa, H. Kameno, R. Takahata, and H. Ueyama, *IEEE Trans. Appl. Supercond.* **9** [1999] 996.
[265] T.A. Coombs, A.M. Campbell, I. Ganney, W. Lo, T. Twardowski, and B. Dawson, *Mater. Sci. Engng.* **B53** [1998] 225.
[266] J.R. Hull and T.M. Mulcahy, *IEEE Trans. Appl. Supercond.* **9** [1999] 390.
[267] S. Hanany, et al. ASC 2002.
[268] R. Decher, P.N. Peters, R.C. Sisk, E.W. Urban, M. Vlasse, and D.K. Rao, *Appl. Supercond.* **1** [1993] 1265.
[269] A. Rivetti, G. Martini, R. Goria, and S. Lorefice, *Cryogenics* **27** [1987] 8.
[270] M. Lamb, K.B. Ma, R. Cooley, D. Mackey, R. Meng, C.W. Chu, W.K. Chu, P.C. Chen, and T. Wilson, *IEEE Trans. Appl. Supercond.* **5** [1995] 638.
[271] K.B. Ma, C.K. McMichael, M.A. Lamb, and W.K. Chu, *IEEE Trans. Appl. Supercond.* **3** [1993] 388.
[272] J.H.. Yu, E. Postrekhin, K.B. Ma, W.K. Chu, and T. Wilson *IEEE Trans. Appl. Supercond.* **9** [1999] 908.
[273] G. Binnig, H. Rohrer, Ch. Gerber, and E. Weibel, *Appl. Phys. Lett.* **40** [1982] 178.

FIELDS AND FORCES FROM SUPERCONDUCTORS AND PERMANENT MAGNETS

A.M. CAMPBELL
IRC in Superconductivity, Madingley Road, Cambridge, UK

1. INTRODUCTION

Superconductors can be used in a similar way to ferromagnetic materials, but there are fewer calculations of field distributions. In this article the techniques common in ferromagnetic materials are extended to superconductors. A harmonic analysis is used which has a number of advantages. Firstly rather simple expressions can be obtained using only the fundamental and this is often sufficiently accurate. Secondly if a more detailed analysis is required the simple solution can be easily extended and usually allows a separation of the contribution of the shape, magnet dimensions and other distances. This is helpful in the design of a system. Ferromagnetic materials are assumed to be either linear, or to have a fixed magnetic moment. Superconductors are assumed to be in the Meissner state, as a sintered permeable (but diamagnetic) material or to carry a fully penetrated critical current. The problems considered are :-

1) Screening by a permeable material
2) Field Smoothing.
3) Screening by a thin sheet in the Meissner state
4) Forces between magnetic arrays
5) Levitation above a conducting plane.

In most cases the method used to is to postulate a potential function and match the boundary conditions. This brings out the similarity between the problems.
It is essentially an application of what the electrical engineers call 'layer theory'. The components can be split into layers with simple potentials in each layer. Starting with a simple known potential and matching boundary conditions at each boundary allows very complex systems to be analysed. For planar systems the potentials are sin and cos terms, for cylinders

and spheres cylindrical and spherical harmonics fulfil the same function but for most practical purposes the layer thickness is small compared with the radius so that the planar theory is good enough.

Although the obvious way to start is with a given external field and current, when there are exponential terms involved this can lead to a large array of very ill conditioned equations. A useful technique to avoid this problem is to assume unit field in a region where there is only one term in the potential, for example at the centre of a cylinder in an applied field. This allows the coefficients of each layer to be calculated in turn until the field to produce the starting assumption is found. Since the systems are linear we can then find the fields for any applied field. The subject is well known for conventional magnetic materials. One of the objects of this article is to see how much of this knowledge can be usefully transferred to superconducting and hybrid devices.

2. PERMEABLE MATERIALS

2.1 Screening

Screening of spaces by a permeable material is a well known technique,see for example reference [1] and references therein. This section is included because there are close parallels between a screened box and the following sections. Well known results for screened boxes can give useful leads for less familiar geometries. For a cylinder parallel to the external field we try magnetostatic potentials of the form $(ar+b/r)\cos\theta$ and make the potential and H_θ continuous. The procedure for multiple shells is made easier if we start at the center with b=0 and a=1. This allows the coefficients to be calculated at each boundary in turn by solving only two equations at each boundary until we reach the outside and can determine the uniform field which gives unit field in the center. The attenuation factor of a uniform external field by a single cylinder of radius r and thickness d is:-

$$1+\left(\frac{d}{2r}+\frac{d^2}{4r^2}\right)\left(2+\mu+\mu^{-1}\right)\left(1+\frac{d}{r}\right)^{-2} \tag{1}$$

It can be seen that the permeability appears only in the factor $(\mu+\mu^{-1})$ so that both paramagnetic and ferromagnetic materials screen the field. A material with μ=0.001 will be as good as one with μ=1000. This also appears to be a general result for multilayer screens. However sintered YBCO has a minimum μ of about 1/3 once the intergrain currents are quenched so it seems unlikely that it can be used in this way. Of course screening with intergrain currents flowing in the bulk is extremely effective, but is limited by the intergrain critical current.

A surprising result of going to a two layer screen is that a single layer screen can be improved by removing material, leaving an air gap between the inner and outer radii. This imposes a higher reluctance path on the flux which prevents it reaching the interior [1,2]. Another way of understanding this is to see that a single layer has a screening factor proportional to its thickness so that screening by two contiguous layers is the sum of the screening factors and proportional to μ. However if one cylinder is inside a larger outer one the final screening factor is the product of the two screening factors which is proportional to μ^2 and much larger.

2.2. Field Smoothing

Inhomogeneous fields are a source of hysteresis loss in bearings and one technique of reducing inhomogeneity is to put a thin sheet of a permeable material in contact with the face of the permanent magnet. This problem can be solved analytically for a two dimensional system. For simplicity we assume a thin magnet represented by a sinusoidal current sheet $I_y \cos(kx)$ and put above it a sheet of material of thickness d and permeability μ.

At all points except the current sheet $\nabla^2 A=0$ so that the z variation is as $\exp(-kz)$. This is a very general result, variations of potential in the x direction lead to variations in the perpendicular direction which decay exponentially over a distance comparable to the wavelength in the x direction.

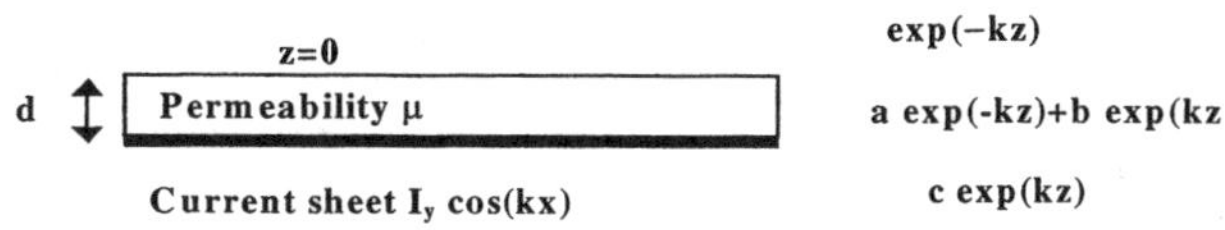

Fig.1. A permeable sheet above a sinusoidal current sheet. Trial solutions for A are on the right.

We put in trial solutions for A of the form:-

$$A_y = \left[a\exp(-kz)+b\exp(kz)\right]\cos(kx) \qquad (2)$$

as shown in Fig.1. A_y is continuous, $\Delta H_x = I$ at the current sheet, and zero elsewhere
Then matching boundary conditions:-:-

$$2I_y k^{-1} = 2\left[1+\exp(-2kd)\right]-\left[\mu+\mu^{-1}\right]\left[1-\exp(-2kd)\right] \qquad (3)$$

This gives the current to produce unit A above the permeable sheet. The factor by which a particular harmonic is reduced is:-

$$0.5\left[1+\exp(-2kd)\right]-0.25\left[\mu+\mu^{-1}\right]\left[1-\exp(-2kd)\right] \qquad (4)$$

Again a very diamagnetic material will have the same smoothing effect as a ferromagnetic material.

For large μ and $kd \ll 1$ the field fluctuation is reduced by a factor $1+\frac{1}{2}\mu kd$ in comparison with the field at the same point without iron. For large kd (thick layers) the factor is $\mu/4$, i.e 250 for $\mu=1000$. Here thick means thick compared with the wavelength of the fluctuation in magnetisation so thinner layers are still very effective if μ is large. However the benefits must be compared with the reduction in the long wavelength fields which provide the levitation and the reduced space available because of the thickness of the iron. In contrast, below the sheet the field oscillations are doubled.

Following on from the results of screening by cylinders we might expect two layers separated by an air-gap to be better than the same thickness of solid and this proves to be the case. The algebra is very similar to that of concentric magnetic screens with the wavelength of the variation in magnetisation taking the place of the radius. However the advantage disappears if the sheets are too thin. For smoothing the field in a magnetic bearing we would like to have little attenuation at the wavelength of the magnet and a large attenuation at the shorter wavelengths and this can be achieved to some extent. Let us take a magnet with $k=1\ \text{cm}^{-1}$, i.e. about 3 cms in width. We put a 3 mm layer of the permeable material with μ=20 on the magnet.

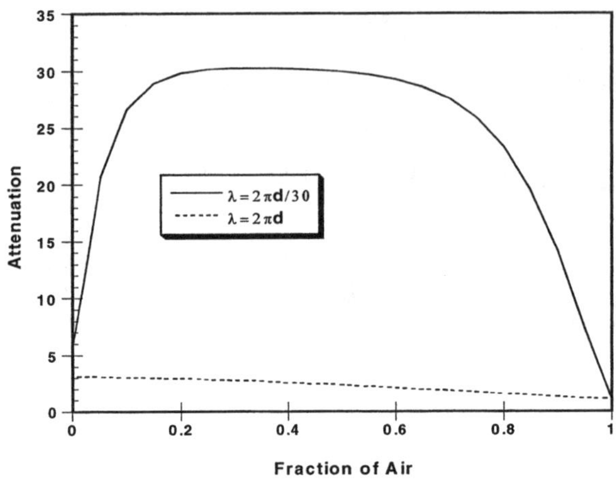

Fig.2 The attenuation of the fundamental, and 1/30 of the wavelength as a function of the air gap between two sheets.

Figure 2 shows how the fundamental and a variation of 1/30th of the wavelength are attenuated as the proportion of air changes from zero to unity. For a solid layer the fundamental is attenuated by a factor of 3 and a 1mm inhomogeneity by a factor 5.5. If we now make a fraction 0.85 of the thickness in the centre air the attenuation of the fundamental goes down to 1.4 while that of the inhomogeneity goes up to a factor of 14.

3. SUPERCONDUCTING SHEETS

3.1. One Dimensional Sinusoidal Currents

The response to a DC field in the Meissner state is very similar to that of a normal conductor in an AC field. This is because in both cases ***A*** is proportional to ***J***, for the superconductor the constant is $-\lambda^2\mu_o$ while for the normal material the constant is $-1/j\omega\sigma = -0.5j\delta^2\mu_o$ where δ is the skin depth. The screening by a cylinder in a parallel field is derived in ref [3].
In both cases flux is excluded from a one dimensional solid when its thickness exceeds the skin depth or penetration depth, but for thin flat sheets and hollow cylinders this occurs when the depth is intermediate between the sheet thickness and the macroscopic dimension. The value is normally close to the geometric mean of the two dimensions and there is a wide regime of almost complete screening where the current density in the sheet is uniform.

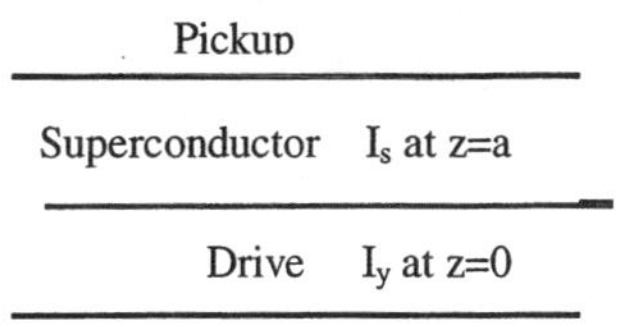

Fig.3. The positions of the drive coil, superconductor, and pickup coil

A number of experiments have used the coupling between two coils on either side of a thin superconducting film to measure the penetration depth, see for example ref [4]. Fiory et. al. gave a computational method [5] and Pippard [6] a simple analytic calculation of the coupling. The following is similar to Pippard's derivation but uses the vector potential instead of fields. We assume a current I_y per unit length varying as cos(kx), fig.3. The paper is the zx plane.

In thin films we can assume the current in the superconductor, I_s, is uniform and from the London equations $\nabla^2 \boldsymbol{A} = -\lambda^2 \boldsymbol{A}$ where $\boldsymbol{A}$ is the vector potential in the y direction. In free space $\nabla^2 \boldsymbol{A} = 0$ so the solution is of the form exp($\pm kz$)cos(kx). We can get the general result with trial solutions but there is a simpler technique applicable to thin films.

For an isolated current sheet I at z=0, to match the boundary condition on B_x :-

$$A_y = \left(\frac{\mu_o I_y}{2k}\right)\exp(-k\,|\,z\,|) \tag{5}$$

If the current in the film is I_s then the vector potential at the film is :-

$$A_y = \left(\frac{\mu_o I_y}{2k}\right)\exp(-k\,|\,z\,|) + \left(\mu_o I_s / 2k\right) \tag{6}$$

To be consistent with the London equations, for a film of thickness d,

$$A_y = -\mu_o \lambda^2 J_y = -\mu_o \lambda^2 I_s d^{-1} \tag{7}$$

Hence

$$I_s = \frac{-I\exp(-ka)}{1 + 2k\lambda^2 d^{-1}} \tag{8}$$

Above the film $\boldsymbol{A}$, and hence also $\boldsymbol{B}$, are reduced by a factor $1+\lambda^{-2}d/2k$.

We note that screening begins when the penetration depth is intermediate between the thickness of the film and the wavelength of the field, i.e. when $\lambda^2 > d/2k$

For films comparable to the penetration depth in thickness we need to try solutions in the film of the form:

$$A_y = a\exp(k'z) + b\exp(-k'z) \tag{9}$$

In the superconductor $\nabla^2 \boldsymbol{A} = -\lambda^2 \boldsymbol{A}$, so:-

$$k'^2 = k^2 + \lambda^{-2} \tag{10}$$

The result is

$$I_y = \exp(-tk)\left[2k\cosh(k't) + \left(k' + k^2/k'\right)\sinh(k't)\right] \tag{11}$$

The attenuation of A_y is by a factor K where:-

$$K = \exp(tk)\left[\cosh(k't) + k'/2k\left(+k/2k'\right)\sinh(k't)\right]^{-1} \tag{12}$$

The attenuation is independent of the position of the superconductor and the height above it, which is reasonable since the ratio of fields on either side of the superconductor cannot depend on the position of the source. This makes it easy to extend the analysis to thick coils on both sides of the superconductor. Since A varies as exp(-kz) integration over a driving coil of thickness w multiplies the result by a factor:-

$$\left[1-\exp(-kw)\right]\left[kw\right]^{-1} \tag{13}$$

If we want the mutual inductance the vector potential must be integrated round the secondary, which will introduce a similar factor containing the secondary thickness. Section 3.3 gives a simple example.

3.2. Two Dimensional Distributions

The main results can be obtained from the very simple current distribution of a harmonic current in the y direction as above. However since most magnets are cubes or cylinders we need to see the effect of a variation in the x direction. In this case if there is a harmonic variation as exp $j(k_x x+k_y y)$ then since $\nabla^2 \boldsymbol{A}=0$ the variation in the z direction must be as exp($\pm kz$) where :-

$$k^2 = k_x^2 + k_y^2 \tag{14}$$

A 2D current distribution in a plane can be derived from a vector potential $\boldsymbol{M}$ such that curl $\boldsymbol{M}=\boldsymbol{I}$ where $\boldsymbol{I}$ is the current per unit length For a planar current $\boldsymbol{M}$ is perpendicular to the plane and is a scalar. M can be interpreted as the magnetisation per unit area, (although only in magnetic materials has it meaning on a local scale), and found from the moment of each current loop. For a permanent magnet $\boldsymbol{M}$ is constant over the magnet, for a superconductor carrying a constant J_c it rises linearly from each edge like a pitched roof.

For a two dimensional current sheet, assumed symmetrical about x=0 and y=0, the fundamental is given by :-

$$M_z = I_o \cos\left(k_x x\right)\cos\left(k_y y\right) \tag{15}$$

where I_o is the peak current. Then the current per unit area is $\boldsymbol{I}$=curl $\boldsymbol{M}$ where:-.

$$I_x = -I_o k_y \cos\left(k_x x\right)\sin\left(k_y y\right) \tag{16}$$

$$I_y = I_o k_x \sin\left(k_x x\right)\cos\left(k_y y\right) \tag{17}$$

Matching boundary conditions for x and y directions leads to the same equations as if only a y current is flowing so we end up with the same attenuation factor K, equation (12) Then :-

$$A_x = -0.5 I_o K k^{-1} k_y \cos\left(k_x x\right)\sin\left(k_y y\right)\exp(-kz) \tag{18}$$

$$A_y = 0.5 I_o K k^{-1} k_x \sin\left(k_x x\right)\cos\left(k_y y\right)\exp(-kz) \tag{19}$$

$\boldsymbol{A}$ can be expressed in terms of $\boldsymbol{M}$ as the curl of an attenuated magnetisation $\boldsymbol{N}$. $\boldsymbol{A}$=curl($\boldsymbol{N}$) where:-

$$N_z = 0.5 K k^{-1} M_z \exp\left(-kz\right) \tag{20}$$

This can be summed over all the Fourier components to give the mutual inductance of coils on either side of the film. A number of Fourier transforms are listed in §4.7.

3.3 Single Coils

For non periodic currents we need integrals instead of sums. As a simple example take a single square coil of side *2a* carrying a current I_o. Equation (57) gives the Fourier integral, and the magnetisation is :-

$$M_z(x,y)=\int_o^\infty\int_o^\infty \frac{4I_o}{\pi^2 k_x k_y}\sin(k_x a)\sin(k_y a)\cos(k_x x)\cos(k_y y)dk_x dk_y \tag{21}$$

Then :-

$$A=0.5\nabla\times\left[MKk^{-1}\exp(-kz)\right] \tag{22}$$

$$A_z(x,y,z)=\int_o^\infty\int_o^\infty \frac{-2I_oK}{\pi^2 k_x}\sin(k_x a)\sin(k_y a)\cos(k_x x)\sin(k_y y)\exp(-kz)\,dk_x dk_y \tag{23}$$

$$A_y(x,y)=\int_o^\infty\int_o^\infty \frac{2I_oK}{\pi^2 k_y}\sin(k_x a)\sin(k_y a)\sin(k_x x)\cos(k_y y)\exp(-kz)\,dk_x dk_y \tag{24}$$

The inductance is the integral of *A* round the second square coil for unit current:_

$$L=\frac{16\mu_o}{\pi^2}\int_o^\infty\int_o^\infty Kk_x^{-2}k^{-1}\sin^2(k_x a)\sin^2(k_y a)\exp(-kz)dk_x dk_y \tag{25}$$

As a check on this, if K=1 and z is large we should have the inductance between two small current loops of area $4a^2$ separated by z, which is $8\mu_o a^4/\pi z^3$. This can be obtained from equation (25) by seeing that the integrand becomes small if $k>>1/z$ and so for all the significant parts $k_x a$ and $k_y a$ are small. We put $k_x=k\sin\theta$ and $k_y=k\cos\theta$ so the element of area becomes $k\,dkd\theta$ and the integral becomes

$$L=16\pi^{-2}\mu_o a^4\int_o^\infty\int_o^{\pi/2} k^2\sin^2(\theta)\exp(-kz)d\theta dk \tag{26}$$

$$=\frac{8\mu_o a^4}{\pi z^3} \tag{27}$$

This is the correct dipole expression.

(In evaluating this type of integral numerically, if there is a positive power of k, it is essential to include the exp($-kz$) term with a finite value for z, since the upper limit of k must be greater than $1/z$ if the integral is to converge).

Since the screening is a function of k it is reasonable to explore the possibility of smoothing fields with superconducting films. However this does not work because the screening increases with increasing k, so that the main field is screened more than the fluctuations.

4. FORCES BETWEEN MAGNETIC ARRAYS

The forces between an array of permanent magnets and its image are important in magnetic bearings and other applications of levitation. To a good approximation rare earth permanent magnets have constant magnetisation and can be modelled either as a surface current sheet or by the electrostatic analogy of a polarised body and a surface charge density. If the critical current density of the superconductor is high the levitation force is that between the magnet and its image. This allows a direct calculation of the force between magnets, but the algebra can be tedious and it is not very easy to see how to optimise the system. For small magnets the force is proportional to the field gradient, but this approach is difficult to apply to large systems close together. There have been a number of calculations of the forces on a single magnet, see for example refs [7,8] and references therein. In practice magnet arrays are used and this brings in a large number of possible geometries. Nagashima et. al. [9] considered arrays of between one and four poles, including the effect of a soft iron back plate to make the magnets appear longer than they really were. Zeisberger et. al. [10] have discussed the optimisation of the forces in a number of arrays with, and without, iron. They conclude that the stiffness can be increased by subdivision of the magnets, but the clearance must be kept small if there is to be a benefit. They also conclude that iron pole pieces are of little benefit. Yang et. al. [11] measured the forces and stiffness of up to nine magnets in different arrangements. They found that the smaller the number of poles the larger the force for a given number of magnets. In these experiments the magnet array was comparable in size to the superconductor, so there are significant edge effects. These are also seen in recent work in connection with a levitated people-carrying vehicle [12].

4.1 General Considerations

The most important parameters are the lift, the stiffness and, for levitated trains, the drag For practical purposes the forces in which we are interested are those between a magnet and its image. Modern magnets have a very high coercive force so that they can be regarded as having a fixed magnetisation M, typically $\mu_o M$ is about 0.5-1.5 Tesla. The corresponding magnetic pressure is $\frac{1}{2}\mu_o M^2$=0.1-1MPa.

The simplest geometry is two long coaxial solenoids with currents in the same sense and separated by a gap small compared with the radius. From the energy in the gap the attractive force per unit area is $B^2/2\mu_o$ where B is the field at the centre of a solenoid. It follows that this is also the repulsive force if we reverse one of the currents, since all forces are proportional to the product of the currents in the coils. Permanent magnets are magnetically equivalent to a surface current sheet of M A/m where M is the magnetisation, so that in principle forces between solenoids in the limit of zero wire diameter can be used to find the forces between magnets. At first sight this looks as if it leads to a singularity, but we can show that this is not the case.

If two solenoids approach to a distance comparable with the wire diameter we must estimate the force that is due to the wires which are almost touching so we need to sum the forces of the turns closest to the gap. If the solenoid radius is a and the wire diameter d the current is M/d. For two coils of length $L<a$ which are just in contact, adding up the forces in the limit of small d gives a force $\mu_o M^2 aL\, 2\ln(2)$ and a stiffness $\mu_o M^2 a \ln(L/d)$. In other words there is no singularity in the force, so that the force can be

calculated from macroscopic terms, but there is a logarithmic infinity in the stiffness. What this means in practice is that magnets in contact will demagnetise where their edges touch, so that it is difficult to get an accurate value for the maximum stiffness. In practice they will not be touching and the singularity is only logarithmic, so that macroscopic calculations give a pretty good estimate of the stiffness. The force between two solenoids falls off significantly only when the gap is comparable to the diameter. Hence the specific stiffness (stiffness/maximum force) is approximately equal to the diameter. We can gain specific stiffness by subdividing a magnet, but in the case of a superconductor with bulk currents this will have the disadvantage of reducing the maximum force and hence also the absolute stiffness. In general passive magnetic bearings are always much less stiff than mechanical bearings.

The main difference between the superconductor and the permanent magnet is that the relevant material parameter is a bulk critical current density, rather than the surface current or magnetisation. This means that the mean magnetisation of a superconductor is proportional to its linear dimensions, and the current is distributed throughout its volume. It makes a harmonic analysis of the current density the most suitable treatment since expressions obtained using only the fundamental will be quite accurate, and higher harmonics can be added as required.

As well as the forces the stability of a magnetic system is very important. Earnshaw's theorem showed that a charge in a fixed potential field cannot be stable, and Maxwell [13] showed that the presence of any movable charges such as appear in conductors or dielectrics, makes the system even more unstable. This is because Earnshaw's theorem applies to a system of fixed charges and if the charges in a dielectric or conductor are allowed to relax this lowers the energy. Since all the energy of a system of charges can be attributed to the mechanical forces which were used to set it up this implies a decrease in mechanical energy if a body is moved and hence more instability. The extension of these results to paramagnetic magnetic systems is straightforward as there is a direct analogy between the electrostatic and magnetic forces. The theorem does not apply to diamagnetic systems as these must have an energy term which comes from the internal energy of the material and is not mechanical.

We can go further than merely state that the system is unstable. The force on a body depends on the gradient in the applied field $\boldsymbol{H_o}$ and since $\nabla \boldsymbol{H_o}=0$ it follows that:-

$$\frac{\partial H_{ox}}{\partial x}+\frac{\partial H_{oy}}{\partial y}+\frac{\partial H_{oz}}{\partial z}=0 \tag{28}$$

Hence in a constant field, if the stiffness in one direction is positive, at least one of the other stiffnesses must be negative and the system is unstable. The more stable we make one direction the more unstable another direction becomes.

A recent theorem due to Hull [14] relates the vertical and lateral stiffness of a magnet above a superconductor which has been cooled in the magnetic field. He showed that the vertical stiffness is equal to the sum of the two lateral stiffnesses. This appears to be a very general result for any system of magnets however complex.

4.2. The Maximum Levitation Height.

Before finding the forces it is of interest to see just how much clearance we might expect between a permanent magnet levitated above a superconducting plane. If we consider a bar magnet the force comes from the poles so that the centre section can be removed which reduces the weight without reducing the force. Similarly if we start with a thin disc parallel to the superconductor we can increase the force per unit volume by making it thicker, and reducing the radius. We conclude that the largest force per unit volume is obtained from a magnet with roughly equal dimensions in all directions and so choose a sphere.

Now a uniformly magnetised sphere behaves as a dipole at its centre so we can calculate the force between two magnetised spheres at any separation. The force between two dipoles of moment m separated by a distance a is

$$F = \frac{3\mu_o m^2}{2\pi a^4} \tag{29}$$

This must equal the weight for stable levitation.

We can lump the material parameters of the magnet into a magnetic length :-

$$L = \frac{\mu_o M^2}{2\rho g} \tag{30}$$

L is the length of magnet which can just support itself. (For Neodymium Iron Boron $\mu_o M$ is 1.3 Tesla, and the specific gravity 7.5 so L is 9.1 metres. These figures will be used as an example). When the levitation force of a sphere of radius r balances the weight the height, h, is given by $h=(0.25Lr^3)^{1/4}$. Since the height increases less rapidly than the radius only small magnets can levitate at heights large compared with the size of the magnet. If the gap between the superconducting plane and the bottom of the sphere is g then $g=h-r$ and substituting for h:-

$$g = \left(0.25Lr^3\right)^{1/4} - r \tag{31}$$

The gap has a maximum value when $r=81L/4^5$ which is 72 cms. The gap is then 24 cms. Therefore even if there is no weight on a bearing, this is the maximum clearance we can expect, (and the magnet must be 1.4 metres in diameter).

Whatever the magnetisation the maximum clearance is 1/6 of the diameter of the magnet. Fortunately the repulsive force increases rapidly with decreasing spacing so that large loads can be carried at spacings which are not too much smaller than the optimum no-load spacing.

4.3. Field from a Sinusoidal Current Sheet.

We now calculate the field due to a periodic array of magnets. We start with a current sheet in the xy plane. As in equation (5) the vector potential is:-

$$A_y = ½\mu_o I_y k^{-1} \exp(-kz)\cos(kx) \tag{32}$$

By direct analogy with the electrostatic potential a similar result holds for the vector potential of the components of the current in the other directions, and also for variations as sin kx. Any given magnetisation $\boldsymbol{M}$ can be replaced by equivalent currents $\boldsymbol{J}$=curl $\boldsymbol{M}$, and the vector potential found by integrating the vector potential due to these currents.

If we have a layer of current of thickness w we integrate through the thickness and obtain:-

$$A_y = ½\mu_o I_y k^{-1} \exp(-kz)\cos(kx)\left[\left(1-\exp(-kw)\right)/kw\right] \tag{33}$$

where z is the distance from the surface.

4.4. Force on a Sinusoidal Current

From equation (33) the x component of the field a distance z from the sheet is

$$B_x = \tfrac{1}{2}\mu_0 I_y \exp(-kz)\cos(kx)\left[(1-\exp(-kw))/kw\right] \tag{34}$$

If the array is a distance g from an ideal superconductor the x component of the field at the surface is doubled, so the mean pressure is :-

$$P = \tfrac{1}{4}\mu_0 I_y^2 \exp(-2kg)\left[(1-\exp(-kw))/kw\right]^2 \tag{35}$$

If we have a magnetisation of amplitude M_z then, the current density is curl$\boldsymbol{M}$, so $I_y = kwM_z$. Hence:-

$$P = \tfrac{1}{4}\mu_0 M_z^2 \exp(-2kg)\left[1-\exp(-kw)\right]^2 \tag{36}$$

The stiffness is the differential with respect to g:-

$$Stiffness = \tfrac{1}{2}\mu_0 k M_z^2 \exp(-2kg)\left[1-\exp(-kw)\right]^2 \tag{37}$$

For more complex magnets we express the magnetisation as a Fourier series:-

$$M_z(x) = \sum_n M f_n \cos(nkx) \tag{38}$$

Here M is the magnitude of the magnetisation and the Fourier coefficients are related to the size and shape. Then the pressure is :-

$$P = \tfrac{1}{4}\mu_0 M_z{}^2 \sum_n f_n^2 \exp(-2nkg)\left[1-\exp(-nkw)\right]^2 \tag{39}$$

We see that Fourier components with a wavelength smaller than the gap are very attenuated, so that the fundamental often gives an excellent approximation. Also for thin magnets the force is proportional the thickness of the magnet, but there is not much point in making the magnet much thicker than its diameter.

The stiffness is the differential with respect to g:-

$$Stiffness = \tfrac{1}{2}\mu_0 M_z{}^2 \sum_n f_n^2 nk \exp(-2nkg)\left[1-\exp(-nkw)\right]^2 \tag{40}$$

4.5 The Halbach Array

A Halbach array of magnets is one where the angle of the magnetisation changes continuously while its magnitude remains constant [15]. Originally proposed for a 'wiggler' it has some quite surprising properties with other applications.

4.5.1 The Planar Array

The simplest example is the linear Halbach array which involves only a sinusoidal magnetisation. It consists of a plane of magnets in which the magnetisation is constant in magnitude but rotates uniformly in the xy plane on moving in the x direction as in Fig.4 i.e.:-

$$M_x = -M_o \sin(kx) \tag{41}$$

$$M_z = M_o \cos(kx) \tag{42}$$

We take the top surface of the magnets at $z=0$ and the bottom at $z=-w$.

The equivalent currents, which are all in the y direction, are :-

Firstly surface currents in the y direction $M_o \sin kx$ at $z=0$ and an equal and opposite surface current at $z=-w$. Secondly there is a bulk current density:-

$$J_y = curl_y M = M_o \sin kx \tag{43}$$

Using equation (32) the vector potential for $z>0$ from the sum of these currents is :-

For $z>0$
$$A_y = \mu_o M_o k^{-1} \exp(-kz)\left[1-\exp(-kw)\right]\sin(kx) \tag{44}$$

For $z<-d$ $A_y=0$

Hence on the top surface

$$B_x = \mu_o M_o \left[1-\exp(-kw)\right]\cos(kx) \tag{45}$$

$$B_z = -\mu_o M_o \left[1-\exp(-kw)\right]\sin(kx) \tag{46}$$

Below the sheet B=0.

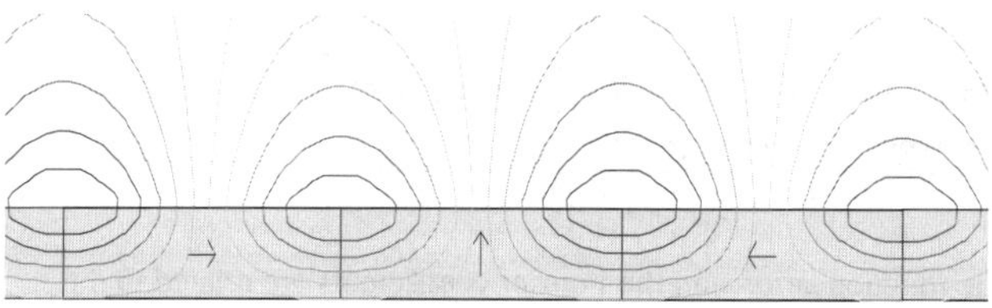

Fig.4. The lines of force round a Halbach array with a continuous variation of magnetisation in the sense shown. (The arrows here indicate the general direction of the magnetisation).

This is illustrated in Fig.4. which shows the contours of ***A***, which are the lines of force of ***B***. Thus for a thick layer the array concentrates all the field on the top surface and gives zero field on the bottom surface. At first sight this is a very surprising result, and seems to violate the symmetry of the problem. However this is not the case since this array rotates the magnetisation anticlockwise as we move in the x direction, if it rotated clockwise the field would appear on the bottom surface. The reason for this asymmetry can be seen by considering only three magnets. If two are horizontal they reinforce the field of the vertical magnet above the layer, and diminish it below. However the point for practical applications is that if we put such an array above a superconductor so that B_z is zero and B_x is doubled, the mean pressure for a thick array is $\mu_o M_o^2$. This is twice what can be achieved by bringing two infinite magnets together ($\frac{1}{2}\mu_o M_o^2$) and although an ideal sinusoidal array is impractical something like it is

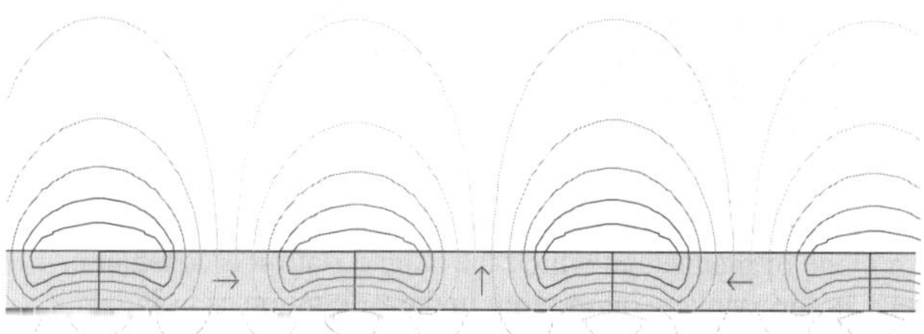

Fig.5. The lines of force for a pseudo Halbach array in which the magnetisation varies discretely in jumps of $\pi/2$. There is still a large concentration of field on the upper surface.

achievable with discrete magnets. The Pseudo-Halbach array is a set of magnets where the magnetisation changes direction by $\pi/2$ between neighbouring magnets. The lines of force are shown in fig.5. and it can be seen that there is still a major asymmetry. This can be realized with superconducting pellets [16])

4.5.2 A Cylindrical array

We can now roll the Halbach sheet into a cylinder with the top surface becoming the interior, see fig.6. If we measure θ anticlockwise from the x axis then $M_r = M_o \sin\theta$ and $M_\theta = -M_o \cos\theta$. which produces equivalent axial current densities of $\pm M_o \cos\theta$ along the exterior and interior surfaces respectively, and a bulk current density curl $M = 2M_o r^{-1} \cos\theta$ The surface currents produce equal and opposite uniform fields in the centre which cancel.

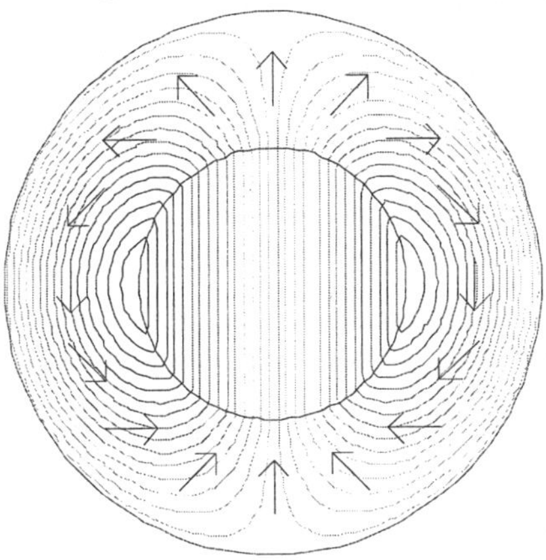

Fig.6. A Cylindrical Halbach Array

Each layer δr of bulk current also produces a uniform field in the bore of $M_o \delta r/r$ and integrating the fields the total is $M_o \ln(r_1/r_2)$ where r_1 and r_2 are the outer and inner radii respectively. In principle we can get an infinite field in the bore, but in practice the field is limited by demagnetisation. A pseudo Halbach array in this form using superconductors has been proposed by Hull [16] as a powerful stator for a synchronous motor or generator.

4.6. Rectangular Magnets

In practice we usually have relatively small blocks of uniform material and the question is how to arrange them to get the desired levitation force and stiffness. As there is a number of variables this is a complex design problem which there is not space to tackle here in any detail. However it is clear that having found the fields due to a sinusoidal current we can find the field due to a more complex array by adding the higher harmonics. For blocks of permanent magnets the Fourier components are those of a rectangular wave, for superconducting blocks the magnetisation is a triangular wave.

For a two dimensional array the mathematics is the same as in §3.2 above. If we have a 2D periodic array of magnets with magnetisation $\boldsymbol{M}(x,y)$ of depth in z direction w, and in terms of Fourier components :-

$$M_z(x,y) = \sum_{n,m} f_{n,m} \cos nk_x x \cos mk_y y \tag{47}$$

then the coefficients f_{nm} are essentially Fourier components of a square wave and of order unity. The fields decay in the z direction as $\exp(-k_{nm}z)$ where:-

$$k_{nm}^2 = (nk_x)^2 + (mk_y)^2 \tag{48}$$

If the array is a distance g from a superconducting plane the pressure P is:-

$$P = (0.125\mu_o M^2)\sum_{n,m} f_{nm}^2 \exp(-2gk_{nm})[1-\exp(-k_{nm}w)]^2 \tag{49}$$

The stiffness is the differential with respect to g, i.e.each component is multiplied by k_{nm}:-

$$Stiffness = (0.25\mu_o M^2)\sum_{n,m} f_{nm}^2 k_{nm} \exp(-2gk_{nm})[1-\exp(-k_{nm}w)]^2 \tag{50}$$

This allows the calculation of the forces and stiffness of a wide range of periodic arrays of magnets. It makes clear the effect of the main design parameters which are:-

(i) *M*:- Forces and stiffness are proportional to the square of the magnetisation. This is inevitable from dimensional arguments.

(ii) *f*:- These are Fourier components of order unity. The number that needs to be taken into account depends on the sizes of *kg* and *kw*. These components will be independent of the direction of magnetisation, so that arrays with magnetisation parallel to z and x will have the same effect as arrays with the magnetisation in the *xy* plane.

(iii) Dimensions:- *k* fixes the fundamental period, *g* the gap to the superconductor, and *w* the depth of the magnet. We need *kg* to be small and *kw* large to get large effects.

(iv) Gap:- The force is proportional to exp(-2$k_{nm}g$). Fourier components with a wavelength much less than the gap will have little effect.

(v) Magnet thickness:- This appears as a factor (1-exp($k_{nm}w$)^2. For thicknesses small compared with the period the forces increase as the square of the thickness. For a magnet of width L in an alternating array $kL=\pi$ so if $L=w$ this is a factor 0.9. There is little benefit in making the magnets much thicker than their width. Note however that in the pseudo Halbach array the period is four times the magnet width, as opposed to double in the alternating array, and the factor is 0.6. These arrays will benefit by being made thicker.

(vi) Stiffness:- This is found by multiplying each component by the wave number so that the ratio of stiffness to pressure is 1/size.

(vii) If all the magnets are in the same direction with no separation between them then they cannot be treated as a periodic array but treated as a single magnet with a surface current round the perimeter. This is also true of superconducting blocks. This arrangement may be necessary if the gap is significantly larger than the size of the magnets. To obtain the full levitation force the thickness must then be comparable to the diameter of the array, rather than that of an individual magnet. Given that the size range of superconducting pellets is limited, the design of an array of superconducting pellets is very dependent on whether we are interested in fields at distances large compared with the pellet size. In this case we need to make all the magnetisations parallel and the thickness of the complete array comparable with its diameter.

4.7. Fourier Transforms

Fourier transforms are as follows.

I) For a 2D array of square magnets of unit magnetization of side W with alternating signs, as in a chess board.

$$M = \sum_n \sum_m f_{nm} \cos\left[(2n-1)\pi x/W\right] \cos\left[(2m-1)\pi y/W\right] \tag{51}$$

where

$$f_{nm} = (-1)^{n+m} \frac{64}{\pi^2 (2n-1)(2m-1)} \tag{52}$$

II) For a similar array of superconducting pellets with uniform J_c.
The peak M is 1 and the current density $2/W$

$$M = \sum_n f_n \cos\left[(2n-1)\pi x/W\right] \cos\left[(2n-1)\pi y/W\right] \tag{53}$$

where

$$f_n = \frac{8}{\pi^2 (2n-1)^2} \tag{54}$$

III) For isolated magnets we use a Fourier integral. For a magnet symmetric about x and y=0 we can put

$$M(x,y) = \int_o^\infty \int_o^\infty f(k_x,k_y) \cos(k_x x) \cos(k_y y)\, dk_x dk_y \tag{55}$$

where

$$f(k_x,k_y) = \frac{4}{\pi^2} \int_o^\infty \int_o^\infty M(x,y) \cos(k_x x) \cos(k_y y)\, dx dy \tag{56}$$

For a rectangular magnet of magnetisation M per unit area with length $2X$ along x and $2Y$ along y :-

$$f(k_x,k_y) = \frac{4M}{\pi^2 k_x k_y} \sin(k_x X) \sin(k_y Y) \tag{57}$$

For a square superconductor of side $2a$, carrying a uniform current density and with a magnetisation M at the centre:-

$$f(k_x,k_y) = \frac{2M\left[k_x \sin(k_y a)\cos(k_x a) - k_y \cos(k_y a)\sin(k_x a)\right]}{\pi^2 k_x k_y \left(k_y^2 - k_x^2\right)} \tag{58}$$

The current density is M/a.

4.8. Results

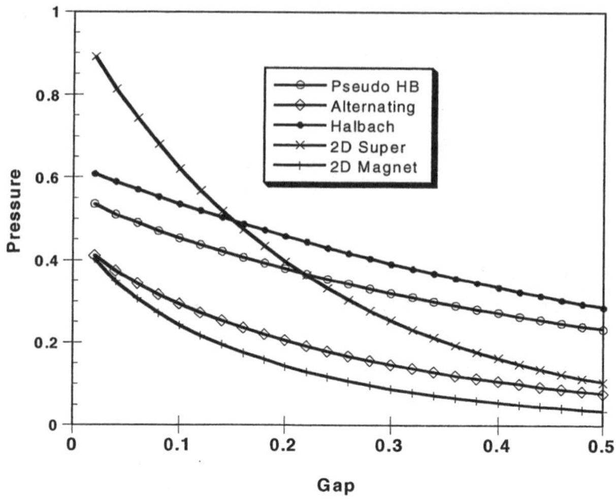

Fig.7. The levitation pressures of a number of magnet arrays.

In Fig.7. are presented the results of calculating the pressure for a number of arrangements of magnets with the depth equal to the diameter. The superconducting ones have the current density chosen somewhat arbitrarily to give unit pressure, defined as $\mu_o M^2$, when in contact. Alternating magnets are line magnets alternating in sense, the Halbach array is of the same thickness but with a continuous change in direction of the magnetisation. The two dimensional arrays magnets arranged to alternate in the x and y directions like a chess board. The superconducting ones have the current density chosen somewhat arbitrarily to give unit pressure, defined as $\mu_o M^2$, when in contact. The unit of pressure is $\mu_o M^2$ and the magnets have unit diameter and depth.

5. LEVITATION ABOVE CONDUCTING SHEETS

5.1. Background

The levitation of a magnet above a conducting plane is well known as the original basis for the Japanese Maglev train. The first relevant work was by Maxwell [17, 13] who showed that the transient currents which appeared when a magnet was placed above a thin conducting sheet could be described by an image pole which receded to infinity with a constant velocity $v_o=2/\mu_o\sigma d$ where σ is the conductivity of the sheet and d its thickness. A paper by Saslow [18] uses this technique to obtain a number of results including the important fact that the ratio of the drag force to the lift force on any magnet moving with velocity v is v_o/v. This follows at once by considering the work done by the forces on the magnet and the receding pole, although it was originally derived by Davis using the Pointing vector [19]. For a 2mm. thick copper sheet this velocity is 13m/sec or 30 mph.

The system can be characterised by a dimensionless parameter $S=\mu_o \sigma v d$ which is $2v/v_o$, sometimes known as the 'Magnetic Reynolds Number'

5.2 Speed Regimes

There are three regimes of speed. At the lowest speed, $S<<1$, the field due to the induced currents is small compared to the applied field and the currents are proportional to the speed. At intermediate speeds, $S>1$, the currents create an image magnet, but are uniform through the thickness. This is the thin sheet approximation which is the regime of most interest. The levitation force can be calculated from magnetostatics as in §4 above, and the drag force obtained by multiplying by v_o/v. The lift force is constant as it is the image force and the drag is inversely proportional to v. This also follows from the fact that the currents are independent of v so the dissipation is constant. The high frequency regime is when $S>1/kd$, where k is the wave number of the magnet as before, and in this , the skin depth regime, the currents decay exponentially within the sheet. This is the situation treated by Reitz and Davis [20] who use a dimensionless parameter $1/kd$ times that used here. In this article the harmonic approach is used to obtain the lift and drag for a number of situations.

5.3. Levitation and Drag Forces.

5.3.1 Levitation by Thin Conducting Sheets

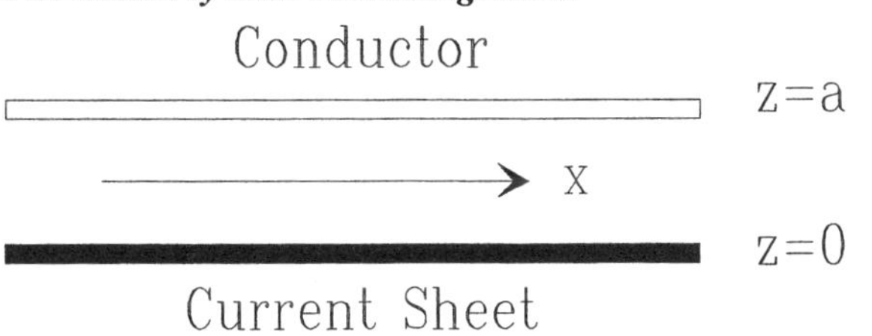

Fig.8. The basic configuration is a sinusoidal current sheet moving at constant velocity v below a conducting sheet.

The case of interest in this section is that of a magnet moving above an infinite sheet. We put a thin current sheet in the *xy* plane at z=0, see fig.8. Above this at $z=a$ there is a thin conducting sheet of thickness d. The current sheet has I_y amps per metre in the y direction which varies sinusoidally with wavelength L in the x direction, is independent of y, and travels with a velocity v. Then all fields are phasors multiplied by $\exp j(kx-\omega t)$ where $k=2\pi/L$ and $\omega=kv$. The peak field density available from two sheets in opposition is $\mu_o I$ with a corresponding magnetic pressure $P_o=\frac{1}{2}\mu_o I^2$. This is the maximum pressure with which we shall compare the lift and drag of real systems. The maximum mean pressure available from a sinusoidal sheet is $\frac{1}{2}P_o$.

Since $\nabla^2 \boldsymbol{A}=0$ fields must vary in the z direction as $\exp(\pm kz)$. To match the boundary condition on B_x at the sheet:-

$$A_y = \tfrac{1}{2}\mu_0 I_y k^{-1} \exp(-k\,|\,z\,|) \tag{59}$$

We now label the superconducting sheet current I_s and add a sheet of copper at z=a carrying a current I_c. For this sheet

$$A_y = \tfrac{1}{2}\mu_0 I_c k^{-1} \exp(-k\,|\,z-a\,|) \tag{60}$$

This must be added to the potential due to the original current sheet, so at the copper sheet (z=a):-

$$A_y = \tfrac{1}{2}\mu_0 k^{-1}\left[I_c + I_s \exp(-ka)\right] \tag{61}$$

The electric field is:-

$$\begin{aligned} E_y &= j\omega A_y \\ &= \tfrac{1}{2}\mu_0 j\omega k^{-1}\left[I_c + I_s \exp(-ka)\right] \end{aligned} \tag{62}$$

This electric field produces the current I_c, so if the current is to be consistent with the electric field which induces it:-

$$I_c = \tfrac{1}{2}\mu_0 j\omega k^{-1}\sigma d\left[I_c + I_s \exp(-ka)\right] \tag{63}$$

Putting $S=\mu_o \sigma v d$ gives the following expression for the induced current:-

$$I_c = I_s \exp(-ka)\frac{jS}{2-jS} \tag{64}$$

At the current sheet (z=0) the fields due to the induced currents in the conductor are:-

$$B_{xc} = \tfrac{1}{2}\mu_0 I_c \exp(-ka) \tag{65}$$

$$B_{zc} = \tfrac{1}{2}\mu_0 I_c j \exp(-ka) \tag{66}$$

The lift force on the current sheet is obtained by multiplying the current I_s by B_x due to the copper, B_{xc}, and averaging over x:-

$$\begin{aligned} \text{Lift} &= \tfrac{1}{2}\text{Re}\left(I_s B_{xc}\right) \\ &= \tfrac{1}{2}P_o \exp(-ka)\frac{S^2}{4+S^2} \end{aligned} \tag{67}$$

and the drag is:

$$\begin{aligned} \text{Drag} &= \tfrac{1}{2}\text{Re}\left(I_s B_{zc}\right) \\ &= \tfrac{1}{2}P_o \exp(-ka)\frac{2S}{4+S^2} \end{aligned} \tag{68}$$

The coefficient of friction at all speeds (up to the skin effect regime) is $2/S$ in agreement with the result of §5.1, and for S>1 the loss per unit area is independent of speed since the current density remains that needed for field exclusion. It can be written $\tfrac{1}{2}P_o v_o$ where:-

$$v_o = \frac{2}{\mu_o \sigma d} \tag{69}$$

the velocity of Maxwell's receding pole.

By analogy with the complex susceptibility, we can write the complex lift as:-

$$Lift + jDrag = \tfrac{1}{2}P_o \exp(-ka)\frac{jS}{2-jS} \tag{70}$$

This is the thin sheet approximation which is the one which applies to most practical situations. It can be used until the skin depth is equal to the sheet thickness, i.e. $S=2/kd$. Since the lift at small S is proportional to S^2 and hence the square of the thickness we see that two conducting sheets will be lifted higher than a single one, so if a conducting sheet is laid on another levitated one the two will rise up. This is because the phases of the two induced currents become closer to that of the driving current.
The vector potential between the sheets is :-

$$A_y = \frac{\mu_o}{2k}\left[I_s \exp(kz) + I_c \exp(-kz)\right] \tag{71}$$

$$= \frac{\mu_o I_s}{2k}\left[\frac{jS}{2-jS}\exp(kz) + \exp(-kz)\right] \tag{72}$$

Note that the exp(kz) term depends only on the induced current and the exp($-kz$) term on the driving current. This result will be used in the next section.

5.3.2. Thick Conducting Sheets

If the conductor is thick then within the conductor:-

$$\nabla^2 \boldsymbol{A} = \mu_0 \boldsymbol{J} = \mu_0 \sigma \boldsymbol{E} = j\omega\sigma\mu_0 \boldsymbol{A} \tag{73}$$

There is a solution to this equation of the form $\exp(k_z z + jkx - j\omega t)$ provided that:-

$$k_z^2 = k^2(1 + jS/kd) \tag{74}$$

In general the vector potential must be written in the form $\alpha \exp(kz) + \beta \exp(-kz)$ in each region with the appropriate k, and we need to match boundary conditions at each face of the conductor and current sheet. However the number of coefficients can be reduced by recognising that in the gap between the current sheets the term in exp(kz) is the potential of the driving current and the term in exp($-kz$) is that of the induced current (see equation (72)).
We put the conducting material between $z=a$ and $z=a+d$. In the gap:-

$$A_y = \alpha \exp(kw) + \beta \exp(-kw) \tag{75}$$

where β is the vector potential due to the current sheet alone which is $\mu_o I_s/2k$. There is a similar expression within the conductor with a different value of k, and below the current sheet a term proportional to exp(kz) only.
Hence given the potential at the conductor surface due to the current sheet (β) we can get the four remaining coefficients by making A and B_x continuous at the two surfaces of the conductor. This gives the fields in terms of the vector potential at the conductor surface, which may be the most convenient for some purposes, and was the approach used in ref [20]. However in what follows we put β in terms of the driving current. The lift and drag can be found from the fields and the Maxwell stress tensor, but in fact they can be found more directly from the coefficients of A in the gap).
Since the first term is that due to the current sheet and the second due to the conductor, using subscript s for the current sheet and c for the copper the lift averaged over a period is:-

$$\begin{aligned}&½I_sB_{xc} = ½I_sk\,\mathrm{Re}(\alpha)\\&= \mu_o^{-1}k^2\beta\,\mathrm{Re}(\alpha)\end{aligned} \quad (76)$$

Similarly the drag is

$$\begin{aligned}&½I_sB_{zc} = ½jI_sk\,\mathrm{Im}(\alpha)\\&= \mu_o^{-1}k^2\beta\,\mathrm{Im}(\alpha)\end{aligned} \quad (77)$$

Hence the complex lift is;-

$$Lift + jDrag = \mu_o^{-1}k^2\beta\alpha \quad (78)$$

This expression also applies to current layers below z=0 since for a layer at $z= -a$ the constant β is multiplied by exp(ka) and α by exp (-ka) so the factors cancel.
We can also show that the result follows from a direct evaluation of the Maxwell stress tensor. From the Maxwell stress tensor the lift is

$$Lift = \frac{1}{2\mu_o}\left[|B_x|^2 + |B_y|^2 - |B_z|^2\right] \quad (79)$$

and the drag

$$Drag = -\frac{1}{2\mu_o}\left[B_xB_y\right] \quad (80)$$

If A_y is a phasor varying as exp j($kx-\omega t$) and :-

$$A_y = \alpha\exp(kz) + \beta\exp(-kz) \quad (81)$$

Then:-

$$\begin{aligned}B_x &= -k\left[\alpha\exp(kz) - \beta\exp(-kz)\right]\\B_y &= jk\left[\alpha\exp(kz) + \beta\exp(-kz)\right]\end{aligned} \quad (82)$$

Averaging over x and t :-

$$Lift + jDrag = \frac{1}{4\mu_o}\left[B_xB_x^* - B_zB_z^* - j\left(B_xB_z^* + B_zB_x^*\right)\right] \quad (83)$$

$$\begin{aligned}&= \frac{1}{4\mu_o}\left[\left(B_x + jB_z\right)\left(B_x^* + jB_z^*\right)\right]\\&= -\mu_o^{-1}k^2\alpha\beta^*\end{aligned} \quad (84)$$

Since β is due to the current sheet and therefore equal to $½k^{-1}\mu_oI$ the complex lift is directly proportional to α. The result for a thin current sheet is :-

$$\alpha = \frac{\mu_oIk^{-1}\exp(-2ka)(k^2-k_z^2)\sinh(k_zd)}{(k-k_z)^2(\exp(-k_zd) - (k+k_z)^2\exp(k_zd))} \quad (85)$$

so that the complex lift is:-

$$Lift + jDrag = \frac{P_o\exp(-2ka)(k^2-k_z^2)\sinh(k_zd)}{(k-k_z)^2(\exp(-k_zd) - (k+k_z)^2\exp(k_zd))} \quad (86)$$

This is the expression derived by Reitz and Davis (20). In the limit of large d, the skin effect regime, the drag is :-

$$Drag = \tfrac{1}{2} P_0 \exp(-2ka)\sqrt{\frac{kd}{2S}} \tag{87}$$

5.3.3. Thick Driving Current Sheets

As for the magnet array we can integrate the effect of each current sheet so that if the driving current is spread over a thickness w the force is multiplied by a factor $[1-\exp(-kw)/kw]^2$ Although in the general case the lift and drag are non-linear functions of the current source, for most practical sources we can add the lift and drag due to individual current elements and harmonics. The condition required is that the current density is a product of a function of z and a function of x, i.e. the variation with x is extruded in the z direction so that all the induced currents have the same phase. This is usually the case.

5.3.4. Stiffness

The ratio of stiffness to lift force is $2k$. From dimensional arguments this ratio must always be the reciprocal of a distance which is characteristic of the size of the magnet, and we see here it is $4\pi/L$ where L is the magnet period. This should be contrasted with iron magnetic circuits where the relevant distance is the gap, and the stiffness at low speeds is negative and large.

5.3.5. Results

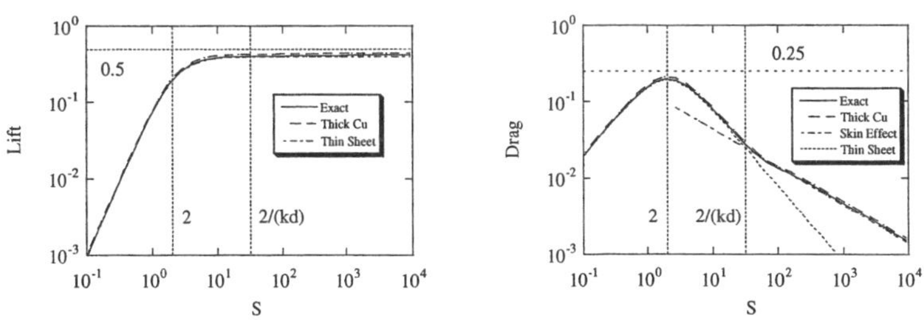

(a) (b)

Fig.9. The Lift and drag for each approximation compared with the exact calculation.

Figure 9 shows the results of the approximate expressions compared with the exact values for a typical arrangement. In this case the thickness of all three layers is 1% of the magnet period and the approximations are very good. The peak values are reduced from the maximum possible by the prefactor exp(-2ka) and the finite current sheet thickness. Unit lift is ½ $\mu_o I^2$.

5.4. Damping and Transients.

To analyse vertical motion superimposed on the horizontal motion we allow the gap, *a*, to vary with time. Since this is a non-linear problem we use the full expressions rather than phasors. As before

$$A_y = ½\mu_o k^{-1}\left[I_c + I_s \exp(-ka)\cos(kx-\omega t)\right] \tag{88}$$

Differentiating with respect to time:-

$$E_y = -½\mu_o k^{-1}\left[\dot{I}_c + I_s \exp(-ka)\left(\omega\sin(kx-\omega t) - \dot{a}k\cos(kx-\omega t)\right)\right] \tag{89}$$

This electric field produces the current I_c so:-

$$I_c = -½\mu_o k^{-1}\left[\dot{I}_c + I_s \exp(-ka)\left(\omega\sin(kx-\omega t) - \dot{a}k\cos(kx-\omega t)\right)\right] \tag{90}$$

Putting $S=\mu_o\sigma vd$ gives the following differential equation for the induced current:-

$$\dot{I}_c + 2\omega S^{-1} I_c = \omega I_s \exp(-ka)\left[v^{-1}\dot{a}\cos(kx-\omega t) - \sin(kx-\omega t)\right] \tag{91}$$

The solution for any variation of the gap a with time is:-

$$I_c = -I_s\omega\exp\left(\frac{-2\omega t}{S} - ka\right)\int\exp\left(\frac{-2\omega t}{S}\right)\left[\frac{a'}{v}\cos(kx-\omega t) - \sin(kx-\omega t)\right]dt \tag{92}$$

To find the lift we multiply this current by the field at the copper due to the current sheet, B_x :-

$$B_z = \mu_0 I_s \exp(-ka)\cos(kx-\omega t) \tag{93}$$

The result for a constant vertical velocity u is :-

$$Lift = -½P_0 \exp(-2ka)\left[\frac{S^2}{4+S^2} - \frac{2uS}{v\left(4+S^2\right)}\right] \tag{94}$$

The first term is the lift and the second is a viscous damping term. Reference [21] gives more detailed calculations of the forward and horizontal damping in a number of geometries. The damping has two characteristics. It is very small compared with the lift since the vertical velocity is much smaller than the horizontal velocity.

For large *S* the ratio of this vertical drag to the lift is $2u/Sv$ and the response to transient forces is very rapid, the image following the driving magnet up to frequencies of order kv.

Low damping can be a major problem in magnetic levitation so we now estimate the effect of a shorted turn round the magnet to increase the damping. If the field of the magnet is $\boldsymbol{B}$, this is also the field of the image, which induces currents in the shorted turn. The current density is $\sigma B^2 u$ so the damping force is $V_s\sigma B^2 a$ where V_s is the volume of the conductor in the shorted turn. Since the magnetic pressure balances the weight it follows that $½B^2 A_m/\mu_o = mg$ where A_m is the bearing area of the magnet. We can get critical damping if:-

$$V_s = A_m\left(\frac{2}{\sigma\mu_o}\right)\sqrt{\frac{\pi}{Lg}} \tag{95}$$

For a square section magnet, long perpendicular to the direction of travel, with the shorted turn the same height as the magnet, this gives a thickness of 1.3cm for L=1m, which seems acceptable, particularly if the shorted turn could also act as the magnet support. Smaller magnets will require a thicker conductor in comparison with the magnet, but a superconducting link with a short resistive section could be used instead of a solid conductor.

5.5. Higher Harmonics

For more complex current distributions we need to add the Fourier series of the harmonics as for the permanent magnets in §4.6. The induced currents are a linear superposition of the contributions from each harmonic and both lift and drag are the sum of the contributions from each harmonic of the driving current distribution. The complex lift can still be written as $P_o\, jS/(2-jS)$ where now P_o is the pressure between the current sheet and its image, which can again be calculated from magnetostatics . Although this only applies strictly to thin driving current sheets, as pointed out above it can still be used for thick driving current sheets, provided that the phase of the currents is the same, i.e. the current is a function of z multiplied by a function of x

5.6. Two Dimensional Current Distributions.

We now allow the current to vary in the y direction, so that we can deal with coils of different shapes. We describe the current by a magnetisation $\boldsymbol{M_A}$ per unit area in the z direction such that I=curl($\boldsymbol{M_A}$). Fields are still phasors varying as $\exp j(k_x x-\omega t)$ with the velocity $v=\omega/k_x$. Then

$$I_x = -k_y M_A \sin\left(k_y y\right) \tag{96}$$

$$I_y = -jk_x M_A \cos\left(k_y y\right) \tag{97}$$

We can then repeat the calculation in §5.3.1. The variation in the z direction is as $\exp(\pm kz)$ where now $k^2=k_x^2+k_y^2$. We put the induced currents in the copper as curl($\boldsymbol{M_c}$) where $\boldsymbol{M_c}$ is the induced magnetisation. In calculating the electric field (equation (62)) we must put $\omega=vk_x$ instead of vk so that where S appears in the 1D case we must put Sk_x/k. The vector potentials A_y and A_x are obtained by multiplying the currents by $\mu_o/2k$ as before. Whether we use the x currents or y currents we get the same equation for the induced magnetisation as for the induced currents, equation (64) with $S' = Sk_x/k$.

$$M_c = M_A \exp\left(-ka)\right)\frac{jS'}{2-jS'} \tag{98}$$

Between the sheets the vector potential due to the induced magnetisation is given by :-

$$A_x = -½\mu_0 k^{-1} k_y M_c \exp(kz-2ka)\sin k_y y \tag{99}$$

$$A_y = -½j\mu_0 k^{-1} k_x M_c \exp(kz-2ka)\cos k_y y \tag{100}$$

From $\boldsymbol{A}$ we can calculate the $\boldsymbol{B}$ at the superconductor (z=0) due to the magnetisation in the copper, multiply by the current in the superconductor and average over x and y. The result is :-

$$\text{Lift}=\tfrac{1}{2}\,\text{Re}\left(I_x^*B_{yc}\right)-\tfrac{1}{2}\,\text{Re}\left(I_y^*B_{xc}\right)$$

$$=\left(-0.125\mu_o M_A^2 k^2\right)\exp(-2ka)\frac{S^2k_x^2}{4k^2+S^2k_x^2} \tag{101}$$

$$\text{Drag}=\tfrac{1}{2}\,\text{Re}\left(I_y^*B_{zc}\right)$$

$$=\left(-0.125\mu_o M_A^2 k^2\right)\exp(-2ka)\frac{2Sk_x^2}{4k^2+S^2k_x^2} \tag{102}$$

If the magnet is of thickness w then M_c is reduced by a factor $\exp(-kz)$ for layers a distance z below $z=0$. This introduces a factor $(1-\exp(-kw))/kw$. The field this produces is then also reduced by a factor $\exp(-kz)$ and this must be integrated through the thickness to give the force. The result is factor $[(1-\exp(-kw))/kw]^2$. If the magnetisation of the magnetic material is M then $M_A=wM$. So in terms of the magnetisation :-

$$Lift=\left(-0.125\mu_o M_A^2 k^2\right)\exp(-2ka)\frac{S^2k_x^2}{4k^2+S^2k_x^2}\left[1-\exp(-kw)\right]^2$$

$$Drag==\left(-0.125\mu_o M_A^2 k^2\right)\exp(-2ka)\frac{2Sk_x^2}{4k^2+S^2k_x^2}\left[1-\exp(-kw)\right]^2 \tag{103}$$

(These are the z and x forces on the superconductor, so since it is below the copper the forces are negative).

If we have a magnet long in the y direction $k_y=0$ and so $k=k_x$. We then reproduce equations (67) and (68), apart from an extra factor ½ due to the sinusoidal variation in the y direction. At high speeds the lift, and hence the drag are independent of the direction of the magnet, the lift is the image force and the drag is $2/S$ times this. However if the magnet is long in the x direction, to reach full flux exclusion S must be increased by a factor of about the aspect ratio. This is consistent with the Maxwell picture since a line of poles in the direction of motion will be closer to image poles of the opposite sense than if the line is perpendicular to the motion. It is also reasonable that a current moving perpendicular to itself will generate more lift than a current moving longitudinally. In general we expect a reasonable approximation if we extend the velocity scale for magnets which are long in the y direction by a factor $(1+k_y^2/k_x^2)^{1/2}$.

We conclude that if speeds are high enough the shape of the magnet makes little difference to the lift and drag. However if there is a problem in reaching sufficient speed to exclude the flux it is important to make the magnet short in the direction of motion compared with the perpendicular direction. Round or square pellets of superconductor for which $k_x=k_y$ will need to reach speeds about twice those indicated for magnets long in the y direction to achieve the same flux exclusion and lift.

The forces due to different harmonics of the current distribution do not interfere and can be added. Currents distributed in the z direction can be easily dealt with provided that they are all in phase, i.e. that the current can be written as the product of a function of z and a function of x and y. In practice we usually have a current uniform in z to a thickness w in which case we just multiply by a factor $((1-\exp(-kw))/kw)^2$ as before.

5.6.1 Thick Conductors

For a thick conductor we follow the procedure of §5.3.2. If:-

$$A_y = \left[\alpha \exp(k_z z) + \beta \exp(-k_z z)\right] \cos(k_y y) \tag{104}$$

Then

$$A_x = \left(\frac{-jk_y}{k_x}\right)\left[\alpha \exp(k_z z) + \beta \exp(-k_z z)\right] \sin(k_y y) \tag{105}$$

Putting the values of $\boldsymbol{B}$=curl($\boldsymbol{A}$) from equations (104) and (105) into the lift and drag from equations (79) and (80), and averaging over x,y and t gives the following expressions:-

$$Lift = \frac{-k^4 \beta \operatorname{Re}(\alpha)}{2\mu_o k_x^2}$$
$$Drag = \frac{-k^3 \beta \operatorname{Im}(\alpha)}{2\mu_o k_x} \tag{106}$$

Since

$$\nabla^2 \boldsymbol{A} = \mu_0 \boldsymbol{J} = \mu_0 \sigma \boldsymbol{E} = j\omega\sigma\mu_0 \boldsymbol{A} \tag{107}$$

It follows that in the copper sheet:-

$$k_z^2 = k_x^2 + k_y^2 + jSk_x d^{-1} \tag{108}$$

As in §5.3.2 we define $k=\sqrt{(k_x^2+k_y^2)}$ then in free space $k_z=k$ and in the conductor

$$k_z = k\sqrt{1 + \frac{jSk_x}{k^2 d}} \tag{109}$$

If we now write down the boundary conditions on the fields and currents it will be seen that they are all satisfied by the same equations as were used for the case where k_y was zero provided we substitute Sk_x/k for S. In particular the values for α and β, the coefficients of A in the gap due to the induced currents, are the same. We now repeat the calculation of the lift and drag for a current with x and y components.
In equation (85) k is the free space wave number in the z direction, which is now $\sqrt{(k_x^2+k_y^2)}$, and k_z the value in the conductor. So as before

$$\alpha = \frac{\tfrac{1}{2}\mu_o I_s k^{-1} \exp(-2ka)(k^2 - k_z^2)\sinh(k_z d)}{(k - k_z)^2 (\exp(-k_z d) - (k + k_z)^2 \exp(k_z d))} \tag{110}$$

and

$$\beta = \tfrac{1}{2}\mu_o I_s k^{-1} \tag{111}$$

Substituting these expressions in equation (106) gives the lift and drag in the skin effect regime.

5.7. Rectangular Coils

For single coils we integrate over the various k vectors of the Fourier components. As an example we take a rectangular magnet of magnetisation M_o, size $2X$ by $2Y$, thickness w, whose transform is equation (57). Then at the superconductor the magnetisation M_s is :-

$$M_s = \int_0^{\infty}\int_0^{\infty} \frac{4M_o}{\pi^2 k_x k_y} \sin(k_x X)\sin(k_y Y)\cos(k_x x)\cos(k_y y)dk_x dk_y \tag{112}$$

The induced magnetisation in the thin sheet approximation is:-

$$M_c = \int_0^{\infty}\int_0^{\infty} \frac{4M_o\left[1-\exp(-kw)\right]}{\pi^2 k_x k_y}\left(\frac{jSk_x\exp(-ka)}{2k-jSk_x}\right)\sin(k_x X)\sin(k_y Y)\cos(k_x x)\cos(k_y y) \tag{113}$$

We then calculate the field at the magnet surfaces, as in §5.6, multiply by the surface current M_o and integrate along each face to get the total force. The result is :-

$$Lift = \int_0^{\infty}\int_0^{\infty} 8M^2\,\mathrm{Re}\left[\frac{-Sk_x\exp(jk_x X)}{2k-jSk_x}\right]\left[1-\exp(-kw)\right]^2\exp(-2ka)\frac{\sin(k_x X)\sin(k_y Y)^2}{\left(\pi k_x k_y\right)^2}dk_x dk_y$$

$$Drag = \int_0^{\infty}\int_0^{\infty} 8M^2\left[\frac{S\sin(k_x X)}{4k^2+S^2k_x^{\,2}}\right]\left[1-\exp(-kw)\right]^2\exp(-2ka)\frac{\sin(k_x X)\sin(k_y Y)^2}{\left(\pi k_y\right)^2}dk_x dk_y \tag{114}$$

For reasons that are not entirely clear these integrals caused some numerical difficulties and a different approach using receding image construction and the Biot Savart law gave more reliable results. These are shown in figs. 10a and 10b.

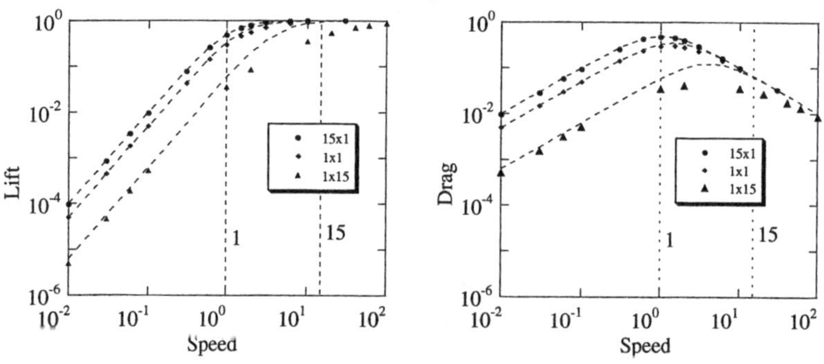

(a) (b)

Fig.10. Lift and drag for several rectangular coils compared with first harmonic expressions. The symbols show the lift and drag for rectangular coils 15x1, 1x1 and 1x15, where the first dimension is perpendicular to the direction of motion. In all cases the coil was $0.4/k_y$ above the conducting plane.

The dotted lines show what we would predict from a harmonic analysis, taking the aspect ratio as the ratio of k vectors. In other words the theory predicts a lift of $v^2/((1+ev^2)$ where e is the aspect ratio. It can be seen that this gives quite a reasonable approximation for all three coils.

5.8. Ferromagnetic Sheets

Following a suggestion by Laithwaite passed on to me by Dr.M.K.Bevir, it was found that lift could be improved by backing the copper with ferromagnetic sheets. At first sight this is very surprising since usually magnetised sheets attract. However what is happening is made clear by considering the effect of putting an iron core in a copper cylinder subjected to an axial oscillating magnetic field, as occurs in an inductive fault current limiter. The first effect is that the iron increases the induced currents so that complete flux exclusion is reached at lower frequencies or speeds. We put laminated iron both below the current sheet and above the copper so that the reluctance is dominated by the air gap, g, which we assume to be small compared with the wavelength (Fig.11.). The currents are uniform in the y direction and vary as $\exp(j(kx-\omega t))$. In practice the conductors would be put in slots.. However to a first approximation this just has the effect of spreading the current into a thicker layer, which reduces Joule losses, but does not change the general behaviour. We call I_T the sum of the supercurrent I_s and the induced current I_c.

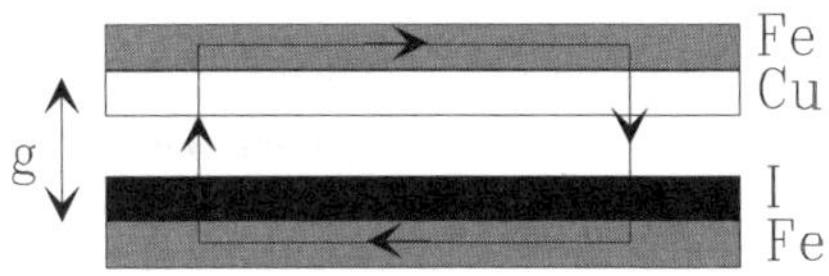

Fig. 11. Iron sheets backing a current sheet and a conductor.

From Ampères circuital theorem:-

$$-g\delta B_z = \mu_o I_T \delta x \tag{115}$$

and

$$B_x = \mu_o I_T \tag{116}$$

Hence

$$B_z = -(I_c + I_s)/jgk \tag{117}$$

As before the induced current must be consistent with the induced electric field so:-

$$\begin{aligned} I_c &= E\sigma d = vB_z\sigma d \\ &= -v\sigma d(I_c + I_s)/jgk \end{aligned} \tag{118}$$

Hence

$$I_c = \frac{jSI_s}{kg - jS} \tag{119}$$

and

$$B_z = \frac{-j\mu_o I_s}{kg - jS} \tag{120}$$

From the Maxwell stress tensor the lift, assuming $kg<<1$, is:-

$$Lift = \tfrac{1}{4}\mu_o^{-1}\left(B_x B_x^* - B_z B_z^*\right)$$

$$= \tfrac{1}{2}P_0 \frac{S^2 - 1}{(kg)^2 + S^2} \tag{121}$$

and the drag is

$$Drag = \tfrac{1}{4}\mu_o^{-1}\left(B_x B_z^* - B_z B_x^*\right)$$

$$= \tfrac{1}{2}P_0 \frac{2S}{(kg)^2 + S^2} \tag{122}$$

At high speeds ($S>2/kd$) the skin effect prevents any flux entering the iron and the skin effect regime is the same as if the iron were not there. In fact as can be seen from the expressions above, the iron has little effect for S greater than about 4, independently of the skin effect. The coefficient of friction at all speeds up to the skin effect regime is $2S/(S^2-1)$.
If we compare the expression for the lift, eq (121), with that obtained without iron, eq (67), we see the ratio is $1+3S^{-2}-4S^{-4}$. This is greater than unity for all $S>1$, so a larger lift is obtained. However the maximum benefit is 56% at $S=\sqrt{(8/3)}$ and the ratio of the coefficients of friction. is $S^2/(S^2-1)$ which is always larger than unity. Hence the greater lift goes with a larger coefficient of friction, which is why the system has more applications as a brake. The effect of iron backing is shown in Fig.12, and fig.13 is a comparison of lift, drag and coefficient of friction with and without iron backing.

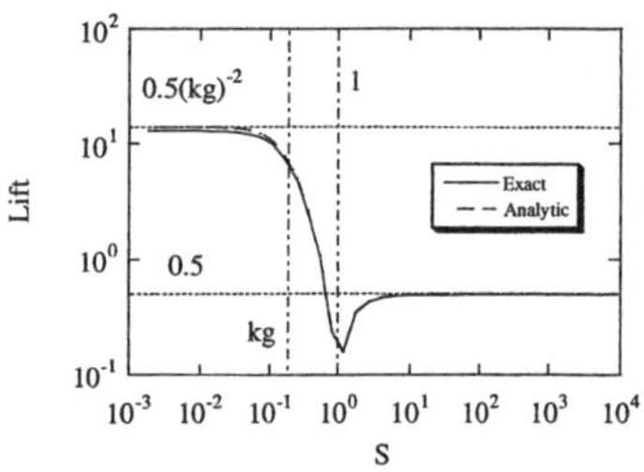

(a)

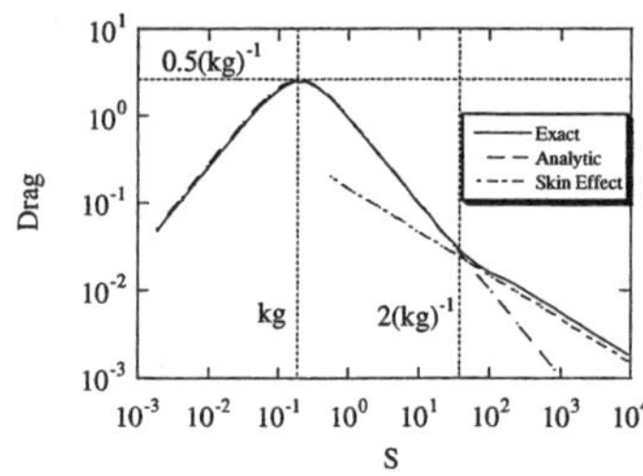

(b)

Fig.12 Lift with iron backing Drag

The dimensions are unit period, gap, ferromagnetic sheets .05 thick. The exact solutions were obtained using the method described in §5.3.2. The conclusion is that a ferromagnetic backing can increase the lift at certain speeds, but will also increase the coefficient of friction and it is not easy to see a useful application in levitation. However with magnets in reinforcement rather than opposition the system has been analysed by a number of authors for use as an eddy superconductor and copper .01 thick, and the current brake, see for example ref [22].

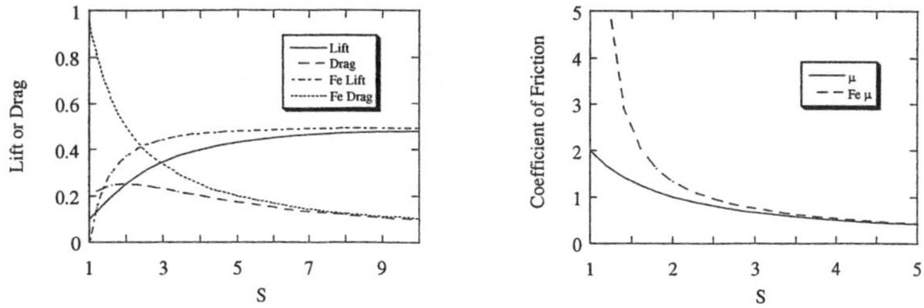

(a) (b)

Fig. 13 Lift and Drag Coefficient of Friction

This system behaves quite differently from one where the conductor is itself a ferromagnetic material. This can be treated by inserting the permeability into the boundary conditions of §5.3.2, and the result is that we do not get lift until the skin depth regime is reached, while the losses are increased. This geometry is therefore not of practical interest.

5.9. Null Flux

When levitation is achieved there is an image magnet in the copper conductor, which means that currents must flow in the conductor to create this image magnet. Since it is difficult to produce a steady field of more than about 0.1 tesla from copper coils it is clear that generating image magnets of even one tesla will involve a lot of heating. This may be acceptable if it is a transient as in a launcher or levitated train. It is much less acceptable in a bearing. Losses can be greatly reduced by using magnets in opposition on either side of a thin conducting sheet, the null flux configuration [23,24]. By symmetry there will be no levitation force in this configuration. However if the magnet is displaced the ratio of lift to drag can be very high, although larger magnets will be needed to generate the same amount of lift. In a bearing the lift force may only be needed during dynamic loads, and even if there is a mean force it can be borne by permanent magnets. The null flux configuration could therefore produce a very low loss non contact bearing in circumstances where the dynamic loads are small (or infrequent) compared with the static load, and stiffness is not important. This is the situation in flywheel energy storage.

In the skin depth regime stiffness and drag are doubled by having a magnet on both sides of the conductor, so only the thin sheet regime is of interest. (Iron sheets are not useful in this configuration either). Let us put one current sheet very close to the conductor and another in antiphase on the other side a distance a away. The induced currents and B_z subtract while the x components of B add. Since a only enters the field and currents expressions in the prefactor $\exp(-ka)$ the lift compared with the single sheet is multiplied by a factor $(1-\exp(-ka))(1+\exp(ka))$ while the drag is multiplied by $(1-\exp(-ka))^2$. Assuming ka is small the coefficient of friction is reduced from $2/S$ to $2ka/S$ while the magnet size to obtain the same lift force is increased by a factor $1/(ka)$. Whether this is useful depends on the relative costs of the loss and the magnet size, but the Japanese

levitated train uses a null flux configuration, albeit in the form of figure of eight coils at the side of the track rather than a continuous sheet under the train.
However in a bearing we can reduce the losses further by taking the weight with a permanent magnet bearing and stabilising this with a null flux bearing in the symmetric position. This hybrid bearing has some very attractive features.

5.10. Null Flux Losses

The thin sheet approximation gives the loss as the sheet is displaced, but extrapolates to zero as the symmetric position is approached although there is inevitably a finite loss in this position. In fact it is not too difficult to obtain the exact expression in any position from the coefficients of A, α and β, as calculated in §5.3.2, equations (85) and (86). We just shift the origin by a so that the conductor is at z=0. We then use symmetry to find the coefficients of A if the current sheet is above the conductor. Finally we add the coefficients for different values of a for the upper and lower current sheets. If in the symmetric position there is an air gap a on each side of the conductor, and the conductor is then displaced a distance h, the result is that the complex lift is :-

$$P_o \exp(-2ka)\Big(\frac{\exp(-2kh)\sinh(k_z d)(k^2-k_z^2)-2kk_z}{(k-k_z)^2\exp(-k_z d)-(k+k_z)^2\exp(k_z d)} - \Big(\frac{\exp(2kh)\sinh(k_z d)(k^2-k_z^2)-2kk_z}{(k-k_z)^2\exp(-k_z d)-(k+k_z)^2\exp(k_z d)}\Big)^*\Big) \tag{123}$$

(Here * denotes complex conjugate)
For design purposes this is a bit complicated so we need to find a second order approximation to the drag, which is not easy from this expression. It proved shorter to find the vector potential in the conductor and from this calculate the electric field and loss. In this approximation the result agrees with that of Richards and Tinkham [24] which is as follows.
The drag is:-

$$P_o \exp(-2ka)\, S\left[\frac{4k^2h^2}{4+S^2}+\frac{k^2d^2}{12}\right] \tag{124}$$

The first term is the thin sheet approximation, which gives zero loss in the symmetric position, and the second term is the loss in the symmetric position to first order in kd. Here h is the displacement from the central position and d the thickness of the conducting sheet. There is therefore a minimum drag of

$$P_o \exp(-2ka)\, Sk^2d^2/12 \tag{125}$$

The stiffness is

$$2P_o k \exp(-2ka)\frac{S^2}{4+S^2} \tag{126}$$

and the maximum restoring force at h=a is

$$½P_o\left(1-\exp(-4ka)\right)\frac{S^2}{4+S^2} \tag{127}$$

There are two modes of operation. In one we have significant displacements caused by the load and the properties are determined by the thin sheet approximation to first order. The other is where the conductor is normally symmetric, the lift force is zero, and the loss is determined by the last term in equation (124). This can be achieved if there is a load to bear by taking it on a permanent magnet and only using the copper as a stabiliser, (an Evershed system). This leads to a very low loss bearing. Such a bearing is completely passive with very low loss and a stiffness typical of passive magnetic bearings. It is maintenance free with an indefinite life. It is therefore a possible replacementfor the bearing of an energy storage flywheel.

Acknowledgements
I would like to thank Dr. M.K.Bevir for a number of useful discussions, and in particular for pointing out to me both the possibility of the iron circuit, and the null flux configuration.

References

1) Wadey, W.G., 'Magnetic Shielding with Multiple Cylindrical Shells', Rev. Sci. Instr, **27**, 910-916, 1956.
2) Rucker, A.W., Phil. Mag. **37**, 95, 1894.
3) Kittel, C., Fahy, S., and Louie, S.G., 'Magnetic Screening by a Thin Superconducting Surface', Phys. Rev. B. **37**, 642-643, (1988).
4) Claassen, J.H., Evetts, J.E., Somekh, R.E., and Barber, Z.H., Phys. Rev.B. **44**, 9605, 1991.
5) Fiory, A.T., Hebard, A.F., Mankiewich, P.M., and Howard, R.E. Appl. Phys. Let., **52**, 2165, 1988.
6).Pippard, A.B., 'Magnetic penetration Through a Superconducting Film', Super. Sci. and Tech., **7**, 696-699, 1994.
7) J.R.Hull A. Cansiz, J.Appl. Phys., **86**, 6396-6404, (1999).
10). Chen, I.G., Hsu, J.C.,and Wu, M.K., IEEE. 'Calculation of Magnetic Levitation Force of Singlr Grain YBCO', Trans Appl. Superconductivity, **11**, 1745-1748, (2001).
9). Nagashima, K. Otani T.and Murakami, M, Physica C, **328**, 137-144,(1994).
10) Zeisberger, M, .Habisreuter,D., Litzkendorf, D., Surzhenko, O., Muller, R., and Gawalek W., IEEE Trans Appl. Superconductivity, **11**, 1741-1744, (2001).
11) Yang, W.M..Zhou, L.Feng Y, Zhang P.X.,.Wang J.R, Zhang, C.P.,.Zu, Z.M,.Tang X.D and Wei, W., Physica C, **354**, 5-12, (2001).
12). Wang,S, Wang, J, Ren, Z,jiang, H, Zhu, M, Wang,X, and Tang, Q, ' Combination Levitation Force of Multiblock YBCO Bulk High Temperature Superconductors, To be published in Physica C.
13) Maxwell, J.C. , 'A Treatise on Electricity and Magnetism', vol1 chap.6, Dover, New York, 1954.
14) Hull J.R 'Effect of Pemanent Magnet Irregularities in on Levitation Force Measurements', Super. Sci. and Tech., **13**, 854-856, 2000
15).Halbach, K, Nuclear Instruments and Methods, **169**, 1-10, (1980).
16) Hull, J.R., 'Trapped flux Internal –dipole Superconducting Motor-generator'. IEEE Trans on Applied Superconductivity, **9**, 1229-1232, 1999.

17) Maxwell,J.C., 'On the Induction of Electric Currents in an Infinite Plane Sheet', Proc. Royal. Soc., **XX**, 160-168, (1872).
18) Saslow, W.M., 'Maxwell's Theory of Eddy currents in Thin Conducting Sheets., Am. J. Phys.,**60**, 693-711, (1992).
19) Davis, L.C., 'Drag force on a Magnet Moving near a Thin Conductor', J. Appl. Phys.,**43**, 4256-4257, 1972
20) Reitz, J.R., and Davis, L.C., 'Force on a Coil Above a Conducting Slab', J. Appl. Phys, **43**, 1547-1553, (1972).
21) Davis L.C., and Wilkie, D.F., 'Analysis of Motion of Magnetic Levitation Systems', J. Apl. Phys., **42**, 4779-4793,1972.
22) Edwards, J.D., Jayawant, B.V.,Dawson,W.R.C., and Wright, D.T, 'Permanent-magnet Linear eddy Current Brake with a Non-magnetic Reaction Plate', IEE Proc. Elec. Power Appl., **146**, 627-631,
23) Powell, J.R. and Danby, G.T., 'Cryogenics and Industrial Gases', **4**, 19, (1969).
24) Richards,P.L. and Tinkham, M, 'Magnetic Suspension and Propulsion Systems for High Speed Transportation', J. Appl. phys. **43**, 2680-2691, (1972).

TREND OF APPLICATIONS TO HIGH-Tc SUPERCONDUCTING POWER TRANSFORMERS IN JAPAN

Kazuo FUNAKI and Masataka IWAKUMA
Research Institute of Superconductor Science and Systems, Kyushu University
6-10-1 Hakozaki, Higashi-ku, Fukuoka 812-8581, JAPAN

1. INTRODUCTION

A positive prospect for the application of superconductors to power transformers was first given in conceptual design of a 1000 MVA low-T_c superconducting (LTS) power transformer [1]. The conceptual design reached conclusions of economic merit and reduction in the weight on the basis of advanced low-AC-loss superconducting wires, which had not yet been made at that time. Since NbTi composites with very fine filaments less than 1 μm in diameter were developed, many researchers studied the basic properties of superconducting transformers by constructing demonstration models with the NbTi wires [2], [3]. In the cases of developing the LTS transformers, nevertheless the severe condition for refrigeration penalty of liquid helium resulted in the main reason for various types of unsolved technical breakthroughs, reliable large-capacity LTS conductors with low AC loss, high-voltage technique for helium cooling, cost saving in small- and middle-size systems, and so on.

For high-T_c superconducting (HTS) materials such as Bi-2223 systems, which can be used in liquid nitrogen, techniques to make long wires with critical current density J_c more than 10^4 A/cm^2 have been also developed. The new superconductors gave a drastic change in the pessimistic view for the application of superconductors to power devices including transformers. First the feasibility of near-future HTS transformers was practically considered from both viewpoints of high performance of the windings and lower cost of the refrigeration.

The authors' group started to study the technical and economic fundamentals for the HTS transformers with the proposal of transposed parallel conductors of Bi-2223 composite tapes as a practical candidate of large-capacity and low-AC-loss windings. This article mainly includes three topics up to developing a field-test model of the HTS transformer cooled by liquid nitrogen as a milestone for the application of high T_c superconductors to electric power devices. The lead is a demonstration model, which is surely the *first* model of the HTS power transformer cooled by liquid nitrogen (subcooled liquid nitrogen) to perform the rated operation at 500 kVA (800 kVA) and 6.6 kV/3.3 kV with high efficiency using the transposed parallel conductors of Bi-2223 tapes. The next is conceptual design of a 20 MVA HTS transformer of 66 kV/6.9 kV for urban underground substations in order to indicate the high performances and the conditions for the economic merit. From viewpoint of basic technology, there existed big differences between the demonstration model and the target in the conceptual design. One of them comes from quantitative aspect in the rated capacity and voltage. The other is in qualitative aspect that is related with reliability in connection with practical electrical power grids. The reliability is provided for conventional oil-filled transformers through internationally standardized specific tests such as AC voltage test, impulse voltage test and so on. In this way, the third one is the design and development of the high-voltage type of field-test model in accordance with the international standard.

2. BACKGROUND

2.1 Practical critical surface

In practical type 2 superconductors, magnetic flux penetrates into them as fluxoids. While the fluxoids are exposed to electromagnetic force due to the transport current, pinning effects of impurities, defects, grain boundaries etc. in the superconductor prevent their motion. If the pinning effects surpass the electromagnetic force, the transport current flows with no dissipation in the superconductor. The upper limit of the transport current density with no dissipation is denoted as critical current density J_c. Practically, J_c is obtained as the current density that corresponds to a threshold level of electrical strength, e.g. 10^{-6} V/cm in the E versus J characteristic. Taking into account thermal fluctuation, flux creep, of the fluxoids held at the pinning sites, we can easily predict that the fluxoids can be released from the pinning sites in a statistical manner even if the electromagnetic force does not attain the maximum level of pinning effects. In the case where high-T_c superconductors are used in a higher-temperature region, the thermal effect on J_c may be much more marked in comparison with metallic superconductors used around 4.2 K. The thermal effect on J_c in the HTS material is schematically shown for temperature T and magnetic induction B by thick lines in Figure 1, where the thin lines indicate intrinsic properties without thermal agitation. The two-dimensional function of J_c for the temperature and the magnetic induction is called a practical critical surface. We have high performance of superconductivity inside the surface. In Figure 1, the cross sections of the surface, J_c-B characteristics, are also shown on three isothermal planes typically classified by the targets of the application. In both cases of LTS and HTS materials, the thermal effect on the J_c-B characteristic is negligible on the plane of the lower-temperature region (4.2-20 K). On the other hand, J_c for the HTS materials has a more severe degradation in the region of high magnetic induction on the planes of higher temperature.

Around liquid nitrogen temperature, for example, J_c of Bi-2223 Ag-sheathed tapes decreases extremely in magnetic induction of more than 1 T even in the direction parallel to the tape surface. However, the efficiency of refrigeration, which is about 50 times as high as that around liquid helium temperature, increases a permissive level for AC loss in HTS wires drastically, taking into account an effective ratio of total AC loss in superconductors to Joule heating in copper, as shown in the next section. This economic merit may result in a prospective application of HTS materials to AC uses especially in the region of lower magnetic induction, such as the windings of transformers, power cables and current limiters. In the intermediate region of temperature (30-40 K), we may also have wider applications

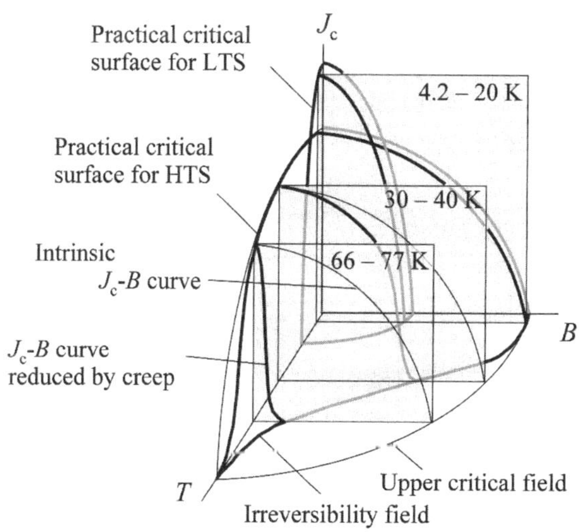

Figure 1 Practical critical surfaces for LTS and HTS materials.

both for higher magnetic field and economic merit. In this region, the temperature of system operation may be given by optimizing the system design under the conditions from properties of superconductor, system sizes, cooling methods and so on.

2.2 An equivalent loss ratio for HTS wires [4]

For prospective applications of the HTS power devices, we will basically need to develop large-capacity HTS windings with low AC loss. While various types of HTS conductors are progressing, we recommend parallel conductors [5] as the large-capacity windings with low AC loss for HTS flat wires such as Bi-2223 multifilamentary tapes, Y-123 coated conductors and so on. We shall estimate AC losses in the windings of large-capacity parallel conductors using the ratio of equivalent AC loss in the HTS windings to that in conventional copper ones.

Permissible level of AC loss in superconducting wires for AC use has been studied with a parameter ξ defined by the loss ratio [6]

$$\xi = \frac{\zeta P_{SC} V_{SC}}{P_{NC} V_{NC}} \quad (1)$$

$$= \frac{\sqrt{2}\zeta P_{SC}}{\rho_{NC} J_{NC} \lambda J_c} \quad (2)$$

where P_{SC} and V_{SC} are the AC loss power density and the volume for the superconducting wire, and P_{NC} and V_{NC} are for the copper wire, respectively. The coefficient ζ is a penalty factor of refrigeration, for example, 500 to 1000 for liquid helium cooling and 10 to 20 for liquid nitrogen cooling. The numerator of the right hand side in Equation (1) is an equivalent AC loss in the superconducting wire taking into account refrigeration power. Equation (2) is an explicit form of the loss ratio with resistivity ρ_{NC} and current density J_{NC} for the copper wire. When ξ is reduced to less than unity, the efficiency of power devices is to be improved by introducing superconducting wires.

First we consider the loss ratio for superconducting transformer windings of n-strand parallel conductors approximately using the expression of total loss power in Equation (A4), where each strand behaves as monofilamentary and its characteristics are given in the left column of Table 1. While the wire has a multifilamentary structure, the filaments are electromagnetically coupled with each other because of no twisting. Figure 2 shows the dependence of ξ upon overall critical current density λJ_c of the strand under the condition of $k = J_{tm} / \lambda J_c = 0.7$ for the amplitude of transport current density J_{tm}. In this figure, we have 2.0×10^{-8} Ωm and 4.0×10^{6} A/m^2 as typical values of ρ_{NC} and J_{NC} for the copper winding. Solid lines show the results under condition of constant current capacity for the conductor.

Table 1 Specifications of test devices and conceptual design

	Demonstration model	Field-test model	Conceptual design
Capacity	0.5 MVA (0.8 MVA)	1.0 MVA	20 MVA
Phase	Single phase	Single phase	Three phases
Prim./Second. voltage	6.6/3.3 kV	22/6.9 kV	66/6.9 kV
Prim./Second. current	76/152 A (121/242 A)	45.5/145 A	175/966 A
Operation temperature	77 K (66 K)	77 K	66 K
Frequency	60 Hz	60 Hz	60 Hz
Magnetic induction	1.7 T at room temp.	1.6 T at room temp.	1.6 T at room temp.
Impedance	0.67 %	5.0 %	7.5 %
Strand	Bi-2223 Ag-sheath tape	Bi-2223 Ag-sheath tape	Bi-2223 Ag-sheath tape
Cross section	0.22 mm × 3.5 mm	0.25 mm × 3.8 mm	←
Twist pitch	No twist	←	10 mm
Filament number	61	←	←
Filament thickness	--	--	10 μm
Matrix/Sheath	Pure silver	Pure silver/Ag-Mn	Ag alloy
Critical current	42 A_{peak} at 0 T	45.5 A_{peak} at 0 T	200 A_{peak} at 0 T
Conductor type	Transposed parallel	←	←
Strand number	3-6	2-6	5-30
Layer number	2-2	4-2	6-2
Transposition	5 times/layer	11 times/layer	(optimized)
Coolant	(Subcooled) Liquid N_2	Liquid N_2	Subcooled N_2
AC test voltage	--	50 kV	140 kV
Impulse test voltage	--	100 kV	350 kV
Thermal load	2.30 kW (3.54 kW) * +	5.14 kW * +	10.8 kW * ++
Iron loss	2.29 kW	3.20 kW	12.3 kW
Efficiency	99.1 % (99.3 %) *	99.2 % *	99.9 % *
In-rush	--	No quench	←
Sudden short	--	No damage	←
Advanced technology	Parallel conductor High efficiency at rated operation	High-voltage type for spot network grid Same high-voltage test as conventional devices Excess-current test for accidental short and inrush	KA-class conductor Low-loss type of strands Tap changers included Same high-voltage test as conventional devices Excess-current test for accidental short and inrush

* Reconfirmed with the penalty factor 20 of refrigeration.

+ Thermal load includes heat leakage and AC loss in windings.

++ Thermal load includes only AC loss in windings.

Dashed lines are for a fixed number of strands. Only cross points between the solid lines and the dashed lines with a positive integer n have a realistic meaning. The solid lines are asymptotically independent of λJ_c in a region of large n, because the total AC loss is nearly proportional to $n\,(\lambda J_c)^2$ for large n as seen in Appendix A. In the case where the total current of the parallel conductor increases up to more than 2000 A, the loss ratio is beyond the upper limit for AC use.

Assuming a remarkable progress of HTS multifilamentary tapes in near future, we shall consider prospects of their parallel conductors as AC windings in the same manner as the previous case mentioned above. Assumed characteristics of the advanced HTS tapes for estimating the loss ratio is given in the right column of Table 1. Here the advanced tapes have isolated and twisted filaments and the sheath material with about 10 times higher resistivity than pure silver. Electromagnetic configuration and corresponding expressions of AC loss for the windings of the advanced tapes are summarized in Appendix B. In this case, the total AC loss of the windings is approximated as a sum of hysteresis loss of filaments and coupling loss among filaments in each strand. The loss ratio calculated is indicated as a function of λJ_c at $k = 0.7$ in Figure 3. Parameters in the figure have the same meanings as

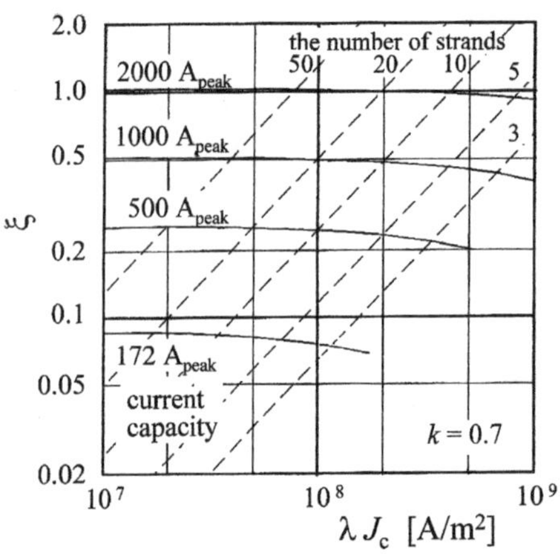

Figure 2 Loss ratio ξ for n-stacked strands of usual monofilamentary HTS tapes as a function of λJ_c with $\zeta = 10$. Solid and dashed lines are given for constant current capacity and fixed strand number, respectively. Characteristics of the strands are listed in the left column of Table 1.

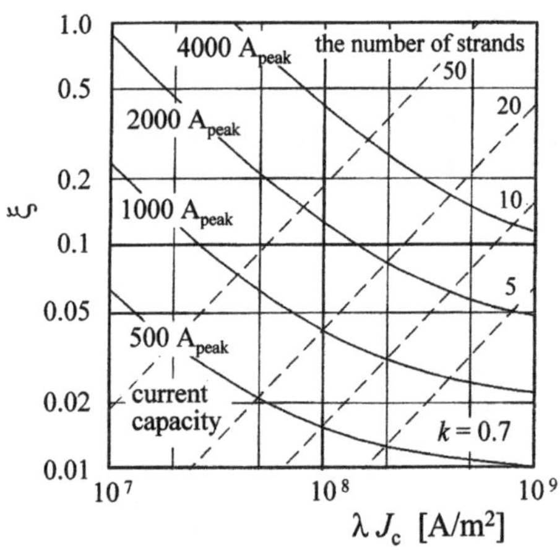

Figure 3 Loss ratio ξ for n-stacked strands of advanced multifilamentary HTS tapes as a function of λJ_c with $\zeta = 10$. Solid and dashed lines are given for constant current capacity and fixed strand number, respectively. Characteristics of the strands are listed in the right column of Table 1.

those in Figure 2. The loss ratio for the conductor with a fixed current capacity decreases with increase in λJ_c . This tendency comes from a decrease in contribution from the coupling loss to the loss ratio. Figure 3 shows that several kA-class parallel conductors with a low ξ can be designed as AC windings in a region of λJ_c more than 1×10^8 A/m^2.

In the advanced tapes exposed to a parallel magnetic field, hysteresis loss decreases, roughly speaking, with a ratio of the filament thickness to the tape in comparison with the present ones like monofilamentary. The decrease in the hysteresis loss results in a reduction in the total AC loss especially in a region of high λJ_c as seen from Figure 3. In a perpendicular magnetic field to the tape surface, on the other hand, the dominant component of AC loss must be a coupling loss among filaments, because the interlinkage area is proportional to the tape width. Changing the definition of β into w / d in Equation (B9), we obtain the coupling time constant in this configuration from Equation (B8). If the coupling time constant is larger than the period of AC magnetic field, the coupling current has saturation in a part of filaments, which may generate the coupling loss comparable with the hysteresis loss in a monofilamentary tape. In this way, even in transformer windings composed of advanced HTS tapes, both edge parts of the windings may generate one of major parts in the total AC loss. We need to estimate AC losses in more detail for practical design of HTS windings.

While we consider only the ideal configuration of parallel magnetic field for the transformer windings, we can find a prospective feasibility of future HTS transformers with current capacity of several kA-class from the above estimation on AC losses in the HTS parallel conductors. In those days when we kept a plan for the application of HTS materials to electric power devices in our mind, practical multifilamentary HTS wires with low AC loss had not yet been fabricated for the advanced windings. Nevertheless we thought that we could construct a high-efficiency model system of the HTS transformer with lower current capacity than 1 kA by means of Bi-2223 multifilamentary Ag-sheathed tapes already developed, which were not twisted and had monofilamentary property, as shown in Figure 2.

2.3 Merits of HTS transformers

In HTS transformers, it is expected from AC loss estimation in the previous section that the load loss generated mainly in the windings decreases drastically in comparison with copper windings even if the refrigeration penalty is accounted for. Furthermore, the coolant of liquid nitrogen or subcooled liquid nitrogen functions as an electrical insulator almost equivalent to oil. These conditions may lead to the following advantageous aspects for the HTS transformers against the conventional oil-filled ones:

(i) small size and light weight;
(ii) non-oil, non-flammable;
(iii) high efficiency even including refrigeration penalty;
(iv) current limiting effect;
(v) reduction in life cycle cost.

Furthermore, it can be also predicted that the above merits could be improved in coexistence with other HTS power devices, cables and current limiters.

3. A 500/800 KVA DEMONSTRATION MODEL [4], [7]

For future HTS transformers with the advantageous properties mentioned in the previous section, we have designed and developed a 500 kVA-800 kVA test device cooled by liquid nitrogen-subcooled liquid nitrogen, where the windings are parallel conductors of Bi-2223 tapes with transposition. The main aims are (a) to develop HTS windings with large capacity and low AC loss, and (b) to construct a demonstration model for high-efficiency HTS transformers cooled by liquid nitrogen and subcooled liquid nitrogen, which *must* lead to high-voltage-type power devices.

3.1 Specifications of strands and windings

Specifications of strands and windings used on basis of the above concepts are listed in the left column of Table 1. Values in the parentheses are for the operation in subcooled liquid nitrogen at 66 K. The strands are Ag-sheathed Bi-2223 multifilamentary tapes that have 61 filaments with no twisting. The rectangular cross-section is 0.22 mm thick and 3.5 mm wide and the ratio of superconductor λ is 0.284. Bi-2223 tapes have anisotropy in transport properties for the direction of external magnetic field to the surface. The critical current was measured with a criterion of 0.1 μV/cm for a typical short sample exposed to external magnetic field both parallel to and perpendicular to the tape surface in the range of temperature between 64 K and 77 K. The results at two levels of operation temperature are shown in Figure 4 (a). Squares and circles show the critical current at 77 K and 66 K, respectively. The latter data are evaluated with an interpolation between experimental results at 70 K and 64 K. We can expect that the transformer capacity at 66 K increases about two times compared with at 77 K due to enhancement in the critical current I_c as shown in Figure 4 (a), where I_c is plotted for magnetic induction. While the critical current in the parallel field, indicated by open symbols, is nearly constant in the range of magnetic field for the horizontal axis, a drastic decrease in the critical current by closed symbols occurs with increase in the perpendicular field. This shows that the rated current of transformer winding results from the properties of strands exposed to the perpendicular field in edge parts of the coils.

We proposed and first used the parallel conductors with a proper transposition as large-capacity windings of tape-shaped strands with a large aspect ratio. The primary

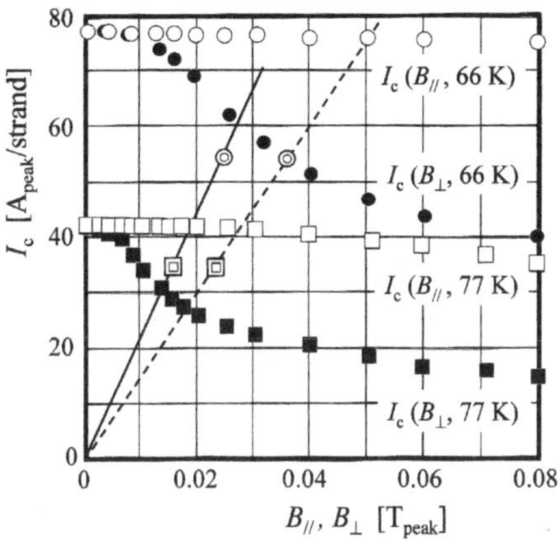

Figure 4 (a) Dependence of critical current of Bi-2223 taped strand on applied magnetic induction $B_{//}$ parallel and $B_\perp$ perpendicular to the tape surface. Solid and dashed lines indicate load lines for the strand in edge and central turns, respectively, in each layer.

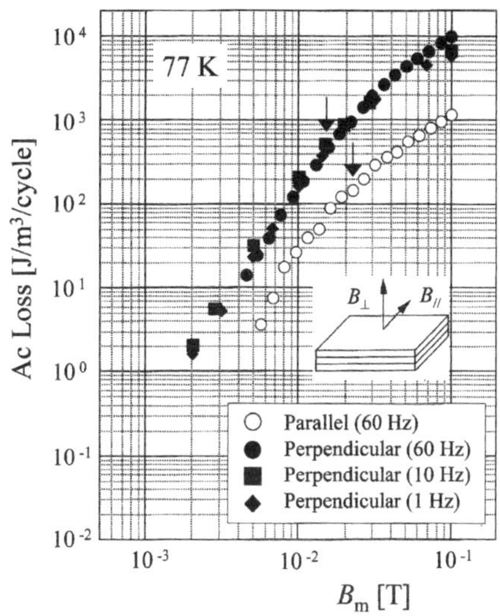

Figure 4 (b) AC losses of Bi-2223 tapes at 77 K in external magnetic field parallel and perpendicular to the tape surface.

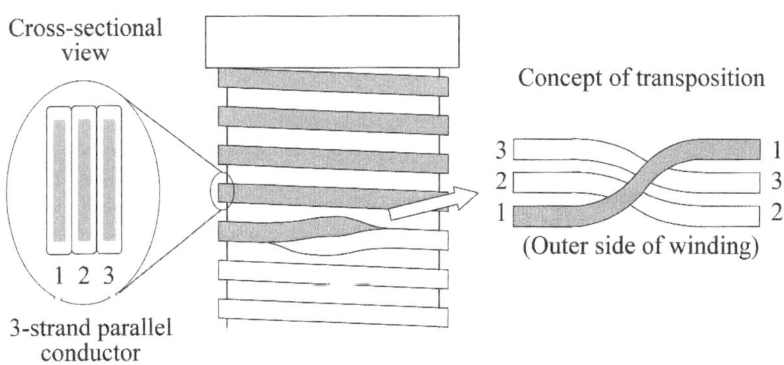

Figure 5 Schematic view of transposition in a layer of solenoid coil for the three-strand parallel conductor of Bi-2223 tapes.

winding is the parallel conductor with three strands, that is a unit of parallel conductor, and the secondary one is composed of two units connected in parallel. Figure 5 shows a schematic view of the transposition in a layer of solenoid coil for the three-strand parallel conductor. In the parallel conductors, the strands are electrically insulated with each other by a glass-fiber tape and transposed five times per layer to cancel out net interlinkage flux between each pair of strands for a uniform current distribution among them. The positions of transposition are set by calculating the distribution of magnetic field in the winding structure as shown in the following. The total length of strands is about 2.2 km.

3.2 Design parameters of transformer

We designed an oxide-superconducting transformer with capacity of 500 kVA and 800 kVA operated in liquid nitrogen at 77 K and subcooled nitrogen at 66 K, respectively, for the primary / secondary voltage of 6600 V / 3300 V. The primary and secondary coils have two layers, respectively, that are sandwiched mutually as shown schematically in Figure 6 (a) for diminishing perpendicular component in the magnetic field and consequently AC loss in the windings. In order to estimate critical current and AC loss of the windings, we numerically calculated the distribution of magnetic field in the winding part. In the calculation, each layer of the three-strand windings is approximated as three superconducting hollow cylinders

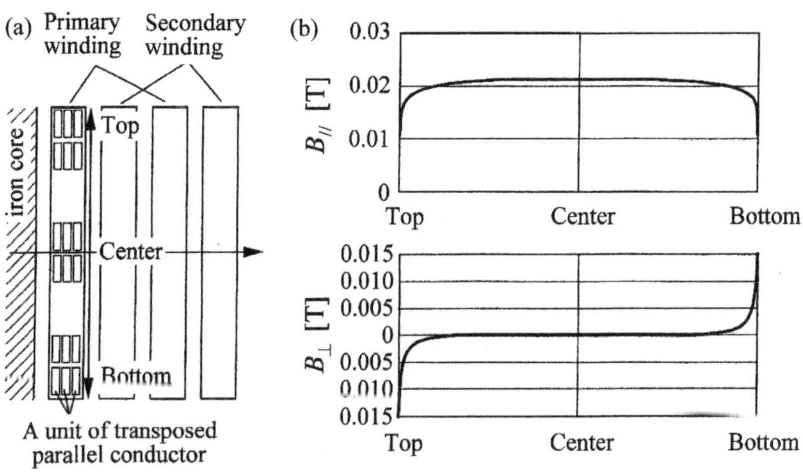

Figure 6 (a) Schematic cross-section of a sandwiched winding structure and (b) Numerically calculated distributions of magnetic induction parallel and perpendicular to the tape surface along the coil axis on the outer surface of an innermost coil.

coaxially arranged. The distributions of magnetic induction along the coil axis on the outer surface of the innermost coil are illustrated in Figure 6 (b) both for the components parallel to and perpendicular to the tape surface. As can be seen in this figure, a relatively large perpendicular component is generated near both edges of the coils, and the ratio of turns exposed to the large perpendicular component is not larger than a few %. In Figure 4 (a), solid and dashed lines give load lines for the strands in edge and central turns, respectively, in each layer. Symbols of double circles and squares on the load lines indicate the peak level of branch currents balanced among the strands at 66 K and 77K, respectively. At the 66 K operation, the peak level of transport current in the rated condition almost attains to critical level by the perpendicular magnetic field. Furthermore, at the 77 K operation, the rated current is set a little over the critical level for the edge turns. Even in this case, we can perform an overload test of the transformer with no remarkable decrease in the efficiency, as seen from the distribution of magnetic field in Figure 6 (b).

Design parameters of the transformer are also summarized in the left column of Table 1. The windings are cooled by liquid nitrogen or subcooled nitrogen in a G-FRP (Glass Fiber Reinforced Plastics) cryostat of 785 mm in diameter and 1210 mm in height, which has a room-temperature space around a columnar axis for an core of silicon steel plates with the diameter of 314 mm. The turn ratio of the primary to the secondary is 2. The primary / secondary rated current is about 76 A / 152 A and 121 A / 242 A, respectively, in 77 K and 66 K operations. Design principle of electrical insulation for the present transformer is

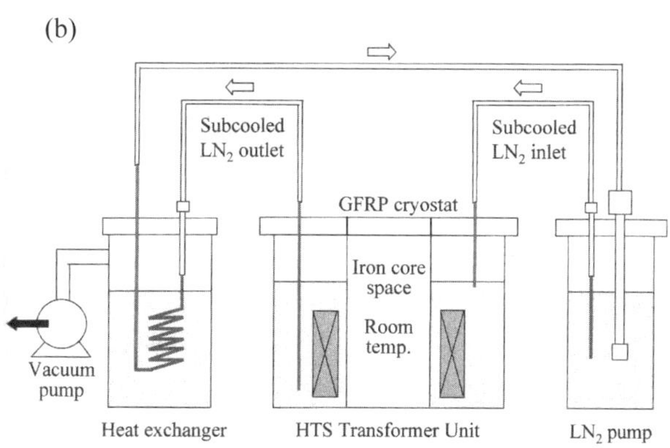

Figure 7 (a) Photograph of a demonstration model of 800 kVA HTS transformer cooled by subcooled liquid nitrogen of 66 K, and (b) the flow diagram of subcooled liquid nitrogen in the cooling system for the demonstration model.

attributed to electrical properties of nitrogen gas in a room temperature. As a result, we may have an excess margin in the design of electrical insulation. Further consideration in the optimum design will be necessary for devices of higher voltage with reduction in the sizes.

Figure 7 (a) and (b) show an overview of the *first* HTS transformer and the flow diagram of subcooled liquid nitrogen in the cooling system for the demonstration model.

3.3 Estimation of AC losses in the windings

As we can see in Figure 6 (b), the windings are divided into two parts with different types of AC losses. One is dominant central turns where the strands are exposed to nearly uniform magnetic field parallel to the surface and the other is a few turns of both edge parts in an external field with larger perpendicular component.

We measured the AC losses of the strands in the external field of 60 Hz parallel to and perpendicular to the tape surface to roughly compare contributions from these two parts to the total AC loss. Figure 4 (b) shows the AC losses per unit volume per cycle measured with a racetrack shaped pickup coil at 77 K. The samples are short Ag-sheathed Bi-2223 tapes with characteristics similar to those in the left column of Table 1. For the measurement of the perpendicular-field loss, the sample is a piled-up three pieces of tapes to simulate electromagnetic environment of the strands in the winding, as shown by the insert in Figure 4 (b). In Figure 4 (b), arrows show the levels of AC loss for the maximum amplitudes of parallel and perpendicular components of magnetic induction in the windings. It can be seen from the comparison that the loss density for the perpendicular component is a few times larger than that for the parallel one. From a viewpoint of the total AC loss, however, the contribution from the central turns exposed to the parallel field is dominant as well.

In this way, we approximately estimated the total AC loss in the windings as that in a sufficiently long solenoid coil of a three-strand parallel conductor with uniform transport current among the strands. Expressions of the AC losses are introduced in APPENDIX A. In the rated operation of 800 kVA at 66K, for example, the theoretical prediction gives the winding loss of 74 W. The calculated level of the AC loss is equivalent to reduction of 0.2 % in transformer efficiency taking into account the penalty factor 20 of refrigeration.

3.4 Characteristic tests of transformer at 77 K

The operation at 77 K included no-load, short-circuit and load tests to obtain steady characteristics in the rated operation in the same manner as the conventional oil-filled transformers. In the no-load test, we excited the transformer from the secondary side at the rated voltage 3300 V for thirty minutes. The exciting current was 2.60 A and the no-load loss was 2289 W in usual procedure with a power meter, which can be attributed to a core loss. The short-circuit test was performed from the primary side at the rated current 75.8 A for 2.5 hours. The primary voltage was 44 V and the impedance voltage was estimated as 0.67 % of the rated level. Steady loss power in the rated condition was estimated first by a calorimetric method with a flow meter of nitrogen gas at room temperature. The total amount of thermal load in coolant, which is composed of AC loss in the windings and heat leakage through the cryostat and current leads, was about 115W. The heat leakage measured separately at no transport current was 51 W. It can be shown from the difference between the two results that the AC loss in the windings is 64 W. We also measured the AC loss directly by an electromagnetic method. Figure 8 simply shows an electrical circuit of the loss measurement with a pair of voltage taps attached to the current terminals on the top plate of the cryostat.

The AC loss power P in both windings was measured from the primary voltage $V_1(t)$ and current $I_1(t)$ in the following numerical procedure by a computer system

$$P = \frac{1}{T}\int_0^T V_1(t)\, I_1(t)\, \mathrm{d}t \qquad (3)$$

where T is a period of alternating signals. For suppressing an error in the numerical procedure, an inductive component was removed in advance from the voltage signal by means of the terminal voltage from a Rogowski coil attached in the primary circuit. In the rated condition, we obtained 60 W as the total amount of AC loss in the windings. Agreement between the results by the two methods shows that we can measure the steady loss power of the windings with a high accuracy in the short-circuit test.

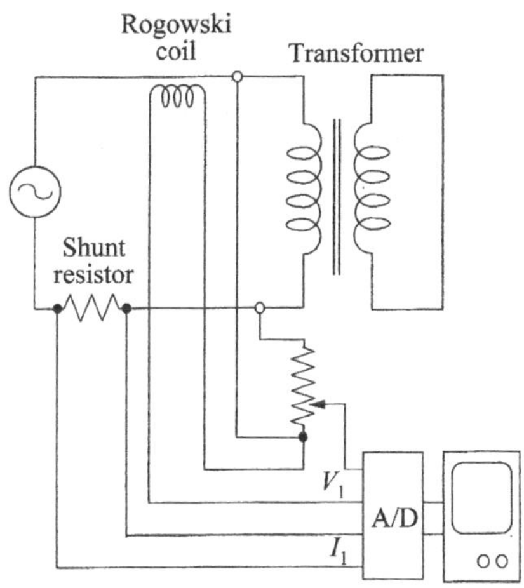

Figure 8 An electrical circuit of AC loss measurement for the primary and secondary windings.

The efficiency of the transformer can be calculated from the iron loss of 2.29 kW and the thermal load of 115 W in the above two tests. For the thermal loads of the AC loss and heat leakage to liquid nitrogen, however, we have to take into account a refrigeration penalty of liquid nitrogen to get an equivalent loss. Even for the penalty factor of 20, we have a high efficiency 99.1 % under the rated condition of 500 kVA.

In the load test, the transformer was connected with the secondary load of a 500-kVA inductor. The transformer was excited up to about 500 kVA and operated steadily at the rated condition for one hour. If we consider the situation of over-current as shown in Figure 4 (a), the results show that the windings of Ag-sheathed Bi-2223 tapes are very stable against thermal disturbance even over the critical current.

3.5 Characteristic tests of transformer at 66 K

We prepared a continuous flow system of subcooled liquid nitrogen to cool down the windings of the transformer. The system is composed of additional two cryostats for heat exchange and forced flow of subcooled liquid nitrogen to the main cryostat of the transformer. The flow diagram of subcooled liquid nitrogen in the cooling system is given in Figure 7 (b). The warmer nitrogen from the main cryostat is forced out from the vessel with a LN_2 pump. The forced-flow nitrogen is cooled down in the vessel of heat exchange that is evacuated by a rotary vacuum pump at a level of 20.6 kPa. The temperature of saturated liquid nitrogen in the heat exchanger was adjusted around 65 K. The cryostats for the transformer and the LN_2 pump were maintained at an atmospheric pressure. The steady flow rate of the nitrogen is about 1 liter/min. The cooling capacity was restricted by the flow rate of subcooled nitrogen. In steady operations of the transformer, the cooling capacity was about 20 ΔT [W] for $\Delta T < 5$ K, where ΔT is the temperature difference between the subcooled liquid nitrogen in the main cryostat and the saturated liquid nitrogen in the heat exchanger.

We shall make use of the results of the no-load test in the saturated liquid nitrogen as those in the subcooled liquid nitrogen. Therefore, we performed only the short-circuit test, where we set the initial temperature of the main cryostat to 66 K. In the short-circuit test, we supplied the primary current

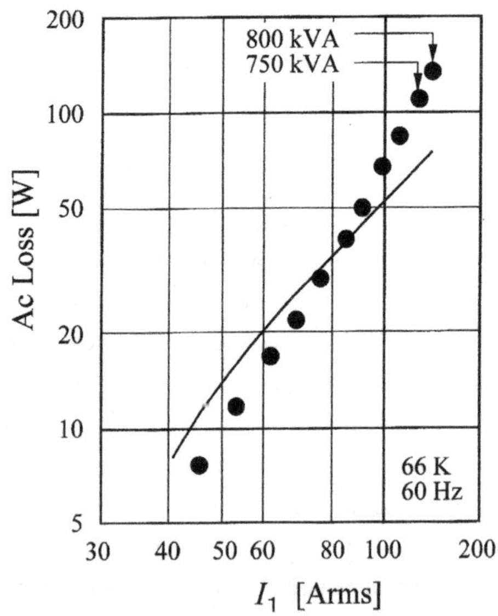

Figure 9 AC loss power in the windings for the primary current I_1 at 66 K and 60 Hz. A solid line is the theoretical prediction using the expression in Appendix A.

up to 121.6 A. The level of the maximum excitation is indicated by double circles on the load lines for the two types of external field in Figure 4 (a). In this case, the transformer has a capacity of 800 kVA for the primary voltage of 6600 V.

We excited the transformer with the secondary short circuit at current levels corresponding to 750 kVA for 60 minutes and 800 kVA for 15 minutes, respectively. Since the total thermal load to the system exceeded the cooling capacity, the thermal situation of subcooled liquid nitrogen was non-steady. These continuous operations were interrupted when the increase in temperature of the windings came up to about 1.5 K. It was, nevertheless, confirmed that the HTS transformer at the rated condition has a maximum capacity of 800 kVA.

We measured the AC losses using the electromagnetic method shown in Figure 8. The AC loss obtained must include an ohmic loss in the current leads in addition to the AC loss in the windings. We confirmed that errors from this ambiguity could be neglected in estimating the AC loss. Figure 9 shows measured AC losses by closed circles for the primary current at 66 K and 60 Hz. Corresponding levels of the transformer capacity are also indicated in Figure 9. During a series of loss measurements at 66 K, we had no increase of the winding temperature more than 0.1 K. The efficiency of the transformer under the rated condition of 800 kVA is estimated as 99.3 %, where the iron loss at room temperature is 2.29 kW and the thermal load (AC loss) is 177 W (126 W), with the refrigeration penalty of 20.

We also calculated the AC loss in the windings approximately using theoretical expressions in APPENDIX A. In Figure 9, a solid line shows the theoretical prediction for the dependence of the AC loss upon the primary current. We used the following values of parameters in the calculation; $n = 3$, $f = 60$ Hz, $d = 0.22$ mm, $\lambda J_c = 1.05 \times 10^8$ A/m^2 and $V = 1.60 \times 10^{-3}$ m^3. We have an agreement between measured and theoretical results except for a region near the rated current. The deviation in the large-current region may come from additional AC losses due to local field of the loosely wound turns and non-uniform distribution of branch current in the parallel conductors [4].

We fabricated the 800 kVA demonstration model of the HTS transformers in higher efficiency than conventional oil-filled ones with equivalent capacity in a practical situation including the refrigeration penalty. While the Bi-2223 tapes used had a monofilamentary property, the transformer possessed a high-efficiency performance at the rated operation. As can be predicted in the design, the AC losses in the strands could be suppressed by reducing the maximum levels of magnetic field, especially perpendicular component, in the winding section.

4. DESIGN OF FUTURE TRANSFORMERS IN URBAN UNDERGROUND SUBSTATIONS [8]

Taking into account advantageous aspects to be expected for the HTS transformers, especially non-flammability, small size and weight, and high efficiency, we can propose that HTS transformers in underground distribution substations are promising targets for HTS applications from the viewpoint of high performance. On the other hand, another point to be considered is the life cycle cost relative to that of the conventional ones. As a typical example for transformers in the underground substation, a 20 MV A HTS transformer has been conceptually designed under some assumptions of future development in HTS technology. The design parameters used are listed in the right-most column of Table 1. In the conceptual design, two levels of AC losses in HTS wires are used. One is the present level of AC loss and the other is reduced to 15 % for hysteresis loss and 20 % for coupling loss. The factor of refrigeration penalty is 20. We shall compare the life cycle cost with the same-size conventional transformer using the results of the conceptual design. In the comparison, the life cycle cost is defined as the sum of initial cost of construction and maintenance cost. The latter is equal to the product of annual loss power rate and lifetime in years. The total loss power is represented as

$$\text{Total loss power} = \text{load loss power} \times (\text{operation rate})^2 + \text{no-load loss power} \quad (4)$$

for conventional oil-filled transformers

$$= \text{load loss power} \times 20 \times (\text{operation rate})^2 + \text{no-load loss power} \quad (5)$$

for HTS transformers

where the load loss term includes the refrigeration penalty 20 for the HTS transformers. The ratio of the life cycle cost of HTS transformers to that of conventional oil-filled ones is illustrated for various load rates in Figure 10. Here, we adopted a load loss of 111 kW and a no-load loss of 11 kW as typical levels for the conventional transformers and an initial cost ratio of 2.6. At an operation rate of 70%, for example, the life cycle cost of HTS transformers is lower than that of the conventional ones for life times of more than 22 years for the present HTS windings and 11 years for the advanced ones. If we consider that the high initial cost ratio results from the costs of HTS wires and cooling system, we can obtain more proper targets by reduction in their costs. Figure 11 (a) shows an overview of the 20 MVA HTS transformer in the conceptual design. Figure 11 (b) also shows a comparison of the weight ratio. The weight of the HTS transformer is about one half that of the conventional one, because of a drastic reduction in the volumes of coolant (oil) and conductor. These results of AC loss and weight also lead to a halving of the installation area of the HTS transformer. Finally, the conceptual design shows that it is feasible to introduce the HTS transformers to the underground substations instead of the conventional oil-filled ones.

We catch a prospective target for the HTS transformers in the underground substations by the conceptual design mentioned above. While the development of advanced HTS conductors is a prerequisite for the future HTS transformers as shown in the present section,

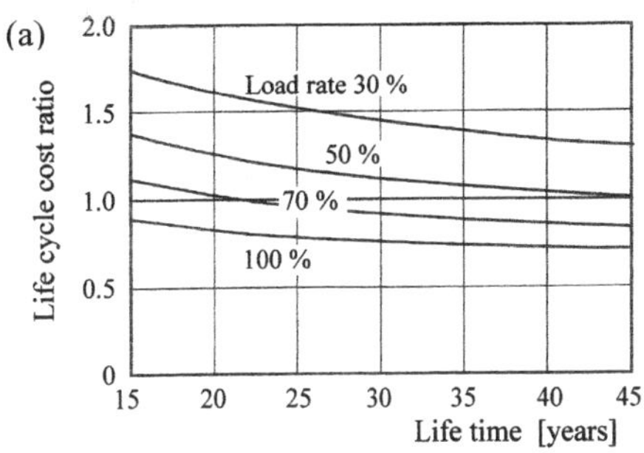

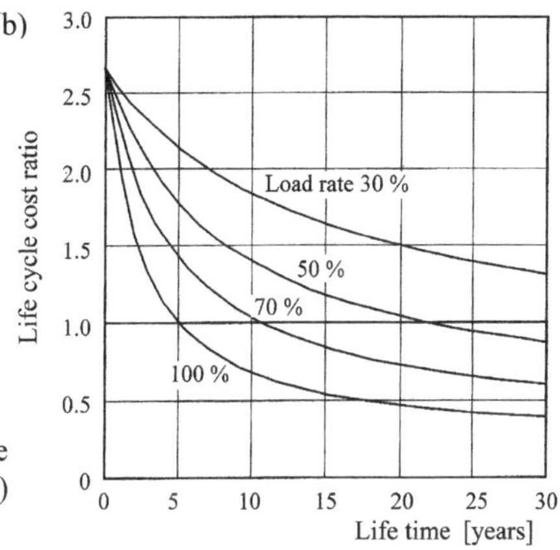

Figure 10 Life cycle cost ration vs. life time (a) for present HTS wires, and (b) for advanced HTS wires.

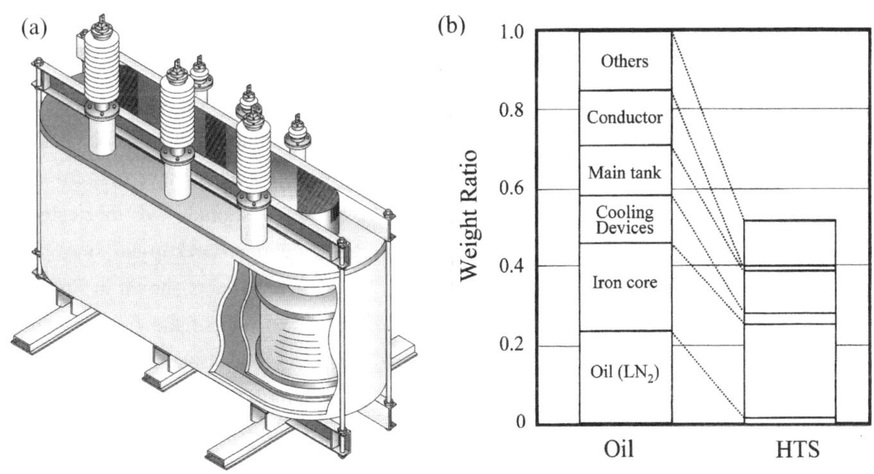

Figure 11 (a) Overview of a 20 MVA HTS transformer and (b) Comparison of weight ratio of HTS transformers to Oil ones in the conceptual design.

the next step is to establish essential bases of novel technology for the application of superconductors to power devices.

5. A 1 MVA FIELD-TEST MODEL [9]

5.1 Model coils for basic technology

We were designing a high-voltage-type single-phase HTS transformer for field test. The primary/secondary rated voltage is 22 kV / 6.9 kV. The rated capacity is 1 MVA in 77 K operation. The main parameters are listed in the middle column of Table 1. In addition to the laboratory-test device mentioned in the section 3, the field-test device is placed as a step for prospective application to HTS transformers in underground distribution substations. The followings are main technical subjects to be imposed on the field-test device in similar way to conventional oil-filled transformers;

i) over-current of five times as high as the rated level
ii) AC withstand voltage of 50 kV
iii) lightning impulse voltage of 100 kV

In order to establish the basic technology for the field-test device, we carried out two types of preliminary tests with model systems of the transformer. One is a model coil for over-current test, which is composed of a HTS inner coil and a copper backup coil. The others are a set of local models with particular winding structures both for turn-to-turn and layer-to-layer withstand voltage tests.

5.1.1 A model coil for over-current tests

In the design of the field-test transformer, we set a level of excess current 5 times as high as the rated one for installation of the transformer into a distribution grid or for accidental short in the grid. The high level of excess current may result in severe thermal and mechanical stresses to the winding. A model coil was designed and fabricated for over-current tests. The model coil has a HTS-coil part and a copper-backup-coil one. Main characteristics of the two coils are listed in Table 2. The overview is shown in Figure 12. The winding of the HTS coil is composed of similar wires to that of the field-test device mentioned in the middle column of Table 1. The winding is a four-strand parallel conductor of Bi-2223 Mn-Ag sheathed tapes with 7-time transposition in a layer of 64 turns. The HTS coil can be connected with the backup coil in different two ways. One is in a normal direction and the other is in a reverse one. The HTS winding in the model coil can be exposed to tensional or compressive excess stress by the over-current similarly to the primary and secondary windings in the transformers. The HTS tapes are supported with glass-fiber tapes mechanically from the outside on the basis of numerical estimation of mechanical stress in the winding for the test of over-current five times as high as the rated level of the field-test device.

Table 2 Specifications of model coil for over-current test.

	HTS coil	Copper backup coil
Inner diameter	200 mm	240 mm
Height	497 mm	496.4 mm
Winding width	1 mm	9.2 mm
Strand	Mn-Ag sheath Bi-2223 tape	Copper tape
Strand cross-section	3.4 mm × 0.24 mm	7.3 mm × 2.3 mm
Parallel number	4	1
Turn number	64	68
Layer number	1	4
Transposition number	7	--
Critical current	118 A at 10^{-13} Ωm	--

Figure 12 Overview of the model coil for over-current test.

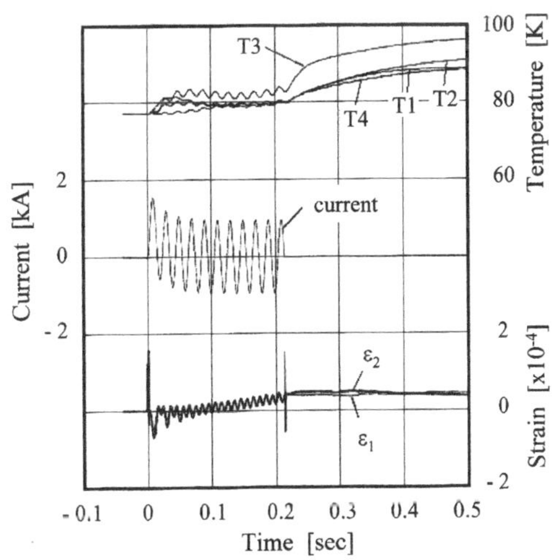

Figure 13 Waveforms of applied current and signals from thermal sensors and strain gauges.

In order to verify the effects of the excess current on *V-I* characteristics of the winding in the field-test transformer, we performed the over-current tests by applying equivalent excess current to the model coil with normal or reverse connections between the HTS coil and the copper-backup one. Figure 13 shows the waveform of the applied current and the signals from the thermal sensors and the strain gauges at the application of five-time over-current with duration of 0.2 sec in the reverse connection, where the HTS winding is exposed to compressive stress. The winding temperature does not attain to 100 K and the levels of strain are also kept permissive. Similar results were also obtained in the normal connection.

The *V-I* characteristics of the HTS winding were scarcely affected by the over-current in both connections. It has been confirmed from these results of the over-current tests that the winding has no damage up to the over-current five times as high as the rated level with the duration of 0.2 sec.

5.1.2 Local model coils for withstand voltage tests

Three-layer local model coils were constructed for the AC voltage test and the impulse voltage test. An example of the local model coil for the layer-to-layer withstand voltage test is schematically shown in Figure 14. The middle layer of a HTS winding was connected with a high-voltage terminal and the other two layers of copper windings were grounded. The distances between the HTS layer and the grounded layers were designed on basis of electrical insulation properties of liquid nitrogen. We performed the two types of withstand

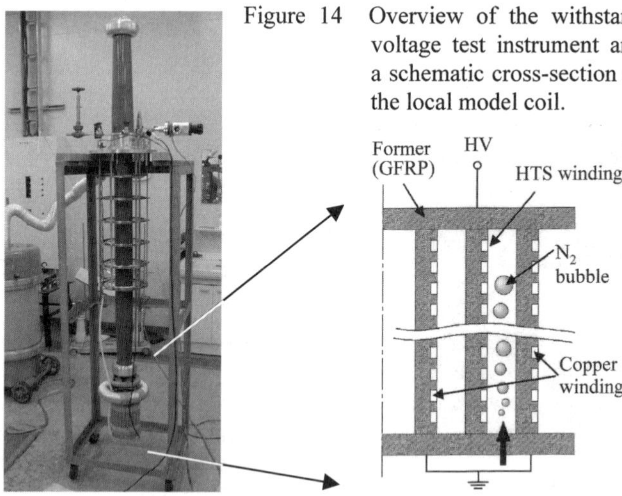

Figure 14 Overview of the withstand voltage test instrument and a schematic cross-section of the local model coil.

voltage tests above-mentioned in coexistence with bubbles for simulating evaporation of liquid nitrogen due to the heat generation in the HTS windings. The AC voltage was 50 kV at 60 Hz and the peak and duration of the impulse voltage were 150 kV and about 0.08 ms. Especially, the level of impulse voltage was determined as a maximum level of local electric potential in the field-test HTS transformer winding for lightning impulse of 100 kV by numerical simulation. No electrical breakdown occurred in the withstand voltage tests under the preliminary consideration on electric insulation by liquid nitrogen. The experimental results were put into practice for the basic design of the field-test model.

5.2 Design and construction

5.2.1 Specifications of strands and windings

The primary and secondary windings of the field-test transformer are composed of 2-strand and 6-strand parallel conductors, respectively. The strands are Bi-2223 Ag/Mn-sheathed multifilamentary tapes. The main specifications of the strands and windings are given in the middle column of Table 1. Figure 15 shows average characteristics of critical current in typical strands for perpendicular and parallel magnetic induction.

The strands are transposed in each layer of the primary and secondary windings, respectively, for uniform current distribution among strands in the same manner as for the laboratory-test device. The total length of strand is 4.7 km.

5.2.2 Specifications of transformer

We designed and fabricated a single-phase oxide superconducting transformer with the maximum capacity of 1 MVA operated in liquid nitrogen of 77 K for the primary / secondary voltage of 22 kV / 6.9 kV. The main parameters are also listed in the middle column of Table 1. The primary and secondary windings are cylindrical coils of 4 and 2 layers, respectively. The primary winding is located coaxially outside the secondary one in a GFRP cryostat around an iron core of room temperature. In Figure 15, the load lines are indicated for the strands exposed to maximum levels of perpendicular and parallel magnetic field in the primary and secondary windings of the designed transformer on assumption of uniform current distribution among strands. The squares show the positions of the rated operation for the strands in the windings. The rated levels are within the critical one on the load lines for the parallel field and a little exceed that for the perpendicular field.

In the design and construction, high performances both for over-currents of inrush and short circuit and for lightning impulse voltage were introduced on the basis of the test results with the model coils. Figure 16 is an overview of the developed HTS transformer installed

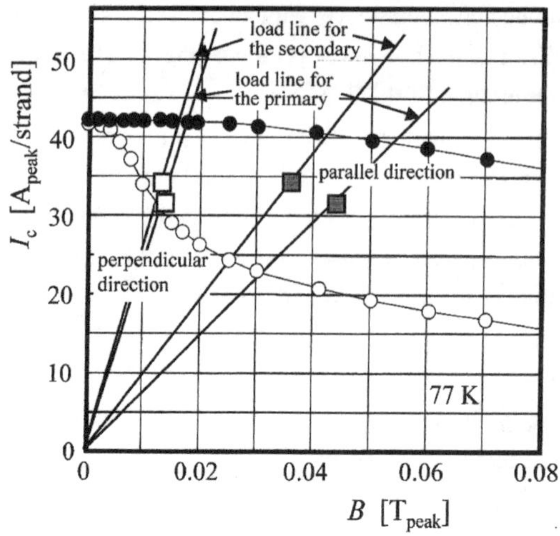

Figure 15 The dependence of critical current (peak) of strand on magnetic induction at 77 K and load lines for strands in the primary and secondary windings. Open and closed symbols are quantities in perpendicular and parallel magnetic field, respectively. Squares show positions of rated condition in case of uniform current distribution among strands.

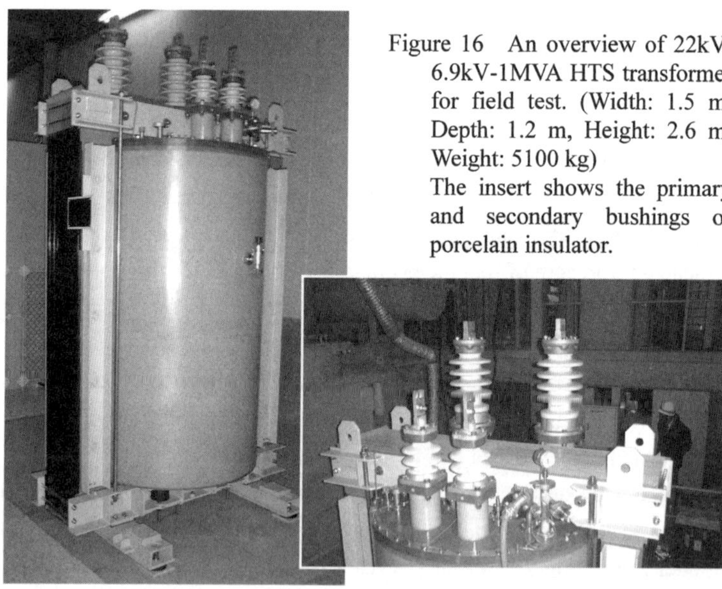

Figure 16 An overview of 22kV/ 6.9kV-1MVA HTS transformer for field test. (Width: 1.5 m, Depth: 1.2 m, Height: 2.6 m, Weight: 5100 kg) The insert shows the primary and secondary bushings of porcelain insulator.

in the GFRP cryostat. The total height of the unit is 2630 mm. The G-FRP cryostat has a height of 1616 mm, an outer diameter of 993 mm and a room temperature bore of 374 mm in diameter through which an iron core is installed. The volume of the liquid nitrogen in the cryostat is 330 liters. The inset shows two sets of bushings of conventional porcelain insulator for the primary and secondary windings.

5.3 Individual tests based on international standards

We had the same test items for the HTS transformer as those for the conventional oil-filled ones, which include AC voltage test and lightning impulse voltage test in accordance with the international standards of test methods.

Main results of the individual tests for the HTS transformer cooled by liquid nitrogen are shown in the middle column of Table 1. AC loss of 0.206 kW in the windings was evaluated by usual electromagnetic method [4] in the short-circuit test at the rated primary and secondary currents. Core loss of 3.20 kW was also estimated in the no-load test at the rated voltages. The rated capacity estimated was 1 MVA. We obtained efficiency of 99.2 % in the rated operation taking into account the refrigeration penalty of 20 for liquid nitrogen cooling.

We also performed the AC voltage test by opening the primary side of the transformer and applying sinusoidal voltage of 12 kV and 60 Hz to the secondary side during 1 min. The primary voltage was 50 kV. We had no electrical breakdown in the test.

In the lightning impulse test, we had two levels of impulse, 50 kV (half wave) and 100 kV (full wave) in peak from the primary. The secondary side was shorted and grounded with the iron core and another terminal of the primary. Figure 17 (a) and (b) show the waveforms (upper) of impulse and the responses (lower) of the primary current for the half and full waves, respectively. Vertical axes of Figure 17 (a) are scaled twice as large as those of Figure 17 (b). The waveforms of the voltage and the responses of the current are equivalent to each other. The results show no breakdown for full wave applied to the transformer.

5.4 Field test [10]

5.4.1 System configuration

Figure 18 shows a flow diagram of subcooled liquid nitrogen in the cooling system for the field test of the HTS transformer. The cooling system consists of two main units, a HTS transformer unit and a pump unit, and a control panel. The two main units are connected with each other by means of flexible transfer-tubes and the subcooled liquid nitrogen is circulated through the transfer tubes. The basic concept in the design is on the previous procedure to develop the demonstration model mentioned in the section 3.

A pair of current leads, a relief valve, a nitrogen gas inlet and transfer-tube ports are located on a top flange. An inlet of subcooled liquid nitrogen is set at the bottom of the cryostat and an outlet of the subcooled liquid nitrogen for return to the pump unit is set at the

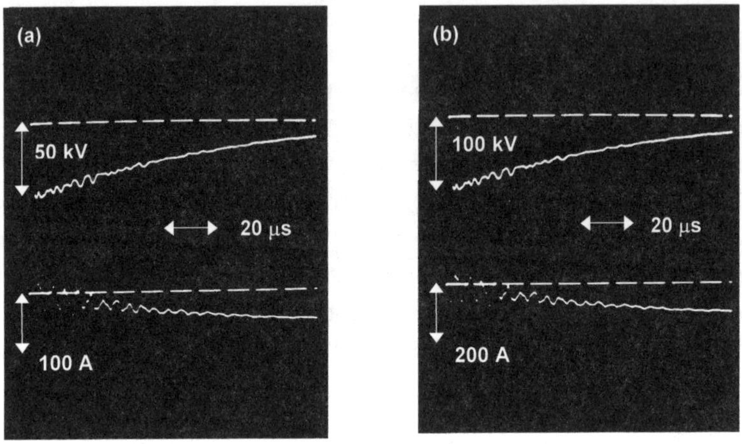

Figure 17 Waveforms of the primary voltage (upper) and current (lower) in lightning impulse voltage tests with peak values of (a) 50 kV and (b) 100 kV. The horizontal axis of time and the vertical axes of voltage and current are scaled as shown in (a) and (b), respectively. Vertical axes for voltage and current in (a) are scaled twice as large as in (b).

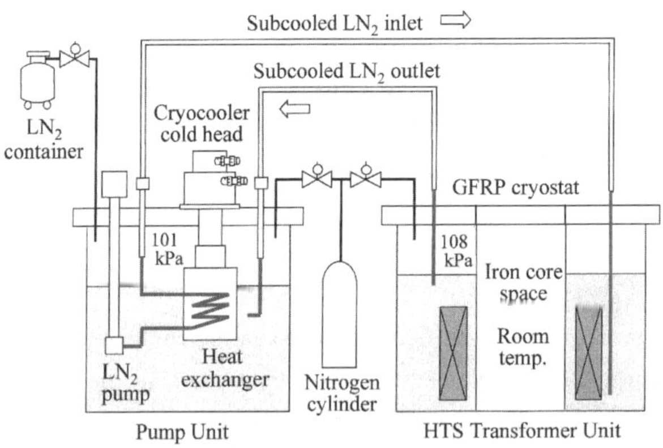

Figure 18 Flow diagram of subcooled liquid nitrogen in the cooling system for the HTS field-test transformer.

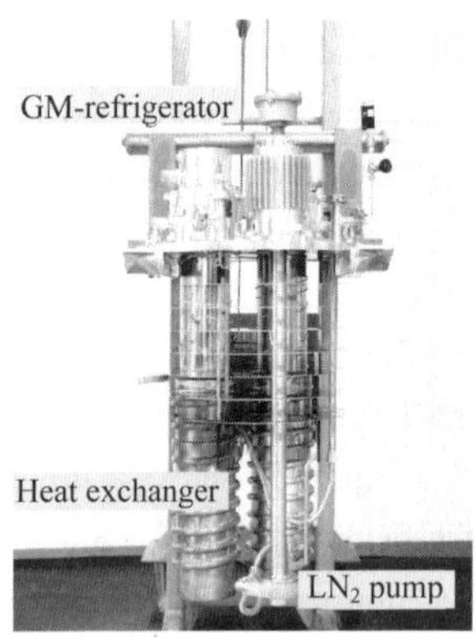

Figure 19 Internal view of the pump unit assembly.

position of 185 mm under the top flange. The 1 atm subcooled liquid nitrogen is transferred to the bottom of the cryostat. Then the warmed-up subcooled liquid nitrogen is returned to the pump unit. The pressure in the HTS transformer unit can be controlled between 101 kPa and 122 kPa by means of nitrogen gas injection and the relief valve. The cryostat pressure is kept 7 kPa higher than the pump unit pressure. This pressure difference makes the subcooled liquid nitrogen in the HTS transformer unit transfer to the pump unit.

The pump unit consists of a stainless steel cryostat, a liquid nitrogen pump and two GM-refrigerators. Figure 19 shows an inside view of the pump unit. The stainless steel cryostat has a height of 1090 mm, an inner diameter of 480 mm and an outer diameter of 580 mm. The liquid nitrogen pump is a centrifugal type and the diameter of the rotor is 68.58 mm. Each GM-refrigerator at 50 Hz operation has a refrigeration power of 145 W at 64 K and a power consumption of 6 kW. Each cold head of two refrigerators is attached with a heat exchanger made of a copper spiral pipe. The cold heads and the heat exchangers are immersed in the liquid nitrogen. The subcooled liquid nitrogen discharged by the pump goes through the copper tubes to be cooled. The subcooled liquid nitrogen is transferred into the HTS transformer unit after the heat exchange. Heaters control the temperatures not to be below 64 K on the cold heads, because liquid nitrogen is solidified at 63 K and 1 atm.

The control panel has a pressure controller, a liquid level indicator, four temperature indicators and a liquid-nitrogen-pump controller. The pressure controller always keeps the pressure in the HTS transformer unit a little higher than an atmospheric pressure. The liquid level indicator shows the level of liquid nitrogen in the HTS transformer unit. The three temperature indicators show temperatures at the top, the middle and the bottom of the HTS transformer, respectively. The other temperature indicator shows the temperature of subcooled liquid nitrogen at the inlet of the pump.

5.4.2 Field test procedure

The HTS transformer system was installed at a distribution power grid of Kyusyu Electric Power Co., Inc. A simplified electrical circuit of the field test is given in Figure 20, where the HTS transformer is connected with the power grid of 6.9 kV through a 6.9 kV/22 kV step-up oil-filled transformer. The over current in a sudden short is restricted at a level five times higher than the rated current by means of the % impedances of the HTS and step-up transformers. We performed short-circuit tests with an individual AC power supply before the connection to the grid. In the connection with the grid, the HTS transformer was operated at 821 kVA for 18 hours and at 586 kVA for 144 hours with the subcooled liquid nitrogen cooling system.

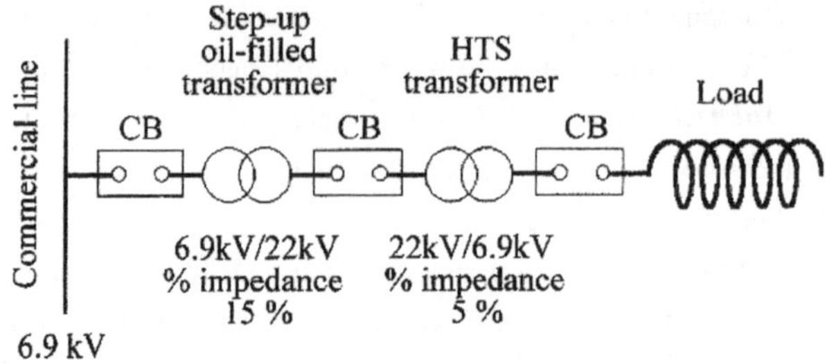

Figure 20 Electrical circuit for field test of the HTS transformer.

The procedure for the field test is in the following way. Both the HTS power transformer unit and the pump unit were initially filled up with 77 K saturated liquid nitrogen, and then the GM-refrigerators and the pump were operated for cooling-down and circulation of liquid nitrogen. Subsequently, subcooled liquid nitrogen was produced in the pump unit and circulated through the system to cool the units down to 66 K. During the system operation, the pressure in the HTS transformer unit was maintained at above the atmospheric pressure by means of the pressure controller. Furthermore, the pressure in the HTS transformer unit was kept at a level 7 kPa higher than the pressure of the pump unit to transfer subcooled liquid nitrogen to the pump unit. In the steady state, the flow rate of the liquid nitrogen in the pump and the HTS power transformer unit was 4 liter/min. In the whole field test, the circulation flow rate was steadily kept constant and there was no evaporation of the circulated subcooled liquid nitrogen in the system.

During cooling down, no current was supplied to the transformer. After the liquid nitrogen temperature in the transformer unit reached 66 K, the transformer was connected with the commercial power grid and the characteristic tests were carried out.

5.4.3 Test results

Figure 21 shows liquid nitrogen temperatures in the HTS transformer unit and the pump unit, and GM-refrigerator cold head temperatures during the field test. It took 12 hours for liquid nitrogen to reach 70 K and 26 hours to reach 66 K, respectively. After 21 hours, the heaters in the refrigerator cold heads started working to keep the cold heads above 64 K.

When the transformer was excited to the rated current equal to 1 MVA for 10 hours in a short-circuit test, the liquid nitrogen temperature increased up to 67.9 K in the HTS transformer unit and to 66.9 K in the pump unit. In Figure 21, it is shown that the HTS transformer was operated at 821 kVA for 18 hours. Liquid nitrogen temperatures in the HTS transformer unit and the pump unit were kept at about 67 K. The HTS transformer was also operated at 586 kVA for 144 hours just after the 821 kVA operation.

The inrush responses were monitored at seven-time connections of the transformer with the grid. While the maximum peak of inrush current attained to a level about 5 times as large as the rated one, the transformer was operated with no quench. An example of the response is given in Figure 22. The transformer was also excited up to an overload current of 1.67 MVA operation only for a short time because the heat generation exceeded the capacity of cooling system.

During the whole field test, the refrigerator cold head temperatures were entirely kept at

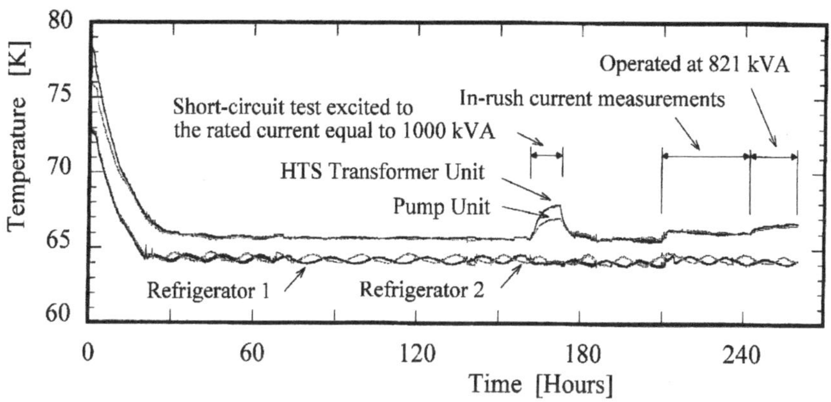

Figure 21 LN_2 temperatures in HTS transformer and pump units and cold head temperatures of two refrigerators at the field test.

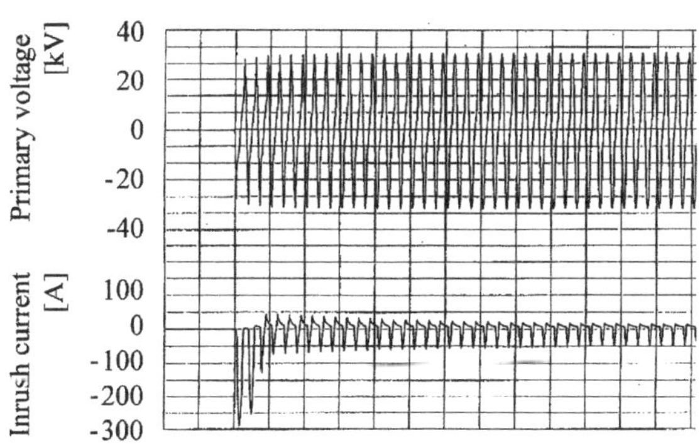

Figure 22 An example of waveforms of the primary voltage (upper) and inrush current (lower) after connection of HTS transformer with commercial line.

64 K except for the over-current operation. Finally, the cooling system was operated continuously for three weeks with no additional liquid nitrogen supply and the liquid level of the subcooled liquid nitrogen in the HTS transformer unit was retained at the position of 145 mm under the top flange.

5.5 Thermal analysis of the cooling system

In advance we performed thermal analysis of the cooling system for the HTS transformer individually. In a set up of heat load tests for the thermal analysis, a stainless steel cryostst, which has an inner diameter of 432 mm, an outer diameter of 476 mm and a height of I 324 mm, was used as a test cryostat instead of the HTS transformer unit. An electric heater set in the test cryostat supplied thermal loads for the cooling system. In the experiments, the temperature distributions of the subcooled liquid nitrogen in the test cryostat and in the pump unit were measured at the heat loadings of 230 W, 250 W, 270 W and 300 W, respectively. Platinum resistance thermometers were used to measure the temperatures and set mainly in a vicinity of the liquid surface.

Figure 23 and Figure 24 show temperature distributions of the liquid nitrogen in the test cryostat and in the pump unit with 50 Hz operation of the GM refrigerators. The liquid nitrogen pump was operated at 35 Hz in Figure 23 and at 50 Hz in Figure 24, respectively. Both in Figures 23 and 24, thermal gradient layers of about 40 mm thick are formed just below the liquid surfaces in the same manner as mentioned in the section 3.5 [7]. The

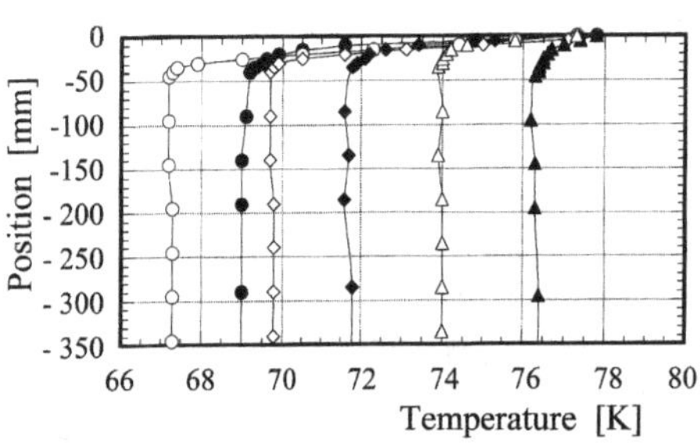

Heat Load	250 W	270 W	300 W
Test Cryostat	—●—	—◆—	—▲—
Pump Unit	—○—	—◇—	—△—

Figure 23 Temperature distribution pattern of liquid nitrogen at 35 Hz LN_2 pump operation. Position 0 mm is the liquid surface.

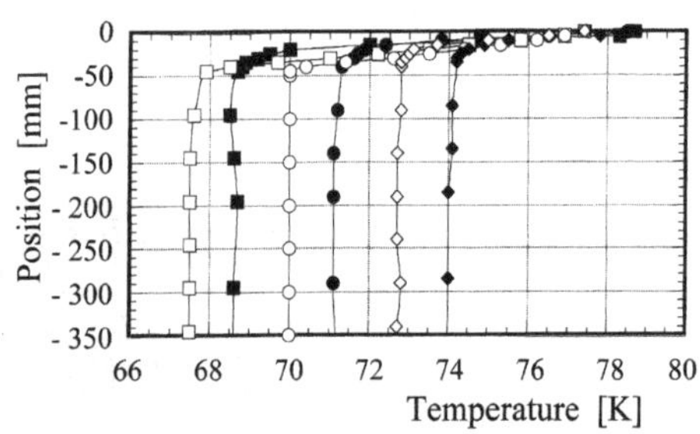

Heat Load	230 W	250 W	270 W
Test Cryostat	—■—	—●—	—◆—
Pump Unit	—□—	—○—	—◇—

Figure 24 Temperature distribution pattern of liquid nitrogen at 50 Hz LN_2 pump operation. Position 0 mm is the liquid surface.

uniform temperature distribution is observed below the thermal gradient layer. It was assured that the HTS transformer could be surrounded by a unified layer of subcooled liquid nitrogen in the cryostat with the cooling system. Practically the HTS transformer was set at a position of more than 150 mm under the liquid surface inside the cryostat in the field test. Temperatures of the bulk liquid nitrogen below the thermal gradient layer depend on the level of heat loads and pump operation frequency. The 50 Hz pump operation makes a larger flow rate of subcooled liquid nitrogen but generates a larger heat. Consequently, the 35 Hz pump operation makes the temperature of the test cryostat lower than the 50 Hz pump operation for the same heat load. Then the 35 Hz pump operation was chosen at the field test.

The preliminary test shows that the cooling system has a refrigeration power of 250 W at 69 K. Considering the individual cooling power 290 W of the two refrigerators, we could know that the thermal load to the pump unit including heat generation by the liquid nitrogen pump was around 40 W. In the field test of the HTS transformer in the connection with 60 Hz power grid in Kyushu Island, the cooling power was enhanced up to more than 300 W, which covered the thermal load in the rated operation of 1 MVA.

5.6 Electromagnetic analysis of the HTS transformer [11]

For the design of the transformer system, we first measured the AC losses in the Bi-2223 strand exposed to AC magnetic field. In practical situation, the strand is exposed to both of alternating magnetic field and alternating transport current. Therefore the AC loss in the condition of the simultaneous application of them should be investigated properly. However the contribution of transport current to the total AC loss is no more than 30% in usual situation, $k' \ll 1$, as shown in Figure A3. Hence we do not take into account the contribution of transport current, for simplicity, in the following. Furthermore we assumed that the additional AC losses due to the formation of parallel conductors are negligible, i.e. the AC loss densities in the windings are the sum of those in the strands, since the transposition should be properly performed for uniform distribution of branch current.

The AC losses in the Bi2223 strand were measured by a saddle-shaped pick-up coil in 77 K saturated and 65 K subcooled LN_2. Straight sample strands with a length of 50 mm were stacked into some layers as shown in ref. 12. The stacked strands of 2 and 6 layers correspond to the situations in the primary and secondary windings, respectively. The sinusoidal magnetic field of 60 Hz was applied in the parallel and perpendicular direction to the wide surface of the stacked strands. Observed AC losses in the strand at 77 K and 65 K are shown for the amplitude B_m of magnetic induction in Figure 25. The breaking points of loss curves in Figure 25 give the penetration field of the strand both for the parallel and perpendicular fields. In the region of larger amplitude than the breaking point, AC losses at 65 K are larger than those at 77 K according to the increment of overall J_c. The AC losses due to perpendicular field are no less than 10 times those due to parallel field for the large amplitude both at 65 K and 77 K. However they decrease with the number of stacked strands for the smaller amplitude than the breaking points. This may result from the effect of demagnetizing field.

The frequency dependence of the AC loss was scarcely observed from 1 Hz to 60 Hz. This means that the AC losses in the strands are almost hysteretic. We compare the measured results with theoretical predictions. The solid and broken lines in Figure 25 represent the theoretically calculated AC losses in a homogeneous infinite superconducting slab with the same thickness as the thickness and the width of the filamentary region of the strand for the parallel and perpendicular configurations, respectively. Here the J_c-B characteristics and the anisotropy of J_c in the strand are taken into account [9]. The observed AC losses in the parallel field are almost explained by the theoretical prediction. On the other hand, while the AC losses in the perpendicular field approach to the theoretical ones in the larger amplitude, it is difficult to predict the AC loss in the smaller amplitude. These results show we need to accumulate the AC losses of strands measured in equivalent local magnetic field to reproduce the entire AC loss in the windings, rather than the theoretical estimation.

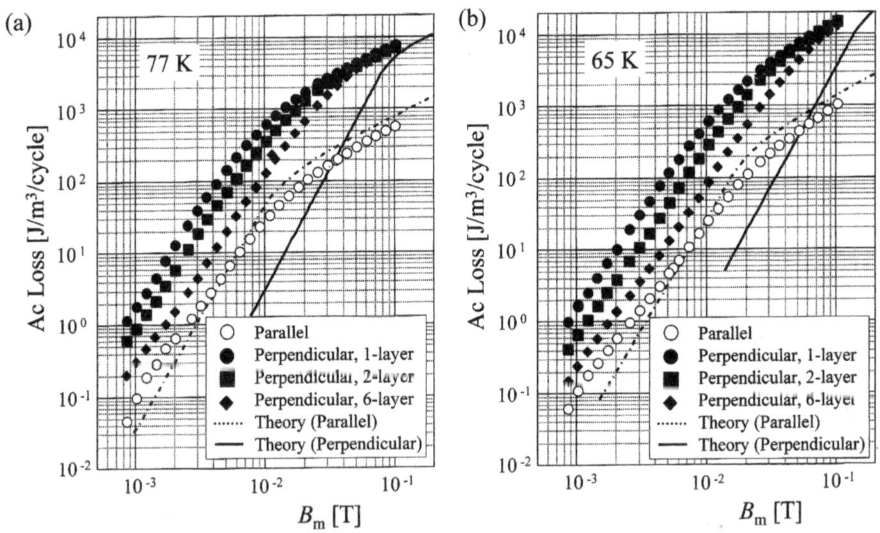

Figure 25 AC losses in the stacked strands in the parallel and perpendicular magnetic field to the wide surface (a) at 77 K and (b) at 65 K.

We numerically calculated the magnetic field distribution in the windings taking into account the effect of an iron core. For the estimation of AC loss, the amplitudes of the parallel and perpendicular components of magnetic field applied to each strand are represented by the maximum ones inside the strand in the windings. The calculated results of both the components in the rated operation are schematically shown in Figure 26 (a) and (b) as contour maps.

The AC loss W in each turn was estimated from the observed results in the stacked straight strands shown in Figure 25 by the following equation

$$W(B_m, \theta, n) = W_{//}(B_m \cos\theta, n) + W_{\perp}(B_m \sin\theta, n) \quad (6)$$

where B_m and θ are the amplitude of local magnetic induction and the angle to the wide surface in the winding of n-stacked strands. In the right side of Equation (6), $W_{//}$ and $W_{\perp}$ are the measured AC losses of n-stacked strands obtained from Figure 25 for the amplitude $B_m \cos\theta$ of the parallel component and the amplitude $B_m \sin\theta$ of the perpendicular one, respectively. The obtained AC loss distribution in the rated operation at 65 K is shown in Figure 27 (a). The AC losses in the rated operation at 77 K and 65 K are listed in Table 3.

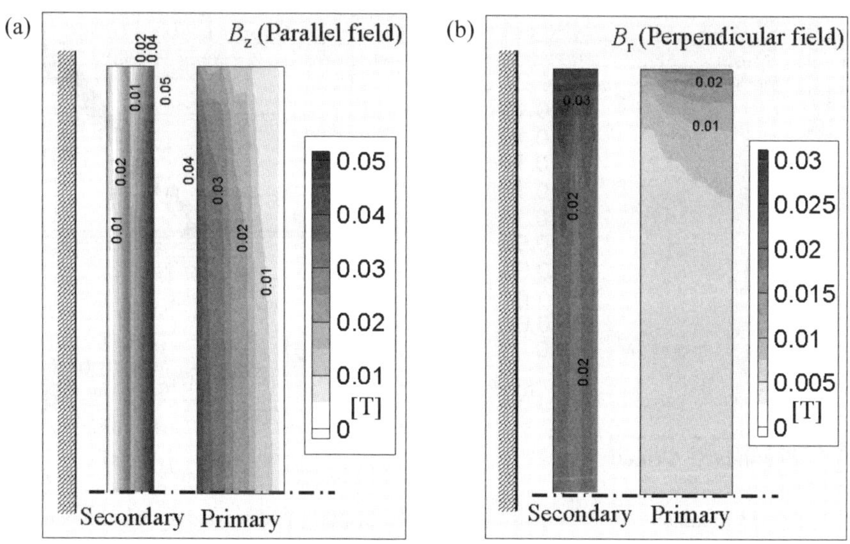

Figure 26 Calculated distributions of (a) parallel- and (b) perpendicular-field components to the wide surface of strands in the primary and secondary windings of the field test transformer at the 65 K rated operation.

We can see that the AC losses due to the perpendicular component are larger than those due to the parallel one. This fact means that the contribution due to the perpendicular component with the smaller amplitude than the breaking point is dominant in the AC loss of the windings.

Table 3 Parallel and perpendicular components of AC losses estimated in the windings of the field-test transformer at the rated operation.

	Calculation			Measurement
	Primary w.	Secondary w.	Total	Total
AC losses at 77 K	48.7 W	93.9 W	142.6 W	205.8 W
parallel component	17.9 W	15.8 W	33.7 W	
perpendicular component	30.8 W	78.1 W	108.9 W	
AC losses at 65 K	43.2 W	79.7 W	122.8 W	158.5 W
parallel component	21.0 W	20.4 W	41.4 W	
perpendicular component	22.2 W	59.3 W	81.5 W	

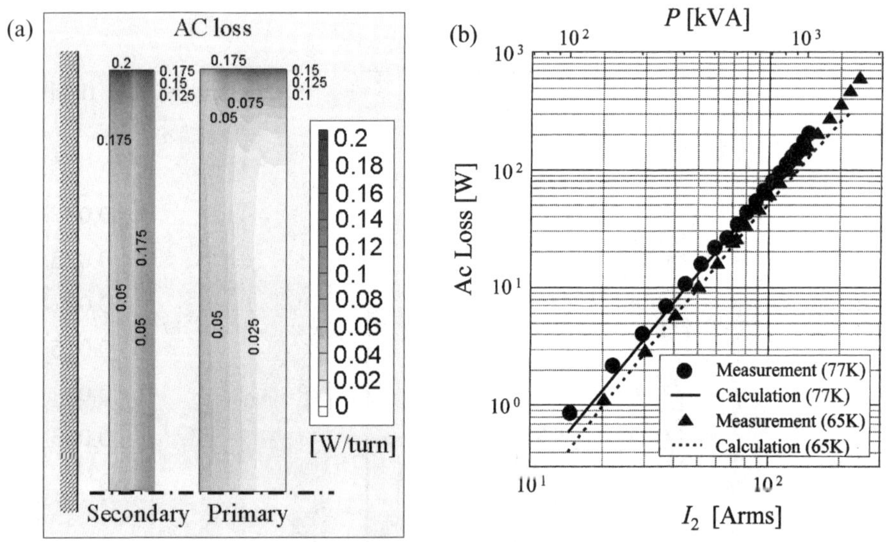

Figure 27 (a) Calculated AC loss distribution in the primary and secondary windings of the field-test transformer at the 65 K rated operation,
(b) The dependence of AC losses in the windings on the amplitude of secondary current at 77 K and 65 K. The upper axis is in the corresponding operation capacity of the field-test transformer.

We also compare the measured AC loss of the windings with the numerical prediction by Equation (6) for the secondary current I_2 and corresponding capacity of the field-test transformer. The results are given in Figure 27 (b). You can see that the observed AC loss in the windings are almost explained by accumulated estimation from the elemental losses measured for the strands except for the region in a vicinity of the rated conditions, where the peak level of the current is comparable to the critical current of the windings. Further considerations on the dynamic resistance loss and some kind of flux-flow loss may clear the discrepancy of 20-30 % around the rated operations.

The above procedure to estimate the AC loss in the windings shows that we can make design of the electromagnetic properties including the AC loss for the HTS transformers from AC losses separately measured in the strands exposed to external magnetic field parallel to and perpendicular to the wide surface.

6. CONCLUDING REMARKS

This chapter shows a series of activities to develop basic technology of future high-T_c superconducting transformers for urban underground substations in Japan by constructing the demonstration and field-test models and conceptually designing the full-scale model. Main topics in the basic technology are related with large-capacity low-AC-loss windings proposed with transposed parallel conductors composed of HTS wires with a large aspect ratio, electrical insulation and continuous cooling methods by (subcooled) liquid nitrogen and positive prospects of the HTS transformers with advanced HTS wires from the viewpoints of the high performances and the economic merits. These activities will lead to the next step for developing prototypes of the HTS transformers with the rated conditions of 66 kV/6.9 kV in the primary/secondary voltage and 10 to 20 MVA in the capacity with views of the following conditions to be established:

i) kA-class windings composed of HTS wires with the low AC losses,
ii) insulation properties of bushings and windings by subcooled liquid nitrogen in accordance with the international standard for insulation tests,
iii) compact cooling systems with subcooled liquid nitrogen, etc.

ACKNOWLEDGMENT

The scientific results summarized in the present chapter were reconstructed with supports by Prof. M. Hara (Kyushu University), Dr T. Uede (Fuji Electric Co., Ltd.), Dr. S. Yoshida

(Taiyo Toyo Sanso Co., Ltd.), Mr. H. Hayashi (Kyushu Electric Power Co., Inc.) and other many coworkers. The authors are grateful for their cooperative efforts.

REFERENCES

1. H. Riemersma et al., IEEE Trans. Power Appar. Syst. 100 [1981] 3398
2. A. Février et al., IEEE Trans. Magn. 24 [1988] 11477
3. K. Funaki et al., IEEE Trans. Magn. 28 [1992] 394
4. K. Funaki et al., Cryogenics 38 [1998] 211
5. M. Iwakuma et al., Proceedings of ICEC16/ICMC Part 2 [1996] 1325
6. I. Hlásnik, Cryogenics 23 [1983] 508
7. K. Funaki et al., IEEE Trans. on Appl. Supercond. 7 [1997] 824,
   S. Yoshida et al., Adv. in Cryog. Eng. 43 [1998] 1191
8. K. Funaki et al., Supercond. Sci. Technol. 13 [2000] 60
9. K. Funaki et al., IEEE Trans. on Appl. Supercond. 11 [2001] 1578
10. S. Yoshida et al., Avd. in Cryog. Eng. 47 [2002] 473
11. M. Iwakuma et al., IEEE Trans. on Appl. Supercond. 11 [2001] 1482
12. M. Iwakuma et al., Adv. in Supercond. X [1997] 833
13. T. Ogasawara et al, Cryogenics 16 [1976] 33, 89

APPENDIX A [4]

While the Bi-2223 tapes listed in the left column of Table 1 have a multifilamentary structure for a react & wind procedure, their electromagnetic properties may be equivalent to those of monofilamentary ones because the filaments are not twisting. We shall consider a sufficiently long single-layer solenoid coil wound by a parallel conductor of the n monofilamentary tapes with a thickness d. All strands have the same average density of transport current as each other. Considering n is the total number of strands stacked in the radial direction of the coil, we can see that the following estimation is also useful for the multi-layer parallel conductors, where n is equal to the product of the strand number of the parallel conductor and the layer number of the coil. We also approximate the parallel conductor in the long coil by n layers of coaxial superconducting hollow cylinders with a thickness d. The radial distribution of axial magnetic induction in the cylinders are indicated in Figure A1, where the alternating transport current is in an increasing half cycle. We describe the distribution of magnetic flux by means of the critical state model with a constant value λJ_c of overall critical current density. We can calculate AC loss per cycle of the transport current by integrating a Poynting's vector both on inner and outer surfaces for each hollow cylinder over the cycle. The AC loss density calculated in each layer is given by

$$W_1 = \frac{2}{3}\mu_0 k^3 (\lambda J_c d)^2 \qquad \text{for the 1st layer} \tag{A1}$$

$$W_i = \frac{\mu_0}{8}\left(\frac{k}{k_i}\right)^2 (\lambda J_c d)^2 \{\frac{16}{3}kk_i + 4(1 - 2kk_i + k_i^2)\alpha_i - (2 - 5kk_i + 2k_i^2 + \frac{k}{k_i})\alpha_i^2$$
$$+\frac{2}{3}(-kk_i + \frac{k}{k_i})\alpha_i^3\} \qquad \text{for the } i\text{-th layer}; \quad i = 2, 3, \ldots, n \tag{A2}$$

$$\alpha_i = 2\frac{k_i}{k}\frac{1-k}{1-k_i}, \; k_2 = \frac{1}{3}, \ldots, k_n = \frac{1}{2n-1} \tag{A3}$$

where k is the ratio $J_{tm} / \lambda J_c$ for the amplitude J_{tm} of transport current density. Especially, Equation (A2) is composed of magnetization loss and dynamic resistance loss [13]. We obtain the total loss power of the windings

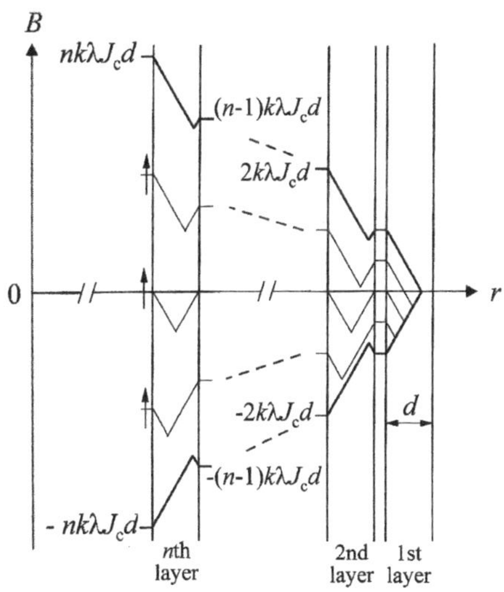

Figure A1 Radial distribution of axial magnetic induction in n-layer coaxial superconducting hollow cylinders with uniform transport current among the layers.

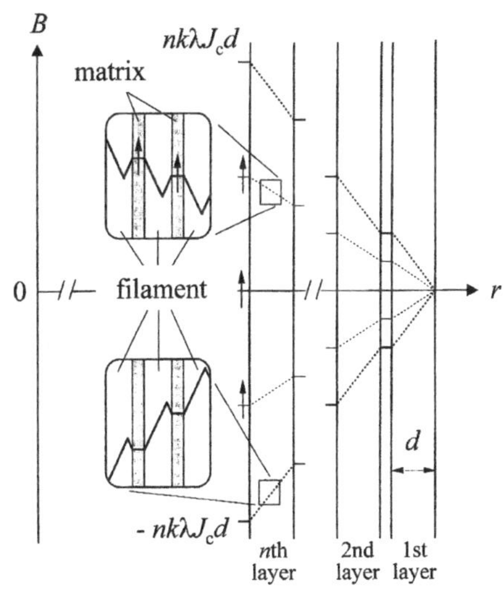

Figure A2 Radial distribution of axial magnetic induction in a single-layer coil wound by a parallel conductor of n multifilamentary tapes with uniform transport current among the strands and among the filaments inside each strand.

$$P = \frac{1}{n} f V \sum_{i=1}^{n} W_i \tag{A4}$$

where f is the frequency and V is the whole volume of the strands.

APPENDIX B [4]

We shall consider a sufficiently long single-layer solenoid coil wound by a parallel conductor of n advanced tape-shaped strands with a thickness d. All strands have the same branch current as each other and for simplicity, the current distribution is also uniform among filaments inside the strand. We also approximate the distribution of magnetic flux in the parallel conductor in the same manner as Appendix A. In this case, however, each strand has multifilamentary behaviors, where the strands are isolated and twisted. The radial distribution of axial magnetic flux in the conductor is illustrated in Figure A2. In this figure, dashed lines in the layers show an average flux distribution and the inserts give the detailed profiles in the multifilamentary structure.

The total AC loss in the conductor may be composed of a pinning loss in the filaments and a coupling loss among the filaments. As can be seen from comparison between the profiles of flux distribution in the inserts and in Figure A1, the pinning loss in the filaments is also a sum of magnetization loss and dynamic resistance loss. Since the exact calculation brings only lengthy expressions of AC loss, we approximate the pinning loss by the magnetization loss that is a dominant component for parallel conductors with many strands, as shown in the following. For simplicity, we shall calculate the ratio p_r of the magnetization loss to the pinning loss in a superconducting slab with a thickness d_s indicated in the insert of Figure A3. In this case, the obtained result is given by

$$p_r = \frac{k}{4k'}\left(1-\frac{2k'}{3k}\right)^{-1}\left\{\frac{16}{3}kk' + 4(1-2kk'+k'^2)\alpha - (2-5kk'+2k'^2+\frac{k}{k'})\alpha^2 + \frac{2}{3}(-kk'+\frac{k}{k'})\alpha^3\right\} \tag{B1}$$

$$\alpha = 2\frac{k'}{k}\frac{1-k}{1-k'}, \quad k' = \frac{\mu_0 J_{tm} d_s}{2B_{em}} \tag{B2}$$

where J_{tm} and B_{em} are the amplitudes of transport current density and magnetic induction outside the superconductors, respectively. Figure A3 shows contours of the ratio p_r for k and k'. In the parallel conductor with many strands, namely in the region of $k' \ll 1$, error from the above approximation is rarely over about 30 % even at $k = 1$.

Consequently, the total AC loss power density in the parallel conductor of multifilamentary strands can be approximated as the sum of the magnetization loss power density P_h and the coupling loss power density P_c as follows;

$$P_{SC} = P_h + P_c \tag{B3}$$

$$P_h = \frac{1}{n}\sum_{j=1}^{n}\frac{2}{\mu_0} f B_{mj} B_{fp}\left(1 - \frac{2B_{fp}}{3B_{mj}}\right) \tag{B4}$$

$$P_c = \frac{1}{n}\sum_{j=1}^{n}\frac{(2\pi f)^2 \tau_s}{1+(2\pi f)^2 \tau_s^{\,2}}\frac{B_{mj}^{\,2}}{\mu_0} \tag{B5}$$

$$B_{mj} = (2j-1)\,k\mu_0\lambda J_c d/2 \tag{B6}$$

$$B_{fp} = \mu_0\lambda J_c d_f/2 \tag{B7}$$

$$\tau_s = \mu_0\sigma\left(\frac{L_s}{2\pi}\right)^2 \frac{1+\lambda}{1-\lambda} f(\beta) \tag{B8}$$

$$f(\beta) = \frac{\pi^2}{120}\frac{\beta^2(7\beta^2+20\beta+35)}{(1+\beta)^2}, \quad \beta = \frac{d}{w} \tag{B9}$$

In the above equations, d and w are the thickness and the width of the tape, and d_f and L_s are the thickness and the twist pitch of the HTS filament, respectively.

In the same manner as in Appendix A, the above result is introduced to the winding of the multi-layer parallel conductor, where n is the product of the strand number of the parallel conductor and the layer number of the coil.

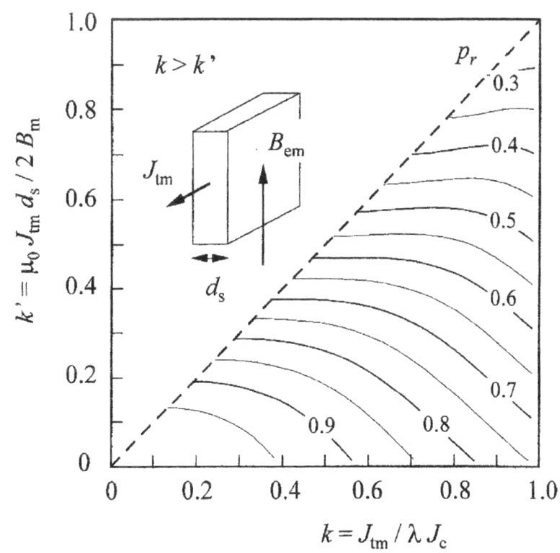

Figure A3 Ratio p_r of magnetization loss density to total pinning loss density in a superconducting slab simultaneously exposed to alternating parallel magnetic field and transport current.

APPLICATIONS OF HIGH-Tc SUPERCONDUCTORS TO SUPERCONDUCTING MAGNETIC ENERGY STORAGE (SMES)

Takakazu SHINTOMI
High Energy Accelerator Research Organization
1)Oho, Tsukuba-shi, Ibaraki 305-0801, JAPAN

1. INTRODUCTION

The basic concept of Superconducting Magnetic Energy Storage (SMES) was proposed by Irie and Yamafuji in 1969 [1]. The concept vs that electric energy col store into a superconducting coil without energy loss because the resistivity .s zero. R. Boom and H. Peterson at University of Wisconsin proposed the SMES system for use in electric power networks by combining a superconducting coil and an ac-to-dc converter in 1972 [2]. Their group had enthusiastically developed the concept of the large scale SMES as an alternative storage system of the hydro-pumped storage stations which are served in practice for diurnal electric energy storage. After their proposal, many researches have been carried out not only on conceptual design [3, 4] but also on development of components. Some of them were proposed to be installed as practical use [5, 6]. The recent plan is a small scale SMES to be installed for a power compensation in Florida [7].

The concept of the SMES system is very simple that the electric energy can be stored in a superconducting coil with small amount of energy losses. Since electric energy can be stored in the system directly without conversion into any other type of energies such as mechanical or potential energies, the SMES system has advantages of relatively high efficiency and fast response in comparison with other energy storage systems. The energy losses occur in keeping a superconductng coil in a cold stage and in the converter which is used for transferring energy between the coil and electric power networks. The efficiency had been estimated as more than 90 % when it is a large SMES system to be used as diurnal storage. However, the scale of the SMES system for diurnal storage is to be hundreds meters in diameter, and a lot of technology breakthroughs have to be established. These technologies have to be developed from a small scale step by step. According to those research activities, the micro-SMES using low Tc superconductors has recently been introduced in practical use [8]. Those micro-SMESs are used as UPS or network stabilization in combination with several units in vol. A. Nagaoperation.

After the discovery and development of high Tc superconductors in 1986, these materials are considered to have a possibility to be applied to SMES and it is one of the interesting research items for SMES. Researches with design works and model coils using high Tc materials have been carried out [9 – 12]. Three possible materials, Bi2212, Bi2223 and YBCO have been investigated, and the materials, Bi2223 and YBCO, have potential to be used for SMES coils. They have merits in operation at 20 K or higher temperatures and higher magnetic field as compact systems. Another advantage will be obtained when they will be operated in a direct-cooling scheme.

In this section, the possible applications of high Tc materials to the SMES system will be discussed and described.

2. SMES AND ITS APPLICATIONS

The concept of the SMES system is as shown in Fig. 1. A superconducting coil is connected to electric power networks through an ac-to-dc converter. During energy charging into the coil, the converter is operated as an ac-to-dc converter to increase the coil current. On the other hand, the converter is operated as a dc-to-ac inverter for energy discharge from the coil. The operation for the converter or the inverter can be much easily controlled by changing phase angles of solid-state switching elements. The converter system is well established and commonly used in dc power transmission systems, the frequency conversion stations and other many equipments. The superconducting coil is kept at low temperature using a refrigerator.

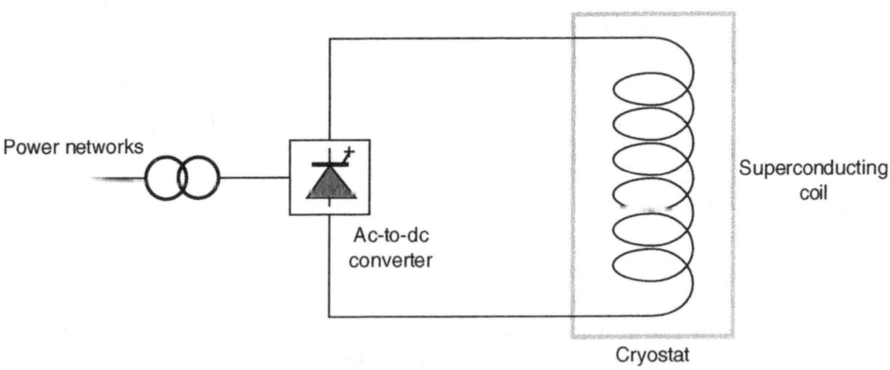

Fig. 1 The basic configuration of the SMES system. An ac-to-dc converter is connected between a superconducting coil and ac power networks. The converter can control electric energy transfer between them.

The stored energy E in the superconducting coil is given as $E = L\,i^2$, where L is the inductance of the coil and i is its current. The stored energy of the coil varies by changing the coil current, and it is easy to control not only the active power but also the reactive power by controlling the ignition phase of the converter since the energy is stored inductively.

The type of SMES is distinguished such as the micro-SMES, the small-scale ones, the medium-scale and the large scale ones by the scale of the storage capacity and the output power as shown in Fig. 2. The micro-SMES has been in practical use for UPS and compensation of voltage fluctuation. The small-scale SMES is to be used for stabilization of power networks, load fluctuation of relatively short duration, and frequency control of power networks. The medium-scale SMES will be installed for the dispersed load leveling such as steel mill factories and multi-purpose use. The large-scale SMES will be developed as a diurnal storage system for an alternative use of the hydro-pumped storage. The development of the SMES system has to be achieved step by step from the smaller-scale to the larger-scale.

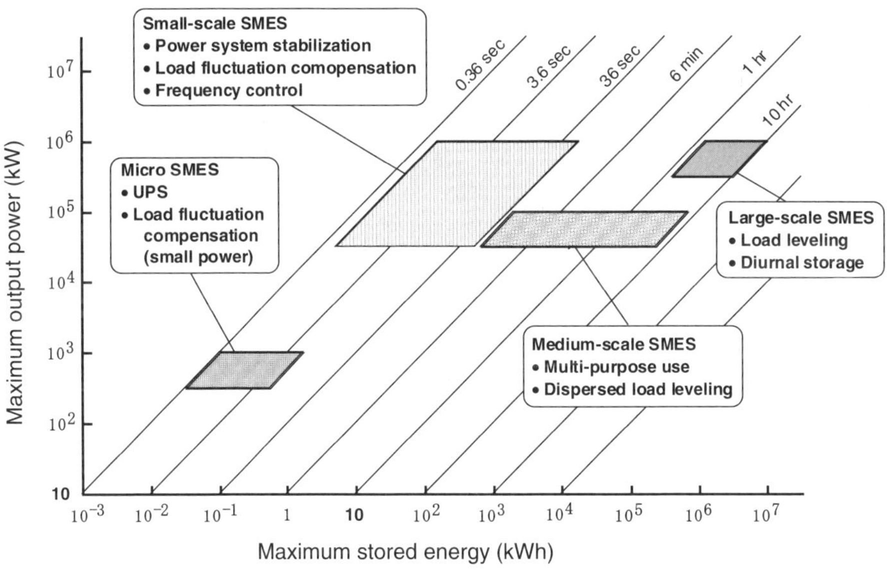

Fig. 2 The type of SMES distinguished by the scale of the storage capacity and the output power.

3. STATUS OF NATIONAL R&D PROJECT FOR HIGH TC SMES

The national project of basic researches to develop the high Tc SMES is being carried out in Japan as a 5-year project from 1999 to 2003 [13]. The project is managed by the New Energy and Industrial Technology Development Organization (NEDO) under control of the Ministry of Economy, Trade and Industry. The researches are feasibility studies on the high Tc SMES by conceptual design study, experimental study on properties of high Tc materials in a view point of practical use, basic researches with model high Tc coils. These investigations are directed toward the high field and compact SMES in consideration of the advantage of high Jc properties of the materials in higher magnetic field regions and also directly-cooled coil systems with refrigerators in higher temperature regions.

The main research items of this national project are as follows:

1) Conceptual design study of high Tc SMES system (1999 - 2000), and
2) Basic technology development with high Tc superconducting model coils (2001 - 2003).

In these research activities the following technologies will be clarified:

1) Optimization of the high Tc SMES coil configurations,
2) Assessment on superconductors for applying to the high Tc SMES,
3) Application of directly-cooled coils to the high Tc SMES,
4) Evaluation of electro-magnetic properties of the high Tc SMES with model coils,
5) Possible development of superconducting cables with large current carrying capacities, and
6) Stability of the high Tc SMES by over-current operation.

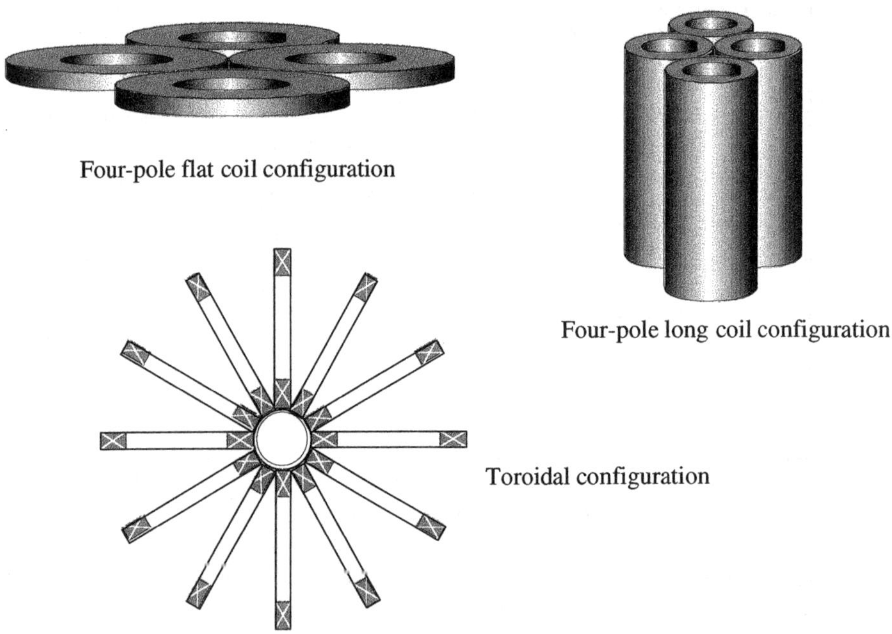

Fig. 3 Evaluation of coil configurations: (a) four-pole flat coil, (b) four-pole long coil, and (c) toroidal coil configurations.

Three types of coil configurations are evaluated taking into considerations of an economical point of view and smaller stray field as shown in Fig. 3. They are a four-pole flat coil configuration, a four-pole long coil one and a toroidal one. In the study, the stored energy of each coil was fixed as 20 kWh. The four-pole long coil configuration has merits of small occupation area and small coil volume. However, in a view point of small magnetic field leakage the toroidal coil configuration has the best advantage. These coil configurations will have a possibility to be adopted in the high Tc SMES system.

Three types of the high Tc superconductors, Bi2212, Bi2223 and YBCO, are taken into consideration [14]. The superconductors, Bi2212 and Bi2223, were experimentally investigated with five small model coils. These coils were focused on targets which were mechanical strength, high stability, small ac-loss, high stability with direct-cooling, and high pulse current operation. By these investigations, the followings were figured: (1) both superconductors, Bi2212 and Bi2223, can be used for direct-cooling coils, (2) Bi2212 should be used below 20 K, and (3) Bi2223 can be used for coils operated at around 50 K. Although the YBCO conductor is still under development to be used for SMES coils, it is a desirable candidate because of its better Jc characteristics as shown in Fig. 4. The critical current density of the superconductor of around 10^4 A/mm^2 was obtained over 20 T [15], and the property is relatively insensitive to magnetic field in operation below 50 K.

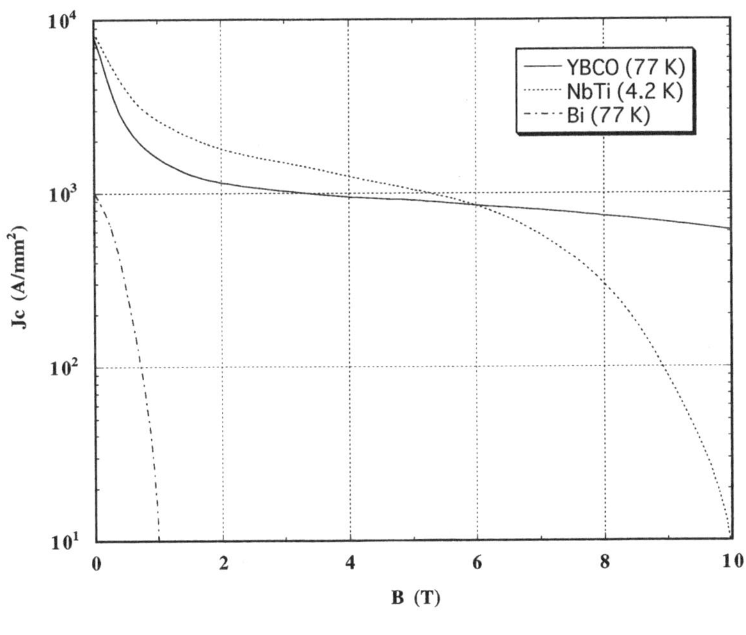

Fig. 4 Characteristics of critical currents for superconductors.

In the operation of the SMES system, thermal stability, ac losses and coil protection at quench are the most concerns. In case of the SMES system with high Tc superconductor, it is to be operated around 20 K or higher by making the most of the characteristics of the superconductors. By the investigations, the following results were obtained: (1) overrated operation, that is with over-current of two times of the critical current, is possible when the operation duration is not long, (2) the directly-cooled SMES is possible, and (3) compact SMES coil systems are available using better Jc performance of the high Tc superconductors in a high magnetic field region.

The high Tc SMES has merit that it can be operated at higher temperatures over 20 K. For SMES with the YBCO superconductors, it is expected to be operated at 50 K to 60 K. In case of operation at much higher temperatures than the temperature of liquid helium, it is favorable for the high Tc SMES to be cooled not with liquid helium but with refrigerators. One of the possible configurations might be a directly-cooled coil with distributed small capacity refrigerators. The small-scale SMES system less than 100 kWh as shown in Fig. 2 may come into practice as the next step of the micro-SMES. The recent development and applications of the small-scale refrigerators in fields such as cryo-pumping equipment will make such configuration in practical use for the small-scale SMES system with high Tc materials. Moreover, many directly-cooled superconducting magnets have been recently developed, and the techniques developed for such magnets will help to develop the directly-cooled high Tc SMES using small refrigerators as shown in Fig. 5.

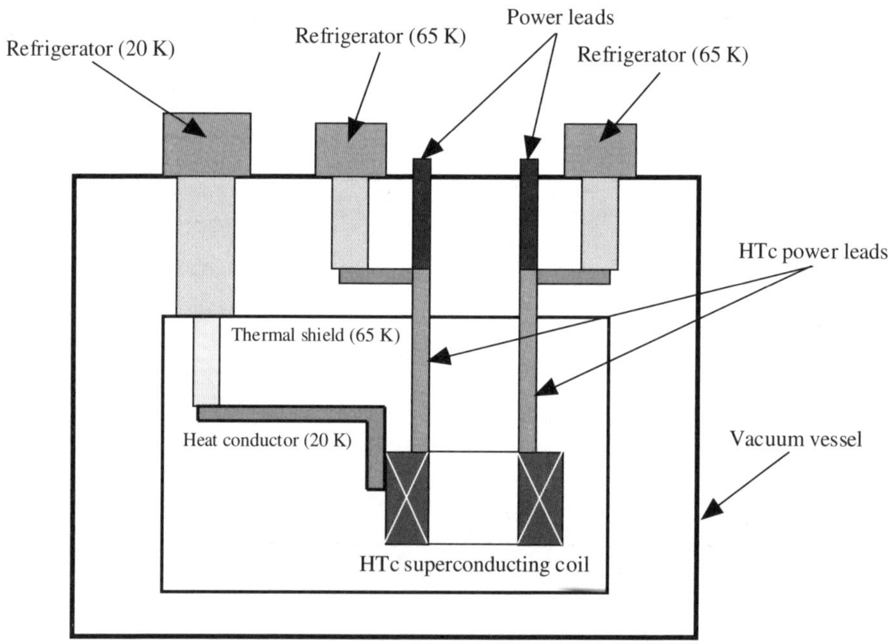

Fig. 5 The concept of directly-cooled SMES coil.

In fiscal year 2002, the model coil as shown in Fig. 6 was made in order to investigate electro-magnetic properties and thermal stability of the high Tc superconducting coil. The main parameters are given in Table 1. By the investigation with the model coil, it is clarified that the high Tc superconducting materials have a potential to be used for the high Tc SMES, the directly-cooled coil can be accepted and the high Tc superconducting coil has relatively large transient current capacity. However, the following investigations are mainly pointed out in order to make the high Tc SMES in practical use: (1) establishment of optimized design method, and (2) development of large current capacity superconductors.

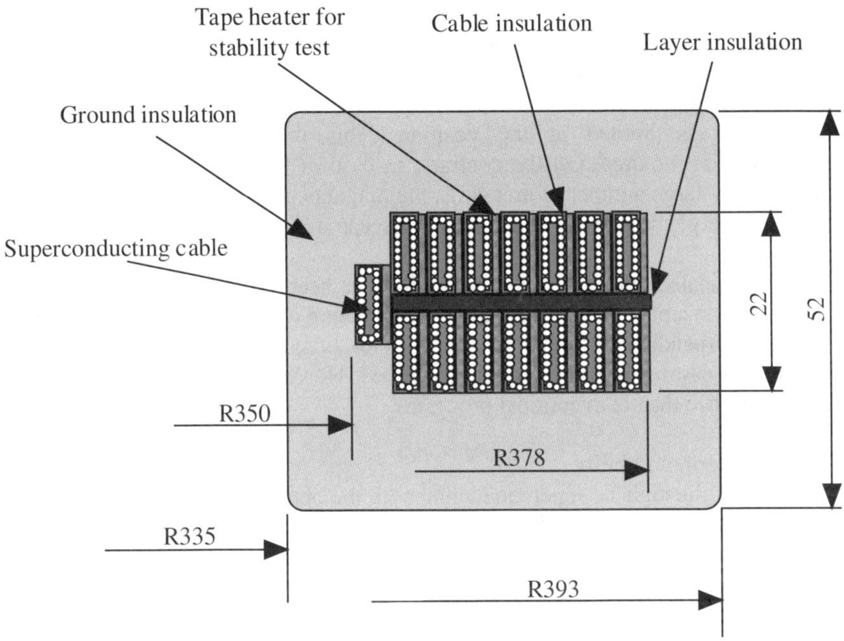

Fig. 6 Model coil for investigating the electro-magnetic properties of high Tc coil for SMES use.

Table 1 Main parameters of model coil to investigate electro-magnetic properties of the high Tc superconducting coil.

Material of superconductor	Bi2212 with NiCr reinforcement
Number of strand	16
Coil inner diameter	670 mm
Coil outer diameter	786 mm
Winding inner diameter	700 mm
Winding outer diameter	756 mm
Number of turns	14
Inductance	0.32 mH
Rated current	3000 A
Magnetic field at coil center	0.073 T

4. NECESSARY TECHNOLOGIES TO BE DEVELOPED

4.1 Anisotropy of superconductor and coil configuration

In case of the high Tc superconductor, the anisotropy of the superconductor has to be considered in the design of the coil system configuration. A few coil configurations, a toroidal type and a multi-pole solenoid, have been proposed for the SMES system. The magnetic field direction is not uniform in the solenoid. The critical current of the coil is limited at the portion of the solenoid where the magnetic field is oriented from the perpendicular direction to the c-axis of the high Tc superconductors. The magnetic field of the solenoid is inclined from the perpendicular direction to the c-axis, and the critical current is limited at the position. This means that more amount of superconductor is required. On the contrary, in case of the toroidal coil configuration with a relatively large number of unit coils, the magnetic field of the coil becomes more uniform and perpendicular to the c-axis. This can reduce the necessary amount of superconductor.

A sophisticated concept of a coil winding has been proposed for a solenoid type coil [16]. The concept is that a high Tc superconducting cable is wound by taking the c-axis always perpendicular to the magnetic field direction of a solenoid. Technical fabrication processes should be further developed. However, in case the coil can give better performance than conventional processes.

4.2 Transient thermal stability

In case of the high Tc superconducting coil, the operation temperatures will be 50 K for Bi2223 and around 50 to 60 K for YBCO. When the high Tc superconducting coil becomes a critical state, its thermal run-away is relatively slow in comparison with the case of conventional low Tc superconducting coils because of the larger heat capacity at their operation temperatures than those of the low Tc coils. In some case, the high Tc superconducting coil can be operated with overrated current. The run-away time constant, of course, depends on the configuration of superconductors used. The logarithmic increase of the heat capacity of the coil increases the transition time of the superconducting coil at quench. It is much easier to protect the coil from damage. When the coil turns to a normal state and it is detected, there is enough time to decrease the current and turn off a power supply.

4.3 Ac losses

The ac losses of superconductor used are a great concern for SMES. In case of the UPS use, ac losses are not severe because a phenomenon is rare and one shot like a delta function. However, when the SMES is used for the network stabilization or for pulse power compensation, ac losses cause not only heat load of a refrigerator but also unstable operation of SMES. The high Tc superconductors have usually a tape shape and the ac losses depend on the orientation of the tape-shape superconductors to alternate magnetic field. The difference of the ac losses between perpendicular and parallel to the wide surface of the tape in the subjected magnetic field becomes sometimes one order. The cautious estimation of ac losses and technique to reduce them are quite important for the high Tc SMES in practice.

4.4 Mechanical properties

The high Tc superconductors are very brittle like A15 superconductors. The Bi2223 superconducting tape, for example, can only stand the mechanical properties under stress of 200 MPa at 77 K without any Jc degradation even though Ag is strengthened with Ag - 0.3 % Sb/Ag - 0.5 % Mn. The careful mechanical analysis and design are required for the high Tc SMES, especially for the high field and compacted SMES.

5 SUMMARY

The SMES using high Tc superconductor is interesting and may have a potential to be used in practice, especially for the micro-SMES and small-scale SMES. In order to develop the high Tc SMES, the following technical items should be investigated:
1) optimization of coil configurations,
2) direct cooling technologies,
3) development of high Tc superconductors with low ac losses,
4) development of large-scale high Tc superconductors, and
5) investigation on thermal stability.

References

1. F. Irie and K. Yamafuji, Inter. Institute of Refrigeration, Low temperatures and Electric Power, March 1969, p. 411
2. R. W. Boom and H. A. Peterson, Superconducting Energy Storage for Power Systems, IEEE Trans. Magnetics, vol. MAG-8, 701 (1972)
3. Conceptual Design and Cost of a Superconducting Magnetic Energy Storage Plant, EPRI EM-3457, April 1984
4. Research on Superconducting Magnetic Energy Storage System, NEDO-P-8408, June 1985 (in Japanese)
5. R. L. Verga, Superconducting Magnetic Energy Storage and Other Large-Scale SDI Cryogenic Applications Program, Adv. Cry. Eng., vol. 35, 555 (1990)
6. X. Huang et al., 30 MWBabcock and Wilcox SMES Program for Utility Applications, IEEE Trans. Applied Superconductivity, vol. 5, 428 (1995)
7. C. A. Luongo, A 100 MJ SMES Demonstration at FSU-CAPS, presented at 2002 Applied Superconductivity Conference, Houston, 2002
8. Web site of American Superconductor Corp.: www.amsuper.com
9. J. H. Kim et al., Design of a 200-kJ HTS SMES System, IEEE Trans. Applied Superconductivity, vol. 12, 774 (2002)
10. H. Kasahara et al., Basic Characteristic Evaluation of Cryocooler-Cooled HTS Coils, IEEE Trans. Applied Superconductivity, vol. 12, 766 (2002)
11. A. Friedman et al., HT-SMES Operating at Liquid Nitrogen Temperatures for Electric Power Quality Improvement Demonstrating, presented at 2002 Applied Superconductivity Conference, Houston, 2002
12. A. Nagaya et al., Study of a High-Tc Superconducting Magnet Made with Bi-2212/Ag Rutherford Cable, IEEE Trans. Applied Superconductivity, vol. 11, 1761 (2001)

13. For example, NEDO report on technical development of SMES system in 2001 (in Japanese), 2001
14. L. J. Masur and J. Kellers, Development of Long Length High Temperature Superconductor Wire for Electrical Power Applications, presented at 2002 Applied Superconductivity Conference, Houston, 2002
15. Y. Iijima et al., IEEE Trans. Applied Superconductivity, vol. 3 (1993) 1510
16. F. Sumiyoshi, private communications

UNRIVALLED SENSITIVITY -
SQUIDS IN NONDESTRUCTIVE TESTING

Marc von KREUTZBRUCK
Institute of Applied Physics, Justus-Liebig-University Giessen
Heinrich-Buff-Ring 16, 35390 Giessen, Germany *

1. INTRODUCTION

For many years extremely sensitive magnetic flux measurements were of little or no concern to the field of nondestructive testing (NDT), where established magnetic testing methods relied on conventional sensors such as induction coils, hall probes or more lately magnetoresistive sensors. But recently, increasing demands in materials testing have led to growing calls among industrial users for advanced, highly sensitive and reliable testing methods. It was the advent of high temperature superconductivity that paved the way for a new niche in the NDT market - and after more than a decade of research, a magnetic flux sensor of superior sensitivity is poised to take its position in this emerging market: The Superconducting Quantum Interference Device - or SQUID.

The SQUID's outstanding sensitivity is based on the physical effects of flux quantization and the Josephson Effect. The device is made up of a loop of superconductor, interrupted by either one Josephson junction (radio-frequency SQUID) or two (dc SQUID). The application of an external magnetic field results in a voltage change across the SQUID loop. In this way, the SQUID acts as a flux-to-voltage transducer. This voltage response is periodic in the flux quantum $\Phi_0 = h/2e = 2.07\cdot10^{-15}$ Wb and usually is linearized by means of a feedback circuit. A resolution of more than a millionth of the flux quantum can then be detected, making it possible to measure a magnetic field as minute as fT/√Hz, i.e. more than 100 billion times weaker than the earth's magnetic field.

* This work is dedicated to the memory of Christoph Heiden, who initiated the German SQUID NDE Consortium, including aircraft industry partners such as EADS Airbus and Lufthansa Technik AG.

The development of the first SQUID in the mid-1960s promised to open the door to a series of ground-breaking applications. The technology entered the field of clinical research as early as the late 1970s, where it proved invaluable in a range of biomedical applications - first and foremost in measuring magnetic fields of the human brain (magnetoencephalography, MEG), as well as those of the heart (magnetocardiology, MCG) and other biomagnetic fields in the human body, including those generated by peripheral nerves and the liver. In the early 1980s, SQUID technology was also introduced in geophysical research. Today, SQUIDs are commonly used in geomagnetic prospection methods such as borehole sounding or magneto-tellurics. With the discovery in 1986 of high-temperature superconductivity, the general focus of global research shifted to developing and implementing SQUIDs made of high-transition-temperature (high-T_c) materials - in most cases YBCO. These high-T_c SQUIDs appeared to solve a problem that had prevented the technology from being embraced by industry: contrary to low-transition-temperature (low-T_c) SQUIDs, which operate at liquid helium temperature (4.2 K), high-T_c SQUIDs are cooled by liquid nitrogen (77 K), a cheaper and widely available coolant. The prospects for wide-scale industrial implementation sparked a boom in high-T_c research activities that led to a series of proposals for the SQUID's use in the field of nondestructive evaluation (NDE). It should, however, be mentioned that two pioneers in SQUID NDE research, H. Weinstock and G.B. Donaldson, first suggested the use of SQUIDs in NDE in 1985, a year before the discovery of high-T_c superconductivity [1, 2]. But the systems and cooling environment used in the early days of SQUID NDE research were still a long way from meeting industrial requirements, and it took several years before the technology developed to the point that it could be used in industry.

The main benefit SQUIDs offer in NDE is their high field sensitivity, which they maintain even at low frequencies of just a few Hz. This makes this field sensing device suitable for all NDE applications in which minute material defects are either directly detected by measuring a sample's magnetic field or indirectly located by measuring the response field of currents generated inside the sample. Possible material damage could include fatigue cracks in conductive samples, deep lying defects, corrosion, small magnetic and non-magnetic inclusions, a variation in magnetic properties due to mechanical stress, etc.. In addition to its high field sensitivity, the SQUID offers two further major advantages; high spatial resolution and outstanding dynamic range. A series of improvements have served to enhance these features: they include a reduction in sensor sizes to boost spatial resolution, as well as advanced read-out electronics which offer better dynamic range, improved slewrate and wider bandwidth. These developments have made the SQUID a candidate for an increasing number of NDE applications. They have also helped to push research beyond the early academic stages, when demonstrations of the technology's potential were limited to the detection of artificial flaws sawed into aluminum plates or measurements conducted in electromagnetically shielded environments.

SQUID NDE has advanced considerably in recent years, both in industry and in the field of materials research, thanks to the achievements of a number of work groups around the world. Their cooperation with industry has been particularly important, not least because it has provided insight into industrial requirements. Because of the bulky cooling equipment in which they are encased, SQUIDs are unwieldy and therefore often not attractive to industrial users. This particular problem has served to widen research from a general focus on the development of the SQUID itself to the search for unobtrusive or adapted cooling techniques. Low-vibration pulse tube coolers, tiny, mobile liquid nitrogen cryostats, and SQUID-microscopes are three examples of developments that have made it possible to implement SQUID sensors in sophisticated industrial NDE applications.

As much as the SQUID's high field sensitivity has advantages, there is an inherent drawback in that it also picks up all types of environmental noise and background clutter. In order to transfer the SQUID's high sensitivity into improved testing quality, several factors need to be taken into account; for example, correct positioning of the sensor in strong excitation field, a low vibration of the system, sophisticated low-noise electronic components as well as additional noise-reducing software when background noise levels are particularly large.

This overview looks at the current prospects for the implementation of SQUID in non-destructive testing. The focus is on SQUID NDE and on its applications in eddy current testing (ECT) in particular. Section 2 outlines SQUID fundamentals and the basics of EC SQUID NDE. Section 3 explores the possibilities presented and limitations posed by the use of SQUIDs in NDE. Section 4 starts with a short review of past NDE activities followed by a detailed description of three recent SQUID NDE applications.

A number of outstanding articles offer valuable insight into the development of SQUIDs in NDE. H. Weinstock [3] and G.B. Donaldson [4] were the first to review progress in the area. For a deeper look at SQUID fabrication and applications, the book "SQUID Sensors" edited by H. Weinstock includes several very informative reviews by leading scientists in the field [5]. The 1997 publication "SQUIDs for Nondestructive Evaluation" by W.G. Jenks and J.P. Wikswo, Jr. [6] is highly recommendable. The many articles by J. Clarke [7, 8] are excellent resources for a broader view of SQUID technology and its different applications.

2. FUNDAMENTALS AND THEORETICAL ASPECTS OF SQUID NDE

We begin with a short review of the principal function, fabrication and operation of a SQUID sensor for readers unacquainted with SQUID technology. For detailed descriptions I refer readers to the comprehensive articles by John Clarke and other experts in the field [9]. There follows a list of the different types of NDE methods in which SQUIDs can be used, taking Eddy Current NDE as an example of one in which a SQUID can be integrated. We will also look at SQUID NDE from a theoretical point of view, shortly describing image processing, modelling techniques as well as inverse methods. A modelling example from the field of EC SQUID NDE is presented to demonstrate how to optimize a NDE system's parameters.

2.1 SQUID Sensor and Read-out

i) An Introductory Look at the SQUID

The SQUID is a superconducting device containing either one (rf SQUID) or two (dc SQUID) Josephson junctions. These junctions can be produced by using an insulating layer a few nanometers thick (tunnel junction) to separate two superconductors. If the layer is thin enough, pairs of electrons (so called cooper pairs) tunnel through the barrier while maintaining phase coherence during the tunneling process. The electrical properties of a Josephson junction are described by the Josephson equations [10, 11]:

$$I = I_C \cdot \sin\delta \quad (2.1) \qquad V = \frac{\hbar}{2e}\left(\frac{d\delta}{dt}\right) \quad (2.2)$$

Equation (2.1) is the current-phase relation in which the phase shift differences between the two superconductors $\delta = \phi_1 - \phi_2$ are governed by the applied current I. I_C describes the critical current, which is the maximum superconducting current the junction can carry. Higher currents $I > I_C$ result in a voltage across the junction according to the voltage-frequency relation, shown in Equ. (2.2), in which $\hbar$ is the Planck's constant and e is the charge of the electron.

In the case of a dc SQUID (see sketch in Fig. 1, left), an external field or flux variation changes the phase difference across the Josephson junction(s). This is due to flux quantization, which describes the fact that in a superconducting loop magnetic flux only appears as multiples of the flux quantum $n\Phi_0$, in which n is an integer and $\Phi_0 = h/2e \cong 2.067 \cdot 10^{-15}$ Vs (equivalent to the amount of magnetic flux from the earth's magnetic field passing through an area the size of one human red blood cell). In order to keep the flux constant, the loop will compensate the external flux variation Φ_{ex} by generating a screening current $j = -\Phi_{ex}/L$, in which L is the inductance of the loop. In so doing, the loop is able to remain in its current flux quantum state until the external flux is increased to the point at which an additional flux quantum Φ_0 can enter the loop and raise it to the next flux quantum state. The screening currents are superimposed on the bias current at which the SQUID is operated, turning the critical current I_C into an oscillating function of the external flux. Assuming that the SQUID is operated with a bias current larger than the sum of the critical currents I_C of the Josephson junctions, the voltage across the SQUID loop oscillates with increasing external flux Φ_{ex}. Figure 1, right, shows the current-voltage (*I-V*) characteristics of a dc SQUID for $\Phi_{ex} = n\Phi_0$ and $\Phi_{ex} = (n + ½)\,\Phi_0$ and its transfer function for two different bias currents applied to the SQUID loop. One can recognize that a linear variation of the external flux leads to a periodic, first order sinusoidal transfer function. The periodicity of the voltage modulation is the magnetic flux quantum Φ_0. The SQUID is operated with bias currents and fields that shift the working point to the steepest parts of the *I-V* curve, at which the transfer coefficient $\partial V/\partial \Phi$ is maximum. This turns the SQUID into an effective flux-to-voltage transducer. Typical values for bias currents are of the order of 10 μA for low-T_c and 100 μA for high-T_c resulting in a modulation depth of a few tens of μV. Finally it should be said that Josephson junctions have a distinct hysteresis in their *I-V* characteristics, which in practice makes it necessary to resistively shunt a low-T_c junction to eliminate its hysteretic behavior [12].

Contrary to the dc SQUID, the rf SQUID is inductively coupled with a resonant circuit which is driven by an rf current source at frequencies above a few 100 MHz. The voltage across the tank circuit also shows a periodic V-Φ transfer function. High-T_c rf SQUIDs have the distinct advantage of not needing a galvanic connection to the read-out electronics. On the other hand they are less suitable for NDE applications in which the SQUID has to be placed in close proximity to a conducting sample. One can observe a shift of the tank circuit's resonance frequency as the SQUID-to-sample separation is changed. Such an inductive and capacitive interaction between circuit and sample can exercise an adverse effect on magnetic field detection because it reduces the transfer function.

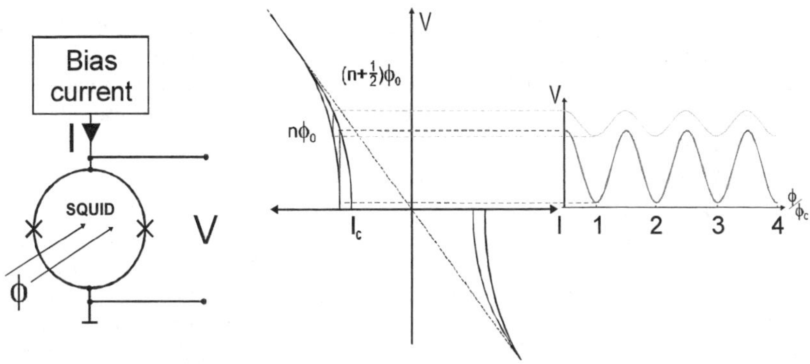

Figure 1. Left: Schematic diagram of a dc SQUID, Right: *I/V*-characteristic and *V-Φ* transfer function of a current biased dc SQUID.

ii) SQUID Fabrication

In low-T_c technology thin film SQUIDs are usually produced through sputtering of niobium thin films. The pattern is generated by techniques such as photolithography or electron beam lithography. The tunnel junction can be made with a few nm-thick layer of Al_2O_3 (S-I-S). The shunts need to be made of a material that does not superconduct at the temperature of operation.

Over the years of research into high temperature superconductors a lot of work has concentrated on the development of high-T_c compounds for SQUID fabrication such as $YBa_2Cu_3O_{7-\delta}$ (YBCO, $T_c \cong 93$ K), $Bi_2Sr_2Ca_nCu_{n+1}O_{2n+6}$ and $Ta_2Ba_2Ca_nCu_{n+1}O_{2n+6}$ (n = 0,1,2). To date YBCO remains the most widely used material. In the early days of high-T_c technology SQUIDs were made of polycrystalline bulk YBCO. But this was quickly replaced by thin-film YBCO because of the high 1/f noise observed in SQUIDs made from polycrystalline material (bulk or film). Common film deposition methods are pulsed laser deposition (PLD) and sputtering. Naturally the junction can be fabricated in the same way as it is in its low-T_c counterpart, e.g. as an S-I-S or S-N-S configuration, where N is a normal metal which is resistive at high-T_c temperatures. But junctions have better noise characteristics when made with grain boundary weak links. One way to produce these is through a bicrystal junction in which the substrate has a grain boundary with a tilt angle in the crystal direction of e.g. 24° [13-15]. During deposition the YBCO film replicates the substrate's grain boundary, leading to well-defined grain boundaries in the epitaxial layer. Another approach is the step edge junction in which the substrate has a small pit. Provided the step edge of the pit is sufficiently sharp, the deposited YBCO film will replicate the pit and the layer will contain well-defined grain boundaries at the top and bottom of the pit's edges [16-18]. This is a state-of-the-art technique that achieves very low noise levels. The substrates generally used for both of the above described techniques are $SrTiO_3$ or $LaAlO_3$.

There are a variety of dc or rf SQUID configurations to choose from, such as single-layer magnetometers or multi-layer techniques. But these devices are impractical for many applications because a bare SQUID's effective flux capture area is small, leading to a low magnetic field resolution. An attractive alternative that increases the SQUID's active area is the use of a flux transformer either in the form of an integrated pattern on the SQUID chip (flip chip configuration) or in the case of low-T_c SQUIDs a simple, wire-wound flux transformer inductively coupled to the SQUID. The latter offers a further advantage because it circumvents the need to mount the SQUID chip close to the dewar's outer wall, thus offering better shielding from environmental magnetic influences.

A flux transformer consists of a pick-up coil at the one end and an input coil inductively coupled to the nearby SQUID loop at the other. The energy transfer from the pick-up loop to the SQUID is maximized by adapting the inductances of the pick-up loop, the input coil and the SQUID itself. A quantitative description can be found in [7]. It has been shown that use of a large diameter pick-up coil results in astounding field sensitivity [19]. But even small pick-up loops of just a few millimeters can achieve field sensitivities of a few fT/√Hz when used in conjunction with flux sensitive SQUIDs [20]. While flux transformers for low-T_c applications can be easily made by hand using niobium wires, the fabrication of flexible high-T_c superconducting wires is rather difficult. For this reason the pick-up coil needs to be integrated on the SQUID chip. This either can be done on a single substrate, or on separate substrates which are then referred to as a flip chip configuration.

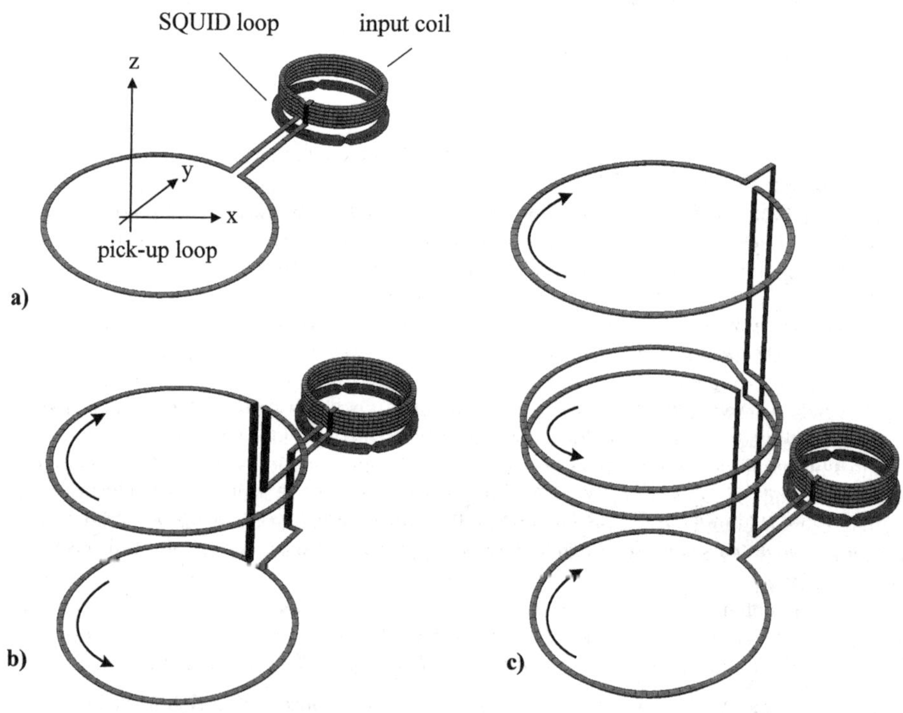

Figure 2. Sketch of flux transformer configurations. a) magnetometer, b) first-derivative gradiometer measuring $\partial B_z/\partial z$, c) second-derivative gradiometer measuring $\partial^2 B_z/\partial z^2$.

A flux transformer is a particularly useful tool when dealing with gradiometer configurations comprising two pick-up loops wound in opposition. Accurately balancing the loops prevents a homogeneous field from coupling flux into the SQUID loop. This helps suppress background noise. Figure 2(b) shows an example of an axial 1st-derivative gradiometer measuring $\partial B_z/\partial z$. Of course higher orders of gradiometer configurations can be used when a specific application calls for a higher suppression of background fields (Fig. 2(c)). A single pick-up coil measures the projection of the magnetic field along the coil normal. Because flux transformers can combine several pick-up coils with different spatial alignments, they can be used to measure any spatial derivative of a magnetic field, e.g. $\partial B_z/\partial y$, $\partial^2 B_z/\partial y \partial x$.

iii) Operating a SQUID

A SQUID is usually operated in a feedback circuit as a null detector to linearize the periodic transfer function. Because a SQUID's output signal generally does not exceed amplitudes of a few tens of μV, it is directly coupled to a low-noise preamplifier and integrator circuit. The amplified output voltage is converted into a current by using a feedback resistor. This current is then fed back to a coil which, positioned close to the SQUID, converts the current into magnetic flux. An applied field is thus fed back and generates an opposing flux which keeps the flux in the SQUID constant. This kind of feedback loop is called a flux-locked loop (FLL) and provides a linear relationship between an applied field and its output voltage. When using flux sensitive SQUIDs it can measure variations in flux ranging from less than a millionth of a single flux quantum to many flux quanta.

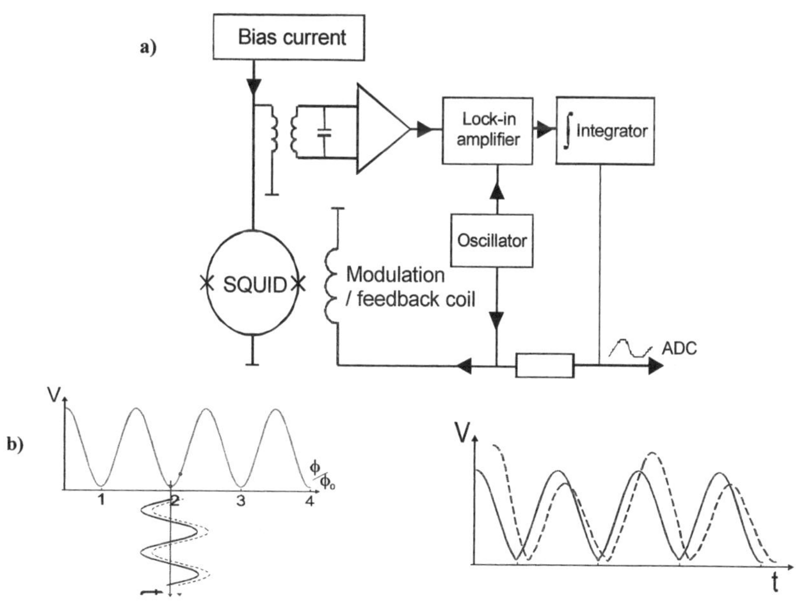

Figure 3. (a) Schematic diagram of a dc SQUID read-out electronics with flux modulation. (b) periodic transfer function and rectified output signal in the case of flux modulation

Apart from a directly coupled feedback, it is also possible to use a flux modulation feedback, an option first proposed by R.L. Forgacs and A.F. Warnick [21] (Fig. 3). In this approach a modulating flux is applied to the SQUID. Its peak-to-peak amplitude is smaller than $\Phi_0/2$ and the frequencies f_m usually range between 10 kHz and a few 100 kHz. If no external magnetic field is applied to the SQUID the output signal is a rectified version of the applied flux modulation, thus containing only a frequency $2f_m$. If this double frequency signal is fed to a lock-in amplifier referenced to the modulation frequency f_m the output signal will be zero. On the other hand, if a small external field is applied to the SQUID the working point will shift slightly to the left or right along the transfer function, leading to an increased output voltage at frequency f_m. Thus the lock-in amplifier will output a dc voltage, which is proportional to the amplitude of the signal at f_m. Analogous to the directly coupled feedback the lock-in amplifier's output is connected to an integrator which in turn is linked via a feedback resistor to the feedback coil. The coil is inductively coupled with the SQUID, so it generates a counterfield at the SQUID which keeps the working point constant at one minimum of the transfer function. Besides directly coupled feedback circuits and flux modulation, SQUIDs can also be operated with additional positive feedback or as a relaxation oscillator, described in detail in [22].

2.2 Integration of a SQUID in a NDE System

2.2.1 Electromagnetic Testing Methods

There are a variety of applications in nondestructive testing in which the SQUID's high field sensitivity can be used to provide detailed information about the integrity of the material under investigation. In conventional NDE electromagnetic testing can be divided into two major fields: Eddy Current Testing (ECT) and magnetics. The latter includes magnetic particle inspection, magnetic flux leakage, Barkhausen noise, etc. While electromagnetic NDE techniques differ in method and procedure, they are all designed to probe local magnetic anomalies in static or alternating stray fields. These can be caused by cracks, corrosion, inclusions, impurities or alternative material damage where the material properties differ from those of the tested sample. The results of electromagnetic NDE methods not only help to assess material damage but also give insight into the grain structure of polycrystalline metals as well as the mechanical strength of a sample.

The following list of testing methods which can be used in SQUID NDE is divided into two categories: passive and active field measurements. In the first case, a magnetic field is generated by the sample or the defect itself which can be used to gauge the mechanical state of integrity. In the second, defect detection depends on some form of excitation, either through the application of a background field or by means of different kinds of current generation.

a) Passive Field Measurement:

i) **Direct field measurement** of *ferromagnetic samples* to determine changes in mechanical stress and magnetic structure.

ii) **Direct field measurement** of ferromagnetic inclusions in *all kinds of samples* (food, wires, hard drives, etc.)

b) Active Field Measurement:

i) Flaw detection in *ferromagnetic samples* by **magnetic flux leakage** techniques (MFL). A magnetization field is applied to the sample. The defect response field is caused by a distortion of the magnetic flux lines in the vicinity of magnetic anomalies in the ferromagnetic structure. This can be a non-ferromagnetic inclusion as well as any kind of damage in the mechanical structure (flaws, corrosion etc.)

ii) Characterization of a *ferromagnetic sample* through analysis of the **Barkhausen noise**. The sample is periodically magnetized and small jumps in the magnetization curve are detected using a wide bandwidth. Mechanical properties are correlated to the characteristics of the Barkhausen emisson.

iii) Detection of material defects such as flaws, corrosion and fatigue cracks in *conductive samples* by **eddy current testing** (ECT). An alternating current is used to induce eddy currents in the conductive sample. The defect response field is caused by perturbation currents flowing around the defect. Most practical and common method used in NDE. Can be applied in a variety of induction techniques, e.g. using several frequencies, pulsed ECT, etc..

iv) Detection of material defects in *conductive samples* by **current injection**. The defect response field is generated by perturbation currents flowing around the defect. The frequency of the injected currents can range between dc or higher frequencies and is injected directly into the sample using two electrodes.

v) Detection of active corrosion in *conductive samples* by measuring **electrochemical currents**.

vi) Detection of material defects in *conductive samples* by **thermocurrents** generated by thermal gradients. Due to the thermoelectric Seebeck effect inclusions in the sample generate currents at their intersection with the surrounding material. Thermocurrent testing is an alternative to ECT if the inclusion has a similar electrical conductivity as the matrix in which it is embedded but differs in its Seebeck coefficient.

vii) Detection of material defects in *nonconductive samples* by determining their **susceptibility**. Flaws or inclusions have a different susceptibility than the material around them. By applying a homogenous background field, the magnetization of the defect will be different from that of the sample. This way a defect acts as a small magnetic dipole.

viii) Determination of material properties in *semiconductor materials* by **photo-generated currents**. It can be used to test photovoltaic devices such as silicon wafers etc.

2.2.2 Integration of a SQUID in an EC NDE System

When using a SQUID for a passive field testing method its integration is rather simple because no additional active components are required to generate fields or currents. All it takes is placement of the SQUID inside the dewar or cryocooler in a position near the sample, and correlation of the scanning stage with the measured field data in order to generate magnetic field maps. The way a SQUID is integrated in a NDE system depends on the application for which it is foreseen. Regardless of the kind of technique being used to test a sample, the SQUID must be thermally shielded so that it can be operated below its critical temperature T_c. The cooling environment into which it is incorporated can be either a cryostat for liquid helium or liquid nitrogen, or a cryocooler. As an example the sketch in Fig. 4 shows a microwave rf-SQUID mounted in a mobile fiberglass cryostat. By contrast, all active field measurements require some type of excitation scheme to generate interacting fields or currents. AC-current methods in particular usually require a demodulating device such as a lock-in amplifier. Further diverse compensation or filtering electronics are often used to fully exploit the SQUID's high field sensitivity.

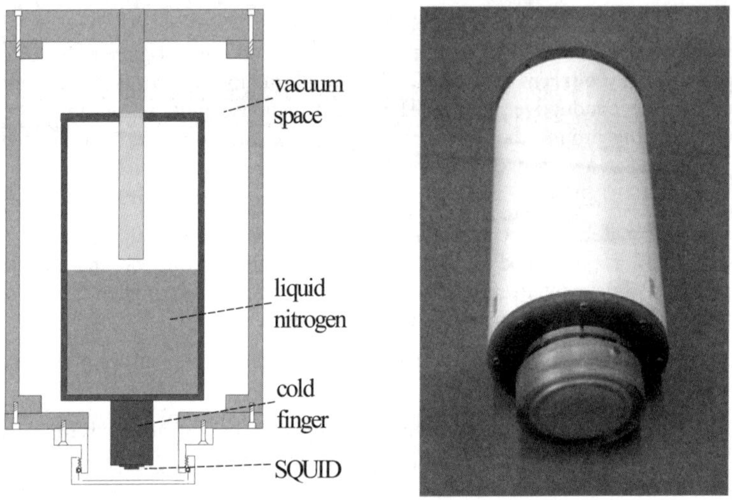

Figure 4. A high-T_c rf SQUID mounted in a mobile liquid nitrogen dewar (11.5 cm in dia.).

i) Basic EC SQUID NDE Setup

Figure 5 shows a basic measurement setup of an ECT system. The SQUID's cooling environment is mounted onto the moveable part of an x/y-translation stage and is thus scanned above a test object. The stage is usually controlled by a PC, which correlates the position (x,y,z) of the stage with the measured field data $B_{x,y,z}\,(x,y,z)$. The excitation unit's signal is fed both to the excitation coil, which is rigidly mounted at the bottom of the cryostat, and to a lock-in amplifier as the reference signal. Depending on the SQUID's alignment, the device will sense a certain component of the magnetic field above the test specimen. It is read out with a FLL electronics, whose output is connected to a lock-in amplifier for demodulation of the measured EC field. Both channels of the demodulation process are fed to a PC. An ADC with sufficient high digital resolution converts the analog signals. Of course several SQUIDs can be used in an array or combined into an electronic gradiometer. In the latter case several read-out electronics are used to subtract the different SQUIDs' signals from each other.

ii) Excitation Scheme Used in EC SQUID NDE

Considering that the field response of technically relevant material defects can be many orders of magnitude lower than the field strength of the excitation field, compensation of the excitation field by geometrical means or electronically is widely practiced. This helps to save the entire dynamic range of a SQUID system for the sample's response field. For this purpose, one usually tries to use an excitation scheme that generates a high field strength at the sample but a low excitation field at the position of the SQUID. This can be done with the help of a specially designed inducer coil. Figure 6 shows five different excitation shapes. When measuring the magnetic field's vertical z-component B_z, for example, a gradiometric coil - a so-called double-D coil - can be used which ideally has zero excitation field in the center of the coil (Fig. 7). When the SQUID is placed exactly between the two halves of the coil, it sees only a fraction of the excitation field which the sample is experiencing and mainly detects the sample's EC response. A single wire can be used in the same way. It also eliminates the z-component if the SQUID is mounted exactly over the wire. A circular excitation coil has no horizontal field component at its center and thus can be used when the SQUID chip is mounted vertically inside the dewar.

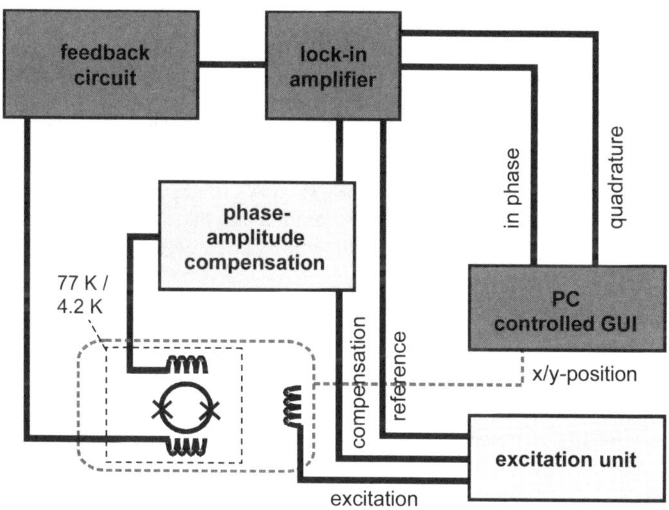

Figure 5. Sketch of a principal measurement setup of a SQUID-based EC NDE system (adapted from ref. [36]).

Alternatively there is the option of an electronically compensated configuration. Here the vertical field component can be measured in combination with a circular excitation coil. A low field at the location of the SQUID is obtained by locally compensating the excitation field at the SQUID with a phase-shifted current, driven in an additional compensation coil which is mounted close to the SQUID. Since this compensation coil is relatively small (<1 mm) with dipole-like fall-off characteristics, it will mainly reduce the excitation field at the location of the SQUID and not at the sample (Fig. 8).

Either way, compensation by excitation geometry or electronics easily achieves a reduction of the excitation field at the location of the SQUID by a factor of up to 1000. A combination of both approaches leads to an improvement of six orders of magnitude in the dynamic range and virtually eliminates the excitation field at the SQUID. This allows one to use large excitation fields without saturating the SQUID read-out.

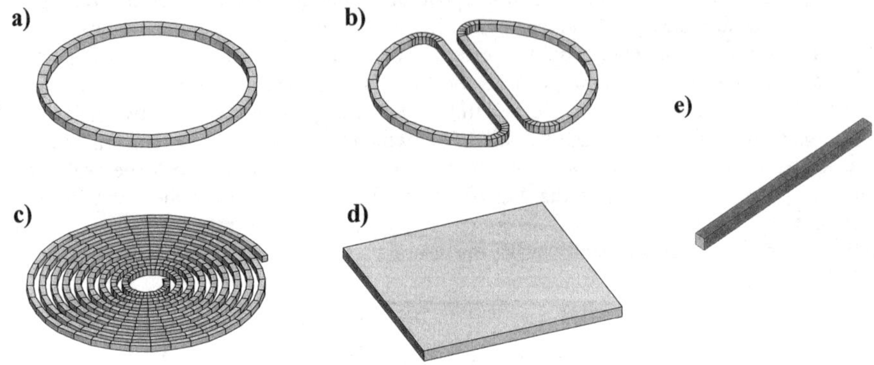

Figure 6. Different types of excitation units for EC generation: a) circular coil, b) gradiometric (double-D) coil, c) spiral coil, d) sheet inducer (homogenous field), e) single wire.

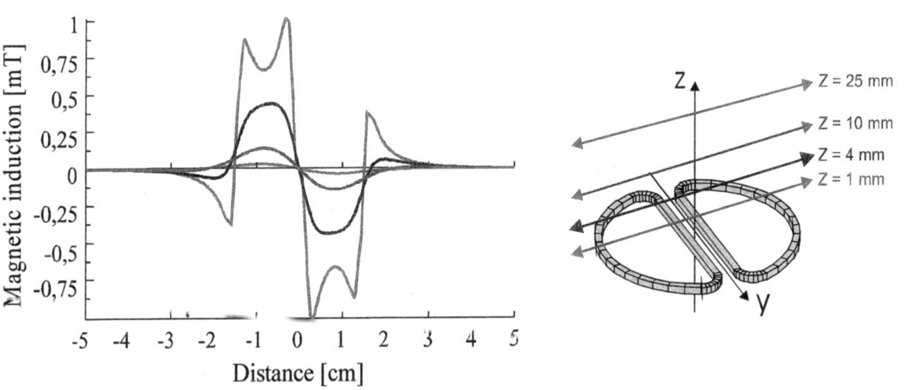

Figure 7. Excitation field distribution of the magnetic z-component B_z above a gradiometric excitation coil indicated for different distances z from the coil. The current flows in one half of the coil clockwise and in the other half anti-clockwise. Coil parameters: 100 turns, 100 mA und $\varnothing = 30$ mm. The SQUID is then positioned at a position $x = 0$ where the excitation field disappears.

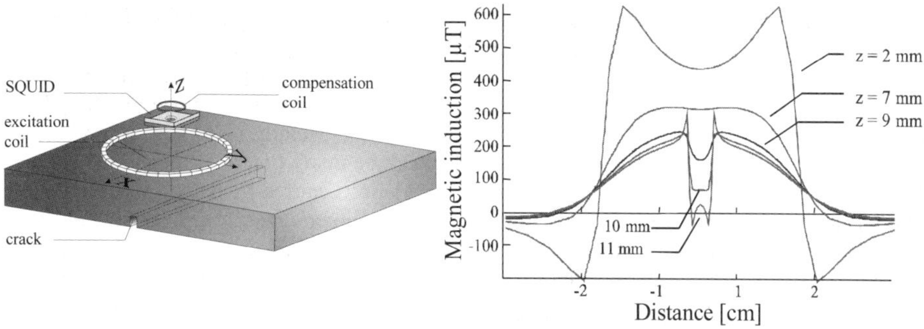

Figure 8. Left: Sketch of a circular excitation coil configuration using a local compensation coil to minimize the excitation field B_z at the location of the SQUID. Right: Field distribution of the magnetic z-component above a circular excitation coil and an additional local compensation coil which was positioned z = 10 mm above the sample. Excitation coil: 100 turns, 100 mA, ∅ = 30 mm, compensation coil: 200 turns, 10 mA, ∅ = 4 mm.

2.3 Theoretical Aspects of SQUID NDE

This section presents an example of Eddy Current Testing (ECT) to highlight the advantages of magnetic field modelling in SQUID NDE. It looks at several types of excitation coils which meet the special requirements for a SQUID-based EC NDE system. To this end, an analytical model first presented in [23] was developed to determine the optimum EC parameters for each type of excitation coil. Optimum excitation frequencies were calculated and the EC response of small material defects estimated by means of finite-element method (FEM). The results showed that high field sensitivity is essential when detecting small or deep-lying material defects.

2.3.1 Magnetic Modelling

In NDE, magnetic modelling is an indispensable tool to gain deeper understanding of the function of different testing methods. It is particularly helpful in cases involving more complicated NDE techniques, such as EC testing, where it can be applied to determine a test system's optimal setting - from sensor design and excitation geometry adaptation to EC test frequencies. Modelling can also be used to determine the expected response field for different types of defects. Because the forward problem is easy to calculate, modelling is an easy way of studying the feasibility of new ideas in SQUID NDE. This is supported by the fact that a SQUID sensor can be easily integrated into a numerical model because it acts like an ideal point sensor.

There are several analysis techniques suitable for tackling electromagnetic problems: simple cases can be solved by analytical solutions [24], the finite-element method (FEM), finite-difference method (FDM), boundary-element method (BEM) and volume-integral method (VIM). Among these, the FEM is the most universal. In contrast to its competitors it can handle complicated 3D-problems and even can deal with non-linear effects in high permeability materials. One can use different mesh arrangements (such as tetrahedral, hexahedral) which are comparably easy to adapt to the geometry of a sample or defect. To achieve a higher degree of accuracy in the model, the nodal density of the finite-element mesh is adapted to the strength of the current gradients and field gradients. By using smaller elements at regions where large gradients occur, e.g. in the vicinity of the material defect, one can calculate accurate current distributions inside the sample and the corresponding field distribution above the sample's surface. High accuracy is especially important in the case of ECT because the fields caused by a material defect can be many orders of magnitude smaller than the excitation field. In order to separate field variations caused by defects, the excitation field has to be subtracted from the defect response. High numerical accuracy will model small field variations generated by deep-lying or small anomalies inside the material. Powerful software packages such as Vector-Fields (OPERA) and ANSYS are commercially available.

FEM's impressive flexibility has won over an increasing number of SQUID NDE researchers since the groups headed by J.P. Wikswo and G.B. Donaldson published intriguing results in 1995 [25-27]. A deeper insight into magnetic modelling by means of FEM can be found in [28]. The important supporting role simulations can play in SQUID NDE is detailed in several examples in section 4. FEM does, however, have a particular disadvantage compared to other methods: It is rather time-consuming, and the computer used for calculations needs a lot of storage capacity - both for memory and its hard drive. Even linear models (no ferromagnetic components) with a mesh size of 160,000 nodes (about 380,000 equations) need about 550 MByte RAM for calculation, and several 100 MByte to store a single model. The calculation time can easily exceed 45 minutes calculated with a 2 GHz PC-workstation. An optimization process often requires the calculation of several models clocking many hours in total calculation time. By contrast, the volume-integral method (VIM) is much faster because the modelling region is confined to the defect, meaning fewer equations need to be solved. But because this method requires a previous determination of the Green's function, it is ill suited for modelling complicated NDE problems and is only advisable when calculating nonferromagnetic materials, isolated defects and simple sample geometry [29]. To overcome the drawbacks presented by FEM's slow numerical calculations and VIM's inflexibility, the giessen group developed an analytical model for EC SQUID NDE that speeds up an NDE system's optimization process [23].

2.3.1.a Dependence of the Penetration Depth on Material Properties and Excitation Coil Geometry

In contrast to a direct injection of dc or ac currents in the sample, the induction of eddy currents by an external excitation coil generates a locally limited current distribution. This reduces the influence of edge effects. A sample's edge can appear as a huge flaw, leading to a strong distortion of currents flowing near an edge. When using injection techniques the effect of distorted currents along the sample's edges are constantly present and mask the actual defects. The edge effect does not occur if the induced currents are confined to a region close to the excitation coil. Induction techniques (EC NDE) don't require an electrical connection to the sample, making them easier to use than injection methods from a practical point of view. But because of the induction law and the skin effect, it is harder to choose optimum measurement parameters such as the excitation frequency and excitation coil geometry. An additional mathematical hurdle is posed by the inverse calculation - from the measured field distribution to the current flow in the sample - which is more difficult than in the case of a direct current injection. While a homogenous field distribution produced by e.g. direct current injection or a sheet inducer [30] allows for easier estimation of the defect geometry, these methods are either unpractical or not sensitive enough to detect technically relevant cracks. This is especially the case when dealing with deep-lying or small cracks.

For a rough estimation of the optimum excitation frequency for a given test object, one can use the well-known expression for the skin penetration depth δ:

$$\delta = \sqrt{\frac{2}{\omega \cdot \sigma \cdot \mu \cdot \mu_0}}\,, \tag{2.3}$$

where ω is the excitation frequency, σ is the conductivity of the material, μ is the magnetic permeability and μ_0 describes the permeability of free space. Here, the wave propagation in a conducting medium is calculated by solving Maxwell's equations, assuming a planar incident electromagnetic wave and a conductor of semi-infinite extent. We can write the time independent solution for the amplitude of eddy current density in the form:

$$j = j_0\, e^{-\alpha z} \quad \text{and} \quad \delta \cdot \alpha = 1, \tag{2.4}$$

where α is the inverse of the skin depth, δ and j_0 the current density on the surface of the conductor. However, when using a small coil rather than a planar incident wave to excite eddy currents, both j_0 and α will be different from the values calculated using Equ. (2.3). To account for this effect, we can define corrected values j_{C0} and α_C, where

$$j_c = j_{c0}\, e^{-\alpha_c z}\,. \tag{2.5}$$

In contrast to the planar wave solution, the penetration depth in the case of a coil can be described - besides the dependence on ω and σ - as a function of the depth z and the coil diameter R:

$$\alpha_c = f(\omega, \sigma, z, R). \tag{2.6}$$

This correction function was calculated for different kinds of excitation coils, like circular coils without a ferrite core, spiral coils, double-D coils and a sheet inducer. For this purpose the eddy current density was determined for frequencies between 10 and 1000 Hz and for depths between 0 and 30 mm. Equation (2.4) and (2.5) can be combined into an equation for α_C in which j_C and j_{C0} (the current density on the surface of the test object) can be calculated by means of FEM. We refer to the ratio of the normalized current densities of Equ. (2.7) as the correction or error function $\varepsilon(\omega, z)$:

$$\alpha_c(\omega, z) = \alpha \cdot \frac{Ln\left(\dfrac{j_c}{j_{c0}}\right)}{Ln\left(\dfrac{j}{j_0}\right)} = \alpha \cdot \varepsilon(\omega, z)\,. \tag{2.7}$$

For each frequency 100 points were taken along a line running from the surface of the conductor into a depth of 30 mm in the region below the coil containing the maximum eddy currents (dashed vertical lines in Fig. 9, middle column). These data are fitted by appropriate polynomials to obtain an analytical expression for $\varepsilon(\omega, z)$ in the frequency and depth interval mentioned above.

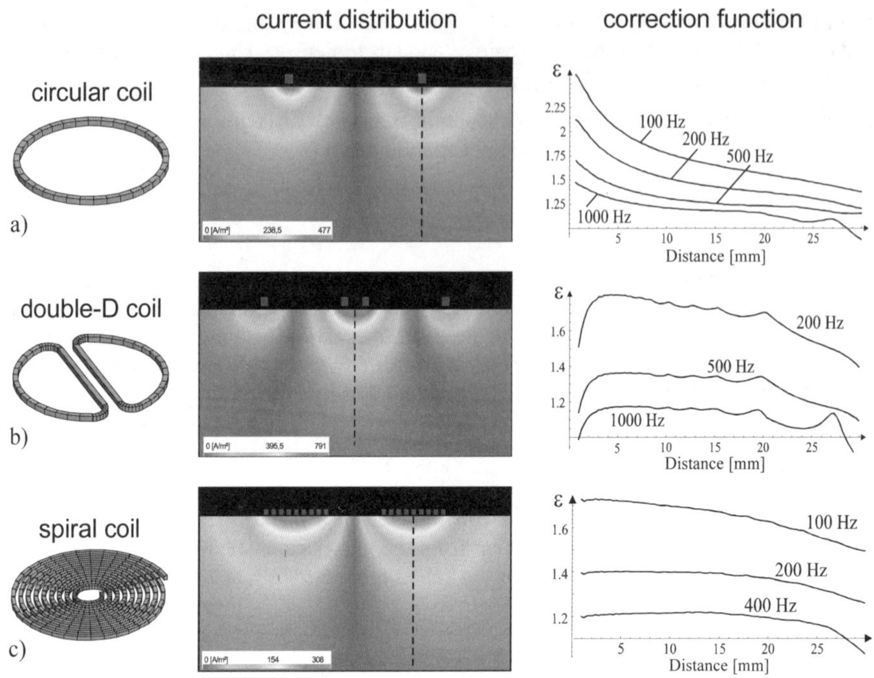

Figure 9. Left column: Different types of excitation coils investigated: a) circular coil (R=15 mm), b) double-D coil (R=15 mm), c) spiral coil (R=18 mm). Middle column: Cross-section of corresponding induced eddy currents inside an aluminum sample. Red color: high EC density; blue color: low EC density. Right column: Corresponding correction function $\varepsilon(\omega, z)$ describing the ratio of real penetration depth and the standard penetration depth - calculated by 3D-FEM for different frequencies.

The right column in Fig. 9 shows the ratio $\varepsilon(\omega, z)$ between the standard penetration depth $1/\alpha$ (for the case of a planar incident wave, see Equ. (2.3)) and the penetration depth in the case of an excitation coil $1/\alpha_c$, for different coils and several excitation frequencies. Three factors let the correction function $\varepsilon(\omega, z)$ converge to 1: large coil diameter, large depth in the conductor, and high frequencies. A value of the correction function close to 1 means that the excitation field is then similar to that of a planar wave. The largest deviations occurred in the case of a small circular coil when using low frequencies, while the field distribution of a sheet inducer barely differed from that of a planar wave. Only in the case of a very low excitation frequency did the finite size of the inducer plate slightly influence the correction function towards higher values. However, it turned out that especially when using a SQUID-based NDE system with its relatively low frequencies for the detection of deep-lying flaws, the estimation of the excitation frequency from the standard penetration depth $1/\alpha$ - instead of $1/\alpha_c$ - is not a wise choice. Note that e.g. for a circular excitation coil (R=15 mm) and an excitation frequency of 100 Hz, the real penetration depth can be more than a factor of two smaller than what is calculated using Equ. (2.3).

2.3.1.b Determination of the Optimum Excitation Frequency

When calculating an optimum excitation frequency one has to consider the following fact: Because of the induction law the current density at the surface of the sample j_0 is frequency-dependant. When frequencies are too high, the current density at the surface - scaling with √Hz - is sufficiently high, but the penetration depth decreases, and this leads to an exponentially reduced response from deep-lying defects. On the other hand, when excitation frequencies are too low, the current density at the surface is too small. Despite a larger penetration depth, this results in a lower field variation at the location of the crack. Therefore in total there is a certain excitation frequency which results in a maximum response field from the crack. We turn now to a discussion of simulations devised to easily find this optimum excitation frequency.

Often the simulation of the actual distortion of the eddy current flow caused by a crack turns out to be too time consuming with any presently available means such as FEM simulations. Let's look at a simple model for calculating the optimum excitation frequencies for cracks in different depths of arbitrary test samples: using Equ. (2.7), we are able to calculate the decrease in eddy current density with increasing depth in the conductor for a given excitation method, taking into account the dependence of the penetration depth α_C on coil geometry and excitation frequency.

Four components contribute to the defect-related variation of the eddy current field ΔB_z: i) the surface current density j_{c0} controlled by the inducer size and shape, the EC frequency f, and coil-to-surface separation, ii) the exponential decay of the excitation field governed by the decay coefficient α_C and the material depth z, iii) the interaction process of the induced currents with the material defect, which can be phenomenologically described by an amplification factor λ, iv) the decay of the return field controlled by the decay coefficient α_{crack}. The decay coefficient α_{crack} depends on the size and shape of the defect. These four factors determine the magnetic EC response above the sample's surface:

$$\Delta B_z = j_{c0}(\omega) \cdot e^{-\alpha_c(\omega,z)\cdot z} \cdot \lambda \cdot e^{-\alpha_{crack}(\omega,z)\cdot z} . \qquad (2.8)$$

The situation is different when detecting small cracks - in other words, when the size d of a crack is smaller than the separation between defect and sensor. Figure 10 shows a sketch of the circular perturbation currents flowing anti-clockwise along the left edge of the crack and clockwise along the right edge. Currents flowing below and above the defect are not taken into account because they have a negligible influence on the magnetic z-component B_z that the SQUID measures. Because of two different current directions at the edges of the defect, there are two magnetic dipoles, one positive and one negative. Therefore, one measures higher magnetic moments in the case of large separations, when d/z is smaller than one, leading to a stronger attenuation of the defect response. This additional attenuation factor - the spatial decay of the quadrupole field, namely $f(d,z)$ - must be added to Equ. (2.8) if the distance z between defect and sensor is larger than the defect size d. The magnetic field variation above the sample can then be summarized by the following expression:

$$\Delta B_z = j_{c0}(\omega) \cdot e^{-\alpha_c(\omega,z)\cdot z} \cdot \lambda \cdot e^{-\alpha_{crack}(\omega,z)\cdot z} \cdot f(\tfrac{d}{z}) , \qquad (2.9)$$

in which the spatial decay function $f\,(d,z)$ can be approximated by:

$$f(\tfrac{d}{z}) \approx -1 + \frac{2}{1+e^{-2.84153\frac{d}{z}}}, \qquad f(\tfrac{d}{z}<1) \sim \frac{d}{z}. \tag{2.10}$$

The function $f\,(d,z)$ is proportional to d/z if the size d of the crack is smaller than its position in a material depth z, and about unity when d is larger than z (Fig. 10, right).

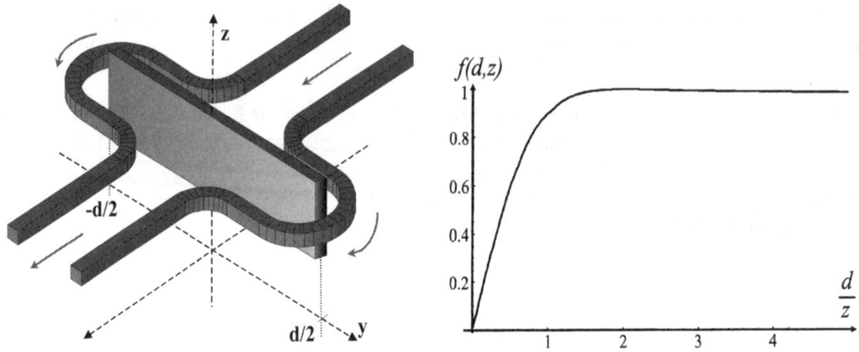

Figure 10. Left: Interference of a homogenous current flow, Right: Attenuation of the EC response field for detecting small defects. The magnetic field's fall-off is increased compared to that of a single magnetic dipole, if the distance between SQUID and defect z is smaller than the size d of the defect.

Now to find the optimum excitation frequency we calculate the first derivative of Equ. (2.9) to determine the maximum response field B_z of cracks in different depths (Fig. 11). For example a crack (20 x 0.6 x 0.2 mm³) in a depth of 9 mm in an aluminum sample (σ= 20 MS/m) can be detected with highest SNR when using a frequency of 260 Hz. This calculation assumes the use of a double-D coil. For cracks located deeper in the conductor, the optimum excitation frequency moves towards lower values. Figure 11 also shows measured data. These measured variations of the magnetic field correspond quite reasonably to the data of our model. At higher frequencies a discrepancy arises due to the small number of nodes in the FEM simulation (30,000). A deviation from the theoretical data is found at low excitation frequencies of the order of 50 Hz (Fig. 11, bottom), caused by the 50 Hz notch filter in our lock-in detector. But this effect has only a minimal influence on the location of the field maximum. Another reason for a deviation between measured and calculated data is the phase shift along the extension of the crack in z direction.

Using Equ. (2.9), we are now able to compute the optimum frequency for cracks in various depths (Fig. 12). For comparison, the optimum excitation frequency for a planar wave or a sheet inducer (300 mm x 160 mm) is also displayed. One can see that a planar excitation source requires a much lower excitation frequency, leading to a significant reduction in the crack's response field.

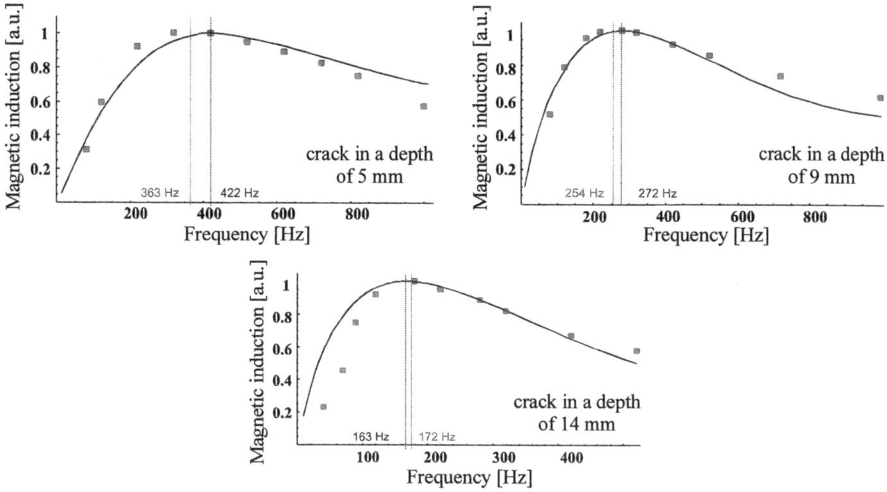

Figure 11. Simulated (full line) and measured data for the field variations caused by cracks, for different excitation frequencies. The size of the crack is 40 x 0.6 x 0.2 mm³.

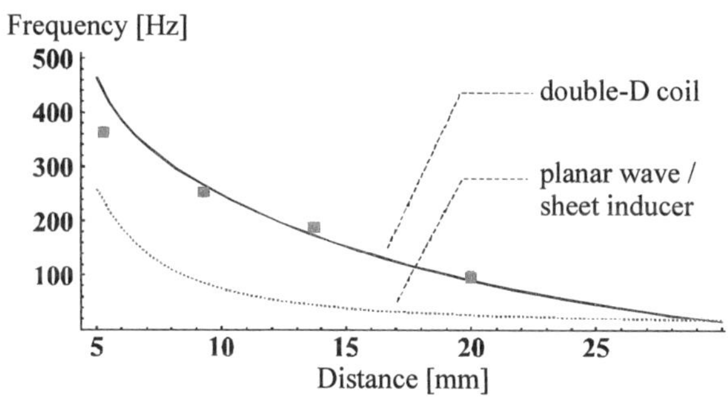

Figure 12. Optimum excitation frequency for cracks in varying depths. The dotted line is for a sheet inducer and the solid one is for a double-D coil. The indicated points represent the measured values.

2.3.1.c Determination of the Response Field Amplitudes

Equ. (2.9) additionally allows one to estimate the magnetic field strength caused by perturbed currents at cracks in different depths, although the signals depend on the heights and lengths of the cracks. For smaller cracks both λ the ratio between eddy current densities with and without cracks, and the mean diameter of the perturbed circular currents are small, which results in a higher attenuation ($\alpha_{flaw} >> \alpha$) of the response field caused by the crack.

Furthermore one has to consider the phase jump effects in cases of cracks with an extension in the z-direction. Here the total response of the crack results in the integral over the decreasing amplitude and the changing phase in the z-interval of the defect. For cracks with no extension in the z-direction, one can assume a constant amplitude and phase shift in the vicinity of the defect.

Figure 13 shows the increasing attenuation for cracks in a depth between 5 and 30 mm, using the optimized excitation frequency for each depth. The coils (circular, double-D) have a current density of 10^6 A/m^2. In the case of circular and double-D coils, this corresponds to an excitation current of 100 mA and 100 turns. The equivalent magnetic field component B_z on the surface of the conductor is about 220 μT_{RMS}. For a 40 mm-long crack ($\lambda \approx 3$) in a depth of 25 mm, we measure a peak-to-peak signal ΔB_z of about 400 pT, which results in a damping factor of $1.8 \cdot 10^{-6}$. The upper curve in Fig. 13 represents the attenuation for a double-D coil. Here we find an attenuation factor of $2.5 \cdot 10^{-6}$.

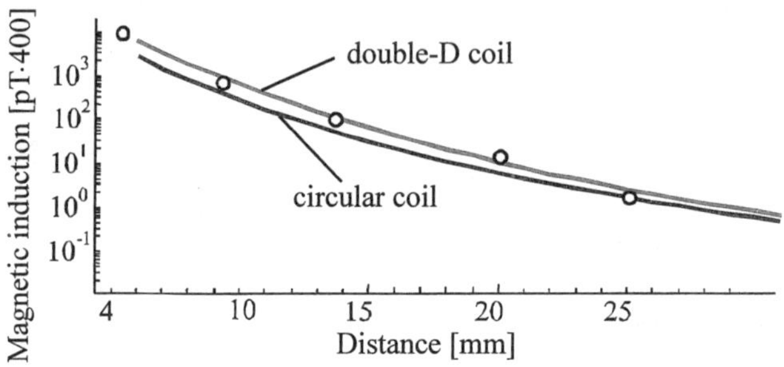

Figure 13. Dependence of the signal strength ΔB_z at the surface (z = 0) on the depth of the cracks. The dots represent the measured values (double-D coil).

Thus the dynamic range of the measuring system for detecting cracks in a depth between $z = 0$ and $z = 30$ mm, depending on the thickness of the test object, should be many orders of magnitude. High values are particularly required when using a non-optimal excitation frequency. The situation is worse in the case of small material defects, resulting in an even lower EC response field. This speaks for the use of a SQUID which can provide both high field sensitivity as well as high dynamic range.

2.3.2 Image Processing

Electromagnetic testing harbors a major disadvantage in that the user can not directly image the sample's mechanical condition - unlike in radiography, for example, where one obtains a clear image of material defects. Electromagnetic NDE is different because field sources generated by either ferromagnetic inclusions or current-distorting material defects create a blurred magnetic field distribution above the sample's surface. This is due both to the divergence of the magnetic field and the placement of the sensing device, which in an exterior investigation is naturally always a certain distance from the field source. What begins as a sharp current distribution inside the sample turns into a broad field distribution when the field sensor is either too large or positioned at too far a distance above the sample's surface. This blurredness makes data interpretation more complicated and begs additional support through signal processing to evaluate the mechanical state of the sample. Fortunately, many NDE applications seek only to determine whether a defect exists at all, rather than investigate its exact position and its size. In these cases the method is referred to as nondestructive testing (NDT) instead of NDE. Signal processing here aims to distinguish between background clutter and the defect signal being measured.

Filter techniques can be applied to reduce any kind of noise influence. A common computational approach is the use of a finite-impulse-response filter (FIR) [6], which reduces unwanted frequencies from a time series data set that do not contain any information about the material defect. C.H. Barbosa et al. [31], for example, presented adapted FIR filtering for two-dimensional field maps to reduce effects caused by strong variations of the SQUID-to-sample separation (lift-off effect) during the scanning process.

Correlation techniques can be used to suppress unwanted signals generated by the sample's geometry or its components. G. Sawade and H.-J. Krause [32] presented an autocorrelation algorithm to test concrete structures with magnetic stray field measurements. They used this technique to separate the signals of mild-steel reinforcements from the signatures of rebars located deeper inside the concrete. A minimum in the autocorrelation function pointed to a broken rebar.

In ECT phase sensitive detection is frequently used to suppress signals with different phase angles. It is also used in EC SQUID NDE to diminish unwanted background [6, 33-35]. The Giessen group showed a correlation technique in conjunction with phase sensitive detection. It suppressed the influence of steel bars inside an aircraft wheel, which also contained tapped holes. An excitation frequency was used so that signals caused by the steel bars appeared only in the in-phase channel of the EC demodulation and signals generated by the tapped holes were mainly detected in the quadrature channel. By contrast, defect signals appeared in both in the in-phase channel as well as in the quadrature channel. A correlation of both channels then suppressed the tapped holes as well as the steel bars and enhanced the defect signal [36].

In ECT the shape of defect signals depends both on the type of material defects present and the sensor-head being used. The latter can be designed to give a specific kind of defect, such as a surface crack, corrosion etc, its own, characteristic signal shape. It is then possible to design matched defect filters which enhance the expected signal structure and separate defects from superimposed background signals. The Giessen group [23] showed such adapted filters for a configuration containing a gradiometric excitation coil and a SQUID magnetometer in 1998. H.-J. Krause [37] later used this technique when he combined a SQUID gradiometer with a gradiometric excitation coil for aircraft wheel testing.

As a last example I would like to mention the work of J. Cadzow and X. Li [38], who showed how the iterative algorithm of blind deconvolution can be used to improve the quality of a two dimensional field map. They investigated a two-layer aircraft lap joint sample. The cracks were adjacent to the rivets and could hardly be seen in the field map. By means of blind deconvolution the crack signal could be clearly seen in a two-dimensional image.

The above-mentioned filters help enhance the signal-to-noise ratio of defect signals but come nowhere near to providing tomographic-like information about a defect's exact position and its shape and size. Yet there are a variety of NDE problems in which the user needs to know about the defect itself. In the aviation industry, for example, precise information on the length of cracks in aircraft parts can be a matter of life and death. Fatigue cracks have a known characteristic exponential growth. This helps inspectors determine when fatigue cracks will exceed critical lengths and cuts down on inspection time. But drawing a conclusion from magnetic field data about the geometry and position of a defect is a rather more complicated process than merely establishing the presence of a crack. This task is addressed by the different inverse modelling approaches shortly described in the next section.

2.3.3 Inverse Modelling

SQUID NDE has numerous NDE competitors, such as X-ray imaging and ultrasound, many of which are capable of giving a direct image of material damage. X-ray investigations that incorporate well-developed inverse modelling techniques usually provide a reliable indication and visualization of the defect's position and geometry. Research in the field of electromagnetic testing is seeking to find the 3D-distribution of a material's conductivity. This would properly describe the position and size of a material defect. But existing calculation methods to invert a magnetic field map into a sample's electrical conductivity do not entirely solve the inverse problem. Until now, the electromagnetic inverse problem has not even been fully solved for 2D-problems and simple 3D-geometry of a sample, let alone for configurations involving technically relevant EC excitation schemes and the measurement of complicated 3D- samples in which the induced currents flow in all three directions.

The assumption that there can be no unique solution to the 3D-inverse problem was first spelled out in Helmholtz's theorem. The expansive work that has nevertheless been undertaken on this mathematical problem has not sought to refute Helmholtz, but rather to simplify the interpretation of magnetic field data. An excellent overview with details of the magnetic inverse problem with respect to SQUID NDE is given by J.P. Wikswo [39].

The most common approach for two-dimensional current distribution $j\ (x,y)$ inverted from a magnetic field map is the Fourier-transform spatial filtering, first developed by W.J. Dallas [40, 41]. It was the group headed by J.P. Wikswo which first introduced it in the field of SQUID NDE. Further methods were investigated by Wikswo and co-workers to determine a 2D current distribution such as dipole fitting, Hosaka Cohen transformation, alternating projections, FEM and blind deconvolution.

Besides these 2D-approaches, many efforts have been undertaken [42-44] to tackle the inverse problem using neural networks to classify the defect geometry. For this purpose as much information as possible is required to successfully extract features and classify a material defect's geometry. In order to obtain more specific data about the sample one can use sensor arrays, sensors with different alignments measuring different field components or several excitation frequencies. The determination of defect parameters by neural network techniques was shown by the Giessen group in [45]. Further activities are in the pipelines to address ill-conditioned 3D-inverse problems by means of statistical methods such as the Bayesian image reconstruction method, which seeks to match the most probable current distribution to the measured field distribution. This method is frequently used to reconstruct x-ray data sets [46].

Regardless of the advantages or disadvantages of the above-mentioned methods, the quality of inverted data always depends on the quality and amount of the measured field values. Noise in the magnetic field data or too large a SQUID-to-source separation can significantly hamper the capability of finding an inverse solution to the measured field distribution.

It is clear that the SQUID in particular has the potential to speed up research into the magnetic inverse problem because of its ability to measure magnetic fields in a point-like manner and with an excellent signal-to-noise ratio. And it is likely that the demand for magnetic imaging in future SQUID NDE applications will soon spur an improvement in available imaging techniques.

3. POSSIBILITIES AND LIMITATIONS OF SQUID NDE

This chapter discusses the possibilities presented and limitations posed by SQUID NDE. Several examples are used to illustrate its promising potential and highlight areas in which the technology is already proving its worth in nondestructive testing applications which rely on the SQUID's high performance. But the technology also poses limitations, and before these extremely sensitive sensors can be implemented in real-life industrial applications, there are often a number of hurdles that need to be overcome in order to fully exploit the SQUID's high field sensitivity.

3.1 Advantages of SQUID in Nondestructive Testing

The main reason for contemplating the use of a SQUID in NDE applications is its excellent flux resolution which offers a higher degree of sensitivity in all aspects of electromagnetic testing than conventional systems. This makes the SQUID suitable for the detection of material defects such as small ferromagnetic inclusions with low magnetization, active corrosion in metals or weak magnetic flux leakage caused by tiny inhomogeneities in a ferromagnetic structure. Because the SQUID's field sensitivity is virtually frequency independent it is also well suited for low frequency ECT to detect small perturbation currents caused by material defects or non-magnetic inclusions hidden in conductive samples. This is in contrast to conventional induction coils measuring the rate of change of the magnetic field, resulting in reduced sensitivity for low frequency applications. The SQUID, therefore, is a promising sensor for low frequency ECT and has already proved its value to several types of electromagnetic NDE applications. There are further advantageous properties, such as dynamic range, linearity and spatial resolution which enable one to measure signals with an outstanding signal-to-noise ratio and which hold out new possibilities for NDE. These includes different kinds of signal processing, pattern recognition and inverse methods.

3.1.1 Field Sensitivity

Early high-T_c SQUID systems provided low field sensitivities at low frequencies. These were far superior to the performance of the low-T_c systems generally available at the time, whose pick-up coils and flux transformer provided field resolutions down to a few $fT/\sqrt{Hz}$. To date high-T_c SQUID magnetometers, fabricated with a flip-chip flux transformer, integrated flux transformer or multiloop configuration, provide similar field resolutions of a few ten $fT/\sqrt{Hz}$ at 1 Hz [47-50]. Modern, fast read-out electronics offer this high sensitivity in a wide bandwidth ranging between 1 Hz and a few MHz.

The following measurement shows an example in which high field resolution in the pT-range is needed to ensure successful defect detection. A deep-lying defect hidden in aircraft aluminum alloy (electrical conductivity of 18 MS/m) was measured by means of ECT-technique in conjunction with a high-T_c SQUID magnetometer [51]. For this measurement a 3 GHz rf high-T_c SQUID was used with a sensitivity of about 1 $pT/\sqrt{Hz}$ and the overall sensitivity of the EC-system was 3 $pT/\sqrt{Hz}$. The currents were induced with a gradiometric coil and an excitation field of 0.5 mT_{pp} which is comparable to the fields used in conventional ECT systems. The EC frequency was adapted to 80 Hz to detect a defect at a deeper level of the sample. The graph in Fig. 14 shows the absolute value of the field distribution above a sample containing a 40 mm-long and 0.6 mm-high crack located 26 mm below the surface of an aluminum plate. One finds increased field values at the left and right edge of the defect, caused by high circular perturbation currents flowing in the x/y-plane around the defect's edge. This leads to an enhanced vertical magnetic field component above the surface. The defect could be detected with a SNR of more than 60, and a field amplitude of approximately 250 pT - six orders of magnitude smaller than the excitation field.

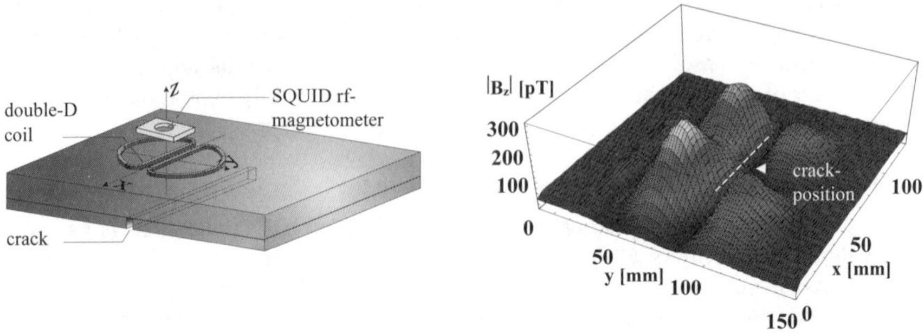

Figure 14. EC measurement using a high-T_c SQUID magnetometer as field sensor. The graph on the right shows the absolute value $|B_z|$ of the measured field above the sample's surface. Crack dimension: 40 x 1.2 x 0.2 mm^3, overall field resolution of about 3 pT/√Hz, artificial crack in a depth of 26 mm, EC frequency: 80 Hz, SNR of defect signal: 60, bandwidth: 30 Hz (from ref. [51]).

It emerged that in some NDE applications the SQUID's high field sensitivity not only enhances the probability of detection (POD) by showing up smaller material defects, it also cuts down on the time needed for taking measurements. One must keep in mind that a given field sensitivity of 1 pT/√Hz always relates to the bandwidth of 1 Hz. This means that magnetic field variations with an amplitude of 1 pT, can be detected with a SNR of only 1 when using a bandwidth of 1 Hz. But to ensure reliable evidence of material defects one generally needs to use a much higher SNR and industrial applications require much faster measurements, using bandwidths of 100 Hz to 10 kHz. This is shown in the following example in which measurements were performed to compare the SQUID system with an ECT system based on conventional induction coils. The experiment used a sensitive conventional ECT system with a pick-up coil without a ferrite core [52].

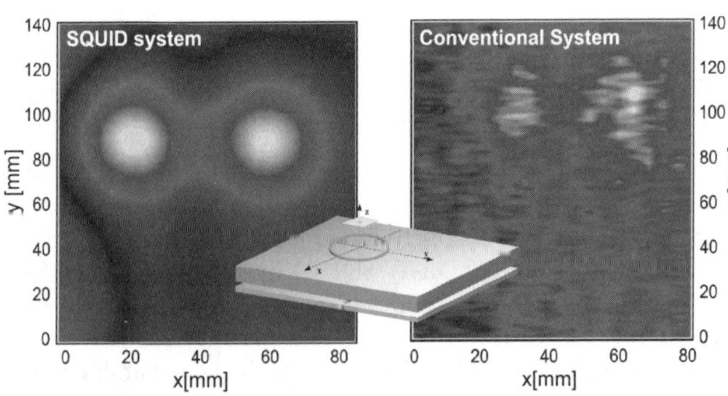

Figure 15. An artificial crack (40 x 1.2 x 0.15 mm^3) below a 13 mm thick plate of aluminium. Left: Microwave rf-SQUID-system, circular excitation coil with a diameter of 20 mm and 90 Hz excitation, bandwidth: 30 Hz, SNR: 1200, Right: Conventional EC system, induction coil with a diameter of 44 mm and 170 Hz excitation, bandwidth: 2 Hz, SNR: < 3 (from ref. [52]).

The test sample had a small flaw at a depth of 13 mm. Although the maximum amplification was used (pre: 24 dB, post: 60 dB), the crack did not show up when using a bandwidth of 30 Hz. So the field distribution above the sample's surface was measured again using a bandwidth of 2 Hz and a corresponding scanning time of 4 minutes, which resulted in a signal-to-noise ratio that was smaller than 3. The optimum excitation frequency was determined to be 170 Hz. Figure 15, left, shows a measurement of the SQUID-based EC system. Although a considerably higher bandwidth of 50 Hz was used, the signal-to-noise ratio was improved by a factor of 400. The highest SNR was achieved with an excitation frequency of 90 Hz. This was somewhat lower than the optimum excitation frequency of the conventional ECT system, which is more sensitive for higher frequencies. Furthermore, detecting deep-lying defects like the one in Fig. 15 with a conventional system not only involves a slow scanning process, it also calls for the use of large pick-up coils. The latter often results in a spatial resolution that is too low.

3.1.2 Spatial Resolution

A further advantage of a SQUID sensor is its potential for high spatial resolution, which is determined by the diameter of the cryogenic pick-up coil or the size of the sensor itself, if no flux transformer is used. However, one has to take into account that a SQUID is exquisitely sensitive to magnetic flux and it will loose field sensitivity if the effective area of the pick-up coil or the SQUID loop is kept small. But because of the fact that SQUID NDE applications generally don't require the maximum achievable field sensitivities of a few fT/√Hz, it suffices to keep the SQUID loop and the field collecting area very small. The question of how much spatial resolution is required largely hinges on the individual application.

Mathai et al. [53] began in 1993 to image static magnetic fields with a combination of a high field sensitivity of 5.2 pT/√Hz at 6 kHz and spatial resolution of 66 μm. In the same year, Vu and Van Harlingen [54] also achieved a field sensitivity of 100 pT/√Hz and 10 μm spatial resolution. These were followed by numerous developments, including the works of Tsuei et al. (1994), Black (1995), Lee et al. (1996) [55-57].

But of course the spatial resolution is not only controlled by the size of the sensor. The space between the sensor and field source is another key determining factor in the spatial resolution. This poses a challenge for the construction of SQUID NDE systems, requiring the development of a cooling environment that allows the SQUID to come as close as possible to the material being tested at room temperature.

It's not just the exploitation of the SQUID's high spatial resolution that speaks for its use in an adapted cooling environment. It also can detect a minute field source under examination at a closer range with higher SNR, because magnetic field sources are inherently dipolar in nature, thus decreasing with the inverse cube of distance ($1/r^3$). Even when measuring current distributions which fall-off as $1/r$ or $1/r^2$ one can achieve a better SNR by placing the SQUID as close as possible to the sample. The latter reasons were a strong motivation for many groups to develop SQUID systems with sophisticated cryogenic devices. These made it possible to greatly reduce the SQUID-to-sample separation. To date, this branch of SQUID technology is called SQUID microscopy, which can attain a SQUID-to-sample separation of less than 30 μm [58].

The distance between SQUID and sample can be minimized by constructing an adapted dewar. The SQUID is placed in the vacuum below the cryogen reservoir, as near as possible to the sample at room temperature. Normally the SQUID is cooled by a high-thermal conductivity link of sapphire or copper. Alternatively the SQUID can be cooled by cryogen flowing in a channel adjacent to the SQUID. For SQUID microscopes using an adapted cryostat the SQUID is separated from room temperature by a very thin (a few tens of μm) window. A known problem is the bowing of the window under the vacuum which enlarges the stand-off distance. Using window materials like sapphire, iridium, tungsten, silicon, or silicon-nitride can minimize this effect. SQUID microscopes with a window to separate the SQUID from the sample typically can reach a spatial resolution of about 50 μm. This approach normally uses various pick-up mechanisms, such as flux transformers or pick-up loops, that guide the magnetic flux into the SQUID loop. In these systems the experiment sample generally is scanned below the fixed SQUID, which unfortunately makes these bulky devices unsuitable for mobile applications.

To demonstrate the SQUID's potential for high spatial resolution, Fig. 16 shows a field measurement of a 20 German mark bill printed with magnetic ink. A segment of the zero was measured with a spatial resolution of about 100 μm. The SQUID magnetometer has an intrinsic flux noise of about $1 \cdot 10^{-6}\ \Phi_0/\sqrt{\text{Hz}}$ and an effective sensor area of 45 μm x 45 μm. The corresponding field resolution of such SQUID microscopes was on the order of 1 pT/√Hz [59]. While early SQUID NDE work was generally performed with SQUID systems that attained a spatial resolution of more than 1 cm, modern SQUID NDE systems have a resolution of about 1-2 mm and it is hoped that this value can be further improved for small standard cryogenic systems.

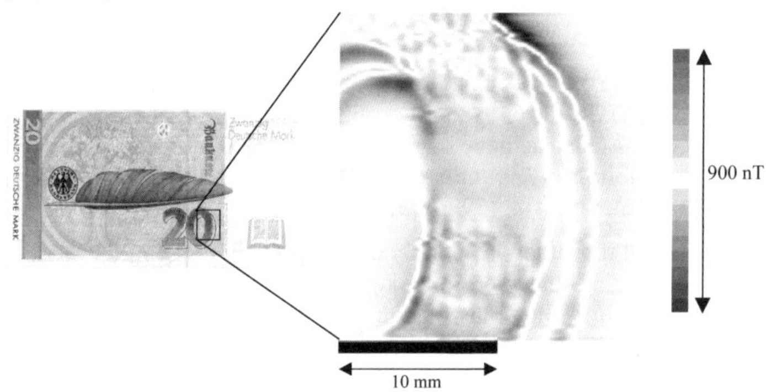

Figure 16. Measurement of a 20 German mark bill printed with magnetic ink. A segment of the zero was measured with a spatial resolution of about 100 μm (Courtesy M. Mück [60]).

3.1.3 Dynamic Range

The dynamic range of a SQUID system is as important a performance parameter in SQUID NDE as field sensitivity. It determines the operational range of SQUID read-out electronics. The importance of dynamic range was long underestimated. But there are two reasons why it is indispensable for measurements in unshielded environments, especially when using a SQUID magnetometer instead of gradiometer configurations. Movement in the earth's magnetic field or in other large background fields, as well as the presence of large excitation fields in the case of EC NDE, all call for the use of a magnetometer system which can detect fields over many orders of magnitude.

Usually the dynamic range is referenced at the SQUID output as the ratio of full scale output Φ_{max} to the flux sensitivity Φ_{noise}. This quantity is frequently represented in decibels (dB) by the relation:

$$20\,Log\frac{\Phi_{max}/\sqrt{Hz}}{\Phi_{noise}} \tag{3.1}$$

For example a field sensitivity of 1 pT/√Hz and a dynamic range of 140 dB/√Hz leads to a maximum measuring field of 10 μT. That is still smaller than the earth's magnetic field (≈ 50 μT) and thus not suitable for mobile applications if magnetometers are being used. The design of the SQUID's flux-locked loop is a key factor in determining dynamic range. Sophisticated read-out electronics can achieve values of more than 150 dB/√Hz in the case of high-T_c systems. Even higher values usually are possible for low-T_c systems. Drung et al. [61] reported on the development of wideband 5 MHz read-out electronics containing a low-noise preamplifier and a two-pole-integrator. A flux noise of $\sqrt{S_{\phi,t}} = 3.4\cdot10^{-6}\ \Phi_0/\sqrt{Hz}$ and a maximum feedback range of ± 620 Φ_0 results in a static dynamic range of 171 dB/√Hz. R.H. Koch [62] developed a wideband feedback electronics which provides an operational range of 2400 Φ_0, leading to a dynamic range of more than 180 dB/√Hz for a SQUID having a white noise level of $2\cdot10^{-6}\ \Phi_0/\sqrt{Hz}$. This means that over 9 orders of magnitude can be measured without having to switch the measuring range.

Usually the digital signal processing is incorporated into the analog electronics. Because of this the dynamic range of an analog-to-digital converter (ADC) must be taken into account and should have an operational range of the same order of magnitude as the analog electronics. For example, a digital resolution of 24 bit corresponds to a dynamic range of 144 dB. 24 bit ADCs are available in current ADC technology. To date, though, they have the drawback of still being too slow to use maximum sample rates of a few kHz. But in the case of EC testing the demodulation of the EC response at the SQUID output requires a much higher sample rate, bearing in mind that the working frequency (10 Hz to 100 kHz) should be sampled at higher rates (at least double, according to the sampling theorem). Demodulation, therefore, is often performed by analog lock-in amplifiers and their output is then digitalized by an ADC. Lock-in amplifiers with a dynamic range of better than 130 dB/√Hz are commercially available.

By contrast, the field data in standard EC testing is usually recorded in a 12 bit to 16 bit format, corresponding to an analog dynamic range of 65 dB to 85 dB. But when measuring samples that either have complicated geometrical structures or are made up of different materials, defect signals are superimposed by additional EC-effects (edge effect, lift-off effect). These often are several magnitudes larger than the field variation caused by any material defect. Furthermore, in standard EC signal processing algorithms, such as subtraction methods and phase rotation, the defect signal must be separated from an artificial background, which frequently requires a dynamic range of more than 16 bit to separate and classify the defect signal, especially when dealing with small or deep-lying defects.

The following example demonstrates a SQUID-based EC measurement in which the high dynamic range was not only used to tackle the problem of background noise but also to perform an EC measurement with an extremely high signal-to-noise ratio in the field under investigation. A simple specimen was used, consisting of two aluminum sheets with a total thickness of 2.6 mm. The upper sheet was 2 mm-thick and covered a 0.6 mm-thick sheet, which in turn contained a 20 mm-long and 0.6 mm-high defect. The probe made use of a circular excitation coil, which was driven at 2 kHz, and a high-T_c microwave SQUID for the detection of the vertical field component B_z of the response field. The measurement was performed in an unshielded environment and the peak-to-peak amplitude of the measured background noise, mainly dominated by the power line at 50 Hz, was in the range of 50 nT. The bandwidth was adjusted to 1 Hz and the field values were recorded in a 21 bit mode. The output voltage of the high-T_c SQUID was demodulated by a lock-in amplifier with a high analog dynamic range (E&G 5210). Figure 17 (a) shows a typical crack signal. It had a maximum amplitude of slightly more than 4 μT. The two right-hand diagrams in Fig. 17 (b,c) magnify the interval of the response field, indicated by the dotted rectangle in Fig. 17 (a,b). The noise amplitude was about 4 pT/√Hz, leading to a signal-to-noise ratio of about 120 dB/√Hz, which approximately corresponds to the digital dynamic range of the 21 bit data recording.

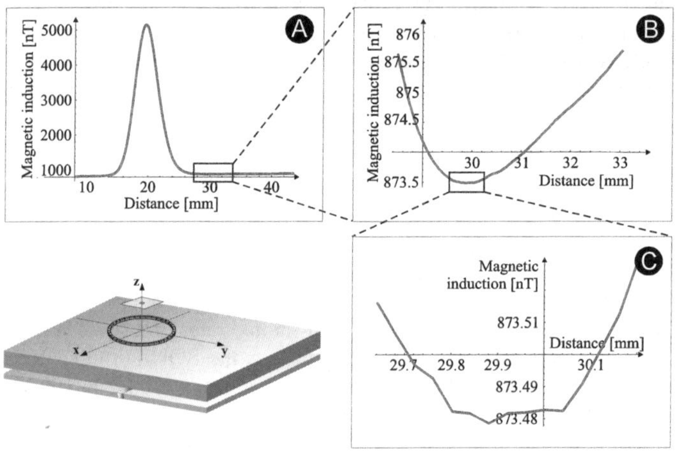

Figure 17. EC measurement with a resolution of 21 bit using a high-T_c SQUID (2 kHz excitation, circular coil) of a defect with 20 mm x 0.6 mm in size located in a depth of 2 mm. a) Crack signal, b) Magnification of the interval indicated by dotted line, c) Noise amplitude of about 4 pT/√Hz.

An example of SQUID NDE that requires such a high dynamic range is the detection of material defects in complicated aircraft samples - ones, for example, which contain bolts or rivets made from different materials, e.g. titanium alloy, surrounded by conventional aircraft aluminum alloy. A problem is presented by the fact that minute, deep-lying defects which cause small field variations are superimposed on field changes, in their turn caused by current distortions in the vicinity of the titanium bolts. Due to the size of the bolts and their different electrical conductivity, the eddy current response here is much larger than the response of the material defect. In order to separate the defect signal from the background, several steps in signal processing have to be performed. As an example Fig. 18 shows a line scan across the rivet and the resulting field after several of these signal processing steps. First, the lift-off effect (change of response field due to varied spacing between SQUID and sample) has to be eliminated by subtracting a linear function (Fig. 18, line 2). The second step contains a phase rotation to minimize the rivet signal and enhance the defect signal on one of the lock-in amplifier's two channels. (Fig. 18, line 3). Nevertheless, the defect signal is still superimposed by a signal from parts of the rivet, located at the same depth as the defect. A third step therefore is needed to further suppress the rivet signal, namely the subtraction of a "reference" rivet signal. The residual signal will then only contain a field variation caused by the defect. This can be seen in the magnification of line 4.

Obviously the SNR of the defect signal will be insufficient if the analog or digital dynamic range of the electronics components are too low. These include read-out electronics, further electronics for signal processing (e.g. filters, lock-in amplifiers), or electronics for data acquisition (e.g. ADC). Figure 19, shows defect signals of a 10 mm-long crack located adjacent to the bolt in a depth of 10 mm. When using an ADC resolution of only 14 bit (Fig. 19, left) the SNR, after the defect signal enhancement algorithm was applied, was low. Data collected with a 21 bit ADC led to a much higher SNR (Fig. 19, right).

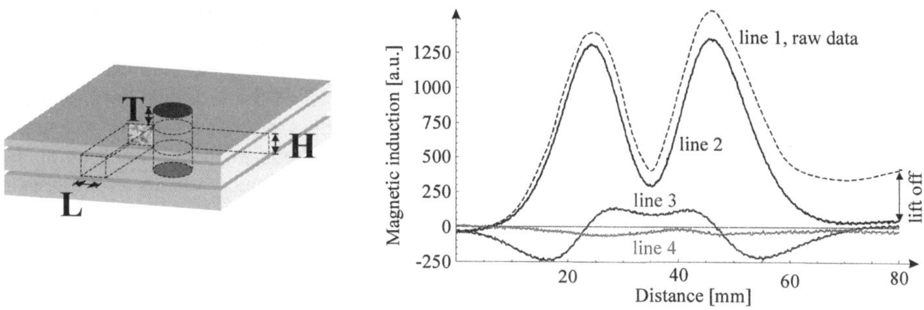

Figure 18. Line scan across a riveted multilayer sample using 180 Hz excitation frequency. Both a 5 mm and 10 mm-long crack were located in a depth of 10 mm. Line 1: raw data containing both rivet and defect signal. Line 2: suppression of the lift-off effect by subtracting a linear function. Line 3: suppression of the rivet signal by rotating the phase. Line 4: subtracting a "reference" rivet signal to minimize the residual rivet signal (from ref. [45]).

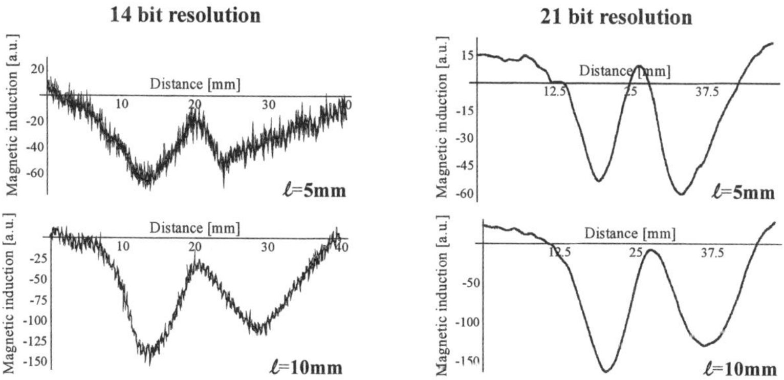

Figure 19. EC measurement of a riveted multi-layer sample, using 180 Hz excitation frequency. A 5 mm and a 10 mm-long crack was located in a depth of 10 mm. Suppression of the rivet signal by rotating the phase and subtracting a "reference" rivet signal to enhance the defect response. Left: 14 bit ADC resolution, Right: 21 bit ADC resolution (from ref. [45]).

3.2 Limitations of SQUID NDE

Turning to the limitations of SQUID NDE, the following section discusses the problems that can arise when taking very low magnetic field measurements, and presents ways of tackling these obstacles - even when the SQUID is exposed to the high-level magnetic noise of a particular environment. In principle, the limitations in SQUID NDE stem both from destabilizing influences on the measurement system, and additional noise-generating influences which affect the SQUID's output signal. These decrease the probability of detecting weak magnetic field variations. A SQUID measurement can be destabilized if the external field changes are larger than the maximum operational range of the read-out electronics. The feedback circuit can then no longer "lock on" external field changes, and the system becomes saturated. Yet an amplitude that is too high is not the only thing that can destabilize the feedback circuit. Very fast field changes also might cause instability if the maximum rate of flux change that the system is able to track without loosing lock is too low. This important property is determined by the slewrate of a system. When measuring in an environment with heavy electromagnetic pollution, SQUID systems in any NDE application are therefore usually adapted to provide very high slewrates and large static dynamic range. But even after a SQUID system has been upgraded to meet all of the technical specifications of a certain NDE application, industrial users are nevertheless put off by a few remaining impractical features of SQUID-operated NDE systems. We therefore begin our look at the limitations of SQUID NDE with the property that is most often held forward as the reason SQUIDs have so far failed to make significant inroads into NDE industry.

3.2.1 Cooling

Providing suitable SQUID cooling techniques has been a challenge ever since SQUIDs debuted on the stage of industrial applications. The requirements of a cooling environment in SQUID NDE call for i) sufficiently low spacing between sensor and room temperature sample, ii) the elimination of influences on the SQUID's field sensitivity caused by electromagnetic interference signals or mechanical vibrations, iii) sufficient temperature stability, iv) high reliability and finally, v) reasonable price. One can use two kinds of cooling environments: a liquid cryogen-filled dewar or a cryocooler.

Research into cryocoolers for SQUID-cooling was initiated by J.E. Zimmerman [63] when he developed a Stirling cooler to operate a low-T_c SQUID. Although high-T_c SQUIDs are much easier to cool via cryocoolers, a problem is presented by the fact that common Stirling and Gifford-McMahon coolers produce too much electromagnetic interference and mechanical vibrations, which are generated by an oscillating displacer in the cold expansions space. In early 1990s C. Heiden and G. Thummes [64-66] started to use a new class of regenerative cryocoolers; the pulse tube refrigerators (PTR), for SQUID-cooling. Their advantage is that they do not have any moving parts in their cold space, which reduces the vibrational noise. For example, C. Lienerth et al. [67] showed a PTR in combination with a high-T_c SQUID with a noise level of 110 fT/√Hz for frequencies higher than 10 Hz and 45 fT/√Hz for frequencies above 100 Hz.

Apart from pulse tube coolers, the Joule-Thompson cooler is also an attractive candidate for high-T_c SQUID operation in SQUID NDE due to the low level of its cold head's mechanical vibrations. For instance R. Hohmann et al. [68, 69] used a Joule-Thompson cooler for a stationary maintenance application in aircraft wheel testing. Here, electricity replaced liquid nitrogen as the cooling agent. In this application a commercial Joule-Thomson cryocooler (APD *Cryotiger*®) was successfully used for high-T_c SQUID cooling. The SQUID was mounted on a sapphire cylinder which was in thermal contact with the cold head. The head, in turn, was inside the thin bottom end of the titanium cooler with a SQUID-to-sample separation of about 5 mm (Fig. 20). A sapphire rod was chosen because of its high heat

Figure 20. Top, left: Mobile liquid nitrogen cryostat ILK6 mounted on a xy-stage, right: Mobile liquid nitrogen cryostat ILK4 (University of Giessen) with flexible adjustment of SQUID-to sample separation ranging from 2 mm to 6 mm. A high-T_c microwave SQUID was used. Bottom, left: Mobile liquid nitrogen cryostat ILK4 (Research Center Jülich) mounted on a scanning stage for the investigations of an aircraft wing. Bottom, right: Joule-Thompson closed-cycle machine cooler used for stationary SQUID cooling requiring only electricity, 500 W at 230 V developed at Research Center Jülich. (Fig. bottom: Courtesy H.-J. Krause)

conductance and, in contrast to metal, because of the absence of magnetic noise such as Nyquist noise and eddy currents. Eddy currents generated in the cylinder by varying background fields in particular can affect the performance of the FLL electronics. This can be avoided by using materials with low electrical conductivity. A mylar insulation foil was used for radiation and high frequency EMI shielding. Here, flexible plastic gas lines enabled displacements of the cold head [37]. Such systems, however, are bulky, and their use restricted to stationary operation. They are also more expensive to acquire and cost more to operate than nitrogen cryostat systems.

A. Binneberg developed a mobile cryostat for SQUID NDE at ILK Dresden. The ILK series contains lightweight nitrogen cryostats [69, 70] constructed for portable high-T_c SQUID operation. The mobile head of the ILK 4 weighs less than 2 kg and has a hold time of 12 hours. Figure 20 shows a smaller version of the so-called ILK 6, which weighs even less and has an operating time of about 6 hours. Since the SQUID is mounted on a sapphire finger in the vacuum space of the cryostat, it can be operated at any orientation.

Regardless, however, of the method one chooses to cool a SQUID, cooling systems have inevitable disadvantages: compared to other field sensors, they weigh more, have greater volume and are more expensive. In addition, because of the warm-cold separation, they cannot make contact with the sample's surface and therefore fail to exploit the full potential of the SQUID's spatial resolution.

The following NDE example shows the loss of spatial resolution if cooling systems are not adapted to the NDE application. Figure 21, left, shows the sketch of an Airbus-300 fuselage skin with its rivets, spant (strutting of the fastener row) and stringer (braces along the flight path). The external sheets are riveted to the stress-bearing framework. The samples have a length of about 0.5 m and have four rivet rows, each containing 21 rivets. The top layer of the three-layer sample has a thickness of 1.4 mm to 2 mm, the second layer is only 0.6 mm to 0.8 mm thick, and the third layer 1.8 mm to 2.5 mm thick. Due to the specific cyclic stress loading, defects occur mainly in the third layer of the fuselage skin and fatigue cracks are most likely to develop along the direction of the fastener row. The rivets have a diameter of 4.8 mm and a distance between each other of 22.5 mm.

The solid line in Fig. 21, right shows a line-scan along the rivet row using a low-T_c SQUID microscope. To achieve EC measurements with a high spatial resolution, a single wire was used, positioned perpendicular to the rivet rows to excite eddy currents in the sample. The wire was placed below the SQUID to minimize the excitation field at the location of the SQUID. This made it unnecessary to perform additional electronic compensation. The excitation frequency was 2 kHz and the separation between sample and SQUID was 600 μm. The defects adjacent to the rivets can be clearly recognized. When using wire excitation, the signal of a flawless rivet shows an increase in magnetic flux density B_z on one side, and a decrease on the other side. This is caused by induced currents which flow clockwise on one side and counter-clockwise on the other. A defect induces an additional current which leads to an increase or decrease of B_z, depending on the position of the defect. The data was raised to the third power. This made the amplitudes of the signals nearly proportional to the crack length and the defect signals were easier to separate from the background. In comparison, the dashed line in Fig. 21 shows the measurement when using a high-T_c SQUID in conjunction with a common liquid nitrogen dewar whose stand-off distance between SQUID and sample was about 6 mm. In the high-T_c system a small ring coil (3-5 mm in diameter) was used for excitation. One can easily see that the conventional cryostat provides much lower spatial resolution. Using systems with higher spatial resolution will completely overcome the drawback of the conventional dewar, which cannot locate cracks shorter than 5 mm. But unlike mobile liquid nitrogen dewars (e.g. Fig. 20), a SQUID microscope cannot be moved across the surface of the fuselage. The problem posed by the lack of mobility in high spatial resolution SQUID systems has yet to be solved.

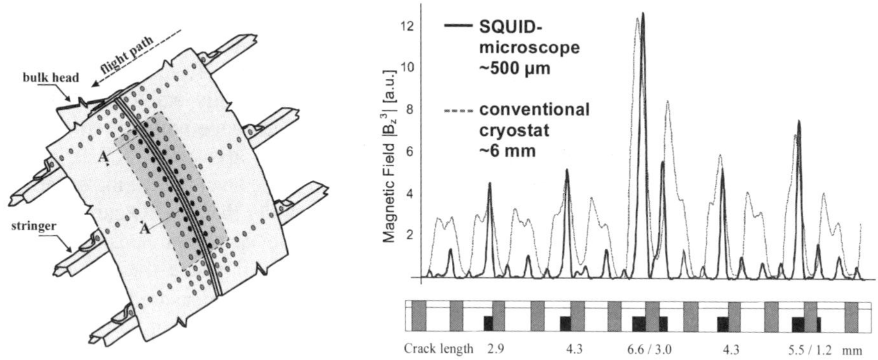

Figure 21. Left: Sketch of an Airbus A-300 fuselage skin. Right: EC measurement using a SQUID microscope compared with a conventional high-T_c system. Solid: SQUID microscope with wire excitation at 2 kHz and a dc low-T_c SQUID. Dashed: high-T_c rf SQUID in a fiberglas cryostat with small circular coil excitation at 2 kHz (from ref. [59]).

Finally one should mention that the inconvenient but necessary cooling of the SQUID is actually an advantage with respect to temperature stability. The SQUID's temperature is kept quite constant by the cryostat or cryocooler. These serve to shield external temperature variations - something that still poses a severe problem to other field sensors such as magnetoresistive sensors (GMR, AMR) or fluxgate probes, whose output signal is highly dependent on temperature.

3.2.2 Noise in SQUID NDE

In the field of shielded biomagnetic SQUID applications much attention has been placed on decreasing the SQUID's low frequency noise level to measure the magnetic fields of the brain, heart or other human organs with amplitudes in the lower fT-range. Usually in SQUID NDE the noise level of the sensor itself doesn't present as much a problem. This is because in any real, unshielded application environmental background clutter usually dominates the measured field of interest. This can be several orders of magnitude larger than the SQUID system's own noise level, which is determined by both the intrinsic noise level of the sensor itself and the electronic noise caused by the read-out electronics. The latter does not play a role in adapted SQUID systems, where the SQUID's intrinsic noise is the only contributing factor to field sensitivity. This includes flicker or 1/f noise caused by motion of flux lines (vortices) which are trapped in the body of the SQUID, or fluctuations in the critical current of the Josephson junctions. Early high-T_c SQUIDs made from polycrystalline YBCO films showed high $1/f$ noise amplitudes at low frequencies of a few Hz [71, 72]. To date, the $1/f$-knee frequency f_c, which describes the frequency at which the extrapolated values of the white noise level and the $1/f$ noise intersect, can be reduced to about 1 Hz.

Nevertheless compared to their low-T_c counterparts, high-T_c materials have the distinct disadvantage of having much lower flux-pinning energies. Flux will penetrate the YBCO film during cooling in the earth's field or movement in the presence of background fields. This leads to an increased noise level at low frequencies due to the thermally activated hopping of weakly pinned vortices. One approach to eliminate this kind of noise is to use a narrow linewidth in the SQUID's design in which vortices are prevented from penetrating the YBCO film [73-75].

But the focus of work in SQUID NDE has been less to increase the sensitivity of SQUID systems, as has been the case in alternative areas. The main thrust of work in unshielded SQUID NDE applications, remains the development of electronically stable and electromagnetically robust systems. The aim is to devise systems that can cope with extremely loud external noise sources, which can lie many orders of magnitude above the SQUID's own noise level. For unshielded, mobile high-T_c systems, common noise levels are in the order of a few pT/√Hz, whereas systems in which the sample is scanned under the SQUID can deal with even higher sensitivities in the range of 100 fT/√Hz. Whenever higher sensitivities are required the following points can be taken into consideration to reduce the influence of external noise: i) One can surround the SQUID and the sample to be measured by a μ-metal shield to screen all external background fields. ii) A SQUID gradiometer can be used so that only magnetic field gradients are measured and homogeneous background fields eliminated. iii) Higher working frequencies will usually reduce the influence of external noise, bringing it closer to the SQUID's white noise level. iv) Finally, noise reducing filter techniques also can be used.

Since SQUIDs are being used frequently in ECT applications at the moment, this section not only looks at the problems posed by external noise sources but also describes the kinds of noise influences that can occur in EC testing. Depending on the NDE application, these additional noise sources can be much higher than external noise. Aside from noise generated by the SQUID itself, one can group all types of interference into four different categories: i) external noise sources, ii) electronical noise from additional components of a NDE system, iii) vibrational noise and EC effects created by the sensor head's artificial movement (e.g. tilting and Lift-Off effect), and iv) EC effects arising from the sample itself.

3.2.2.a External Noise

External noise sources include all environmental noise, such as noise from power lines (50 Hz and its harmonics), PC-monitors (75 Hz to 150 Hz), any moving particle such as a passing vehicle, radio frequency interferences or static background fields etc.. Figure 22 shows a frequency spectrum of the environment in a common lab. One can recognize the power line noise. The amplitude$_{rms}$ of the 50 Hz peak is about 12 nT and the higher harmonics occur with decreased amplitudes. The 550 Hz peak, for example, has a field amplitude of about 100 pT$_{rms}$. The white noise level of the microwave SQUIDs is in the order of somewhat higher than pT/√Hz.

In the case of dc measurements, e.g. for the detection of tiny magnetic particles, the amplitudes of mono-frequent interferences are often higher than the field of interest to be measured. That means that even higher frequency signatures, such as the 50 Hz field variation of the power line, are superimposed on the detected dc signatures. Naturally one can also opt for one or more of a number of different filter types. Depending on the chosen bandwidth, one can use low or high pass filters, notch filters, or comb filters to suppress mono-frequent interference. The degree to which such external sources of interference can affect a measurement depends i) on the amplitude and frequency of background clutter, ii) on the chosen working frequency as well as, iii) the bandwidth, and iv) the kind of filter being used. The impact of external noise very clearly can be minimized through the use of ECT technique. Choosing a narrow bandwidth for demodulation of the working frequency can shut out surrounding interfering frequencies. To this end it is important to select an EC excitation frequency which provides the largest possible frequency distance to the nearest interfering frequency.

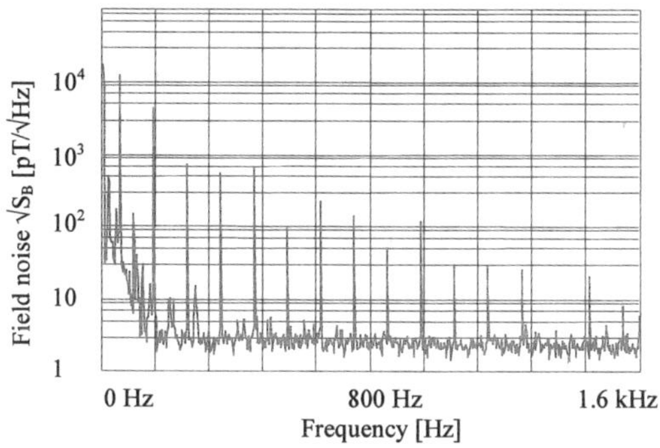

Figure 22. Frequency spectrum of a 3 GHz rf-SQUID at 77 K operated without magnetic shields. One can see the 50 Hz peak and its harmonics.

If a measurement is still seeing too much interference by surrounding noise one should try using a gradiometer instead of a magnetometer. The degree to which the external noise can be suppressed depends on the spatial homogeneity of the interfering fields. The choice of an appropriate gradiometer and the length of the baseline should be determined by the kind of external noise present as well as the type of test sample and its distance from the SQUID. Details of how to determine an adaptive gradiometer configuration can be found in [76]. Two types of gradiometer configuration are available for the subtraction of two SQUID signals - the axial gradiometer and the planar gradiometer. The axial gradiometer consists either of a flux transformer with two pick-up coils or can be operated as an electronic gradiometer in which two single SQUIDs are positioned vertically over one another and subtracted with two single read-out electronics. Like a single SQUID, an axial gradiometer with a large baseline also produces a simple response signal, simplifying data interpretation. However when it comes to voluminous samples, relatively high frequency interference (>100 Hz) can only be minimally suppressed. This is because the sample itself serves as a shield against external interference, leading to the induction of eddy currents inside the sample. The degree of shielding is a function of the vertical separation between the SQUID and the sample's surface. This results in different interference amplitudes and phase shifts at the two SQUIDs' positions, so that subtracting both SQUID signals from each other is no longer an efficient option.

This problem is not as significant when using a planar gradiometer. It can be operated as an integrated gradiometer, where a single chip contains both SQUIDs which are then positioned with equal separation to the sample. But planar gradiometers have a different drawback; they yield more complicated signatures because of their derivative $\partial B/\partial y$ along the gradiometer axis. In ECT this handicap is augmented because one usually applies a gradiometric excitation scheme which results in a further derivative $\partial B/\partial x$ orthogonal to the gradiometer axis. This complicates data interpretation [37, 77].

Both types of gradiometers have still another disadvantage. If the baseline is smaller than the SQUID-to-field source-separation, they measure the field source with a reduced signal-to-noise ratio. This particular disadvantage is shown in Fig. 23 with respect to a planar gradiometer. The detected field strength of a magnetic dipole is measured as a function of different baselines and SQUID-to-source-separations. The gradiometer's field values have been normalized to the amplitude that a single magnetometer would measure. For instance, for the detection of deep-lying defects with separations between gradiometer and defect of about 20 mm, a gradiometer with a baseline of 2 mm would only measure 14% of the magnetometer's output signal. For ratios of baseline and SQUID-to-source separation which are between 0.25 and 13.25 this loss of sensitivity can be described with an accuracy of better than 0.6 % by the following polynomial function:

$$g(z,b) = \sum_{i=1}^{8} a_i \cdot \left(\frac{b}{z}\right)^i , \tag{3.2}$$

in which the coefficients a_i are: a_1=1.35782, a_2=0.16647, a_3=1.29093, a_4=0.9721, a_5=0.34088, a_6=6.22408·10^{-2}, a_7=5.59081·10^{-3}, a_8=1.89159·10^{-4}.

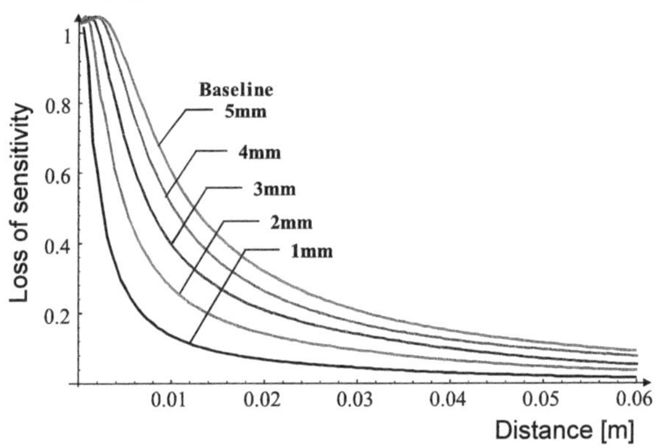

Figure 23. Relative loss of sensitivity of a planar gradiometer with respect to an equivalent magnetometer as a function of spacing between gradiometer and magnetic dipole.

Finally the earth's magnetic field is not at all static, so it also acts as an external noise source. In the course of a day one will usually measure a temporal field variation of a few 10 nT_{pp}. During magnetic storms this value can rise tremendously to more than several 100 nT_{pp}. This geomagnetic noise is caused by ocean waves or ionospheric currents [78, 79]. It mainly occurs in low frequency field variations at levels ranging from 1 pT/√Hz to 100 pT/√Hz at 0.1 Hz. In summer the activities are higher than in winter due to increased solar radiation and for frequencies above 100 Hz the geomagnetic noise is about 100 fT/√Hz. This high frequency contribution of the earth field's temporal variation will be further shielded by the building in which a NDE measurement is performed. Therefore an eddy current measurement is not affected by this kind of noise source and other background clutter is by far more significant.

3.2.2.b Challenge of Strong Magnetic Background Fields

Besides the SQUID's high field sensitivity, its dynamic behavior is another important criterion for SQUID NDE measurements. While the field sensitivity of a SQUID system is usually dominated by the SQUID itself and its intrinsic noise, the dynamic behavior is mainly limited by read-out electronics and their flux-locked loop. This includes system parameters such as maximum dynamic range, for dc-fields as well as for ac-fields, maximum slewrate and bandwidth. These properties play a key role for SQUID NDE, determining the level of permitted environmental noise, the amplitude of the EC excitation field, etc.

i) Static Dynamic Range

As already described in section 3.1.3, SQUID systems provide an outstanding dynamic range. However, sometimes the SQUID system can be destabilized by background fields, such as the remanent fields of ferromagnetic components in the order of several 100 μT to 1 mT. If the SQUID is exposed to such strong background dc-field gradients and its dynamic range is too low, the read-out electronics will saturate during sample scanning. To prevent this, a higher transfer function $\partial U/\partial B$ should be used by adapting the feedback loop. This will increase the static full-scale output. It's done by using a small feedback resistance or a feedback coil with large mutual inductance. The integrator and its maximum admissible drive current then results in large feedback fields. But one has to keep in mind that the SQUID's intrinsic noise then might be exceeded by the integrator's voltage noise, which means a loss of sensitivity.

Usually the dynamic range of a read-out electronics is specified in flux quantums Φ_0. Typical values are in the range of between $\pm 10^2$ and a few $\pm 10^3$. Therefore it is important to use SQUID sensors whose field-to-flux coefficient $\partial B/\partial \Phi$ is adapted to the specific SQUID NDE application. For instance the microwave high-T_c SQUID used in [80] has values of about 45 nT/Φ_0. Using an electronics with a static dynamic range of ± 1100 Φ_0 leads to a corresponding field value of almost ±50 μT, which enables movement in the earth's magnetic field. Hohmann et al. [77] reported on NDE measurements of aircraft wheels using a high-T_c rf washer magnetometer with a field-to-flux coefficient of 15 nT/Φ_0. Using a read-out electronics with a dynamics of ± 2500 Φ_0 achieved a maximum measurable field range of ±37.5 μT. A further example is reported by M.E. Espy et al. [81] who used a high-T_c dc SQUID with a field sensitivity of 105 nT/Φ_0 for weld inspection. Commercial high-T_c SQUID magnetometers are available from such manufacturers as Star Cryoelectronics (Mr. SQUID ™ and mini MAG ™). The "Mr. SQUID" chip has a large field-to-flux coefficient of 700 nT/Φ_0, leading to a comfortable static dynamic range in the order of some 100 μT. But one has to consider that typical noise levels of $5\cdot 10^{-5}\,\Phi_0/\sqrt{\text{Hz}}$ then lead to poor field sensitivities of a few ten pT/√Hz.

The employment of a field reset is a common technique to increase the dynamic range of a FLL when the maximum detectable field value has been reached [21, 82]. The number of resets are counted and added to the measured instantaneous field value. Nevertheless, this method has the strong disadvantage of a time delay of up to a few ms for the reset procedure. For applications requiring frequencies of more than 100 Hz, as in the case of ECT, the detected field values are affected by the time delay of the field reset and therefore inadmissible. There are other ways of getting rid of the influence of background dc fields' when measuring higher frequency fields in EC SQUID NDE and SQUID NDE applications:

To prevent the sensors from becoming saturated, we have proposed a new background field compensation scheme. Each time the dc field approaches the read-out electronics' saturation value, an additional compensation electronics generates a step-like field change. This lowers the total magnetic field seen by the SQUID and keeps the field at the SQUID within the dynamic range of the FLL. With regard to the limited slewrate this step-like counterfield is low-pass-filtered and controlled by an 8 bit counter, thus enhancing the dynamic range by a factor of 256 to 184 dB/√Hz without increasing the noise at the eddy current frequency. With this system, eddy current measurements were carried out on both aircraft wheels containing steel keys with high remanent magnetization, and aircraft lap joints with steel bolts that had a remanent field of up to 1 mT. The system is described in more detail in [83, 84].

Beyer et al. [85] developed a direct coupled FLL electronics with a frequency-dependent feedback circuit, resulting in a transfer function V_{out}/B with a high-pass characteristic. Therefore large dc fields result in lower voltage output and thus cannot saturate the read-out. Section 4.4 takes a closer look at this kind of dc compensation in conjunction with SQUID photo scanning. It is important to mention that these two solutions to the problem of background noise - step-like dc field compensation and frequency-dependent feedback - do not really increase the dynamic range of the measured field of interest. Instead, they reduce the influence of background noise in a certain frequency range of the feedback circuit's output voltage.

ii) Dynamic Range for AC Fields

In addition to the static dynamic range, the dynamic range for ac fields plays an important role either when scanning fast in the presence of static background field gradients, or when the SQUID is exposed to higher frequency signals. This is because the SQUID is operated in a feedback loop.

SQUID flux-locked loop electronics operate by "locking on" to an operating point on the SQUID V-Φ transfer function. As a result, the read-out electronics attempt to drive the flux error signal at the SQUID to zero by means of negative feedback. Because the integrator includes a capacitor, the circuit requires a finite time to react to a changing applied flux in the SQUID loop and generate the appropriate feedback. As long as changes in the flux error signal are less than $\Phi_0/4$ in magnitude, the circuit will be able to track the input signals. If the rate of change of external fields is large enough to create a flux error signal with magnitudes larger than $\Phi_0/4$, the SQUID current will exceed the critical current of a junction, and at least one flux quanta will enter or leave the SQUID. In this way the feedback electronics will "lose lock" and all information on the measured field of interest will be lost. The slewrate is the parameter that quantifies the feedback circuit's ability to keep the system locked. It measures the rate at which the SQUID system can track an incoming signal and is given by the relation:

$$\frac{d\Phi_{error}}{dt} = \frac{\omega \cdot \Phi_0}{4} \tag{3.3}$$

Because the rate of flux change is proportional to the frequency, high-frequency fields can cause even greater problems. If the amplitude of an external alternating field is ΔB and its frequency is f then the following minimum slewrate performance S_{min} is required:

$$S_{\Phi,\min} = \frac{1}{\alpha} \cdot 2\pi \cdot f \cdot \Delta B \qquad [\Phi_0 / s], \tag{3.4}$$

where $1/\alpha$ is the inverse of the field-to-flux coefficient $\alpha = \partial B/\partial \Phi$. This shows that for example power line noise is no challenge for a SQUID system's slewrate. Power line field amplitudes are usually in the order of 100 nT and common field-to-flux coefficients are about 10 nT/Φ_0. This requires a slewrate performance of slightly more than $3{\cdot}10^3$ Φ_0/s, which is about three orders of magnitude below a SQUID's average slewrate performance.

To attain a large signal bandwidth and high slewrate, the bandwidth of the flux-locked loop must be increased. But this results in more noise being fed back into the SQUID and will therefore increase the tendency for the feedback loop to "lose lock". R.H. Koch [86] theoretically determined how feedback loop noise effects the SQUID. He numerically estimated that bandwidths of a few ten MHz and slewrates of a few 10^7 Φ_0/s are achievable. There are several examples in which sophisticated wideband read-out electronics confirm such predictions to handle fast, external field variations. R.P. Giffard (1980) and F.C. Wellstood (1984) [87, 20] showed that the slewrate can be improved by using a two-pole integrator. They reached a slewrate of $3{\cdot}10^6$ Φ_0/s. R.H. Koch et al. [62] (1996) described a wideband electronics using a modulation frequency of 16 MHz. They reported on a slewrate of better than 10^6 Φ_0/s and a bandwidth of about 2.5 MHz. D. Drung [88] showed a direct coupled dc SQUID read-out electronics with a slewrate of $2.3{\cdot}10^6$ Φ_0/s (for 77 K-operation) and a field-to-flux coefficient of 5.1 nT/Φ_0, corresponding to more than 10 mT/s.

Equation (3.4) shows that slewrate requirements can be reduced by using a high enough field-to-flux coefficient $\alpha = \partial B/\partial \Phi$ to measure fast-changing and strong magnetic fields. Systems with an optimized slewrate of more than 10 mT/s are particularly advantageous when it comes to measuring configurations in which the magnetometer is moved in the presence of the earth's magnetic field and/or other background clutter. A.N. Matlashov et al. [89] developed a fast feedback-electronics using different high-T_c SQUIDs. They achieve slewrates of better than $5{\cdot}10^6$ Φ_0/s, leading to an extremely high corresponding field value of almost 1 T/s at 10 kHz.

Due to lower intrinsic noise levels, low-T_c technology is often able to achieve better slewrate performance. D. Drung et al. [22] reported on a 5 MHz feedback circuit achieving $8{\cdot}10^5$ Φ_0/s for frequencies higher than 480 kHz and a maximum slewrate of $3{\cdot}10^7$ Φ_0/s (corresponding to approximately 60 mT/s) for frequencies below 7.6 kHz. M. Mück et al. [90] developed a fast low-T_c SQUID NDE system with maximum slewrates of better than approximately 350 mT/s, capable of detecting 15 fluxquanta (525 nT_{pp}) at 100 kHz. This is also described in more detail in section 4.3.

3.2.2.c Electronical Noise

The second type of noise comes from the electronically generated field noise of additional components of the EC system. This can comprise the noise of the sinusoidal EC excitation, the lock-in amplifier or even the digital noise of the ADC at the measurement board in the PC. As an example Fig. 24 shows the frequency spectrum of an EC measurement using a high-T_c microwave SQUID and a low noise EC excitation generator (Kenwood AG204 & amplifier TDA 2003). The excitation field strength was about 400 μT_{pp} at 170 Hz using a gradiometric coil (double D shape). To compensate the strong excitation field at the location of the SQUID, the SQUID was placed exactly between the two halves of the gradiometric coil. The SQUID thereby only sees an EC excitation peak-to-peak field of approximately 4 μT, and the noise influence of the excitation unit is also suppressed. With respect to the white noise level of 2 pT/√Hz, which is not affected by the current excitation, this results in a corresponding SNR of 128 dB/√Hz of the measured working frequency.

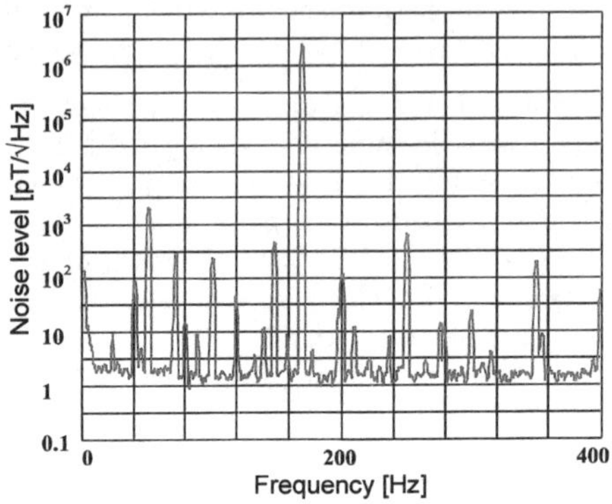

Figure 24. Frequency spectrum of a 3 GHz rf-SQUID at 77 K during an EC measurement in an unshielded environment. One can see the 170 Hz peak of the EC excitation.

3.2.2.d EC Effects Created by the Sensor Head's Artificial Movement

ECT is a popular method of getting rid of the influence of background clutter, especially external noise at lower frequencies. However, further noise sources can occur that are caused by the sample itself and its interaction with the excited eddy currents. This third kind of noise source stems from the EC effects created by the sensor head's artificial movement - such as vibration, tilting and a variation of the sensor-to-sample separation. In the NDE community these effects are referred to as the "tilting-effect" or "lift-off-effect". They depend highly on the design of the excitation coil's geometry which should be adapted to any application [91].

i) Lift-off Effect

When the excitation coil is lifted some distance *d* from the surface of the conducting sample, a portion of the magnetic flux created by the excitation coil's current fails to reach the test material. For example, if the coil is lifted too far above the sample's surface, then none of its magnetic flux reaches the test sample. If the coil then is moved closer to the sample's surface again, more and more of its magnetic flux lines intercept the test material. This leads to more induced eddy currents and an increase of the eddy current response field above the surface. The induced currents reach a maximum density when the face of the excitation coil is placed in firm contact with the sample's surface. Lift-off effect refers to the changes in the EC output signal that occur when the excitation coil is lifted away from direct contact with the sample or the coil-to-sample separation changes. Due to the skin effect, rising excitation frequencies redouble the lift-off effect. Even conventional ECT systems can detect small changes in the coil-to-sample separation of just a few 10 μm. So to attain reproducible and avoid artificial signals, it is essential to control the lift-off. As an example, the lift-off effect is shown in Fig. 25 when using a SQUID-based ECT system.

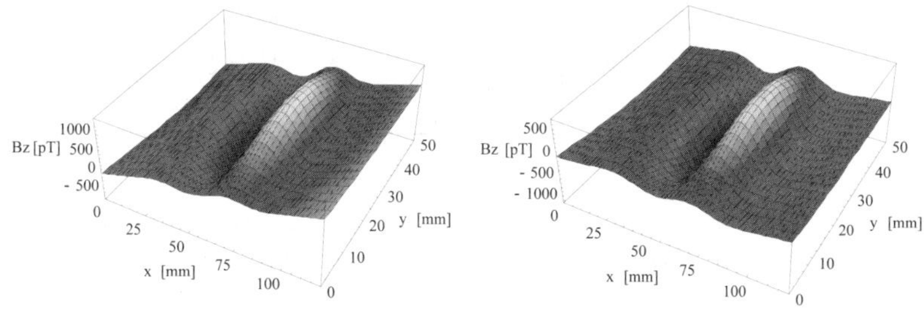

Figure 25. Lift-off effect, left: EC measurement with an overlapping lift-off signal. Right: Measurement after eliminating the lift-off signal using a high pass filter.

ii) Tilting Effect

Tilting of the excitation coil moves at least a portion of the coil-winding away from the sample's surface, thus increasing the lift-off of some of the winding. This, in turn, reduces the coupling between excitation coil and test material.

iii) Edge Effect

A third kind of EC effect is the edge effect. It occurs when measuring in the presence of edges (edges, grooves etc.). When the excitation coil moves towards an edge of the sample a portion of the magnetic flux created by the excitation coil's current fails to remain inside the test material. Because the induced currents cannot leave the sample, they are distorted at the edge. In this way the edge acts as a defect, infinite in size, leading to a strong variation of the EC response signal (Fig. 26).

These three EC effects, especially the lift-off, occur in all types of excitation coils. They can easily be measured or numerically determined by 3D FEM computations. Consequently, their lift-off effect can be corrected fairly easily if the position or the separation of the coil is determined during a measurement. This allows one to reduce the sensor handling during an EC measurement. If the lift-off is small during a measurement it is often sufficient to correct it afterwards by subtracting a linear function from the EC response or by using a highpass filter. Furthermore, these effects depend highly on the type of excitation coil and can be suppressed by using adapted excitation schemes.

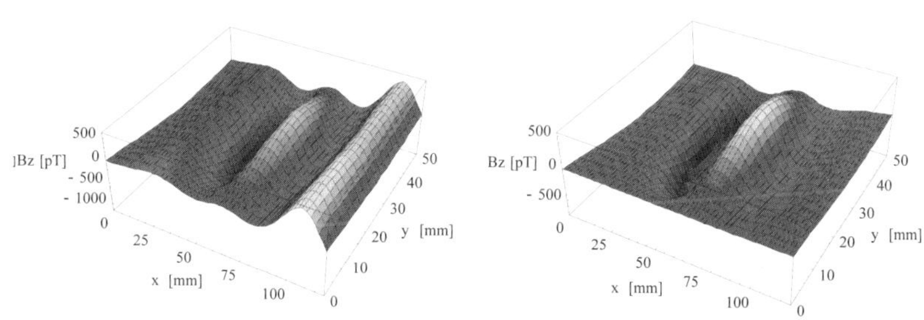

Figure 26. Egde effect, left: EC measurement with a high signal from the sample's edge covering parts of the measurement signal. Right: Measurement after eliminating the edge signal.

iv) Vibrations and Inadequate Mounting

When measuring in an unshielded environment with a SQUID magnetometer, AC background noise is not the only source of disturbance to SQUID sensors. SQUIDs also pick up all dc fields, gradients from the earth and local magnetic anomalies. In principle, the dc field and its gradients should not exercise any influence at all on a stationary NDE system, and the SQUID should only measure the field of interest. In practice, however, even stationary NDE systems move slightly because of unavoidable vibrations. In the presence of background field gradients, these natural movements create additional vibrational noise. The influence of vibrational noise on a SQUID measurement is described in [76].

In the case of eddy current testing vibrational noise which is directly measured by the SQUID's movement in the earth's field usually does not disturb the EC measurement. This is because the measured EC response is demodulated, and the detected additional noise doesn't affect the EC response if the vibrational noise occurs in a low frequency range. On the other hand, the system's vibration can cause severe artifacts due to the movement of the sensorhead. As described above, the lift-off and tilting effects cause a slight change in the separation between the sensorhead which houses the SQUID and excitation coil, and the sample. In this way the vibration directly results in a change in the EC response. Although a user can choose from a variety of filters to achieve maximum sensitivity in many SQUID NDE applications, it is nevertheless advisable to avoid moving the SQUID in an unshielded environment. It is much easier to simply move the sample being tested - particularly when measuring in the fT range. Movement in the earth's magnetic field is a viable option for SQUID NDE systems that don't require ultra-high field sensitivity. Field sensitivities of a few pT have been achieved when scanning the SQUID above the sample [51, 84]. But attention must be paid to the fact that the scanning system on which a cryostat is mounted causes its own vibrations. This is particularly prevalent in low-cost systems, where the vibrational noise just about always exceeds the SQUID's intrinsic noise level. Regardless of the type of scanning system, the use of a short moment arm to cut down on the vibration is therefore always advisable when moving a SQUID above a sample's surface. Failure to do so will result in an acceleration of the mounted SQUID system, leading to large EC effects particularly at both the beginning and end of the scanning process.

3.2.2.e EC Effects Arising from the Sample Itself

Finally, the fourth type of noise in an EC measurement is caused by EC effects arising from the sample itself. In the case of complex samples, this kind of noise is the most significant noise factor.

i) EC Effects Arising from Complex Samples

Due to the skin effect, difficulties crop up when testing complex samples, which can include edges, holes or a number of various materials close to their surface. The induced EC density is higher at the sample's surface than it is at deeper levels, giving rise to significant distortion currents at the surface near components with differing electrical conductivity. This in turn results in strong variations of the magnetic field distribution above the sample's surface. Thus in some EC NDE applications the SQUID's high field sensitivity can only be exploited through the use of additional signal processing, whereby defect-induced field variations are separated from any type of superimposed background signal. To prevent EC interference caused by the make-up of the sample itself, the detection of deep-lying cracks requires the additional use of sophisticated excitation schemes with adapted coil geometries to decrease the EC density at the sample's surface. A new approach is described in section 4.2, in which an excitation coil system is used to reduce surface currents and enhance EC density in deeper material levels by adapting superimposed excitation fields. This reduces the influence of any bolts or rivets inside the sample and allows for easier defect signal extraction from superimposed background. This structure-generated noise can also be caused by changes in the sample's thickness, a rough sample surface, a relative change of the electrical conductivity or, when testing ferromagnetic samples, a relative change in magnetic permeability.

ii) Relative Change in the Electrical Conductivity

Eddy current testing is based on material properties like electrical conductivity and magnetic permeability. A defect, such as a crack, porosities or corrosion, appears as a local step-like change in the electrical conductivity. This leads to a variation of the induced current distribution. Unfortunately, each material also shows a relative change in its spatial conductivity contribution, which also influences the current distribution even though the mechanical structure is intact. Figure 27 shows an example of an aircraft aluminum plate with a high relative change in conductivity $\Delta\sigma/\sigma$ of about 3% and the field distribution above the 10 mm-thick plate. A high excitation field amplitude of 4 mT was used, to make visible the effect of the local change of conductivity. The EC response peak-to-peak field amplitude, measured with a microwave high-T_c SQUID, was a few 100 pT. This showed that the possibility of exploiting the SQUID's high-field sensitivity also depends on the quality of the aluminum alloy being tested.

A worse problem of the same kind usually arises when testing ferromagnetic samples, because the relative change in permeability is generally a fair degree larger than that of the electrical conductivity that leads to a significant local change in penetration depth. For this reason it is common in ECT to saturate a ferromagnetic sample during a measurement, so as to reduce the local magnetic inhomogeneities. But attempting to operate a super sensitive SQUID sensor in the face of saturation fields in the order of 1 T is a futile undertaking.

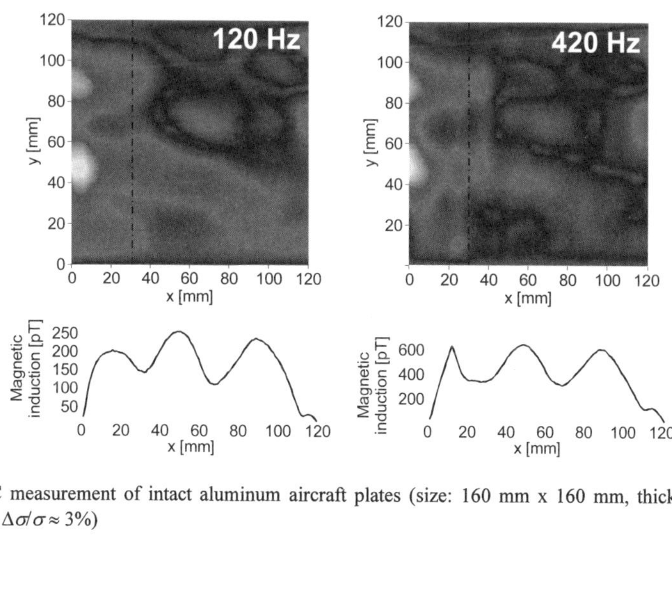

Figure 27. EC measurement of intact aluminum aircraft plates (size: 160 mm x 160 mm, thickness: 10 mm, $\sigma = 20$ MS/m, $\Delta\sigma/\sigma \approx 3\%$)

iii) Nyquist Noise

Finally one should mention that the sample actively generates yet another small noise source. It's the so-called "Nyquist noise", created by the thermal motion of charge carriers in a conducting material. Although it does not play a role in conventional ECT it can be a limiting factor in sensitive SQUID NDE applications. The peak-to-peak magnetic noise amplitude can achieve values of the order of 1 pT/√Hz, depending on the sample's conductivity and of course on the SQUID-to-sample separation. A model for the Nyquist noise of conducting samples was developed by Varpula and Poutanen [92]. They calculated the thermal magnetic noise for infinite conducting slabs and found the following approximation for the rms-noise spectrum of the z-component:

$$B_{n,z} \cong \frac{\mu_0}{2} \cdot \sqrt{\frac{\sigma k T t}{2\pi}} \frac{1}{z} \frac{1}{\sqrt{1+\left(\frac{2}{\pi} f / f_c\right)^2}}, \tag{3.5}$$

where k, t, T, σ and μ_0 are Boltzmann constant, thickness of the slab, temperature, electrical conductivity of the metallic conductor and permeability of the free space respectively. The sensor's distance from the slab's surface is described by z. Equation (3.5) was calculated by assuming that the frequency is sufficiently lower or higher than $f_c = 1/(4\mu_0 \cdot \sigma \cdot z \cdot t)$ which describes the frequency at which the rms-noise is reduced by three decibles from its zero-frequency value. For example if a 3 mm-thick copper plate is measured with a SQUID-to-sample separation of 10 mm, the frequency f_c is 457 Hz. For frequencies lower than that the noise amplitude is quite frequency-independent and in the limit of zero-frequency one yields the following approximation.

$$B_{n,z} \cong \frac{\mu_0}{2} \cdot \sqrt{\frac{\sigma k T t}{2\pi}} \frac{1}{z} \tag{3.6}$$

On the other hand Equ. (3.5) shows that for frequencies significantly higher than f_c the noise amplitude scales with $1/f$. This is because the sample itself acts as a shield, screening the high frequency fields generated by charge motion. This is good news for sensitive SQUID NDE applications that use higher frequencies.

Thermal magnetic noise should not only be taken into consideration with regard to a sample but also for the fabrication of adapted cooling systems. The thermal magnetic noise influences of cryostats containing electrically conductive components is described by N. Kasai et al [93].

3.3. Competition from Alternative Magnetic Field Sensors

SQUID sensors face some stiff competition, not just from other well established NDE methods such as conventional ECT, ultrasonics and x-ray imaging, but also from a number of alternative magnetic field sensors. These include hall-effect sensors, inductive sensors (coils), all kinds of magnetoresistive sensors (XMR; X=A-Anisotropic, X=G Giant, X=T-Tunneling), fluxgate probes, Giant Magneto-Impedance Sensors (GMI), and magnetostrictive sensors. Though none of these have sensor properties that come close to matching the SQUID's performance, some of them have advantages over the induction coils conventionally used in EC applications. In addition, they are considerably cheaper to use than SQUIDs.

The main competitors for SQUID NDE applications are the GMR sensor and the fluxgate probe. The latter can reach field resolutions of almost a few pT/√Hz, whereas to date GMR sensors, optimized for high field resolution, achieve a field sensitivity somewhere in the

100 pT-range in 1 Hz bandwidth. Both of them have the advantage of not needing to be cooled and thus being lighter, smaller and more flexible to move. On the other hand, they suffer from a significant decrease in magnetic field sensitivity if the sensing volume is reduced with the aim of achieving better spatial resolution. Sensitive fluxgates therefore have large volumes which make them ill suited for many NDE applications.

Fluxgates in general though have several key advantages over other sensors: high field sensitivity, high linearity, large dynamic range and ruggedness. As a result they are frequently used today in a wide range of applications such as terrestrial applications (e.g. geomagnetic observation), measurements in outer space (satellites) and naval applications (e.g. submarine detection). Fluxgate sensors measure the absolute strength of a surrounding magnetic field in a frequency range usually between DC and a few kHz. That means they also differ from induction coils. In contrast to coils which respond to the magnetic field strength's time derivative, leading to a decreased field sensitivity for low frequencies, the fluxgate's field sensitivity is virtually independent of frequency. In most cases, the measuring principle of a fluxgate sensor is based on the "second harmonic principle". The sensor consists of a ferromagnetic core surrounded by a modulation coil which periodically saturates the probe core with the working frequency *f*. The response field is detected by a further pick-up coil and - for a vanishing external field - does not contain a 2nd harmonic of the reference frequency. If an external magnetic field is applied to the sensor, the symmetry of the cyclic saturation process is disturbed and a 2nd harmonic of the reference signal appears, its amplitude corresponding to the external field strength.

Sensitive fluxgate probes are relatively large. They usually have a core length of more than 20 mm. Although a room-temperature sensor can be positioned in close proximity to a field source, the large sensing volume will measure an averaged magnetic field. The effective spatial resolution of such an averaged field distribution is, therefore, less than that of a point-like sensor positioned at the same distance to the field source under detection. Bartington Instruments Inc., for example, offers a commercial fluxgate probe with high field sensitivity. It contains three single probes (each 25 x 8 x 8 mm³), each of which is connected to the read-out electronics using a 0.75 m-long cable. The maximum output voltage of the sensor is limited to ±12 V, corresponding to a maximum detectable field strength of ±120 μT. The field noise of the fluxgate sensor is about 6 pT/√Hz for frequencies larger than 1 Hz. The probe is also operated with a feedback electronics which results in an excellent dynamic range of approximately 150 dB/√Hz. Besides its large sensing volume and bandwidth of only 3 kHz, a further disadvantage is the fluxgate's high temperature drift, which has been determined to be 2 nT/°C. This makes it necessary to prevent temperature fluctuations during measurements in the lower nT-range.

A second candidate for NDE application are magneto-resistive sensors such as AMR, GMR and TMR sensors. The GMR sensor (Giant Magneto Resistance) in particular seems to be an attractive alternative for the measurement of small structures. They are widely available in different configurations. Due to the high temperature dependence of their output signal, the GMR magnetic field sensor usually employs a Wheatstone bridge of four resistors and various forms of flux shields and flux concentrators. In this configuration two resistors (at opposite legs of the bridge) can be exposed to the magnetic field. The other two are located below a flux shield. When the external field is applied, the electrical resistance of the exposed resistors decreases, while the other resistor pair remains unchanged, causing a signal output at the bridge terminals. The temperature-generated signal output is thus suppressed.

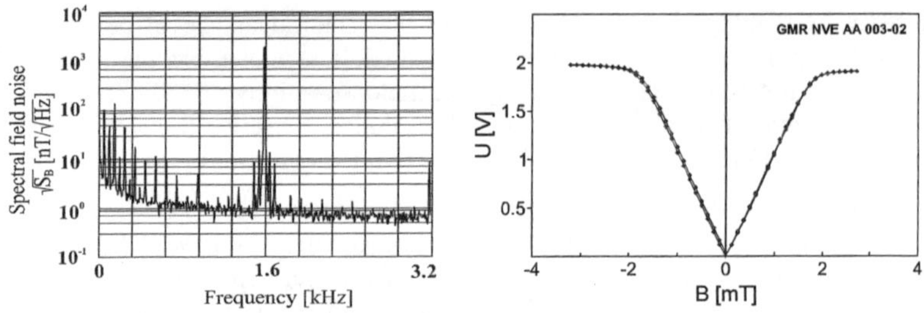

Figure 28. Left: Spectral field noise of a GMR sensor (NVE AA003-02) during an eddy current measurement with 1.6 kHz. Right: Transfer function of a GMR sensor (40 V constant Voltage supply, NVE AA003-02).

Figure 28, right shows the V-shaped transfer function of a commercial GMR sensor. A linearization of the output voltage can be realized with the help of either a small permanent magnet attached in close vicinity to the GMR sensor or the magnetic field of an additional adjusting current flowing in an adjacent wire. In this example, the working point of a GMR-NVE AA-003 sensor was then shifted to a field value of about 0.7 mT. The field noise of this GMR sensor was about 1 nT/√Hz at signal frequencies above 1 kHz (Fig. 28, left). The sensor saturated at 2 mT and provided a linear range of 1.4 mT, leading to a dynamic range of approximately 120 dB/√Hz. The non-linearity of about 1% could be reduced by using further feedback electronics. The magnetically active area of the active resistor pair was about 200 x 200 μm². However, when measuring the magnetic *z*-component, the effective attainable spatial resolution of the sensor was only about 2 mm, due to package dimensions of 4 x 5 mm².

To date, alternative sensors are still some way from offering the combined benefits of a field resolution better than 1 pT/√Hz and a spatial resolution smaller than 1 mm. Although the SQUID sensor offers these desirable properties, the high cost of an entire SQUID system and its bulky, unwieldy cooling environment serve to discourage its implementation in some applications.

4. CURRENT SQUID NDE APPLICATIONS

4.1 SQUID NDE – a Short Review

A short review follows of past and current SQUID NDE research with respect to different kinds of SQUID NDE techniques such as the inspection of ferromagnetic components and detection of magnetic inclusions, as well as the examination of conducting samples by means of different electromagnetic testing techniques.

i) Ferromagnetic Samples and the Detection of Magnetic Inclusions:

There are several examples in which SQUIDs have been used to investigate ferromagnetic samples. J. Banchet et al. and H. Weinstock [94, 95] employed them to study the magneto-mechanical hysteresis curve of different kinds of steel. G.B. Donaldson et al. [96] also measured the effect of aging at high temperatures in samples of duplex stainless steel. An example of the detection of dislocation in magnetic materials was given by N. Kasai et al.

in 1996 [97]. They used a high-T_c gradiometer to measure magnetic flux images of ferromagnetic specimen and corresponding, fine deformations caused by stress loading.

That same year, T. Nagaishi et al. [98] became the first to show the principal detection of fine ferromagnetic particles in wires by using different types of high-T_c dc SQUID magnetometers with a field resolution of about 140 fT/√Hz or 48 fT/√Hz. The read-out electronics provided a slewrate of 7.8 mT/s. The wire ran about 15 mm below the SQUID through the side holes of a cylindrical permalloy magnetic shield. Iron particles as small as 40 μm in diameter could be detected with a scanning speed of 10 m/min. The detection of defects in superconducting NbTi/Cu wires was shown two years later by H. Weinstock et al. [99] when they used a low frequency current injection technique.

A further example of the successful use of SQUID in NDE applications was the detection in 1998 of ferrous inclusions in aircraft turbine discs by Y. Tavrin *et al.* [100]. A high-T_c SQUID second-order gradiometer was used to locate such inclusions, which can lead to cracks that eventually may cause engine failure. The magnetic inclusions in the non-magnetic alloy of the turbine discs were pre-magnetized and their remanent field was measured. Lateral localization of the inclusions was possible by disc rotation and a radial scan, while the depth was determined from the $1/r^3$-dipole fall-off by sequential measurements at various depths. The approximate particle mass was then determined from its dipole strength. A mass of 1 mg could be reliably measured in depths ranging up to 70 mm.

The jülich group [82, 101, 102] used a high-T_c dc SQUID to conduct magnetic flux leakage testing in 1998. The SQUID was used to detect ruptures in pre-stressed steel tendons encased inside concrete bridges. A yoke magnet was attached to the bridge's surface to magnetize the embedded tendon with a longitudinal excitation field. Local disturbances in the magnetic field distribution, e.g. from ruptures, caused the emanation of a magnetic leakage flux from the tendon. A high-T_c SQUID was then used to detect the corresponding stray field within the yoke. The probe containing the yoke magnet and the sensors was moved along the direction of the pre-stressed tendon outside the concrete surface. For this application a high-T_c SQUID was developed made up of ramp-type junctions and a slitted washer [103]. This SQUID NDE application aimed to achieve a large dynamic range with a system that could deal with maximum excitation fields of up to 15 mT. Other works on the testing of ferromagnetic samples can be found in other publications [104, 105].

As a final example of ferromagnetic materials testing there is the work of H. Weinstock and co-workers [106] who used a SQUID to detect Barkhausen noise. They measured the jumps in the magnetization curve of polycrystalline iron pills and wires. The combined benefits of the SQUID's high field sensitivity and its large bandwidth made the sensor suitable for detecting the large frequency range of Barkhausen emission.

ii) Testing Non-Ferromagnetic Materials

Non-ferromagnetic and electrically conductive materials usually are tested by alternating fields. In 1985 H. Weinstock and M. Nisenoff [1] became the first to demonstrate that SQUIDs could be used for testing conductive structures such as copper pipes by injecting an alternating current (1 A and 4.6 Hz). This was followed by numerous works by the group headed by J.P. Wikswo, Jr. [107-109].

Conventional electromagnetic testing normally uses inductive techniques, such as ECT, for the investigation of conductive samples. This is because providing constant electrical contact when using current injection techniques can be difficult or even impossible if the structure is covered by paint or passivation. By contrast, ECT does not require direct contact with the structure under investigation. Early works on EC SQUID NDE were presented by F.R. Fickett, J.C. Moulder, and T.E. Capobianco, followed by W.N. Podney and J.P. Wikswo, Jr. [110-115].

Over the past decade work in the field of EC SQUID NDE has increasingly concentrated on defect detection in aircraft parts such as wheels, riveted fuselage lap-joints, wing splices with bolted components etc., in the hope of establishing a niche in the NDE market. The first attempt to use a SQUID for testing aircraft lap-joints was made back in 1993 by Y.P. Ma and J.P. Wikswo, Jr. [116, 117]. This demonstration used a low-T_c system. Two years later Wikswo's group [25] presented a series of examples in which current induction techniques (ECT) or current injection techniques were used to investigate sub-surface cracks, caused by corrosion or fatigue, in aging aircraft and reinforcing rods of concrete structures. In the same year, W.N. Podney [118] employed a cryo-cooled low-T_c system with a remote differential sensing coil to investigate airframes. By employing drive currents of a few amperes with frequencies up to 300 Hz, he was able to identify 1 mm size defects through several millimeters of aluminum. In addition, the system was able to detect areas where corrosion had caused a material loss of between 5 to 10 percent in an aluminum sample between 10 and 15 mm thick.

These early applications helped spark intense interest in SQUID NDE of aircraft components, and groups around the world turned their attention to this promising field. Since 1996 the giessen group has been carrying out various measurements with a high-T_c microwave SQUID on original aircraft parts. These include riveted multi-layer samples from an Airbus A320 and bolted aircraft structures containing deep-lying defects. The group was the first to measure fast-rotating aircraft wheels and developed a system to compensate large, slow varying background fields generated by ferromagnetic components in the wheel's interior [36, 45, 80, 83].

During the same period, the University of Strathclyde group [119] showed measurements of a flawed multi-layer aircraft section. H.-J. Krause et al. [120] presented a mobile, high-T_c SQUID system for ECT of aircraft using an adapted cryostat equipped with wheels for easy movement above the surface of an aircraft wing. R. Hohmann et al. [77, 37] reported on tests conducted on aircraft parts employing a high-T_c SQUID system cooled by a Joule-Thomson cryocooler. R. Hohmann et al. also gave details of a test on aircraft wheels using a high-T_c SQUID in conjunction with a remote EC technique. The SQUID, encased in either a Joule-Thompson cooler or a mobile liquid nitrogen dewar, was mounted onto a robot and integrated in a fully automated wheel-testing system. The system was able to detect inner flaws that had penetrated just 10% of the interior wall's thickness (between 8 mm and 12 mm).

We cannot end this overview of past NDE activities without mentioning two different methods that have been used to generate currents in the sample. The first is the detection of material defects using thermocurrents based on the Seebeck effect. Y. Tavrin and J.H. Hinken [121] were the first to demonstrate that thermoelectric measurements can be made using high-sensitivity SQUID magnetometers. These sense the weak thermoelectric currents around inclusions and other types of inhomogeneities when the specimen is subjected to directional heating or cooling. P.B. Nagy and A.H. Nayfeh presented the first theoretical prediction on the expected signal strength of thermocurrents in different metals containing spherical inclusions [122]. The material investigated was a 7075-T6 aluminum alloy specimen containing an inclusion with a size of 1 mm in diameter. Neglecting the SQUID-to-sample separation, they found a generated field strength of 80 nT. For their calculation they assumed a temperature gradient of 1°C/cm, a thermoelectric power of the inclusion with respect to the surrounding host of 0.1 μV/°C. Naturally the detected field is distinctly lower due to the unavoidable spacing between SQUID and sample. A second, different approach of current generation was shown in the early days of SQUID NDE by J.G. Bellingham et al. and M. Misra et al. [123, 124]. They used a SQUID to detect electrochemical currents generated by an active corrosion process in non-voltaic cells.

iii) Scanning SQUID Microscopes in NDE

The range of possible applications for a room temperature SQUID microscope is wide and not limited to NDE applications alone. SQUID microscopes are widely used in the field of biomagnetism to study the motion of magnetotactic bacteria [125], to detect an antibody labelled with tiny magnetic particles [126] or to study the propagation of action currents and potentials in cardiac tissue [127-129]. In SQUID NDE microscopy the most promising application is the visualization of weak electric currents in semiconductors or the inspection of traces on a circuit board. The latter includes the investigation of multi-chip modules as well as the fault detection of integrated circuits. To this end F.C. Wellstood and coworkers [130] developed a high-T_c scanning SQUID microscope in which a high-T_c SQUID was separated from a room temperature sample through a 25 mm thick sapphire window. They used an additional optical microscope and a *z*-stage for vertical movement to properly position the sample close to the window, leading to a SQUID-to-sample separation of up to 30 μm. Unfortunately the packaging of a microelectronic device limits the SQUID-to-sample separation to a few 100 μm. Wellstood showed that the spatial resolution can be improved by using a 2D Fourier transform which inverts the magnetic field map into a current density distribution. They showed the detection of a current path with a resolution of 72 μm (FWHM) for a SQUID-to-sample separation of 150 μm.

Naturally SQUID microscopes can be used for several other NDE purposes. Y.P. Ma et al. [131] used a high resolution SQUID system for magnetic susceptibility imaging to investigate the composition of diamagnetic or paramagnetic objects. T. Shaw et al. [132] also used a SQUID microscope to investigate effects of deformation on the local remanent magnetization of steel. They used a SQUID system which allowed for a distance between SQUID and sample of approximately 500 μm. They found a significant correlation between the measured 2D field images and different microstructures produced by controlled heat treatments and different levels of mechanical deformation.

iv) Current Applications

SQUID NDE is still an exciting research area. Each of the next three examples details a recent application of SQUID NDE to show how the SQUID's potential in NDE applications is transferred into reliable material testing. The first section looks at ECT of thick aircraft components from the planned Airbus A380 - a joint project conducted by German researchers in collaboration with the European aviation industry. Measurements are presented by the author's group at the University of Giessen. The next section examines an EC SQUID NDE system designed by M. Mück et al. to detect tantalum inclusions in niobium sheets. Finally, an example of SQUID photoscanning will be shown, developed by T. Schurig et al., in which silicon wafer or photovoltaic devices are investigated by magnetic fields caused by photo-generated currents.

Besides the above mentioned applications there are currently several other interesting high-T_c SQUID NDE projects in the pipelines. C. Carr et al. [133] has developed a fully portable cryocooler-based high-T_c SQUID NDE system. It incorporates a closed-cycle cooler which is connected to a copper thermal battery. Last year (2002) C. Carr et al. and Y. Hatsukate et al. [134, 135] reported on measurements of carbon fiber reinforced polymer, a test material first proposed by A. Ruosi in 2000 [136]. H.-J. Krause et al. [102, 137] have proposed a pulsed EC- testing method in which a calculation technique originally developed for geophysical data interpretation is used to yield quasi-tomographic EC images. There have been used several approaches to speed up the measurement process or to simplify data interpretation, such as to use several SQUIDs combined into a sensor array, to use several EC channels [45] or to use adapted excitation schemes [138]. Among those who have used a SQUID array are the group headed by H.-J. Krause and the author's group [139]. Three SQUID sensors were read-out simultaneously with a multiplexed read-out scheme, leading to a reduction in the time needed to perform a measurement of aircraft rivets and bolts. The use of several SQUIDs also has been shown by D.D. Clark et al. [140]. He performed weld quality evaluation using a high-T_c SQUID array, consisting of 11 SQUIDs evenly distributed along an 8 mm long baseline. The weld was tested by inducing EC inside the conducting weld. Preliminary data showed magnetic field variations in places with varying weld quality. J. Matthews et al. [141] recently constructed a multi-channel scanning SQUID microscope. The array consisted of 8 high-T_c SQUIDs used to image defects in wires and short circuits in computer chips.

4.2 Testing Thick Aircraft Parts of the Airbus A380

This example implemented a high-T_c SQUID magnetometer in an ECT system to measure very thick structures of the Airbus A380. The work was a German joint project uniting the European Aeronautic Defence and Space Company's EADS Airbus, national air carrier Lufthansa AG, ECT systems manufacturer Rohmann GmbH, ILK Dresden, the Research Center Jülich and the Institute of Applied Physics at the University of Giessen. The following measurements were taken by the authors group of a 62 mm-thick, bolted aluminum sample supplied by EADS-Airbus. It resembles the material structure of the planned, three-layer outer wing splice that is being proposed for the Airbus A380. The task was to detect material defects at a depth of up to 40 mm in the second layer. Previous SQUID NDE work has already shown that a SQUID can detect such deep-lying defects through the combination of field sensitivities of a few pT/√Hz and a large dynamic range of about 150 dB/√Hz [51]. But in this case the NDE challenge was far more complicated. A problem was presented by the fact that deep-lying defects which caused small field variations were superimposed onto large field changes, in their turn caused by current distortions in the vicinity of the sample's titanium bolts. To prevent EC interference caused by the sample itself, the detection of deep-lying cracks requires the additional use of adapted excitation schemes which reduces the influence of bolts or rivets inside the sample and allows for easier defect signal extraction from the superimposed background [138].

4.2.1 Simulations of Adapted Excitation Frequencies

The Airbus A380 wing splice connects the outer wing (CFR P-compound) with the inner wing (aluminum alloy) and consists of 3 layers of aluminum with a total thickness of 62 mm (16, 30, 16 mm). The layers are bolted together with 25 mm dia. titanium bolts, whose upper part is conically shaped and has a diameter of 42 mm at the surface of the sample (Fig. 29). The second layer contains an exchangeable sheet, in which defects of different lengths have been introduced adjacent to the bolt in two different depths; 16-31 mm ("upper" defect) and 31-46 mm ("lower" defect). The conductivity of the aluminum was about 20 MS/m, whereas the conductivity of the bolt, made from titanium alloy, was more than one order of magnitude smaller, leading to strong current distortions in the vicinity of the bolt.

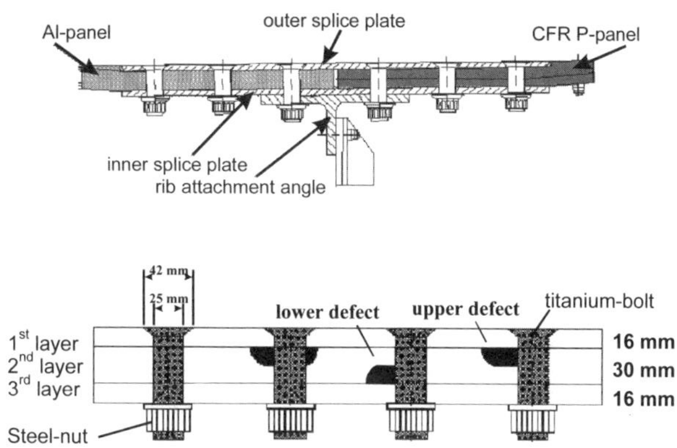

Figure 29. Sketch of a projected A380 wing splice (from ref. [35]).

Due to the induction law and skin effect, the excitation frequency must be chosen correctly in order to obtain a sufficiently high eddy current density at the defect [45]. However, when testing real samples, e.g. multi-layered samples containing titanium bolts, the aim is to provide an easy distinction between a defect and a bolt. The required frequency is not always the one that creates a maximum eddy current density at the defect. For this purpose one can make use of the depth-dependent phase shift between the magnetic fields generated by defects and bolts in different depths. In order to suppress the disturbing signal of the bolt, the excitation frequency can be chosen so that there is a phase shift between defect and bolt signal of $(2n-1)\cdot\pi/2$. In this case, the in-phase output of the demodulating lock-in amplifier mainly contains the defect signal, whereas the quadrature output mainly contains the bolt signal. In order to achieve an optimum signal-to-noise ratio of the defect response, calculations were carried out by means of 3D-FEM modeling.

In Fig. 30 the normalized field amplitude caused by "upper" and "lower" defects in the second layer is shown on the assumption that the bolt's signal is simultaneously minimized by phase rotation. The simulation shows that compared to the bolt signal, the field amplitude variation caused by the "upper" defect is smaller by a factor of 18 (5.5 %). The "lower" defect only gives rise to a field strength of about 1.2 % of the bolt's signal amplitude. Optimum excitation frequencies are 38 Hz for the "upper" defect and 22 Hz for the "lower" defect.

This simulation was made on the assumption that a circular coil excited the eddy currents. When using different excitation schemes the optimum excitation frequencies are shifted towards lower or higher frequencies and the amplitude of the ratio of bolt signal and defect signal appears to be different.

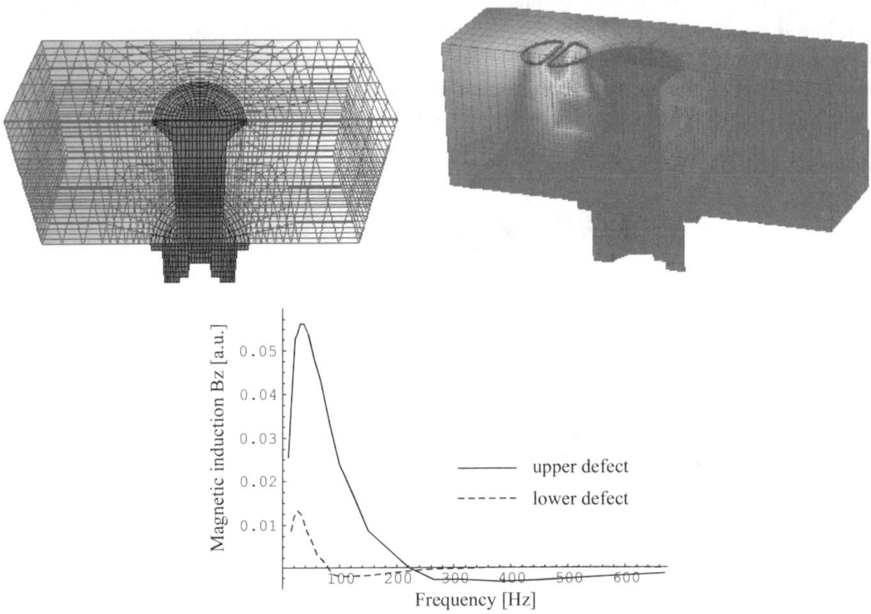

Figure 30. Left: FE-mesh of A380 wing splice bolt structure, Right: EC distribution in the A380 wing splice sample using a double-D excitation. Bottom: Field amplitude of the in-phase channel of the projected A380 wing splice calculated for different excitation frequencies by 3D FEM-simulations. The field amplitude is normalized to the bolt signal's amplitude using a circular excitation coil with 30 mm in diameter (in part from ref. [35]).

4.2.2 Measurement Configuration

This measurement used two excitation frequencies. One of them was chosen to provide depth-selective information about the defect, the second to localize the position of the bolt for easier distinction between bolt and defect signals. Figure 31 shows the principle block diagram of the measurement configuration. The excitation unit generates alternating signals of frequency f_1 (<100 Hz) and f_2 (>3 kHz), each of which is fed to an individual inductor to excite eddy currents in the sample. Two single wires were used as excitation coils for frequency f_1 and f_2. The wires were positioned perpendicular to each other, their crossover located below the center of the cryostat. The SQUID was located right above the crossover of the two wires, which were attached to a small x/y-stage in between the cryostat and the sample and positioned so that the excitation field at the SQUID was as low as possible. Two lock-in amplifiers measured amplitude and phase of the SQUID read-out output at the two excitation frequencies.

The measurements were carried out with both high-T_c rf-washer SQUID magnetometers biased at 630 MHz [142], and high-T_c microwave rf SQUIDs [143]. The field sensitivity of both kinds of SQUIDs was in the range of 1 pT/$\sqrt{Hz}$ at signal frequencies above 10 Hz, and the field to flux transfer coefficient was about 15 nT/Φ_0 for the rf-washer SQUID and 45 nT/Φ_0 for the microwave SQUID. Equation (3.6) shows that for thick samples ($t > 20$ mm) with an electrical conductivity of 20 MS/m and a SQUID-to-sample separation of about 5 mm the Nyquist noise of the sample is in the order of 2 pT/$\sqrt{Hz}$. The use, therefore, of higher sensitivity is not required.

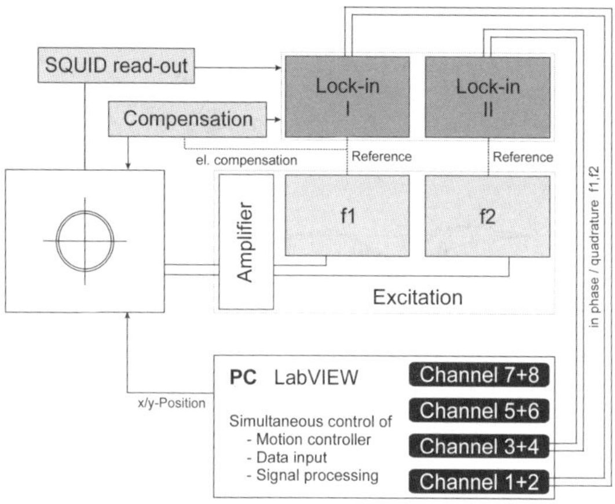

Figure 31. Schematic diagram of the dual-frequency measurement configuration (from ref. [35]).

The SQUID could be cooled using either a Joule-Thomson cooler or a liquid nitrogen dewar. The spacing between SQUID and sample was determined to be 5.5 mm or 3 mm. The size of the SQUID washer was 2.5 mm. In this application, neither the SQUID-to-sample separation nor the size of the washer affected the system's spatial resolution because the defects and thus the perturbation currents as field sources were located in a depth of more than 30 mm. More information on the cooling, SQUID integration and setup of this application can be found elsewhere [77].

4.2.3 Measurements

Figure 32 shows the measured field distribution above the sample, when using wire excitation. The wire was aligned perpendicular to the *x*-axis and the excitation frequency was 38 Hz. No adapted phase rotation was performed, so only the field variation caused by the bolt can be recognized. The defect signal is much smaller than the superimposed bolt signal and thus cannot be seen in the raw data. Several steps in signal processing had to be performed in order to separate the defect signal from background interference. First, the lift-off effect was eliminated by subtracting a linear function which resulted in curve 2 in Fig. 33. A phase rotation was then performed to minimize the bolt signal and enhance the defect signal. Curve 3 shows the result: the main bolt signal was suppressed, and the only remaining field variations stemmed from the depth in which the bolt and the adjacent crack were located.

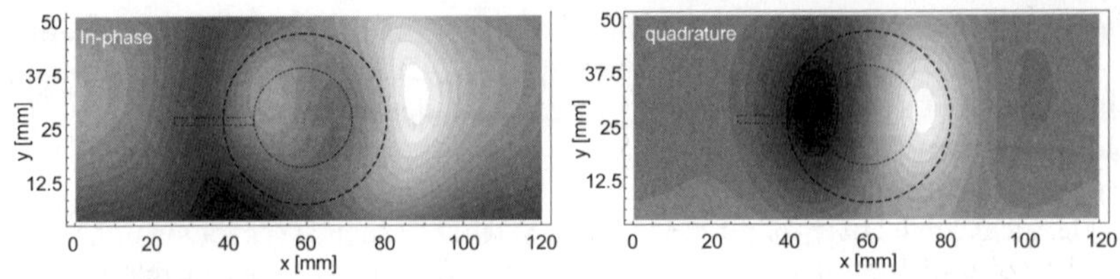

Figure 32. Field distribution above the bolted wing splice sample using wire excitation at 38 Hz. Raw data with arbitrary phase rotation, (black: low values; white: high values). Bolt and adjacent defect are indicated by the dashed lines (from ref. [35]).

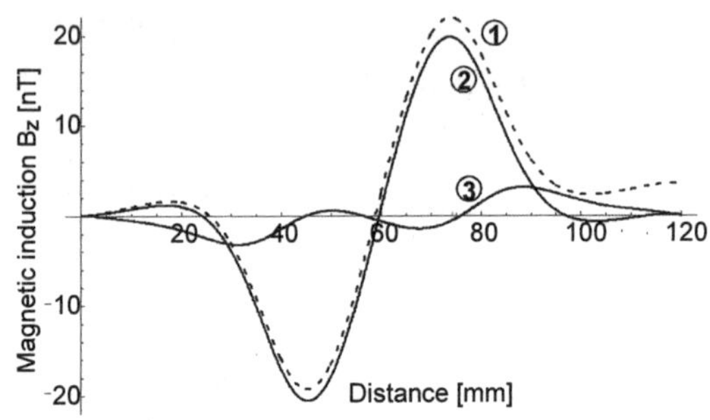

Figure 33. Line scan, extracted from Fig. 32, across a bolted multi-layer sample using 38 Hz excitation frequency. Line 1: raw data containing both bolt and defect signal. Line 2: suppression of the lift-off effect by subtracting a linear function. Line 3: suppression of the bolt signal by phase rotation (from ref. [35]).

In conventional EC signal processing both channels usually are combined into a complex plane-like diagram, a move that simplifies the interpretation of eddy current trajectories. Due to adapted phase rotation the defect signal then mainly occurs in changes along the *x*-axis. This is shown in Fig. 34, where 10 trajectories of intact bolts and bolts with 30 mm long "lower" defects are represented in a complex plane-like diagram.

In practice the reduction of the bolt signal in the in-phase channel (*x*-axis) of the lock-in amplifier is limited to a factor of between 5 and 30, depending on the excitation frequency and the shape of the excitation coil. It is worth mentioning that the axes in the complex plane-like diagram above have different scales. Because of the performed phase rotation the bolt signal gave rise to a field amplitude of slightly more than 50 nT_{rms} along the *y*-axis, whereas the field amplitude along the *x*-axis was only about 6 nT_{rms}. Despite adapted phase rotation therefore, in the case of lower defects it is the bolt that contributes the major part of the in-phase response (*x* axis). These superimposed bolt signals are caused by current distortions from the parts of the bolt located in the same depth as the defect and thus have an equal phase shift. Nevertheless, defects with a length of about 10 mm located in a depth of 16-31 mm could still be detected. The difference in the response field between the flawed and intact bolts was a few 100 pT_{rms} when using an excitation frequency of 32 Hz. This means the bolt signal was about 100 times larger than the defect signal, when using single wire excitation.

The measurements were carried out with different excitation frequencies. Because of the induction law, the eddy current response was reduced when using lower frequencies. It turned out that in practice the optimum excitation frequency was somewhat lower than the calculated values when using wire excitation instead of excitation with a circular coil. The maximum defect signal for the "lower" defect was obtained by using frequencies of about 20 Hz. But sometimes it proved sensible to use smaller frequencies, which give rise to a larger ratio of defect signal and bolt signal, to enable an easier extraction of the defect signal (Fig. 33, 15 Hz). As an example, the bolt signal was only about 20 times larger than the defect signal when using an excitation frequency of 15 Hz. However, other eddy current effects have to be taken into consideration, e.g., the edge effect, which could cause inadmissible interference with the EC measurements if frequencies are too low.

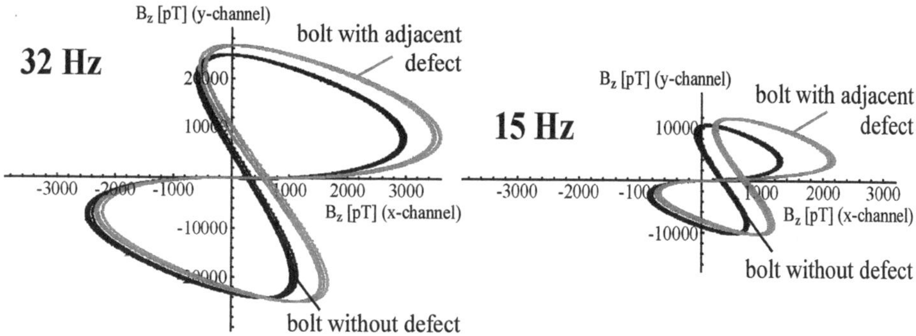

Figure 34. 10 Trajectories of line scans across an intact bolt and one with a "lower" defect, y-axis: quadrature channel; x-axis: in-phase channel. Frequency: 32 Hz, 15 Hz; defect length: 30 mm; HTS SQUID, using a bandwidth of 30 Hz (from ref. [138]).

A comparison with a conventional EC system shows that the employment of extremely low frequencies of 20 Hz to detect deep-lying defects results in an inapplicably low SNR of the EC response (Fig. 35). In order to obtain higher field resolution and sufficient penetration depth, a large coil with 50 mm in dia. and many windings was chosen for the EC measurement. However, no difference between trajectories of bolt signals could be observed regardless of whether they had a defect or not.

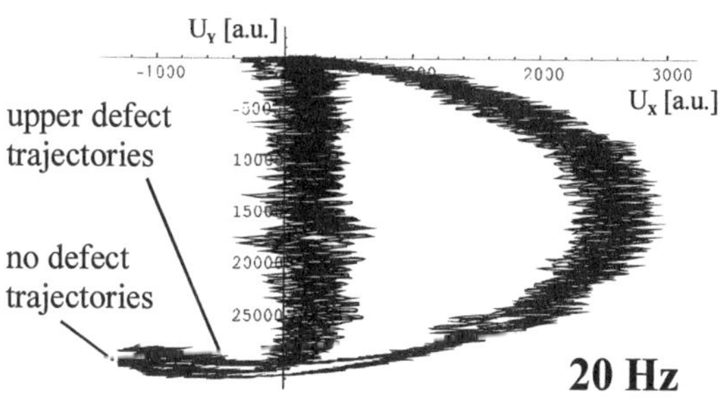

Figure 35. Trajectories of the "lower" defect signal (defect length 30 mm) using a conventional EC system with 50 mm dia. coils and a bandwidth of 30 Hz (from ref. [35]).

4.2.4 Example of Adapted Eddy Current Excitation

Due to the skin effect, the induced current density at the surface of the sample is higher than at deeper levels, thus currents flowing close to the surface generate most of the background signals caused by a bolt. One way of minimizing EC interference caused by the sample's own components is to reduce the surface currents and enhance the current density at the depth in which the crack is located. Unfortunately, magnetic fields cannot be focused to one point in the way rays of light can be concentrated in optics. Magnetic fields are always divergent - as quantified by Maxwell's Equations. To tackle this problem we superimposed two different excitation fields generated by a dual gradiometric coil excitation scheme (DGC). It consisted of two double-D coils (Fig. 36). Both double-D coils were applied with the same frequency but driven with two different excitation currents (J_1, J_2) and phase shifts (φ_1, φ_2). The use of 3D finite-element simulations determined values for J_1/J_2 and the phase difference $\Delta\varphi=\varphi_1-\varphi_2$ that compensated the induced eddy current density at the sample's surface (x = -20 mm to +20 mm, z = 0 mm). The outer coil's current J_2 was 3.7 times higher than J_1 when using a distance *d* of 28 mm and an excitation frequency of 10 Hz. Due to an adapted phase setting $\Delta\varphi$ the surface currents were even significantly reduced below the inner part of the inner double-D coil (Fig. 37).

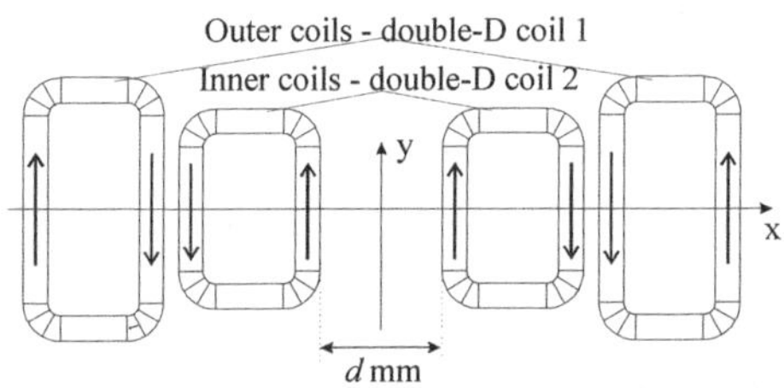

Figure 36. Dual-gradiometric coil system (DGC) consisting of 2 double-D-coils. The direction of the excitation current is indicated by the arrows along the coil winding (from ref. [138]).

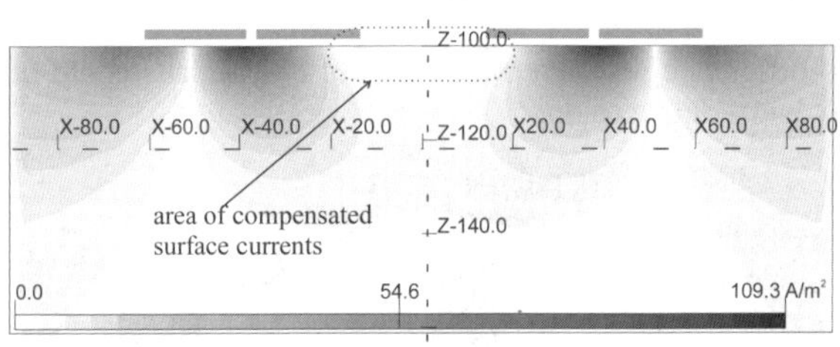

Figure 37. Induced eddy current density in the sample at an excitation frequency of 10 Hz and a distance *d* of 28 mm, calculated by FEM-simulations (from ref. [138]).

Figure 38 shows the induced EC density at different depths along the z-axis in the sample. The greater the depth, the larger the EC density, which peaks at a specific depth (here at 22 mm). The depth at which maximum induced currents occur depends on the applied frequency and the distance *d*. In comparison, the double-D coil's graph is plotted showing a maximum eddy current density at the surface of the sample and a continual attenuation for deeper levels. The induced EC density reaches its peak along the x-axis after a certain distance from the center where EC density has practically vanished (Fig. 39 a). The curve shape is similar at a depth of z = 21 mm, but through the use of the DGC the EC density in regions close to the center only dropped to half the values of the peaks to the left and right (Fig. 39 b). Because of these peaks, a defect located adjacent to the bolt gives rise to large distortion currents and is thus easier to detect. At the same time, the minimum EC density at the center reduces the influence of the rivet or bolt itself.

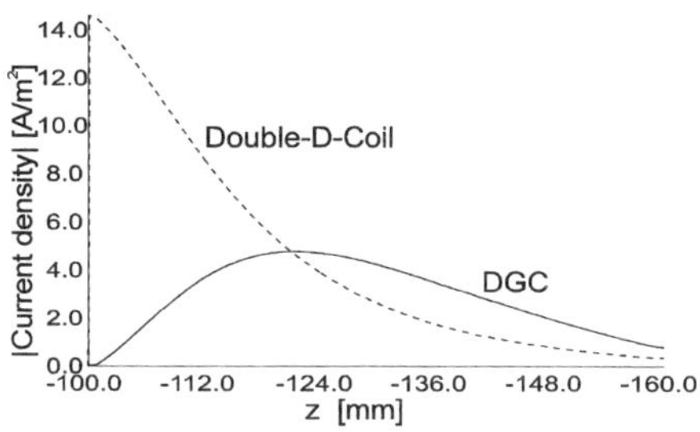

Figure 38. Induced EC density on the z-axis of the sample (x=0 mm, y=0 mm) (from ref. [35]).

Measurements on the A380 wing splice samples were carried out using the optimal settings for J_1/J_2 and $\Delta\varphi$. The "upper" defect at a depth of 16-31 mm was measured with an adapted excitation frequency of 38 Hz, the "lower" defects at a depth of 31-46 mm were detected with a frequency of 15 Hz. Because of adapted phase rotation inside the complex plane-like diagram, the defect signal mainly occurred in changes along the *x*-axis. Compared to measurements with a single double-D coil, the shape of the trajectories was more complicated. However, a higher deviation from the reference bolt signal could be recognized when using an excitation with the DGC.

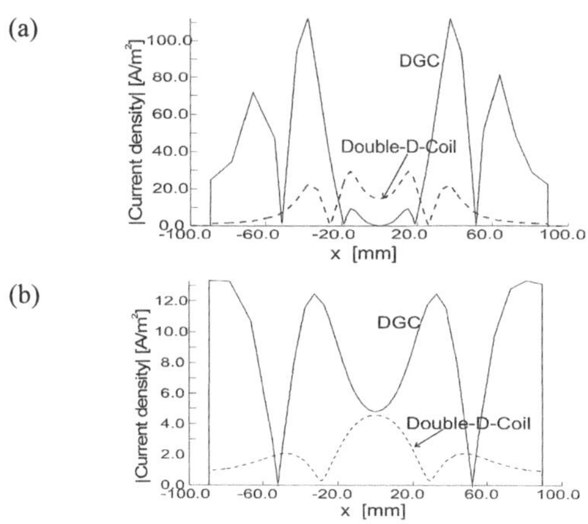

Figure 39. Induced EC density on the x-axis on the surface of the sample (a) and at a depth of z=21 mm (b) (from ref. [35]).

The "upper" defect could be detected easily with both excitation frequencies (Fig. 40 a,b). The deepest-lying defect at the depth of 31-46 mm caused only a relatively small field variation (Fig. 40 c), due to too small a distance *d* inside the inner double-D coil. The maximum value of the induced EC arose at a depth of about 20 mm, while the defect was located at a depth of 31-46 mm. However, the defect signal easily could be separated from the reference bolt signal through, for example, the extraction of different signal features using pattern recognition [45]. Compared to other excitation methods, the DGC offers two key advantages: a higher amplitude in the in-phase channel and signal shapes that are easier to differentiate from the bolt signal.

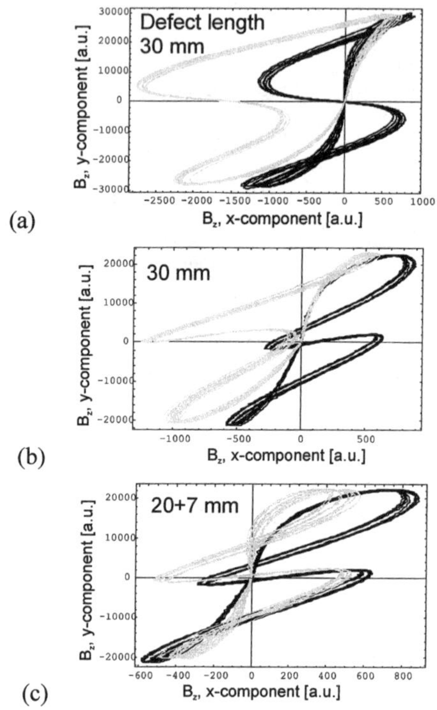

Figure 40. Measurements of bolts with adjacent cracks of different lengths at different depths (gray: bolt with adjacent crack, black: intact bolt), 10 measurements for each bolt. a) 38 Hz, defect depth 16-31 mm; b) 15 Hz, defect depth 16-31 mm; c) 15 Hz, defect depth 31-46 mm (from ref. [35]).

The successful detection of defects in thick aircraft parts, such as the planned wing splice of the Airbus A380, is just one example in which SQUID sensors can be applied in industrial NDE applications. In this application it is particularly the SQUID's ability to remain highly sensitive at an extremely low working frequency which makes it superior to conventional induction coils that are used as field sensors. The SQUID's field sensitivity of slightly more than 1 pT/√Hz was sufficient to distinguish between intact bolts and bolts with adjacent defects. Differences in the field variation above the sample's surface of a few 100 pT between intact bolt signals and bolts with adjacent defects could be easily detected with sufficiently high SNR.

Because airplanes don't earn money on the ground, inspection time is money for airlines. But every maintenance interval involves the inspection of numerous bolts. Airlines want both reliable and fast defect detection. To meet this demand, large bandwidths in the order of up to

100 Hz were used in this example of an EC SQUID system. Sufficient high SNR of the detected defect signal could be achieved when using high-T_c SQUIDs with field sensitivities of 1 pT/√Hz at frequencies ranging between 10 Hz and 40 Hz. This example also showed that coupling SQUIDs with adapted excitation schemes can help solve a range of NDE problems by providing a "deeper" look into a sample's structure.

4.3 Testing Niobium Sheets for Superconducting Resonators

A more than promising application for SQUID NDE is the inspection of niobium sheets for superconducting resonators [90, 144]. One example where these resonators are needed is the international accelerator project TESLA (Tera Electron Volts Superconducting Linear Accelerator) in Hamburg, Germany. An international collaboration at the DESY synchrotron in Hamburg is currently developing the project. It is a 500 GeV center-of-mass linear collider using high gradient superconducting accelerating cavities (Fig. 41). Two 15 km-long linear accelerators are required to achieve such high energies. The two linear accelerators consist of more than 20,000 microwave cavities each of which, in their turn, needs to provide a minimum accelerating field larger than 25 MV/m. A problem is presented by the fact that small inclusions in niobium, such as tantalum, can be an obstacle for the obtainable electric field strength of these cavities.

Fig. 41: Left: Accelerating microwave cavity made from high purity niobium. The superconducting cavity will be operated at a temperature of 2 K. Right: A section of the linear accelerator in clean room environment; the total length of each linear accelerator is about 15 km; approx. 20,000 accelerating units are needed for the linear accelerator. (Courtesy W. Singer, DESY)

Because niobium sheets are expensive to produce and because of the large number of accelerating units required, the sheets need to be inspected before the manufacturing of a microwave cavity can begin. To date the use of EC testing is the common technique for such inspections. Unfortunately, niobium and tantalum have a similar electrical conductivity, leading to only slight changes of the EC distribution when detecting small inclusions. Because even small tantalum inclusions with a size of 100 μm in diameter can significantly influence the superconducting properties by reducing the minimum required field strength of the cavities, a sensitive magnetometer is required for the detection of low field variation.

A German research collaboration is investigating the feasibility of using SQUIDs to test the quality of niobium samples. Its partners are DESY, Heraeus AG, measurement systems manufacturer WSK Messtechnik, and the Institute of Applied Physics at the University of Giessen.

4.3.1 Simulations for ECT Tantalum Inclusions Detection

In the following ECT simulation the author's group in giessen estimated the field amplitudes generated by small tantalum inclusions in a niobium matrix using 3D FEM calculations. Figure 42 shows the eddy current distortion of a circular tantalum inclusion embedded in a niobium sheet in the *xy*-plane (parallel to the sheet) and located in a depth of $z = 2$ mm. The center line of the *y*-axis, indicated by the dashed line (Fig. 42, right) shows the current distribution along the *y*-direction when a homogenous excitation current flows in the *y*-direction. Tantalum has a slightly higher electrical conductivity σ than niobium ($\sigma = 7610$ S/mm and $\sigma = 6930$ S/mm, respectively), which results in a somewhat higher current density inside the inclusion. This means that the perturbation currents do not flow around the defect, as in the case of cracks or inclusions with low conductivity, but tend to flow towards the inclusion's center. One therefore can observe a reduced EC density at the edges along the *x*-axis of the inclusion (Fig. 42, left). Although the different direction of this current flow causes a 180 degree turn in the direction of the magnetic moment, it has only a negligible effect on the generated field amplitude above the sample's surface. The increase of the EC density inside a tantalum inclusion, therefore, is relatively low compared to the induced current density inside the surrounding niobium matrix - namely just 6%. As a result the niobium and tantalum's similar electrical conductivity naturally also results in a field response above the sample's surface which is not as high as it is in measurements of alternative inclusion materials.

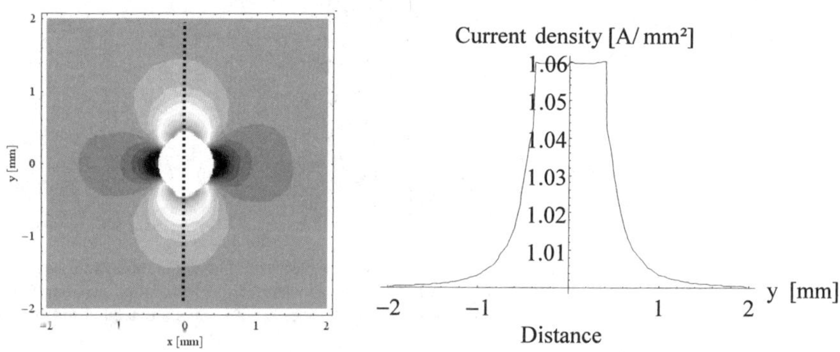

Figure 42. Left: Eddy current distortion of a 2 mm-deep circular tantalum inclusion (dia. of 800 μm) in a planar niobium sheet (FEM simulation), Right: Eddy current distortion along the dashed line in the left figure.

Figure 43 shows the magnetic field response (B_z-component - perpendicular to the sample's surface) 1 mm above a planar niobium sample for different inclusion materials. The inclusion has a diameter of 200 μm and is positioned in a depth of 1 mm. It is self-evident that inclusions of materials such as aluminum or titanium would result in a high field response due to niobium's very different electrical conductivity. Unfortunately, inclusions occur more frequently in tantalum than in other elements during the manufacturing process by means of electron-beam melting. The simulation shows that a 200 μm tantalum inclusion in a depth of 1 mm leads to a magnetic field amplitude 1 mm above the surface of 1700 pT when using a large excitation field of 0.8 mT. The response field amplitude scales with the inverse cube of the inclusion's diameter. A tantalum inclusion with a size of only 100 μm therefore results in field variations of only a few ten pT, especially when the SQUID-to-sample separation is larger than 1 mm.

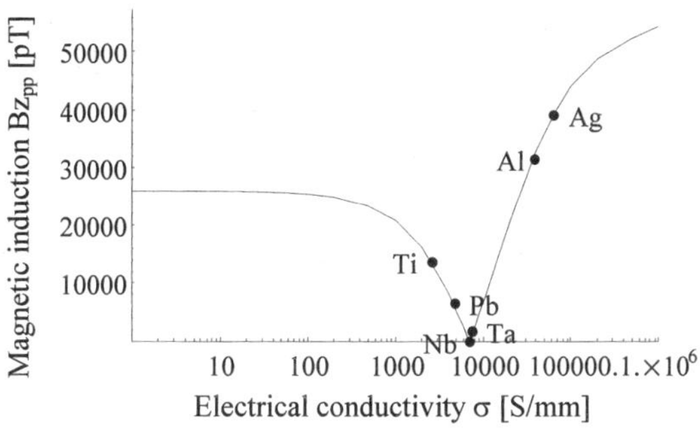

Figure 43. Eddy current response of a 1 mm-deep, 200 μm dia. circular inclusion with varying conductivity in a planar niobium sheet (FEM simulations, SQUID-to-sample separation: 1 mm). The maximum achievable magnetic field (0.8 mT excitation field) 1 mm above the sample is about 1700 pT for a tantalum inclusion.

The above simulation took an excitation frequency adapted to the skin depth to obtain a maximum inclusion-generated field response. The dependence of the magnetic field variation on different excitation frequencies is shown in Fig. 44. Such a finite-element simulation shows that the maximum achievable field variation shifts towards lower frequencies if the defect is located at deeper levels - 7 kHz (330 pT) for inclusions in a depth of 2 mm, and 23 kHz (1700 pT) for a depth of only 1 mm. Fortunately the fall-off of the frequency characteristic is not too narrow, giving one the freedom to choose a somewhat lower or higher excitation frequency than the optimum. This allows an adjustment of EC phase angles to suppress artifacts caused by the lift-off effect or surface roughness.

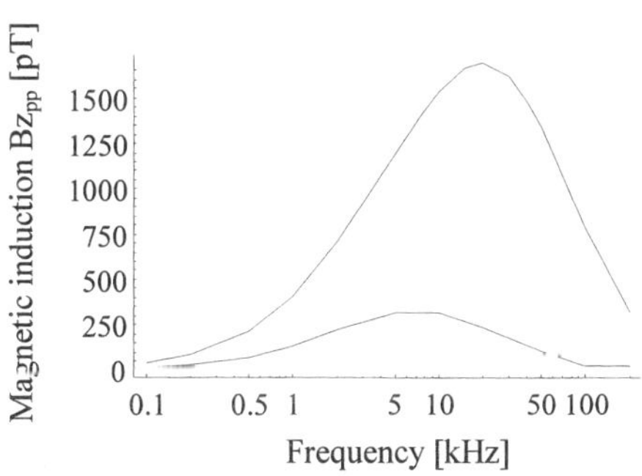

Figure 44. 3D FEM simulations of EC response of a circular tantalum inclusion with 200 μm dia. in a planar niobium sheet for varying EC frequency, SQUID-to-sample separation: 1 mm, depth of the inclusions: 1 mm and 2 mm, excitation field strength: 0.8 mT. Maximum signal 1 mm above the sample is about 1700 pT for a depth of 1 mm and 330 pT for a depth of 2 mm.

4.3.2 Measurement Setup

Figure 45 shows the principle measurement configuration for the EC testing of niobium sheets. The system, developed by M. Mück [90], used a low-T_c SQUID magnetometer with a field sensitivity of about 50 fT/√Hz and a field-to-flux transfer coefficient of 35 nT/Φ_0. The SQUID was cooled in a low-noise fiber-glass dewar for biomagnetic measurements. The cryostat had a lifetime of about one day and was filled with 1 liter of liquid helium. A circular eddy current excitation coil was placed between the cryostat and the niobium sheet and driven with excitation field values of up to 1 mT. For this reason a small compensation coil (driven with the same excitation frequency but at an adapted phase) was placed close to the SQUID to reduce the excitation field seen by the SQUID sensor. For the purpose of following fast rf interference, the system used a conventional flux-modulated FLL with a modulation frequency of 4 MHz. The voltage noise of the electronics was about 90 pV/√Hz and the dynamic range about 15 Φ_0 at 100 kHz. Finally, the scanning process was performed by a PC-controlled, high-quality *xy*-table (Newport Inc., better than 1 μm accuracy in positioning) which moved the sample under the fixed cryostat containing the SQUID. The SQUID-to-sample separation was about 6 mm. The system's high dynamic range and high slewrate make it an attractive choice for scanning in an unshielded environment. In addition, high excitation frequencies of some tens of kHz and the use of lock-in detection allow one to perform EC measurements with sensitivities comparable to the SQUID's white noise level of 50 fT/√Hz. The thermal magnetic noise of a niobium sheet at room temperature is in the same order of magnitude. By using Equ. (3.6) in section 3 the zero-frequency noise amplitude of a 2 mm thick niobium sheet is calculated to be approximately 300 fT/√Hz. For higher frequencies used in this application, of say 40 kHz, the thermal magnetic noise amplitude is reduced to about 60 fT/√Hz, making it unnecessary to choose a SQUID with a higher field sensitivity.

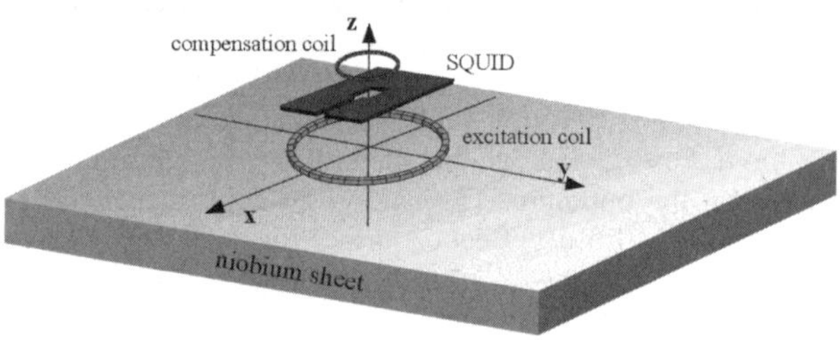

Figure 45. EC measurement configuration for ECT of niobium sheets. The excitation field is eliminated at the position of the SQUID by a minute compensation coil located near the SQUID (Courtesy M. Mück from [90]).

4.3.3 Measurements

As shown in the above simulation, the optimum excitation frequency for this NDE problem is given by the niobium's skin depth and the expected depth of the inclusions. Superconducting rf currents will only flow in the resonator in a depth of a few μm from the surface, but more than 0.5 mm will be etched away from the surface during fabrication, so the search for defects has to take place in levels deeper than 0.5 mm. To maximize the eddy current density in this depth, an excitation frequency of about 40 kHz was required. Maximum excitation currents of up to 2 A were used to achieve the highest possible excitation field in the range of 1 mT_{pp}. Due to dissipation the coil temperature was then about 60 °C.

The test sample provided by DESY for quantitative measurements had tantalum inclusions whose size and positions were known. A sketch of this sample is shown in Fig. 46. It was made of a high purity, 2 mm-thick niobium sheet. Arranged in the shape of a cross, small tantalum spheres were embedded into the niobium in eleven places by carefully melting the surface of the sample in a vacuum with an electron beam..

High purity niobium sheets used for superconducting resonators don't have a significant relative change in their electrical conductivity. This means variations in the EC response field during sample scanning are caused only by the lift-off effect and the roughness of the surface. Even small variations in the SQUID-to-sample separation of the order of 10 μm produce significant field changes which can be larger than responses from tantalum inclusions. Pressing the excitation coil firmly onto the sample during scanning can reduce the lift-off effect caused by the roughness and unevenness of the sample.

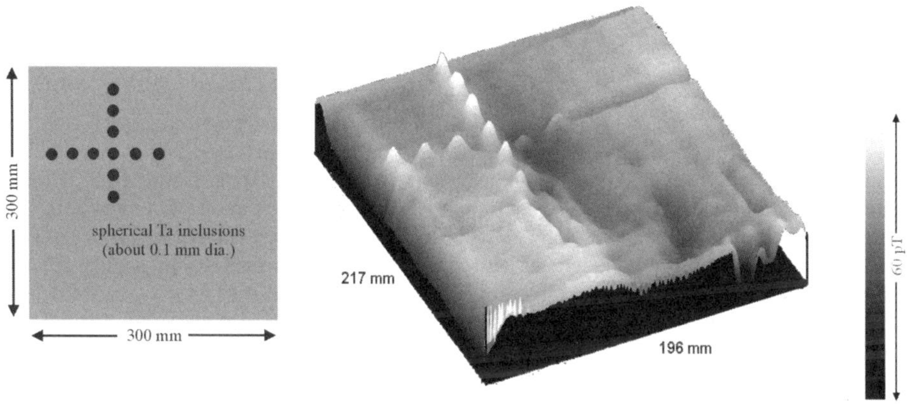

Figure 46. Left: Sketch of the test sample. By electron beam melting of the surface, 11 Tantalum spheres (each about 100 μm in dia.) were embedded into a 30 cm x 30 cm niobium sheet. Right: EC field distribution above the niobium plate. The eleven tantalum inclusions can be clearly detected with a field amplitude of about 20 pT. A small excitation coil 3 mm in diameter was used to generate an excitation field of about 0.6 mT (Courtesy M. Mück from ref. [90]).

The lift-off effect is increased when higher frequencies (100 kHz) are used. And even if the excitation coil is pressed firmly against the sample, the signals caused by the slightest tilt of the coil during scanning can obscure the inclusion-generated field response. To solve this problem the niobium sheet can be measured from the other (back) side. The phase of eddy currents is a function of the depth in which the currents flow. Therefore one can maximize eddy currents in an adjusted depth just by choosing the right reference phase for the lock-in amplifier. Eddy currents flowing simultaneously at the sheet's surface thus are suppressed in the EC signal, leading to a smaller lift-off and tilting effect.

Using this technique, all tantalum inclusions in the test sample were detected with a sufficient SNR. Figure 46, right, shows the two-dimensional distribution of the EC response field above the niobium sheet. The EC frequency was 10 kHz, the excitation field about 0.6 mT, the scanning speed 7 cm/s and the post-detection bandwidth 100 Hz. The tantalum inclusions could now be clearly detected. Some additional structures (from the rolling of the sheet) could also be seen.

Despite such large excitation fields, which cannot be increased due to dissipation of the excitation coil, the field variation caused by the small tantalum inclusions was only about 20 pT - as predicted by the simulation. In this case, fast scanning is of the essence - ultimately DESY will need to test some 30 km of niobium sheets. And because this calls for the use of large bandwidths, a field resolution of better than 100 fT/√Hz is needed to achieve sufficiently high SNR of the EC response field caused by tantalum inclusions.

This SQUID NDE example showed an application in which the SQUID's high field sensitivity can be transferred into higher detection reliability. Although a low-T_c system was used, the measurement in principal should be reproducible with high-T_c systems using fast read-out electronics. By contrast, field sensors such as induction coils, GMR sensors or fluxgate-probes cannot provide field sensitivities in the fT-range with comparable spatial resolution, making the SQUID the only candidate in the running that can measure minute tantalum inclusions in a niobium matrix.

4.4 SQUID Photoscanning

A further promising application for the employment of SQUIDs in nondestructive evaluation is photoscanning. Developed by Th. Schurig and coworkers [85] at Germany's national institute of natural and engineering sciences, the PTB in Berlin, the application was designed to inspect semiconductor samples such as silicon wafer or photovoltaic devices by detecting the magnetic field of photo-generated currents. Possible imperfections in semiconductor wafers are mainly local variations of the doping level but they can also appear as electrically active defects like grain boundaries. The recognition of artifacts in photovoltaic devices is another potential task of this application. The strength of the photo-generated magnetic field depends on whether the fields are caused by slight fluctuations of the doping level in high purity silicon wafers or by photovoltaic artifacts. In the first case, typical values lie in the range of 100 fT, in the latter they are in the order of several nT.

4.4.1 Photoscanning Measurement Setup

SQUID photoscanning sets out to detect photo-generated current distributions by measuring their magnetic field. For this purpose, the bottom side of the sample was illuminated by a (commercial diode) laser light with a wavelength of 675 nm or 980 nm focussed by an optics to a spot size of about 20 µm in diameter. This led to a power consumption on the sample's surface of several mW. It is worth mentioning that in contrast to common NDE techniques, the spatial resolution in this application is not limited by the size of the SQUID itself but by the excitation spot size. Two SQUIDs were arranged orthogonally (sensitive to the horizontal field component) to measure the B_x- and B_y-component of photo induced magnetic distribution. The scanning process involved moving the sample between the stationary optics and SQUID. Figure 48 shows the photoscanning measurement system.

The SQUID could be operated with several types of cryogenic environments: a liquid helium bath cryostat with a cold-warm distance of 7 mm, a liquid nitrogen bath cryostat with a reduced cold-warm distance of 2 mm, or a stirling cryocooler (see also Fig. 47, right). A magnetic field sensitivity of 2.35 nT/Φ_0 and a white system magnetic flux density noise of 8 fT/√Hz was achieved when using multiloop low-T_c SQUID magnetometers. When direct coupled high-T_c SQUID magnetometers were used, the field sensitivity was about 5 nT/Φ_0 whereas the white system magnetic flux density noise was about 42 fT/√Hz. As shown in Fig. 47, the SQUID output after the FLL was band-pass filtered and phase-sensitively analyzed by a DSP, which also controlled the modulation of the laser diodes. The upper limit for this modulation frequency f_{mod} - a square-wave amplitude - depends on the excess charge carrier lifetime $\tau_{ne} : f_{mod} < 1/\tau_{ne}$. For semiconductor samples τ_{ne} is in the range of 5 μs to 500 μs leading to useful values for f_{mod} of less than some tens of kHz.

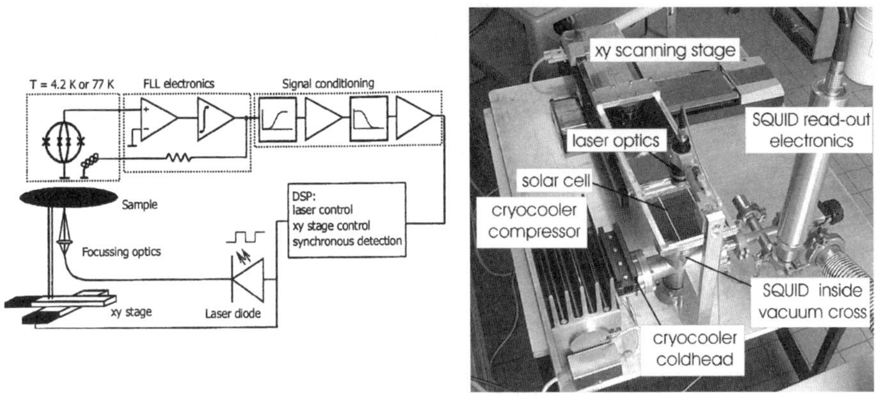

Figure 47. Left: Sketch of a SQUID photoscanning system. Right: SQUID photo scanning system using a stirling cryocooler (Courtesy Th. Schurig from ref. [85]).

The system was designed to provide high field sensitivity and thus has a relatively low field-to-flux coefficient. Despite the read-out electronics' large dynamic range, this results in a static dynamic range of only a few μT. So to use the photoscanning system in an unshielded ambience, special care should be taken to reduce the influence of external noise. Instead of expensive μ-metal shielding, HF shielding can also be used to suppress noise for frequencies above 1 kHz. But it is not effective for frequencies below about 100 Hz. External interference in this low-frequency spectrum can reach values of several 100 nT for the power line fields (50-60 Hz), and several μT at frequencies of below 1 Hz. To compensate low frequency interference, the group headed by Th. Schurig [85] in 2000 proposed the implementation of a frequency-dependent feedback.

The idea behind using a frequency-dependent feedback circuit is that it will increase the system's operational range in the lower frequency spectrum without losing field sensitivity at the modulation frequency f_{mod}. Replacing the feedback resistor with a RC network achieved a transfer function V_{out}/B with high-pass characteristics. After adjusting the RC network, the transfer function V_{out}/B became nearly constant at frequencies smaller than 100 Hz (1.6 V/μT) whereas V_{out}/B was increased by about one order of magnitude for frequencies higher than 1 kHz (Fig. 48). (The calculation of adapted resistor values is detailed in [85].)

This approach prevented low frequency interference from causing a decline in the high-T_c SQUID's SNR in the relevant laser modulation frequency range f_{mod}. Such an adapted transfer function will in all probability reduce the field sensitivity at lower frequencies. The field sensitivity then is governed by the integrator's electronical noise which is inductively coupled to the SQUID. However, it is not low frequency field sensitivity which is required in this application, but rather maximum sensitivity at frequencies around the modulation frequency f_{mod}. Generally, frequency-dependent feedback is a convenient approach when measuring higher frequencies in unshielded environments as it ensures a sufficiently high full-scale-output at low frequencies and a larger system transfer function for higher frequencies.

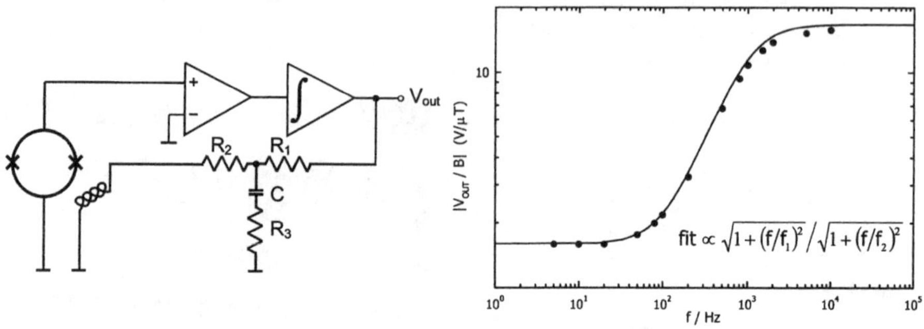

Figure 48. Left: FLL electronics with a RC network replacing the single feedback resistor. Right: System transfer function V_{out}/B of a high-T_c SQUID working in a frequency dependent feedback circuit with $R_{1,2}$ = 514 Ω, R_3 =27 Ω, C=6 µF. Points indicate the measured transfer function (Courtesy Th. Schurig from ref. [85]).

4.4.2 Measurements

i) Electronic Grade Silicon

High purity silicon is produced via the Czochralski method (CZ) or the floatzone method (FZ). In the latter case a poly-SI bar is moved through an induction coil (inert gas atmosphere). After melting, the bar rearranges into a single, high purity silicon crystal. The silicon can be additionally doped by adding dopands in the gaseous atmosphere. The FZ-method fabricates wafers with a maximal diameter of about 5" (12.5 cm).

Figure 49 a) shows a measurement (B_y-component, horizontal) of a commercial 3" n-type phosphorus-doped (n-Si:P) wafer. There are various concentric rings which can be associated with growth-related inhomogeneities of the doping level. The maximum magnetic signal is in the order of 1 pT. The results were verified through resistivity mapping and photovoltage scanning. Figure 50 b) shows a photoscanning measurement (B_x and B_y-component, horizontal) of a p-Si:B sample. The sample was cut from the core of the crystal parallel to the direction of growth. The maximum magnetic signal was 1 pT for the B_x-component and 5 pT for the B_y-component. Meniscus-like shaped striations appeared which visualized the shape of the crystal-melt phase boundary.

Visualizing inhomogeneities at the doping level – as demonstrated in this section is a way of optimizing the floatzone process by determining suitable input parameters (e.g. the temperature field of the induction heating).

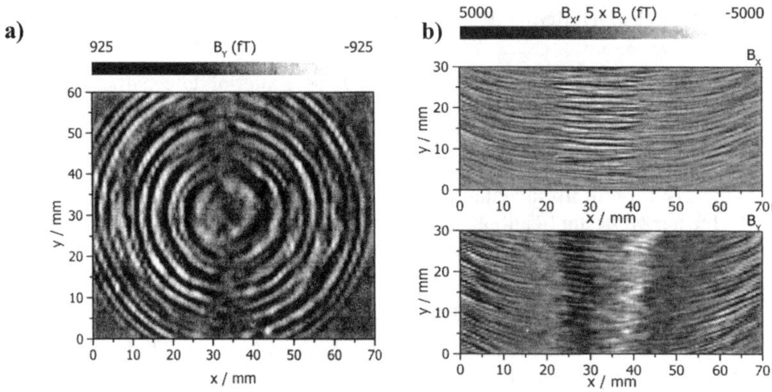

Figure 49. a) Photoscanning measurement (B_y) of a thermally oxidized 3" n-Si:P wafer (resistivity of ~70 Ωcm). The rings can be associated with growth-related inhomogeneities of the doping level. b) Photoscanning measurement (B_x and B_y) of a vertical cut FZ p-Si:B sample (resistivity of ~0.3 Ωcm). The meniscus-like striation pattern visualizes the shape of the crystallization phase boundary. (Courtesy Th. Schurig from ref. [85]).

ii) Solar Silicon

Silicon is the most widely used material for the production of solar cells. Because of the large demand for cheap solar cells, its fabrication process differs from that of monocrystalline silicon in that it is optimized to achieve a high throughput rate. This results in poorer quality solar silicon with a particularly high impurity content.

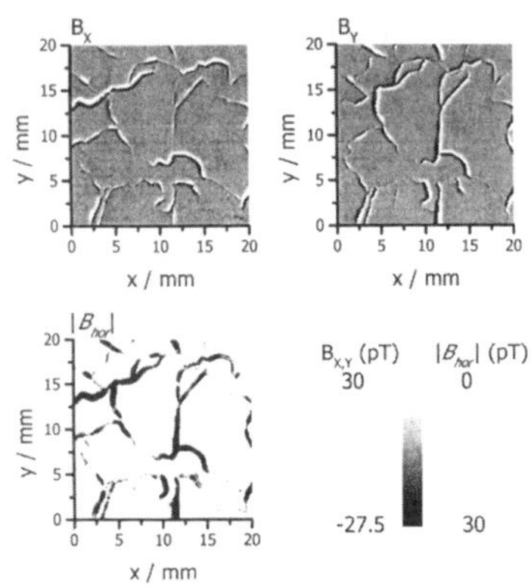

Figure 50. SQUID photoscanning B_x and B_y images of a p-Si:B sample with a resistivity of about 2 Ωcm. The signals are obtained from electrically active grain boundaries. $|B_{hor}|$ shows the morphology of the grains. (Courtesy Th. Schurig from ref. [85]).

Electrically active grain boundaries are one of the loss mechanisms that reduce the efficiency of solar cells. Figure 50 shows a measurement (B_x and B_y-component, horizontal) of a p-SI:B sample visualizing these grain boundaries. Bipolar (black-white) lines occur in the image when the laser spot is positioned in the vicinity of a grain boundary. Figure 50 also visualizes the morphology of the sample by the absolute value $|B_{hor}|$.

5. CONCLUDING REMARKS

The examples in section 4 show that a SQUID can detect material defects that remain hidden to other NDE devices. Without a doubt, its high field sensitivity translates into improved testing reliability in certain applications. No other measuring technique can at once match the SQUID's sensitivity and offer its excellent dynamic range, providing wide scope for data signal processing. But for all its advantages, the SQUID's future in NDE depends on evidence of its value to further NDE applications, and on the success of efforts to win over new users. The SQUID's unparalleled sensitivity ultimately will be of little use if the market doesn't actually put it to use.

The global market for nondestructive testing equipment and services is currently estimated at approximately 2 billion euros a year. The market is expected to see further rapid growth, fuelled by increasing quality-control demands in manufacturing and maintenance. Ultrasonics make up the lion's share of the NDE market, followed by further NDE techniques such as radiography, penetrants, magnetics, eddy currents and acoustic emission, used to test the integrity of a structure, component or material. They are employed in such fields as aeronautics and aerospace, civil engineering and the nuclear industry. The two testing methods in which SQUIDs can be applied, magnetics and eddy current testing, account for about 1/6th of the total NDE market. In choosing a particular NDE method, a user will naturally opt for the one that strikes a balance between quality control and cost-effectiveness. Because of its higher costs and inconvenient handling, SQUID technology is not always the most attractive option and might only be used where standard NDE techniques fail to ensure sufficiently high reliability.

The SQUID is by no means a guarantee for increased probability of detection in every NDE application. There are several NDE problems in which the difficulty of finding small defects stems from the sample itself rather than the sensor's noise. In section 3 we saw that a SQUID NDE measurement can be disturbed by several different kinds of noise sources. Most of them can be eliminated. But there are a series NDE applications in which the sample generates noise due to spatial variation of the material properties. For example, ECT measurements of many types of steel can be difficult due to the metal's intrinsic permeability heterogeneities. Spatial variation of the permeability can be much higher than that of the electrical conductivity. When dealing with very sensitive field measurements the conductivity heterogeneities can even become a problem in non-ferromagnetic materials and small material defects can disappear in background noise. There then is no point in using super-sensitive and expensive sensors.

Quite apart from the competition posed by alternative testing methods, the SQUID has several of its own competitors among magnetic field sensors such as flux gate probes, and magnetoresistive sensors which are already used in some NDE applications. The magnetoresistive layers of a spin valve sensor, based on the giant magnetoresistive effect (GMR), can be kept small and therefore have the distinct advantage of easy positioning in close proximity to a sample's surface. But their field sensitivity is still too low when measuring weak field sources which are far away from the sensor position. Flux gates can achieve astounding field sensitivities in the lower pT-range if their ferromagnetic core size is chosen to be more than a few cm. But this leads to poor spatial resolution, often making flux gates too large for use in many types of high field sensitivity applications. Although a lot of work is being done in the field of magnetic field sensors, room temperature sensors are still a long way from matching the SQUID's combined assets of high field sensitivity and practical spatial resolution. For this reason there is no sensor that can compete with the SQUID in NDE applications that succeed in turning high field sensitivity into improved probability of detection. The goal now is to find further applications in which SQUIDs can be used for enhanced testing quality.

Much of recent SQUID NDE work has moved beyond the stage of basic research. Many groups are now seeking to make SQUID systems easier to handle. As a result, engineering work is featuring more prominently in their activities. This not only includes the use of hardware components with adapted cryogenic devices but also easy-to-handle software for SQUID operation to control the scanning systems and provide guidance in data interpretation and defect classification. Increasing requirements in the area of quality control have generated greater demand for more reliable techniques for material inspection. It is no surprise then, that many recent SQUID NDE applications have been conducted in cooperation with industry. Such partnerships are serving to erode the skepticism and reluctance on behalf of industry to accept SQUID NDE as a reliable tool for solving sophisticated NDE problems, which remains in place despite years of successful demonstrations. It is hoped that further partnerships will end years of struggle by researchers to convince industry and NDT societies – who have yet to accredit the technology – that SQUID NDE is a sound testing method.

While the SQUID itself is commercially available from a number of SQUID manufacturers (e.g. Quantum Design, Conductus and several small German companies such as Fino AG (FIT-Meßtechnik), Jülicher SQUID GmbH (JSQ)), an integrated SQUID NDE system was long offered by only two U.S. companies, Neocera or Tristan Inc.. Tristan Technologies sells SQUID microscopes using low-T_c SQUIDs as well as high-T_c sensors. These microscopes can image a sample's magnetic field with frequencies ranging between dc and 10 kHz with a spatial resolution of a few tens of μm. But these devices are rather expensive and only suitable for small samples. While more flexible prototypes were developed within the framework of scientific and industrial research collaborations, these NDE systems were not commercially available for a long time. That is changing. Over the past couple of years, for example, the Jülich SQUID Company (JSQ) [145] has been offering SQUID systems assembled for certain NDE applications. In total they have sold more than ten NDE high-T_c SQUID systems to research institutes as well as industrial users. A new German company can be added to this small list of SQUID NDE manufacturers: ez-SQUID [146] sells low-T_c SQUID NDE systems suitable for different NDE applications and counts companies like German industrial giant WC-Heraeus GmbH among its customers. These are small but significant indications that the SQUID is gradually carving out a niche in the NDE market.

Because there is no alternative to SQUIDs when extreme field sensitivity is a priority, and because there are several types of problems that cannot be solved through rival testing techniques, the SQUID appears to have a real and lasting chance of survival in the field of nondestructive testing. Both the results published to date and the growing interest among industrial users justify hopes that the SQUID will establish itself as a reliable testing tool for ambitious and challenging NDE applications.

ACKNOWLEDGEMENTS

I would like to thank the following experts for their kind permission to publish illustrations in this chapter: H.-J. Krause (Fig. 20), Th. Schurig (Fig. 48-51), W. Singer (Fig. 41), and my colleague M. Mück for Fig. 16 and Fig. 45-47 as well as his invaluable comments. I am also indebted to G. Gierelt, F. Gruhl, and K. Allweins in particular, for their assistance in drawing up some of the illustrations. My thanks also go to the German Ministry of Education and Research for funding parts of the work presented here.

REFERENCES

[1] H. Weinstock and M. Nisenoff, in SQUID '85, Proc. 3rd International conference on Superconducting Quantum Devices, H.D. Hahlbohm and H. Lübbig, Eds., Berlin: deGruyter , pp. 843-847 (1985)

[2] R.J.P. Bain, G.B. Donaldson, S. Evanson and G. Hayward, in SQUID '85, Proc. 3rd International Conference on Superconducting Quantum Devices, H. D. Hahlbohm and H. Lübbig, Eds., Berlin: deGruyter , pp. 841-846 (1985)

[3] H. Weinstock, *IEEE Trans. Mag.*, **27**(2), 3231-3236 (1991)

[4] G.B. Donaldson in NATO ASI: Superconducting electronics, H. Weinstock and M. Nisenoff, Eds., New York: Springer Verlag, pp. 175-207 (1989)

[5] H. Weinstock, in SQUID Sensors: Fundamentals, Fabrication and Applications, Ed. Dordrecht: Kluwer, (1996)

[6] W.G. Jenks, S.S.H. Sadeghi, J.P. Wikswo Jr., *J. Phys D: Appl. Phys.* **30**, 293 (1997)

[7] J. Clarke: SQUID Fundamentals, in Weinstock, H.: *SQUID Sensors: Fundamentals, Fabrication and Applications*, H. Weinstock Ed. Dordrecht: Kluwer, pp. 1-62 (1996)

[8] J. Clarke, SQUIDs: theory and practice, in H. Weinstock and R.W. Ralston (eds.), the New Superconducting Electronics, Kluwer Academic Publishers, dordrecht, pp. 123-180 (1993)

[9] D. Koelle, R. Kleiner, F. Ludwig, E. Dantsker and J. Clarke, *Rev. Mod. Phys.,* **71**, 631-686 (1999)

[10] B.D. Josephson, *Phys. Lett.* **1**, 251-253 (1962)

[11] B.D. Josephson, *Adv. Phys.* **14**, 419-451 (1965)

[12] W.C. Stewart, *Appl. Phys. Lett.* **12**, 277-280 (1968)

[13] D. Dimos, P. Chaudhari, J. Mannhart and F.K. LeGoues, *Phys. Rev. Lett.* **61,** 219-222 (1988)

[14] D. Dimos, P. Chaudhari and J. Mannhart, *Phys. Rev. B.* **41,** 4038-4049 (1990)

[15] R. Gross, P. Chaudhari, M. Kawasaki and A. Gupta, *IEEE Trans. Magn.* **27,** 3227-3230 (1991)

[16] R.W. Simon, J.F. Burch, K.P. Daly, W.D. Dozier, R. Hu, A.E. Lee, J.A. Luine, H.M. Manasevit, C.E. Platt, S.M. Schwarzbeck, D. St John, M.S. Wire and M.J. Zani, Progress towards a YBCO circuit process, in R.D. McConnell and R. Noufi (eds.), Science and Technology of Thin Film Superconductors 2, Plenum Press, New York, 549-558 (1990)

[17] C.L. Jia, B. Kabius, K. Herrmann, G.J. Cui, J. Schubert, W. Zander, A.I. Braginski and C. Heiden, *Physica C* **175**, 545-554 (1991)

[18] C.L. Jia, B. Kabius, K. Herrmann, G.J. Cui, J. Schubert, W. Zander, and A.I. Braginski, *Physica C* **196**, 211-226 (1992)

[19] D. Shoenberg, Superconductivity, Cambridge University Press, Cambridge (1962)

[20] F.C. Wellstood, C. Heiden and J. Clarke, *Rev. Sci. Inst.* **55**, 952-957 (1984)

[21] R.L. Forgacs and A.F. Warnick, *Rev. Sci. Inst.* **38**, 214-220 (1967)

[22] D. Drung: Advanced SQUID Read-out electronics, in Weinstock, H.: *SQUID Sensors: Fundamentals, Fabrication and Applications*, H. Weinstock Ed. Dordrecht: Kluwer, pp. 63-116 (1996)

[23] M. v.Kreutzbruck, M. Mück, and C. Heiden, in: B. Larsen (Ed.), Proceedings of 7th European conference on Nondestructive Testing ECNDT, Vol.3, pp. 2513-2520 (1998)

[24] N.G. Sepulveda and J.P. Wikswo, Jr., *J. Appl. Phys.* **79 (4)**, 2122-2135 (1996)

[25] J.P. Wikswo, Jr., *IEEE Trans. Magn.* **5(2),** 74-120 (1995)

[26] S. Cochran, G.B. Donaldson, C. Carr, D. McA. McKirdy, M.E. Walker, U. Klein, A. McNab and J. Kuznik, in Proc. 2nd Eur. Conf. On Appl. Supercond., IOS Press, Amsterdam, pp. 53-64 (1995)

[27] L.N.C. Morgan, C. Carr, A. Cochran, D. McA. McKirdy and G. B. Donaldson, *IEEE Trans Appl. Supercond.* **5**(2), 3127-3130 (1995)

[28] A. Kost, *Numerische Methoden in der Berechnung elektromagnetischer Felder*, Springer Verlag (Berlin) (1994)

[29] D. McA. McKirdy, S. Cochran, G.B. Donaldson, and A. McNab, in Proc. 2nd Eur. Conf. On Appl. Supercond., IOS Press, Amsterdam, pp. 185-194 (1995)

[30] R.J. Bradley, N.G. Sepulveda and J.P. Wikswo, Jr., *J. Appl. Phys.* **65 (1)**, 361-372 (1989)

[31] C.H. Barbosa, A.C. Bruno, L.F. Scavarda, E.A. Lima, P.C. Ribeiro, C. Kelber, *IEEE Trans Appl. Supercond.* **5**(2), 2478-2481 (1995)

[32] G. Sawade, J. Straub, H.-J. Krause, H. Bousack, G. Neudert, R. Ehrlich, in: Proc. Int. Symp. NDT Civil Engng. (NDT-CE), vol. 2, DGZFP, Berlin, pp. 1077-1084 (1995)

[33] Y.P. Ma and J.P. Wikswo, Jr., *J. Nondestr. Eval.* **14**, no.3, 149-167, 15 401-408 (1995)

[34] Y.P. Ma and J.P. Wikswo, Jr., *Review of Progress in QNDE*, **15**, 401-408 (1996)

[35] M. v. Kreutzbruck, K. Allweins, G. Gierelt, H.-J. Krause, S. Gärtner, W. Wolf, *Physica C* **368**, 85-90 (2002)

[36] M. Mück, M. v. Kreutzbruck, U. Baby, J. Tröll und C. Heiden, *Physica C*, **407**, 282-287 (1997)

[37] R. Hohmann, M. Maus, D. Lombarski, M. Grüneklee, Y. Zhang, H.-J. Krause, H. Bousack and A.I. Braginski, *IEEE Trans. Appl. Supercond.* **9**, 3801-3804 (1999)

[38] J.A. Cadzow, X. Li, *Digital Signal Processing*, **5(1)**, 3-20 (1995)

[39] J.P. Wikswo, Jr.: The Magnetic Inverse Problem, in Weinstock, H: *SQUID Sensors: Fundamentals, Fabrication and Applications*, H. Weinstock Ed. Dordrecht: Kluwer, pp. 629-695 (1996)

[40] W.J. Dallas, *Applied Optics*, **24**, no. 24 (1985)

[41] W. Kullmann, W.J. Dallas, *IEEE Trans. Biomed. Eng.*, **34**, no. 11, 837-842 (1987)

[42] M. Kurokawa and I. Takeuchi, *Nondestructive Testing of Materials*, R. Collins et al. (eds.), IOS Press, pp. 321 (1995).

[43] D.S. Forsyth, A. Fahr and C.E. Chapman, *Review of Progress in QNDE*, Vol 13, 879-886 (1994)

[44] M. Negley, M.R. Govindaraju and D.C. Jiles, *Review of Progress in QNDE*, Vol 13, 1817-1824 (1994)

[45] M. v. Kreutzbruck, K. Allweins, T. Rühl, M. Mück, C. Heiden, H.-J. Krause and R. Hohmann, *IEEE Trans. Appl. Supercond.* **11**, 1032-1037 (2001)

[46] V.L. Vengrinovich, Yu.B. Denkevich, G.R. Tillack, *J. Phys D: Appl. Phys* **32**, 2505-2514 (1999)

[47] L.P. Lee, J. Longo, V. Vinetskiy, R. Cantor, *Appl. Phys. Lett.* **66**, 1539-1541 (1995)

[48] F. Ludwig, E. Dantsker, D. Koelle, R. Kleiner, A.H. Miklich, D.T. Nemeth, J. Clarke, D. Drung, S. Knappe, H. Koch, *IEEE Trans. Appl. Supercond.* **5**, 2919-2922 (1995)

[49] J. Beyer, D. Drung, F. Ludwig, T. Minotani, K. Enpuku, *Appl. Phys. Lett.* **72**, 203 (1998)

[50] Y. Zhang, U. Krüger, R. Kutzner, R. Wördenweber, J. Schubert, W. Zander, M. Strupp, E. Sodtke and A.I. Braginski, *Appl. Phys. Lett.* **65**, 3380-3382 (1994)

[51] M. v. Kreutzbruck, M. Mück and C. Heiden, Nondestructive Testing of Aircraft Parts by SQUID based Eddy Current Systems, *Studies in Applied Electromagnetics and Mechanics*, Vol.18, IOS Press (Amsterdam), 257-260 (2000)

[52] M. v. Kreutzbruck, M. Mück, U. Baby and C. Heiden, Experiments on eddy current NDE with HTS rf SQUIDs, in *Studies in Applied Electromagnetics and Mechanics*, Vol.13, IOS Press (Amsterdam), 165-168 (1998)

[53] A. Mathai, D. Song, Y. Gim, F.C. Wellstood, *IEEE Trans. Appl. Supercond.* **3**, 2609-2612 (1993)

[54] L.N. Vu, M.S. Wistrom and D.J. Van Harlingen, *IEEE Trans. Appl. Supercond.* **3**, 1918-1921 (1993)

[55] C.C. Tsuei, J.R. Kirtley, C.C. Chi, L.S. Yu-Jahnes, A. Gupta, T. Shaw, J.Z. Sun, and M.B. Ketchen, *Phys. Rev. Lett.* **73**, 593-596 (1994)

[56] R.C. Black, F.C. Wellstood, E. Dantsker, A.H. Miklich, D. Koelle, F. Ludwig and J. Clarke, *IEEE Trans. Appl. Supercond.* **5**, 2137-2140 (1995)

[57] L.P. Lee, J. Longo, V. Vinetskiy and R. Cantor, *Appl. Phys. Lett.* **66** (12), 1539-1542 (1995)

[58] F.C. Wellstood, Y. Gim, A. Amar, R.C. Black and A. Mathai, *IEEE Trans. Appl. Supercond.* **7** (2), 3134-3138 (1997)

[59] F. Gruhl, M. Mück, M. v. Kreutzbruck, J. Dechert, *Rev. Sci. Instrum.* **72** (4), 2090-2096 (2001).

[60] J. Dechert, M. Mück and C. Heiden, *IEEE Trans. Appl. Supercond.* **9**, 4111-4114 (1999)

[61] D. Drung, H. Matz and H. Koch, *Rev. Sci. Instrum.* **66** (4), 3008-3015 (1995)

[62] R.H. Koch, J.R. Rozen, P. Wöltgens, T. Picunko, W.J. Goss, D. Gambrel, D. Lathrop, R. Wiegert and D. Overway, *Rev. Sci. Instrum.* **67** (8), 2968-2976 (1996)

[63] D.B. Sullivan, J.E. Zimmerman, and J.T. Ives, Operation of a Practical SQUID Gradiometer in a Low-Power Stirling Cooler, in J.E. Zimmerman, D.B. Sullivan and S.E. McCathy (eds.), Refrigeration for Cryogenic Sensors and Electronic Systems, NBS Special Publication SP-607, Superindendent of Documents, U.S. Government Printing office, Washington, D.C. 20234, pp. 186-194 (1981)

[64] G. Thummes, F. Giebeler and C. Heiden, Cryocoolers 8, R.G. Ross, Jr. (ed), Plenum Press, New York, pp. 383-393 (1995)

[65] C. Wang, G. Thummes and C. Heiden, *Cryogenics* **37**, 159-164 (1997)

[66] C. Wang, G. Thummes and C. Heiden, *Cryogenics* **38**, 689-695 (1998)

[67] C. Lienerth, G. Thummes and C. Heiden, *IEEE Trans. Appl. Supercond.* **11** (1), 812-815 (2001)

[68] R. Hohmann, H.-J. Krause, H. Soltner, Y. Zhang, C.A. Copetti, H. Bousack, A.I. Braginsky, *IEEE Trans. Appl. Supercond.* 7, 2860-2865 (1997).

[69] R. Hohmann, M.L. Lucía, H. Soltner, H.-J. Krause, W. Wolf, H. Bousack, M.I. Faley, G. Spörl and A. Binneberg, in: Proc. Cryocoolers **9**, Plenum, New York, pp. 925-934 (1997).

[70] M.L. Lucía, R. Hohmann, H. Soltner, H.-J. Krause, W. Wolf, H. Bousack, M.I. Faley, G. Spörl and A. Binneberg, *IEEE Trans. Appl. Supercond.* **7**, 2878-2881 (1997).

[71] M.J. Ferrari, M. Johnson, F.C. Wellstood, J. Clarke, P.A. Rosenthal, R.H. Hammond, and M.R. Beasley, *Appl. Phys. Lett.* **53**, 695 (1988)

[72] M.J. Ferrari, M. Johnson, F.C. Wellstood, J. Clarke, A. Inam, X.D. Wu, L. Nazar, and T. Venkatesan, *Nature* (London) **341**, 723 (1989)

[73] R.H. Koch, J. Z. Sun, V. Foglietti, and W.G. Gallagher, *Appl. Phys. Lett.* **67**, 709 (1995)

[74] E. Dantsker, S. Tanaka, P.A. Nilsson, R. Kleiner, and J. Clarke, *Appl. Phys. Lett.* **69**, 4099 (1996)

[75] E. Dantsker, S. Tanaka, and J. Clarke, *Appl. Phys. Lett.* **70**, 2037 (1997)

[76] J. Vrba: SQUID Gradiometers in Real Environments, in Weinstock, H: *SQUID Sensors: Fundamentals, Fabrication and Applications*, H. Weinstock Ed. Dordrecht: Kluwer, pp. 117-178 (1996)

[77] R. Hohmann, D. Lombarski, H.-J. Krause, M. v. Kreutzbruck, and W. Becker, *IEEE Trans. Appl. Supercond.* **11**, 1279-1282 (2001)

[78] J.B. Cladis, G.T. Davidson, L.L. Newkirk, The Trapped Radiation Handbook, DNA 2542H (1971)

[79] W.H. Campbell, *Ann. Geophys.* **22**, 492 (1966)

[80] M. v. Kreutzbruck, M. Mück und C. Heiden, *IEEE Trans. Appl. Supercond.* **7**, 3279-3282 (1997)

[81] M.A. Espy, A.N. Matlashov, J.C. Mosher and R.H. Kraus, Jr., *IEEE Trans. Appl. Supercond.* **11**, 1303-1306 (2001)

[82] H.-J. Krause, W. Wolf, W. Glaas, E. Zimmermann, M.I. Faley, G. Sawade, R. Mattheus, G. Neudert, U. Gampe, J. Krieger, *Physica C*, **368,** 91-95 (2002)

[83] M. v. Kreutzbruck, U. Baby, A. Theiss, M. Mück and C. Heiden, *IEEE Trans. Appl. Supercond.* **9**, 3805-3808 (1999)

[84] M. v. Kreutzbruck, A. Theiss, M. Mück and C. Heiden, *Rev. Sci. Instrum.* **70** (9), 3714-3718 (1999)

[85] J. Beyer, D. Drung and Th. Schurig, *IEEE Trans. Appl. Supercond.* **11**, 1162- 1167 (2001)

[86] R.H. Koch, *IEEE Trans. Appl. Supercond.* **7**, 3259-3262 (1997)

[87] R.P. Giffard, in Superconducting Quantum Interference Devices and Their Applications, H.D. Hahlbohm and H. Lübbig (eds.), de Gruyter Berlin, pp. 445 (1980)

[88] D. Drung, S. Bechstein, K.-P. Franke, M. Schreiner and Th. Schurig, *IEEE Trans. Appl. Supercond.* **11**, 880-883 (2001)

[89] A. Matlashov, M. Espy, R.H. Kraus, Jr., K.R. Ganther, Jr., L.D. Snapp, *IEEE Trans. Appl. Supercond.* **11**, 876-879 (2001)

[90] M. Mück, C. Welzel, F. Gruhl, M.v. Kreutzbruck, A. Farr, F. Schölz, *Physica C*, **368,** 96-99 (2002)

[91] R.C. McMaster, P. McIntire, P., M.L. Mester, Nondestructive Testing Handbook, Vol. 4: Electromagnetic Testing, American Society for Nondestructive Testing, (1986).

[92] T. Varpula and T. Poutanen, *J. Appl. Phys.* **55** (11), 4015-4021 (1984)

[93] N. Kasai, K. Sasaki, S. Kiryu and Y. Suzuki, *Cryogenics* **33**, 175-179 (1993)

[94] J. Banchet, J. Jouglar, P.L. Vuillermoz, P. Waltz and H. Weinstock, *IEEE Trans. Appl. Supercond.* **5**, 2486-2489 (1995)

[95] J. Banchet, J. Jouglar, P.L. Vuillermoz, P. Waltz and H. Weinstock, *Review of Progress in QNDE,* **14**, 1675-1682 (1995)

[96] G.B. Donaldson, S. Evanson, M. Otaka, T. Shimzu, K. Takaku, *British J. NDT*, **32**, 238-240 (1990)

[97] N. Kasai, N. Ishikawa, H. Yamakawa, K. Chinone, S. Kakayama and A. Odawara, , *IEEE Trans. Appl. Supercond.* 7, 2315-2318 (1997)

[98] T. Nagaishi, H. Kugai, H. Toyoda, H. Itozaki, *IEEE Trans. Appl. Supercond.* **7**, 2886-2889 (1997)

[99] H. Weinstock, N. Tralshawala, J.R. Claycomb, *IEEE Trans. Appl. Supercond.* **9**, 3797-3800 (1999)

[100] Y. Tavrin, M. Siegel, J. Hinken, *IEEE Trans. Appl. Supercond.* **9**, 3809-3812 (1999)

[101] J. Krieger, H.-J. Krause, U. Gampe, G. Sawade, in Proc. Intl. Conf. on NDE Tech. f. Aging Infrastr. NewPort Beach, USA, pp. 229-239 (1999)

[102] H.-J. Krause and M. v. Kreutzbruck, *Physica C*, **368,** 70-79 (2002)

[103] M.I. Faley, U. Poppe, and K. Urban, E. Zimmermann, W. Glaas, H. Halling, M. Bick, H.-J. Krause, D.N. Paulson, T. Starr and R.L. Fagaly, *IEEE Trans. Appl. Supercond.* **9**, 3386-3391 (1999)

[104] F. Schmidl, S. Wunderlich, L. Dörrer, H. Specht, S. Linzen, H. Schneidewind, P. Seidel, *IEEE Trans. Appl. Supercond.* **7**, 2756-2759 (1997).

[105] S. Wunderlich, F. Schmidl, H. Specht, L. Dörrer, H. Schneidewind, U. Hübner and P. Seidel, *Supercond. Sci. Technol.* **11**, 315-321 (1998)

[106] H. Weinstock, T. Erber and N. Nisenoff, *Physical Review B*, **31**(3), 1535-1553 (1985)

[107] Y. Ma, D.J. Staton, N.G. Sepulveda and J.P. Wikswo, Jr., *Review of Progress in QNDE*, **10 A**, 979-986 (1991)

[108] Y. Ma and J.P. Wikswo, Jr., *Review of Progress in QNDE*, **11**, 1153-1159 (1992)

[109] D.C. Hurley, Y.P. Ma, S.Tan and J.P. Wikswo, *Jr., Res. Nondestr. Eval.* **5**, 1-29 (1993)

[110] F.R. Fickett and T.E. Capobianco, *Review of Progress in QNDE*, **4**, 401-406 (1985)

[111] T.E. Capobianco, F.R. Fickett and J.C. Moulder, *Review of Progress in QNDE*, **5**, 705-711 (1985)

[112] W.N. Podney, *IEEE Trans. Mag.*. **27** (2), 3241-3244 (1991)

[113] W.N. Podney, *Review of Progress in QNDE* **13**, 1947-1954 (1993)

[114] Y. Ma and J.P. Wikswo, Jr., *Review of Progress in QNDE*, **13**, 303-309 (1994)

[115] Y. Ma and J.P. Wikswo, Jr., *Journal of NDE*, **14**, 149-167 (1995)

[116] J.P. Wikswo, Jr., Y.P. Ma, N.G. Sepulveda, D.J. Staton, S. Tan, I.M. Thomas in M.T. Valley, N.K. Grande, A.S. Kobayashi (eds.), Nondestructive Testing of Aging Aircraft, SPIE Proc. vol. 2001, pp. 164-190 (1993)

[117] Y. Ma and J.P. Wikswo, Jr., *Review of Progress in QNDE*, **12**, 1137-1143 (1993)

[118] W.N. Podney, *IEEE Trans. Appl. Supercond.* **5**, 2490-2492 (1995)

[119] C. Carr, D.McA. Kirdy, E.J. Romans, G.B. Donaldson and A. Cochran, *IEEE Trans. Appl. Supercond.* **7**, 3275-3278 (1997)

[120] H.-J. Krause, R. Hohmann, Y. Zhang, M. Grüneklee, M. Banzet, J. Schubert, W. Zander, A.I. Braginski, M.L. Lucía, G. Spörl, A. Binneberg, M. Junger, *Review of Progress in QNDE*, **16**, 1053-1060 (1997)

[121] J.H. Hinken and Y. Tavrin, *Review of Progress in QNDE*. **19**, 2085-2092 (2000).

[122] P.B. Nagy and A.H. Nayfeh, *J. Appl. Phys.* **87**, 7481-7483 (2000).

[123] J.G. Bellingham, M.L.A. McVicar, M. Nisenoff and P.C. Searson, *J. electrochem. Soc.*, **133** (8), 1753-1754 (1986)

[124] M. Misra, S. Lordi, M.L.A. McVicar, *IEEE Trans. Mag.* **27**, 3245-3248 (1991)

[125] T.S. Lee, Y.R. Chemla, E. Dantsker and J. Clarke, *IEEE Trans. Appl. Supercond.* **7**, 3147-3150 (1997)

[126] R. Kötitz, H. Matz, L. Trahms, H. Koch, W. Weitschies, T. Rheinländer, W. Semmler and T. Bunte, *IEEE Trans. Appl. Supercond.* **7**, 3678-3681 (1997)

[127] F. Baudenbacher, N. T. Peters, P. Baudenbacher, J. P. Wikswo, *Physica C* **368**, 24-31 (2002).

[128] A. Moya, F. Baudenbacher, J.P. Wikswo and F.C. Wellstood, *IEEE Trans. Appl. Supercond.* **9**, 3511-3514 (1999).

[129] F. Baudenbacher, N.T. Peters, and J.P. Wikswo, Jr., *Rev. Sci. Instrum.* **73** (3): 1247-1254 (2002).

[130] E.F. Fleet, S. Chatraphorn, F.C. Wellstood, and L.A. Knauss, *IEEE Trans. Appl. Supercond.* **9**, 4103-4106 (1999).

[131] Y.P. Ma, I.M. Thomas, A. Lauder, J.P. Wikswo, Jr., *IEEE Trans. Appl. Supercond.* **3**, 1941-1944 (1993).

[132] T. Shaw, K. Schlenge, R.Mc Dermott, J. Clarke, J.W. Chan, S.-H. Kang and J.W. Morris, Jr., *IEEE Trans. Appl. Supercond.* **9**, 4107-4110 (1999).

[133] C. Carr, J.C. Macfarlane, and G.B. Donaldson, to be published in *IEEE Trans. Appl. Supercond.* **13** (2003)

[134] C. Carr, D. Grahman, J.C. Macfarlane and G.B. Donaldson to be published in *IEEE Trans. Appl. Supercond.* **13** (2003)

[135] Y. Hatsukade, A. Ishiyama and N. Kasai to be published in *IEEE Trans. Appl. Supercond.* **13** (2003)

[136] A. Ruosi, M. Valentino, G. Peluso and G. Pepe, *IEEE Trans. Appl. Supercond.* **11**, 1172-1175 (2001)

[137] H.-J. Krause, G.I. Panaitov, and Y.Zhang to be published in *IEEE Trans. Appl. Supercond.* **13** (2003)

[138] K. Allweins, G. Gierelt, M. v. Kreutzbruck, H.-J. Krause to be published in *IEEE Trans. Appl. Supercond.* **13** (2003)

[139] H.-J. Krause, S. Gaertner, N. Wolters, R. Hohmann, W. Wolf, J. Schubert, W. Zander, Y. Zhang, M. v. Kreutzbruck and M. Mück, *IEEE Trans. Appl. Supercond.* **11**, 1168-1171 (2001)

[140] D.D. Clark, M.A. Espy, P.D. Beinke, R.H. Kraus and A.N. Matlachov, to be published in *IEEE Trans. Appl. Supercond.* **13** (2003)

[141] J. Matthews, S.Y. Lee, F.C. Wellstood, A.F. Gilbertson, G.E. Moore and S. Chatraphorn, to be published in *IEEE Trans. Appl. Supercond.* **13** (2003)

[142] Y. Zhang, M. Mück, K. Hermann, J. Schubert, W. Zander, A.I. Braginski and C. Heiden, *IEEE Trans. Appl. Supercond.* **3**, 2465-2468 (1993)

[143] M. Mück, *Rev. Sci. Instrum.*, **63**, 4, (1992)

[144] M. Mück, C. Welzel, A. Farr and F. Schölz, to be published in *IEEE Trans. Appl. Supercond.* **13** (2003)

[145] http://www.jsquid.com

[146] http://www.ez-squid.de

DC SQUID MAGNETOMETERS FROM $YBa_2Cu_3O_{7-\delta}$ FOR BIOMAGNETIC APPLICATIONS

Henry-Jobes BARTHELMESS[a], Frank LUDWIG[b], Meinhard SCHILLING[b], Dietmar DRUNG[c], and Thomas SCHURIG[c]

[a]Magnicon GbR, Wachtelweg 7, 22869 Schenefeld, Germany
[b]Technische Universität Braunschweig, Institut für Elektrische Messtechnik und Grundlagen der Elektrotechnik, Hans-Sommer-Str. 66, 38106 Braunschweig, Germany
[c]Physikalisch-Technische Bundesanstalt, Abbestrasse 2-12, 10587 Berlin, Germany

1. INTRODUCTION

Superconducting quantum interference devices (SQUIDs) are the most sensitive detectors of magnetic flux. This sensitivity can be exploited for the measurement of any physical quantity that can be converted into magnetic flux, for example, magnetic field, magnetic field gradient, voltage, current etc. Excellent introductions into the SQUID fundamentals can be found in [1,2]. One of the potential fields of application of SQUID devices is biomagnetism, especially the detection of biomagnetic signals from the human brain and heart. Today, the largest number of SQUID devices – fabricated from the conventional, so-called low-T_c superconductor (LTS) Nb – is accommodated in multichannel systems for magnetoencephalography (MEG). Compared to their electrical counterparts, electrocardiogram (ECG) and electroencephalogram (EEG) systems, the magnetic sensor systems are more expensive and suffer from the need to cool the sensors with liquid helium (T =4.2 K). With the discovery of superconductivity in ceramic oxides, especially of $YBa_2Cu_3O_{7-\delta}$ with a transition temperature of about 90 K, the expectation arose that devices based on these so-called high-T_c (HTS) superconductors will quickly become more widely used than LTS ones. The reasons for this are that liquid nitrogen with a boiling temperature of 77 K is much cheaper than liquid He and that – for a given heat load

– liquid nitrogen boils off much more slowly. In addition, the cryogenic technology considerably simplifies so that much smaller cold-warm distances can be realized with HTS systems. The smaller separation between sensor and source provides a larger signal which partly compensates the intrinsically higher noise of HTS devices.

Considering the biomagnetic fields from the human heart, the peak signal amplitude of a magnetocardiogram (MCG) amounts to several tens of picotesla with a fine structure of clinical interest of about 2 pT. The base-to-peak amplitude of a biomagnetic signal to be identified in a single measurement in a bandwidth Δf is given by $B_p = c_R\sqrt{S_B \cdot \Delta f}$ where the crest factor c_R is determined by the probability that an observed magnetic field is signal rather than noise. Assuming that the noise is white throughout the measurement bandwidth, a value of c_R =3 to 4 is used [3]. Thus, for typical values of $\Delta f = 150$ Hz and $B_p = 2$ pT, the sensor should have a noise of about 50 fT/$\sqrt{Hz}$ down to low frequencies. Such noise levels have been achieved with directly coupled SQUID magnetometers on 10 x 10 mm^2 substrates by a number of groups. Clearly, to record even smaller signals from the human brain more sensitive multilayer SQUID magnetometers are required. HTS multilayer SQUID magnetometers with intrinsic white noise levels below 10 fT/$\sqrt{Hz}$ down to frequencies below 10 Hz were reported by a few groups [4–6], however, in contrast to single-layer devices the fabrication technology is much more demanding and the yield of such devices is correspondingly poor.

MCGs recorded with HTS SQUID magnetometers were reported by numerous groups, mainly to illustrate the good noise performance achieved with their devices. However, most practical applications require a multichannel system to enable one to map fields from the heart at different locations. The most elaborate HTS SQUID systems have been built at the former Superconducting Sensor Laboratory in the middle 90s in Japan [7,8]. The noise of the sensors in their 32-channel system was 70-250 fT/$\sqrt{Hz}$ at 1 kHz and 300-800 fT/$\sqrt{Hz}$ at 1 Hz. Although the noise specifications did not fulfil the requirements listed above, Itozaki and co-workers were able to obtain magnetocardiac isofield contour maps in a magnetically well shielded environment.

Clearly, if one wants to see a widespread use of SQUIDs in biomagnetism, it is not sufficient to replace the liquid He with liquid N_2 but also to eliminate the need for heavily magnetically shielded rooms which significantly contribute to the total system costs. In a real, magnetically unshielded environment the sensor is exposed to various noise sources. The largest environmental magnetic field is generally the static magnetic field of the earth. As will be described in Section 2, static magnetic fields can indirectly influence the intrinsic low-frequency noise of the SQUID devices by the number of vortices trapped in the superconducting films. This effect can be avoided or minimized by appropriate device layouts. On the other hand, at urban locations, the sensors are exposed to time-varying environmental fields like those originating from power line cables or railway systems. For example, the peak-to-peak amplitude of the 50 Hz power line interference at typical PTB or University of Hamburg measurement sites amounts to between a few 100 nT and about 1 μT. This is about a factor of 10^5 to 10^6 above the desired noise amplitude of a clinically relevant MCG measurement.

The simplest way of performing background noise cancellation is to subtract the background noise signal measured by a reference sensor B_{ref} from the signal measured by the source signal sensor B_{src}, i.e., to assume that the background noise signal at the source signal sensor equals B_{ref}. The signal corrected for the superimposed background noise is then given as

$$B_{corr} = B_{src} - B_{ref}. \tag{1}$$

Considering, e.g., a MCG measurement, the source signal sensor is positioned directly above the chest. Choosing an appropriate baseline, B_{corr} measures how the magnetic field of the heart changes from the source signal to the reference sensor whereas background magnetic fields originating from far away environmental sources, neglecting gradients, will cancel. Clearly, the baseline (distance between source signal and reference sensor) should be not too small since the measurement signal B_{corr} will then considerably be suppressed. On the other hand, the baseline should not be too large since then deviations from a spatially constant background field will deteriorate the background noise suppression. Assuming a source sensor-heart distance of 7 cm, a current dipole moment for the human heart of 15 μAm and a balance of the gradiometer of 1 part in 10^4, an optimum signal-to-noise ratio is found for a gradiometer baseline of about 7 cm [9].

Gradiometers can be realized in different ways. In LTS technology one generally connects an axial wire-wound gradiometric pickup loop to the spiral input coil integrated on the washer-type SQUID. The main advantages of such hardware gradiometers are that one can achieve quite large baselines and that the subtraction $B_{src} - B_{ref}$ is performed in the gradiometer coils providing a high dynamic range. Unfortunately, there does not exist such HTS wire bonding technology. Thus, first-order HTS hardware gradiometers are fabricated as planar ones using thin-film technology. Due to the expensive substrates, such gradiometers are mostly fabricated on 10 x 10 mm^2 chips yielding baselines of a few mm. An exception is the large-baseline planar gradiometer developed at Berkeley [10]. Here an asymmetric large-baseline single-layer flux transformer was inductively coupled to a directly coupled SQUID magnetometer. Properly aligning the flux transformer over the magnetometer, the authors achieved a balance of 1 part in 3000. Taking two such first-order gradiometers and a gradiometric large-baseline flux transformer, respectively, the concept could be extended to form a second-order planar gradiometer, allowing them to record high-quality MCGs in magnetically unshielded environment [11]. The main disadvantage of this approach is the limited mechanical stability of the flip-chip configuration.

Other HTS approaches to form gradiometers were electronic ones. Here the electrical output of the reference sensor is subtracted from the output of the source signal sensor. In contrast to the hardware gradiometer were the subtraction is performed in the gradiometer coil, here it is carried out electronically. The adjustment of the field-to-voltage conversion factors can be performed using trimming resistors.

Generally, the two magnetometers are not perfectly aligned in one direction. If there is an angle between the two sensor normals, one can certainly adjust the field-to-voltage conversion factors such that a homogeneous magnetic field in one direction is perfectly cancelled but a sensitivity to in-plane magnetic fields remains. In the first HTS approach to form a first-order electronic gradiometer, Tavrin et al. [12] developed a complex cardanic suspension system where one sensor was mounted rigidly while the plane inclination of the second one could be adjusted from outside the cryostat. Adding a third, again mechanically adjustable sensor the authors extended the approach to form a second-order gradiometer [13]. However, this approach is not applicable for more complex systems, especially involving many sensing channels. The same holds for the approach by Borgmann et al. [14] who used – in analogy to early LTS gradiometer work – superconducting plates for tuning the common mode rejection to better than 10^4 for uniform background fields. To get rid of mechanical balancing which is impractical and unreliable, He et al. [15] developed electronic first-order HTS SQUID gradiometers where additional "x" and "y" reference sensors were implemented to perform the balancing. The balancing was generally performed with a 3-axis Helmholtz coil system achieving a balance of better than 1 part in $5 \cdot 10^3$ for all three directions. Zhang et al. [16] reported that the optimum balance obtained in a homogeneous magnetic field (provided by the Helmholtz coil system) does not always result in the lowest background noise when operated unshielded. The best signal-to-noise ratio was often obtained when the gradiometer was slightly off-balanced. The reason for this is that, in an urban real-world environment, it is not sufficient to suppress homogenous background magnetic fields since also gradient fields contribute to background noise. Consequently, the first-order gradiometer, optimally balanced in a Helmholtz coil system, does not eliminate any gradient fields whereas the slightly unbalanced system partly compensates for both background fields and gradients so as to minimize the total background noise. As shown by Vrba [17], for high-performance MEG measurements in unshielded urban environment third-order gradiometry is required.

So far, there has been only a few attempts to set up HTS multichannel systems for unshielded operation [18–20]. All three systems were based on directly coupled SQUID magnetometers. Despite enormous efforts to reduce the environmental noise, the system noise at 1 Hz remained typically around 1 $pT/\sqrt{Hz}$ and only poor quality MCGs could be recorded.

In this paper we give a summary of the recent work towards the development of HTS systems for biomagnetic applications performed by our groups at the PTB in Berlin and at the University of Hamburg. Section 2 starts with a brief overview about general design considerations of highly sensitive directly coupled SQUID magnetometers. This is followed by a description of special refinements of the pickup loop structure performed by the two groups to make the intrinsic magnetometer noise less susceptible to environmental dc and ac magnetic fields. In Section 4 we give a description of the two SQUID systems developed for biomagnetic measurements. On the one hand, a single-channel second-order gradiometer system for operation in magnetically unshielded environment was developed at the PTB which allowed the recording of high-quality real-time MCGs. This work mainly focused on the suppression of environmental noise. On the other hand, a small, easy-to-handle and flexible 8-channel system was set up at the University of Hamburg which allowed one to record magnetic field maps of human heart signals as well as electrically evoked signals of the human brain in a magnetically shielded room. We will conclude with a short outlook on what further improvements should be done in the near future.

2. DESIGN RULES FOR DIRECTLY COUPLED DC SQUID MAGNETOMETERS

The directly coupled SQUID magnetometer as schematically shown in Figure 1(a) consists of a large pickup loop of inductance L_p and area A_p directly connected to the body

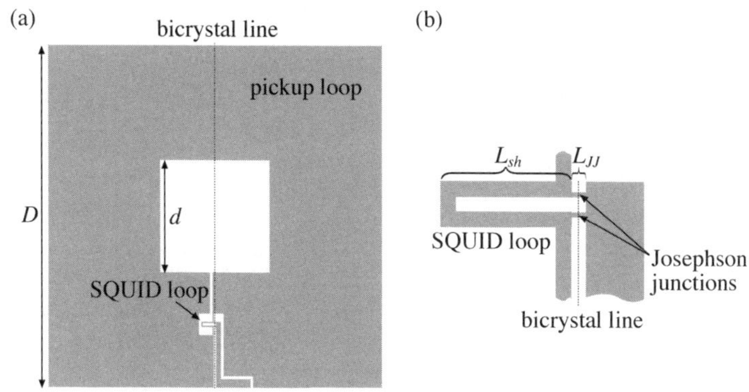

Figure 1. (a) Schematic of a directly coupled dc SQUID magnetometer with bicrystal Josephson junctions. (b) SQUID loop with Josephson junctions in detail.

of the SQUID loop of inductance $L = L_{sh} + L_{JJ}$ which is generally much smaller than L_p. Here L_{sh} is the shunt inductance of the SQUID, i.e., the inductance of the SQUID to which the screening current couples. L_{JJ} is the inductance of the Josephson junctions area (see Figure 1(b)). As can be seen the pickup loop and the shunt inductance of the SQUID loop form a closed superconducting loop. Obeying flux quantization, a magnetic field change ΔB applied to the pickup loop induces a screening current $\Delta I = \Delta B A_p/(L_p + L_{sh})$ which in turn links a flux $L_{sh}\Delta I$ to the SQUID. The effective area is thus given by

$$A_{eff} = A_p \frac{L_{sh}}{L_p + L_{sh}} \pm A_s, \tag{2}$$

where A_s is the effective area of the bare SQUID. The sign of A_s depends on the relative orientation of the SQUID and the pickup loop. If $A_{eff} >> A_s$ and $L_p >> L_{sh}$, the expression for the effective area simplifies to

$$A_{eff} = A_p \frac{L_{sh}}{L_p}. \tag{3}$$

As shown by Lee et al. [21] and Cantor et al. [22] the large mismatch between L_{sh} and L_p, which is the main disadvantage of this magnetometer approach compared to the multiturn flux transformer coupled and multiloop SQUID magnetometers, can be reduced by maximizing the ratio A_p/L_p. This can be achieved by using a pickup loop with a large linewidth. In the limit of a square washer the area is approximately given by $A_p = dD$ [23] and its inductance by $L_p = 1.25\ \mu_0 d$ [24], where d and D are the inner and outer dimensions of the washer-type pickup loop. Kinetic contributions to the inductance are neglected. Using these expressions for A_p and L_p and their simulated dependence of the flux noise power S_Φ on the SQUID inductance, Koelle et al. [2] have found that a minimum magnetic field noise can be reached for SQUID inductances around 100 pH. Here the effect of parasitic resonances which can considerably influence the SQUID performance [25] has been neglected.

Applying these design criteria, various groups have demonstrated that directly coupled HTS SQUID magnetometers can be fabricated on 10 x 10 mm^2 $SrTiO_3$ bicrystal substrates with magnetic field noise values down to 24 fT/$\sqrt{Hz}$ at 1 kHz with $1/f$ corners

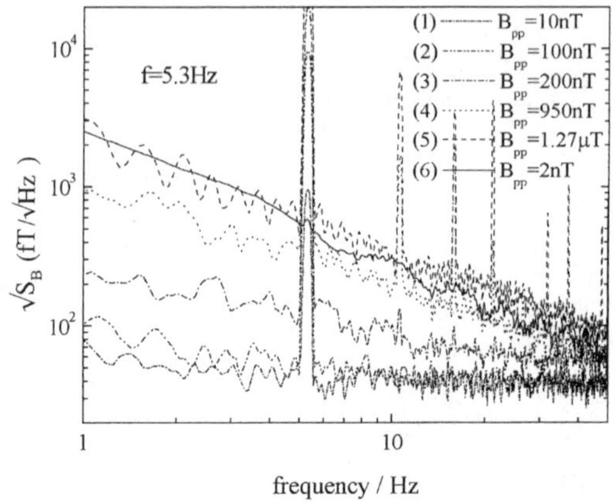

Figure 2. Magnetic field noise versus frequency of a directly coupled HTS SQUID magnetometer with wide pickup loop for different amplitudes of an ac magnetic field with frequency $f_s = 5.3$ Hz. The measurements were performed successively from (1) to (6).

of a few Hz [26]. To achieve such low noise levels also in magnetically disturbed environment, it is important to avoid thermally activated hopping of flux vortices trapped in the $YBa_2Cu_3O_{7-\delta}$ films. As found in various studies, the low-frequency noise of a $YBa_2Cu_3O_{7-\delta}$ SQUID increases above a certain threshold in proportion to the magnetic field in which the device was cooled. The increased noise is attributed to the number of vortices which are trapped in the film and which can thermally activated hop around. Dantsker et al. [27] could demonstrate that this threshold cooling field can be increased by decreasing the linewidth of the superconducting structures. Clem [28] showed theoretically that it is energetically unfavorable for vortices to enter a film strip of width w for cooling fields $B_0 < \pi\Phi_0/4w^2$. Reducing the linewidth of their SQUID structure to a value of 4 μm, Dantsker et al. [29] could show that no increase of the noise at 1 Hz was observed for cooling fields up to about 100 μT.

The standard directly coupled SQUID magnetometer design developed and tested at Conductus and adapted by various groups includes a coplanar SQUID loop with a linewidth of 4 μm coupled to a wide pickup loop (typically 3-4 mm wide). As outlined by Cho et al. [30], regarding the case that the device is cooled in the static magnetic field of the earth, the width of the pickup loop structure must not necessarily be reduced to linewidths of 4 μm. The reason for this is that – due to the large inductance mismatch of the directly coupled SQUID magnetometer – the excess flux noise in the pickup loop is rather ineffectively coupled to the SQUID. Taking the flux noise values of an unpatterned $YBa_2Cu_3O_{7-\delta}$ film cooled in a static magnetic field of 50 μT and the design values of a typical directly coupled SQUID magnetometer, they found that a rms flux noise at 1 Hz well below 1 $\mu\Phi_0/\sqrt{Hz}$ is coupled to the SQUID which is well below the typical flux noise of a HTS SQUID.

In Figure 2 the magnetic field noise spectra of such a wide-pickup-loop directly coupled dc SQUID magnetometer, fabricated at PTB Berlin, exposed to sinusoidal ac magnetic fields of varying amplitude are shown. The device was cooled down in well shielded environment ($B_0 \approx 10$ nT). As can be seen, no increase of the low-frequency noise for amplitudes B_{pp} up to about 100 nT was measured whereas for higher signal field amplitudes excess noise due to the motion of flux vortices arises. In addition, for $B_{pp} > 500$ nT flux jumps with amplitudes of typically (0.015-0.020) Φ_0 were observed. These flux jumps correspond to a change of the magnetic flux in the pickup loop of about 1 Φ_0. The excess noise did not change after lowering the field amplitude from 1.27 μT to 2 nT, indicating that the vortices remain in the film. The observed increase of the low-frequency noise with increasing signal amplitude is well understood [31]. A changing magnetic field induces a circulating current flowing through the pickup loop and the shunt inductance of the SQUID. If the Lorentz force related to this current exceeds a certain value, vortices can enter the superconducting film and produce excess noise via their thermally activated motion. As directly shown by Hirano et al. [32] by local flux noise measurements vortices most probably enter the $YBa_2Cu_3O_{7-\delta}$ film where the pickup loop crosses the bicrystal substrate grain boundary (see also section 4.2).

3. DIRECTLY COUPLED DC SQUID MAGNETOMETERS IN EXTERNAL MAGNETIC FIELDS

In this section, two different concepts of the pickup loop layout are investigated in static and changing magnetic fields with respect to noise and linearity of the devices. First, magnetometers with slotted pickup loops are introduced. Second, magnetometers with a perforated pickup loop are investigated.

3.1. Directly coupled dc SQUID magnetometers with slotted pickup loop

A photograph of a directly coupled SQUID magnetometer design with slotted pickup loop as developed in the PTB is shown in Figure 3. As can be seen, in contrast to other approaches to replace a washer-like structure by several narrow parallel lines [29,33,34], the parallel loops are not equidistant. Since both, A_p and L_p, are determined by the current distribution, it was intended to arrange the 16 parallel loops to reproduce the distribution of a circulating current in a wide, solid pickup loop. Consequently, although drastically reducing the area of superconducting thin film, the area A_p and inductance L_p of the pickup structure and thus the effective area of the magnetometer should remain almost unchanged. The second idea behind the line distribution and linewidth was that all 16 parallel loops should carry about the same current induced by an external magnetic field. In addition, since all loops twice cross the substrate grain boundary, each loop contains two flux dams. A flux dam as introduced by Koch et al. [35] is a junction in a closed superconducting loop which restricts the maximum circulating current. Since the induced current in each loop is expected to be about the same and each loop contains a flux dam with nominally the same critical current, all flux dams should open simultaneously. With a critical current density of 1 x 10^4 A/cm^2, which is a typical value for 30° grain boundary junctions, this provides a critical current of about 1 mA for each flux dam. Taking the design parameters of the SQUID magnetometer this means that they remain closed for field amplitudes up to a few μT. Such power line amplitudes can easily be found in a laboratory or clinical environment. The SQUID loop has a linewidth of 4 μm and an inner area of 4 x 82 μm^2, and the two bicrystal junctions are 1.5 μm wide, resulting in an

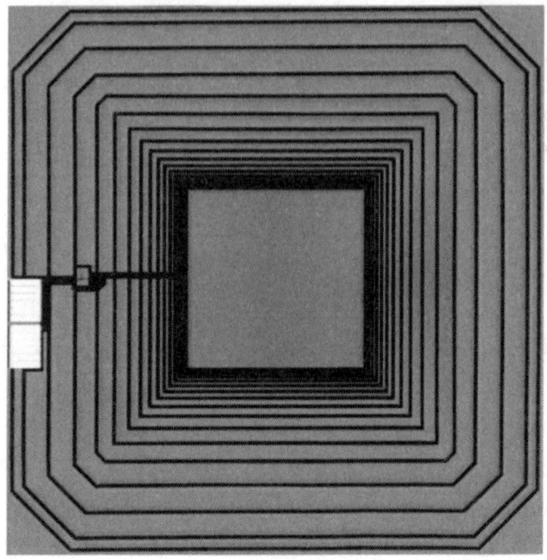

Figure 3. Photograph of directly coupled SQUID magnetometer with slotted pickup loop

estimated inductance of about 100 pH (including kinetic contributions).

Mistakenly, for the layout shown in Figure 3 the current distribution of an open square washer was used yielding a maximum line density at the inner edge. However, in a directly coupled SQUID magnetometer, the pickup loop is shorted by the SQUID loop and the signal current flowing into the SQUID loop due to an applied magnetic field is concentrated near the outer pickup loop edge [36]. Detailed numerical simulations, using the model developed by Jansman et al. [37], indicated that the pickup loop line distribution depicted in Figure 3 is indeed suboptimal but the loss in effective area is about 3% compared to the solid pickup loop design and even less compared to layouts with optimized pickup line distribution [38].

In spite of the suboptimum line distribution regarding the effective area, a number of directly coupled SQUID magnetometers of the layout shown in Figure 3 was fabricated, tested and successfully operated in systems (e.g., in the second-order HTS SQUID gradiometer system described in section 4.1). The slotted magnetometers were fabricated from typically 200 nm thick $YBa_2Cu_3O_{7-\delta}$ films deposited by hollow-cathode discharge sputtering on 10 x 10 mm^2 $SrTiO_3$ bicrystals with 30° misorientation angle [39]. To prevent the $YBa_2Cu_3O_{7-\delta}$ films from moisture and to allow one to heat them above T_c, the SQUID chips are encapsulated in a ceramic housing with integrated thick-film heater. All sensors were operated in the Berlin magnetically shielded room (BMSR) using a directly coupled flux-locked loop (FLL) electronics [40] with 100 kHz bias reversal. The magnetometers were mounted in the center of a 100-turn coil with a diameter of 38 mm and a coil constant of 3.4 μT/mA. A low-noise current source was developed consisting of two 9-V battery blocks buffered by operational amplifiers. For a static field as high as 80 μT the current source contributes to the system noise only 17 fT/$\sqrt{Hz}$ at 1 kHz and 24 fT$\sqrt{Hz}$ at 1 Hz, respectively.

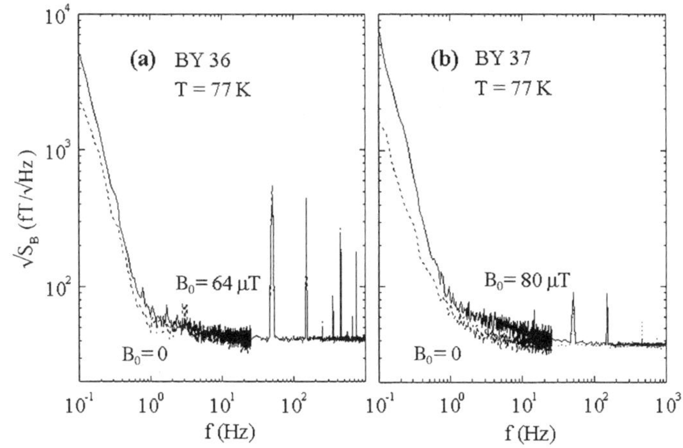

Figure 4. Magnetic field noise spectra of two directly coupled SQUID magnetometers cooled in zero magnetic field (dashed lines) and in a static magnetic field (solid lines) of (a) $B_0 = 64\mu$T and (b) 80 μT, respectively. Measurements were performed in BMSR with 100 kHz bias reversal FLL electronics. The noise includes the contributions from the directly coupled preamplifier of the FLL, from the background noise inside the BMSR, and from the current source driving the field coil.

3.1.1. Directly coupled dc SQUID magnetometers with slotted pickup loop in static magnetic cooling fields

The magnetic field noise of two directly coupled SQUID magnetometers with 16 parallel pickup loop lines in nominally zero magnetic field is shown as dashed lines in Figure 4. The total white noise amounts to about 40 fT/$\sqrt{Hz}$ for both devices, increasing to 54 fT/$\sqrt{Hz}$ and 65 fT/$\sqrt{Hz}$ at 1 Hz for devices BY 36 and BY 37, respectively. The field sensitivity B/Φ of the magnetometers, measured with a Helmholtz coil, was 5.2 nT/Φ_0 which is in good agreement with the simulated value [38].

To measure the noise when cooled in a static magnetic field, the SQUIDs were cycled above T_c by means of the integrated heater after the static magnetic field was switched on and had stabilized. Figure 4(a) shows the field noise spectrum of magnetometer BY 36 cooled in nominally zero magnetic field and in a static magnetic field of $B_0 = 64$ μT, respectively. As can be seen, the magnetic field noise is almost the same down to about 1 Hz. Device BY 37 could be cooled without significantly increasing the 1 Hz noise even in a static magnetic field of 80 μT (Figure 4(b)). Here, the noise at 1 Hz was about 80 fT/$\sqrt{Hz}$. Details of the measurements and especially of the origin of the noise below 1 Hz can be found in [41].

3.1.2. Directly coupled SQUID magnetometers with slotted pickup loop in changing magnetic fields

To investigate the influence of ac magnetic fields on the SQUID noise, after cooling the device in zero magnetic field, a sinusoidal signal with a constant frequency $f \leq 50$ Hz was

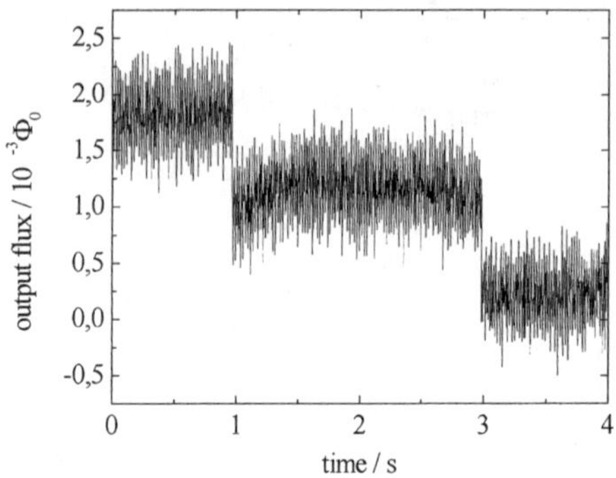

Figure 5. Time trace of directly coupled magnetometer with flux jumps which occurred after applying a transient magnetic field signal. Amplitude of flux jumps is about 0.9 mΦ_0.

applied to the field coil while the SQUID was in the unlocked mode. In this case, the ac field induces a shielding current in the closed superconducting loop formed by the pickup loop and the SQUID shunt inductance, in contrast to the FLL mode where the feedback coil should compensate all external field changes and the SQUID sensor should ideally remain field-free (this is of course only the case if a Helmholtz coil is used as feedback coil as done in [42], otherwise considerable local deviations can occur for conventionally used feedback coils). In the experiments, the function generator was connected to the field coil at a low voltage level – typically providing an ac magnetic field amplitude of about 17 nT –, increased the amplitude to the given maximum value, decreased it back again to the initial value and disconnected the generator from the field coil. After this, the FLL was closed and the noise was measured. When the generator voltage was gradually changed, peak-to-peak amplitudes of at least 54 μT, limited by the experimental setup, could be applied without increasing the noise of the SQUIDs cooled in zero magnetic field. However, when transients were present (e.g., when a square-wave signal was applied or when the voltage range of the generator was changed), peak-to-peak amplitudes of a few 100 nT were sufficient to produce flux jumps with amplitudes of typically 0.9 mΦ_0. Simulations indicated that a flux change of 0.8 mΦ_0 is produced in the SQUID by a flux jump of 1 Φ_0 across the outermost pickup loop line, in good agreement with the experimental findings. A typical time trace with flux jumps is depicted in Figure 5. Note that in between these jumps the SQUID low-frequency noise was unchanged indicating that the induced shielding currents did not drive vortices into the superconducting films but that just single flux quanta jumped across the individual lines.

The same experiment was also performed for the case that the device was – before applying ac magnetic fields – cooled in a static magnetic field above that of the earth. Flux jumps of the same amplitude as for the above mentioned case were now observed when peak-to-peak amplitudes above a few μT were applied. As for the zero-field cooling case, a sinusoidal signal was gradually increased from a value of a few tens of nT to a maximum value and then decreased again to the starting value, and after disconnecting the generator from the field coil the FLL was closed and the noise measured. One possible reason for

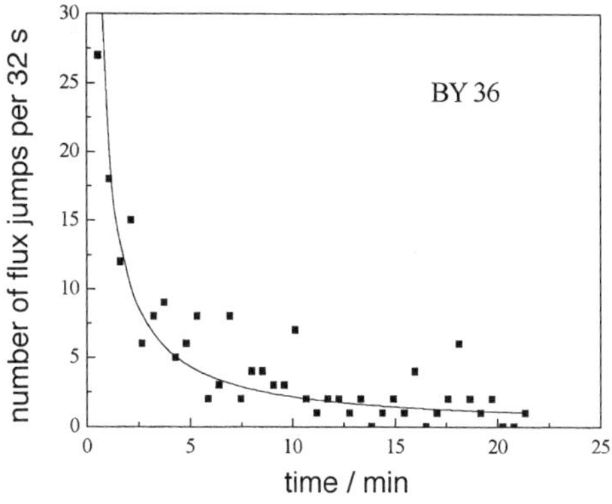

Figure 6. Number of flux jumps per 32 s versus time after cooling SQUID magnetometer BY 36 in a static magnetic field of 64 μT and exposing it to an ac magnetic field with an amplitude of 2.5 μT. The solid line is calculated using a model from Enpuku, described in section 3.3.

the reduced threshold ac field amplitude of field-cooled magnetometers could be that shielding currents are induced during the cool-down which add to the currents induced by ac fields. When a directly coupled magnetometer is cooled in a static magnetic field, due to flux conservation in a closed superconducting loop, the change of the pickup loop area with temperature produces an apparent magnetic field change which is compensated by a shielding current. A rough estimate yields that the total shielding current in all 16 loops is of the order of 2 mA when the device is cooled in $B_{cool} = 64$ μT assuming that the T_c of the flux dams is 88 K, i.e., 2 K below that of the films.

Figure 6 shows the number of flux jumps per 32 s over time after the magnetometer had been cooled in a static magnetic field of 64 μT and a sinusoidal field with an amplitude of 2.5 μT had been applied. As can be seen, the number of flux jumps per time interval decays with time. It took more than 20 min until no further flux jumps occurred. A more detailed analysis of this experiment can be found in section 3.3.

3.2. Directly coupled dc SQUID magnetometers with perforated pickup loop

In this section directly coupled dc SQUID magnetometers on 1 x 1 cm^2 $SrTiO_3$ bicrystal substrates are investigated where the complete pickup loop is patterned with holes to reduce the maximum linewidth [43]. These devices where realized at the University of Hamburg. The idea behind this concept in comparison to the slotted pickup loop is to replace flux focussing via Meissner effect by flux focussing via flux quantization of the holes. Then one does not have to take care about the given current distribution within the pickup loop. The $YBa_2Cu_3O_{7-\delta}$ films used in this chapter were made by pulsed laser deposition and are patterned by standard photolithography as described elsewhere [44]. All measurements were performed in liquid nitrogen inside of a triple MUMETAL shield.

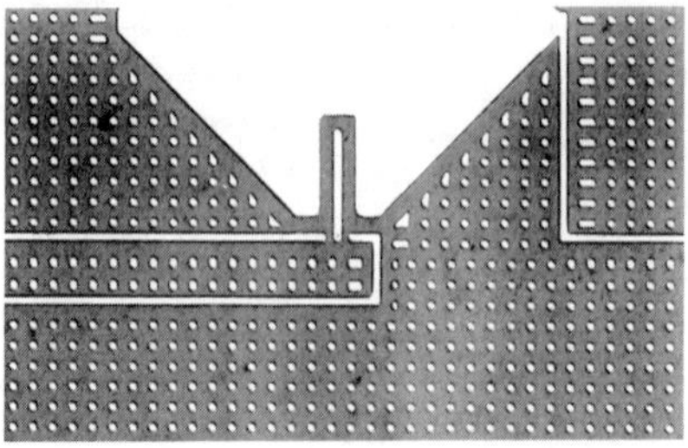

Figure 7. Micrograph of magnetometer part around the SQUID loop region after it was perforated.

3.2.1. Directly coupled dc SQUID magnetometers with perforated pickup loop in static magnetic cooling fields

The effects of this kind of patterning were directly compared for the same device, since the comparison of different devices may be misleading because of a variation in the microstructural or epitaxial quality of the involved superconducting thin films. So after characterizing the magnetometer with a solid pickup loop this device was patterned again to perforate all superconducting areas, and the measurements were repeated.

The magnetometer layout is similar to the one introduced in section 2 with a washer-type pickup loop of an outer dimension of 8.3 mm and an inner hole size of 3 x 3 mm^2. The SQUID loop has an outer area of 16 x 56 μm^2 and an inner hole of 4 x 50 μm^2, from which a SQUID inductance of 60 pH is calculated, including the kinetic contribution. The nominal width of the bicrystal junctions is 3 μm. After the perforation a structure width of 7.1 μm between the circular holes with a diameter of about 4 μm was obtained. An optical micrograph of the region around the SQUID loop is depicted in Figure 7.

The magnetic field during cooling B_0 was applied with Helmholtz coils inside the shielding can. The coil was supplied by a lead-acid battery and an appropriate resistor. The flux density noise resulting from the current noise in the coil is less than 85 $fT/\sqrt{Hz}$ at 3 Hz and B_0=100 μT, and does not significantly contribute to the measured noise values. The noise measurements were made in flux-locked-loop mode with bias current reversal to reject the low-frequency noise from critical-current fluctuations in the junctions. In all field dependent measurements the magnetic field was applied while the magnetometer was heated above T_c, and it remained on during the cooling and the measurements. Hence, no large shielding currents were generated in contrast to a field change process. The currents induced during cool-down due to the change of the effective area of the pickup loop were neglected. The cooling process from 100 K to 77 K has lasted about 15 minutes.

The magnetometer characteristics for the zero field case before and after the second patterning are listed in Table 1. After patterning, I_0 and R_n were somewhat changed, presumably due to a slight oxygen loss in the Josephson junctions. The patterning had only a minor effect on the sensitivity of the device although the character of the input coupling changed. With a solid pickup loop, flux is focused into the loop by the Meissner effect. In the patterned device the flux in the holes is quantized at integer flux quanta Φ_0 [45]. In the zero cooling field case it is quantized at 0 Φ_0. After the device is cooled below the critical temperature of the superconducting film, a change in an external magnetic

Table 1. Properties of the investigated magnetometer without external magnetic field before and after the patterning.

			Before	After
Voltage swing	ΔV	(μV)	26	27
Critical current	I_0	(μA)	107	76.5
Normal resistance	R_n	(Ω)	2.1	2.8
Sensitivity	B/Φ	(nT/Φ_0)	9.20	8.98
Effective area	A_{eff}	(mm^2)	0.224	0.230

field leads to shielding currents around each hole which keeps the total flux in each hole at zero. So the magnetic field produced by the shielding currents around the holes seems to couple as much field into the pickup loop as the Meissner effect in the unpatterned device for zero field cooling. However, a difference is found for larger cooling fields as shown in Figure 8.

For the solid device an increase of the effective area above $B_0 = 100$ μT was found, whereas it decreases with nearly the same slope for the patterned device above 35 μT. Hence, the calibration and therewith the field noise of the sensor is affected by larger cooling fields. The change of the effective area at different cooling fields is not limited to the used magnetometer design. In [46], an integrated multiloop pickup coil (IMPUC) magnetometer with a flux transformer was investigated at different cooling fields. The IMPUC magnetometer shows a change in the effective area for different cooling fields quite similar to the perforated device in this experiment. The reason for this layout dependent behavior is not yet clarified.

Figure 9(a) depicts several flux density noise spectra for the unpatterned magnetometer in different cooling fields B_0. An increasing low-frequency noise with B_0 due to the increasing number of vortices trapped in the film can be found. The spectra for $B_0 = 71$ and 198 μT include a Lorentzian contribution, caused by the random telegraph signal (RTS) from a predominant two-level fluctuator. Figure 9(b) shows the noise values $\sqrt{S_B}$ at the frequency 3 Hz versus B_0, before and after the holes were patterned into the pickup loop. In the unpatterned case, the low-frequency noise increases with B_0 even

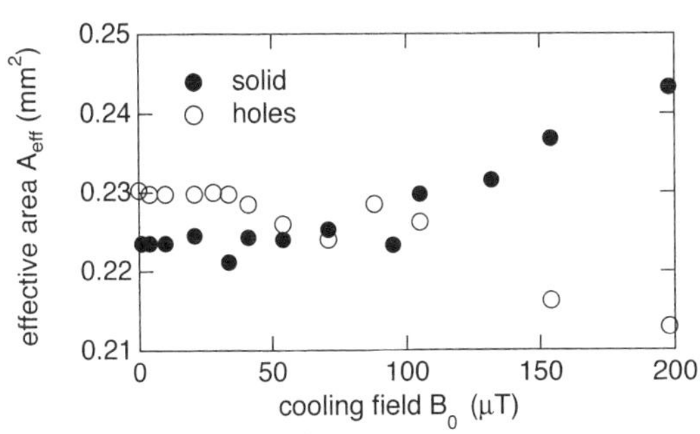

Figure 8. Effective area of magnetometer depending on the cooling field before (solid) and after (holes) the second patterning.

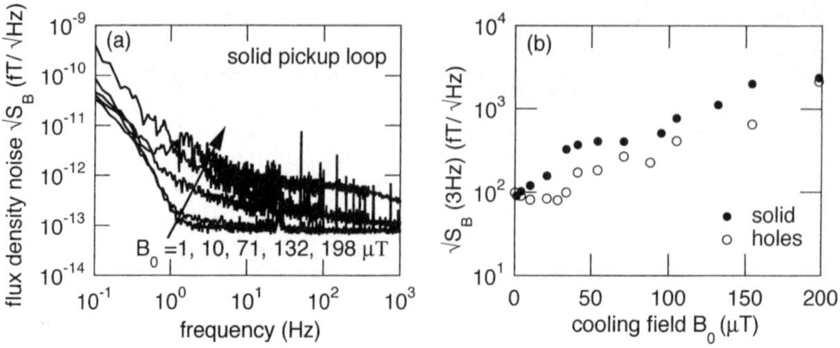

Figure 9. (a) Flux density noise spectra of the unpatterned magnetometer for different cooling fields B_0. (b) Flux density noise at 3 Hz vs cooling field B_0 before (solid) and after (holes) the patterning.

for the lowest cooling fields. This noise below $B_0 = 35\ \mu$T presumably originates from the motion of vortices in large superconducting structures near the SQUID loop (e.g., the 20 μm wide lead to the Josephson junctions), since the indirect component from noise currents in the pickup loop is strongly suppressed by the poor coupling between the inductances of the pickup loop and the SQUID as mentioned in section 2.

In the patterned case, the noise remains approximately constant below a threshold field $B_t = 35\ \mu$T. This value corresponds to a maximum linewidth of $w = \sqrt{\pi\Phi_0/4B_t} = 6.8\ \mu$m and is in good agreement with the maximum structure width of 7.1 μm. For all values of B_0, the magnetometer with perforated pickup loop remains markedly less noisy than that with solid pickup loop. At cooling fields above 100 μT a $1/f$ like dependence of the low frequency noise was found which – as shown by Ferrari et al. [48] – results from the incoherent superposition of many thermally activated hopping processes.

3.2.2. Directly coupled dc SQUID magnetometers with perforated pickup loop in changing magnetic fields

In this section, the field sensitivity of our dc SQUID magnetometers with holes in external ac magnetic fields is investigated. The design of the used magnetometers is quite similar to the one mentioned in the above section besides some optimizations according the sensitivity and the maximum linewidth. The outer dimension of the pickup loop is now 9 mm with a pickup loop width of 3 mm from which an effective area A_p of 30 mm^2 and an inductance L_p of 5 nH was calculated. The linewidth between the holes was reduced to 4 μm. The SQUID loop is now 60 μm long with a linewidth of 4 μm. The gap between the SQUID loop striplines is also 4 μm wide. For the SQUID loop, an inductance of about 100 pH was calculated. The bicrystal Josephson junctions are 1.5 μm wide. These devices can be cooled in static magnetic fields of up to 60 μT without increasing the noise down to 3 Hz.

All the measurements were performed at 77 K inside the shielding can consisting of three layers of MUMETAL and two layers of aluminum. The external magnetic fields were generated using Helmholtz coils inside the shielding can. The devices are biased

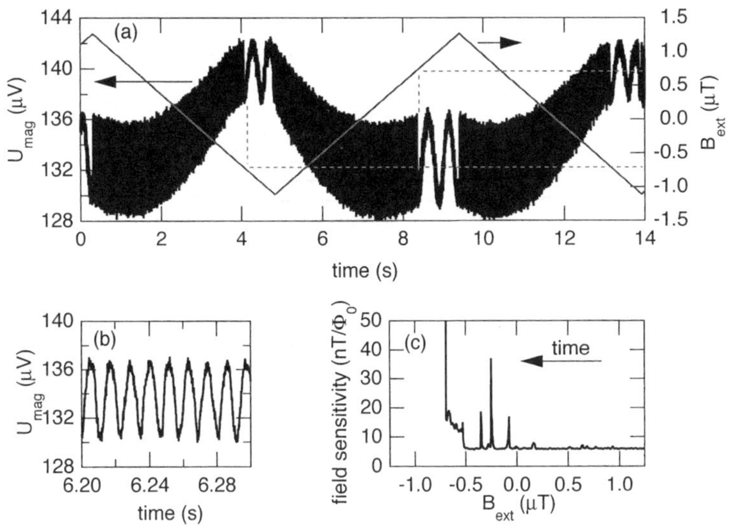

Figure 10. (a) Magnetometer voltage U_{mag} and external magnetic field B_{ext} versus time. The dashed lines mark the threshold fields and corresponding times where the magnetometer sensitivity increases by a factor of about 44 to 260 nT/Φ_0. (b) Magnification of the magnetometer modulation with regular field sensitivity. (c) Magnetometer field sensitivity for the interval between 1 s and 5 s. The jump to 260 nT/Φ_0 at $B_{ext} = -0.7\mu$T is cut-off.

with a constant current I_b at which the modulation depth is near its maximum. The voltage over the SQUID is measured directly. No flux-locked-loop or feedback electronics were used.

An encapsulated directly coupled dc SQUID magnetometer from our multichannel system, which is described in detail in section 4.2, was used to investigate the behavior in ac fields of about 1 μT amplitude. As mentioned in section 2, the current change ΔI_p which is induced in the pickup loop by an external field change ΔB_{ext} can be estimated with $\Delta I_p/\Delta B_{ext} = A_p/L_p = 6$ mA/μT. The shunt inductance L_{sh} of the SQUID loop can be neglected. So, with the given layout, every external field change of 1 μT induces a current of 6 mA in the pickup loop and therewith in the directly coupled SQUID loop. Figure 10(a) shows the magnetometer voltage U_{mag} versus time if a triangular-shaped magnetic field of about ± 1.1 μT is applied. The data of the applied magnetic field B_{ext} and the magnetometer output signal U_{mag} were recorded about one minute after starting the external magnetic field change at $B_{ext} = 0$ μT.

The effect of a Fraunhofer pattern can clearly be seen at the magnetometer output voltage in an increasing voltage and a decreasing modulation depth. The magnetic field change ΔB_J at the Josephson junctions caused by the induced current can be estimated using the Biot-Savart law of a linear conductor for each part of the SQUID loop and the leads between pickup loop and SQUID loop. An external field change of 1 μT results in a

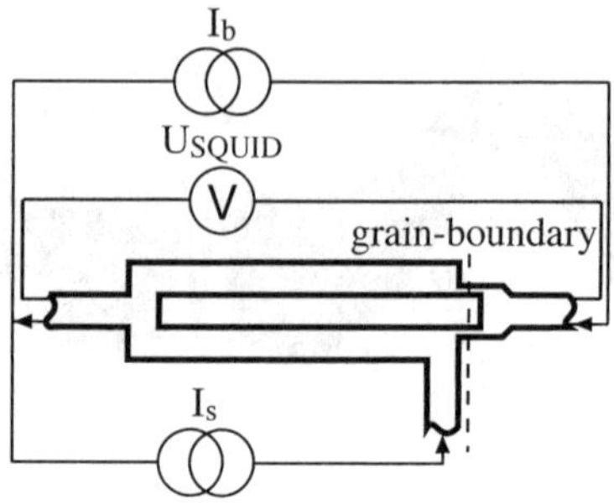

Figure 11. Schematic of the SQUID layout to inject the current I_s into half of the SQUID loop.

field change of about 165 μT at the site of each Josephson junction. From the size of the Josephson junction, the first minimum in the Fraunhofer pattern is expected to occur at a magnetic field of about 2 mT at the site of the junction. As can be seen from Figure 10(a), the maximum magnetometer modulation is not at zero external field but shifted by about -0.8 μT and 0.3 μT for decreasing and increasing external fields, respectively. This may be due to inhomogeneous flux distributions within the Josephson junctions. In Figure 10(b) a magnification of the magnetometer modulation is shown for external fields near zero. Here, the field sensitivity calculated from the measured modulation equals the expected value of about 5.9 nT/Φ_0 for the magnetometer. At higher magnetic fields, the field sensitivity and therewith the effective area A_{eff} of the magnetometer changes with increasing external field. In Figure 10(a), large magnetometer modulations with a corresponding field sensitivity of 260 nT/Φ_0 can be observed at external fields about $\pm$0.7 μT (see dashed lines). The results of a detailed analysis of the field sensitivity in the measuring interval between 1 s and 5 s are depicted in Figure 10(c). The field sensitivity values are calculated by doing a sinusoidal fit over at least 2 magnetometer modulations. At $B_{ext} = -0.5$ μT, the field sensitivity changes to about 11 nT/Φ_0 before the large sensitivity change to 260 nT/Φ_0 occurs. At the turning points of the triangular-shaped field change the field sensitivity of the magnetometer changes back to the regular value. So, the process that is responsible for the large field sensitivities seems to depend not only on the absolute values of the external field, but also on the dynamics of the field change. During one measurement with a fixed field amplitude and frequency, the behavior of the magnetometer was reproducible, i.e., the changes of the field sensitivity did always appear exactly at the same external fields.

As the magnetometer becomes less sensitive at some threshold fields, flux must have entered the device, depending on the induced shielding currents. There are two weak parts, which could be sensitive to these currents. First, the SQUID loop itself is a narrow 4 μm wide $YBa_2Cu_3O_{7-\delta}$ line with a fixed critical current density. While flux vortices are entering the SQUID loop, driven by the shielding current, a reduced effective area should be observed. Second, as mentioned above, the pickup loop has to pass the bicrystal line two times if, as usual, the bicrystal line is in the middle of the substrate and the whole substrate area is utilized. In the used design there are two solid crossings over the bicrystal line, one 3 mm wide and one about 1 mm wide. If B_{ext} exceeds a critical value so that

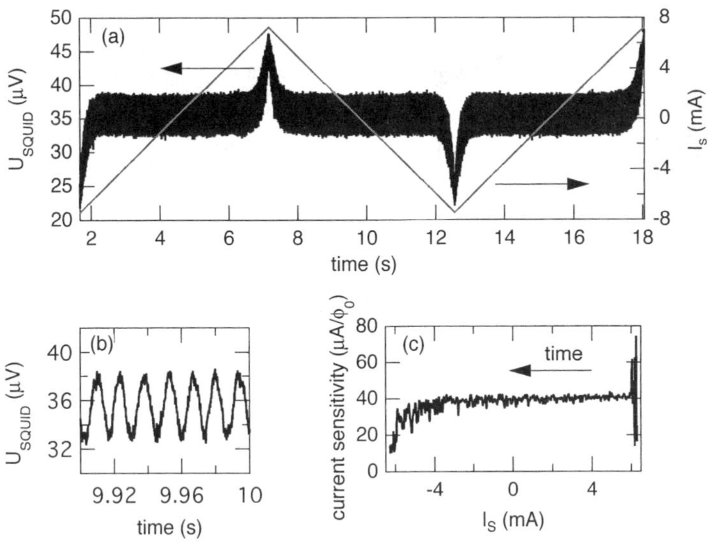

Figure 12. (a) SQUID voltage U_{SQUID} and induced current I_s versus time. The frequency of the ac current is about 0.2 Hz. (b) Regular SQUID modulation at small impressed currents. (c) Current sensitivity for the interval between 8 s and 12 s.

vortices start to penetrate the grain boundary crossings of the pickup loop due to the Lorentz force of the induced current, the vortices will be pulled in one direction along the crossing [49]. With every fluxon that enters (or leaves) the pickup loop, the shielding current is correspondingly decreased and can not couple flux into the SQUID loop.

To clarify the reason for the sensitivity change at large field changes in the devices, the shielding current in a simple SQUID loop was simulated to see what happens when the critical current of the $YBa_2Cu_3O_{7-\delta}$ line is reached. A layout was used as shown in Figure 11. The 4 μm wide SQUID loop has a length of 56 μm and a film thickness of 130 nm. The slit is 4 μm wide. A current $I_s = \pm 7.5$ mA was injected into half of the SQUID loop, the bias current I_b was fed symmetrically to the SQUID. The SQUID voltage U_{SQUID} was also measured symmetrically.

If a triangular-shaped current was applied to the SQUID loop, a regular modulation can be seen at low currents due to the flux generated by I_s (Figure 12(b)). The average current sensitivity for the regular SQUID modulation of 40 $\mu A/\Phi_0$ leads to an inductance per unit length of 1.9 pH/μm for the SQUID loop. At a critical current I_c the modulation decreases and the averaged SQUID voltage rises or decreases, depending on the sign of I_s (see Figure 12(a)). At an I_c of about 6 mA when the critical current of the weakest point of the SQUID loop is reached vortices start to penetrate the film, the modulation vanishes, fractions of I_s add to the bias current I_b and cause the voltage over the SQUID to increase or decrease for positive or negative I_s, respectively. A critical current of 6 mA for the 130 nm thick, and 4 μm wide $YBa_2Cu_3O_{7-\delta}$ film is in good agreement to

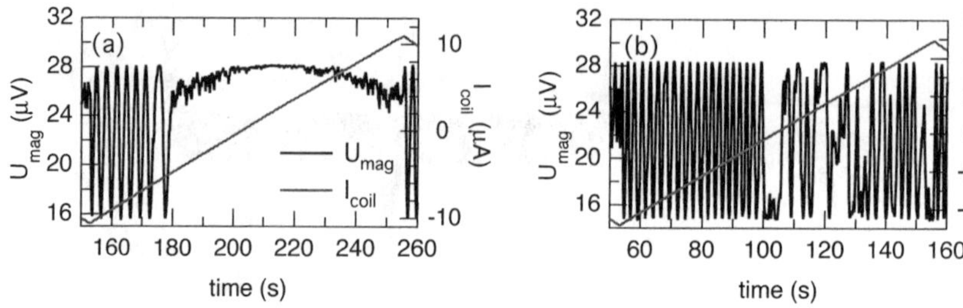

Figure 13. Magnetometer modulation and coil current at different coil positions (a) pickup loop center, and (b) above SQUID loop.

critical current density measurements of $J_c = 1.1 \cdot 10^6$ A/cm^2 on this chip. The current sensitivity shown in Figure 12(c) was calculated the same way as in the magnetometer experiment. At positive currents above 6 mA, Figure 12(c) shows large fluctuations of the current sensitivity, whereas the current sensitivity decreases for impressed currents below -3 mA more or less uniformly. Assuming, that parts of the SQUID loop are in a resistive state at currents $I_s > 6$ mA, the observed current sensitivity fluctuations can be explained with vortices entering and leaving the SQUID loop.

The decreasing current sensitivity at increasing negative currents I_s could result from a change in the SQUID inductance due to current distribution changes in the part of the SQUID loop where the current is impressed. Enpuku et al. [50] found a nonlinear increase of the kinetic inductance in superconducting thin films at large bias currents. This would increase the total inductance of the SQUID loop and thus decrease the current sensitivity. It is not yet clarified why no effect of a Fraunhofer pattern of the Josephson junctions was observed in this experiment although the impressed current produces a comparable flux at the site of the junctions as in the magnetometer experiment mentioned above. It has to be taken into account that the voltage changes due to a reduced critical current strongly depend on the characteristics of the $I(V)$ curve of the SQUID and on the bias current where the maximum modulation depth of the device is found.

However, the change in sensitivity in the SQUID experiment is different from the change the magnetometer shows. So it can be concluded, that not the SQUID loop is the limiting factor at large shielding currents, but the grain boundary crossings of the pickup loop.

To prove this assumption, local fields at different magnetometer positions were applied to investigate the modulation properties. A small 30 turn coil with an outer diameter of about 1.5 mm and an inner diameter of about 0.5 mm was positioned at the center of the magnetometer and above the SQUID loop, respectively. The coupling of an applied field at the central position should induce a shielding current in the pickup loop in the usual way, i.e., most of the flux will be indirectly coupled into the SQUID loop via shielding currents induced in the pickup loop. This experiment is comparable to the magnetometer experiment mentioned above. In the position above the SQUID loop, high local fields can be applied to the SQUID loop itself. A similar behavior to the SQUID experiment is expected, i.e., most of the flux is directly coupled to the SQUID loop. A magnetometer with the above described layout, but with a high sensitivity to flux penetration was used.

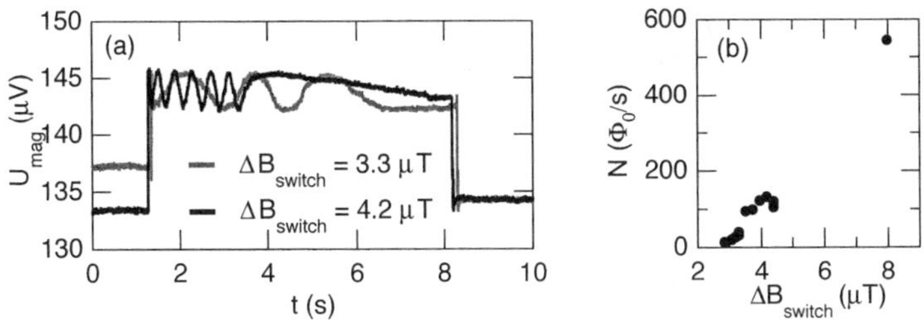

Figure 14. (a) Magnetometer response at two different switching fields ΔB_{switch}. (b) Number of vortices entering the pickup loop per second N versus switching field.

Measurements in homogenous external fields showed that flux enters this magnetometer at fields as low as about 30 nT. With a magnetometer field sensitivity of 5.6 nT/Φ_0, only five to six magnetometer modulations could be observed until flux entered the device in homogeneous magnetic fields.

The magnetometer modulation and the coil current versus time are displayed in Figure 13(a) and (b) for measurements with the coil at the central position and above the SQUID loop, respectively. Figure 13(a) shows a behavior comparable to the magnetometer measurement discussed above. At the turning point of the field changes, regular magnetometer modulations can be observed (150 s to 180 s). Above a threshold field change, flux starts to penetrate the grain boundary crossings of the pickup loop driven by the induced shielding current leading to a large sensitivity change (180 s to 255 s). The number of observed regular magnetometer modulations agrees well with the value measured in homogeneous field. The situation for the position above the SQUID loop looks quite different (Figure 13(b)). A total of 19 regular modulation periods can be seen due to less induced current in the pickup loop and larger fractions of field coupled directly to the SQUID loop. However, in contrast to the magnetometer experiment (Figure 10) and the experiment with the coil in the central position (Figure 13(a)), for larger applied field changes irregular modulations are observed, comparable to the behavior at large currents in the SQUID experiment (Figure 12(c)). A quantitative analysis for this case is difficult since the various direct and indirect contributions to the flux coupled into the SQUID loop depend very sensitively on the exact position of the coil.

To investigate the dynamics of the vortex movement along the grain boundary crossings, homogeneous switching fields ΔB_{switch} of different amplitude were applied to the magnetometers, using a Helmholtz coil. Figure 14(a) shows the magnetometer voltage U_{mag} versus time for two different switching fields $\Delta B_{switch} = 3.3$ μT and $\Delta B_{switch} = 4.2$ μT. The external field was switched on at about 1.4 s and turned off at about 8.3 s. Although there is no change of the external field during that time period, a magnetometer modulation can be observed with a frequency depending on the amplitude of the switching field. From the modulation frequency, one can calculate the number of vortices per second N that have to enter the pickup loop to produce the observed modulation. Figure 14(b) shows that there is an increase in N for higher switching fields as expected for higher induced currents.

3.3. Discussion of directly coupled dc SQUID magnetometers and flux dams in changing magnetic fields

The observed flux jumps of directly coupled SQUID magnetometers with slotted pickup loops (Figure 6) and the irregularities of the field sensitivity measured on directly coupled SQUID magnetometers with perforated pickup loops (Figures 11, 14 and 15) when exposed to large time-varying magnetic fields are – as discussed in the corresponding sections – most probably related to the fact that the pickup loop lines in both cases twice cross the grain boundary of the bicrystal substrate, thereby containing flux dams. The main advantage of flux dams incorporated in the pickup loop of a directly coupled or flux transformer coupled SQUID magnetometer (i.e., in a closed superconducting loop) is that they limit the maximum screening current which is induced by changes of an applied magnetic field [35]. As shown by Sun et al. [51], too high critical currents flowing through the pickup loop can cause vortices to penetrate the superconducting films and produce excess low-frequency noise. The usefulness of flux dams has been demonstrated by a number of groups [31,34,41,52,53]. For example, Milliken et al. [31] directly compared the noise of directly coupled SQUID magnetometers where the 3 mm wide pickup loop crosses the substrate grain boundary with devices where a flux dam was incorporated. In the latter case the linewidth of the pickup loop where it crosses the substrate grain boundary was reduced to 20 μm. When the devices were cooled in zero magnetic field and – being cold – after a magnetic field was switched on, the devices without pickup loop restriction exhibited considerable levels of low-frequency noise for magnetic field values below 1 μT. For comparison, the devices with flux dams showed no increase of the low-frequency noise for field values above 20 μT. However, as discussed in Section 3.1.2 and also pointed out by Oyama et al. [52], when the flux dam opens, it often requires a long waiting time until the flux dam completely closes.

The properties of a flux dam incorporated in a closed superconducting loop were studied in more detail by Enpuku et al. [54,55]. For a quantitative description, he considers the flux dam as a Josephson junction with critical current I_c and normal-state resistance R_s. The circuit equation of a closed superconducting loop with a flux dam is thus given by [55]

$$\frac{d\Phi_{ext}}{dt} = L_p \frac{d(\Delta I)}{dt} + V_{JJ}(\Delta I) \tag{4}$$

where Φ_{ext} is the applied flux, L_p the pickup loop inductance, and V_{JJ} the voltage across the flux dam. Taking into account the effect of thermal noise, the voltage across the junction for $\Delta I < I_c$ is given by [56]

$$V_{JJ} = I_c R_s \exp\left[-\frac{\Delta U(\Delta I)}{k_B T}\right] \tag{5}$$

with the potential barrier of thermally activated flux entry

$$\Delta U(\Delta I) = \frac{I_c \Phi_0}{\pi}\left[\sqrt{1-(\Delta I/I_c)^2} + \frac{\Delta I}{I_c}\left[\arcsin\left(\frac{\Delta I}{I_c}\right) - \frac{\pi}{2}\right]\right]. \tag{6}$$

Applying, e.g., a constantly rising magnetic field (as in the experiments described in Section 3.2.2), thus producing a constant $d\Phi_{ext}/dt$, a screening current ΔI is induced to conserve the magnetic flux inside the loop. In this case the junction voltage V_{JJ} is zero. Further increasing the field and neglecting thermal noise, the flux dam will open if the screening current exceeds the critical current I_c of the flux dam. With thermal noise, a voltage V_{JJ} appears already for ΔI values below I_c. Since the voltage across the flux dam is not zero, flux can enter the pickup loop through the flux dam and ΔI decreases correspondingly. If there is no further increase of the magnetic field, i.e., $d\Phi_{ext}/dt = 0$, the screening current decreases logarithmically, as analytically shown by Enpuku et al. [55]. Although the value ΔI is below its maximum value, it is still close to I_c and the potential barrier against the thermal activation of flux entry is small and consequently vortices can continue penetrating the loop. Due to the logarithmic decay of ΔI, it takes a long waiting time until ΔI decreases to a value where no further flux jumps occur during a typical measuring time.

Enpuku's model can also be applied to describe the experimentally observed time dependence of flux jumps per time interval (Figure 6). According to Enpuku [57], the number of thermally activated flux entries per second is given by $n = 1/T = V_{JJ}/\Phi_0$ where T is the average time between two flux jumps. With equation (A4) from [55] and assuming that $V_{JJ} < V_0$ for $t > 1$ s (V_0 is the value of V_{JJ} at $t = 0$) [57], one finds for the number of flux jumps per 32 s

$$N = 32n = \frac{32(0.65 L_p I_c \Gamma)}{\Phi_0} \cdot \frac{1}{t} \tag{7}$$

where t is the time and $\Gamma = 2\pi k_B T/(I_c\Phi_0)$ the noise parameter. Using an estimated inductance of $L_{p,out} = 40$ nH for the outermost of the 16 parallel loops of the directly coupled SQUID magnetometer with slotted pickup loop and $T = 77$ K, one finds $N = 1300/t$ which is shown as solid line in Figure 6. The agreement is very good supporting the validity of Enpuku's model.

There are various options to solve the problem of uncontrolled flux entry into the pickup loop. Assuming that a flux dam is a useful component, especially if the devices are exposed to large magnetic field changes, Enpuku et al. [55] proposed the use of a compensation field. The compensation field is applied after the flux dam has opened so that the screening current is quickly decreased to zero. Alternatively, the flux dam could be replaced by a switch including two [55] or four junctions [58]. These two approaches may be helpful in applications where the device is just once exposed to a large magnetic field change, e.g., when a system is moved in the earth's magnetic field or when the measurement system is exposed to a magnetic field pulse (e.g., for the premagnetization of magnetic nanoparticles [59–61]).

For certain applications, one does not necessarily need to incorporate a flux dam. One of the main problems of using SQUID devices based on bicrystal junctions is that the grain boundary of commercially available bicrystal substrates passes directly through the substrate middle. Consequently, exploiting the whole substrate as pickup area, the pickup loop must necessarily cross the substrate grain boundary twice. To avoid this problem, Yokosawa et al. [62] proposed a directly coupled SQUID magnetometer design with two pickup loops, located on the two opposite sides of the bicrystal substrate, coupled to the same SQUID. Though exploiting the whole substrate area, the effective area of this design is about the same as if just one of the two pickup loops is used. Thus, as far as there is no competitive HTS junction technology, bicrystal substrates with off-centric grain boundary might be desirable to fabricate directly coupled SQUID magnetometers with a pickup loop of about 9 x 9 mm^2 size which does not not cross the substrate grain boundary.

4. BIOMAGNETIC MEASUREMENT SYSTEMS

In this section, two systems for biomagnetic applications are reviewed, equipped with the above mentioned magnetometers. The systems are built, following two different concepts. The first, equipped with slotted pickup loop devices, has a single measuring channel consisting of a total of five magnetometers. It is made out of a fiberglass dewar and works stably in unshielded environment. The second system, equipped with the perforated pickup loop magnetometers, is an 8-channel system, with a variable sensor configuration. It is made out of a handy low cost glass dewar and was used to illustrate the potential of HTS SQUID systems for biomagnetic applications.

4.1. System for use in unshielded environment

In the following section a single-channel second-order gradiometer system developed at the PTB Berlin is described. The system concept takes into account the following criteria:

1. To make the system compact and mobile, a 45° inclined LN_2 dewar was chosen so that MCGs can be measured in a sitting position, i.e., the signal sensor detects horizontal magnetic fields, and no complicated bed construction is needed.

2. Second-order gradiometry is used for the environmental noise suppression. In contrast to the gradiometer work summarized in section 1, the balancing of the system is performed adaptively at the measurement site using sophisticated software algorithms.

3. To relax the requirements on the A/D conversion which is required to do the software data processing, gradiometry is performed combining electronic (first-order) and software gradiometers. Balancing the electronic first-order gradiometers to about 1 part in 100, 16-bit A/D converters should be sufficient to suppress, e. g., the 50 Hz power line signal with peak-to-peak amplitudes up to 1 μT to the required level of a few pT.

4.1.1. System setup

The system includes five directly coupled SQUID magnetometers of the design described in Section 3.1 based on 30° $SrTiO_3$ bicrystal junctions. As a result of the 45° inclined dewar, the three "z" magnetometers do not form a purely axial gradiometer but a combination of an axial and a planar one (Figure 15(a)). The magnetic field noise of the three "z" sensors is about 50 $fT/\sqrt{Hz}$ with a 1/f corner around a few Hz; the noise of the other two devices used as x and y reference sensors was slightly higher. The magnetometers are encapsulated with an integrated thick-film heater which allows one to easily heat the devices above T_c to release trapped magnetic flux.

A crucial point of electronic (or software) gradiometers in magnetically unshielded environment is the FLL electronics. For unshielded operation in urban environment, closed-loop bandwidths of the order of 1 MHz and slew rates of up to 10^6 Φ_0/s are required.

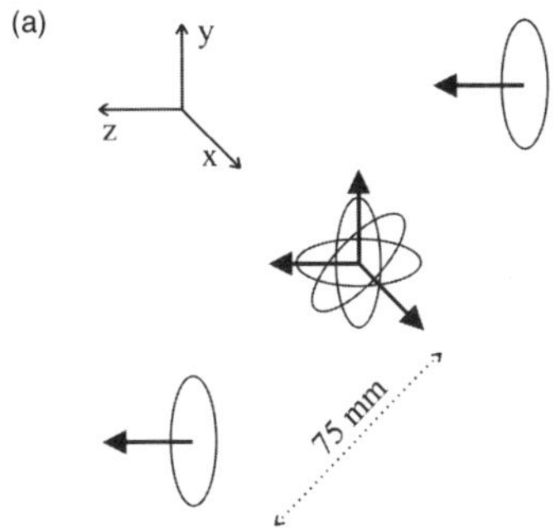

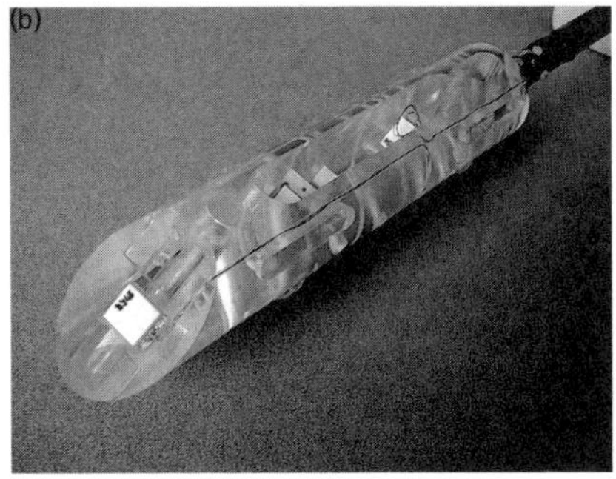

Figure 15. (a) Scheme of the sensor arrangement. Arrows indicate directions of field sensitivities. (b) Photo of acrylic glass probe head.

To achieve these margins, a directly coupled FLL electronics was used, developed at the PTB Berlin and extensively described in [40]. For LTS SQUIDs having a voltage noise of typically 0.1 $nV/\sqrt{Hz}$ in the white region, the preamplifier voltage noise generally dominates the total system noise. Generally, one uses a flux modulation FLL electronics with a transformer, however, it is hard to achieve the required dynamic performance with such electronics, although Koch et al. [63] presented a flux modulation FLL electronics with 16 MHz modulation frequency, resulting in a closed-loop bandwidth of about 1 MHz. Another solution to the problem is the application of additional positive feedback (APF), as proposed by Drung et al. [64]. For HTS SQUID devices the situation is slightly different. Good HTS SQUIDs exhibit voltage noise levels of about 0.5 $nV/\sqrt{Hz}$ in the white region. Utilizing 4 and 8 bipolar transistors in parallel, the preamplifier voltage noise at 1 Hz and 1 kHz can be reduced to 0.44 $nV/\sqrt{Hz}$ and 0.4 $nV/\sqrt{Hz}$, respectively. Taking these numbers, using such directly coupled electronics to read-out HTS SQUIDs, the preamplifier voltage noise typically increases the system noise by 20-30%. The circuit diagram of the used FLL electronics is depicted in Figure 16. As can be seen, the electronics allows one to characterize the SQUID devices in open-loop (reset) and closed loop (flux-locked loop) mode. To suppress low-frequency excess noise due to critical current fluctuations it is equipped with a bias current reversal option with a bias reversal frequency of typically 100 kHz. This high bias reversal frequency is required since both in-phase and out-of-phase critical current fluctuations have to be suppressed by the bias reversal. In contrast, in a flux-modulated FLL electronics the in-phase critical current fluctuations are suppressed by the flux modulation, typically having a modulation frequency of some 100 kHz, so that a bias current reversal frequency of a few kHz is sufficient. The maximum closed-loop 3 dB bandwidth was about 1 MHz resulting in a slew rate of about 10^6 Φ_0/s. For the system measurements in unshielded environment, a typical system bandwidth of about 300 kHz, depending on the given environmental noise level, is chosen. Recently, Drung et al. [65] reported a further improvement of their directly coupled FLL electronics implementing automatic bias voltage tuning. An in-system programmable microcontroller is used to control all functions of the SQUID system.

The five encapsulated SQUID magnetometers are accommodated in a probe head machined from 48 mm diameter acrylic glass, shown in Figure 15(b). The baseline of each

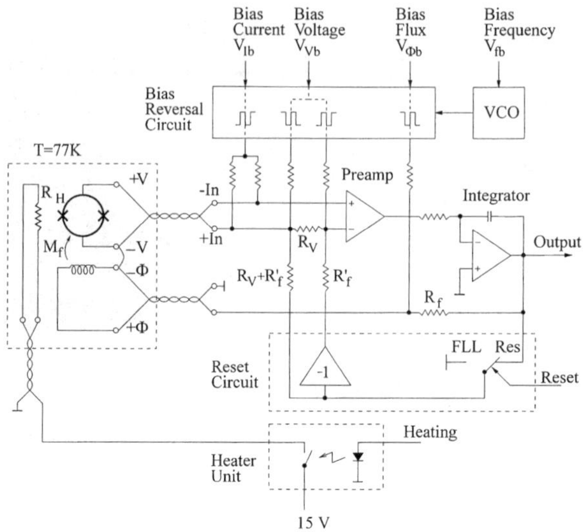

Figure 16. Circuit diagram of directly coupled FLL electronics of PTB Berlin.

two "z" gradiometers amounts to 75 mm (Figure 15(a)). The three magnetometers forming a vector magnetometer in the middle of the probe head are arranged in such a way that the crosstalk is minimized. Separate measurements yielded crosstalk values of a few tenth of a percent for signal frequencies below 5 kHz in spite of the close proximity of the magnetometer sensors (center-to-center distance of $\geq$ 13 mm) [66]. The probe stick is inserted in a liquid nitrogen dewar with a holding time of about one week. The dewar has an outer diameter of 180 mm and a total length of 800 mm. For rf shielding, 10 layers of superinsulation foil are placed in the vacuum space between the two shells of the fiberglass dewar in addition to the thermal insulation foils. The cold-warm distance of the dewar amounts to about 10 mm. A photograph of the complete system including the dewar with plugged-in FLL electronics and the dewar support as well as the control unit and industrial computer is shown in Figure 17.

4.1.2. Gradiometry

As mentioned before, a combination of electronic and software gradiometry is chosen for the environmental noise suppression. For this the five analog signals from the FLL electronics are fed into a so-called signal conditioner, representing the interface between SQUID electronics and A/D converter. Using DIP switches, resistor cascades can be adjusted to balance the two first-order gradiometers, formed by the upper and center as well as by the center and lower sensor, to 1:100. All five analog output signals (i.e., two first-order gradiometers and the three magnetometer channels of the center vector triple) contain anti-aliasing filters to adjust the bandwidth to the clock frequency of the 16 bit A/D converter. The further signal processing is performed using a Pentium industrial computer with integrated A/D converter cards from National Instruments.

To adjust the system to the given noise environment, adaptive noise cancellation is applied. In the first approach, the adaptive balancing was performed in the time domain

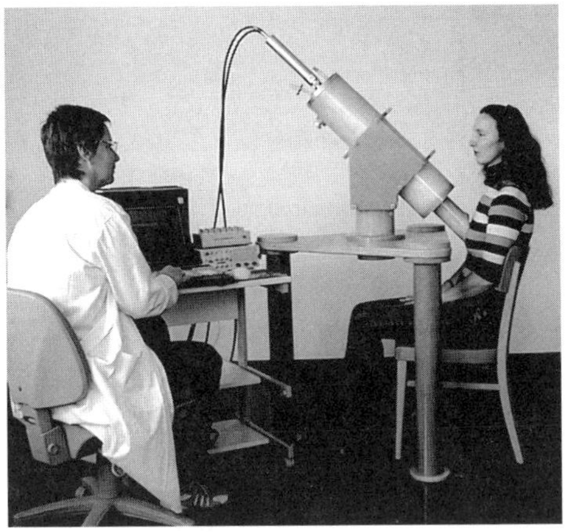

Figure 17. Photo of the complete system including fiberglass dewar, dewar support, plugged-in FLL electronics, control unit, and industrial computer. Each set of electronics with cylindrical lid houses the FLL electronics for three SQUID sensors.

(TD) using the algorithm described by Woeltgens and Koch [67]. The background signal is given by

$$B_{back}(t) = \sum_i a_i B_{ref,i}(t). \tag{8}$$

Here $B_{ref,i}(t)$ is the signal from the i-th reference sensor. The calibration coefficients a_i are obtained from an algorithm which is based on the minimization of the residual sum of squares, i.e.,

$$\frac{\partial}{\partial a_i} \sum_t (B_{src}(t) - B_{back}(t)) = 0. \tag{9}$$

Note that the coefficients a_i are independent of time and frequency. The algorithm was implemented in LABVIEW software.

Figure 18 shows the magnetic field noise spectra of the center "z" magnetometer, the lower first-order electronic gradiometer, containing the signal sensor, and the TD balanced second-order gradiometer measured in a typical PTB laboratory environment. The sampling rate of the A/D converter was 1 kS/s, the record time was 60 s. As one can see the first-order gradiometer provides a low-frequency noise suppression of about two orders of magnitude, in agreement with the projected value. The TD balanced second-order gradiometer suppresses the 50 Hz power line signal for this particular measurement position by about a factor of 20000. It should be emphasized that the adaptive TD balancing procedure is based on a minimization of the total noise energy and not directed on the maximum suppression of a certain signal. Consequently, the corrected time signal $B_{grad}(t)$ is still dominated by certain environmental frequency components (16 2/3 Hz, 50 Hz, or 150 Hz), as can be seen from the corresponding spectrum in Figure 18. As will be discussed later, these frequency components can be further reduced without using notch filters performing adaptive balancing in the frequency domain (FD).

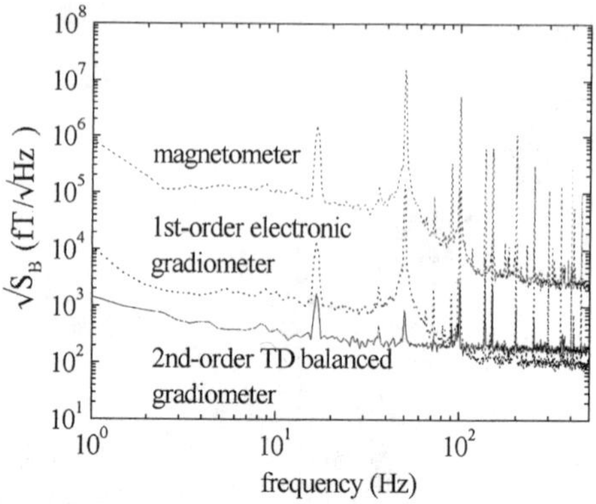

Figure 18. Magnetic field noise spectra for "z" magnetometer, electronic first-order gradiometer, and TD balanced second-order gradiometer measured in magnetically unshielded environment. All SQUID magnetometers were operated with PTB directly coupled FLL electronics with 100 kHz bias reversal.

To record MCGs at a given measurement site the following procedure was pursued: First, with no patient, we recorded data for typically 60 s at a sampling rate of 1 kS/s. Taking these data, the coefficients a_i were calculated. Now, knowing the weights for the individual channels for the given noise environment and after reducing the measurement bandwidth to 100 or 200 Hz, MCGs can be recorded and the corrected time traces can be calculated. Figure 19 depicts a real-time MCG recorded in a 100 Hz bandwidth in a quiet PTB laboratory environment. One can clearly see the QRS complex and the T-wave. The peak-to-peak noise amounts to about 10 pT. Another example is depicted in Figure 20 showing a single heart beat recorded on a volunteer at a meeting in Garmisch-Patenkirchen in December 2000. The measurement site was in the city center about 100 m away from the railway station. The noise amplitude of the real-time trace (lower curve) of

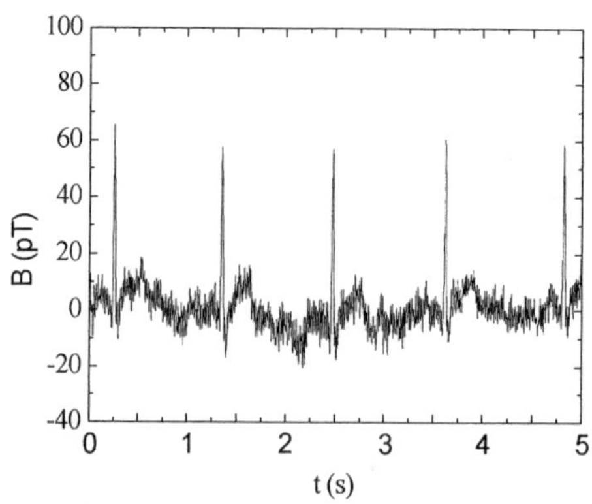

Figure 19. Real-time magnetocardiogram recorded in magnetically unshielded environment. The measurement bandwidth was 100 Hz, no notch filters were used.

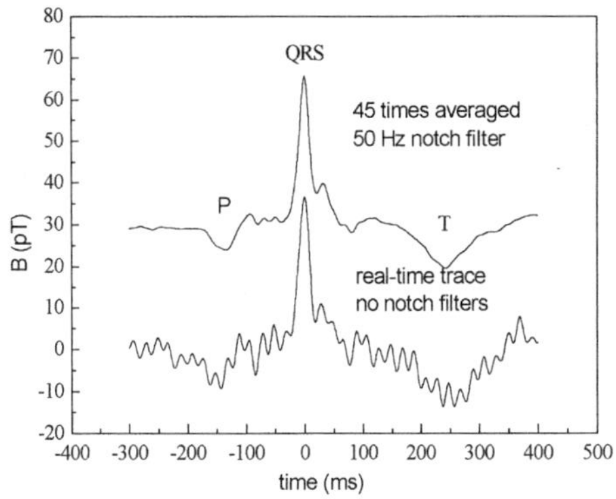

Figure 20. Time trace of a single heart beat recorded from volunteer at a meeting Garmisch-Patenkirchen (Germany). The measurement bandwidth was (0.3-100) Hz. For clarity the upper curve was shifted up by 30 pT.

less than 10 pT for a measurement bandwidth of 100 Hz is dominated by the remaining 50 Hz disturbances which could not be completely suppressed by the TD balancing method. The upper curve shows the 45 times averaged and 50 Hz notch filtered time trace of a single heart beat with an excellent signal-to-noise ratio. Finite impulse response (FIR) software filters were used.

To further improve the noise suppression without using additional notch filters to cancel the remaining levels of 16 2/3 Hz, 50 Hz and 150 Hz disturbances, adaptive balancing in the FD was tested. The FD balancing algorithm is again adapted from [67]. Here the background signal is given by

$$B_{back}(f) = \sum_i h_i(f) B_{ref,i}(f) \tag{10}$$

where $B_{ref,i}(f)$ is the Fourier transform of $B_{ref,i}(t)$. In contrast to the TD balancing procedure, each channel is not weighted by a single, time and frequency independent number a_i but by a frequency dependent vector $h_i(f)$. The vectors $h_i(f)$ are determined such that the spectral density of $B_{corr}(f) = B_{src}(f) - B_{back}(f)$ is minimized, i.e.,

$$\frac{\partial S_{corr}(f)}{\partial h_i(f)} = 0 \qquad \text{and} \qquad \frac{\partial S_{corr}(f)}{\partial h_i^*(f)} = 0 \tag{11}$$

where S_{corr} is the power spectral density of $B_{corr}(f)$ and $h_i^*(f)$ is the complex conjugate of $h_i(f)$. As for the TD case, the algorithm was implemented in LABVIEW software. To compare the FD corrected noise spectra with the TD corrected ones, the Hanning window function was applied to the data.

Figure 21 shows the magnetic field noise spectra of the center "z" magnetometer, the TD balanced and the FD balanced second-order gradiometers measured in a PTB laboratory room. As can be seen, the noise of the TD and FD balanced gradiometers is about the same except that the FD balancing procedure much more effectively suppresses the noise peaks at 16 2/3 Hz, 50 Hz etc. The remaining 50 Hz peak in the FD balancing spectrum is presumably caused by imbalances of the second-order gradiometer system which can not

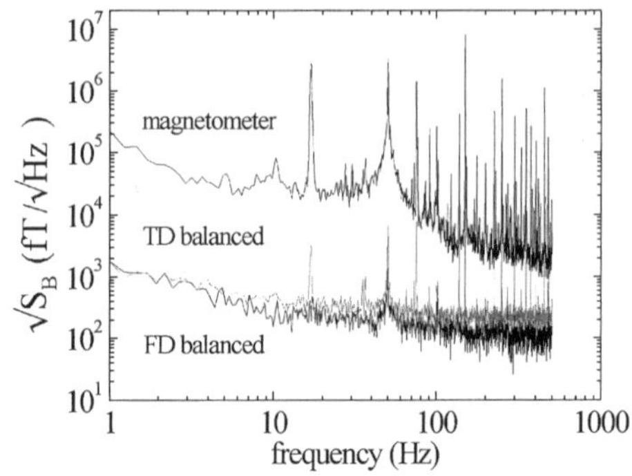

Figure 21. Magnetic field noise spectra of "z" magnetometer, TD balanced and FD balanced second-order gradiometers measured in typical magnetically unshielded PTB laboratory. Record time was 60 s at a sampling rate of 1 kS/s.

be corrected with the given two reference sensors. It is straight forward to show that the "x" and "y" reference sensors in the center vector triple are sufficient to perfectly balance a first-order gradiometer but not the given second-order gradiometer. Assuming that the three "z" magnetometers are perfectly aligned on the gradiometer axis, one would e.g. have to replace the upper "z" sensor by a second vector triple. Whereas the "x" and "y" reference sensors allow one to compensate parasitic sensitivities of the gradiometer system to magnetic fields in the "xy" plane, the two additional reference sensors can be used to also compensate parasitic sensitivities to magnetic field gradients other than $G_{zz} = \partial B_z/\partial z$.

One can also see from Figure 21 that the white noise level of the FD balanced gradiometer is somewhat lower than that of the TD balanced one. For the latter one a value of about 200 $\mathrm{fT}/\sqrt{Hz}$ is found. This is about a factor of four above the intrinsic magnetometer noise. Ideally one would expect a noise increase of $\sqrt{1^2+2^2+1^2} = \sqrt{6}$ for the given second-order gradiometer configuration. There are two possible reasons for the increased system noise: First, the intrinsic magnetometer noise, which was independently determined in the BMSR, was increased by the presence of the various environmental fields. Here it would be sufficient if one of the three "z" sensors exhibited an increased "white" noise (for the sampling rate of 1 kS/s the maximum bandwidth is 500 Hz) due to, e.g., some RTS signal caused by trapped flux. Second, the residual noise may be environmental since the second-order gradiometer in its given configuration can not be perfectly balanced. The "white" noise of the FD balanced gradiometer is about 110 $\mathrm{fT}/\sqrt{Hz}$. A careful analysis of the weighting vector $h(f)$ of the upper first-order gradiometer indicated that it had values of around 1 for frequencies like 50 Hz, as to be expected for a second-order gradiometer, but only of about 0.5 for frequencies above 100 Hz. Assuming that the weight of the upper first-order gradiometer amounts to 0.5 for all frequencies and neglecting the noise contribution of the other reference sensors, one obtains that the noise is increased by a factor of $\sqrt{1^2+(1+0.5)^2+0.5^2} = \sqrt{3.5}$ compared to the intrinsic noise of a single magnetometer. Again, the measured value is slightly higher.

MCG measurements were performed analogously to the TD case: First, data were

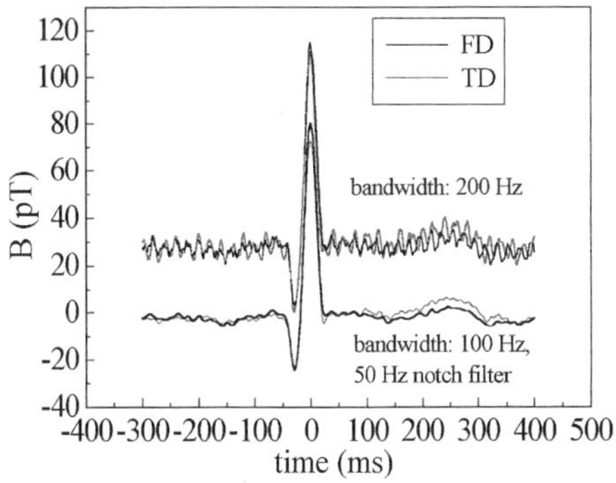

Figure 22. Time trace of TD and FD corrected single heart beat recorded in typical PTB laboratory. Upper two curves are shifted by 30 pT for clarity.

recorded without patient and used to determine the vector weights $h_i(f)$. Knowing the frequency dependent weights, MCGs can be recorded and FD corrected. To get a direct comparison between TD and FD balancing, one set of time traces without patient and one with patient was recorded. The former set was used to determine the weighting coefficients a_i for the TD case and the weighting vectors $h_i(f)$ for the FD case. Knowing these, the real-time MCG data were processed and the corrected MCGs calculated. Figure 22 shows the 10 times averaged time trace of a single heart beat recorded in a PTB laboratory. The upper two curves were obtained for a bandwidth of 200 Hz without any notch filters. The peak-to-peak noise amplitude of both traces amounts to less than 10 pT and is more or less dominated by 50 Hz and 150 Hz. For the TD trace this is to be expected since a considerable level of the 16 2/3 Hz, 50 Hz and harmonics peaks remained in the noise spectrum. In contrast, these discrete peaks are much more effectively suppressed in the FD spectrum. As can be seen from Figure 21, the 50 Hz peak is more than a factor of four smaller than in the TD spectrum and the background noise amplitude is expected to be correspondingly lower. In the FD curve of Figure 22 it is also discernible that the character of the background noise changes with time: in the interval 50 to 150 ms before the R peak there is only a low level of 50 Hz and the background signal is rather dominated by 150 Hz interference, whereas in the interval 200 to 400 ms after the R peak the 50 Hz interference is strongest. For comparison, the background noise character of the TD corrected curve does not vary with time. With the TD routine it was often possible to stably record – on a given measurement site – MCGs over hours after just once performing the adaptive balancing algorithm. In contrast, applying the FD routine to determine the weighting vectors, the background noise often significantly changed with time and exceeded the corresponding TD values although the initial noise suppression was superior to the TD routine. This is most likely due to variations in the background signal which violate the requirement that the background noise should be stationary. The FD routine seems to be much more sensitive to the nonstationarity than the TD procedure. This is not surprising since the weighting coefficients determined with the TD balancing algorithm represent an average value of all signal components. In the FD routine, one determines a single weight for each frequency bin (usually we choose 2048 frequency bins at a sampling rate of 1 kS/s) and the corrected signal is much more susceptible to slight variations of the individual frequency components. One obvious problem of any kind of gradiometry is that – proposed that the reference sensor/sensor system also sees a certain level from the signal source – the signal amplitude and shape are influenced by the gradiometric measurement configuration. Both in the TD case or for a gradiometer with fixed weights, one subtracts from the signal measured with the source signal sensor the signal from the reference sensor multiplied by some factor. In the ideal case, this results simply in a reduced signal amplitude. The problem is more severe in the FD case since the weight of the reference signal is a frequency-dependent vector and consequently the gradiometrically corrected signal is distorted by the character of the noise environment. However, as can be seen from the lower two traces in Figure 22, which were obtained after reducing the bandwidth to 100 Hz and applying a 50 Hz notch filter, there are only slight differences between the FD and the TD corrected traces. This is mainly due to the fact that the weighting coefficients for the upper first-order gradiometer were found to range around 1 for frequencies below 100 Hz in the given magnetic environment.

4.1.3. Summary

A single-channel HTS second-order gradiometer system has been successfully developed and tested at various magnetically unshielded locations. Real-time magnetocardiograms with peak-to-peak noise amplitudes of typically 10 pT were recorded. It could be demonstrated that, as long as the noise environment can be considered as stationary, adaptive balancing represents a very effective method to operate HTS gradiometer systems in unshielded environment. It was also found that – since there will be always slight deviations from a strict stationarity – balancing in the TD is more reliable since it is not as susceptible to nonstationarity as in the FD routine. In addition, the TD algorithm requires by far less computing effort. Combining the described second-order gradiometer system with proper notch filters to get rid of the remaining levels of 16 2/3 Hz, 50 Hz and harmonics, background noise levels can be achieved in real-time that are adequate for clinical diagnostics. It was also shown that the white system noise is still well above the noise which is expected from the intrinsic noise of the individual magnetometers. As pointed out, one reason might be that the intrinsic noise of the magnetometers is enhanced by the presence of the various environmental fields. It must also be kept in mind that the measurements of the intrinsic magnetometer noise, when the devices are exposed to dc or ac magnetic fields, are always performed in idealized and well-defined conditions. To minimize the influence especially of time-varying magnetic fields, the use of encapsulations with integrated field compensation, as proposed by Drung and Schurig [68] might be helpful (see also section 5). The second reason for the increased system noise could be that five magnetometers are not sufficient to set up a perfectly balancable second-order gradiometer system. A further improvement of the background noise suppression could be achieved by adding each one "x" and one "y" channel to the upper "z" sensor. The presented approach of combining electronic and software gradiometry is also adaptable to multichannel systems.

4.2. System for use in a magnetically shielded room

The goal of this system is to demonstrate a simultaneous multichannel measurement for spatial mapping of a magnetic field distribution with a small, easy-to-handle, flexible

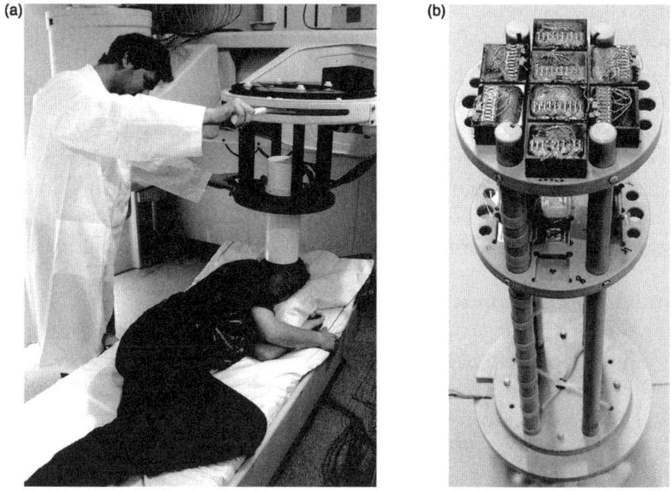

Figure 23. (a) System, mounted in a gantry inside the magnetically shielded room. The plastic shell containing the glass cryostat is about 400 mm high with an outer diameter of 140 mm. (b) Photograph of the magnetometer holder with eight encapsulated magnetometers, mounted on the plane closest to the chest.

low-cost system to take use of the advantages of liquid N_2 cooling in comparison to liquid He cooling. The system dewar is depicted in Figure 23(a), mounted in a gantry inside a magnetically shielded room.

4.2.1. System setup

As sensors the directly coupled dc SQUID magnetometers with perforated pickup loop based on 24° and 30° $SrTiO_3$ bicrystal substrates are used. These HTS dc SQUIDs achieve a magnetic noise level below 50 $fT/\sqrt{Hz}$ at 10 Hz in shielded environment. The difference between 24° and 30° $SrTiO_3$ bicrystal substrates is rather small, with a tendency of Josephson junctions on 30° substrates to have a higher normal state resistance and in proportion a smaller critical current. The noise performance of dc SQUID magnetometers of both kinds of bicrystals is comparable. The sensors are encapsulated for protection against moisture during warm up. The housings also contain a heater and the feedback coil for flux-locked loop operation. The heating procedure for one sensor takes less than 2 minutes with a power of about 1 W. The encapsulated sensors mounted on the sensor holder are depicted in Figure 23(b).

The system can be equipped with up to 24 sensors in three planes to realize different gradiometer and magnetometer configurations. The quite large cold warm distance of about 30 mm results from using a simple glass cryostat which reduces the overall cryogenic system costs to below 400 Euro. The measurement time for the 1.5 l cryostat is about 8 h with eight encapsulated sensors in the lowest plane. An eight channel commercial electronics from Philips with ac-bias and flux modulation at 250 kHz [69] was used. The analog output signals were digitized and evaluated with a commercial data acquisition system from Neuroscan. The MCG mapping (see below) was calculated using the CURRY software [70]. All measurements were performed inside the magnetically shielded room of the physiological institute of the University Hospital Eppendorf, Hamburg, where a LTS multichannel system based on niobium and operated with liquid helium at 4.2 K is used routinely for MEG measurements. This has allowed the comparison of our HTS system to a commercial LTS system from Philips on the same day and the same volunteers [71]. Our cryostat was mounted in a gantry where it can be shifted horizontally from the center position by 60 mm to four further positions. Thereby a sensing area of about 160 x 144 mm^2 was covered. With eight magnetometer channels per measurement position 40 different sensor positions were realized for magnetic field mapping measurements.

4.2.2. Biomagnetic measurements

The unfiltered time traces of two channels in axial first-order gradiometer configuration are shown in Figure 24. Channel 1 is inverted for better overview. For these measurements sensors in the two lowest planes of the sample holder were used in a distance of 7 cm. The measurements were performed inside the magnetically shielded room with a system bandwidth of 500 Hz. In the difference signal, the low-frequency noise can be suppressed by a factor of about 100 with a gradiometer noise of about 10 $fT/cm\sqrt{Hz}$ at 70 Hz. The signal-to-noise ratio of the unfiltered gradiometer signal is better than 8:1. The maximum signal amplitude of the QRS complex is about 57 pT measured at channel 1. The heart signal is also present at reference sensor channel 2 with an amplitude of about 19 pT. So despite the large gradiometer distance of 7 cm, about 30 % of the measuring signal is lost due to the subtraction of both sensor signals.

The following MCG measurements were made using eight magnetometer channels at the lowest plane without gradiometry. An ECG trigger was recorded to correlate the relative time scales of the five system positions. After measuring for 120 s with a bandwidth from 0.3 Hz to 70 Hz in every position the time traces were averaged and for every time step a magnetic field map was calculated. Figure 25(a) shows the averaged time traces of all magnetometers at the five measuring positions. Maximum signal amplitudes of about 25 pT were found in these measurements. The corresponding field map at the time of the S-wave is depicted in Figure 25(b). Here, the dipole character of the heart signal can be recognized well.

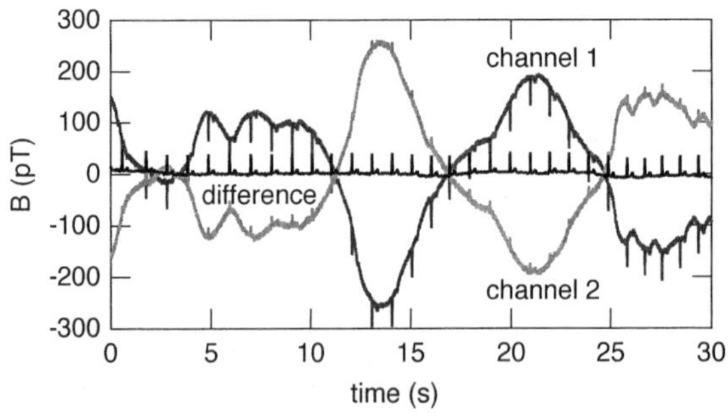

Figure 24. Measured time traces of two magnetometers and their electronic difference signal. Channel 1 is inverted for better overview.

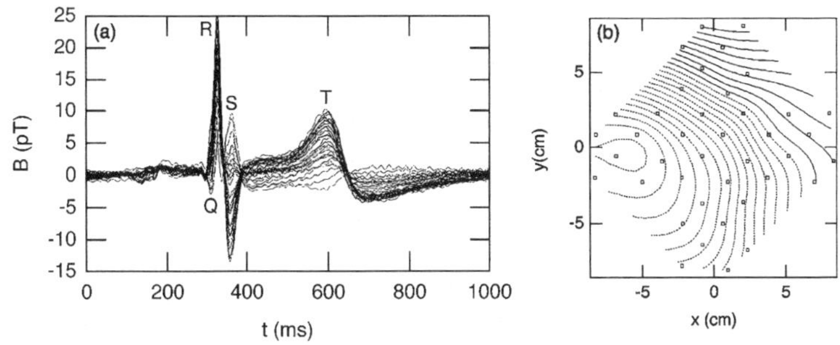

Figure 25. (a) Time traces of the 40 filtered channels after averaging for 120s. (b) Corresponding field map at the time of the S wave. The dotted lines mark positive field direction, the solid lines negative direction.

For the MEG measurements, evoked magnetic fields of an electric stimulus of the median nerve at the left hand of 4 mA at 2 Hz was recorded. Figure 26(a) shows the 2000 times averaged signals of two channels corresponding to different measurement positions (dotted and solid lines) with a software low-pass filter of 200 Hz bandwidth. The two signals show the positive and negative field components of the stimulated activity of the right hemisphere of the brain. The stimulation took place at 0 ms and the first response from the brain can be found about 20 ms later. The maximum amplitude of 150 fT is at about 100 ms. Reference measurements with the LTS multichannel system with average white system noise below 10 $fT/\sqrt{Hz}$ of the same volunteer at the same day are depicted in Figure 26(b) and show almost the same temporal evolution of the signal, but with a better signal to noise ratio. The differences between the systems signals – especially in the signal amplitudes – can be explained by different measurement positions, relative to the signal source.

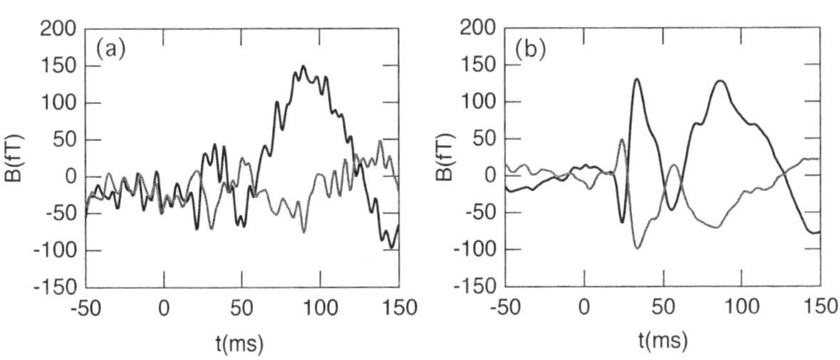

Figure 26. (a) MEG signals of the electrically stimulated median nerve at two different positions, recorded with the HTS system. (b) Reference measurement with the LTS system.

4.2.3. Summary

A handy eight-channel HTS system with encapsulated sensors which gives a great flexibility for sensor arrangement was built. The use of a low-cost glass cryostat and a quite simple magnetometer holder is suitable for systems working in shielded environment. MCG field maps could be recorded as well as evoked MEG signals.

5. CONCLUSION AND OUTLOOK

In this paper we have shown that HTS SQUID magnetometers indeed have the potential to be used for biomagnetic applications. As demonstrated with an eight-channel system exploiting directly coupled SQUID magnetometers with a white noise level around 50 fT/$\sqrt{Hz}$, the signal-to-noise ratio is sufficient to map magnetocardiac activity and even to detect evoked signals from the human brain. As pointed out in section 1, a widespread application of SQUID systems requires systems without expensive magnetic shielding. A single-channel system based on second-order gradiometry was presented which allows the recording of real-time MCGs of sufficient quality at various magnetically unshielded locations. A prerequisite is the availability of SQUID magnetometers with a low level of low-frequency noise even when cooled in the static magnetic field of the earth and exposed to various time-varying environmental magnetic fields. Two different approaches to optimize the directly coupled SQUID magnetometer layout for unshielded operation were presented. Both had in common that the pickup loop had to cross the grain boundary. Various experimental findings indicate that problems of flux penetration and vortex motion in and through the superconducting structures are often caused by the weakened

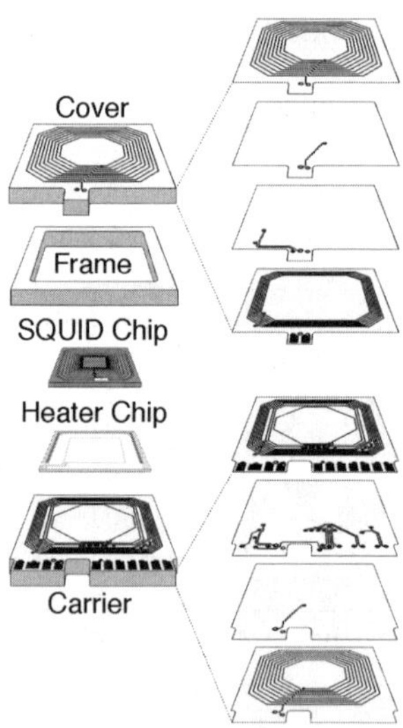

Figure 27. Components of a SQUID capsule with integrated field compensation coils and feedback coil [68]. Carrier, cover and frame are made of standard 1.6 mm thick printed-circuit board material.

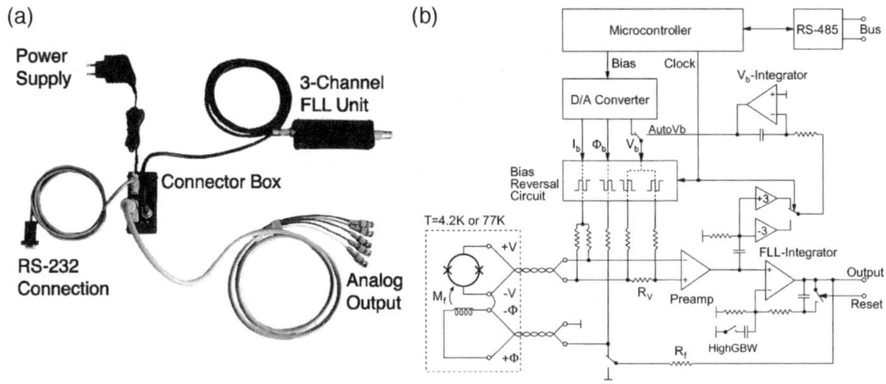

Figure 28. Components (a) and schematic (b) of the new fully computer controlled PTB SQUID electronics with automatic voltage bias (AVB) and integrated low noise current source [65,68]. The electronics is commercially available via MAGNICON [72].

superconductivity in the grain boundary area. Furthermore, it was observed that environmental magnetic fields, which are generally much larger than the biomagnetic fields of interest, can not only cause low-frequency excess noise but can also influence the magnetic field sensitivity B/Φ of the magnetometer.

There are various approaches to solve the observed problems, especially in a harsh magnetic environment. One promising approach is to use Helmholtz or Helmholtz-like feedback or compensation coils. Helmholtz coils have been successfully used as compensation coils in the 3-SQUID gradiometer system presented by Clem et al. [42] for mobile magnetic anomaly detection. Such large coils are, however, not acceptable in a multichannel biomagnetic system. Recently, Drung and Schurig [68] introduced a SQUID encapsulation with carefully designed integrated field compensation coils (Figure 27). In spite of the small capsule size of 18 x 16 x 5 mm^3, the compensation field was estimated to be homogeneous to about ± 2 % over the area of a typical SQUID magnetometer pickup coil of 9 x 9 mm^2. A low noise current source was integrated into the SQUID electronics to allow the generation of compensation fields of up to ± 71 μT with only about 20 fT/$\sqrt{Hz}$ excess noise above a few Hz. The compensation field can be set manually or automatically using a feature called dynamic field compensation (DFC). For this, the FLL output is supervised permanently by reading out the analog-to-digital converter. Each time the FLL output exceeds a specified threshold, the compensation field is changed by a certain amount such that the FLL output goes to about zero. It is important to note, that the whole time the FLL keeps locked, i.e. the change in the compensation field is slow enough that the slew rate of the FLL is not exceeded. The usefulness of DFC was demonstrated using an encapsulated HTS SQUID magnetometer as introduced in section 3.1 [73]. This SQUID was cooled down in liquid nitrogen inside the Berlin magnetically shielded room (BMSR), carried out of the shielded room by hand, set to the floor, and carried back into the BMSR without flux entering the device. The SQUID noise did not change even though the magnetometer was exposed to 60 μT dc and 1.5 μT peak-to-peak ac magnetic fields while being outside the BMSR. In a practical biomagnetic system, the

SQUID devices might be shielded by a simple MUMETAL can while being cooled. After the activation of DFC the shielding can is removed and the system keeps locked all the time even if the dewar is moved in the earth magnetic field. Such a system together with high quality gradiometry should be sufficient to allow stable operation and low noise MCG recordings with a mobile cryostat even in harsh magnetic environment.

The authors would like to thank Sylke Bechstein, Bettina Schiefenhövel, Jörn Beyer and Erik Heim for their contribution.

REFERENCES

[1] J. Clarke, in: *SQUID Sensors: Fundamentals, Fabrication and Applications*, NATO ASI Series, edited by H. Weinstock (Kluwer Academic, Dordrecht 1996), p.1

[2] D. Koelle, R. Kleiner, F.Ludwig, E. Dantsker, and J. Clarke, Rev. Mod. Phys. **71**, 631 (1999)

[3] H. W. Ott, *Noise Reduction Techniques in Electronic Systems*, 2nd edition (Wiley, New York 1988), p.233

[4] E. Dantsker, F. Ludwig, R. Kleiner, J. Clarke, M. Teepe, L. P. Lee, N. McN. Alford, and T. Button, Appl. Phys. Lett. **67**, 725 (1995)

[5] D. Drung, F. Ludwig, W. Müller, U. Steinhoff, L. Trahms, Y. Q. Shen, M. B. Jensen, P. Vase, T. Holst, T. Freltoft, and G. Curio, Appl. Phys. Lett. **68**, 1421 (1996)

[6] M. I. Faley, U. Poppe, K. Urban, D. N. Paulson, T. Starr, and R. L. Fagaly, IEEE Trans. Appl. Supercond. **11**, 1383 (2001)

[7] H. Itozaki, S. Tanaka, T. Nagaishi, and H. Kado, IEICE Trans. Electron. **E77-C**, 1185 (1994)

[8] H. Itozaki, S. Tanaka, H. Toyoda, T. Hirano, Y. Haruta, M. Nomura, T. Saijou, and H. Kado, Supercond. Sci. Technol. **9**, A38 (1996)

[9] P. R. E. Petersen, PhD Thesis, Technical University of Denmark, 2001

[10] E. Dantsker, O. Froehlich, S. Tanaka, K. Kousnetsov, J. Clarke, Z. Lu, V. Matijasevic, and K. Char, Appl. Phys. Lett. **71**, 1712 (1997)

[11] K. A. Kousnetsov, J. Borgmann, and J. Clarke, Appl. Phys. Lett. **75**, 1979 (1999)

[12] Y. Tavrin, Y. Zhang, M. Mück, A. I. Braginski, and C. Heiden, IEEE Trans. Appl. Supercond. **3**, 2477 (1993)

[13] Y. Tavrin, Y. Zhang, W. Wolf, and A. I. Braginski, Supercon. Sci. Technol. **7**, 265 (1994)

[14] J. Borgmann, P. David, G. Ockenfuss, R. Otto, J. Schubert, W. Zander, and A. I. Braginski, Rev. Sci. Instrum. **68**, 2730 (1997)

[15] D. F. He, H.-J. Krause, Y. Zhang, M. Bick, H. Soltner, N. Wolters, W. Wolf, and H. Bousack, IEEE Trans Appl. Supercond. **9**, 3684 (1999)

[16] Y. Zhang, G. Panaitov, S. G. Wang, N. Wolters, R. Otto, J. Schubert, W. Zander, H.-J. Krause, H. Soltner, H. Bousack, and A. I. Braginski, Appl. Phys. Lett. **76**, 906 (2000)

[17] J. Vrba, in: *SQUID Sensors: Fundamentals, Fabrication and Applications*, NATO ASI Series, edited by H. Weinstock (Kluwer Academic, Dordrecht 1996), p.117

[18] P. J. M. Woeltgens, R. H. Koch, R. Matthews, S. L. Brown, R. A. Altman, W. J. Gallagher, S. G. Haupt, and D. K. Lathrop, Applied Superconductivity Conference (ASC96), Pittsburgh (unpublished)

[19] B. David, O. Dössel, V. Doormann, R. Eckart, W. Hoppe, J. Krger, H. Laudan, and G. Rabe, IEEE Trans. Appl. Supercond. **7**, 287 (1997)

[20] H. J. M. ter Brake, N. Janssen, J.Flokstra, D. Veldhuis, and H. Rogalla, IEEE Trans. Appl.

Supercond. **7**, 2545 (1997); H. J. M. ter Brake, W. A. M. Aarnink, P. J. van den Bosch, J. W. M. Hilgenkamp, J. Flokstra, and H. Rogalla, Supercond. Sci. Technol. **10**, 512 (1997)
[21] L. P. Lee, J. Longo, V. Vinetskiy, and R. Cantor, Appl. Phys. Lett. **66**, 1539 (1995)
[22] R. Cantor, L. P. Lee, M. Teepe, V. Vinetskiy, and J. Longo, IEEE Trans. Appl.Supercond. **5**, 2927 (1995)
[23] M. B. Ketchen, W. J. Gallagher, A. W. Kleinsasser, S. Murphy, and J. R. Clem, in: *SQUID85, Superconducting Quantum Interference Devices and Applications*, edited by H. D. Hahlbohm and H. Lbbig (Walter de Gruyter, Berlin 1985), p. 865
[24] J. M. Jayxox and M. B. Ketchen, IEEE Trans. Magn. **17**, 400 (1981)
[25] K. Enpuku, T. Maruo, and T. Minotani, J. Appl. Phys. **80**, 1207 (1996)
[26] J. Beyer, D. Drung, F. Ludwig, T. Minotani, and K. Enpuku, Appl. Phys. Lett. **72**, 203 (1998)
[27] E. Dantsker, S. Tanaka, P.-A. Nilsson, R. Kleiner, and J. Clarke, Appl. Phys. Lett. **69**, 4099 (1996)
[28] J. R. Clem (unpublished)
[29] E. Dantsker, S. Tanaka, and J. Clarke, Appl. Phys. Lett. **70**, 2037 (1997)
[30] H.-M. Cho, R. McDermott, B. Oh, K. A. Kousnetsov, A. Kittel, J. H. Miller Jr., and J. Clarke, IEEE Trans. Appl. Supercond. **9**, 3294 (1999)
[31] F. P. Milliken, S. L. Brown, and R. H. Koch, Appl Phys. Lett. **71** 1857 (1997)
[32] S. Hirano, H. Oyama, S. Kuriki, T. Morooka, and S. Nakayama, Appl. Phys. Lett. **78**, 1715 (2001)
[33] F. Ludwig, J. Beyer, D. Drung, S. Bechstein, and Th. Schurig, IEEE Trans. Appl. Supercond. **9**, 3793 (1999)
[34] M. S. DiIorio, K.-Y. Yang, S. Yoshizumi, S. G. Haupt, D. Haran, R. H. Koch, F. P. Milliken, J. R. Rozen, D. K. Lathrop, S. Kumar, and H. S. Trammell III, IEEE Trans. Appl. Supercond. **9**, 4428 (1999)
[35] R. H. Koch, J. Z. Sun, V. Foglietti, and W. J. Gallagher, Appl. Phys. Lett. **67**, 709 (1995)
[36] L. Saury, C. Gunther, and C. Dubuc, IEEE Trans. Appl. Supercond. **9**, 50 (1999)
[37] A. B. M. Jansman, M. Izquierdo, A. Eiguren, J. Flokstra, and H. Rogalla, Appl. Phys. Lett. **72**, 3515 (1998)
[38] F. Ludwig, A. B. M. Jansman, D. Drung, M. O. Lindström, S. Bechstein, J. Beyer, J. Flokstra, and Th. Schurig, IEEE Trans. Appl. Supercond. **11**, 1315 (2001)
[39] F. Ludwig, J. Beyer, D. Drung, S. Bechstein, and Th. Schurig, Appl. Supercond. **5**, 345 (1998)
[40] D. Drung, Rev. Sci. Instrum. **68**, 4066 (1997)
[41] F. Ludwig and D. Drung, Appl. Phys. Lett. **75**, 2821 (1999)
[42] T. R. Clem, D. J. Overway, J. W. Purpura, J. T. Bono, R. H. Koch, J. R. Rozen, G. A. Keefe, S. Wilson, and R. A. Mohling, IEEE Trans. Appl. Supercond. **11**, 871 (2001)
[43] S. Krey, H.-J. Barthelmess, and M. Schilling, J. Appl. Phys. **86**, 6602 (1999)
[44] J. K. Heinsohn, D. Reimer, A. Richter, K.-O. Subke, and M. Schilling, Physica C **299**, 98 (1998)
[45] H.-J. Barthelmess, S. Krey, S. Ostertun, and M. Schilling, Appl. Phys. Lett. **77**, 1882 (2000)
[46] K. O. Subke, C. Hinnrichs, H.-J. Barthelmess, M. Halverscheid, C. Pels, and M. Schilling, EUCAS1999, Inst. Phys. Conf. Ser. **167**, 573 (1999)
[47] T. Dahm and D. J. Scalapino, Phys. Rev. B **60**, 13125 (1999)
[48] M. J. Ferrari, M. Johnson, F. C. Wellstood, J. J. Kingston, T. J. Shaw, and J. Clarke, J. Low Temp. Phys. **94**, 15 (1994)

[49] Y. M. Zhang, D. Winkler, P.-Å. Nielsson, and T. Claeson, Phys. Rev. B, **51**, 8684 (1995)
[50] K. Enpuku, M. Hoashi, H. Doi, and T. Kisu, Jpn. J. Appl. Phys. **32**, 3804 (1993)
[51] J. Z. Sun, W. J. Gallagher, and R. H. Koch, Phys. Rev. B **50**, 13664 (1994)
[52] H. Oyama, S. Kuriki, and M. Matsuda, IEEE Trans. Appl. Supercond. **11**, 1331 (2001)
[53] H.-M. Cho, Y. T. Andresen, J. Clarke, M. S. DiIorio, K.-Y. Yang, and S. Yoshizumi, Appl. Phys. Lett. **79**, 2438 (2001)
[54] K. Enpuku, A. Nakahodo, M. Hotta, S. Hijiya, D. Tokimizu, and D. Kuroda, Jpn. J. Appl. Phys. **40**, 4013 (2001)
[55] K. Enpuku, D. Kuroda, D. Tokimizu, and T. Q. Yang, J. Appl. Phys. **92**, 4751 (2002)
[56] K. Likharev, *Dynamics of Josephson Junctions and Circuits* (Gordon and Breach, New York 1986)
[57] K. Enpuku (private communication)
[58] K. Enpuku, D. Tokimizu, D. Kuroda, and S. Hijiya, Jpn. J. Appl. Phys. **40**, L869 (2001)
[59] A. Haller, H. Matz, S. Hartwig, Th. Kerberger, H. Atzpodien, and L. Trahms, IEEE Trans. Appl. Supercond. **11**,1371 (2001)
[60] Y. R. Chemla, H. L. Grossman, T. S. Lee, J. Clarke, M. Adamkiewicz, and B. B. Buchanan, Proc. Natl. Acad. Sci. U.S.A. **97**, 14268 (2000)
[61] K. Enpuku, T. Minotani, T. Gima, Y. Kuroki, Y. Itoh,M. Yamashita, Y. Katakura, and S. Kuhara, Jpn. J. Appl. Phys. **38**, L1102 (1999); K. Enpuku and T. Minotani, IEICE Trans. Electron. **E84-C**, 43 (2001)
[62] K. Yokosawa, S. Kuriki, S. Hrano, H. Oyama, D. Suzuki, and K. Tsukada, J. Appl. Phys. **90**, 4049 (2001)
[63] K. H. Koch, J. R. Rozen, P. Wöltgens, T. Picunko, W. J. Goss, D. Gambrel, D. Lathrop, R. Wiegert, and D. Overway, Rev. Sci. Instrum. **67**, 2968 (1996)
[64] D. Drung, R. Cantor, M. Peters, H. J. Scheer, and H. Koch, Appl. Phys. Lett. **57**, 406 (1990)
[65] D. Drung, S. Bechstein, K.-P. Franke, M. Scheiner, and Th. Schurig, IEEE Trans. Appl. Supercond. **11**, 880 (2001)
[66] S. Bechstein, D. Drung, M. Scheiner, F. Ludwig, and Th. Schurig, Inst. Phys. Conf. Ser. **167**, 597 (2000)
[67] P. Wöltgens and R. H. Koch, Rev. Sci. Instrum. **71**, 1529 (2000)
[68] D. Drung and Th. Schurig, IEEE Trans. Appl. Supercond. (Proc. ASC02, in print)
[69] B. David, O. Dössel, V. Doormann, R. Eckart, W. Hoppe, J. Krüger H. Laudan, and G. Rabe, IEEE Trans. Appl. Supercond. **69**, 3267 (1997)
[70] The CURRY software is available from NEUROSOFT Inc.
[71] H.-J. Barthelmess, M. Halverscheid, B. Schiefenhövel, E. Heim, M. Schilling, and R. Zimmermann, Proc. 12th Int. Conf. on Biomagnetism, J. Nenonen, R.J. Ilmoniemi, and T. Katila, eds., 927 (2000)
[72] Magnicon GbR, Wachtelweg 7, 22869 Schenefeld, Germany, web: www.magnicon.com
[73] D. Drung et al., unpublished

HTS APPLICATIONS: PROGRESS IN SQUID MICROSCOPY AND HIGH RESOLUTION NON-DESTRUCTIVE EVALUATION

S.A. GUDOSHNIKOV and L.V. MATVEETS

Institute of Terrestrial Magnetism, Ionosphere and Radio Wave Propagation RAS (IZMIRAN), 142190 Troitsk, Moscow region, Russia

1. INTRODUCTION

For more, than the 30-year's period of development there were many applications of SQUIDs due to its high sensitivity and broad frequency band. One of the most recently developed research SQUID-based instrument is a scanning SQUID microscope (SSM). The first SSMs were based on low transition temperature (LTS) SQUIDs, operated at liquid helium temperature (~ 4.2 K) [1-5], the appearance and successful development of high transition temperature (HTS) SQUIDs operated at liquid nitrogen temperature (~77 K) [6] considerably has expanded area of SQUID microscope researches and has opened a road to commercial applications. At present time SQUID microscopes are used for imaging superconducting and magnetic materials, currents flowing in circuits, imaging of high frequency electric and magnetic fields, biomagnetic research. Their capabilities are shown in the field of a nondestructive evaluation (NDE) by search of different defects in materials, metallic structures and hardware products of microelectronics.

In comparison with tiny Hall-sensors, magneto-optical methods and magnetic force microscopy, the SQUID-microscopy offer quantitative measurements of magnetic fields with substantially higher sensitivity, minimal back action on object under study, capability of detecting subsurface and buried magnetic sources.

There are a few papers, including overview the scanning SQUID microscopes based both HTS and LTS SQUID and its application [6-9]. It is difficult to refrain from repetition, but we make an attempt to give a brief description of successful designed HTS SSM, functioning currently in a few research groups, and progress in their applications. It is reasonable that we place sufficiently high emphasis on results attained by our own group in the field of scanning SQUID microscopy. It should be noted that in this paper we limit consideration of applications of HTS SSM for non-destructive evaluation (NDE) only by cases requiring a high spatial resolution magnetic imaging (≤ 500 μm). The SQUID-NDE area now is rather vast as it was shown in several reviews [10-12].

Section 2 describes briefly the operating principles and capabilities of SQUID microscopes and gives descriptions of various types of the practical HTS SSM instruments. In Section 3 a brief review of several application areas of the HTS SSM is considered. Section 4 is concerned with various current research engineering and development activity.

2. SCANNING SQUID MICROSCOPES BASED ON HTS DC SQUIDS

In a scanning SQUID microscope, a sample is scanned in close proximity to a SQUID in two-dimensional raster (X-Y plane). The SQUID is oriented parallel to the plane of the sample and is therefore sensitive to the perpendicular component of magnetic field, B_Z. A computer records the SQUID response as a function of sample position to produce two- or three dimensional magnetic field image.

Basic characteristics of the SQUID microscope are: a magnetic field sensitivity and spatial resolution. They are interdependent and determined by a size of the SQUID and a distance between a SQUID and a sample. The optimum ratio of parameters is reached in that case, when the SQUID-to-sample distance, h, is approximately equal to the linear size of the SQUID loop, a, (Fig. 1a). Most of the SSM instruments are capable to detect magnetic fields less than 100 pT and to produce magnetic images in a millimeter and sub-millimeter scale.

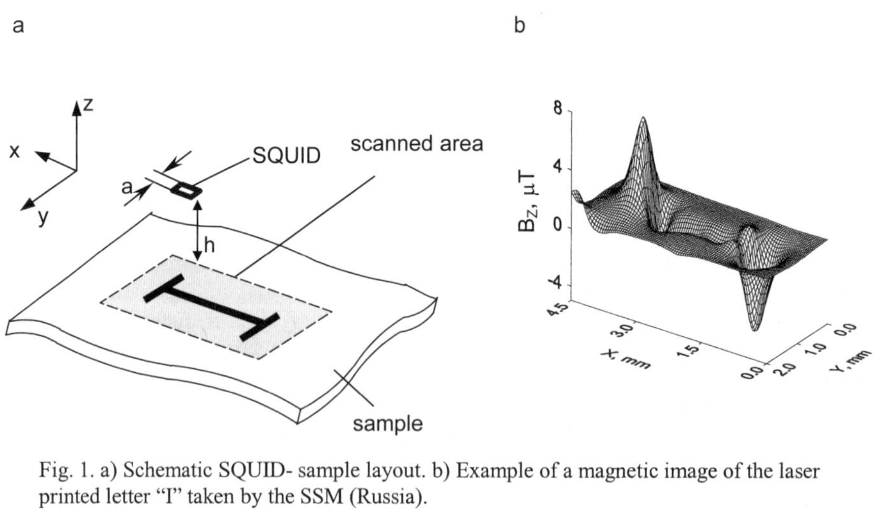

Fig. 1. a) Schematic SQUID- sample layout. b) Example of a magnetic image of the laser printed letter "I" taken by the SSM (Russia).

The combination of a high sensitivity with rather high spatial resolution allows to research small-sized and low magnetic objects by SSM. As an example, 3D image of magnetic field spatial distribution taken at 50 μm over a laser printed letter "I" is shown in Fig. 1b.

In generally, a SQUID microscope consists of a few stages: a cryostat or cryogenic dewar, a small SQUID, a mechanism for micro-positioning the sample, a set of SQUID electronics, and a computer for registration SQUID output signal and controlling the positioning of the sample.

High-temperature scanning SQUID microscopes as well as microscopes with LTS SQUID, can be divided into two types: SSM with a cold sample, in which the sample as well as the SQUID immerses in liquid nitrogen (briefly SSM-77), SSM with a warm sample located at a room temperature (briefly SSM-300). Each of the types has the design features and capabilities. Several successful designs of HTS SSM instruments have been developed in recent years. Their brief overview is given below.

A. The liquid nitrogen sample temperature SSM (SSM-77)

The main feature of this SSM is that both a SQUID and a sample are placed together in a cryogenic cryostat so a SQUID loop and surface of a sample can be brought in a direct physical contact, i.e. a minimum SQUID-to-sample distance about a few micrometers can be reached. Usually, the SQUID is placed near the corner of the substrate and the substrate is tilted so that only the corner is brought close to the sample. The second feature - the main part of positioning mechanism is placed in liquid nitrogen too. As a rule, the stage for a sample is actuated by a system of thrusting rods, sliders and wedges put in action by stepper motors, posed outside of the cryostat. The obvious but not crucial disadvantages of cold sample microscopes are that the sample size is constrained and that the time required for thermal cycling is necessary long.

The first HTS microscope for cold sample was developed at University of Maryland in 1993 [13]. A dc SQUID made of $YBa_2Cu_3O_{7-x}$ (YBCO) film laser-deposited on a $SrTiO_3$ 23° bicrystal substrate was used as a magnetic field sensor. The equivalent flux noise ranged from 0.1 $m\Phi_0/Hz^{1/2}$ at 1 kHz to 0.4 $m\Phi_0/Hz^{1/2}$ at 1 Hz, that corresponds to a field sensitivity of 20-80 $pT/Hz^{1/2}$. The magnetic images of a current distribution, a permanently magnetized sample and a superconducting sample were taken with spatial resolution of 60 –200 μm.

The similar SSM-77 based on a bicrystal YBCO dc SQUID was built at Moscow State University in 1994 [14]. It is functioning up to present time. By the example of the Russian system let's briefly consider design of a liquid nitrogen sample temperature SSM. The functional scheme of SSM-77 is shown in Fig. 2. A cryogenic stick containing a SQUID, a sample and a scanning stage is immersed into a liquid nitrogen cryostat placed inside of a two-layer μ-metal cylindrical shield. Stepper motors, SQUID electronics and stepper motor drive electronics are placed at the top of the cryogenic stick. The SQUID electronics and the stepper motors electronics are connected with the base block. It contains power supplies for the electronics blocks, filters, additional amplifiers, calibration sources, and a sixteen-bit analog-to-digital converter The system is operated automatically under computer driven mode.

To make raster scanning of a sample two stepper motors are used, which turn the micrometer screws of a scanning mechanism. The connecting rods translate a forward movement to X-, Y-rods of the cryogenic scanning stage moving a sample table in a plane X-

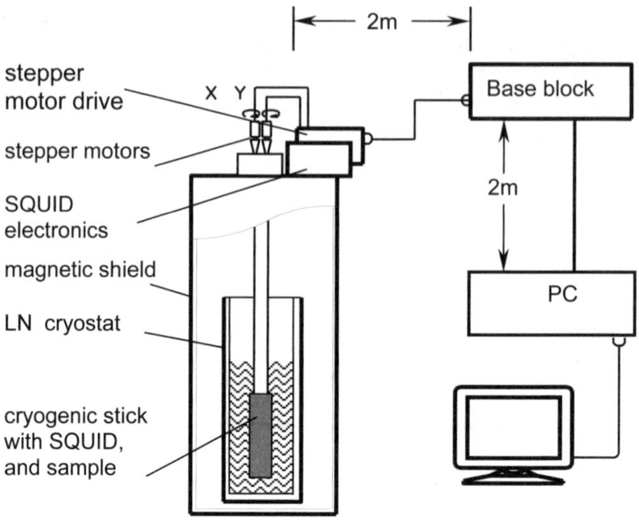

Fig. 2. Schematic of the scanning HTS SQUID microscope system for liquid nitrogen cooled sample (SSM-77).

Y with system of sliders and wedges (Fig. 3). To adjust the SQUID-sample separation, the SQUID is mounted on a spring-loaded holder that can be moved in Z-direction relatively to a sample by a Z-wedge. To minimize magnetic noise the scanning stage is made of nonmagnetic composite materials.

The square-washer shaped SQUID used in the SSM-77 is made of thin YBCO film laser-deposited on a $SrTiO_3$ 36° bicrystal substrate. The Josephson junctions of width 3 μm are formed at grain boundary of a film where the two crystal orientation of a bicrystal meet. The

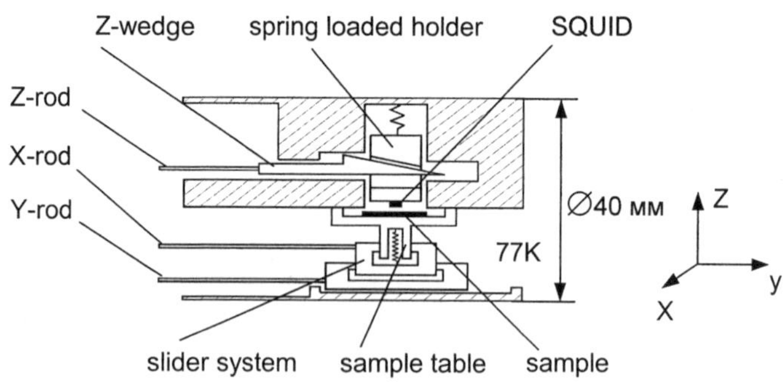

Fig. 3. Schematic drawing of the measuring head of the SSM-77 [585]. The X-wedge and Y-wedge are not visible.

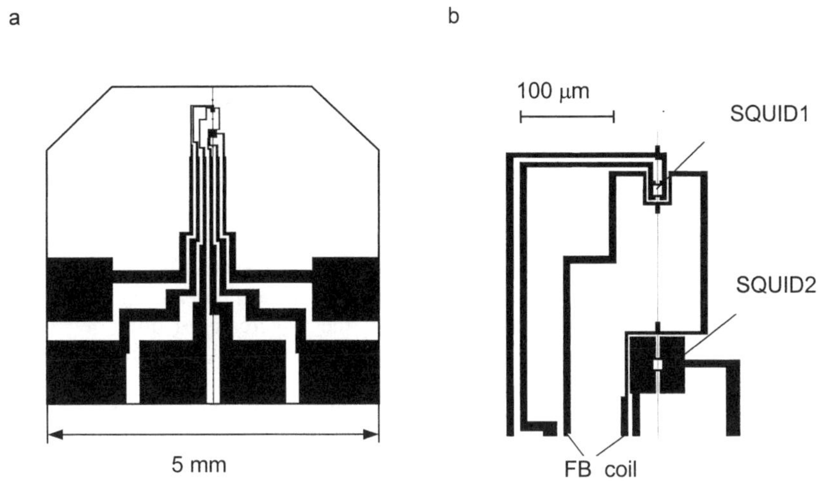

Fig. 4. a) General view of the chip with two HTS SQUIDs. b) Enlarged view of small and large SQUIDs. Dashed line indicates bicrystal grain boundary.

general view of the chip, on which two SQUIDs 1 and 2 and feedback loop, FD, are formed, is shown in Fig. 4a. The grain boundary passes upright on center of the chip, the SQUIDs are posed near the pointed edge of the substrate, and the bonding contact pads are moved to the opposite edge of the chip. If the substrate is under a small angle to a sample such design allows maximum SQUID approaching to a surface of a sample. The enlarged view of the SQUIDs is shown in Fig. 4b. The SQUID 1 with an outer size of 20 μm and internal hole side of 10 μm is posed closer to an edge of the substrate and used at measurements requiring of the best spatial resolution. The SQUID 2 with the corresponding sizes 60 and 10 μm has more large effective area and is used, when increased sensitivity is required.

The SQUID is operated in a simple direct coupled flux locked loop mode. A feature of this electronics is a liquid-nitrogen-cooled amplifier used as the first stage. The bandwidth of the SQUID electronics is about 200 kHz. The equivalent flux noise changes as 1/f in frequencies range of 0,1 - 200 Hz and ranges from $5\times10^{-4}\Phi_0/\mathrm{Hz}^{1/2}$ at $f = 1$ Hz to 10^{-5} $\Phi_0/\mathrm{Hz}^{1/2}$ at f = 1 kHz. A flux-locked range of the system is $\pm$ 50 Φ_0. In Sec. 4 this electronics will be considered in detail.

This SSM-77 was used to examine the magnetic properties of superconducting films and srtuctures [15-17], ultrathin magnetic films [18, 19], giant magnetoimpedance elements [20], Langmuir-Blodgett films [21], amorphous ferromagnetic microwires [22] and others. These applications will be described in detail in Sec. 3.

B. The room sample temperature SSM (SSM-300)

In scanning SQUID microscopes intended for imaging room temperature sample the SQUID must be thermally isolated from the sample. The main technological method used for designing both LTS, and HTS SQUID microscopes with room-temperature sample, is a

SQUID location in vacuum in close proximity to a thin window, separating evacuated space from a room-temperature sample. The cooling of a SQUID is implemented due to a high thermal conductive link with the cryogen can, situated in the same vacuum enclosure. LTS SQUID microscopes with this design have achieved separation between SQUID and room-temperature sample of the order of units of millimeters [1, 23], whereas the spatial resolution of HTS SQUID based SSM-300 is less than 0.1 mm [24, 25]. Nevertheless recently John Wikswo's group at Vanderbilt has developed LTS SQUID based microscope for imaging biological room-temperature samples with sub-millimeter resolution [26] but undoubtedly, its design more complicate than that of HTS SSM-300.

A few years ago Fred Wellstood's group at the University of Maryland, and John Clarke's group at U.C. Berkley have built the room temperature sample HTS SQUID microscopes using this technological method, but their microscopes differ by the design solution. Both these systems are successfully functioning up to now, and can be considered as classical.

In Fig. 5 a schematic drawing of the lower section of SSM-300 developed at Maryland [7] is shown. A SQUID, a sample and a scanning stage of this system are placed in a bottom of the cold-finger type liquid nitrogen cryostat. A sapphire rod is plugged into the end of a cryostat cold finger and projects into the dewar's vacuum space. The SQUID is mounted on the pointed end of the rod and faces a 25 μm thick and 1 mm diameter sapphire window that separates air from vacuum. An adjusting mechanism allows the window to be moved in X, Y, and Z and enables the cold SQUID to be brought within about 30 μm of a room-temperature sample. For imaging, a sample is mounted on a computer controlled X-Y-Z translation stage and raster scanned under the SQUID. The YBCO bicrystal SQUID with square inner hole of 20 μm on a side and outer body of 60 μm on a side, is operated in a flux-locked loop mode. The sensitivity of the system is about 20 $\mu\Phi_0/Hz^{1/2}$ (~35 $pT/Hz^{1/2}$) at frequencies above 1 kHz.

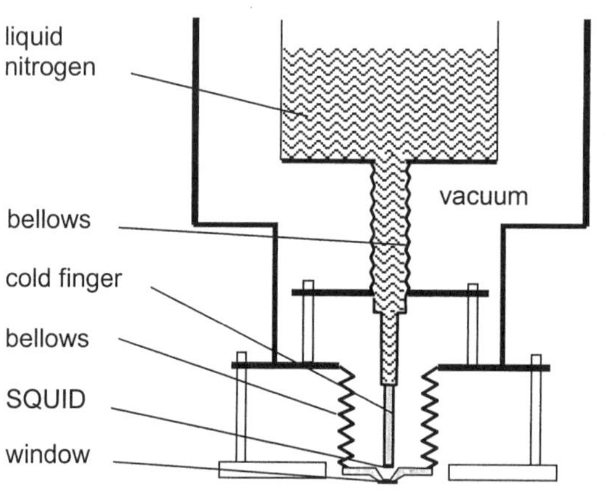

Fig. 5. Schematic of the scanning HTS SQUID microscope for room temperature sample. (From [25], with permission.)

Over the last few years these system was used for imaging high-frequency magnetic and electric fields [27], and for testing integrated circuits [28].The spatial resolution of 50 μm is achieved due to a magnetic inverse technique [29]. Because of the working frequency f > 1 kHz and using lock-in detection the system can be operated in an unshielded environment. Recently a closed-cycle refrigerator was incorporated in this system for cooling SQUID to 77 K. Design description was reported in [30] in which authors have analyzed in detail each stage of the system. Usage of the closed cycle cryocoolers instead of cryogen liquids will be considered below in Sec. 4.

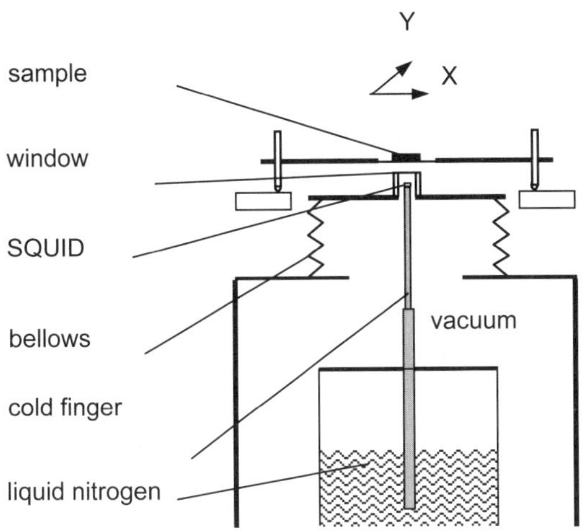

Fig. 6. Schematic of the scanning HTS SQUID microscope for room temperature sample. (From [7], with permission.)

The second system, built in Berkley [25], has the inverted design (sample above the SQUID), in which a SQUID-sample measuring stage is set at the top of a cryostat (Fig.6). SQUID is placed in vacuum beneath a thin vacuum window that separates the SQUID and the sample. It mounted at the upper end of a sapphire rod, the lower end of which is attached to a can with liquid nitrogen. Super-insulation surrounded the rod ensure that the temperature gradient along it is negligible. The position of the window can be adjusted to reduce its separation from the SQUID to a few tens of micrometers. The sample is either placed directly on the window in the air or raster scanned over the window by a non-magnetic, non-metallic scanning mechanism mounted on the cryostat top. In their basic article [25] authors have analyzed in detail each stage of the system. Particularly, they discuss a choice of material for the vacuum window, which is dictated by the need for high elastic modulus, a low electrical conductivity to eliminate Jonson noise, and optical transparency to enable one to center the SQUID chip under the window. In addition a choice of the window shape is determined by an experimental task. The window may be either a 75-μm-thick sapphire disk or a 3-μm-thick Si_xN_y window fabricated on a Si chip. In the first case, the SQUID-to-sample separation is typically 150 μm , whereas in the latter, the separation can be as low as 15 μm [6, 32].

To eliminate external noise the entire microscope is surrounded by three layer of μ-metal shield. Stepper motors driving the translation stage are located outside the shielding enclosure and are linked to the stage by fiberglass rods. The YBCO 24° bicrystal SQUIDs with different sensing effective area are used as a sensor. The sensitivity of the SQUID with inner hole side of 30 μm and outer dimension of about 50 μm is 36 $pT/Hz^{1/2}$ at f = 1 Hz and 22 $pT/Hz^{1/2}$ at 1 kHz. The SQUID electronics operate at a flux modulation frequency of 100 kHz and are designed to run the SQUID in a bias current reversal mode to reduce the 1/f noise due to critical current fluctuations. A Macintosh computer operating with custom software controls measurements and data acquisition.

The system was applied for biomagnetic imaging: both detection of magnetic fields produced by the motion of magnetotactic bacteria [32, 33], and rapid detection of biological targets by using superparamagnetic nanoparticles [34]. A remanent magnetization of ferromagnetic steels in dependence on its microstructure and deformation [35, 36] was examined with this high resolution SSM-300. These SSM applications will be considered in detail in Sec. 3.

A new type of HTS SQUID microscope head was designed and constructed in Sumitomo Electric Industries, Ltd. [37]. It is a simple structure including mutually screwed inner and outer enclosures made of glass fiber reinforced plastic without a bellows. The SQUID is placed in the vacuum space between an inner and an outer enclosure, above the sapphire window. It is mounted on a sapphire rod glued through a bottom of the inner enclosure and partially immersed in the liquid nitrogen. The separation between SQUID and the window is adjusted with the screw depth of the inner and outer enclosure and can be set at about 110 μm. This design makes the HTS SQUID microscope to be compact and simple: the diameter of the head is 55 mm and the total length is 221 mm. For measurements the microscope is placed inside the magnetic shield. Test measurements have shown that iron particles less than 20 μm can be detected with this SQUID microscope.

Saburo Tanaka et al. [38] have presented design of a windowless HTS SQUID microscope intended for imaging samples at both room temperature and 77 K. The main feature of the design is a sample chamber isolated by a shutter gate. After pumping both the cryostat and the sample chamber, the shutter is opened and the sample is positioned as close as possible to the SQUID. The SQUID-sample separation can be less than several micrometers, in principle. The other advantage of this system is reduction of the thermal cycling time without icing.

Neocera Inc. has developed a commercial version of SSM with a room temperature sample and HTS SQUID sensor (MAGMA-C10) [39]. This system uses a small SQUID (about 30 μm wide) with a sensitivity of 20 $pT/Hz^{1/2}$. A minimal SQUID-to-sample separation is 50 μm. This microscope can image buried current-carrying wires by measuring the magnetic fields produced by the currents, or it can be used to image fields produced by magnetic materials. This microscope is intended for nondestructive imaging currents from both the front and back side of the integrated circuits, making failure analysis of multi-chip modules, for localizating of buried defects with a precision of better than 100 μm.

C. SSM based on the SQUID–flux guide sensor

There are two ways to achieve high resolution. The first way is to place a special small SQUID chip in close proximity to a very thin window of a small area (like the foregoing systems at Maryland and Berkeley). The second way is to use a special magnetic flux guide (like the pioneer systems from the University of Saarbrucken and the Julich Research Center), ferromagnetic needle with sharpened tip to transmit magnetic flux from the sample surface to the SQUID. A spatial resolution of the order of 0.1 μm for liquid-nitrogen-temperature samples was achieved in the first one [40] and resolution of 50 μm was demonstrated for the room temperature sample HTS SSM [41, 42]. Later this idea was developed by Japanese and Russian researchers [43-46].

A prototype of the SSM-300 based on a sensor in which the HTS dc SQUID is combined with a needle have been developed in Russia [47]. The needle made of soft magnetic material serves as a magnetic flux guide (MFG) to couple the magnetic flux from the sample into a SQUID. The schematic drawing of the measuring head is shown in Fig. 7. The SQUID is placed inside of vacuum enclosure on the top of the cold sapphire rod 150 mm in length that is cooled due to good thermo-contact with a 2-liter liquid nitrogen can. The non-sharpened MFG 3-mm length made of 50-μm CoFeCrSiB amorphous ferromagnetic wire is glued into vacuum plexiglass window 0.7 mm thick and 25 mm in diameter. In order to the

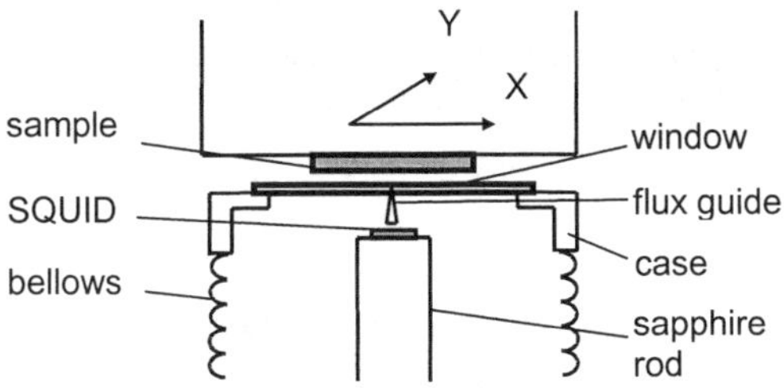

Fig.7. Schematic of a measuring stage of a room sample temperature SSM with SQUID-flux guide sensor. Flux guide is glued through the vacuum window.

MFG does not damage a sample the room-temperature side of the window is polished together with a drop of epoxy with the MFG tip outcrop. The window is glued on a quartz case, beneath of which the SQUID is placed. The case was attached to a stainless steel bellows with using ribbon O-ring and can be moved by the adjusting mechanisms to bring the MFG into a coincidence with the SQUID.

The washer-type bicrystal YBCO SQUID has the effective area (55 × 55) μm^2 of the order of MFG cross section area. For estimation of spatial resolution, a line magnetic field image of two double-wire current-carrying lines with 150-μm center-to-center separation was taken. At minimum sample-MFG separation of about 50 μm the signals from each of DW lines are clearly observed.

An original design of HTS microscope with SQUID-fluxguide sensor has been made in Sumitomo Electric Hightechs [43]. The 10-mm long permalloy rod of 5-mm diameter was glued in the center of the 5-mm thick glass plate with its flat end such that the sharpened end was at the room temperature. The curvature of the rod tip was less than 20 μm. The HTS SQUID made of $HoBa_2Cu_3O_{7-x}$ thin film was leaded to the flat end of the permalloy rod in vacuum space of the cryostat. The SQUID-rod separation was about 0.1 mm. A 0.3-mm line and space meander line on the printed circuit board carrying current was used to investigate the performance of the SSM with and without the sharpened rod. A comparison of the data showed the superiority of using the rod. Less than 0.2 mm spatial resolution was proved. However, the sensitivity of the system with the rod degraded by a factor of 15 due to coupling loss between the rod and the SQUID and the magnetic loss in the permalloy. Spatial resolution with various tip angles and diameters of curvature of the rod has been investigated [44]. The magnetic field spatial distribution was resolved better with smaller tip angle and smaller diameter of tip curvature.

Tanaka and coworkers at Toyohashi University investigated the feasibility of the SQUID-needle system using computer simulation [45]. They supposed a perfect conductor as the material for the SQUID loop and needle permeability as high as 60000. One set of small-field coils as a local sources of magnetic field was used. Using an FEED electrodynamic simulation they found that when the needle penetrated the SQUID loop, the magnetic field density at the loop was one order of magnitude larger than that in the case of no penetration. The second result shows that the needle with a sharper tip gives better space resolution.

They have designed and constructed SSM using HTS SQUID-needle sensor. The needle made of high-μ metal based on Fe-Ni has 7 mm in length and 300 μm × 300 μm bottom cross-section. The size of the needle oval-shaped top edge was 50 μm × 10 μm. A direct-coupled YBCO SQUID magnetometer was used and the substrate was machined to create a dimple for the needle at the center of the pick-up loop. One end of the needle penetrates a superconducting pick-up loop in vacuum; the sharpened end of the needle is glued through the hole in vacuum window. About 100 μm from the top of the needle is outside the window at room temperature. Laser-printed bar lines were used as a sample to test the microscope. Line bars with a line width of 100 μm and spacing between lines of 200 μm were clearly imaged. This result confirms the advantage of such way. The space resolution can be improved by using more sharper needle tip.

D. Laser-SQUID microscopy

A novel technique combined LASER and SQUID technologies has been developed and demonstrated by the researchers of NEC Electron Device (Japan) [48] and PTB (Germany) [49]. This technique is designed for nondestructive evaluation of semiconductor wafers and electrical circuit failure analysis. The principle of operation of the laser-SQUID microscope is schematically shown in Fig. 8. One side of a sample is locally illuminated by the focused laser beam, and a SQUID located above an opposite side of a sample detects the magnetic field produced by optical beam induced currents, or photocurrents. The scanning of a magnetic field with a SQUID magnetometer is made synchronously with scanning of a sample surface with the laser. The intensity of the laser beam is modulated and the resulting modulated SQUID output signal is detected by a lock-in amplifier. The output of lock-in amplifier is used for imaging. The spatial resolution that can be achieved with laser-SQUID technique is determined by the size of the laser spot, typically of the order of units of micrometers. So relatively large SQUIDs can be used, and SQUID-sample separation is not so crucial.

The scanning laser-SQUID microscope designed in Japan intends for nondestructive diagnostics of large scale integration chips (LSI chip). A HTS SQUID, holder of a sample, lens for focusing of a laser beam are inside a shield in a vacuum chamber. The sample can be replaced with the new one through the specialized exchange chamber. A sample is scanned using a piezo scanner (scanning area of 100 μm × 100 μm) or stepper motors (7 mm × 7 mm). The SQUID-sample separation can be made as small as 100 μm. The DC SQUID is made on the basis of $HoBa_2Cu_3O_{7-x}$ film and step-edge Josephson junctions, sensitivity of the SQUID is 30-60 $\mu\Phi_0/Hz^{1/2}$ at frequency range 1–8.3 kHz. The laser beam with wavelengths of 488 nm or 1064 nm is introduced from a bottom side of the cryostat. The beam intensity is modulated with an optical-acoustic modulator or a mechanical chopper. The modulated (8,3 kHz) output signal of the SQUID is detected by a lock-in amplifier.

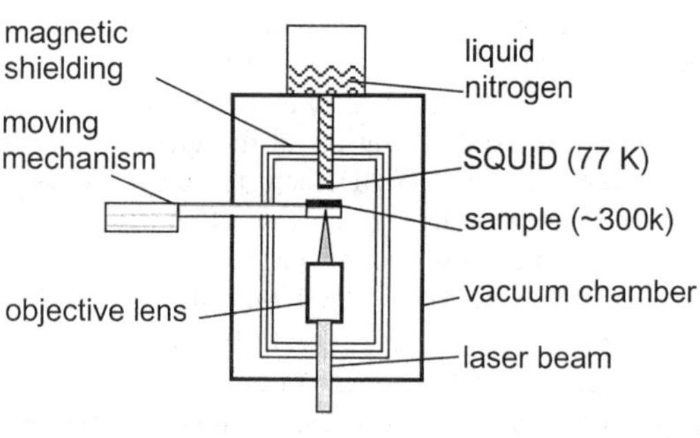

Fig. 8. Schematic view of the Laser-SQUID microscope. (From [48], with permission.)

For demonstrating capabilities of the laser-SQUID microscope a test structure consisting of Ti-Si/poly-Si lines was used. Gate oxide shorts caused by electrical static discharges produce current paths from lines to a silicon substrate through p-n transitions. The obtained magnetic map of p-n transition area has allowed to determine the spatial resolution of the laser-SQUID system, which is 1.3 micrometers, that is record outcome for scanning

SQUID microscopes. Future prospect of the laser-SQUID microscopy is in-line inspection of semiconductor devices, the failure analysis of LSIs.

The PTB laser-SQUID photoscanning system was developed as noninvasive method for investigation of doping inhomogeneity in semiconductor wafers. As the magnetic signals arising from the induced photocurrents in some cases are very small ranging from several tens fT to 1 pT, the high sensitive LTS and HTS SQUIDs were used [49-52]. A photoscanning system with HTS dc SQUID cooled by a Split-Stirling cryocooler has been developed [49]. Because the system is intended for unshielded operation, a small washer single-layer bicrystal HTS SQUID with effective area of 0.028 mm^2 is used. The SQUID is mounted on a sapphire rod that is directly assembled to the cold head of the cooler. By using a 250 μm thick sapphire window a sensor-to-sample distance of 0.5 mm is achieved. The SQUID is oriented as to detect the magnetic field component in parallel to the sample surface. A frequency dependent feedback scheme was applied for stable SQUID operation in the presence of the cooler's magnetic interference. A spatial resolution that is determined by a spot diameter is about 10 μm. These system has been used to analyze current distribution in photovoltaic devices.

E. The variable sample temperature SSM

The described systems is limited to a fixed operation sample temperatures that is either $T \approx 77$ K or $T \approx 300$ K. It is interesting to carry out some of SSM investigations in wide temperature range. It is, for example, observation of a magnetic phase transition, or variation of vortex configurations in superconductors with temperature and so on. There are a few LTS SQUID-based SSM with variable sample temperature in which the sample can be heated to over 100 K [53, 54]. HTS SQUID-based SSM, which allows to investigate magnetic related phenomena in a temperature range 5 – 100 K [55] was designed and fabricated at Chalmers University of Technology. The temperature regulation is realize by maintaining the balance of cold and hot fluxes of He gas taken directly from the LHe bath.

3. SSM APPLICATIONS

There are two vast area of the SQUID imaging system applications. The first one is a scanning HTS SQUID microscopy for fundamental scientific research of magnetic properties of surfaces, ultrathin magnetic films and low dimensional structures. The second one is connected to industrial application such as a nondestructive testing materials and current-carrying structures. Depending on a specific problem the requirement to SSM design and performances can be different. The first direction develops by the way of improvement of a spatial resolution of SSM, keeping high sensitivity. As a rule, the scientific researches are carrying out in laboratory conditions, using the electromagnetic shielding. The second direction develops by the way of designing the compact devices working in unshielded real-world environment, which are simple to make and operate..

In this Section we concentrate on the scientific SSM applications, outlining SQUID-NDE applications for revealing defects of submillimeter size only.

3.1. SCANNING SQUID MICROSCOPY

A. Magnetic measurements of physical objects

Most present SQUID microscopes have been applied to studies of superconducting film systems. From the fundamental point of view these investigations were important for clarification of the mechanism of superconductivity in the cuprates. On the other hand, behavior of vortices in HTS thin-film devices is a key feature that determines their technological applicability.

Now it is undoubtedly that the most important fundamental result achieved by means of SSM is the images of the vortices in HTS films obtained by John Kirtley's group (IBM) with use of the high spatial resolution LTS SQUID based SSM [56, 9]. Analysis of these images has allowed to reveal a half flux quantum which is the best evidence for d-wave pairing in these materials.

Mainly, the LTS SSM instruments were used for imaging magnetic vortex configuration and motion in superconducting samples for different applied fields and temperatures [9, 57-59]. The HTS SSM (Russia, Sec. II) was applied to examine the location of trapped fluxes in a HTS films at 77 K [16]. A 160-nm thick YBCO film was deposited by eximer pulsed laser ablation on SrTiO3 substrates, and square-shaped specimens with side of 1 mm were patterned. The critical temperature of the film was about 90 K. The magnetic field images of HTS films were taken in a low external magnetic field ranging from 0.1 to 50 A/m for zero field-cooled and field-cooled cases. SSM magnetic image for zero field cooled specimen taken with SQUID-sample separation of 20 μm is shown in Fig. 9. A few magnetic field peaks are located randomly in the middle part of the specimen. The intensity of each magnetic peak was about 0.2 μT that corresponds to one magnetic flux quantum [5]. Magnetic image of the same specimen cooled in perpendicular magnetic field ~50 A/m shows much more chaotically distributed peaks.

One more example of magnetic field imaging by HTS SSM is visualization of domain structure for a giant magnetic impedance films [20], which was undertaken with the help of the same instrumentation. The fabrication of high-sensitivity giant magnetic impedance (GMI) elements on the basis of ferromagnetic/conductor/ferromagnetic strip-sandwiches requires the creation of a magnetic anisotropy in a direction perpendicular to a flowing current. Magneto-optic techniques are capable to control only the surface of top layer of such multilayer samples

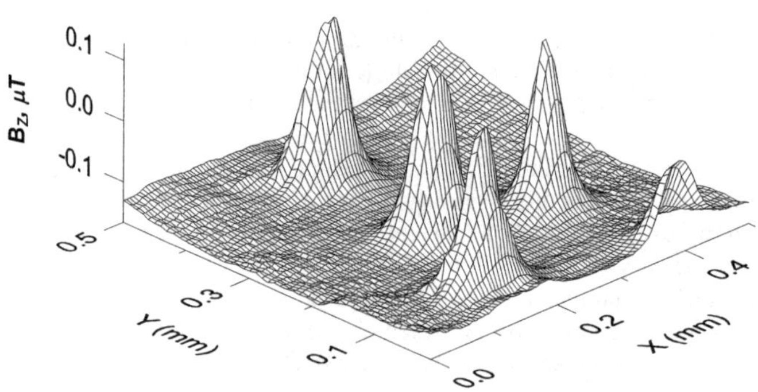

Fig. 9. Image of magnetic flux vortices trapped in the YBCO film at 77 K taken by SSM-77.

without any opaque protective layer. The attempt to apply a SQUID microscope for visualization of magnetic structure in the giant magnetic impedance (GMI) elements was undertaken.

The multi-layer samples with following structures: F/SiO2/Ti/Al, F/SiO2 or F/SiO2/Ti were studied. Here F denotes the ferromagnetic alloy. The Ti layer was deposited in order to increase adhesion and improve the morphology of the magnetic layer surface. The SiO2 layer serves as an insulator between the ferromagnetic layer and Al layer. The samples measured had a shape of 1-μm thick permalloy strips with a size of 6 mm × 0.3 mm. The topography of the vertical component of the magnetic field was imaged with a high spatial resolution of about 30 μm over a small region in the central part of the strip.

A gray-scale magnetic image of the GMI strip in the shield remnant magnetic field is shown in Fig. 10. The positive and negative peaks of magnetic field along the strip boundaries with a period close to 0.08 mm and the average amplitude of about ± 18 μT are clearly seen. In this image the magnetic field peaks near one boundary correlate well with that of the opposite sign near the other boundary. This field configuration agrees with the stripe domains of arbitrary magnetization orientations aligned perpendicular to longitudinal axis of the GMI strip. The areas occupied by both magnetization orientations are approximately equal. The field configuration is stable and reversible. The images taken in magnetic field applied in the strip plane perpendicular to a strip longitudinal axis reveal the domain structure with anisotropy perpendicular to longitudinal sample axis.

Relatively new and very promising aspect of solid state physics now is ability to fabricate ultra-thin films and clusters down to atomic layers, called as nanostructures, and to investigate their physical properties. The production of magnetic particle assembly with well-known periodicity has been a major challenge in the area of magnetic recording media and magnetic devices. Study of the integral magnetic properties of the ferromagnetic nano-patterned thin films and assembly of nanoparticles is actively developed now.

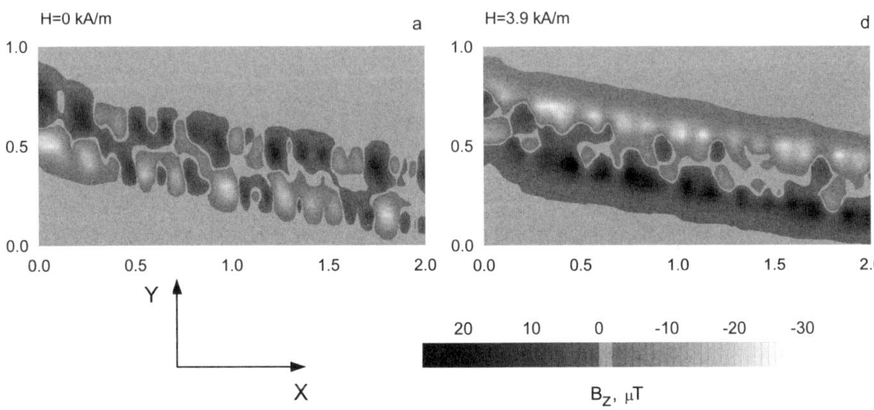

Fig. 10. Magnetic field images of the giant magnetic impedance (GMI) element; on the left hand – in a residual magnetic field of a shield (less than 0.1 μT), on the right hand – in 3.9-kA/m field applied in the film plane perpendicular to a strip longitudinal axis.

As known, the magnetic moments in Ni films exhibit a strong dependence on the number of monolayers, material of substrate, and a structure of a covering layer. The magnetic properties of Au/Ni/Si(100) films of different thickness have been examined at T = 77 K with the SSM-77 [18] described in Sec. 2, A. The Ni films examined were deposited on a monocrystalline Si(100) substrate and covered for preservation with 300-Å layer of Au. The thickness of the square-shaped (600 × 600 μm^2) specimens was ranged from 200 to 8 Å. A SQUID-sample separation was varied in experiments from 100 to 400 μm.

The SSM images of Ni films with thickness ranged from 43 Å to 200 Å have a clearly recorded pair of the symmetric positive and negative magnetic field peaks, placed in opposite corners of the film square. Such images correspond to single domain magnetic structure with an in-plane easy-axis. After applying a parallel to Y-axis 4-kA/m field, the peaks shifted to the midline position, pointing that the magnetic moment rotated to the field direction. The 26-Å film produced a weak magnetic image with in-plane magnetization. When a magnetizing field of 4 kA/m was applied parallel to Y-axis, a clear magnetic peaks had appeared. This magnetic field image is shown in Fig. 11a. Attempts to take images of the films with the thickness 15 and 8 Å in the same parallel magnetic field haven't shown any magnetic signal. This result indicates that the value of magnetization was below the SSM sensitivity threshold and volume magnetization decreases by more than two orders of magnitudes.

The saturation magnetization of the films was calculated from these magnetic measurements. As a result, thickness dependence of the saturation magnetization in these films is shown in Fig. 11b.

a b

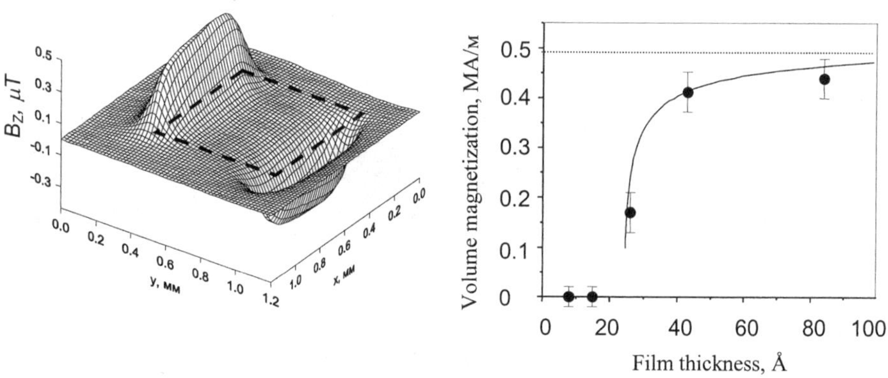

Fig. 11. a) Magnetic field image of 26 Å thick Ni film in magnetizing field equal to 4 kA/m, applied to axis Y. b) Dependence of the saturation magnetization on Ni film thickness.

B. Biomagnetic measurements

Electric currents and magnetic fields play a great role in a wide range of biological phenomena. In order to image and measure magnetic fields from small sized biological objects one needs a room sample temperature SQUID microscope with small SQUID - sample distance.The room-temperature HTS SSM (Sec. 2, B, Berkeley), was used to investigate the dynamics of magnetotactic bacteria, which have a permanent magnetic dipole moments produced by a chain of single domain, ferrimagnetic magnetite (Fe_3O_4) nanoparticles [32, 33]. In this instrument the sample can be brought to within 15 μm of the SQUID that allows to resolve a dipole moment less than 10^{-17} A·m^2. A series of measurements of magnetic field fluctuations produced by ensembles of motile and nonmotile bacteria have been performed. Authors determined the average rotation drag coefficient, magnetic moment, and frequency and amplitude of the vibration and rotation of magnetotactic bacteria. The effects of external disturbances were studied, too. The presence of high levels of oxygen causes the ensemble to slow over time. The application of static magnetic field causes a large-scale migrations of the bacteria. Time trace of single magnetotactic bacterium can be measured by this instrument. The authors plan future studies of the dynamic and migration of magnetotactic bacteria in opaque porous media.

Magnetic nanoparticles are wide used in biology and medicine as diagnostic technique. The SQUID microscope can be applied to study any cell to which a suitably coated magnetic particle can be attached. Bound to a suitable antibody, nanoparticles are used to label specific molecules, structures, or microorganisms. Detection of small particles by a SQUID for immunoassay applications is performed in several groups [61, 62].

Researchers from Berkeley were developed this technique for sensitive, quantitative and rapid detection of biological targets by using superparamagnetic nanoparticles [33]. A pulse magnetic field are applied to a suspension of magnetic nanoparticles carrying antibodies directed against the target. In the presence of this aligning field the nanoparticles develop a net magnetization, which relaxes when the field is turned off. Unbound nanoparticles relax rapidly by Brownian rotation and contribute no measurable signal. Nanoparticles that are bounded to the target produce a slowly decay magnetic flux which is detected by the SQUID. It was shown that SSM is able to detect magnetic field from $5 \cdot 10^4$ superparamagnetic nanoparticles bound to a suitably chosen target. Extremely small sample volumes less than 1 ml are required to perform homogeneous assays with this technique. The magnetic assay also may have advantages in its speed. With the modified version of instrument the authors expect to improve their detection limit to 50-500 magnetic particles.

3.2. NON-DESTRUCTIVE EVALUATION

A. High spatial resolution eddy current technique

Over the past few years a number of groups has demonstrated that the use of SQUID sensors increases the sensitivity of the eddy current technique by several orders of magnitude, allowing to reveal the much deeper defects than by traditional induction coil sensors, to detect and monitor the micro-flaws, the hidden damage in aircraft structures, heterogeneity in electrical conductivity, and the flow of steady state corrosion currents [for example, 63, 64]. HTS SQUID based systems have a preference because their cryogenic enclosure is more simple and cheap. The overview of activities in the HTS SQUID NDE area is given in [6, 12], so we will limit ourselves with a consideration of SQUID systems for NDE allowing to take images of defects with spatial resolution of the order sub-millimeters.

At fabrication of superconducting microwave cavities there is a problem of existence of inclusions of metals different from material of a resonator. A few submillimeter inclusions per resonator could already lead to a substantial reduction in the projected maximum electric field strength of the cavity. An eddy current NDE system based on the LTS SQUID which is able to detect tantalum inclusions in a test sample made from high-purity niobium has been developed by M. Mück et al. [64].

Detection of small-sized artificial inclusions was made using the SSM-77 (Sec. 2,A , Russia) [65]. As a sample a copper plate of 1.5 mm thick with drilled holes of various diameters and depths have been used. A drawing of a scanned area of the copper plate is shown in Fig. 12a. There are five holes drilled on half-depth of the plate and fully filled with solder and two empty holes drilled on half-depth of plate. Diameters of the holes are ranged from 250 to 600 μm. A SQUID-sample separation is installed close to 100 μm. An alternative magnetic field at frequency f = 1779 Hz and amplitude 80 μT is applied in parallel to the plate and the SQUID with the Helmholtz coils. Magnetic field excites eddy currents at the plate surface within the skin depth. Due to difference of conductivity a distortion of current lines is appeared nearby the hole likewise a current dipole. A normal component of the magnetic field induced by these currents is measured by SSM-77. A lock-in amplifier detects the output signal of the SQUID. The magnetic image of this plate are shown in Fig. 12b. An image from each of the holes is two oppositely directed peaks that corresponds to a signal from current dipole. A distance between peaks is of the order of the hole depth. The peak signs depend on the sign of a difference of conductivity. In our case of the less conductivity inclusion the sign

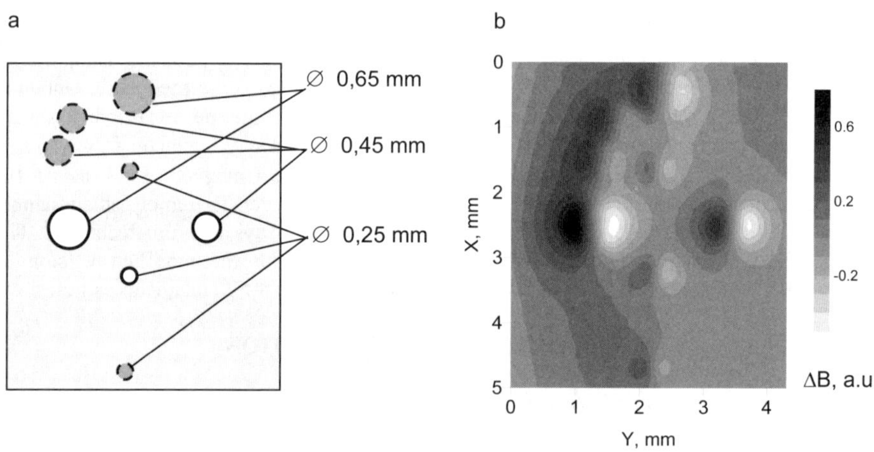

Fig. 12. a) Drawing of a copper plate of 1.5 mm thick with drilled holes of various diameters and depths. Empty holes are marked by white circles, holes filled with solder are denoted by shadowy circles. b) Eddy current response of the copper plate with holes to a 1779-Hz applied field directed in parallel to the plate surface.

of a current dipole moment is oriented oppositely to direction of exciting current.

In a current dipole approximation an inclusion produces the disturbance of magnetic field at a distance *d*:

$$b \sim B_0 \cdot \frac{V}{\delta \cdot d^2} \cdot \frac{\Delta\sigma}{\sigma} \tag{3.1}$$

where: B_0 is an external field on the surface, δ - skin depth, $\Delta\sigma/\sigma$ - relative difference of conductivity of the defect and material, V – a defect volume. Thus, the useful signal is proportional to an exciting field magnitude, inclusion volume, relative difference of conductivity, and inversely as the square of distance to point of observation and skin depth. For characteristic inclusion size ~ 0.1 mm, d ~ 1 mm, magnetizing field of 10^{-5} T and relative difference of conductivity ~ 0.1, the signal magnitude can reach 10^{-9} T. This value can be measured reliably by a HTS SQUID microscope.

B. Current –carrying structures

One possible and promising application for scanning SQUID microscopes, on the smallest scale, is the detection of the defects in microelectronic devices, such as integrated circuits (IC), flip-chip packages and multichip modules (MCM). The capability of high-resolution SQUIDs to localize a current-carrying structure was demonstrated in an early experiments with LTS MicroSQUID [10], and was developed with high-resolution HTS SQUID-based microscopes. As an example, Fig. 13 shows a spatial distribution of magnetic field produced by a direct current of 500 μA flowing through a test well-calculated thin-film current-carrying structure: a square-shaped coil with a side of 400 μm and a conductors width of 20 μm [65]. This image was taken with SSM-77 at SQUID-sample distance of 20 μm.

Important contribution to development of SQUID-based techniques for testing a more complicated circuits was made by Wellstood's group at Maryland University in cooperation

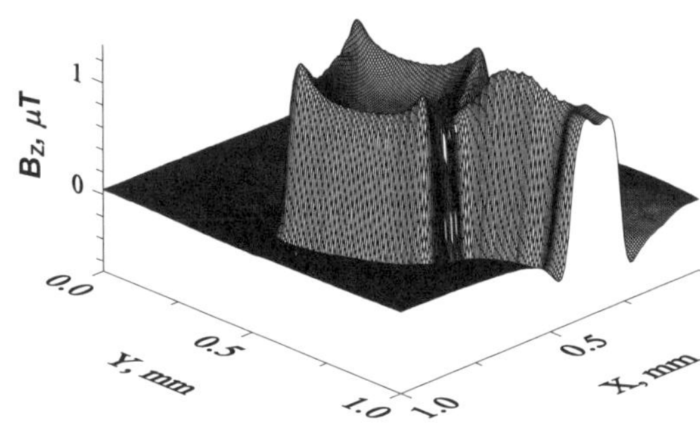

Fig. 13. Image of magnetic field produced by a 500-μA direct current flowing trough the gold thin-film square-shaped coil with a side of 400 μm. Image was made at T = 77 K and SQUID–sample separation of 20 μm [65].

with Neocera Inc. [28,29, 39, 66]. For their experiments they utilized the room sample temperature SSM combined with a closed-cycle refrigerator to cool YBCO dc SQUID to 77 K [30]. To image short circuits in commercial integrated circuit, or MSM, ac current of 0.5-1 mA at frequency 3-4 kHz is applied to a sample and the signal from the SQUID is lock-in detected at the same frequency and phase as the applied current. In examining the image the location of a circuit-short is revealed. An inversion technique was used to convert the magnetic field data to a two-dimensional current density distribution that allowed to locate current paths. A spatial resolution was found less than 50 μm in the converted current density images at SQUID-sample separation of 150 μm. This is three times better than the standard near-field microscopy limit, and about 10 times sharper than the raw magnetic field images.

C. SSM study of the remanent steel magnetization

It is well-known that mechanical or thermal stress can cause a change of magnetic state of ferromagnetic materials due to rotating the magnetization of magnetic domains or moving the magnetic domain wall at microscopic levels. Ability of the SSM to measure the local magnetic fields of the sample with applying a negligible magnetizing field and without perturbing the magnetic state of the sample material makes SSM of a valuable tool for study of magnetic properties of the ferromagnetic materials. Applicability of SQUID magnetometry for examination of stress-related deformations in steels was demonstrated still with LTS SQUID systems [10, 11, 67].

Researchers from Berkley attempted to infer an information about local microstructures of the ferromagnetic steels from measurement of their local magnetic properties [35, 36]. The room sample temperature SSM (Sec. 2, B, Berkley,) with the additional advantage of having a rather large (50 mm × 50 mm) scanning range was applied to study the remanent magnetization of low carbon steel with different microstructures. The samples made of AISI 1040 steel were prepared using different heat treatments, which produced varied microstructures, and different rolling treatments, which produced varied levels of deformation. Effect of microstructure changes in steel on the magnetic field distribution near the sample surface was examined. A few series of samples with different treatment were prepared. Each of the samples shaped 5×12×12 mm^3 block was preliminary demagnetized in order to erase its magnetic history and was subsequently magnetized in with a 50-mT field in thin direction to induce a remanent magnetization. With a SQUID-sample separation of 400-500 μm the sample is scanned over the SQUID. Two-dimensional magnetic image produced by the local remnant magnetization of the sample was taken. In addition, information about the sample microstructure was obtained with optical imaging, permitting the magnetic images to be correlated with the inherent microstructure. Through these measurements it was shown that different microstructures, produced by control treatment, exhibit drastically different magnetic behaviors.

Since the strength and integrity of steels, widely employed in industry, can be well predicted from microstructure information, the found correlation could lead to development of effective method for non-destructive testing of ferromagnetic steels through a simple measurement of remnant magnetization at applying a negligible magnetic field to the sample to be studied.

4. CURRENT RESEARCH ENGINEERING AND DEVELOPMENT ACTIVITY

Research activities directed on optimization of the SSM design, development of new methods, improvement of the characteristics of SSM units are kept on. In this Section we consider in detail the SQUID – fluxguide sensor operation, the new capabilities of SQUID electronics engineering and the use of a close-cycle refrigerator to cool a HTS SQUID.

A. SQUID-fluxguide sensor

As it was shown in Sec. 2, C, several research groups actively develop SQUID-magnetic flux guide systems to improve a spatial resolution of SSM. The SQUID microscope based on such sensor is able to achieve a spatial resolution close to that obtained with magnetic force microscopes [40], exceeding the last in magnetic field sensitivity. This new promising way requires detail theoretical and experimental studying properties of the SQUID-magnetic flux guide system, such as the sensitivity, spatial resolution, optimal flux guide material, dimensions and so on. Some of this questions were made clear in works of Japanese and Russian researchers.

On basis of the theory of thin linear antennas the mathematical model of a sensor which a SQUID is combined with a soft magnetic needle was developed by the researchers of Russian group [73]. In this model is supposed that the thin needle served like magnetic flux guide (MFG) is executed from an ideal magnetic material with magnetic permeability $\mu \to \infty$. During measurements the magnetic field of researched object produces inside MFG a magnetic flux Φ, which is transfer to the SQUID, Fig. 14. The cooperative magnetic charge localized near the measuring end of the MFG is coupled with this flux. The magnetic charge of the opposite sign, neutralizing it, is uniformly distributed on a remaining part MFG, linearly decreasing to the opposite end. The losses of a signal are determined by a position of the SQUID relative to MFG. The maximum value of Φ near the opposite end of the MFG (near the SQUID position) is:

$$\Phi \approx \frac{2\pi}{\ln\frac{L}{a}} \cdot S, \tag{4.1}$$

where $S = \int_L \varphi \mathrm{d}l$, φ - magnetic potential, L, a – length and thickness of the antenna, respectively. Thus, a basic measured value is an integral S from a potential φ of an external magnetic field along the MFG.

Let's consider a case of measurement of a magnetic field above solitary nanoparticle with a magnetic moment M. In accordance to (1) the minimum magnetic flux in MFG is determined as:

$$\Phi_{\min} \approx \frac{2\pi}{\ln\frac{L}{a}} \cdot \frac{M_{\min}}{d} \approx \frac{M_{\min}}{d}. \tag{4.2}$$

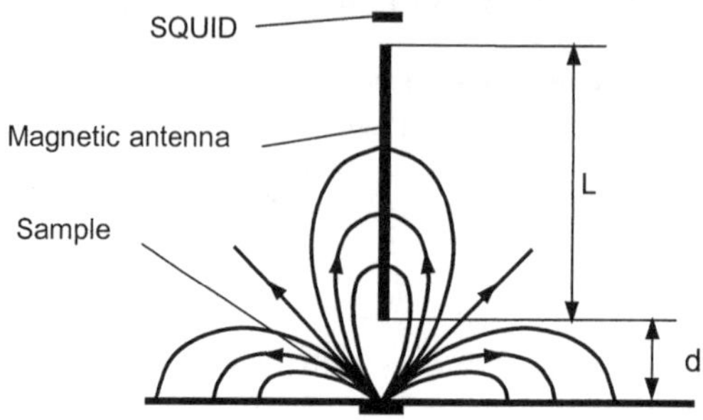

Fig. 14. Layout of a SQUID-flux-guide sensor and a sample.

Here $S = \frac{M_{\min}}{d}$ and d – distance from the MFG to a measured magnetic moment. If to assume Φ_{min}~$10^{-5}\Phi_0$ and d ~ 1 μm, the corresponding minimum magnetic moment will be approximately 2×10^{-16} emu (2×10^{-19} A·m^2). If to assume a magnetization of nanoparticle of about 200 G, its volume will be $V_{\min} \approx 10^{-18} cm^3$, that corresponds to the linear size of a particle ~ 10 nm. The analysis that has been carried out with the help of developed mathematical model, displays that for a case of nanoparticle the spatial resolution is determined, first of all, by a distance *d* from the end of MFG up to the particle measured, i.e. two magnetic nanoparticles will be distinguishable if distance between them is of about *d*. Correspondingly, a radius of curvature of the sharpened needle end must be equal or less than *d*.

It was shown experimentally that real characteristics of SQUID MFG sensors depend on a needle material, MFG geometry, radius of curvature and angle of a needle tip [43-46]. The first results showed that MFG didn't produce an additional noise in weak applied magnetic fields (Reyleigh region). In more strong fields a hysteresis of MFG is observed. Another design of the SQUID-MFG sensor including the dB_Z/dz gradiometer with two MFGs were presented in [68, 69].

B. Direct-coupled readout SQUID electronics

In most cases the HTS SQUID-based systems were designed analogous to the LTS systems, i.e. its operate with the linearity electronics using flux modulation-demodulation techniques and cooled impedance matching circuits. The ability of the HTS SQUIDs to operate at 77 K makes it possible to replace a passive cooled matching circuit on a semi-conducting cooled preamplifier operating at the same temperature [70]. Recently, a simple direct-coupled SQUID electronics free from a flux modulation - demodulation procedure, and corresponding hardware have been developed for HTS dc SQUIDs ([71, 72]. This electronics uses a liquid-nitrogen-cooled amplifier (LNA) as the first stage of the direct-coupled readout electronics and can operate with both direct current (dc) or and alternating current (ac) bias [72].

The basis of the LNA [70] is a differential stage with parallel-connected bipolar transistors. It has an input impedance larger than 70 Ω, a gain 140, a bandwidth more than 2 MHz and power dissipation of about 40 mw. The intrinsic white noise voltage spectral density, $S_V^{1/2}$, is close to 0.18 nV/Hz$^{1/2}$ and the corner frequency of the excess noise is less than 8 Hz. The input current noise density, S_I, estimated from noise measurements under open-circuited amplifier input is close to 3 pA/Hz$^{1/2}$ at frequencies higher than 8 Hz. The preamplifier is packed in a rectangular box (25 ×12 ×6) mm^3 and can be placed near the SQUID. The outline

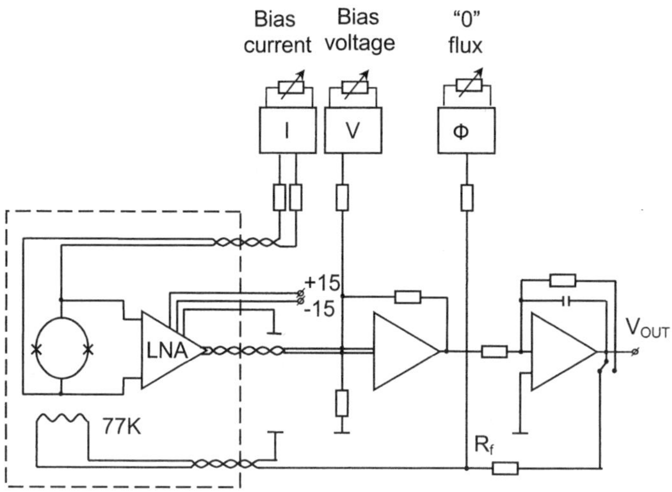

Fig. 15. Equivalent circuit of the direct readout HTS SQUID flux-locked loop feedback electronics based on a liquid-nitrogen-cooled pre-amplifier (LNA).

of SQUID electronics is shown in Fig. 15. The SQUID is directly connected to the LNA. The output signal of the LNA is fed to a room-temperature FLL electronics, which consists of a differential amplifier, an integrator and a feedback circuit. Such SQUID electronics is used now in the SSM-77 (Sec. 2, A). A frequency bandwidth for small signal is ~ 200 kHz, a dynamic range is more than 140 dB, a slew rate is $7{\times}10^5\Phi_0/c$. The electronics is simple and reliable in operation.

C. Cryocoolers for SSM

To the present time HTS SQUIDs and the devices on their basis, including SSM, have reached such stage of the development, when there is possible their commercial realization and broad usage in an industry. More all it concerns SQUID systems intended for nondestructive evaluation. Problems of long-time storage of HTS SQUIDs, elimination of their excess noise, ability to work in unshielded environment and so on were solved sequentially. A main disadvantage for SSM industrial applications is the necessity to cool SQUIDs to below the critical temperatures, that is to utilize liquid nitrogen. Usage of cryostats considerably enlarges the sizes of systems and demands special training for the personnel. Close-cycle refrigeration is increasingly popular alternative to liquid cryogenics, but only recently has been success in using them with SQUID system. There are a several problems relating to a simultaneous operation of a SQUID and cryocooler. The motion of metallic parts inside a cooler compressor produces electromagnetic interference that influence on the SQUID operation. Mechanical vibrations and the movement of metallic parts introduces an excess noise into the measurement process due to Johnson noise and eddy currents induced in cooper constituents near the SQUID.

The are two ways to solve these problems. In [30] authors describe in details the design and operation of a scanning SQUID microscope for room temperature samples that uses a close-cycle refrigerator to cool a HTS SQUID. They selected ADP CryoTiger, a Joul-Thomson refrigerator. A significant advantage of the CryoTiger is its relatively small size. The refrigerator is 6 in. long by 3 in. in diameter, with a 33.3 mm diameter oxigen-free high-conductivity copper mounting plate at the bottom. To prevent the transmission of vibrations from compressor to the cold end, authors connect them with 8 m long flexible houses. This physical separation also reduces the stray magnetic fields from the moving parts in compressor. To prevent undesirable magnetic fields or eddy currents from affecting the SQUID they use a nonmagnetic and nonmetallic sapphire rod on the end of which the SQUID is placed. By maintaining an adequate vacuum in vacuum housing where the cold finger is mounted the thermal conduction by gas between the vacuum walls and cold finger is minimized. The radiation shielding around the cold head also decreases the operating temperature at the sapphire tip up to 74 K.

To start the system, the following steps are need: evacuation of the chamber with turbopumping station requires about 1 h., cooling the system from 300 to 85 K takes about 1.5 h. Usually the refrigerator cools to 72-23 K and then gradually warms up to the final operating point of around 76 – 78 K over a period a few hours. Once cooled, the temperature varies by ±0.1 K over a couple of hours. This level of temperature stability is sufficient for regular operation of the SQUID.

Examining of the system performance shows the spectral density of white noise of 10.5 $\mu\Phi_0/Hz^{1/2}$ at frequencies > 500 Hz that corresponded to 17.5 $pT/Hz^{1/2}$. This system is intended for nondestructive evaluation of microelectronic devices, such as flip-chip packages and multichip modules.

The other way of solving the problem of cooling HTS SQUID without the accompanying increased noise level is usage a system with a cryobattery. As in the previous case the first prototype of such system is intended for SQUID-NDE application [75, 39]. The array of HTS SQUIDs is mounted on the end of a sapphire rod that is connected to a 700-g copper core (alloy 101, OFHC), both held in vacuum chamber with sapphire window. A turnbuckle at the rear attaches a miniature cryogenic refrigerator to the battery for cooling and detaches it during measurements. Cooling the battery from room temperature to 50 K requires 50 hours. The cold stored by the cryobattery ensures 4-6 hours of SQUID operation without refrigeration. Recharging after 6 h. of operation requires 8 h. The microcooler provides 100 mW of cooling at 65 K and 180 mW at 77 K, from 3W of power.

The moderated version the above system using a two-stage cool-down/warm-up process, is currently under development at Strathclyde [76]. In this system cryobattery made of OFHC consists of two separate stages: a 100-g core-type inner battery and a 300-g outer guard ring, surrounding the inner battery. This design allows extending the SQUID operating time because the inner battery loses cold primarily to the outer mass of cooper rather than to the vacuum enclosure which sits at 300 K. In this system the cryocooler used is a Leybold closed-cycle (ROK 10-300) which supply ≈ 10 W of cooling power at 80 K. A copper coldfinger connects the cryocooler with the inner and outer copper simultaneously. To engaged and detach the cooler, a set of bellows is incorporated into a system. Up to now it takes ≈ 5 h for the inner battery to reach 45 K, while the outer guard-ring reaches a relatively poor temperature of ≈ 145 K. By improving the quality of the coldfinger-battery contact points, authors expect that as intended, both inner and outer will reach the same base temperature. The final stage will be to incorporate a sapphire rod onto which will be mounted the HTS SQUID gradiometers.

5. SUMMARY

Recently developed scanning magnetic HTS SQUID microscopy is at the joint of different scientific disciplines. It includes HTS SQUID manufacturing, designing of instrumentation, development of techniques, interpretation of the obtained data. At present, high resolution HTS SSM microscopes with a spatial resolution of 15-500 μm and field sensitivity of the order of 1-100 $pT/Hz^{1/2}$ are real scientific instruments, which operate in industrial and scientific laboratories. Now there is much interest in achieving higher spatial resolution. The HTS SQUID microscopes for room temperature sample are developed and applied both for imaging very weak magnetic fields of biological objects and non-destructive evaluation of different materials and devices. A few examples of their present and potential applications are given in this overview.

The design of the HTS SSM step by step is advanced to development of the commercial realizable SQUID microscopes. The number of HTS SSM instruments combined with small cryocoolers is risen. The first commercial version of HTS SSM intended for imaging currents in integrated circuits and printed circuit boards already exists [39]. These devices will be applied to problems, for which quantitative measurements of magnetic fields with substantially higher sensitivity and minimal back action on study object is needed.

ACKNOWLEDGEMENTS

The authors would like to thank O. V. Snigirev, and B. Ya. Liubimov for helpful discussions, Yu. V. Derioujkina for help in preparing the manuscript. This work is supported in part by ISTC grant No. 1991.

REFERENCES

[1] A. D. Hibbs, R. E. Sager, D. W. Cox et al., Rev. Sci. Instrum. 63 [1992] 3652.
[2] L. N. Vu, D. J. Van Harlingen, IEEE Trans. Appl. Supercond. 3 [1993] 1918.
[3] A. Mathai, D. Song, Y. Gim, F. C. Wellstood, Appl. Phys. Lett. 61 [1992] 598.
[4] A. Mathai, D. Song, Y. Gim, F. C. Wellstood, IEEE Trans. Appl. Supercond., 3 [1993] 2609.
[5] J. R. Kirtley, M. B. Ketchen, K. G. Stawiasz et al., Appl. Phys. Lett. 66 [1995] 1138.
[6] D. Koelle, R. Kleiner, F. Ludwig et al., Reviews of Modern Physics 71 [1999] 631.
[7] F. C. Wellstood, Y.Gim, A.Amar et al., IEEE Trans. Appl. Supercond. 7 [1997] 3134.
[8] K. Enpuku, T.Minotani, IEICE Trans. Electron. E83-C N1 [2000].
[9] J. R. Kirtley, Physica C. 368 [2002] 55.
[10] W. G. Jenks, S. S. Sadeghi, J. P. Wikswo, Jr., J. Phys. D. Appl. Phys. 30 [1997] 293.
[11] J. P. Wikswo, Jr., IEEE Trans. Appl. Supercon. 5 [1995] 74.
[12] H.-J. Krause, M. V. Kreutzbruck, Physica C 368 [2002] 70.
[13] R. C. Black, A. Mathai, F. C. Wellstood et al., Appl. Phys. Lett. 62 [1993] 2128.

[14] S. A. Gudoshnikov, O. V. Snigirev, I. I. Vengrus et al., Cryogenics 34 [1994] 883.
[15] K. E. Andreev, A. V. Bobyl, S. A. Gudoshnikov et al., Supercond. Sci. Technol. 10 [1997] 366.
[16] S. A. Gudoshnikov, M. I. Koshelev et al., Proc. of Europ. Conf. on Appl. Supercond. (EUCAS'99), Barselona, Sept. 1999: Inst. Phys. Conf. Ser. 167 [2000] 541.
[17] S. A. Gudoshnikov, K. E. Andreev, L. V. Matveets et al., Ext. Abst. of 6th Inter. Supercond. Elect. Conf, (ISEC'97), Berlin. Inst. Phys. Conf. Ser. 158 [1997] 755.
[18] O. V. Snigirev, K. E. Andreev, A. M. Tishin et al., Phys. Rew. B. 55 [1997] 14429.
[19] O. V. Snigirev, K. E. Andreev, A. M. Tishin et al., in Rus.: Fizika tverdogo tela 40 [1998] 1681.
[20] S. A. Gudoshnikov, P. E. Rudenchik, L. V. Matveets et al., IEEE Trans. Appl. Supercond. 11 [2001] 223.
[21] S. A. Gudoshnikov, K. E. Andreev, G. B. Khomutov et al., Appl. Supercond. 2 [1995] 1521.
[22] S. A. Gudoshnikov, B. Ya. Liubimov, L. V. Matveets et al., Physica C. 372-376 [2002] 271.
[23] J. Dechert, M. Mueck, C. Heiden, IEEE Trans. Appl. Supercon. 9 [1999] 4111.
[24] R. C. Black, Ph. D. Thesis, University of Maryland, College Park, MD, [1995].
[25] T. S. Lee, E. Dantsker, J. Clarke, Rev. Sci. Instrum. 67 [1996] 4208.
[26] F. Baudenbacher , N. T. Peters, J. P. Wikswo, Rev. Sci. Instrum. 73 [2002].
[27] S. Chatraphorn, E. F. Fleet, F. C. Wellstood et al., IEEE Trans. Appl. Supercond. 9 [1999] 4381.
[28] S. Chatraphorn, E. F. Fleet, F. C. Wellstood et al., Applied physics letters. 76 [2000] 2304.
[29] F. C. Wellstood, E. Fleet, S. Chatraphorn et al., Ext. Abst. of 8th Int. Supercond. Electron. Conf. (ISEC'01) SA-1 [2001] 57.
[30] E. F. Fleet, S. Chatraphorn, F. C. Wellstood et al, Rev. Sci. Instrum. 72 [2001].
[31] T. S. Lee, E. Dantsker, J. Clarke, Rev. Sci. Instrum. 67 [1996] 4208.
[32] T. S. Lee, Y. R. Chemla, E. Dantsker et al., IEEE Trans. Appl. Supercond. 7 [1997] 3147.
[33] Y. R. Chemla, H. L. Grosman, T. S. Lee et al., Biophysical Journal 76 [1999] 3323.
[34] Y. R. Chemla, H. L. Grosman, Y. Poon et al., PNAS 97 [2000].
[35] T. J. Shaw, K. Schlenga, R. McDermott et al., IEEE Trans. Appl. Supercond. 9 [1999] 4107.
[36] T. J. Shaw, J. W. Chan, S.-H. Kang et al., Acta mater. 48 [2000] 2655.
[37] T. Nagaishi, H. Itozaki, Supercond. Sci. Technol. 12 [1999] 1039.
[38] S. Tanaka, O. Yamazaki, R. Shimizu et al., Jpn. J. Appl. Phys. 38 [1999] L505.
[39] www.neocera.com
[40] P. Pitzius, V. Dworak, U. Hartmann, Ext. Abst. of 6th Int. Supercond. Electron. Conf. (ISEC'97), 3 [1997] 395.
[41] Yu. Tavrin, M. Seigel, Ext. Abst. of 6th Int. Supercond. Electron. Conf. (ISEC'97) 3 [1997] 389.
[42] Yu. Tavrin, M. Seigel, Proc. of 3d Europ. Conf. on Appl. Supercond. (EUCAS'97) Inst. Phys. Conf. The Netherlands 158 [1997] 719.
[43] T. Nagaishi, K. Minamimura, H. Itozaki, IEEE Trans. Appl. Supercond. 11 [2001] 226.
[44] K. Minamimura, T. Nagaishi, H. Itozaki, Ext. Abst. of 8th Int. Supercond. Electron. Conf., (ISEC'01), P1-B12 [2001] 201.

[45] S. Tanaka, K. Matsuda, O. Yamazaki et al., Supercond. Sci. Technol. 14 [2001] 1.
[46] S. A. Gudoshnikov, Yu. V. Deryuzhkina, P. E. Rudenchik et al., IEEE Trans. Appl. Supercond. 11 [2001] 219.
[47] S. A. Gudoshnikov, B. Ya. Liubimov, Yu. V. Deryuzhkina et al., Physica C 372-376 [2002] 166.
[48] K. Nikawa, IEICE Trans. Electron. E85-C [2002].
[49] T. Schurig, J. Beyer, D. Drung et al., IEICE Trans. Electron. E85-C [2002] 665.
[50] J. Beyer, Th. Schurig, A. Ludge et al., Supercond. Sci. Technol. 13 [2000] 532.
[51] J. Beyer, D. Drung, T. Schurig, IEEE Trans. Appl. Supercond. 11 [2001] 1162.
[52] T. Schurig, J. Beyer, D. Drung et al., Ext. Abst. of 8th Int. Supercond. Electron. Conf. (ISEC'01) [2001] 12.
[53] J .R. Kirtley, C. C. Tsuei, K. A. Moler et al., Appl. Phys. Lett. 74 [1999] 4011.
[54] T. Morooka, S. Nakayama, A. Odawara et al., IEEE Trans. Appl. Supercond. 9 [1999] 3491.
[55] A. Ya. Tzelenchuk, Z. G. Ivanov, S. Pehrson et al., IEEE Trans. Appl. Supercond. 9 [1999] 4115.
[56] Tsuei C.C., J. R. Kirtly, Physica C 367 [2002] 1.
[57] L. N. Vu, M. S. Wistrom, D. J. Van Harlingen, Appl. Phys. Lett. 63 [1993] 1693.
[58] K. Tanaka, T. Morooka, A. Odawara et al., IEEE Trans. Appl. Supercond. 11 [2001] 230.
[59] A. Sugimoto, T. Yamaguchi, I. Iguchi, Physica C 367 [2002] 28.
[60] T. Morooka, S. Nakayama, A. Odawara, Jpn. J. Appl. Phys. 38 [1999] L119.
[61] K. Enpuku, T. Minotani, M. Hotta et al., IEEE Trans. Appl. Supercond. 11 [2001] 661.
[62] S. Katsura, T. Yasuda, K. Hirano et al., Ext. Abst. of 8th Int. Supercond. Electron. Conf., (ISEC'01), [2001] 483.
[63] R. Hohmann, D. Lomparski, H.-J. Krause et al., IEEE Trans. Applied Supercond. 11 [2001] 1279.
[64] M. Muck, C. Welzel, F. Gruhl et al., Physica C 368 [2002] 96.
[65] http://top.izmiran.rssi.ru/eng/cryo
[66] E. F. Fleet, S. Chatraphorn, F. C. Wellstood et al., IEEE Trans. Appl. Supercond. 9 [1999] 4103.
[67] J. Banchet, J. Jouglar, P.-L. Vuillermoz et al., IEEE Trans. Appl. Supercond. 5 [1995] 2486.
[68] S. I. Bondarenko, A. A. Shablo, P. P. Pavlov, et al., Physica C 372-376 [2002] 158.
[69] S. I. Bondarenko, A. A. Shablo, P. P. Pavlov, Cryogenics 40 [2002] 313.
[70] S. A. Gudoshnikov, I. I. Vengrus O. V. Snigirev et al., Ext. Abst. of 5th Int. Supercond. Electron. Conf. (ISEC'95), Sept. 1995, Nagoya, Japan [1995] 346.
[71] S. A. Gudoshnikov, I. I. Vengrus, L. V. Matveets et al. IEEE Trans. Instrum. Measur., 46 [1997] 624.
[72] S. A. Gudoshnikov, L. V. Matveets, O. V. Snigirev et al., IEEE Trans. Appl. Supercond. 9 [1999] 4397.
[73] S. A. Gudoshnikov, B. Ya. Liubimov, L. V. Matveets et al., Physica C 368 [2002] 66
[74] S. A. Gudoshnikov, A. S. Kalabukhov, L. V. Matveets et al., Proc. of Int. Cryogen. Electron. Conf. (ICEC'98), England [1998] 423.
[75] W. N. Podney, IEEE Trans. Appl. Supercond. 9 [1999] 3483.
[76] C. Carr., J. C. Macfarlane, G. B. Donaldson. Physica C 372-376 [2002] 259.

VORTEX MATTER AND SUPERCONDUCTING ELECTRONIC DEVICES

Roger WÖRDENWEBER
Institut für Schichten und Grenzflächen (ISG),
Forschungszentrum Jülich, D-52425 Jülich, Germany

1. INTRODUCTION

Superconductivity represents an extremely complex and interesting quantum-mechanical phenomenon. The macroscopic quantum state of the Cooper pairs results in phenomena like *perfect conductivity* (zero-resistance ρ=0 at DC current [1] and small microwave surface resistance at high frequencies, e.g. R_s=2-3μΩ for $YBa_2Cu_3O_{7-\delta}$ at 1GHz and 77K [2]), *perfect diamagmetism* (Meissner-Ochsenfeld effect [3]), *quantization of trapped magnetic flux* and *phase correlation* of the charge-carrier wavefunction in weak-link structures leading to Josephson effects [4]. Due to these extraordinary properties, superconductivity owns a high potential for improvement of existing applications or even novel applications among others in the field of cryoelectronics, sensor technology, information distribution or energy management. For these applications the zero-resistance (cables, magnets, etc.), the dramatic change of the resistance at the transition (fault current limiter), the small high-frequency surface resistance (filters, antennas, mixers, etc.) or the Josephson effects (Superconducting Quantum Interference Device (SQUID), single flux quantum logic (SFQ), etc.) are utilized and, thus, carefully studied and analyzed.

Less attention has been paid so far to the impact of quantized magnetic flux (so called flux lines or vortices) in electronic devices, although it is known that (i) most superconducting devices will (or have to) operate in magnetic fields strong enough to create vortices within the device and that (ii) motion of vortices will lead to dissipative processes or flux noise. Thus, flux and flux motion will either corrupt the performance of superconducting device or, as will be shown in this article, it might be taken advantage of in applications. In this article I will try to give an overview on different aspects of vortex matter in superconducting devices.

Most examples refer to thin film devices and high-T_c superconductors, which are the most promising candidates for electronic devices. A brief introduction on vortices in superconductors and the relevance of their consideration in electronic application is provided. I will show, that in applications vortex motion should either be suppressed or can be utilized. In the first case *flux pinning* by material defects is necessary, whereas a controlled *guided motion* of the vortices or *controlled vortex creation* is necessary in the latter case. I will demonstrate, that artificial defects can be used in the cases of flux pinning and guided motion.

2. Vortex Matter in Patterned 2D Systems

2.1 Vortices and Vortex Penetration in Superconducting Thin Films

A superconductor can be characterized by the London penetration depth [5]

$$\lambda(T) = \sqrt{\frac{m}{\mu_o n_s(T) e^2}} \tag{1}$$

and the Ginzburg-Landau coherence length [6]

$$\xi(T) = \sqrt{\frac{\Phi_o}{2\pi B_{c2}(T)}} \tag{2}$$

Here n_s, e, and m denote the number density, charge and mass of the superconducting charge carriers, respectively, B_{c2} is the upper critical field and Φ_o is the flux quantum. Whereas, the penetration depth λ mainly depends on the number density of the superconducting charge carriers (Cooper pairs), the coherence length determines the upper critical field, up to which superconductivity persists. Thus, technical superconductors with high critical fields and high transition temperatures (due to the correlations $n_s(0)\Delta^2(0)=-B_c^2(0)/\mu_o$ and $\Delta(0)=1.76kT_c$ between the transition temperature, energy gap Δ, and thermodynamic critical field B_c, a high T_c implies a high critical field and vice visa) generally possess a small coherence length. For instance, $YBa_2Cu_3O_{7-\delta}$ (YBCO) with a transition temperature T_c = 92K is characterized by a coherence length of the size of the atomic distance ($\xi_c \approx 0.3$nm, $\xi_{ab} \approx 3$nm where the subscript denotes the orientation of the coherence length with respect to the crystallographic orientation).

As a consequence of the small coherence length, technical superconductors posses large Ginzburg-Landau parameters [6]

$$\kappa \equiv \frac{\lambda}{\xi} \tag{3}$$

It can be shown (see for instance [7, 8]), that $\kappa=1/\sqrt{2}$ separates two types of superconductors with different behaviors:

$\kappa<1/\sqrt{2}$ type-I superconductor and

$\kappa>1/\sqrt{2}$ type-II superconductor .

One of the major differences between both types of superconductors consists in the energy of the boundary between normal regions and superconducting regions. Type-I superconductors posses a positive normal-superconducting (NS) surface energy, whereas type-II superconductors are characterized by a negative NS surface energy. As a consequence, magnetic flux is expelled (Meissner-Ochsenfeld effect) from type-I superconductors and type-II superconductors at $B < B_{c1}$ (lower critical field), whereas for type-II superconductors at magnetic fields $B > B_{c1}$ magnetic flux penetrates into the superconductor. Furthermore, due to the negative NS boundary energy of type-II superconductors (at $B > B_{c1}$) the NS surface is maximized by the generation of separated flux lines (vortices) that contain the minimum magnetic flux, i.e. one flux quantum [9-11]

$$\Phi_o = \frac{h}{2e} = 2.068 \cdot 10^{-15} Wb \tag{4}$$

Neglecting the interaction between flux lines and material defects (flux pinning, see section 2.4) the mutual repulsive interaction between the vortices will lead to the formation of a flux-line lattice with lattice parameter [12]

$$a_o = b_o \sqrt{\frac{\Phi_o}{B}} \tag{5}$$

with b_o=1.075 for a hexagonal lattice with the lowest energy and b_o=1.0 for a square lattice with slightly larger energy.

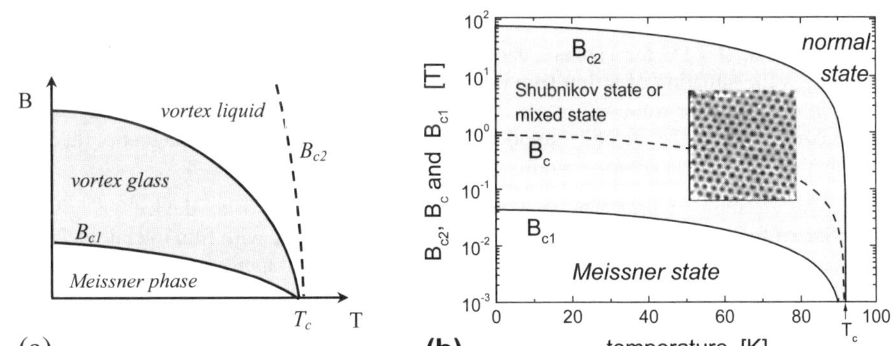

Fig1: (a) Sketch of the phase diagram of a HTS material. The different phases (vortex liquid, vortex glass) observed at high magnetic fields are subject of many research projects and are detailed in various reviews (see e.g. [13]). Electronic devices are usually exposed to small magnetic fields of the order of the earth field of about 50 μT. The situation at small fields is given in fig. (b) where the estimated critical fields B_{c1}, B_c and B_{c2} of YBCO thin films are given according to [14]. The data are derived from experimental data [15] for the field orientation B parallel to the crystallographic c-axis of the YBCO film and theoretical predictions for the temperature dependence of the critical field in highly anisotropic superconductors [16, 17]. The inset shows an image (600nm x 600nm) of the vortex lattice in weak-pinning $NbSe_2$ obtained by magnetic force microscopy at 1.8K and 1T [18].

The critical fields B_{c1} and B_{c2} that confine the so called Shubnikov state (mixed state) are correlated via the Ginzburg-Landau parameter for large κ:

$$B_c = \frac{\Phi_o}{\sqrt{8}\pi\lambda\xi}$$

$$B_{c1} = \frac{\Phi_o}{4\pi\lambda^2}(\ell n\kappa + \gamma) \cong \frac{B_c}{\sqrt{2}\kappa}(\ell n\kappa + \gamma) \tag{6}$$

$$B_{c2} = \frac{\Phi_o}{2\pi\xi^2} = \sqrt{2}\kappa B_c$$

with a small constant for which $\gamma = 0.08 - 0.12$ are given in the literature [7,8,19]. Fig. 1b represents an estimate of the temperature dependent critical field of YBCO thin films. It demonstrates that for large-κ material (i.e. technical superconductors) penetration of vortices is expected at very small magnetic fields. For instance at liquid nitrogen temperatures values of B_{c1}(77K)≈8mT are expected for typical YBCO thin films and B normal to the film surface.

Furthermore, the demagnetization effect will lead to a considerable enhancement of the magnetic field at the edge of a superconductor sample. In the Meissner state (i.e. complete field expulsion) the effective field at the edge of a superconductor according to the classical approach is given by [20]:

$$B_{eff} = \frac{B}{(1-D)} \tag{7}$$

with a demagnetizing factor D. This factor is $D = 0$ for a long rod parallel to the applied magnetic field B, $D = 1/3$ for a sphere, and $D = 1/2$ for a rod perpendicular to the field. $D \approx 1$ characterizes the situation of a thin film perpendicular to B. Thus, superconducting thin film devices in the Meissner state will experience a much larger magnetic field at the edge of the sample compared to the actually applied field or, alternatively, the magnetic flux will penetrate the device. The latter is usually the case.

Moreover, a more appropriate description of patterned thin film devices is given a description of the geometry sample by a rectangular cross-section with film thickness d much larger than the lateral dimension w (in contrast to the elliptic approximation of the cross-section used in the classical picture). In this case the first tunneling of single vortices (e.g. via thermal activated penetration) through the geometrical barrier at the edge of the patterned superconductor is expected at a field [21]:

$$B_T = B_{c1}\frac{d}{2w} \tag{8}$$

Collective penetration occurs at the field of vanishing geometrical barrier at

$$B_P = B_{c1}\sqrt{\frac{d}{w}} \tag{9}$$

The predictions eqs. (8) and (9) are valid for $w >> d > \lambda$.

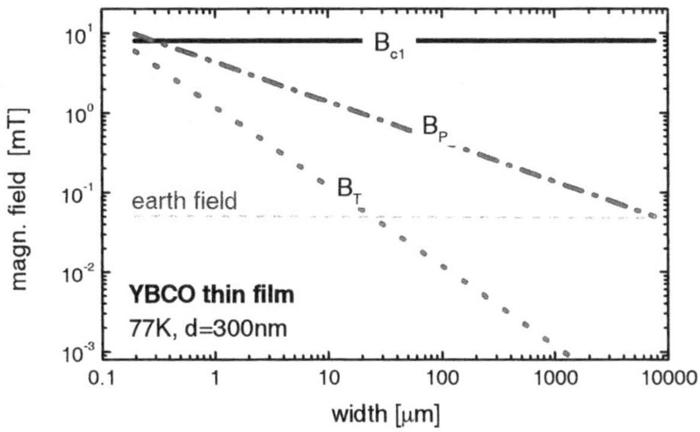

Fig. 2: Comparison of the expected lower critical field, tunneling field (eq.(8)) and penetration field (eq.(9)) as function of the width of a patterned YBCO film (d = 300nm, T = 77K). The field direction is normal to the film surface.

Fig. 2 provides a comparison of characteristic fields for patterned thin films of YBCO. The width dependence of B_{c1}, B_T and B_P is given for a typical film thickness of 300nm and a temperature of 77K. For extended structures, tunneling and penetration of vortices occur at fields far below the lower critical field B_{c1}. Due to this fact, small structures are used for instance for SQUIDs with extremely small *1/f* low-frequency noise performance (see discussion in chapter 3) [22]. However, for applications usually a high sensitivity and, thus, larger sensor areas are required. Typical SQUID-washer dimensions are of millimeter size resulting in B_T(77K) < 1μT, i.e. more than 50 times smaller than the earth field $B_{earth} \approx$ 50μT. Moreover, optimized rf designs for filters or striplines require a 50Ω impedance match. Thus, the use of 0.5-1mm thick $LaAlO_3$ or MgO substrates with a dielectric permittivity of $\varepsilon \approx 24$ or $\varepsilon \approx 10$, respectively, requires a line width in the range of 160μm up to 1mm for thin-film microwave devices [23]. Due to technical reasons often even larger superconductor geometries are utilized in rf applications yielding small values of B_T(77K) << 1μT.

In conclusion, flux penetration has to be expected in most superconducting electronic devices designed for real applications. These devices usually contain extended superconducting areas in the mm-range and have to function in unshielded or barely shielded environment. In some cases (e.g. non destructive testing (NDE) of metallic material) even larger fields are applied in order to monitor defects. As a result of the size, extremely small penetration field values B_T and B_P are expected, e.g. B_T(77K,w=1mm) = 1μT << B_{earth}.

2.2 Vortices in Superconducting Devices

In this section we will discuss some of the effects of vortices in superconducting devices. Usually the magnetic fields in which a device is operating will be quite small, i.e. comparable to or smaller than B_{earth}. As a result, the lattice parameter will be quite large. In case of YBCO at 77K this would result in a vortex-vortex distance $a_o(B_{earth})$ > 7μm ≈ 30λ(T=77K).

Therefore, in a first approach we can describe the effect of the vortices in a single-vortex picture, i.e. vortex-vortex interaction is neglected. Later we will see, that this is not correct and that even at very small fields vortex lattices can be favorable (see section 3.1.1). In this section, I will provide a brief overview on different impacts of vortices (due to vortex motion at moderate and high velocity and induced magnetic flux) on the properties of devices.

2.2.1 Vortex Motion at Moderate Velocity:

If a vortex in a perfect superconductor is exposed to a driving force it will start to move under the action of the force. The driving force can be provided an applied current (Lorentz force),

$$\vec{f}_{vortex} = \vec{J} \times \vec{\Phi}_o \quad , \tag{10}$$

a temperature gradient, or a magnetic field gradient. Furthermore, it can be assisted by a finite temperature (thermally activated or assisted vortex motion). Within a perfectly homogeneous system the driving force is counteracted only by a friction force:

$$\vec{f}_{vortex} = -\eta\vec{v} \quad , \tag{11}$$

with v denoting the steady-state velocity of the vortex. A consequence of the flux motion is a finite electrical field:

$$\vec{E} = n_v \vec{\Phi}_o \times \vec{v} = \vec{B} \times \vec{v} \quad , \tag{12}$$

in the superconductor. Here n_v represents the sheet density of vortices. A steady flow of vortices in a device will result in a finite flux-flow resistance:

$$\rho_{ff} = \frac{\partial E}{\partial J} = \frac{\Phi_o B}{\eta} \quad . \tag{13}$$

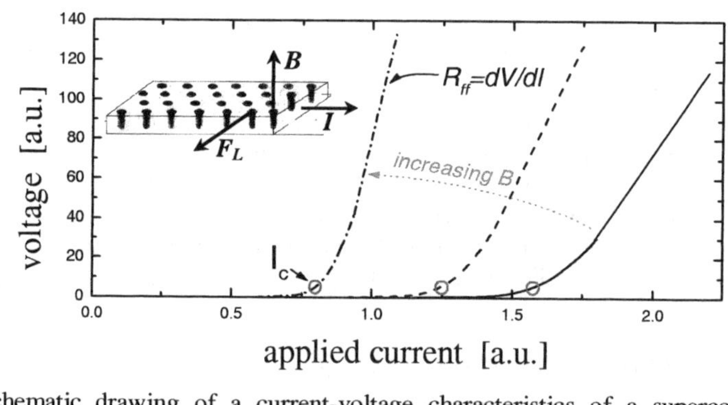

Fig. 3: Schematic drawing of a current-voltage characteristics of a superconductor for different applied fields. The critical current I_c (usually defined by a voltage criterion of 1μV/cm) and the flux-flow resistance are indicated. The inset shows the vortex lattice and the directions of current, applied field, and resulting Lorentz force for a superconducting thin film with B normal to the film surface.

On the other hand, statistic fluctuations (i.e. statistic motion of single vortices) would manifests itself by (flux or voltage) noise signals, e.g. telegraph noise in SQUIDs.

In order to retain the dissipation-free dc current flow or reduce the voltage noises due to vortex motion, the flux lines have to be pinned by defects. The pinning force of the defect compensates the driving force up to a critical value. In case of the Lorentz force this defines the maximum dissipation-free current density, i.e., the critical current density J_c given by:

$$\vec{F}_p = -\vec{F}_L = \vec{B} \times \vec{J}_c, \qquad (14)$$

where $F_L = n_v f_L$ and F_P represents the volume pinning force which is obtained via summation of the elementary pinning force f_P. The elementary vortex-pin interaction will be discussed later. An overview of different summation theories is given in ref. [14].

Two different classes of defects can be used, i.e. natural defects (such as vacancies, precipitates, dislocation loops, stacking faults, grain boundaries, etc.) or artificial defects. In the latter case typical candidates for thin film applications are irradiation defects or specially patterned defects like moats or channels [24,25] or small holes (so-called antidots) [26,27].

2.2.2 Vortex Motion at High Velocity

Energy dissipation due to vortex motion in the mixed state has been studied intensively during the past years. These studies focus predominantly on the limit of small driving forces, i.e., the onset of vortex dynamics in the limit of small vortex velocities. However, a number of applications like HTS wires, coated conductors, fault-current limiters or even microwave devices operate at high power levels and, therefore, potentially in the regime of high vortex velocities. For this regime, Larking and Ovchinikov (LO) [28] predicted that a nonequilibrium distribution of the normal charge carriers (treated as quasiparticles) develops during the motion of vortices, which is based upon the relaxation rate of the quasiparticles. As a result, they expected a discontinuity in the current-voltage characterictic (IVC).

According to the LO theory, the viscous damping coefficient at a vortex velocity v is [28]:

$$\eta(v) = \frac{\eta(0)}{1+\left(\frac{v}{v^*}\right)^2}, \qquad (15)$$

with a critical vortex-velocity v^* :

$$v^* = \left[\frac{D\cdot\sqrt{14\varsigma(3)}\cdot\sqrt{1-\frac{T}{T_c}}}{\pi\tau_{in}}\right]^{1/2}, \qquad (16)$$

τ_{in} denoting the inelastic quasiparticle scattering time, $\zeta(x)$ the Riemann-Zeta function, $D=v_F l_o/3$ the quasiparticle diffusion coefficient, v_F the Fermi velocity, and l_o the electron mean free path. According to LO, a nonlinear IVC is expected for a critical electric field $E^*=v^*B$ according to:

$$J=\frac{E}{\rho_{ff}}\left\{\frac{1}{1+\left(\frac{E}{E^*}\right)^2}+\gamma\cdot\left(1-\frac{T}{T_c}\right)^{1/2}\right\}+J_c \tag{17}$$

with $\gamma \approx 1$. This behavior is depicted in fig. 4a and is observed for conventional superconductors [29,30] as well as for HTS thin films [31-33]. Furthermore, it appears when the current limit is exceeded in superconducting fault-current limiters. In this case the IVC shows an extremely sharp and sudden voltage peak at the critical electric field which is passed during the quench of the superconductor at high power [34].

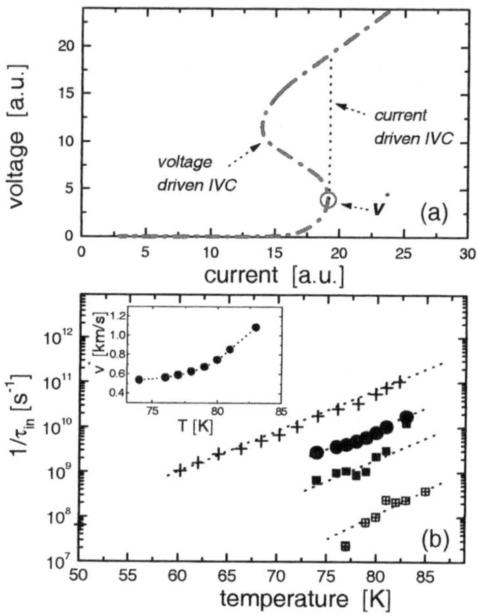

Fig. 4: (a) Schematic drawing of a voltage-controlled IVC and a current-controlled IVC up to voltages $V > V^*$. (b) Inelastic scattering rate and critical vortex velocity as function of temperature for different thin-film YBCO samples obtained from IVC measurements via rapid current ramps (crosses) [31] and current pulses (solid symbols) [33]. The inset shows the resulting critical velocity as function of the temperature for YBCO thin films [33].

2.2.3 *Vortex Motion in rf Field*

Large vortex velocities are also expected when the vortices are exposed to a driving force of higher frequency. Experimental results of the response of a vortex lattice in an applied rf electromagnetic field are given in fig. 5. The difference in the transmission coefficient S_{21} of a transmission line measured at –20dBm and –40dBm represents a measure for the losses due to vortex motion at given frequency. The plot of the maximum loss (minimum in ΔS_{21} in fig. 5b) shows a clear tendency. The losses decrease with increasing frequency. At about 5GHz no additional loss due to vortex motion is recorded in the system. This indicates, that vortex motion at higher frequencies seems not to be possible.

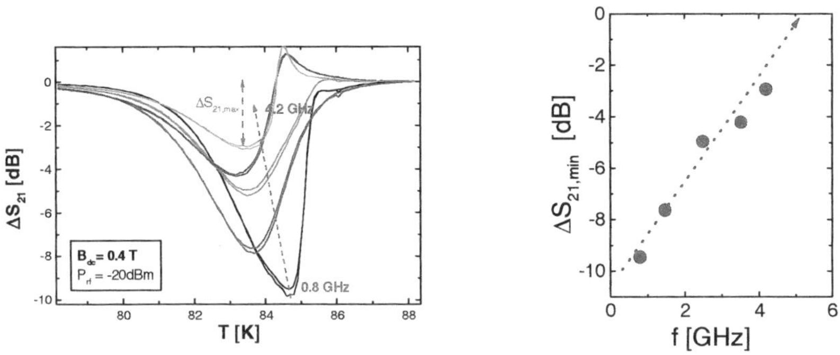

Fig. 5: (a) Change of the microwave losses $\Delta S_{21} = S_{21}(\text{-20dBm}) - S_{21}(\text{-40dBm})$ due to vortex motion (B=0.4T) for a YBCO microstrip line in a microwave field of different frequencies as function of temperature according to [35]. (b) Minimum value of ΔS_{21} at B=0.4T as function of applied frequency.

2.3 *Induced Magnetic Flux*

Next to the effect of vortex motion, the magnetic flux coupled into a superconducting device can affect the performance of the device. A typical example for a device, that can strongly be affected by the magnetic flux due to vortices in the device, is the SQUID. SQUIDs are the up to now most sensitive magnetic field sensors or flux converters. They consist of two components, i.e. the Josephson contact(s) for the read-out of the signal and the washer, which concentrates the signal into the sensor (for a complete description of the SQUID see for instance ref. [22]). In case of commercial SQUIDs the washer can be quite large. Fig. 6 represents a schematic drawing of a washer (omitting the slit) of a commercial rf-SQUID.

The magnetic flux $\phi_v(r)$ coupled by a vortex into a SQUID obviously depends upon the radial position r of the vortex. It can be calculated in analogy to the method of image charges in two-dimensional electrostatics [36] (see fig. 6). Two different expressions for $\phi_v(r)$ are given in the literature. In ref. [37] inner and outer conductor (charge analogy) are assumed to be grounded. This yields:

$$\phi_V(r) = \frac{\Phi_o}{\ln(a/c)}\left[\ln\left(\frac{r-a}{c-r}\right) + \sum_{i=1}^{\infty}(-1)^i \ln\left(\frac{a-a_i}{c-a_i}\right) + \sum_{i=1}^{\infty}(-1)^i \ln\left(\frac{a-c_i}{c-c_i}\right)\right] \qquad (18)$$

with a and c the inner and outer radius of the SQUID washer, and $a_i = \beta(a/c)^i$ and $c_i = \beta(c/a)^i$ with $\beta = r$ (i even) and $\beta = ca/r$ (i odd). In contrast, isolated conductors are assumed in ref. [38]. Using an inversion method [36] a slightly different solution is obtained:

$$\phi_V(r) = \Phi_o \frac{2\left(\arcsin\dfrac{a}{r} - \arcsin\dfrac{r}{c}\right)}{\pi\left(1 - \dfrac{4a}{\pi^2 c}\right)} \tag{19}$$

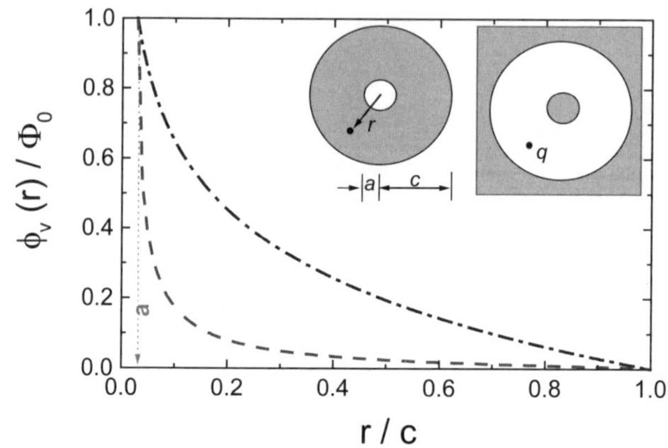

Fig. 6: Comparison of the induced flux as function of reduced radial position according to [39] obtained from eq. (18) (dashed-dotted line) [37] and eq. (19) (dashed line) [38] for SQUID dimensions a=50μm and c=1.75mm [40]. The simplified design used for the calculations is given in the inset. The left figure shows the SQUID with outer radius c and inner radius a and vortex at radial distance r. The right hand figure represents the electrostatic analogy, i.e. a charge in a cylindrical hollow conductor.

A comparison of the resulting coupling strength of the flux is plotted in fig. 6 for a SQUID design used in applications [40]. In both models, $\phi_v(r)$ increases strongly with decreasing distance between vortex and the SQUID hole. Depending on the underlying model a more continuous increase is observed [37] or the effect of flux induction is more or less restricted to vortex positions close to the inner radius of the SQUID [38]. In refs. [39-41] a good agreement with experimental results is obtained via eq. (18). In one of the cases the agreement has been ascribed to the rectangular SQUID geometry, which seems to be better described by eq. (18).

Generally, vortex motion will contribute stronger to the low frequency noise if the moving vortex (fluctuator) is positioned close to SQUID hole. As a consequence vortices close to the SQUID hole should be hindered from moving (e.g. by flux pinning or trapping). An example is described in section 3.1.2.

2.4 *Flux Pinning in Thin Film Devices*

In the previous section, it has been shown, that vortex motion leads to an electric field or changes of the induced magnetic flux in superconducting devices. Both effects can affect the performance of the device. In order to retain the desired device properties (e.g. dissipation-free dc current or low voltage noises due to vortex motion), the flux lines in devices have to be avoided, which is usually not possible, or they have to be pinned by defects. The pinning force of the defect compensates the driving force up to a critical values. In case of Lorentz force the maximum dissipation-free current density (i.e., critical current density J_c) is simply given by F_p=-F_L (eq. (14)). F_P represents the volume pinning force which is obtained via summation of the elementary pinning force f_P. An overview of the theory of flux pinning especially in HTS material is given in ref. [13], different summation theories are summarized in ref. [14].

Two classes of defects have to be distinguished, i.e. natural defects (such as vacancies, precipitates, dislocation loops, stacking faults, grain boundaries, etc.) or artificial defects. Natural defects are generally present in real superconducting material. For instance, YBCO thin films with natural defects can carry large critical currents J_c>10^{10}A/m^2 at 77K and zero magnetic field. It can be demonstrated, that oxygen vacancies in the CuO_2 double layer of only 3-7% of the YBCO unit cells are capable of such strong flux pinning [14].

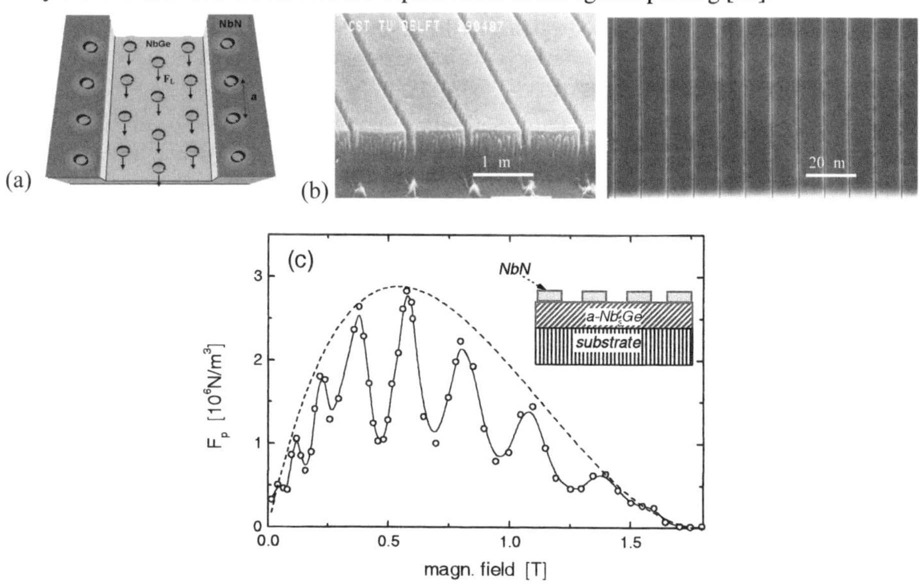

Fig. 7: (a) Schematic drawing of the vortex dynamic at a channel, (b) SEM images of different channel structures in NbN/$NbGe_3$, and (c) critical shear force density F_s=J_sB as function of the applied field for a NbN/a-$NbGe_3$ sample with channel patterned into the strong-pinning NbN (channel width w=290nm, orientation perperdicular to the applied current) on top of the weak-pinning amorphous $NbGe_3$ according to [24,50,51]. The dashed line represents the continuum result of the FLS model, eq.(21). The inset sketches the cross-section of the underlying structure.

However, for a controlled vortex pinning (e.g., controlled pinning of fluctuators in a SQUID) or guided vortex motion, artificial pinning sites should be chosen. In case of thin film applications typical candidates are irradiation defects [42,43] or specially patterned defects like grooves, moats or channels [24,25] or small holes, so called antidotes [26,27,44-49]. Due to the fact, that patterning is one of the essential preparation steps for devices anyway, the patterned defects are the obvious candidates for superconducting thin film devices. Furthermore, these defects can be placed arbitrarily (and, especially, strategically) in superconducting thin film devices.

2.4.1 Guided Vortex Motion via Artificial Channels

An obvious method to provide controlled vortex motion (guided vortex motion) is given by the pattering of narrow channel into superconducting material. Furthermore, it allows to study the microscopic mechanisms of vortex lattice shear and tearing. An intriguing model system of easy vortex flow channels was proposed and first fabricated more than a decade ago [24,25]. The samples consist of weak-pinning amorphous $NbGe_3$ bottom layer and a strong-pinning NbN top layer, into which the small (w=90-500nm) channels are etched (see fig.7). A detailed report of experiments results obtained from resistive measurements on these channels is given in ref. [24,25,52], simulation of the vortex motion in the channels is given in ref. [50]. In the following a brief introduction into the mechanism of flux-line shear is given, that describes vortex motion in this model system as well as vortex motion in most real (i.e., inhomogeneous) and strong-pinning superconductors [14].

The vortices in the channels predominantly experience the interaction with the row of pinned vortices at the edge of the channel within the NbN (see fig.7a). The commensurability between the periodicity of vortex lattice $a_o \propto (\phi_o/B)^{1/2}$ (eq.(5)) can be modified by variation of the applied field B. This can be seen in form of peaks in the field dependence of the critical current or volume pinning force (see fig.7b). A similar situation is present in most real superconductors, where channels for flux flow are defined by a percolative path of weak pinning (e.g. weak links, grain boundaries) across the sample. In this case obviously the matching peak in the critical properties should not be expected since the channel width, channel orientation and local pinning strength will not be uniform over the sample.

In 1973 Kramer first suggested a mechanism of flux motion characterized by synchronous shear of the vortices around flux-lines, which are too strongly pinned to be broken [53]. The resulting flux-line shear (FLS) model was further developed in ref. [25] owing to new insight into the physics of flux pinning obtained from 2D collective flux pinning [54]. In order to describe the FLS mechanism in real (i.e., inhomogeneous) systems further modifications have been suggested [14,55,56] that take into account variations of the superconducting and morphological properties of the sample and effects of thermal activation or assistance of vortex motion.

In the FLS model the vortex lattice is considered on a macroscopic level (continuum). Vortex flow in the channels starts when the shear force density per channel $wF_L = JBw$ (Lorentz force with current perpendicular to the applied field) exceeds the plastic shear stress at the channel edges $2\tau_{max}$. Here w represents the (effective) width of the channel, $\tau_{max} = Ac_{66}$ with c_{66} the shear modulus. For large κ values the isotropic shear modulus of the vortex lattice is given by [57-61]:

$$c_{66} \approx \frac{\Phi_o B_{c2}}{16\pi\mu_o\lambda^2} b(1-0.58b+0.29b^2)(1-b)^2 \tag{20}$$

with the reduced field $b=B/B_{c2}$. The factor A is of microscopic origin. It depends on the vortex-lattice orientation, the (an)harmonicity of the shear potential, the detailed vortex structure at the channel edge and the microstructure inside the channels [50]. In case of the clearly defined channels (edge and width), the effect of the change of the vortex lattice microstructure oscillations upon the factor A is reflected by oscillations of the critical current (see fig.7). In case of a real (i.e. inhomogeneous) superconductor the channels are less well defined and the factor A=0.047 is a constant [54]. The resulting critical shear forces for systems with artificial and random channels, respectively, is given by:

$$\begin{aligned} F_{S,random} &= J_S B = G\frac{2Ac_{66}}{w} \cong \frac{\Phi_o B_{c2}}{64\pi\mu_o\lambda^2 w} b(1-b)^2 \\ F_{S,channels} &= J_S B = G\frac{2A(B)c_{66}}{w} \cong \frac{A(B)G\Phi_o B_{c2}}{8\pi\mu_o\lambda^2 w} b(1-b)^2 \end{aligned} \tag{21}$$

with the critical shear current density J_s and the geometrical factor G that accounts for the orientation of the channels with respect to the Lorentz force, which is about $G \approx 2.5$ for a statistical distribution of channel orientations [24]. The resulting field dependence of the volume pinning force $F_{s,random}$ is characterized by a broad maximum at $b=1/3$, that is observed for a large number of strong pinning superconductors. The field dependence is given in fig. 7. It coincides with the maxima of $F_{s,channels}$ observed for the artificial channel system (see fig. 7c).

More complex artificial NbN/a-$NbGe_3$ channel patterns (e.g. honeycomb or brick-wall patterns) have been fabricated [24]. It could be demonstrated for theses patterns that vortices move within the channels if the total driving force on the vortices in the channel exceeds the shear forces at the channel edges. To my knowledge artificial channels for vortex motion have not been fabricated in HTS thin films. However, a similar mechanism of vortex dynamic is expected for vortex motion along grain boundaries.

2.4.2 Vortex pinning and guided vortex motion via antidots and arrays of antidots

One of the most effective ways to create artificial pinning sites in thin film devices is provided by the preparation of (networks of) small (submicrometer) holes, so called antidots. These defects (i) can be prepared simultaneously with the patterning of the device, (ii) can be placed arbitrarily in superconducting thin film devices, and (iii), in contrast to other pinning defects, which have to be of the size of the superconducting coherence length ξ (core interaction [62]), holes with sizes much larger than ξ will trap magnetic flux very effectively [27]. For a periodic array of holes (antidots) with radius r much smaller than the period d of the antidot array, a well-defined pinning potential is formed. The advances in lithography techniques and the possible use of antidots in applications (e.g. SQUIDs [40, 41]) have lead to a renewed interest in the research of superconducting films containing antidots or antidot lattices. Antidots are successfully prepared in films of conventional superconductors (typically weak pinning Pb, V, or a-WGe thin films, Pb/Ge multilayers, Pb/Cu bilayers or Nb foils) [26,44-48,63-69] as well as HTS material (YBCO) [27,40,41,49,70]. Intensive studies have been performed of (i) the occurrence of commensurability effects (matching effects) between the antidot lattice and vortex lattice, (ii) multiquanta formation in the antidots, (iii) guided motion along rows of antidots, (iv) visualization (e.g. via Hall-probe measurements, magnetooptical imaging or Lorentz microscopy) or (v) in connection with a potential application of antidots (e.g. for the noise reduction in SQUIDs or the improvement of the microwave properties of HTS resonators and filters). An overview of these activities is for instance given in the conference reports [71,72].

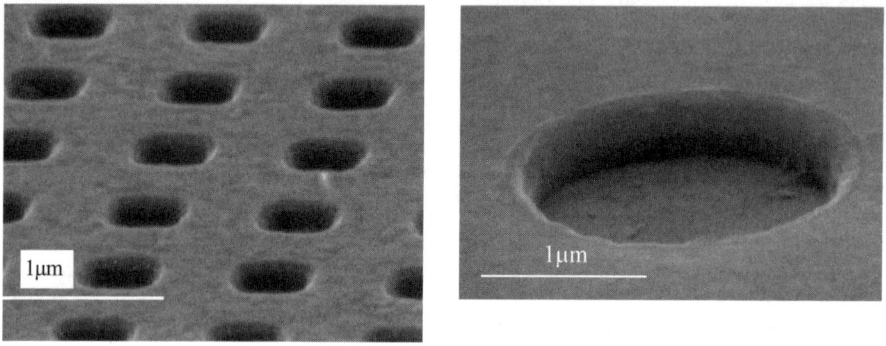

Fig. 8: SEM images of (a) an antidot lattice ($d \approx 1\mu$m and radius $r_o \approx 0.2\ \mu$m) and (b) an antidot ($r_o \approx 1\mu$m) in YBCO films on sapphire [27, 49].

A. *Vortex-Antidot Interaction*

The interaction energy between a vortex and a small insulating cylindrical cavity ('antidot') in a type-II superconductor has been calculated using the London approximation with κ>>1 by Mkrtchyan and Shmidt [73]. An alternative approach (using the analogy between a vortex close to an antidot and a charge line in an infinite dielectric close to a cylindrical cavity of different dielectric permittivity) is given in ref [74]. Recently, the calculations by Mkrtchyan and Shmidt have been extended to arbitrarily large cavities [75]. It is demonstrated that the interaction energy becomes the same as the one between a vortex and the straight edge of a superconductor (Bean-Livingston barrier [76]) when the radius of the antidot goes to infinity. Although the precise form of the interaction potential between a vortex and a cylindrical antidot is slightly different in the three studies [73-75], the main conclusions are identical.

Based upon a description of the free energy F in form of a series expansion of Bessel functions K_o of the second kind, the free energy of a vortex at a radial distance r of an antidot with radius r_o is given by [73,74]:

$$F(r)=\frac{\Phi_o}{4\pi\mu_o\lambda^2}\left[K_o\left(\frac{\xi}{\lambda}\right)+n^2K_o\left(\frac{r_o}{\lambda}\right)+2nK_o\left(\frac{r}{\lambda}\right)+\ln\left(1-\frac{r_o^2}{r^2}\right)\right] \quad (22)$$

for $\xi < (r, r_o) << \lambda$ and n representing the number of flux quanta that are already trapped in the antidot. The elementary interaction between a vortex and an antidot is given by $f_{VA} = -\partial F/\partial r$.

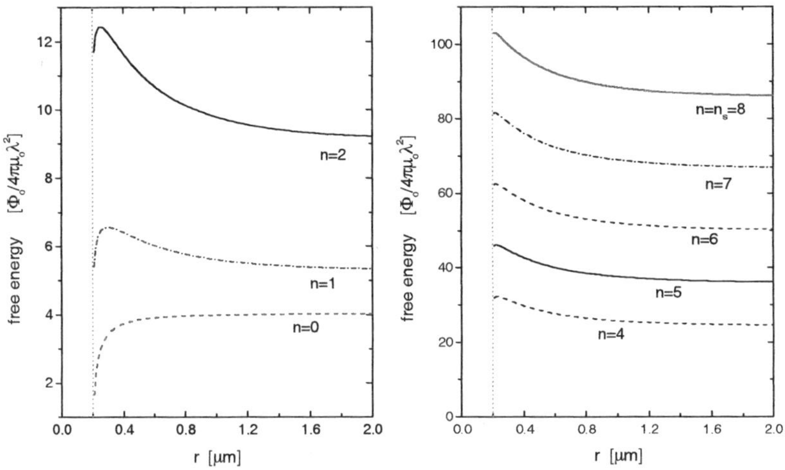

Fig. 9: Elementary vortex-antidot interaction as function of radial distance for different occupation numbers n of the antidot and r_o =200nm, λ=600nm and κ=50.

The vortex-antidot interaction depends upon the magnetic flux $\Phi = n\Phi_o$ trapped in the antidot. The interaction between a vortex and an 'empty' antidot is attractive for all vortex-antidot distances. As soon as one vortex is trapped in the antidot, the interaction potential changes and a surface barrier emerges at the edge of the antidot. The height of the barrier increases with increasing number of trapped vortices. This automatically defines a saturation number n_s, i.e. for $n=n_s$ the interaction becomes repulsive, no additional flux can be trapped by the antidot. The saturation number depends upon the size of the antidot. For small antidots (single antidot with $r_o << \lambda$) the saturation number can be approximated by [73]

$$n_s \cong \frac{r_o}{2\xi(T)} \quad , \tag{23}$$

For a quadratic and hexagonal arrays (periodicity d_{AA}) of extended antidots ($r_o \geq \lambda$) it is given by [27]:

$$n_{s,square} = \frac{B_{c1} d_{AA}^2}{\Phi_o} \sqrt{\frac{d}{\sqrt{2} d_{AA} - 2r_o}} \quad ,$$
$$n_{s,hex} = \frac{B_{c1} d_{AA}^2}{\Phi_o} \sqrt{\frac{d}{d_{AA} - 2r_o}} \tag{24}$$

Here d represents the thickness of the film. In both cases (eqs. (23) and (24)) the saturation number depends upon temperature, i.e. it increases with decreasing temperature. Furthermore, a consequence of a saturation number $n_s > 1$ is the formation of so called *multi-quanta vortices*. Experimental proofs of the existence of multi-quanta vortices will be given below.

The pinning force per unit length depends on the radial distance between vortex and antidot and has a maximum value close to the antidot edge given by

$$f_{p,\max} \approx \frac{\Phi_0^2}{4\sqrt{2}\pi\mu_o\lambda^2\xi}\left(1-\frac{n}{n_s}\right) \quad , \tag{25}$$

i.e., the maximum vortex-antidot interaction decreases with increasing occupation number and $f_{p,max}=0$ for $n=n_s$.

We will see later, that in devices 'strategically positioned' antidots are of importance. However, for most fundamental experiments (e.g. analysis of commensurability effects, formation of multi-quanta or interstitial vortices) regular array of antidots are used. The 2D potential experienced by a vortex moving in a superconductor with a periodic antidot array, has been calculated by direct summation of the elementary interactions between the vortex and the antidot (without and with flux $n\Phi_o$). Fig. 10 shows a comparison of the calculated potential for a square array of antidots with occupation $n = 0$ and $n = n_s$. For $n = 0$ all incoming vortices are attracted towards the 'empty' antidots. As soon as flux is trapped in the antidots, a potential barrier develops at the edge of the antidots. This barrier forms additional shallow local minima between the antidot positions. Thus, new stable positions between the antidots are formed, which are occupied by the so called *interstitial vortices*. Since these interstitial vortices experience a different (more shallow) potential compared to the vortices in the antidot, they usually posses a much higher mobility compared to the vortices trapped in antidots. Experimentally the existence of interstitial vortices has been demonstrated by Lorentz microscopy [66], Bitter-decoration images [67], or resistive measurements [64].

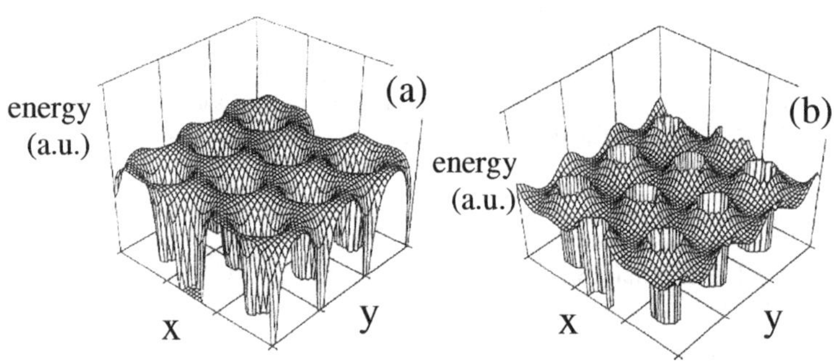

Fig. 10: Free energy of a vortex in a triangular array of antidots ($\kappa = 50$, $d_{AA} = \lambda$, $r_o = 200$nm) for (a) $n = 0$ and (b) $n = n_s$ according to [64].

B. Demonstration of the Attractive Vortex-Antidot Interaction:

The vortex-antidot interaction has been demonstrated among others via microscopy (scanning hall experiments, magneto-optic imaging or Lorentz microscopy) or analysis of commensurability effects for instance in magnetization, resistive or inductive measurements. Here, two examples will be given, i.e. Lorentz microscopy of vortices in Nb foils with an antidot lattice and resistive measurements of the critical current density in HTS films with an antidot lattice.

Ideal proofs of attractive vortex-antidot interaction and the formation of interstitials are given by vortex visualization for instance via Lorentz microscopy. Fig. 11 shows two examples of the vortex distribution in a Nb foil with a square lattice of antidots. Since the vortex density can be changed by changing the applied field according to eq. (5), commensurability (matching) between the antidot lattice and vortex lattice is obtained for

$$B_n = nB_1 = \frac{n\Phi_o}{d_{AA}^2} \quad . \tag{26}$$

Fig. 11a shows the vortex distribution for the first matching field B_1. All vortices are trapped in the antidots. Fig. 11b shows the distribution for the second matching field. Since the saturation number $n_s = 1$ for this temperature (close to T_c), the additional vortices occupy interstitial positions for $B=B_2$. Further arrangements are given in ref. [66].

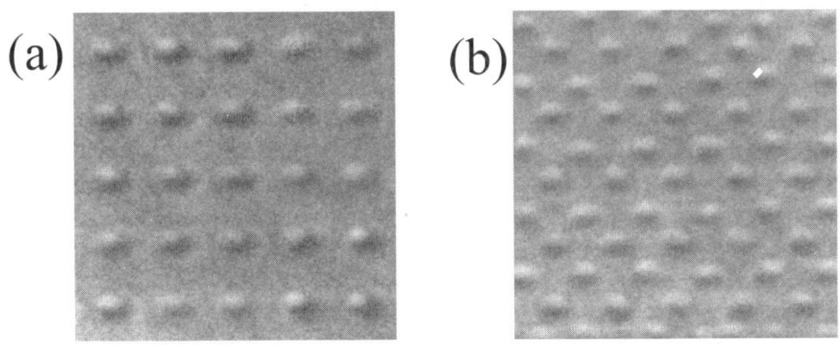

Fig. 11: Vortex positions observed via Lorentz micrograph of a 3 x 3 μm^2 area of a Nb foil with a square lattice of antidots at (a) the first matching field and (b) the second matching field according to [66]. In (b) a composite vortex lattice is formed, vortices are trapped in the antidots and vortices are located at the interstitial positions.

Most of the experiments on flux pinning by antidots has been performed for weak pinning superconducting systems. However, for applications strong-pinning and, especially, high-T_c material is of interest. First reports of a successful preparation and a clear indication of a attractive interaction between antidots and vortices in HTS are given in ref. [27]. Based upon this work antidots in HTS thin films have been successfully been introduced into HTS devices. For the antidot preparation YBCO films have been deposited via a high-pressure on-axis magnetron-sputter technique on CeO_2 buffered sapphire or $LaAlO_3$. High structural perfection and, especially, extremely small surface roughness are required for an optimized patterning process for micrometer and, especially, for submicrometer structures in YBCO films. A typical image of an antidot and an antidot array in YBCO films is given in fig.8.

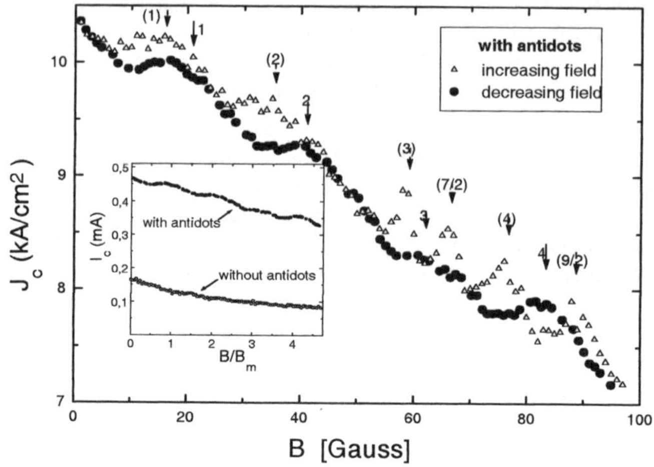

Fig. 12: Critical current density and critical current (inset) as a function of magnetic induction for YBCO with ($r_o \approx 0.22\mu$m, $d \approx 1\mu$m) and without antidots according to [27]. Both samples are obtained from the same film and posses identical dimensions.

Fig. 12 represents a comparison the magnetic field dependence of critical current density J_c and critical current I_c as a function of the magnetic field for a YBCO thin films with and without antidots. A reference YBCO bridge (patterned simultaneously into the same film but without antidots) shows the usually observed monotonic field dependence of J_c. In contrast, the perforated YBCO bridge shows peaks or cusps in the magnetic field dependence of the critical current at the matching fields $B_n = nB_1 = n \times 20.7$G. Matching peaks could be resolved at n=1, 2, 3, 4, but also at rational positions n=7/2 and 9/2 (in conventional thin film material matching peaks were also observed at $1/8 \times B_1$, $¼ \times B_1$, $½ \times B_1$, $3/2 \times B_1$, and $5/2 \times B_1$ [26,46]). The shifts of the position of these peaks for increasing and decreasing magnetic field has to be ascribed to the field gradient in the sample, which is more developed for strong-pinning material [27]. Finally, the increase of the volume pinning force of the sample with antidots with respect the sample without antidots (inset of fig.12) demonstrates the additional flux pinning provided by the antidots. The experiments are executed at temperatures close to T_c. The matching effects (cusps in $J_c(B)$) could be resolved only down to temperatures of 50K. At smaller temperatures the natural pinning dominates the vortex-antidot interaction.

C. Multi-Quanta Vortices

One of the interesting aspects of vortex pinning by antidots is given by the formation of multi-quanta vortices. Already in the first theoretical considerations, Mkrtchyan and Shmidt [73] considered a flux occupation of the antidot of $\Phi = n\Phi_o$ with $1 \leq n \leq n_s$. Due to their specific character, antidots trap quantized flux and not Abrikosov vortices with a normal core. However, by producing so called blind holes (not completely etched holes, see sketch in fig. 13), the vortex character is preserved and the formation of multi-flux-quanta trapping by antidots is visible in form of flux splitting into single-flux-quanta vortices within each hole. This phenomenon has been demonstrated via Bitter-decoration experiments on Nb films with a regular (hexagonal) array of antidots [67]. The multi-quanta formation is clearly visible in these experiments (fig. 13). The saturation number for the experiment shown in fig. 13 seems to be $n_s = 2$, i.e. 2 vortices per antidot. By applying a magnetic field $B_2 < B < B_3$ even vortices at interstitial positions in between the antidots are visible.

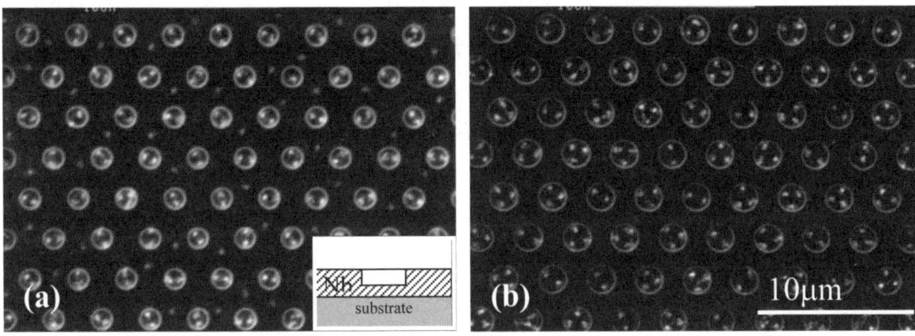

Fig. 13: Bitter-decoration images of Nb films with a hexagonal arrays (d=3.16 μm) of blind holes decorated simultaneously at a field of H=6.38Oe [67]. In contrast to antidots the blind holes are not completely etched (see inset in (a)), thus leaving a small superconducting bottom layer (here a 65nm thick Nb layer is left over), that causes the flux separation into single-quanta vortices within the blind holes. The only difference between the samples is given by the radius of the blind holes, i.e. : r_o=0.6μm (a) and r_o=0.83 μm (b). The small increase of the radius leads to a change of the saturation number from $n_s = 2$ (2 vortices per hole and an additional interstitial vortex per antidot in (a)) to $n_s = 3$ (3 vortices per hole and no interstitial vortices in (b)).

A proof of the existence of multi-quanta vortices in antidots in HTS thin films is given by inductive measurements. These experiments have been performed on YBCO films with regular square lattices of antidots [Woe00]. The experimental technique is based on a lock-in detection of the induced signal produced in a secondary coil due to a small ac field (oriented parallel to the c-axis of the film) in the primary coil (see inset fig. 14a). The in-phase and out-of-phase components of the fundamental of the pick-up coils voltage are proportional to the out-of-phase and in-phase components of the complex susceptibility $\chi(T,B) = \chi'(T,B) + i\chi''(T,B)$. The components represent the stored ($W_m = \chi' B_a^2/2\mu_0$) and dissipated ($W_q - 2\pi\chi''$) energy, respectively. By varying an additional dc magnetic field component B (oriented parallel to the c-axis of the film) the dissipated energy due to vortex motion in the film can be recorded as a function of temperature and field.

Fig.14 shows the variation of χ'' as a function of the bias field at temperatures close to T_c (high mobility of the vortices) for YBCO thin films without antidots and with arrays of antidots of different sizes. For the measurements on the reference films without antidots, χ'' increases monotonically (except for the fields below the magnet's remnant field of about 5G ≈ $0.2B_1$). In contrast, the out-of-phase component for the film with small antidots is constant up to the matching field B_1 = 20.7 G. Above B_1 the data curves rise with a slope similar to that observed for the reference sample. Small curvatures are observed at the matching fields $2B_1$ and $3B_1$. The inductive measurements shown in fig. 14a confirm the presence of matching effects and indicate a saturation number n_s = 1 for the antidot configuration at the examined temperature. Multi-quanta vortices are expected at lower temperatures. Unfortunately a high mobility of the vortices is needed for this experiment, thus, experiments at lower temperatures are not possible. However, by increasing the size of the antidots the presence of multi-quanta vortices can be demonstrated. Fig. 14b shows results, that indicate a temperature dependent saturation number, i.e. n_s changes from 2 to 3 for a small temperature reduction for larger antidots (r_o≈1.75μm).

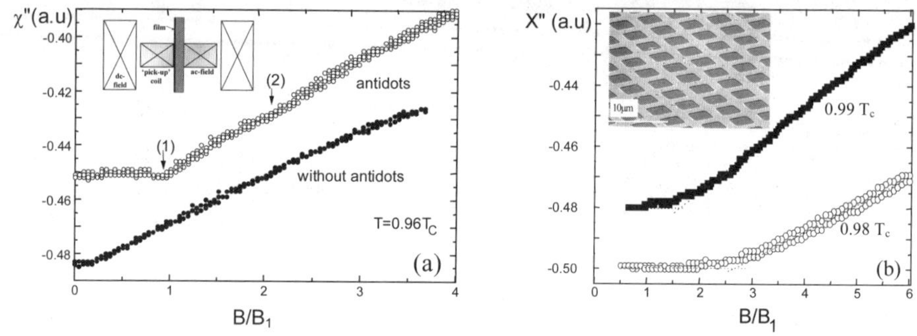

Fig.14: Out-of-phase susceptibility χ'' as a function of dc bias field for YBCO films without and with square lattice of antidots of different sizes according to [49]: (a) comparison of the signals obtained for a reference film (without antidots) and a film with an antidot lattice (d_{AA}=1μm, r_o≈0.34μm) at T/T_c=0.96 (n_s=1) (the inset displays principle of the measurement) and (b) signals for a film with an antidot lattice (d_{AA}=5μm, r_o≈1.75μm) at T/T_c=0.99 (n_s≈2) and T/T_c=0.98 (n_s≈3). The inserts shows (a) a sketch of the principal of the measurement (a) and a SEM picture of the square lattice of the antidots.

Finally, vortex dynamics (i.e. guided vortex motion, see next section) and multi-quanta formation can be visualized via high-resolution magneto-optic (MO) technique [77,78]. The MO set-up produces a two-dimensional image of the local magnetic field, where high fields regions or low-field regions are imaged as brighter or dark regions, respectively.

Fig. 15 shows a 3D picture constructed from the image of trapped flux in a hexagonal antidot lattice in a YBCO film, which was recorded at T = 7K and zero magnetic field after a field sweep up to B = 1T. The scale for the height of the peaks represents the amount of trapped flux quanta. No flux (vortices) could be detected in between the antidots, which indicates the higher mobility (small pinning potential) of the interstitial vortices compared to flux in the antidots. Multi-quanta vortices with flux up to Φ = 280Φ_o are formed in the antidots and are trapped at low temperatures after removal of the field. Using the above given expressions for the saturation number, a value of $n_s \approx 280$ would yield either $\xi(7K) \approx 1.8nm$ for eq. (23) [73] or $B_{c1}(7K) \approx 41mT$ for eq.(24) [27]. Both values are reasonable for YBCO and the orientation of the magnetic field ($B \parallel$ c-axis of YBCO).

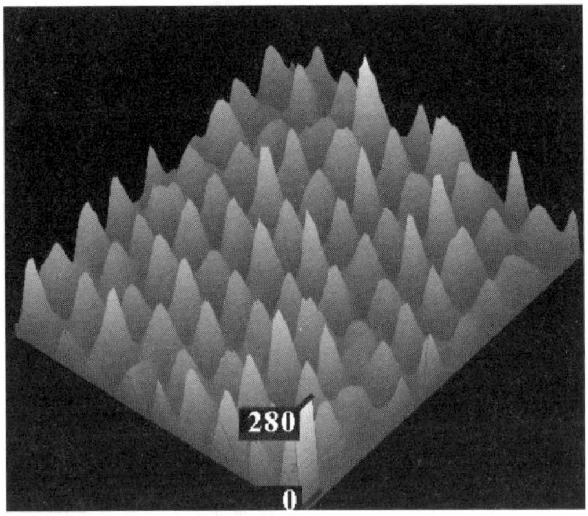

Fig.15. Distribution of the trapped magnetic flux obtained from a MO image that was recorded at 7 K in zero-field after applying at field of 1T according to [77]. The picture is constructed from the field image, the shape of the peaks indicates the local field profile in the sample. The local field shows clear maxima at the position of the antidots (triangular lattice, $d_{AA} = 10\mu m$, $r_o \approx 1\mu m$). The scale for the height of the peaks is given in number of trapped vortices.

D. Guided Vortex Motion via Rows of Antidots

Guided motion of vortices might be interesting for a number of potential applications of HTS thin films (flux transformer, flux pump, or other flux-flow devices). Guided motion has been demonstrated in conventional weak pinning superconductors with channels (see section 2.4.1). The preparation of channels proves to be difficult in HTS material. However, it has been shown, that vortices preferentially move along rows of antidots in HTS thin films [77,78].

Fig. 16 represents a magnetic-field map of a disc-shaped YBCO thin film after cooling in zero field to 4.2K and successive increase of the field up to 30mT. There are two important features: (i) The flux penetration pattern shows a very clear 4-fold (D_4) symmetry, i.e. the penetration is highly anisotropic, compared to the cylindrical symmetry that is observed for non-perforated disc samples [98]. (ii) The flux penetrates along channels that have the same orientation as the antidot lattice vectors. In fact, the flux penetrates along the lattice vector that is closest to the normal to the edge of the sample (direction of the Lorentz force due to the screening current). The measurements demonstrate the effect of guidance of the vortices by adjacent antidots. A simple model for this effect is developed in ref. [78].

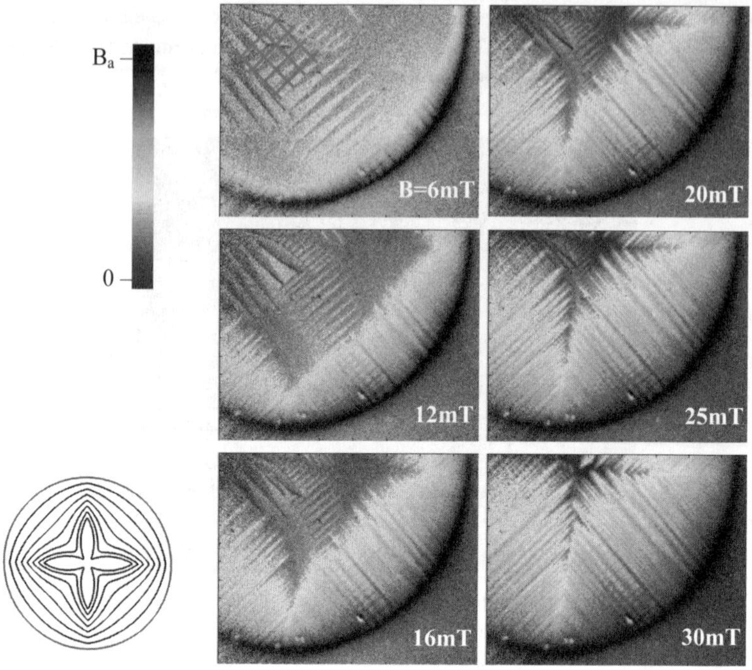

Fig.16: MO visualization of the anisotropic (guided) flux penetration in a disk-shaped YBCO film with a square lattice of antidot ($d_{AA} = 10\mu$m, $r_o \approx 1\mu$m) according to ref. [78]. To enhance spatial resolution only one quarter of the disk is shown. The images are obtained at 4.2K after zero field cooling, the field is gradually increased to B_a = 6mT, 12mT, 16mT, 20mT, 25mT and 30mT, respectively. The scale indicates the magnitude of the local field, the orientation of the antidot lattice is indicated in the image observed for 6mT. Furthermore, the schematic drawing shows the flux front expected for the motion preferential from antidot to antidot (guided motion) [78].

3. Vortex Matter in Cryoelectronic Applications

In the previous sections it has been shown, that (i) vortex penetration into superconducting electronic thin film devices working in little or unshielded environment has to be taken into account and (ii) that vortex motion will lead to dissipative processes due to the presence or motion of vortices. Vortex motion in active devices (e.g. SQUIDs, flux quantum logic, etc.) leads among others to increased low-frequency noise, reduced sensitivity or increased bit error rate [22,79]. In microwave devices (e.g. high-Q resonators, filters, antennas) vortices cause a reduction of the quality factor and power handling capability [80].

In this chapter two examples will be given that illustrate the role that vortices play in superconducting devices, i.e., one representative of active devices and one representative of passive microwave devices:

(a) First, an efficient method for the reduction of low-frequency noise in SQUIDs (active devices) will be described in which pinning of vortices by artificial defects are employed. It will be demonstrated, that (i) vortex motion is responsible for the low-frequency noise in SQUIDs, (ii) that the vortex motion and, thus, low-frequency noise can be affected by means of flux pinning, and (iii) that an effective and field-independent noise reduction can be achieved by means of 'strategically positioned' (thus, artificial) defects. Furthermore, it is shown as a side effect, that SQUIDs can be used as very sensitive tools the analysis of vortex matter in the superconducting thin films.

(b) Second, coplanar microwave resonators are examined in small magnetic fields. It is shown, that (i) the power handling capability of these devices is strongly suppressed by vortices in the device and that (ii) a careful analysis of the local distribution of vortices in the devices can be used for the understanding and optimization of the microwave properties of these devices.

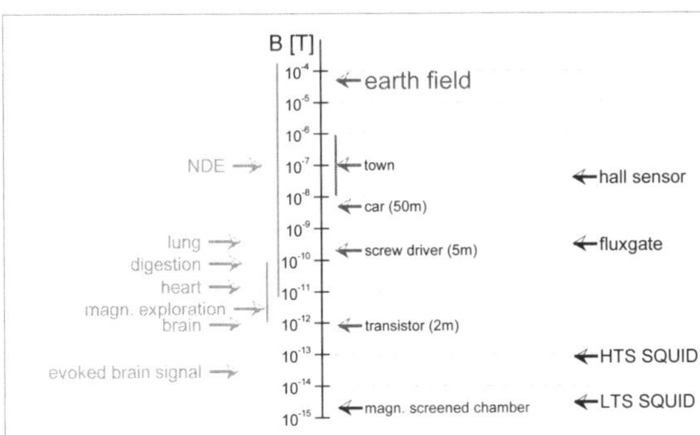

Fig.17: Comparison of the magnetic-field sensitivity of conventional and superconducting field sensors, level of magnetic-field noise, and field sensitivity required for various SQUID applications (i.e. nondestructive testing (NDE), geomagnetic exploration, medical applications).

3.1 Low-frequency noise in SQUID

Superconducting quantum interference devices (SQUIDs) represent the most sensitive sensors of magnetic flux or small physical quantities, that can be transformed into magnetic flux. They are of special interest in applications where extremely small magnetic fields have to be detected, e.g. non destructive testing (NDE), geomagnetic applications and, especially, medical or biomagnetic applications. For an overview on the working principle and the application of SQUIDs, the reader should refer to refs. [22,79]. Here only a brief introduction into the concept of SQUIDs and the problem of vortex motion in SQUIDs is given.

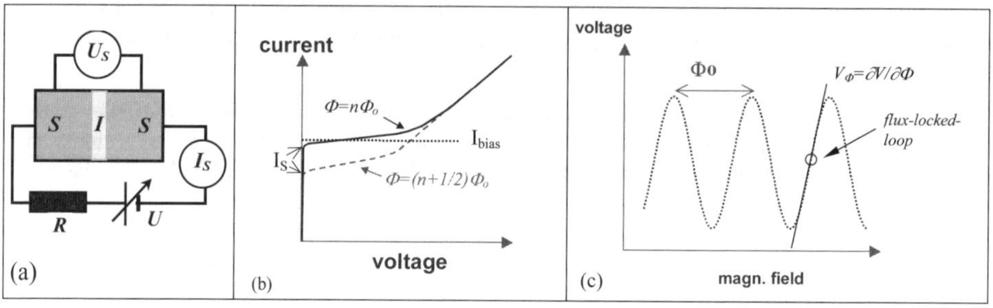

Fig.18: Schematic drawing of the working principle of SQUIDs: The nonlinear IVC (b) of a Josephson junction (a) depends strongly on the magnetic flux. The superconducting tunnel current I_s changes periodically with the flux that threats the junction (eq.(27)). This periodical change is recorded in the SQUID for instance by applying an adequate bias current (c). The sensitivity of the SQUID can be enhanced by using a feedback electronic (FLL) to fractions of the magnetic flux quantum.

The SQUID is a deceptively simple device, consisting of (i) a small superconducting loop with (ii) one (rf-SQUID) or two (dc-SQUID) damped (resistively shunted) Josephson junctions, each characterized by a critical current I_c. Ad (i): magnetic flux threading the superconducting loop is obviously always quantized in units of Φ_o. Thus, the total flux in the SQUID consists of the measured flux Φ plus the flux Φ_{ind} that is induced by current in the loop in order to compensate according to: $\Phi_{tot} = n\Phi_o = \Phi + \Phi_{ind}$. Ad (ii): the phase difference $\Delta\varphi$ across the Josephson junction is correlated with the superconducting tunnel current I_s according to the 1. Josephson equation [4]:

$$I_s = I_c \cdot \sin(\Delta\varphi) = I_c \cdot \sin\left(\frac{\pi\Phi}{\Phi_o}\right) \quad . \qquad (27)$$

The SQUID combines these effects. By reading out the tunnel current (or related quantities), the magnetic flux is given in units of Φ_o. E.g., at a constant bias current the voltage across the Josephson junction is periodically modified in a dc SQUID (see fig. 18) and the flux quanta in the SQUID loop can be counted. Moreover, in conjunction with suitable feedback electronics bucking the measured flux to maintain a stable operating point, the device can be used as a linear null detector. In this so called flux-lock-loop (FLL) mode it becomes the most sensitive fluxmeter known, with a capability of resolving better than 10^{-6} Φ_o. The ranges of measured signal frequencies and intensity are wide and principally only limited by the electronic used. Typical room temperature electronics offer a frequency range extending from dc to kHz, and a dynamic range of 120dB or more [79].

In order to couple sufficient magnetic flux into a SQUID, either flux transformers are added or the superconducting loop (washer) is extended in order to provide sufficient flux focusing. Without going into details (see e.g. [22,79]), in both cases the inductance of the different components plays an important role and determines the final design of the device. In the first case, the large inductance of the transformer has to be matched to the SQUID inductance, i.e., large transformers with large inductance require SQUIDs with comparable inductance (i.e., SQUID of larger size). In the second case, the inductance of the extended washer has to be minimized in order to minimize the SQUID noise. The inductance of the SQUID is dominated by the diameter of the SQUID hole [81]. In both cases the resulting SQUID possesses a large outer dimension (e.g., large focusing effect) and a small hole (small inductance).

In conclusions, SQUIDs for technical applications generally posses designs with rather extended washers. Thus, vortex penetration into the washer is very likely (see section 2.1) and, furthermore, vortex motion within the washer plays a major role for sensitivity of the device at low-frequencies.

The sensitivity of SQUIDs is limited by the frequency dependent noise level of the device. Additional to the contribution of the electronics, in superconducting active devices two different sources are considered to be responsible for the noise. These are the contribution of the active part of the device, which usually consists of Josephson junctions, and the noise of the passive component, the superconducting thin film. The noise mechanisms in Josephson junctions are well understood [82] and a reduction of this noise contribution by simple electronic means has successfully been demonstrated [83]. The contributions of the superconducting thin film (mainly washer or flux transformer) to the noise of active devices are basically understood. They are attributed to the low-frequency noise and are ascribed to motion of quantized flux (vortices) in the superconducting thin films. A nice illustration of this contribution is given by the so called telegraph noise, that occurs when a vortex hops between two pinning sites (see fig. 19a). In case of a statistical motion of many vortices, a scaling of the spectral noise density S_Φ with frequencies f and the applied magnetic field B is expected:

$$\sqrt{S_\Phi(f,B)} \propto \frac{B^n}{f^m} \tag{28}$$

with n=m=0.5. Thus, this so called *1/f* noise-spectrum is ascribed to an incoherent superposition of many thermally activated microscopic fluctuators, which are given by moving or hopping vortices. The characteristic *1/f* frequency-dependence is evidence for a distribution of activation energies for the vortex hopping [84] . Since most measurements of *1/f* noise are sensitive only to small activation energies, usually only the low-energy part of this distribution is probed [85].

Typical examples of the spectral noise density of SQUIDs for applications are given in fig. 19b and c. In fig. 19b, the *1/f* low-frequency noise of the rf SQUID is limited by the corner frequency (10-20Hz at zero-field) above which the frequency-independent so called white noise sets in. The low-frequency noise increases with increasing field in these large SQUID designed for application. The field dependence of this increase is given for two different types of SQUIDs in fig. 19c. Both SQUIDs posses the same geometrical design (see fig. 19c) but are prepared with a different technique (see discussion in section 3.1.1). For magnetic fields larger than the penetration field, the noise increases according to the theoretical expectation, i.e., $S_\Phi \propto B$ (eq. (28)). The low-frequency noise represents a serious limitation for a number of applications of SQUIDs especially if they are used in unshielded or little-shielded environment and in low frequencies applications (e.g., medical applications, NDE, etc.).

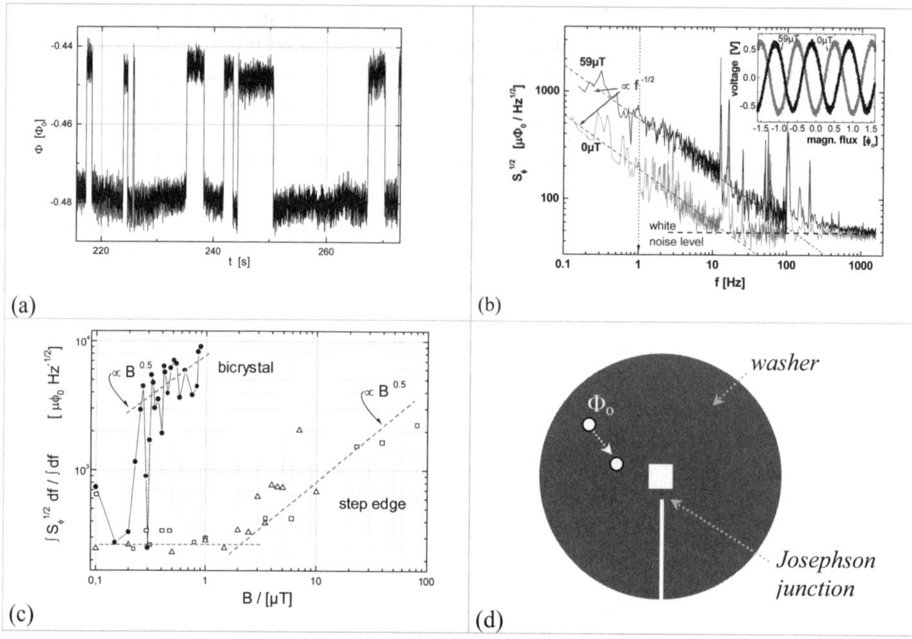

Fig.19: (a) Telegraph noise due to vortex motion, (b) spectral noise density for different magnetic fields measured in a HTS rf-SQUID (SES), and (c) magnetic field dependencies of the noise of SES (open symbols) and BS (solid symbols) according to [41]. The noise data in (c) is characterized by the normalized integration of the noise spectra in the low-frequency regime between 0.5 and 10Hz. The dotted lines represent the theoretical field dependence of the low-frequency noise, eq. (28). The lines in (b) idealize the different noise contributions, i.e. the field-dependent 1/f-low frequency noise and the field- and frequency-independent white noise. The schematic drawing in (d) sketches the geometry of the rf-SQUIDs (with moving vortex in the washer) that is used for the experiments described in this section. The typical dimensions of the planar washer-type rf HTS-SQUIDs used in the following experiments are: outer diameter of 3.5mm, SQUID hole of 100μm x 100μm and 3μm wide Josephson junctions (step-edge type or bicrystal type) [41].

Various remedies to reduce the low-frequency noise have been suggested and tested [40,42,86,87], which in principle can be classified into two categories: (i) either vortex penetration of the superconductor has to be avoided [86] or (ii) vortices have to be pinned by sufficiently strong pinning sites in the superconductor [40,42,87]. In case of flux avoidance, extremely good screening or, alternatively, relatively small superconducting geometrical structures are required. Both solutions are technically difficult to realize and/or very costly. For instance the patterning of the complete washer into a 'grid' of striplines of linewidths $w<6\mu m$ [88] would be necessary for application of these devices in earth field ($50\mu T$). This turns out to be technically quite complicated considering the requirements of the patterning. The alternative is given by vortex pinning by strong pinning sites. In the following I will describe (i) that the low-frequency noise in SQUIDs can be modified via flux pinning due to artificial pinning sites (e.g., antidots) and (ii) that only a few pinning sites very effectively suppress the low-frequency noise if they are strategically positioned in the device.

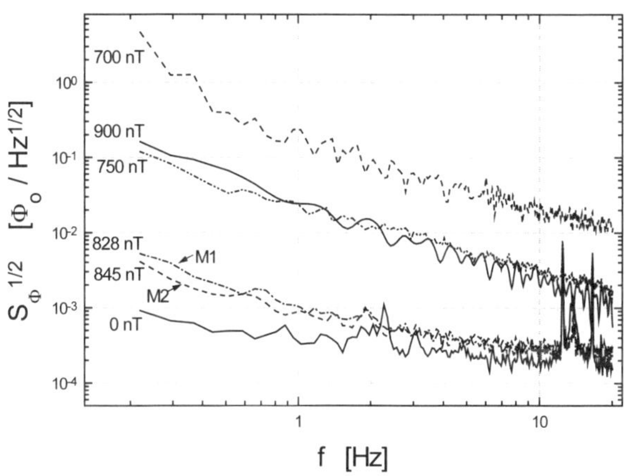

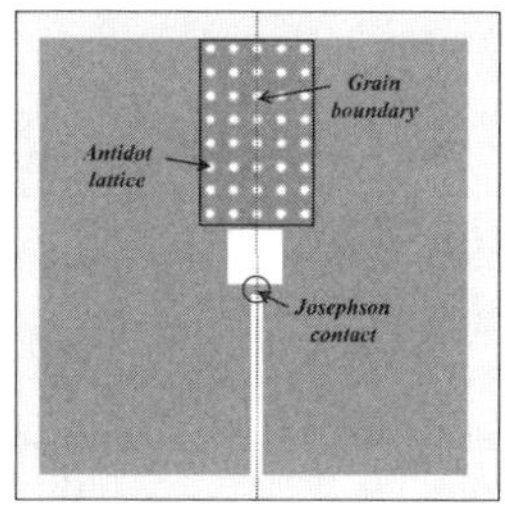

Fig. 20: (a) Low-frequency noise spectra of a BS with a quadratic antidots lattice on top of the grain boundary of the washer for different magnetic fields according to [87]. M_1 and M_2 label two matching conditions (see Table 1). (b): Sketch of the experimental arrangement. A YBCO film with quadratic antidot lattice (d_{AA}=5μm) is mounted in flip-chip configuration on top of the grain boundary of a BS. One of the main axes of the antidot lattice is oriented parallel to the grain boundary.

3.1.1 Manipulation of the Low-Frequency Noise via Antidot Arrays

In order to demonstrate the effect of vortex-antidot interaction upon the low-frequency noise in HTS devices, a characteristic property of bicrystal rf-SQUIDs (BS) is exploited. At extremely small magnetic fields ($B_{p,GB}\approx$200nT for the BS used in the following experiment, see fig. 19c) far below the penetration field ($B_{p,SE}\approx$ 1-2 μT in fig.19c), at which flux penetrates the washer in step-edge SQUIDs (SES) of identical geometry, the grain boundary in the washer of the BS serves as a channel for flux motion (see also sketch of a BS in fig.

20b). This flux motion causes a drastic increase of 1/f noise as shown in fig. 19c. The further increase of the low frequency noise agrees with the expected field dependence $S_\Phi \propto B$, eq.(28).

In order to examine the interaction between an antidot array and vortex lattices a YBCO film with a quadratic lattice (d_{AA} =5μm) of antidots was mounted in flip-chip configuration on top of the grain boundary of the BS washer (see fig. 20). For these measurements, the quadratic antidot lattice is oriented with one axis along the grain boundary (fig. 20b).

The resulting low-frequency noise-spectra of the arrangement (BS with antidot lattice) are given in fig. 20a. As expected, the lowest flux noise is recorded for zero field. At non-zero field, the low frequency noise strongly depends upon the exact values of the applied magnetic induction. In contrast to the standard field dependence $S_\Phi \propto B$ (eq.(28)), it varies non-monotonically over several orders of magnitude. For example, the noise at fields of 750nT or 900nT is more than 2 orders of magnitude larger than for the matching fields M_1 and M_2 (828nT and 845 nT, respectively) at which the noise level of the zero-field spectrum is recorded.

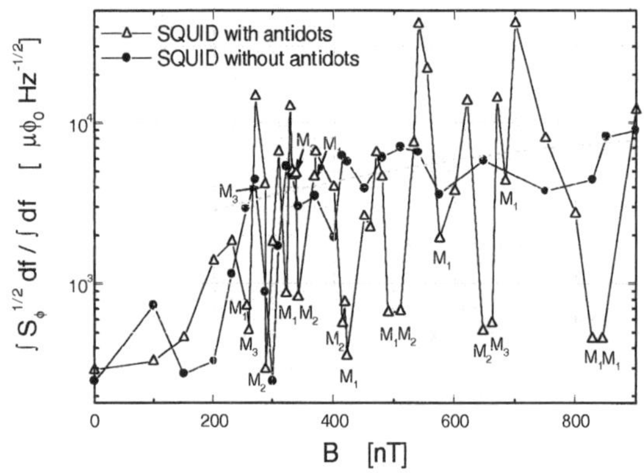

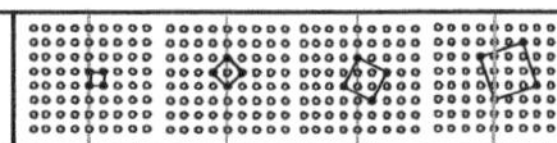

n	M_1	M_2	M_3	M_4
	a_0=d	a_0= n√2d	a_0= n√5d	a_0= n√10d
4				518 nT
5			662nT	331 nT
6			460nT	
7		845nT	338nT	
8		647nT	259nT	
9		511nT		
10	828nT	414nT		
11	684nT	342nT		
12	575nT	288nT		
13	490nT			
14	422nT			
15	368nT			
16	323nT			
17	287nT			
18	256nT			

Fig. 21: (a) Magnetic field dependence of the normalized and integrated low-frequency noise (0.5-10Hz) of a bicrystal rf-SQUID with and without an additional YBCO layer with antidots according to [87]. Each data point has been measured three times in different measuring sessions, the noise data could be reproduced within the experimental error. The data given in the plot are obtained by averaging the experimental results for each magnetic field. The dashed-dotted line represents the theoretical field dependence $S_\Phi \propto B$ (eq.(28)) of the SQUID without antidot lattice. (b) Table of the magnetic fields at which matching conditions are expected in the field range between 255 and 900 nT and schematic sketch of different orientations of a quadratic vortex lattice (solid circles) with respect to the quadratic antidot array (open circles) (d_{AA} = 5μm) and the grain boundary in the washer (grey line). M_i label the matching fields for different orientations of the vortex lattice.

A comparison of the field dependence of the low-frequency noise of the bare SQUID and the same SQUID with the antidot lattice is given in fig. 21. In contrast to the reference measurements on the bare SQUID, which show the usually observed behavior (i.e., constant noise level below $B_{p,GB}$ and linear increase $S_\Phi \propto B$ above $B_{p,GB}$, see also fig. 19), the data display clear minima and maxima. The minima are observed exactly at matching conditions. Matching or commensurability occur when the vortex lattice matches the antidot lattice and the grain boundary, along which the 'noisy' motion of the vortices takes place (see fig. 19b). Since the lattice parameter varies with field, the resulting matching conditions can easily be recorded. They are listed in the table in fig. 21b. It should be noted, that extremely small matching fields down to 247nT = $B_1/324$ have been observed in this experiments. The noise level at matching fields is comparable to the noise level recorded at zero field. This indicated that vortex motion within the grain boundary of the washer is completely suppressed at matching condition. In contrast, at fields in between the matching fields even larger noise levels (up to 10 times larger) are observed for the SQUID with antidot lattice with respect to the bare SQUID. Thus, vortex motion is even enhanced in case of incommensurability between vortex lattice, antidot array and grain boundary.

Thus, we can conclude:

(i) Low-frequency noise reduction via antidots is possible. The experiments shown in fig. 19 and 20 are a proof of this principle.

(ii) However, vortex-vortex interaction leads to vortex lattice formation and, thus, commensurability effects in case of the use of regular antidot array even at extremely small magnetic fields (fractions of the first matching field, here: $\sim B_1/324$).

(iii) Therefore, a field-independent reduction of the low frequency noise can not be achieved via a regular lattice of antidots. In contrary, minima in the noise level at matching fields alternate with excessively high noise-levels for non-commensurability.

Different solutions to this problem have been discussed, such as non-regular antidot lattices or extended holes or moats. In the following, I will show that the use of a few, *'strategically positioned'* antidots is the ideal solution of this problem.

3.1.2 Noise Reduction via Strategically Positioned Antidots

In the previous section it was demonstrated that the low-frequency noise in SQUIDs can be manipulated via antidots. However, it can be seen in figs. 20 and 21 and it is obvious from theoretical considerations, that regular arrays of antidots lead to noise reduction only at discrete values of the magnetic field (at matching fields) whereas in case of non-commensurability between the vortex and antidot lattice even an increase of the low-frequency noise is observed. Therefore it is important to use only a few, 'strategically positioned' antidots in the superconducting device, which trap only those vortices, that attribute strongly to the low frequency noise (vortices close to the SQUID hole) and leave the vortex lattice free to arrange itself within the device. Thus, the important demand is to allocate those strategic positions. Different methods can be applied to find out the best positions for antidots in devices:

(i) *Theoretical considerations*: In case of the SQUID the larges fluctuations are expected for vortex motion close to the SQUID hole. This is the basic message of eq.(18) and (19), that are illustrated in fig. 6. Furthermore, additional to the thermal activation, vortex motion is supported by the currents (Lorentz force). The reading current in the examined SQUID crowds at the constriction at the Josephson junction. Thus, obviously the positions for the first antidots are close to the Josephson junction and close to the SQUID hole (see conf.#1 in fig 22).

The resulting noise reduction can be seen in field-cooled (fc) experiments. In these experiments, the SQUID is cooled from the normal to the superconducting state in an applied field B (typically perpendicular to the plane of the SQUID). A comparison of the spectral noise density in the 1/f regime for one SQUID with different successively patterned antidot arrangements is given in fig. 23. It reveals:

(a) At low fields $B<B_{on}$ the spectral noise density is field independent and the same for all configurations (note, that always the same SQUID has been used).

(b) At higher fields $B>B_{on}$ the spectral noise densities increase linear with increasing field according to the theoretical expectation $S_\Phi \propto B$.

(c) However, the transition from field-independent to field-dependent spectral noise density is strongly shifted from $B_{on}\approx 8\mu$T without antidots (conf. #0, SQUID without antidots) to $B_{on}\approx 40\mu$T with antidots (conf. #1 and #2).

The significant increase of the onset field B_{on} by the arrangement of only 2 strategically positioned antidots is sufficient for most applications of SQUIDs in unshielded environment (earth field $B_{earth}\approx 50\mu$T).

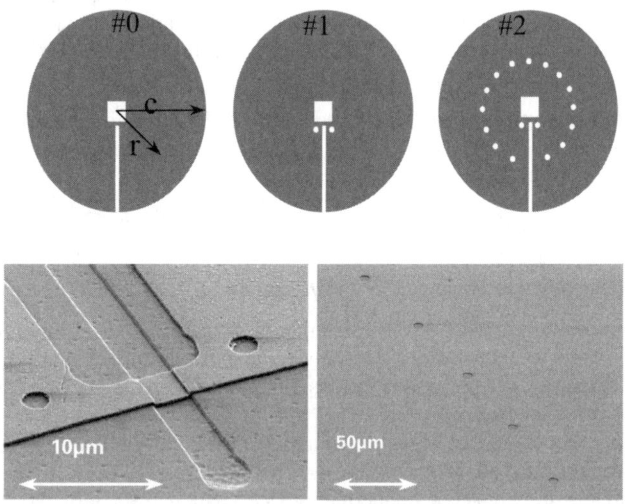

Fig. 22: Sketch of a rf SQUID with different antidot configurations (conf.#0: without antidots; conf. #1: with two antidots in the vicinity of the junction; conf. #2: two antidots at the junction and additional ring of antidots at $r=c/2$) and SEM images of the two antidots in the vicinity of the junction (left, antidots radius: 750nm) and the ring of antidots (right, $d_{AA}=50\mu$m, antidot radius: ~1.25μm).

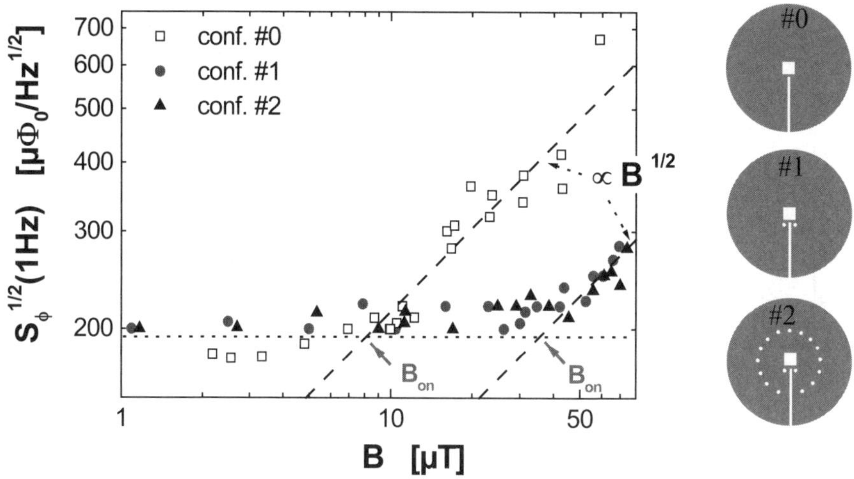

Fig. 23: Spectral noise density at 1Hz as function of the magnetic field for fc measurements and different configurations of strategically positioned antidots. The onset of the increase of the low-frequency noise is characterized by the onset field B_{on}. The proportionality $S_{\Phi} \propto B$ of the low-frequency noise for $B>B_{on}$ (dashed lines) agrees with the theoretical expectation in eq.(28). The onset field for conf. #1 and #2 are comparable to the earth field $B_{earth}\approx 50\mu T$.

(ii) *Experimental detection of penetrating vortices*: The SQUID can be used to experimentally determine the penetration and the position of penetrating vortices [40]. As described in sec. 2.3, a vortex in the washer will induce magnetic flux into the SQUID. Thus, the motion of vortices will lead to a change of the SQUID signal. In case of a gradual movement of a vortex within the washer a gradual change of the flux in the SQUID is observed (see fig. 24a). In contrast, a sudden penetration of a vortex into the washer is accompanied by a sudden change of the SQUID signal (see fig. 24b). The change of the phase of the sinusoidal signal represents a change of the magnetic flux. Thus, zero-field cooled (zfc) measurements can reveal the field and position of vortices penetrating into the washer [40]. In these experiments, the SQUID is cooled in zero-field into the superconducting state (vortex-free state), subsequently the field is ramped. By careful analysis of the SQUID signal (see example in fig. 24b) the magnetic field of individual vortex penetration and the resulting flux change in the SQUID can be recorded. In case of a circular design of the SQUID, the flux change $\Delta\Phi$ depends only on the radial position of the vortex (see eqs. (18) and (19)). It has been shown [40,41] that the vortex penetration happens to occur in a very reproducible way for a given SQUID design. I.e., in zfc experiments, the first vortex penetrations occur reproducible at the same fields and, especially, the resulting radial vortex positions after penetration are identical for the first vortices. Penetration field and vortex position of the first penetrating vortices depend on the chosen design. For the circular rf-SQUID design (see fig.

22) the resulting radial vortex position of the first three vortices is given by the radial position $r_p \approx c/2$ [40].

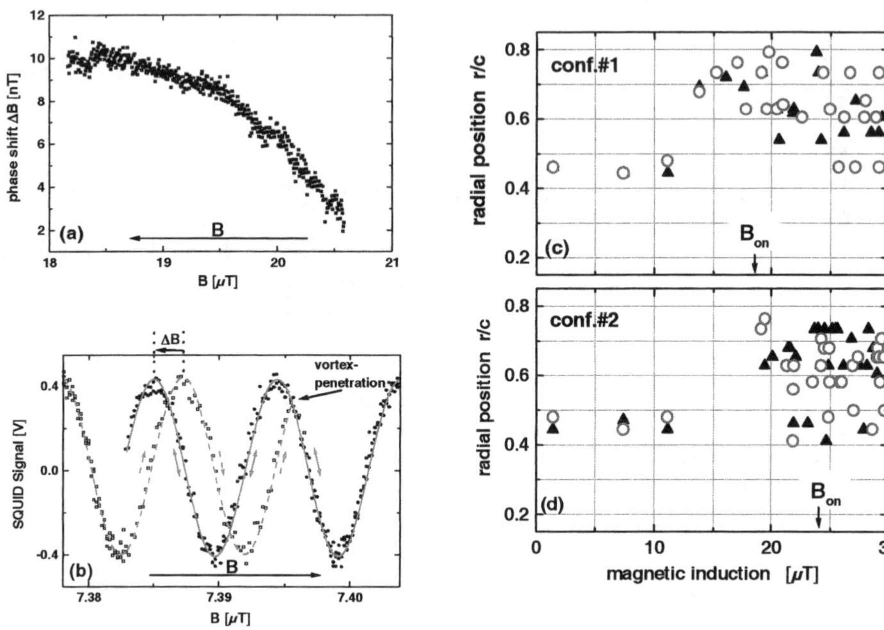

Fig. 24(a): Change of the unlocked SQUID signal due to a continuous movement of a vortex within the washer, (b) discrete phase shift in the SQUID signal due a sudden penetration of a vortex into the washer, and (c) and (d) resulting radial position of vortices as function of the field of penetration according to [40]. The radial positions in (c) and (d) are evaluated according to eq.(18), the different symbols are measurements for different SQUIDs of the same design and configuration.

Introducing a ring of antidots at this radial position leads to a reduction of the low-frequency noise due to the first penetrating vortices. Thus, the onset field B_{on} is not affected in fc experiments (see fig. 23) but increased in zfc measurements (see fig. 25). It also leads to a reduced number of penetrating vortices for magnetic fields $B < B_{on}$, i.e. the trapping of the first penetrating vortices seems to hinder further vortex penetration.

(iii) *Simulation of vortex penetration:* The analytical method is restricted to the first penetrating vortices (small fields) and relatively simple device geometries. In order to optimize the antidot arrangement for complex designs or larger fields, simulation of the vortex penetration and relaxation in the device seem to be more appropriate. Different simulation concepts are possible. Vortex motion in devices could be simulated in terms of molecular dynamics [89-94] or, alternatively, via Monte Carlo method [95]. Both methods are based on a phenomenological equation of the motion for vortices, first introduced in ref [99]:

$$\eta\vec{v} = \vec{F}_L + \vec{F}_{VV} + \vec{F}_P + \vec{F}_T + \vec{F}_M + \vec{F}_B \qquad (29)$$

The velocity v of a single vortex is determined by the balance of forces: the viscous damping force ηv, the vortex-vortex interaction $F_{VV} = \sum f_{VV}$ (summation of vortex-vortex interactions), the vortex-pin interaction $F_P = \sum f_P$ (direct summation of elementary interactions in case of pinning due to few defects, e.g., by antidots), the Lorentz force F_L, the random thermal force F_T, the Magnus force F_M, that tends to move the vortex in a direction perpendicular to the Lorentz force (it is usually negligibly small), and the interaction with boundaries F_B. Numerical simulations have proven to be a very powerful technique to predict the response of the vortices in the pinning potential of a small number of pin sites [89-94]. However, in most of the simulations, the effect of boundaries has not been taken into account. Thus, the real situation encountered in a superconducting device has not been properly simulated up to now. First test of a simulation of vortex penetration and relaxation in a superconducting device via Monte Carlo simulation are in preparation [95].

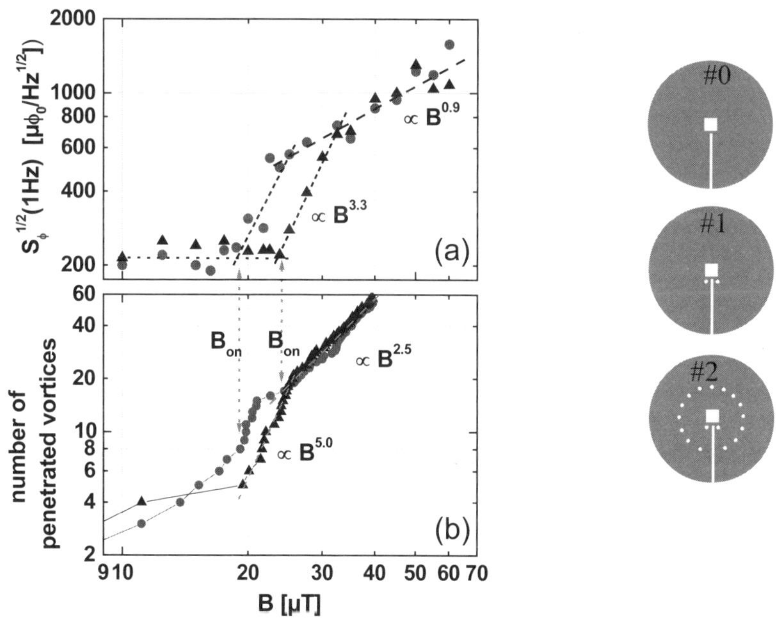

Fig. 25: (a) Spectral noise density at 1Hz and (b) number of penetrating vortices as a function of the applied magnetic induction for zfc measurements of conf. #1 (circles) and conf. #2 (triangles) according to [70]. The arrows mark the onset field B_{on} for both configurations. A sketch of the SQUID design and antidot configurations is added.

3.2 Vortex Matter in Microwave Devices

The behavior of a conductor in a microwave field can generally be characterized by the complex surface impedance $Z_s=R_s+iX_s$ with surface resistance R_s and surface reactance X_s. Z_s describes the electrodynamics at the interface between vacuum and the (super)conductor. The surface resistance represents the dissipated energy, whereas the surface reactance characterizes the energy stored in the conductor. The surface resistance is given by:

$$R_S = X_S = \sqrt{\frac{2\pi\mu_0 f}{\sigma_n}} \qquad \text{and} \qquad R_S \approx 2\pi^2\mu_o^2 f^2\sigma_n\lambda^3 \qquad (30)$$

for the normal conductor and superconductor in the two-fluid model, respectively. Here f represents the frequency and σ_n is the conductivity due to normal charge carriers. The comparison of the surface resistance of a type-II superconductor and a normal conductor shows the difference in frequency dependence and absolute value. At frequencies up to a few GHz, technical superconductors (e.g. Nb, NbN, HTS) possess much smaller R_s values and, thus, expose significantly smaller losses in rf fields.

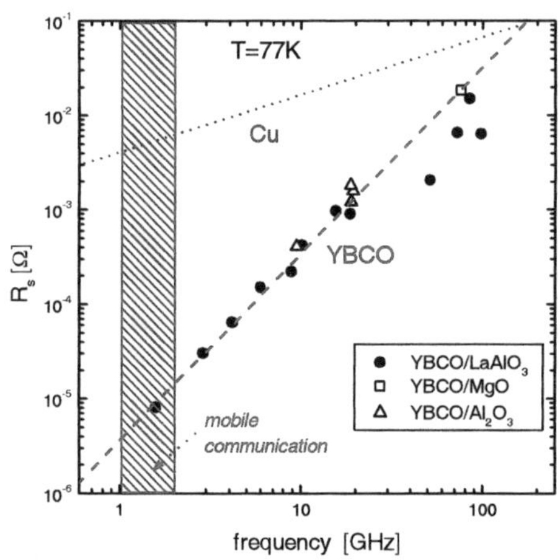

Fig. 26: Comparison of the frequency dependence of the microwave surface resistance R_s at T=77K for YBCO films on different substrates and Cu. In (b) the frequency regime used for mobile communication is marked.

From this point of view, the superconductor offers some benefit in rf applications. The smaller losses can be used to improve the quality factors $Q=f_o/\Delta f$ of resonators or filters (f_o represents the center frequency and Δf is the full width of the resonance at –3dB), or lead to miniaturization of existing devices. For instance the use of HTS thin films in microwave devices allows a dramatic reduction of the device dimension at comparable or higher performance with respect to conventional non-superconducting devices [2]. As a consequence, larger power densities are encountered in these devices. However, in contrast to their normal-conducting competitors, the power handling capability of superconducting microwave devices is limited by the nonlinearity of the surface resistance. At high rf power R_s

increases strongly (see fig. 27). This leads not only to an increase of dissipated rf energy that finally causes a heating of the superconductor, it will also cause a degradation of Q values (fig. 27) and the development of intermodulation signals.

Although various physical origins are discussed (e.g. weak links, thermal or magnetic effects), the physical mechanism, that is responsible for the nonlinear surface resistance and that represents a serious restriction for a number of rf applications, is not fully understood. In the following sections, I demonstrate that the combination of vortex matter and microwave properties leads to new insight into the nature of the nonlinear microwave properties of superconducting material. Using flux and particularly flux lines (vortices) in the superconductor as extremely sensitive probes for the local characterization of the complex microwave conductivity adds to the basic understanding of the magnetic component which seems to be responsible for the nonlinearity. A detailed discussion of this topic is given in ref. [80].

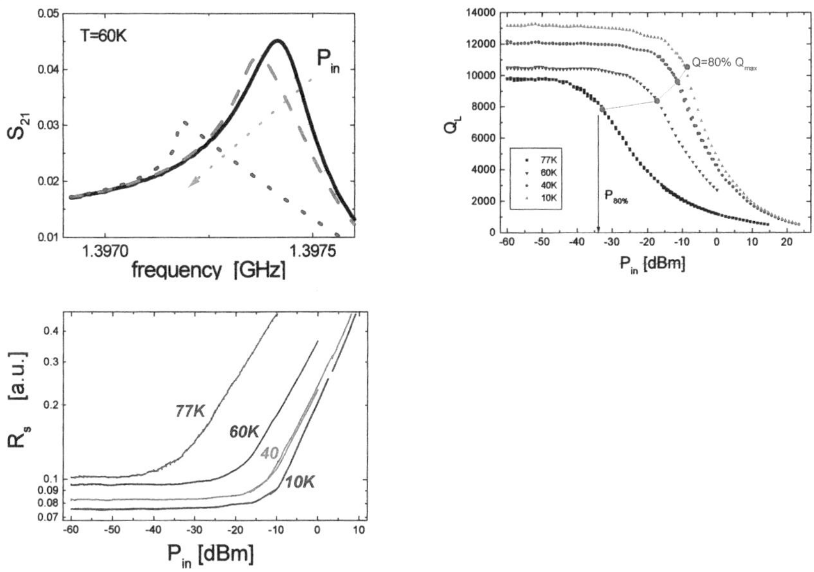

Fig. 27: Nonlinear microwave surface resistance leads to a degradation of superconducting resonators. This visible in the change of the resonance characteristic (a), the increase of the surface resistance (b) or the degradation of the quality factor (c) at large rf power.

3.3 Probing Microwave Properties with Vortices

The microwave measurements at high rf-power are performed on coplanar striplines and resonators. In these measurements, the usual disadvantage of coplanar microwave designs (i.e. the small power handling capability) turns into an advantage. Since the rf current is strongly peaked at the edge of the central conductor (fig.28), (i) the total energy dissipation (and, thus, heating) of the device due to the rf field can be neglected [80] compared to devices with a homogeneous current distribution, and (ii) these devices turn out to be highly sensitive to dc field or vortex penetration at the edge of the superconductor. The latter property provides the basis for the following experiments and their analysis. Experimental determined current densities in the central conductor of a coplanar microwave devices can be approximated by [23]

$$J_{rf}(x)=\frac{I}{w\cdot K\left(\frac{w}{a}\right)}\cdot\begin{cases}\left[\left(1-\left(\frac{2x}{w}\right)^{2}\right)\cdot\left(1-\left(\frac{2x}{a}\right)^{2}\right)\right]^{-1/2} & |x|\leq\frac{w}{2}-\lambda\\ \left[\left(1-\left(\frac{w}{a}\right)^{2}\right)\cdot\frac{\lambda}{w}\right]^{-1/2} & \frac{w}{2}-\lambda\leq|x|\leq\frac{w}{2}\\ -\left[\left(\left(\frac{a}{w}\right)^{2}-1\right)\cdot\frac{\lambda}{w}\right]^{-1/2} & \text{for}\quad \frac{a}{2}\leq|x|\leq\frac{a}{2}-\lambda\\ -\left[\left(\left(\frac{2x}{w}\right)^{2}-1\right)\cdot\left(\left(\frac{2x}{a}\right)^{2}-1\right)\right]^{-1/2} & |x|\geq\frac{a}{2}+\lambda\end{cases}\qquad (31)$$

Due to this characteristic current distribution the nonlinear behavior sets in at extremely small microwave power. For the resonator design used in the following experiments (YBCO coplanar resonators on sapphire with w=100μm, a=182μm) the maximum dissipated power is P_{diss}<0.2mW. According to the balance between dissipated power and cooling rate, a negligibly small temperature rise of ΔT < 16μK is expected [80]. The power handling is characterized by the degradation of the loaded quality factor $Q_L= f_o/\Delta f$. In the following a 80% criterion has been chosen, i.e. P_{max} is defined by the condition $Q_L(P_{max})=0.8\ Q_L(P_o)$ (see fig. 27).

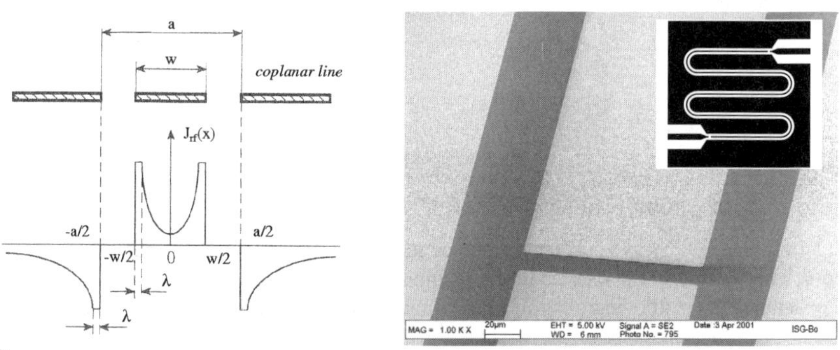

(a) **(b)**

Fig. 28: (a) Microwave current distribution J_{rf} in coplanar superconducting lines according to eq.(31). (b) Design and SEM image of a coplanar HTS thin film resonators according to [80]. Using typical design dimensions of w=100μm, a=182μm (for YBCO on sapphire), and a=500μm (YBCO on LaAlO3) an impedance of Z=50Ω is obtained. The resulting loaded quality factors for $\lambda/2$ resonators in the GHz regime are Q_L(77K)≈10.000-15.000 and Q_L(4K)≈20.000-30.000, respectively.

Experimental procedure: In order to probe the microwave properties with vortices, different measuring procedures are applied that yield different vortex distribution in the microwave devices. These different methods are sketched in fig. 29. In field-cooled (fc) and zero-field cooled (zfc) experiments the resonator is cooled to the superconducting state in an applied magnetic field or in zero-field, respectively. In field-cooled sweep (fcs) experiments, the sample is first cooled to the superconducting state in an applied field B_{fc} (where the field-cooled state is established) and subsequently the field is modified by ΔB.

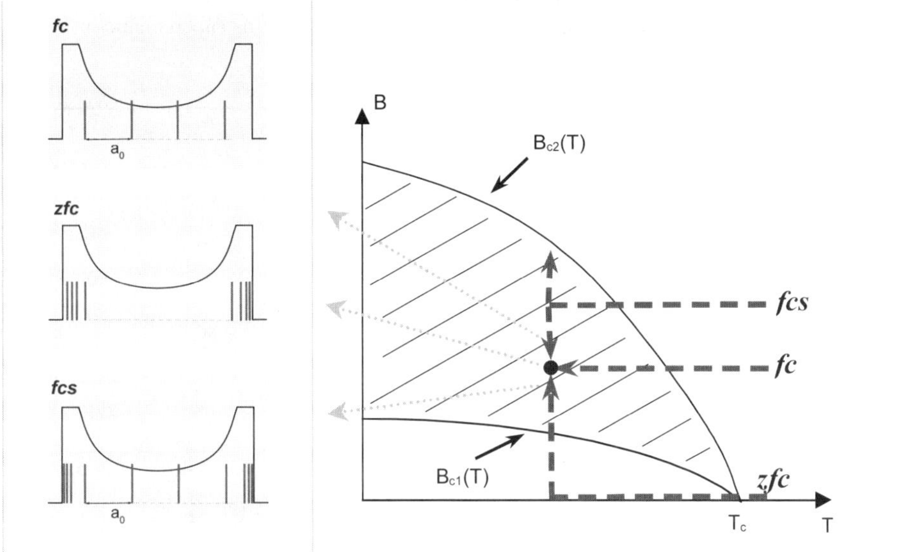

Fig. 29: Schematic drawing of the vortex distribution in field cooled (fc), zero-field cooled (zfc) and field-cooled sweep (fcs) experiments. On the right-hand side the different experimental procedures are sketched in the BT phase-diagram of a type-II superconductors. On the left-hand side the rf current distribution in the central conductor of the coplanar microwave device and the expected vortex distribution is given for the different experimental procedures.

Experimental results and discussion: The power handling capability P_{max} depends strongly upon the way the magnetic field is established. Whereas fc measurements show a gradual decrease of P_{max}, in zfc experiments a strong decrease of the power handling capability within a few μT is followed by an almost constant P_{max} (see fig. 30). The different behaviors are a consequence of the vortex distribution in the resonator. A homogeneous vortex distribution in the superconductor is expected for fc experiments (see fig. 29). The linear increase of the vortex density with increasing field seems to lead to the gradual decrease of P_{max}. In contrast to the fc experiments, an inhomogeneous vortex distribution is expected for the zfc case (see fig. 29). Due to the flux penetration into the field-free superconductor in the zfc case, the position of the largest vortex density coincides with the maximum of the rf current at the edge of the central conductor. This explains the strong reduction of P_{max} already at extremely small fields of a few μT. A further increase of the field in the zfc case does not affect the power-handling capability seriously. This is consistent with other observations of flux penetration at small fields, where an increase of the magnetic field first leads only to a shift of the flux front into the superconductor. Since the interior of the superconductor does not carry large rf currents, this shift does not affect the power handling capability seriously. Furthermore, the fc and zfc values of P_{max} seem to merge at large fields. This is expected for the field, for which the zfc flux front approaches the center of the sample. For these fields similar vortex densities are expected at the edge of the superconductor for the fc and zfc case.

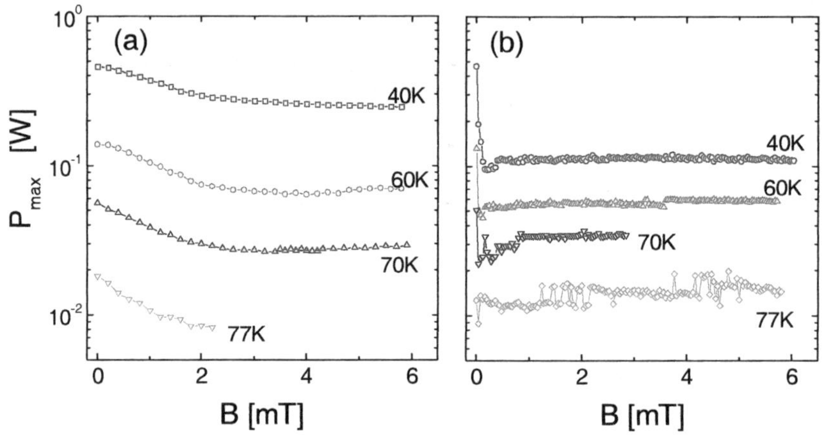

Fig. 30: Field dependence of the power handling capability of a coplanar resonator for (a) fc and (b) zfc experiments according to [96].

The most interesting features, that also confirm the interpretation of the fc and zfc data given above, are observed for fcs measurements (see fig. 31). Starting with the fc value at B_{fc}, fcs experiments show a very characteristic variation of P_{max} with increasing and decreasing field, which provides an ideal fingerprint of flux penetration (at the edge) and vortex penetration. The field dependence is characterized by two clear maxima. For non-zero starting fields B_{fc}, a nonsymmetrical ΔB-dependence of P_{max} is obtained. The position of the main maximum of $P_{max}(\Delta B)$ decreases with increasing field B_{fc} and represents the field at which the frozen in field B_{fc} is compensated by the flux penetrating at the edge of the superconductor, i.e. $B_{eff} = 0$ at the edge of the central conductor [80]. The second smaller peak indicates the field, where vortices leave the edge and penetrate into the central conductor, thus, reducing the effective field at the edge of the conductor [80].

The different underlying mechanisms can be proven by field cycles with different amplitude (fig. 32). For small field amplitudes ΔB, a reversible field penetration of flux (ΔB>0) or 'antiflux' (ΔB<0 compensates the existing frozen-in field) is observed. The flux penetration leads to a decrease of P_{max}, whereas the flux compensation (ΔB<0) causes an increase of P_{max}. Modification of P_{max} by up to 50% take place within a few μT. Only for larger amplitudes $/\Delta B/>10\mu T$ vortices (for $\Delta B>0$) or antivortices (vortices with field direction opposing the direction of the applied field B, i.e. $\Delta B<0$) start to nucleate and penetrate the resonator structure. This leads to an irreversible modification of P_{max} (see fig. 32b). The penetration of the first vortices is accompanied by a small increase of the P_{max} for the positive branch, that can be interpreted by a relaxation of the field at the edge of the superconducting central conductor. Obviously this effect is not possible for the negative branch, here P_{max} decreases when the first antivortices penetrate. Finally, for larger values $/\Delta B/>30\text{-}40\mu T$ the curve levels off and P_{max} approaches the zfc value for large fields. This is an indication for the fact, that the vortex density at the edge of the central conductor is constant.

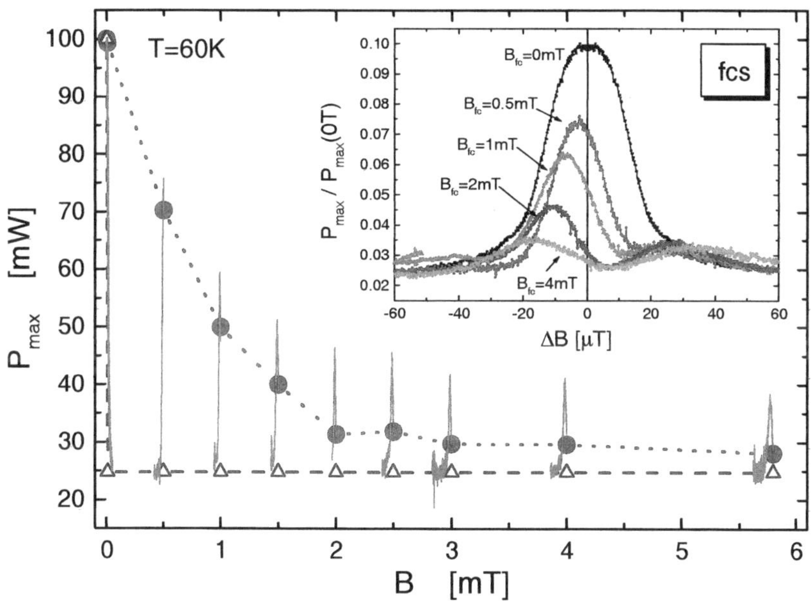

Fig. 31: Comparison of the magnetic field dependence of the power handling capability of a coplanar resonator for fc (solid circles), zfc (open triangles) and fcs (grey lines) experiments. The inset shows the power handling capability in fcs measurements for different starting fields B_{fc} on a normalized scale $\Delta B=B-B_{fc}$.

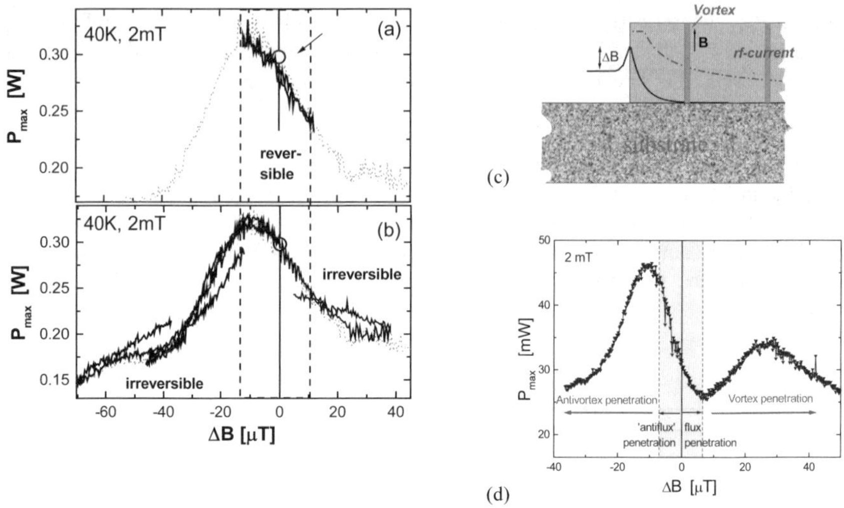

Fig. 32: Magnetic field cycles with different amplitudes for fcs experiments reveal the underlying mechanism of (a) reversible flux penetration and (b) vortex penetration. The dotted gray line represents the power-handling data of a single scan. Figure (c) sketches the different contributions that affect the power-handling capability at the edge of the central strip line, i.e. the rf-current, the penetrating flux (penetration on the scale of λ) and the penetrated vortices. In (d) the field regimes for different mechanisms of flux penetration are sketched in the plot of the fcs experiments for 2mT.

Model for the mechanism of limited rf-power handling capability: The fcs experiments clearly demonstrate, that not only flux and vortices themselves have a large impact upon the power handling capability of the resonator, but also the sign and especially the correlation between the spatial distributions of vortices and rf current density are of importance. The largest impact of vortices upon the microwave properties is observed for vortex positions at the edge of the central conductor. This is that part of the resonator which carries the highest rf current density.

The field-dependence of P_{max} can be explained in a model that considers (i) a superposition and (ii) a limitation of total current density in the superconductor [80]. Ambient magnetic fields lead to screening currents in the superconductor. The resulting screening current densities (J_M due to a magnetization and J_V due to vortices) superimpose to the rf current density J_{rf}, thus, yield a total current density $J_{tot} = J_{rf} + J_M + J_V$ in the coplanar resonator. The nonlinearity implies that the total current density is limited by a critical value, i.e. $J_{tot} \leq J_{max}$. The field dependence of the limiting current is given by [80]

$$J_{\max}(B,x) = J_{\max,B=0}(x) - J_M(B,x) - J_V(B,x) \qquad (32)$$

This model automatically yields a power handling capability that strongly depends on the kind of experiment. It nicely explains the experimentally observed differences in the fc, zfc, and fcs experiments given in figs. 30 and 31. A detailed discussion of this model that consistently explains the physical origin of the power handling capability is given in ref. [80]. Here, I will restrict the discussion to the most simple case, the experiments in zero-field.

Zero-field experiment: Generally the rf current density profile shows a maximum at the edges of the central line (see fig. 28). According to eq.(32) the maximum current density at this position is

$$J_{rf,\max}(B=0)=\frac{\sqrt{2P_{\max}}}{K\left(\frac{w}{a}\right)\sqrt{Z\cdot\lambda_{eff}wd^{2}\left(1-\left(\frac{w}{a}\right)^{2}\right)}}\quad, \tag{33}$$

where d represents the thickness of the superconducting layer and λ_{eff} is the temperature dependent London penetration length corrected for the finite thickness. Thus, the maximum rf current density in the superconductor can be derived from experimental values of P_{max}. Fig. 33 shows the resulting temperature dependence of $j_{rf,max}$ for two resonators patterned from YBCO films of different quality. Two unmasking aspects can be noticed: (i) $J_{rf,max}$ increases linearly with temperature and (ii) the orders of magnitude of $J_{rf,max}$ are about 10^{10}Am^{-2} and 10^{11}Am^{-2} for 77K and 4K, respectively. Both, the linear temperature dependence and the magnitude of the maximum rf current density agree with the behavior and magnitude of the dc critical current for typical YBCO thin films (see fig. 33). This suggests, that the nonlinear microwave behavior is strongly related to the dc critical current of the superconductor. A relations between microwave power handling and pair-breaking mechanisms [97] can not be verified by these experiments [80].

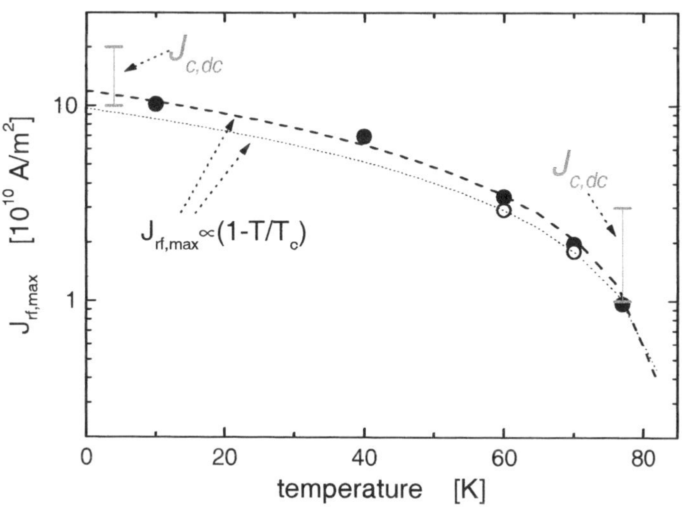

Fig. 33: Comparison of the maximum rf-current for two different coplanar $\lambda/2$ resonators made from YBCO thin films and the dc critical current density in reference lines as function of temperature according to [96]. The dashed lines represent the linear fit of the temperature dependence of $J_{rf,max}$.

In conclusion, it has been demonstrated that the use of vortex matter in microwave experiments yields new insight into the microwave properties of superconducting thin films. The comparison of the microwave power handling capability of coplanar HTS thin film resonators in fc, zfc and fcs experiments shows, that (i) the microwave properties depend on magnetization and vortex distribution, that (ii) power handling and quality factor of the devices are strongly affected by extremely small fields and the way the magnetic field is established (fc, zfc, or fcs), and that (iii) the microwave power handling can be understood in terms of a simple model, that considers all contributions to the current in the superconductor (i.e. rf and screening currents) and limitation of the total current by the dc critical current. The model explains the microwave behavior in non-zero and zero magnetic field. Thus, flux and vortex penetration can be used to probe the local microwave properties, and reveal the mechanism of nonlinear microwave properties. However, it also bears new and interesting aspects of vortex physics in patterned superconducting thin film devices.

4. Conclusions

It is the intention of this work to articulate the following points:

(i) Vortex matter is in general not only an interesting topic for basic research but plays a substantial role for application of superconductivity. In most electronic applications, magnetic flux will penetrate the superconductor. Flux motion and even the presence of magnetic flux will change - in a lot of cases reduce - the performance of the device. If the flux is not taken into account, reduction of the sensitivity, power-handling capability, or an increase of the error rate will often be the consequence. However, vortices could also be used to improve the performance of a device. For instance, various concepts of using a guided motion of vortices exist, or vortex trapping can be used to prevent further flux entry in superconducting devices [40,41].

(ii) The analysis of the local distribution of vortices in the devices can be used for optimization of device properties. In this article it is demonstrated for one example (HTS SQUID) that via quite simple arrangements of a few 'strategically positioned' artificial defects (here only two antidots were used in one example) a significant improvement of the device could be obtained. More careful or more sophisticated analysis (e.g. via computational simulation) could provide a further improvement or could extend this method to the improvement of more complex superconducting devices.

(iii) Active as well as passive superconducting thin-film devices can be used as very sensitive tools for detecting motion and penetration of flux and vortices into superconducting material. In combination with adequate computational simulation this yields new insight into vortex physics in real (e.g. 2D patterned) superconducting systems.

(iv) Finally, the combination of vortex matter with other areas of research of superconductivity might lead to improvement of the basic understanding of the mechanisms of superconductivity. In this article, this has been demonstrated for the combination of vortex matter and microwave properties of HTS material. The careful analysis of the microwave properties of coplanar resonators in small magnetic fields leads to new insight into the physical mechanism that limits the rf power handling capability and is the cause of the nonlinear microwave surface resistance.

Thus, in applications of superconductivity the vortex state of matter appears to be very important and requires further investigation. A better understanding will clearly lead to an improvement of the performance of HTS components, such as reduced noise, better power handling capability, improved 'electronic' homogeneity, or reliability. Furthermore, it promises deeper insight into the basic physics of vortices and vortex matter.

Acknowledgment

The authors like to acknowledge the experimental work of A. M. Castellanos, P. Selders, M. Pannetier, and P. Lahl, which are the basis of this publication. Furthermore A. I Braginski, N. Klein, R. Gross, V.V. Moshchalkov, R. Wijngaarden, R. Griessen, R. Kleiner, D. Koelle, P. Kes, R. Besseling, V. Yurchenkov, M.H.P. Bochem, S. Bunte, W. Hofer, E. Hollmann, R. Kutzner, R. Ott and P. Dymachevski are acknowledged their support, cooperation and discussion. This work was supported by DFG Grant No. WO549/3-1 and ESF scientific program VORTEX.

References

1. H. Kamerlingh Onnes, Leiden Comm. 120b, 122b, 124c (1911)
2. H.J. Chaloupka , Microwave Application of High Temperature Superconductors, NATO ASI Series, H. Weinstock (edt.), Kluwer, Dordrecht (1999); H.J. Chaloupka, M. Jeck, B. Gurzinski, S. Kolesov, Electron.Lett. 32 (1996) 1735
3. W. Meissner, R. Ochsenfeld, Naturwissenschaften 21 (1933) 787
4. B.D. Josephson, Phys. Lett. 1 (1962) 251
5. F. London, H. London, Proc. Roy. Soc (London) A 149 (1935) 72 ; F. London, H. London, Physica 2 (1935) 341
6. V.L. Ginzburg, L.D. Landau, Zh. Eksp. Teor. Fiz. (in Russian) 20 (1950) 1064
7. M. Tinkham, Introduction to Superconductivity, McGraw-Hill, New York (1975)
8. J.R. Waldram, Superconductivity of Metals and Cuprates, IOP Publishing Ltd., London (1996)
9. F. London, Superfluids: Macroscopic Theory of Superconductivity, Dover Publications, New York (1961)
10. R. Doll, M. Näbauer, Phys. Rev. Lett. 7 (1961) 51
11. B.S. Deaver Jr., W.M. Fairbank, Phys. Rev. Lett. 7 (1961) 43
12. A.A. Abrikosov, Sov. Phys. JETP 5 (1957) 1174
13. G. Blatter, M.V. Feigel'man, V.B. Geshkenbein, A.I. Larkin, V.M. Vinokur, Rev. of Mod. Phys. 66 (1994) 1125-1388
14. R. Wördenweber, Rep. Prog. Phys. 62 (1999) 187-236
15. R. Wördenweber, M.O. Abd-El-Hamed, J. Schneider, O. Laborde, J. Appl. Phys. 70 (1991) 2230
16. W. Pint, E. Langmann, E. Schachinger, Physica C 157 (1989) 415
17. E. Langmann, Physica B 165&166 (1990) 1061
18. H. Hess, R.B. Robinson, J.V. Waszczak, Physica B 169 (1991) 422
19. M. Tinkham M, Introduction to Superconductivity (2nd edition) , McGraw-Hill, New York (1996)
20. L.D. Landau, E.M. Lifshitz, Electronics of Continuous Media (Oxford: Pergamon) (1960)
21. E. Zeldov, A.I. Larkin, V.B. Geshkenbein, M. Konczykowski, D. Majer, B. Khaykovich, V.M. Vinokur, H. Shtrikman, Phys. Rev. Lett. 73 (1994) 1428
22. D. Koelle, R. Kleiner, F. Ludwig, E. Dantsker, J. Clarke, Rev. of Mod. Phys. 71 (1999) 631-686
23. I.B. Vendik, O.G. Vendik, D.I. Kaparkov, Superconductor Microwave Technology, Part II: Superconducting Microwave Circuits, Chalmers University of Technology, Report No 24 (ed. Kollberg E) ISSN 1103-4599, ISRN CTH-MVT-R—24--SE (1996)
24. A. Pruymboom, thesis, University of Leiden (1988)
25. A. Pruymboom, P.H. Kes, E. van der Drift, S. Radelaar, Phys. Rev. Lett. 60 (1988) 1430
26. M. Baert, V.V. Metlusko, R. Jonckheere, V.V. Moshchalkov, Y. Bruynseraede, Phys. Rev. Lett. 74 (1995) 3269
27. A.M. Castellanos, R. Wördenweber, G. Ockenfuss, A. v.d. Hart, K. Keck Appl. Phys. Lett. 71 (1997) 962
28. A.I. Larkin, Yu.N. Ovchinnikov, Sov. Phys.-JETP 41 (5) (1976) 960

29. L.E. Musienko, I.M. Dmitrenko und V.G. Volotskaya, Pis'ma Zh. Eksp. Teor. Fiz. 31 (1980) 603 and JETP Lett. 31 (1980)567
30. W. Klein, R.P. Huebener, S. Gauss und J. Parisi, J. Low. Temp. Phys. 61 (1985) 413
31. S.G. Doettinger, R.P. Huebener, R. Gerdemann, A. Kühle, S. Anders, T.G. Träuble und J.C. Villégier, Phys. Rev. Lett. 73 (1994) 1691; S.G. Doettinger, R.P. Huebener, R. Gerdemann, A. Kühle, S. Anders, T.G. Träuble und J.C. Villégier, Physica C 235-240 (1994) 3179
32. S.G. Doettinger, R.P. Huebener, A. Kühle, Physica C 251 (1995) 285
33. P. Lahl, R. Wördenweber, Inst. Phys. Conf. Ser. No 158 (1997) 77
34. R. Wördenweber, U. Krüger, J. Schneider, R. Kutzner, G. Ockenfuß, Inst. of Phys. Conference Series 148 (1995) 619
35. P. Lahl, R. Wördenweber, to be published
36. J. Jackson, Classical Electrodynamics, John Wiley&Sons, Inc. New York (1999)
37. M.J. Ferrari, J.J. Kingston, F.C. Wellstood, J. Clarke, Appl. Phys. Lett. 58 (1991) 1106
38. H. Humphrey, IEEE Trans. Appl. Supercond. 9 (1999) 3741
39. Dönitz D, Thesis, University Tübingen (2002)
40. P. Selders, R. Wördenweber, Appl. Phys. Lett. 76 (2000) 3277
41. P. Selders P, Bericht des Forschungszentrums Jülich, 3826, ISSN 0944-2952 (2000)
42. T.J. Shaw, J. Clarke, R.B. van Dover, L.F. Schneemeyer, A.E. White, Phys. Rev. B54 (1996) 15411
43. R.B. van Dover, E.M. Gyorgy, A.E. White, L.F. Schneemeyer, R.J. Felder, J.V. Waszcak, Appl. Phys. Lett. 56 (1990) 2681
44. A.N. Lykov, Sol. St. Comm. 86 (1993)531
45. E. Rosseel, M. van Bael, M. Baert, R. Jonckheere, V.V. Moshchalkov, Y. Bruynserade, Phys. Rev. B53 (1996) R2983
46. V.V. Metlushko, M. Baert, R. Jonckheere, V.V. Moshchalkov, Y. Bruynserade, Sol. St. Comm. 91 (1994) 331
47. V.V. Moshchalkov, M. Baert, E. Rosseel, V.V. Metlushko, M.J. van Bael, Y. Bruynserade, Physica C 282-287 (1997) 379
48. V.V. Moshchalkov, M. Baert, V.V. Metlushko, E. Rosseel, M.J. van Bael, K. Temst, Y. Bruynserade, R. Jonckheere, Phy. Rev. B57 (1998) 3615
49. R. Wördenweber, A.M. Castellanos, P. Selders, Physica C 332 (2000) 27
50. R. Besseling, Thesis, University of Leiden (2001)
51. N. Kokubo, Phys. Rev. Lett. 88 (2002) 247004
52. A. Pruymboom, P.H. Kes, E. van der Drift, S. Radelaar, Appl. Phys. Lett. 52 (1988) 662
53. E.J. Kramer, J. Nucl. Mater. 72 (1973) 5
54. R. Wördenweber, P.H. Kes, C.C. Tsuei, Phys. Rev. B 33 (1986) 3172
55. R. Wördenweber, Phys. Rev. B46 (1992) 3076
56. D.O. Welch, IEEE Transactions on Applied Superconductivity 3 (1993) 1476
57. E.H. Brandt, Phys.Status Solidi B 77 (1976) 551
58. E.H. Brandt , J. Low Temp. Phys. 26 (1977) 709
59. E.H. Brandt, J. Low Temp. Phys. 26 (1977) 735
60. E.H. Brandt, J. Low Temp. Phys. 28 (1977) 263
61. E.H. Brandt, J. Low Temp. Phys. 28 (1977) 291

62. H. Kronmüller, M. Fähnle, M. Domann, G. Grimm, R. Grimm, B. Gröger, J. Magn. Mater. 13 (1979) 53
63. A. Bezryadin, B. Pannetier, J. Low Temp. Phys. 98 (1995) 251
64. V.V. Moshchalkov, M. Baert, V.V. Metlushko, E. Rosseel, M.J. van Bael, K. Temst, R. Jonckheere, Y. Bruynserade, Phys. Rev. B 54 (1996) 7385
65. L. van Look, E. Rosseel, M.J. van Bael, K. Temst, V.V. Moshchalkov, Y. Bruynseraede, Phys. Rev. B 60 (1999) R6998
66. K. Harada, O. Kamimura, H. Kasai, T. Matsuda, A. Tonomura, V.V. Moshchalkov, Science 274 (1996) 1167
67. A. Bezryadin, Y.N. Ovchinikov, B. Pannetier, Phys. Rev. B53 (1996) 8553; A. Bezryadin, B. Pannetier, J. Low Temp. Phys. 102 (1996) 73; A. Bezryadin, B. Pannetier, Physica Scripta T66 (1996) 225
68. V.V. Metlushko, U. Welp, G.W. Crabtree, Zhao Zhang, S.R.J. Brueck, B. Watkins, L.E. DeLong, B. Ilic, K. Chung, P.J. Hesketh, Phys. Rev. B 59 (1999) 603
69. V.V. Metlushko, U. Welp, G.W. Crabtree, R. Osgood, S.D. Bader, L.E. DeLong, Zhao Zhang, S.R.J. Brueck, B. Ilic, K. Chung, P.J. Hesketh, Phys. Rev. B 60 (1999) R12585
70. R. Wördenweber, P. Selders, Physica C 366 (2002) 135
71. see for instance conference reports in Physica C 332 (2000)
72. see for instance conference reports in Physica C 369 (2002)
73. G.S. Mkrtchyan, V.V. Shmidt, Sov. Phys. JETP 34 (1972) 195
74. A.I. Buzdin, D. Feinberg, Physica C 256 (1996) 303
75. H. Nordborg, CV.M. Vinokur, Phys. Rev. B 62 (2000) 12408
76. C.P. Bean, J.D. Livingston, Phys. Rev. Lett. 12 (1964) 14
77. R. Surdeanu, R.J. Wijngaarden, R. Griessen, J. Einfeld, R. Wördenweber, Europhys. Lett. 54 (2001) 682
78. M. Pannetier, R.J. Wijngaarden, I. Fløan, J. Rector, R. Griessen, P. Lahl, R. Wördenweber, Observation of unexpected D_4 symmetry in patterned superconductors, to be published.
79. A.I. Braginski, H.J. Krause, J. Vrba, Handbook of thin film devices, Vol 3, Superconducting Film Devices, Academic Press, ISBN 0-12-408952-6 (2000) 149
80. P. Lahl, R. Wördenweber, Appl. Phys. Lett. 81 (2002) 505
81. M.B. Ketchen, W.J. Gallagher, A.W. Kleinsasser, S. Murphy, J.R. Clem, SQUID'85 – Superconducting Quantum Interference Devices and Their Applications, Eds. Hahlbohm H D and Lübbig H, deGruyter (Berlin) (1985) 865
82. R. Gross, B. Mayer, Physica C 180 (1991) 235; M. Kawasaki, P. Chaudhari, A. Gupta, Phys. Rev.. Lett. 68 (1992) 1065; A. Marx, R. Gross, Appl. Phys. Lett. 70 (1997) 120; A. H. Miklich, J. Clarke, M. S. Colclough, K. Char, Appl. Phys. Lett. 60 (1992)1899
83. R. L. Forgacs, A. F. Warwick, Rev. Sci. Instrum. 38 (1967) 214; V. Foglietti et al., Appl. Phys. Lett. 49 (1986) 1393; R. H. Koch et al., J. Low Temp. Phys. 51 (1983) 207; R. H. Koch et al., Appl. Phys. Lett. 60 (1992) 507; A. H. Miklich et al., IEEE Trans. Appl. Supercond. 3 (1993) 2434; M. Mück, C. Heiden, J. Clarke, J. Appl. Phys. 75 (1994) 4588
84. P. Dutta, P. Dimon, P.M. Horn, Phys. Rev. Lett. 60 (1979) 646; P. Dutta, P.M. Horn, Rev. Mod. Phys. 53 (1981) 497
85. M.J. Ferrari, M. Johnson, F.C. Wellstood, J.J. Kingston,T.J. Shaw, J. Clarke, J. Low Temp. Phys. 94 (1994) 15

86. E. Dantsker, S. Tanaka, J. Clarke, Appl. Phys. Lett. 70 (1997) 2037; R.H. Koch, J. Z. Sun, V. Foglietta, W. J. Gallagher, Appl. Phys. Lett. 67 (1995) 709
87. P. Selders, A. M. Castellanos, M. Vaupel, R. Wördenweber, Appl. Supercond. 5 (1998) 269
88. J. R. Clem, "Vortex exclusion from superconducting strips and SQUIDs in weak perpendicular ambient magnetic fields," unpublished (1998)
89. C. Reichhardt, J. Groth, C.J. Olson, S.B. Field, F. Nori F, Phys. Rev. B 54 (1996) 16108
90. C. Reichhardt, C.J. Olson, F. Nori, Phys. Rev. Lett. 78 (1997) 2648
91. C. Reichhardt, C. J. Olson, F. Nori, Phys. Rev. B 57 (1998) 7937
92. C. Reichhardt, R.T. Scalettar, G.T. Zimányi, N. Grønbech-Jensen, Phys. Rev. B 57 (1998) 7937
93. G. Carneiro, J. Low Temp. Phys. 117 (1999) 1323
94. C. Reichhardt, N. Grønbech-Jensen, Phys. Rev. B 63 (2001) 54510
95. R. Wördenweber, to be published
96. P. Lahl, Bericht des Forschungszentrums Jülich; 3923, ISSN 0944-2952 (2001)
97. T. Dahm, D.J. Scalapino, Appl. Phys. Lett. 69 (1996) 4248; T. Dahm, D.J. Scalapino, B.A. Willemsen, J. Supercond. 12 (1999) 339
98. Jooss, Ch., et al. , Physica C 299 (1998) 215-230
99. A. Schmid, W. Hauger, J. Low Temp. Phys. 11 (1973) 667

HTS MICROWAVE FILTERS: PROPERTIES, DESIGN AND SYSTEM APPLICATIONS

Heinz J. CHALOUPKA[1)] and Tobias KAESSER[2)]
[1)] University of Wuppertal, Dept. Electrical & Information Engineering, D-42097 Wuppertal, GERMANY
[2)] Tesat-Spacecom GmbH & Co KG, Gerberstr. 49 D-71522 Backnang, GERMANY

1. INTRODUCTION

High-temperature superconductor (HTS) technology has proved to provide a means for the realisation of passive microwave components with superior performance. These components - with microwave filters being the most prominent example- exploit the extremely low conductor losses in planar HTS circuits at cryogenic temperatures. A comprehensive review on this topic is provided by three books [1-3], published so far. Moreover, special issues of microwave journals, e.g. [4] and numerous review articles - the most recent one [5] - are dealing with this particular application of HTS technology.

Having this already available literature in mind, this book chapter aims primarily to stress the very close interdependency and inseparability between HTS technology (materials development, thin-film-technology, assembling and integration technologies), creative filter design and suitable system implementation. Taking the relatively high implementation burden (necessity to work at cryogenic temperatures) into account, applications of HTS filter technology will be restricted to the cases where

(i) desired system properties (including cost) can otherwise (without HTS filters) not be achieved at all,

(ii) novel applications can be created by this new technology.

However, it is important to note that superior filter performance is achieved only if HTS technology is combined with innovative filter topologies and design principles which fully exploit the unique material properties. Furthermore, filters are no end in itself. Hence, the superior filter performance has to be converted into substantially improved sub-system properties, which in turn form a potential of performance enhancements in future systems for communications, sensing etc. It is the specific purpose of this report to stress these two issues, namely (i) the necessary synergy between HTS technology and advanced filter design and (ii) the correspondence of a high performance on the component level with enhancements on the sub-system and system level.

This book chapter is divided into 5 sections. Section 2 deals with material and technological issues. Required knowledge about RF properties of HTS material and thin-film technology is briefly recalled. This is followed by a presentation of some relevant aspects with regard to assembling and integration problems. Section 3 provides fundamentals for the design of high-performance HTS filters. Basic definitions and general properties of filters – with emphasis on the degradation of filter performance by dissipation- are introduced. This is followed by a relatively comprehensive discussion of different filter topologies and related design issues. These aspects turn out to be crucial for the realisation of high-performance filters for challenging applications. In section 4 characteristic examples for HTS filters realised by several groups are presented. Section 5 discusses already existing as well as potential system applications of HTS filters on a sub-system and a system level. Due to the large diversity of the various system requirements this section is focused on sub-system considerations.

Throughout this report the term radio-frequency (RF) is used in a generalised sense to refer to the entire frequency regime between about 30 MHz to 300 GHz which includes the very high frequency (VHF, 30 –300 MHz), the microwave (0.3 to 30 GHz) and the millimetre waves (30-300 GHz) frequency ranges. The operational frequency range of HTS filters considered in this chapter is typically in the regime between about 30 MHz and 10 GHz.

2. MATERIAL AND MANUFACTURING ISSUES

This section starts by recalling basic properties of HTS materials (subsection 2.1) which are relevant to RF-applications and are therefore essential for an understanding of the following sections of this chapter. Since these RF-properties are intimately related to the structure of the HTS material, a further sub-section (2.2) is devoted to thin film technology. Furthermore, sub-section 2.3 covers structurisation as well as assembly and integration technologies which are required to progress from the bare wafer step-by-step towards an operating device such as a RF filter.

2.1 RF properties of HTS materials

Superconducting materials display a number of properties which are unique and which are fundamentally different from those of normal conducting conventional metals (see e.g. [6,7]). The most remarkable one is an electrical DC-resistance which completely vanishes below the so-called critical temperature T_c. This temperature is characteristic for the material, and for HTS materials typically used in RF-applications T_c is about 90 – 110 K. All superconducting properties are only shown below T_c.

In the following, the idealised structure of a conducting (normal or superconducting) half-plane is used to introduce some basic RF-properties. Assuming a homogeneous time-harmonic magnetic field of frequency ω to exist at the surface of this half-plane, an electromagnetic field penetrates the conductor and its magnitude decays, due to the finite (in general complex-valued) conductivity σ, with the penetration depth δ given by

$$\delta = \frac{1}{\mathrm{Re}\left[\sqrt{j\omega\mu_0\sigma}\right]} \tag{2.1}$$

with μ_0 denoting the permeability of free space. Furthermore, at the surface of the conductor the phasors of the tangential electric and magnetic fields E_t and H_t are related by the complex valued surface impedance Z_s:

$$Z_s = R_s + jX_s = \frac{E_t}{H_t} = \sqrt{\frac{j\omega\mu_0}{\sigma}} \tag{2.2}$$

In conventional metals, the RF conductivity σ is generally real-valued and frequency-independent (classical skin effect). Consequently, according to eqs. (2.1/2) the penetration depth (skin depth) δ decreases with the square root of the frequency and real and imaginary part of the surface impedance are identical and possess a $\omega^{0.5}$ dependence.

The result for superconductors is substantially different. For frequencies in the microwave regime and for temperatures between about $0.6T_c$ and T_c, further and sufficient insight is provided by the phenomenological two-fluid model: The set of all electrons carrying the current is subdivided into a subset (fraction f_n) of normal conducting and a subset (f_s) of superconducting electrons, with $f_n + f_s = 1$. The superconducting electrons are bound to pairs, the so-called „Cooper pairs“, and it is this part of the electrons which carries the superconducting current. The distribution between the normal and superconducting parts depends on the temperature. Current transport by the normal conducting electrons (part f_n) is governed by the same effects as current transport in conventional metals: The motion of electrons under the influence of an electric field is dominated by scattering processes with a relaxation time τ. The superconducting part is -due to the finite mass of the electrons-subject to inertia. Both parts contribute to the conductivity which therefore becomes:

$$\sigma = \sigma' - j\sigma'' = \frac{ne^2}{m}\left(f_n\tau - j\frac{f_s}{\omega}\right) \tag{2.3}$$

with n being the density of electrons, e their charge and m their mass, respectively.

Some important conclusions can be deduced from eq. (2.3): A time-independent current (ω=0) implies an infinite conductivity of the superconducting charge carriers, and therefore DC-currents are carried with zero resistance. The imaginary part of the complex-valued conductivity displays a frequency-response characteristic for an inductive component.

For HTS, the temperature dependence of f_n and f_s is approximated by

$$f_n = 1 - f_s = \left(\frac{T}{T_c}\right)^{\gamma} = t^{\gamma} \qquad (2.4)$$

with t denoting the so-called reduced temperature. Often, γ is used as a fit parameter when fitting experimental data, and typically $\gamma \cong 2$.

After inserting the expression for the conductivity into the eqs. (2.1/2.2) for the penetration depth and surface impedance, one can use $\sigma`` >> \sigma`$ to perform a series expansion and to neglect higher-order terms. Thus, the following results are obtained:

$$\delta(T) = \sqrt{\frac{m}{ne^2 \mu_0}} \frac{1}{\sqrt{1-t^{\gamma}}} \equiv \lambda_L(0) \frac{1}{\sqrt{1-t^{\gamma}}} = \lambda_L(T) \qquad (2.5)$$

$$R_S(\omega, T) = \frac{1}{2} \omega^2 \mu_0 \tau \lambda_L(0) \frac{t^{\gamma}}{\left(1 - t^{\gamma}\right)^{\frac{3}{2}}} \qquad (2.6)$$

$$X_S(\omega, T) = \omega \mu_0 \lambda_L(T) = \omega L(T) \qquad (2.7)$$

$L(T)$ is called „kinetic inductance“ since it accounts for both stored magnetic field energy as well as kinetic energy of the accelerated electrons. In case of superconductors the penetration depth becomes identical with the so-called London penetration depth λ_L.

All these quantities depend in very characteristic ways on frequency and temperature, as will be detailed now:

In contrast to conventional metals, the penetration depth λ_L in HTS becomes to a first order independent of the frequency. The London penetration depth is typically 250 nm at a temperature of 77 K.

The surface resistance R_S of HTS scales with the frequency squared. Comparing the HTS material YBCO to copper (both at 77 K), at 4 GHz the surface resistance of YBCO is lower by two orders of magnitude, whereas at about 100 GHz it becomes equal:

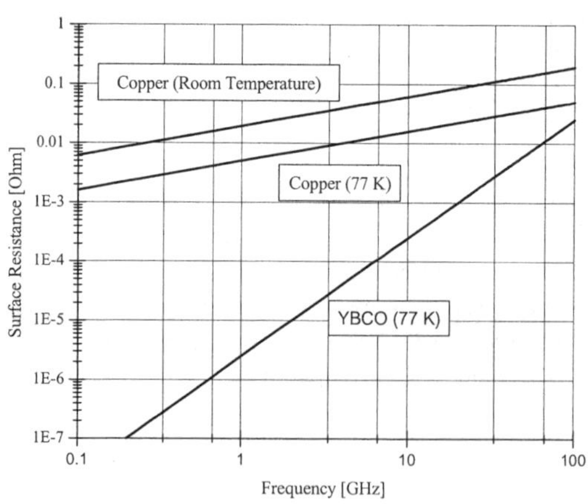

Fig. 2.1: Frequency dependence of the surface resistance of HTS (here: YBCO at 77 K) compared to conventional metal (here: copper).

For HTS the surface reactance X_S is for HTS much larger than the surface resistance R_s (in contrast to conventional metals where they are equal).

The operational temperature for HTS devices should be chosen sufficiently lower than T_c in order to take advantage of the fact that with lower temperature a reduction is obtained for both (i) the surface resistance (Fig. 2.2, left) and (ii) the steepness $\mathrm{d}L(T)/\mathrm{d}T < 0$ of the temperature dependence of the kinetic surface inductance $L(T) = \mu_0 \lambda_L(T)$.
Fact (i) results in a higher unloaded Q of the device, whereas (ii) results in a lower sensitivity of the frequency response of a device to changes in temperature (see Fig. 2.2, right).
Devices based on YBCO are generally operated at 77 K or below. A resonator oscillating at 10 GHz then has a shift in resonance frequency of about 1 MHz when the temperature is varied by 1 K.

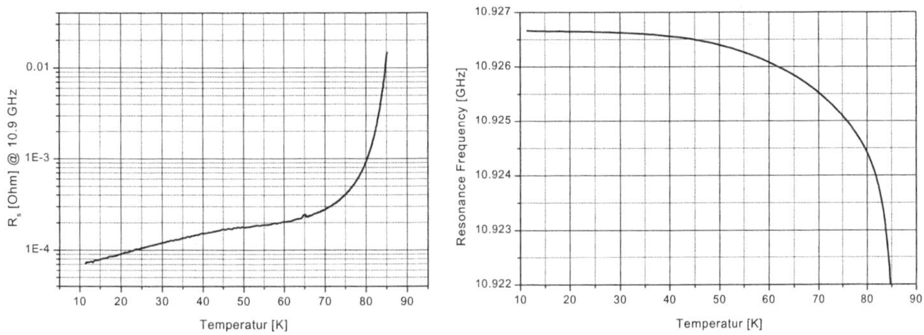

Fig. 2.2: Influence of temperature on HTS properties; left: Surface resistance of YBCO at 10.9 GHz; right: resonant frequency (determined by surface reactance) of an HTS resonator with $T_c = 88.5$ K (data are from the commercial HTS film supplier THEVA [8]).

The used superconducting thin-films often have a thickness on the order of the London penetration depth. Eqs. (2.6/7) do not account for this finite thickness and therefore give slightly too low values for the surface resistance. However, a correction for finite film thickness is straight-forward [9].

The two-fluid model introduced above represents a linear model with respect to the relationship between electric field strength and current density at a given temperature. However, in reality a strong non-linear response occurs if the RF magnetic field strength at the surface of the HTS conductor exceeds a certain temperature-dependent limit (see e.g. [7] and [10]). Above this limit the conductivity becomes a function of current density. Hence, the surface impedance becomes a function of the RF surface magnetic field strength (further increase of the field strength does finally suppress all superconductivity). This limit is, for the material YBCO as an example, for 85 K at about 3.2 mT, for 77 K at 6.4 mT and for 60 K at 23 mT. This fact further supports the statement to chose the operational temperature not too close to T_c.

2.2 Thin film technology

2.2.1 Motivation

In the preceding section it was shown that the RF-properties of HTS materials are fundamentally different from those of conventional metals and that the major advantage of HTS is a drastically reduced surface resistance in the microwave regime. However, this advantage applies only if the HTS material possesses a sufficient crystalline purity.
Single crystals cannot be produced in technologically relevant sizes. However, sufficient crystalline properties are obtained from epitaxially grown thin films on suitable dielectric wafers. Therefore HTS microwave devices are realized as planar circuits.

HTS materials exhibit an anisotropic conductivity. The extreme low value for the surface resistance applies only for currents flowing in a particular ("optimum") plane (the value provided in Fig. 2.1 corresponds to this plane). Currents flowing along the axis perpendicular to this optimum plane experience significant higher dissipation.. It is a lucky circumstance that the crystal orientation of HTS films on dielectric wafers is

(i) such that the plane parallel to the wafer surface coincides with the "optimum plane", and
(ii) the current flow in planar circuits is in parallel to the wafer surface.

2.2.2 Wafer materials

Dielectric wafers suitable for coating with HTS must provide a crystal lattice at the surface similar to that of the HTS film in order to allow for epitaxial growth of the HTS. Furthermore, the wafer must exhibit low electrical losses (small loss-tangent). A high thermal conductivity is desirable to ease cooling and thermal feedback circuitry of the HTS device. The wafer must also be able to withstand the high process temperatures during growth of the HTS film (800° C - 900° C), and of course a high mechanical stability is advantageous when handling the device.

In nearly all realised HTS microwave circuits either sapphire, lanthanum aluminate or magnesium oxide was used as wafer material.

Sapphire has a dielectric constant of about 10, but unfortunately it is dielectrically anisotropic (ε_r-tensor) and, furthermore, HTS can only be grown on a certain crystal plane which does not coincide with the main axes of the ε_r-tensor. The design of a circuit is rendered difficult by the fact that presently available simulation tools cannot cope with this anisotropy and that therefore an approximation by some average value for the dielectric constant must be employed. Furthermore sapphire still possesses some lattice mismatch to the HTS film and despite the fact that a layer sandwiched between the wafer and the HTS (usually ceroxide) compensates partly for the mechanical stress, the film thickness of the HTS is limited to about 400 nm which is disadvantageous with respect to the dissipation and to the onset of non-linear effects (power handling behaviour). Despite these drawbacks sapphire is mechanically very stable, cheap, has superior loss behaviour and high thermal conductivity.

Lanthanum aluminate has a dielectric constant on the order of 24.5 allowing for a higher degree of miniaturisation, and therefore it is often employed for devices at lower operating frequency. However, lanthanum aluminate exhibits twin boundaries with different and random orientation of the anisotropic dielectric tensor between twins, and the resulting uncertainty makes it difficult to use for circuits requiring high precision such as filters with very small bandwidth.

Magnesium oxide again has a lower dielectric constant of 9.8, and since it is isotropic circuits can be designed to a high degree of accuracy. HTS films can be grown on magnesium oxide up to more than twice of the thickness achievable with sapphire. The latter fact results in a lower surface resistance. Compared to sapphire, magnesium oxide is mechanically brittle. But with increasing knowledge in handling and assembly and integration techniques and also with increasing demand for high precision circuits, it seems today that magnesium oxide is the first choice for many groups. All wafers are readily available in standard thicknesses (15, 17 and 25 mil) and diameters (2, 3 and 4 inch) as typically required for HTS circuits.

2.2.4 HTS materials

Next, the choice between different HTS materials is discussed. The most widely used material is $YBa_2Cu_3O_{7-\delta}$ (YBCO) with a critical temperature of 88 - 90 K. At a temperature of 77 K, the surface resistance at 10 GHz is 500 - 600 μΩ for films on sapphire and about 300 μΩ for films on magnesium oxide. There are numerous methods known to produce high-quality films [11]. All methods are typical thin film technologies such as sputtering, thermal evaporation and pulsed laser deposition. Thus, a vacuum chamber is needed to provide ultra clean conditions. YBCO can be grown on all wafer materials and on both sides with similar film quality on each side. This material is readily commercially available [8] with all the associated advantages such as data sheets, test reports, quality assurance and traceability.

It is possible to replace the Yttrium in YBCO by rare earth ions as, for example, Gadolinium, Dysprosium, Erbium, or Holmium [12]. Due to similar chemical properties the same fabrication procedures can be used, and the resulting films differ only slightly (they are generally inferior to YBCO, but this can be attributed to a lack of experience with respect to the optimisation of process parameters). At this point it should be mentioned that the avoidance of patent fees for YBCO represents a main motivation for investigations in these new materials.

Increasing interest is given to $Tl_2Ba_2CaCu_2O_{7-\delta}$ (TBCCO) mainly because it exhibits a higher critical temperature of 102 - 105 K. In applications where cooling power is a rare resource (such as in satellite applications) and the HTS device is therefore operated at a temperature as high as possible, the use of TBCCO relaxes the power balance. This is demonstrated in Fig. 2.3 where it is shown that, for example, the surface resistance R_S for YBCO at 77 K is obtained with TBCCO at already 85 K. If, on the other hand, ultimate performance is the primary design goal the device is operated at lower temperature to reduce R_S. Then it is seen that below about 67 K YBCO is superior to TBCCO. The fabrication process for TBCCO is inherently more complicated than that for YBCO and also much less experience has been gained so far. Consequently, although double-sided high quality films on wafers up to 3 inch diameter have been produced [13,14,15], the development is ongoing with focus on further increase of film quality and reproducibility.

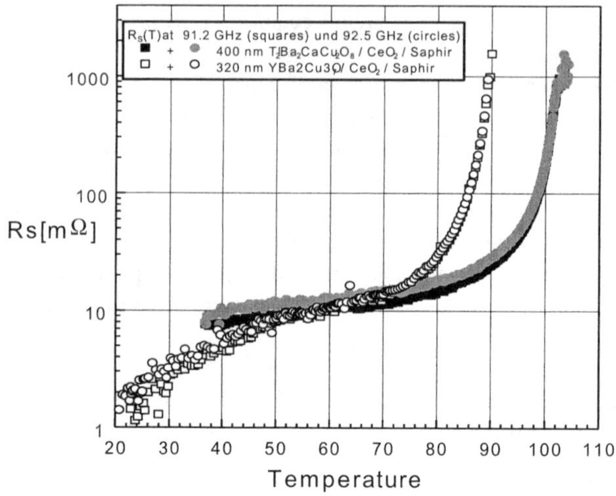

Fig. 2.3: Surface resistance for YBCO and for Thallium-based thin HTS films

Still higher operating temperatures are envisaged by the use of films based on mercury ($HgBa_2CaCu_2O_{7-\delta}$ / HBCCO) which have a critical temperature of 120 - 125 K. However, the fabrication process is again even more involved, and up to now it was only possible to realize films on small areas of a few mm^2 [16,17,18]. It is expected that still a significant amount of development effort is needed until HBCCO films can be employed in microwave devices.

When a HTS film is exposed to the ambient environment, a thin oxide layer is immediately formed at the surface of the HTS. This layer is electrically isolating and is a barrier for the realization of good electrical contacts with low contact resistance. Therefore a thin gold layer is often added on top of the HTS film, either in-situ (that is immediately after the fabrication of the HTS itself, without breaking the vacuum) or after an additional cleaning step removing the oxide layer.

2.3 Structurisation, assembly and integration technologies

The outcome of sub-section 2.2 is a wafer coated at least on one and more often on both sides with HTS material, and often with a gold layer on top of the HTS. The next step is to structurise the RF-circuitry, that is to remove all HTS material except for the conducting parts of the resonators or lumped elements and transmission line sections of the filter. This is achieved by standard etching techniques. First, the gold coating is etched away from all parts except those places where electrical contacts to the outer (non-HTS) circuitry are to be built. For HTS structures both chemical wet etching and more sophisticated techniques such as ion-beam etching have been employed successfully. It is generally reported that chemical wet etching, though being the cheapest technique, is fully sufficient for HTS microwave devices. The smallest dimensions for both line width and spacing between lines are chosen not smaller than about 10 μm.

After the etching process, the wafer is mounted into a suitable housing and electrical contacts are made. The following picture depicts such a setup:

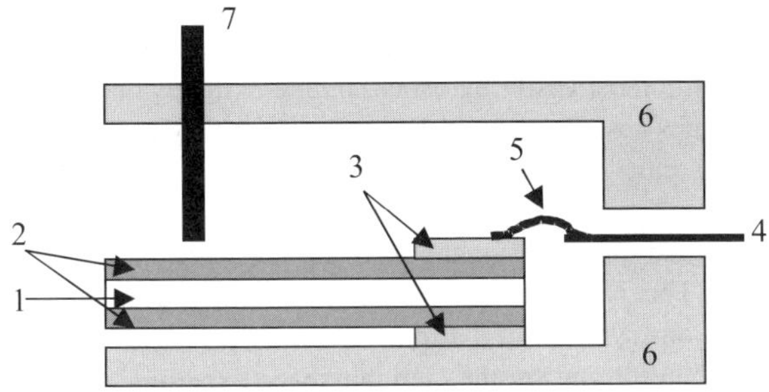

Fig. 2.4: Assembly and integration technologies for planar HTS circuits: The wafer (1) is coated on both sides with HTS (2) and gold (3) on top where necessary for contacts. The backside of the wafer is mounted into the housing (6), and on the front side the feed line of the circuit is connected via a bond wire or ribbon (5) to an HF-connector (4). Many circuits need tuning which can be well achieved by tuning screws (7).

The choice of a suitable material for the housing is highly important. Here, one has to bear in mind the large temperature range, reaching from the cryogenic temperature at which the device is operated to the ambient temperature where the device is set up and stored before system implementation. The best fit in thermal expansion between the wafer and the housing is achieved with steel, however, at the expense of maximum weight. The other extreme is aluminium which has very low weight, but a difference in thermal expansion to the wafer of about a factor two; besides, aluminium has the advantage of better heat conduction compared to steel. There are furthermore alloys allowing for a compromise between tolerable thermal expansion and favourable weight [19]. The best choice depends on the operating temperature, on the absolute size of the wafer and housing and also on the maximum rate of cooling (and heating). The housing is often plated with silver or gold, which reduces the electrical losses, prevents rusting of the housing and which facilitates mounting of the wafer into the housing. (Whereas silver has better electrical properties than gold, gold allows for a wider choice of mounting technologies.)

For mounting the wafer, there is the choice between gluing and soldering. Since most RF-circuits involve current flow on both sides of the wafer, the choice is generally limited to electrically conducting glues. Soldering is superior to most glues with respect to electrical conductivity, but can hardly be used with aluminium housings. Furthermore, the glue or solder can behave in an elastic way, and can thus compensate (to some extent) for thermal mismatch between the housing and the wafer.

After the wafer is mounted in the housing, electrical contacts between the feed lines on the wafer and either feed lines of neighboured wafers or connectors have to be set up. Standard techniques are bonding of thin gold wires or welding of gold ribbons, which can both also be applied to HTS circuits.

As in conventional technology (e.g. metallic cavity-filters and dielectric filters), due to errors in the numerical modelling as well as in the reproducibility of mechanical manufacturing and material properties, the instantaneous (directly after fabrication) filter response will fail to meet the design goals (specifications). Therefore, the device needs to be tuned, which can be done most easily by using commercially available tuning screws. Some more elaborate techniques have been published, e.g. selectively removing HTS material with a laser [20] or coating parts of the circuit with a dielectric layer [21], but none of these approaches can compete with tuning screws with respect to simplicity.

All these technological problems mentioned so far have been well solved by many companies and groups. This is well proved by the HTS front-ends in operational base stations in cellular communications systems, and also by the American space experiment HTSSE [22, and many references therein] and by the German space experiment where these techniques have been fully qualified during the preparation phase according to standard procedures in space industries [23]. However, it must be noted that a newcomer to this field will most likely need a number of own trials to find his own solution for the assembly and integration problems in every detail.

A further point deserves attention in this section: HTS material is sensitive to storage at elevated temperatures combined with elevated levels of humidity. This can be well circumvented in mobile communications, where it is easy to provide a dry atmosphere during fabrication and storage, and where the devices are placed in a vacuum dewar with zero humidity in the subsequent operational phase. However, it represents a more severe problem for application in space communications, where typical requirements demand a storage time of three weeks at a temperature of 40° Celsius and a relative humidity of 93 % with no degradation (as it can appear, for example, when the satellite is mounted on the spacecraft and awaits launch in the hot and humid environment e.g. of Florida or of Guayana). Bare HTS material degrades in such an atmosphere within less than an hour. Despite considerable efforts, up to now it was not possible to find a passivation layer which can be applied to the HTS film and which prevents the HTS film from degradation without negative impact on the high-frequency behaviour [24]. Therefore it is necessary to mount the HTS device in a hermetically sealed housing with dry atmosphere inside. Such hermetically sealed housings are common for other devices in satellite communications, but they are obviously associated with a penalty in mass and also in complexity during setup and testing.

3. FILTERS: GENERAL CONSIDERATION

3.1 Definitions, general properties and different filter types

From a general point of view we consider a passive RF filter to be a linear two-port which changes amplitude and phase of an output signal at port 2 relative to amplitude and phase of an incident signal at port 1 in a prescribed manner as a function of the frequency *f*. It is this desired frequency dependence which makes a two-port to be a filter. The complex-valued transmission coefficient

$$S_{21}(f)=|S_{21}(f)|\exp\left(j\Phi_{21}(f)\right) \tag{3.1}$$

relates the wave-coefficient of the time-harmonic wave ($\propto \exp\left(j2\pi ft\right)$) at port 2 to the wave-coefficient of the wave incident at port 1. Thus, the complex-valued frequency response of the transmission coefficient can be described by 2 real-valued functions, namely the amplitude response $|S_{21}(f)|$ related to the *insertion loss* response via

$$L(f)/\mathrm{dB}=-10\log\left(|S_{21}(f)|^{2}\right), \tag{3.2}$$

and the phase response $\Phi_{21}(f)$ related to the *group delay* response via

$$\tau(f)=-\frac{1}{2\pi}\frac{\mathrm{d}\Phi_{21}(f)}{\mathrm{d}f}. \tag{3.3}$$

Since the insertion loss equals the negative of the magnitude of S_{21} in dB, results in section 4 taken from other authors and displayed as a graph sometimes mark the axis by "S21" to refer to the insertion loss in dB.

If the passive RF filter structure does not contain gyrotropic material (like e.g. magnetised ferrites) it is reciprocal. This means that the transmission coefficient $S_{12}(f)$ from port 2 to port 1 equals $S_{21}(f)$. For a complete description of the two-port the reflection coefficients $S_{11}(f)=|S_{11}(f)|\exp\left(j\Phi_{11}(f)\right)$ and $S_{22}(f)=|S_{22}(f)|\exp\left(j\Phi_{22}(f)\right)$ for port 1 and 2, respectively, are also to be known. Return loss for port 1 is defined as

$$L_{r}(f)/\mathrm{dB}=-10\log\left(|S_{11}|^{2}\right), \tag{3.4}$$

and an analogue definition holds for port 2. The four parameters S_{mn} are also known as S-parameters which represent the elements of the 2 x 2 scattering matrix.

Whereas reciprocity and passivity are properties of the real filter structures, non-dissipative filters do not exist in reality. However, non-dissipative filters are often used as (idealised) reference filters and nearly all methods for filter synthesis are based on the assumption of a vanishing dissipation. Dissipative losses in real filter structures are considered as "perturbation" of an ideal non-dissipative filter structure. Because of this, it is important to know that the scattering matrix of an ideal non-dissipative linear and reciprocal filter is unitary. Thus, the different S-parameters are related by the following equations

$$|S_{22}(f)| = |S_{11}(f)| \tag{3.5a}$$

$$|S_{11}(f)|^2 + |S_{12}(f)|^2 = 1 \tag{3.5b}$$

$$\Phi_{21}(f) = \frac{1}{2}\left(\Phi_{11}(f) + \Phi_{22}(f) \pm \pi\right). \tag{3.5c}$$

The frequency response (amplitude and phase) of a filter is based on the storage of electromagnetic field energy in the filter structure. It can be shown that for time-harmonic signals and the ideal case of a non-dissipative filter the mean field energy W stored in the filter can be related rigorously to the power P_{inc} of the incident wave:

$$W(f) = \tau(f) P_{inc}(f) \tag{3.6}$$

Hence, for given incident power the group delay completely determines the stored field energy.

From a general point of view (without taking practical limits into account) any prescribed amplitude response $|S_{21}(f)| \leq 1$ can be closely approximated by a non-dissipative filter. However, it is important to note that once the amplitude response is fixed the group delay at any arbitrary frequency f must equal or exceed a minimum value (derived from minimum phase) determined by the amplitude response [25]

$$\tau(f) \geq \tau_{\min}(f) \text{ with } \tau_{\min}(f) = 0.0117 \int_{-\infty}^{\infty} \frac{f'^2 + f^2}{\left(f'^2 - f^2\right)^2} L(f')\, df' \tag{3.7}$$

The desired function of a RF filter can primarily rely (a) on the frequency dependence $L(f)$ of its insertion loss or (b) on the frequency dependence $\tau(f)$ of its group delay.

The class (a) of filters where the insertion loss frequency dependence determines the filter function can be subdivided into lowpass, bandpass, highpass and bandstop filters. In an ideal bandpass all frequency components are rejected, except they fall into a pre-specified passband. In an ideal bandstop filter all frequency components are passed through, except they fall into the stopband. A bandstop filter with a very small fractional bandwidth is also called a notch-filter. In the most general case a filter may possess multiple pass- and stopbands.

In the class (b) of filters where the group delay response determines the desired filter function, the insertion loss is required to be as low as possible in the operational frequency range, but the group delay is required to follow a prescribed function of the frequency. Examples for this class of filters ("allpass") are (b1) non-dispersive delay lines with $\tau(f) = \tau_0 = const$, (b2) chirp filters (pulse-compression filters) (b3) RF equalisers for external group delay equalisation.

In case of the chirp filter (b2), a group delay being a linear function of the frequency, e.g. $\tau(f) = 0.5(\tau_{\min} + \tau_{\max}) + (\tau_{\max} - \tau_{\min})(f - f_0)/\Delta f$ in the operational band $|f - f_0| < \Delta f/2$ is to be realised. This filter can, e.g., be used to compress a linearly frequency-modulated pulse of time duration Δt and with $\Delta t \cdot \Delta f = \Lambda >> 1$ into a very short pulse of time-duration $\Delta t / \Lambda$ [26].

For all types of filters mentioned above "optimum" performance would be achieved with non-dissipative structures. Unavoidable presence of dissipative losses leads to a degradation of the performance. A general measure for the amount of dissipation in a filter structure is provided by the (generalised) *unloaded quality factor* Q_0 which relates the dissipated power P_{diss} via

$$P_{diss} = \frac{\omega W}{Q_0} \tag{3.8}$$

to the mean stored field energy W. Combining eqs. (3.6) and (3.8) an estimate for the degradation due to dissipation can be derived. The power transmission with dissipation is via

$$\left|\tilde{S}_{21}(f)\right|^2 \approx \frac{\left|S_{21}(f)\right|^2}{1+2\pi f\tau(f)/Q_0} \tag{3.9}$$

approximately related to the power transmission $\left|S_{21}\right|^2$ without dissipation. Therefore, the dissipation causes the insertion loss to decrease by $\delta L(f)$:

$$\tilde{L}(f) \approx L(f)+\delta L(f) \text{ with } \delta L(f) \approx 10\log\left(1+2\pi f\tau(f)/Q_0\right) \approx 27.2\cdot f\tau(f)/Q_0 \tag{3.10}$$

From eq.(3.10) very important conclusions for the usage of HTS technology can be drawn: Depending on the system application an upper limit for the maximally tolerable degradation (with $\delta L \le \delta L_{\max}$) must always be introduced. This in turn limits (for a given value of Q_0) the maximum achievable "delay-frequency product" via

$$f\tau \le 0.037\cdot\frac{\delta L_{\max}}{\text{dB}}\cdot Q_0 \tag{3.11}$$

This is the rational of HTS applications to filters. Because HTS technology offers much higher Q_0 values than conventional technology, significantly higher group delay values become available. For example, if $\delta L_{\max} = 1$ dB and $f = 4$ GHz are assumed, the achievable group delay is increased from about 20 ns to 1 μs if a normal conducting coaxial structure with $Q_0 \approx 2000$ is replaced with a HTS structure with $Q_0 \approx 100{,}000$.

The importance of higher values of τ is directly obvious in case of filters where the group delay response determines their function (allpass filters). However, it is of equal importance for filters where the function is related to the frequency dependence of the insertion loss (bandpass, bandstop etc.). This comes from the fact that via eq.(3.7) an increase in the steepness of filter skirts always leads to higher values in the associated group delay. This in turn means that the achievable filter steepness (selectivity) is limited by Q_o, too.

All types of filters characterised by a relatively high value of the group delay, namely bandpass and bandstop (including notch) filters with very sharp skirts as well as dispersive (including chirp filters) and non-dispersive delay lines have been successfully realised and tested in HTS technology. In the remaining part of this review we will focus on narrowband bandpass filters. For more details on chirp filters and their applications, see e.g. [26,27].

3.2 Bandpass filters: Specifications and their approximation by a suitable pole-zero scheme. Impact of dissipation.

This subsection is restricted to the topic of narrowband bandpass filters which are required for the majority of applications (see section 5).

Filtering requirements derived from system demands are typically indicated by means of a tolerance scheme for insertion loss, return loss and group delay. Some key parameters for insertion loss specifications are depicted in Fig. 3.1. The passband is specified by its width Δf and centre frequency f_0 as well as (not shown in Fig. 3.1) by the maximum passband insertion loss L_0 und minimum passband reflection loss $L_{r,\min}$. The upper and lower stopbands are specified by a minimum stopband attenuation $L_{s,\min}$ as depicted in Fig. 3.1. These stopband attenuations may be different in both stopbands. The width of the frequency range in which the insertion loss increases from the passband value L_0 to the stopband value $L_{s,\min}$ is denoted by $\delta f_{1,2}$ and called the width of the transition region(s). Within the passband there are often restrictions for the group delay variations. The tolerance scheme can be much more complex as those shown in Fig. 3.1, e.g. it can even include upper values on the frequency derivatives of insertion loss and group delay.

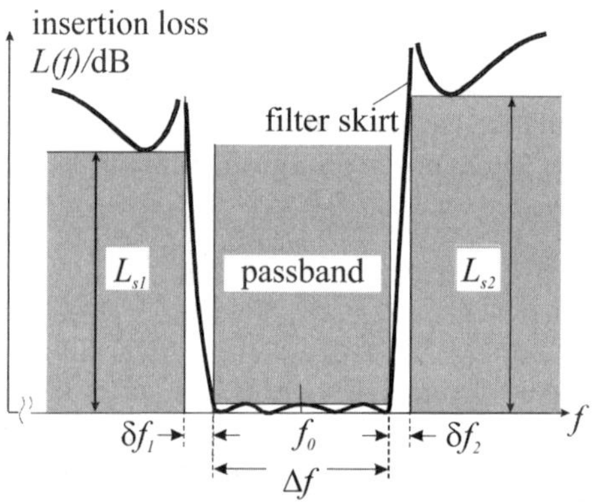

Fig. 3.1: Example for the insertion loss response specifications for a bandpass. Key parameters for the passband, the lower and upper stopbands as well as for the transition regions.

Once the specifications are known, the approximation problem has to be solved. That means, that a function for $S_{21}(f)$ has to be found which (I) satisfies all filtering requirements discussed above, and (II) which is realisable within the considered class of filter structures. Here, two classes of HTS filter structures are taken into account, namely (A) structures comprising coupled resonators, and (B) lumped-element filters composed of planar capacitors and inductors. The resonators in class (A) can in principle be distributed planar resonators or composed of lumped elements. However, in class B there are LC-circuits included which cannot be modelled as coupled-resonator circuits.

In case of lumped-element bandpass circuits (class B) it is well known that $S_{21}(s)$ is a real rational function of the complex frequency $s = \sigma + j\omega$

$$S_{21}(s) = K\frac{\prod_{\mu}(s - s_{0,\mu})}{\prod_{\nu}(s - s_{p,\mu})}, \tag{3.12}$$

which is uniquely determined by its poles, zeros and the scale factor K. The *poles* $s_{p,\nu} = \sigma_{p,\nu} + j\omega_{p,\nu}$ with $\sigma_{p,\nu} < 0$ coincide with the complex-valued eigen-frequencies of the filter between resistive terminations. The zeros $s_{0,\mu} = \sigma_{0,\mu} + j\omega_{0,\mu}$ are the *transmission zeros (TZ)*. Since with each pole in the upper complex plane the conjugated value in the lower complex plane is also a pole, it is (for a bandpass) sufficient to consider the poles in the upper halfplane, only. The same holds for the zeros. An N-pole bandpass has N poles in the upper half plane. The numerator degree in eq. (3.12) is less than or equal to the denominator degree. The difference between denominator and nominator degree determines the number of TZs at infinity. The number of transmission zeros at finite frequency in the upper half-plane is denoted by M.

In case of coupled-resonator bandpass filters (class A), the narrowband approximation allows the transmission coefficient $S_{21}(s)$ to be represented by expression (3.12), too. This approximation is valid only in the regime of the upper s-plane which is "close" to the centre frequency $j\omega_0$. Generally, the narrowband approximation works well for filters with a fractional bandwidth of less than about 5%.

It is convenient to transform the bandpass response according to eq. (3.12) via the BP-LP transformation

$$\tilde{s} = \frac{f_0}{\Delta f}\left(\frac{s}{2\pi f_0} + \frac{2\pi f_0}{s}\right) \tag{3.13}$$

into a lowpass (LP) representation. This means that N poles and M zeros (upper half-plane) from the bandpass regime are represented by N LP-poles and M LP-zeros related to the BP-poles and zeros via eq.(3.13). It should be noted that in this equivalent LP representation the transmission coefficient is still a rational function in $\tilde{s}$, but the coefficients of the polynomials need not necessarily to be real. Hence, poles and zeros in the LP-domain need not to be symmetric to the real axis.

With this representation of the filter response via poles and zeros in the LP-domain, the above mentioned approximation problem is transformed into the problem to find the locations of a suitable number N (filter-order) of (LP)-poles in the left half-plane and a suitable number M of zeros at finite frequencies, such that the corresponding function $S_{21}(s = j\omega)$ meets all filter specifications.

The following general considerations hold for the choice of pole and zero locations:

(i) If no transmission zeros are placed in the $\tilde{s}$-plane (LP) at finite frequencies (M=0), and the locations of N poles are chosen symmetrically to the real axis and such that equal-ripple passband response for the insertion loss is obtained, one ends up with a standard N-pole *Chebyshev response.*

(ii) If the locations of N poles and $M = N$-2 zeros are chosen symmetrically to the real axis with all zeros placed outside of the passband at the imaginary axis, such that equal ripple response is achieved for both the (small) passband insertion loss and the (large) stopband attenuation, one ends up with a standard N-pole *elliptic response*.

(iii) If the locations of N poles and of $0<M<N-2$ zeros are chosen symmetrically to the real axis with all zeros placed outside of the passband at the imaginary axis, such that equal ripple response is achieved for both the (small) passband insertion loss and the (large) stopband attenuation, one ends up with a N-pole *quasi-elliptic response*.

(iv) In case of (i) to (iii) the group delay response is completely determined by the approximation for the insertion loss response. In order to gain additional degrees of freedom for the group delay response (for given insertion loss response) pairs of TZs with $s_{0,\mu} = \pm\sigma_{0,\mu} + j\omega_{0,\mu}$ with $|\sigma_{0,\mu}| > 0$ can be introduced, e.g. for to equalise the *group delay response*.

(v) As long as poles and zeros are located symmetrically to the real axis, like in (i) to (iv), the corresponding bandpass response becomes (see eq. (3.13) symmetrically in frequency with respect to the centre frequency f_0. However, if poles and zeros are no more located symmetrical to the real axis, an *asymmetric frequency response* in the bandpass domain is obtained.

The advantage of *elliptic and quasi-elliptic response* (TZs at imaginary axis) in comparison to the *Chebyshev response* (no TZs at finite frequency) is that the same skirt steepness can be achieved with a lower order N. This is illustrated in Fig. 3.2 where the required order N is shown as function of the ratio $\Delta f / \delta f$. The second advantage of elliptic filters versus Chebyshev filters is that for the same skirt steepness the group delay close to the band edge becomes lower than for Chebyshev filters.

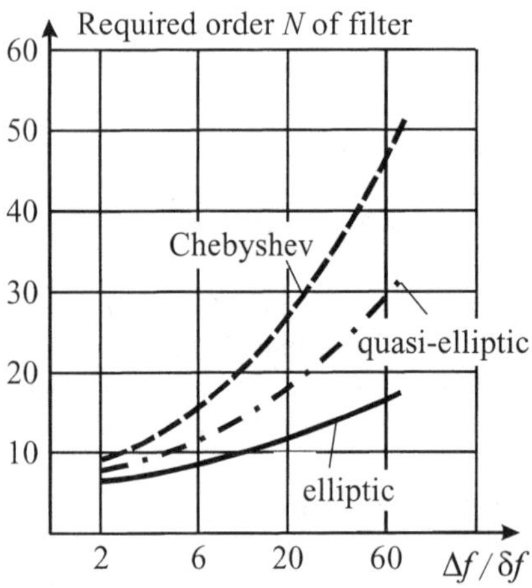

Fig.3.2: Required order N of bandpass as function of $\Delta f/\delta f$ for the following specifications: Minimum passband return loss $L_{r,min}$= 18 dB, minimum stopband attenuation $L_{s,min}$= 90 dB.

In Fig. 3.2 a non-dissipative structure is assumed. Taking losses (described by Q_0) into account the insertion loss is increased according to eq.(3.10) as a function of f und proportional to the value of the group delay. This means that the passband insertion loss can (due to the dissipative insertion loss) become significantly higher than the insertion loss predicted from the non-dissipative filter model. This fact is illustrated in Fig. 3.3.

For an upper limit for the dissipative passband insertion loss of $\delta L_{\max}$ the required unloaded quality factor becomes [28]

$$Q_{0,\min} = \beta \frac{L_s / \mathrm{dB}}{\delta f / \mathrm{MHz}} \cdot \frac{f_0 / \mathrm{GHz}}{\delta L_{\max} / \mathrm{dB}}, \quad (3.14)$$

with constant β being on the order of 250 for elliptic and 750 for Chebyshev response, respectively.

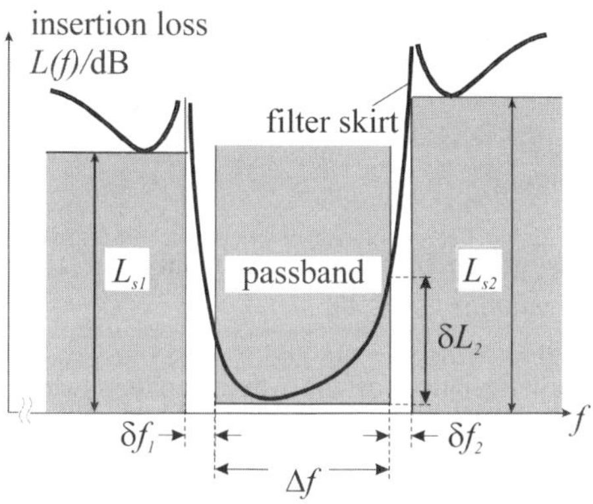

Fig. 3.3: Degradation of insertion loss response due to dissipation

Details on the design of lumped element HTS bandpass filters can e.g. be found in [29] and an example is presented in section 4. We will now focus on the class of coupled-resonator bandpass filters [30]. Here the order N of the filter is determined by the number of coupled resonators, or the number of coupled resonant modes in case of multi-mode resonators. For filters with symmetric frequency response all resonators (resonant modes) are tuned to the bandpass centre frequency and the locations of poles and zeros are determined by the strength and sign of the coupling between the resonators. In case of asymmetric frequency response individual de-tuning of the resonators from the centre frequency can become necessary.

3.3 Topology of bandpass filters composed of coupled resonators

In sub-section 3.2 the solution of the approximation problem via pole and zero location in the complex frequency plane was outlined. The subsequent steps are to find a filter topology and the corresponding values of the filter parameters from the knowledge of the poles and zeros. This step-referred to as *synthesis*- is discussed in this sub-section for the class of bandpass filters composed of coupled resonators. Although this sub-section is still on some fundamentals of filter design, the specific impact of HTS technology and its characteristic properties are strongly emphasized.

A *lowpass (LP) filter prototype* is represented by the network depicted in Fig. 3.4.

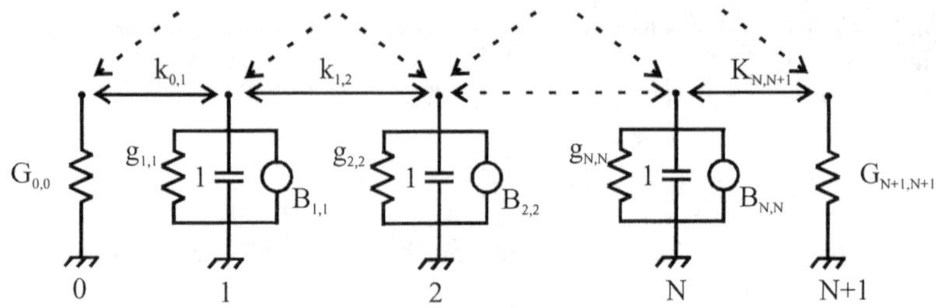

Fig. 3.4: Lowpass prototype network.

The meaning of the symbols in Fig. 3.4 is as follows:

- $G_{0,0}$ is the normalised resistive termination of the input port (commonly, the resistive termination equals 50 Ohm, and the normalization is $G_{0,0} = 1$),
- $G_{N+1,N+1}$ is the normalized output termination,
- $B_{i,i}$ (i = 1...N) is the normalized frequency deviation of the i-th resonator (when the resonator is at centre frequency of the bandpass filter, $B_i = 0$),
- $g_{i,i}$ (i = 1...N) represents resistive loss in the i-th resonator (in the lossless case $g_i = 0$),
- $k_{i,j}$ (i = 0...N+1, j = 0...N+1; $i \neq j$) represents the strength of coupling (or, in other words, the coupling coefficient) between resonator i and j.

The *narrowband approximation* is reflected by the fact that all the above quantities are considered as independent of frequency. Furthermore, in the network above the resonators have only one resonant frequency, while in practice many forms of resonators (e.g. pieces of a transmission line) have an infinite number of resonant frequencies. Generally, the narrowband approximation works well for filters with a relative bandwidth (that is absolute bandwidth divided by centre frequency) of less than about 5%.

Now the low-pass prototype network has to be related to the pole-zero scheme which corresponds to a required insertion loss and group delay response. Before doing this, some essential points must be clarified.

As discussed above, within the approximation stage zeros and poles and the scale factor K are determined such that all requirements (specifications) are met. For a given choice for the frequency dependence of S_{21}, the corresponding zeros and poles and the scale factor are uniquely determined. In the next design step, a lowpass prototype network representing this selected pole-zero-configuration has to be found. It is very important to note that in this step there is in general *no unique choice* of the corresponding network. In many cases, there is even an infinite number of different lowpass prototype networks representing precisely the same set of poles and zeros. This is exactly the point where the unique properties of HTS technology come into play. From the set of all lowpass prototype networks producing the same pole-zero-configuration and thus the same frequency response, the designer must choose the one which is best suited for to be implemented in HTS technology.

One must also recall another general conclusion when realizing a filter with N resonators (N resonant modes in case of multi-mode resonators). The maximum number of inter-resonator coupling paths (described by non-vanishing coupling coefficients) is given by $0.5 \cdot (N-1) \cdot N$. In a real network only some coupling coefficients are finite and the remaining are equal to zero. This corresponds to the situation that not all resonators are linked to each other and to the ports, but that only some are linked. Such a scheme comprising resonators (resonant modes) and coupling paths is called a topology. It is obvious that for a given number of resonators there is a finite set of different topologies. Furthermore, within some topologies there are different values for the coupling coefficients representing the same filter response, and, furthermore, a set of coupling coefficients within one topology can be infinitely large.

There are different mathematical approaches to relate the set of poles, zeros and the scaling factor to the topology and coupling coefficients of the lowpass prototype network. The most general way covers all prototype networks: It calculates the S-parameters of a prototype network as follows [31]. With the quantities introduced above, the so-called nodal admittance matrix is set up:

$$\mathbf{Y} = \begin{pmatrix} G_{0,0} & jk_{0,1} & \cdot & jk_{0,N} & jk_{0,N+1} \\ jk_{0,1} & g_{1,1} + j(\varpi + B_{1,1}) & jk_{1,2} & \cdot & \cdot \\ \cdot & jk_{1,2} & g_{2,2} + j(\varpi + B_{2,2}) & \cdot & \cdot \\ jk_{0,N} & \cdot & \cdot & \cdot & jk_{N,N+1} \\ jk_{0,N+1} & \cdot & \cdot & jk_{N,N+1} & G_{N+1,N+1} \end{pmatrix} \tag{3.15}$$

where ϖ represents the normalised frequency. Furthermore, the current vector $\mathbf{i}$ in the network is

$$\mathbf{i} = \left(i_0, i_1, \cdots, i_{N+1}\right)^t \tag{3.16}$$

and similarly the voltage vector $\mathbf{v}$ is

$$\mathbf{v} = \left(v_0, v_1, \cdots, v_{N+1}\right)^t. \tag{3.17}$$

Current and voltage are related via

$$\mathbf{i} = \mathbf{Y} \cdot \mathbf{v}$$

For a source at the input there is

$$\mathbf{i} = \left(2, 0, \cdots, 0\right)^t \tag{3.18}$$

and the S-Parameters are calculated from $S_{1,1}(\omega) = v_0(\omega) - 1$ and $S_{21}(\omega) = v_{N+1}(\omega)$ (similarly for S_{22} and S_{12} and a source at the output). It is important to understand that the mathematical functions for the frequency dependence of the S-parameters resulting from this matrix inversion are indeed rational functions.

In the envisaged synthesis process one starts from the exact locations of poles and zeros and aims to arrive at a suitable set of coupling coefficients. Unfortunately, the representation given by eqs. (3.15) to (3.18) starts with known coupling coefficients and ends up with the S-parameter frequency response and therefore with the pole-zero locations. This difficulty can be overcome by means of a numerical optimisation procedure. Here, an algorithm based on equations (3.15) to (3.18) is implemented, and the set of coupling coefficients is varied (optimised) until the resulting S-parameters have their zeros and poles located as desired. It is obvious that this procedure requires deep knowledge of available filter topologies, since the optimisation procedure will only converge at all and in a reasonable computing time when suitable starting values for the coupling coefficients are given. This also implies that coupling coefficients not needed are set to zero. Furthermore, restrictions in the available coupling coefficient range have to be taken into account. Moreover, numerical precision becomes an issue in case of high-order filters; e.g. for a 12-pole filter the Y-matrix is of order 14 and the polynomials in the S-parameters are of order 12. It turns out that many standard PC-based routines can no longer be used in this case.

Despite the drawbacks inherent to this method of matrix inversion, it is worth to point out again that this approach can handle all topologies. Hence, it is the most universal method. In the following, alternative methods are presented which are much more straightforward, but which only allow to arrive at certain special filter topologies. Due to the complexity of the subject, an example is used for illustration; the example is arbitrarily chosen such that it displays all relevant characteristics.

A 6-pole filter with insertion loss and group delay response as depicted in Fig. 3.5 serves as an example. It possesses a passband width of $\Delta f = 50\,\text{MHz}$ and a minimum passband return loss of –23 dB. Furthermore, three transmission zeros are used, one on the imaginary axis at $\overline{s}_{0,1} = j1.8$ (for increased skirt steepness) and two at $\overline{s}_{0,2/3} = \pm 1.6248 - j0.1696$ (for partial group delay equalisation). Note that although the prototype network is on the normalised frequency scale, the absolute value for the bandwidth is needed to calculate the absolute values of the group delay. Without further information, the full set of poles and zeros of the S-parameters and the corresponding frequency response (Fig. 3.5) can readily be calculated [32, 33,34,35].

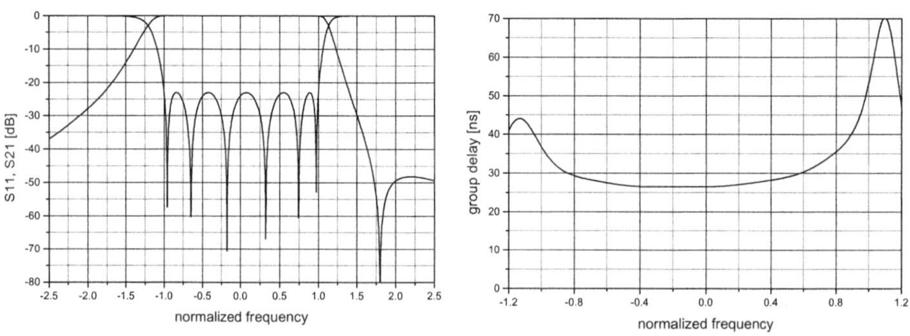

Fig. 3.5: Example for 6-pole filter, used to illustrate how different topologies and corresponding synthesis procedures lead to the same insertion loss and group delay response. Left: Insertion loss and return loss as function of frequency; Right: Group delay response.

Some important properties regarding the zeros of S_{21} can be seen directly:

- The zero on the imaginary axis at $j1.8$ is clearly reflected in the insertion loss; setting zeros on the imaginary axis is therefore straightforward in the design process.
- Since there is only one zero on the imaginary axis, the filter curves are asymmetric (a further zero at $-j1.8$ would lead to a symmetric curve). This asymmetry is also revealed in the passband where it is seen that the return loss, though its ripple remain at –23 dB, has its zeros spaced asymmetrically.
- The two zeros apart from the imaginary axis lead to the characteristic group delay response which is not minimum phase, but rather constant in the passband over a certain bandwidth. Again, as with the return loss, the group delay follows a frequency dependence which is asymmetric with respect to the centre frequency. There is no direct and obvious relationship between the locations of these transmission zeros on one side and the location of minima or maxima of the group delay on the other side. This fact renders the design of such group-delay equalised filters difficult (generally an optimisation algorithm is used to find the optimum values for these zeros).
- The two zeros with finite real part are spaced symmetrically to the imaginary axis; this is a must for zeros with finite real part, otherwise there would be no prototype network as given in Fig. 3.4 realising such response. The finite imaginary value in the zero location reflects the fact that the filter is asymmetric. For a symmetric filter, there would be either such zeros with vanishing imaginary part, or there would be two further zeros with imaginary parts with opposite sign, thus representing a quadruple altogether.

Now the so-called *canonical filter topology* is presented as a first topology for realising the described example filter. This is the most prominent topology, and the coupling coefficients can readily be calculated with analytical formulae [36]. Fig. 3.6 shows the corresponding topology.

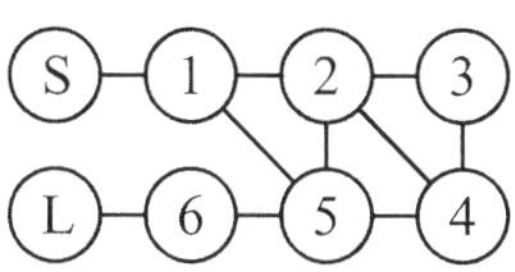

Fig. 3.6: 6-pole filter example with frequency response according to Fig. 3.5 in canonical topology.

Each circle represents a resonator, and a straight line between two circles represents the coupling between the respective resonators. The two ports are denoted by „S“ for source and „L“ for load; in the nodal admittance matrix Y (eq. 3.15) these are at the entries $i = 0$ and $i = N+1 = 7$. The following table 3.1 gives the values for the coupling coefficients $k_{n,m}$ as well as for the normalised frequency deviations $B_{n,m}$ (see Fig. 3.4)

k0,1	1.06622	B1,1	.01728
k1,2	0.90412	B2,2	.01816
k2,3	0.60462	B3,3	-.33810
k3,4	0.52780	B4,4	.00907
k4,5	0.63582	B5,5	.022967
k5,6	0.90506	B6,6	.01728
k6,7	1.06622		
k1,5	0.04133		
k2,4	0.16979		
k2,5	0.05256		

Table 3.1: Coupling coefficients and normalised frequency deviations for the example filter with response shown in Fig. 3.5 in canonical topology (Fig. 3.6).

The following important properties are to be noted:

- In the canonical design, each resonator n is coupled to its direct neighbours n-1 and n+1 („*direct couplings*" of adjacent resonators). If no further couplings would be present, no transmission zeros at finite frequencies would occur. Consequently, further couplings-also called *cross-couplings*- between non-adjacent resonators (or resonant modes) are needed.
- All possible paths from the source S to the load L are touching both the link from the source to the first resonator (here n =1) and the link from the last resonator (here n = 6) to the load. This makes the coefficients $k_{0,1}$ and $k_{6,7}$ to become equal.
- In a sub-structure comprising 4 directly coupled resonators i, i+1, i+2 and i+3, an additional cross-coupling between resonators i and (i + 3) produces a pair of transmission zeros. Such a sub-structure of four resonators is also called a *CQ-section*.
- In a substructure with 3 directly coupled resonators, a cross-coupling linking resonators i and (i + 2) produces a single transmission zero. This sub-structure is called a *CT-section*. When a CT-section interlaces a CQ-section it creates no further zero, but causes the pair of zeros (already present from the CQ section) to become asymmetric.
- These two rules do not necessarily apply to other topologies than the canonical one. In general, adding cross-couplings can, but need not necessarily add the zeros. Furthermore, in some cases the zeros cannot be placed arbitrarily.
- In the canonical design, the cross couplings are filled up by starting from the middle resonators. In our example from Fig. 3.6 the section with the resonators 1-2-5 is also regarded as a CT-section.
- All coupling coefficients related to direct couplings are on the order of 1, whereas the coefficients for cross couplings can become much smaller (this statement does not only hold for the canonical design).
- Coupling coefficients can also (though not in this example) become negative. In the canonical design it is always possible to choose all direct couplings with equal sign (plus or minus), and the couplings with the other sign (if at all present) to be amongst the cross-couplings.

- The canonical design is (with the exception of the choice of signs) unique; there is no other set of coupling coefficients in canonical design representing the same pole-zero configuration.
- For filters possessing a symmetric frequency response, all resonators in the canonical design are tuned to centre frequency. In contrast, an asymmetric response requires the resonators to be detuned from centre frequency ($B_{i,i}$ unequal to zero). Since the considered filter example exhibits an asymmetric response, the $B_{i,i}$ in table 3.1 are unequal to zero.
- A general and helpful rule states that for a given topology, the maximum number of realizable zeros of S_{21} is given by the total number N of resonators minus the number Ns of those resonators which are touched by the shortest path from the source S to the load L. In the considered example (see Fig. 3.6) the shortest path is via resonators 1, 5 and 6 and therefore the number of zeros becomes N-Ns = 6-3 = 3. In some cases less than N-Ns zeros can also be realized, but in these cases the filter is generally very sensitive to parameter deviations.
- Far apart from the passband the insertion loss decreases proportional to $1/\omega^{Ns}$, where Ns is again the number of resonators which are touched by the shortest path from the source S to the load L (The rule above can be deduced from this fact). It also follows that when there is a direct path from source to load, the insertion loss will never become lower than some threshold value given by the coupling coefficient describing this direct path. This rule is especially important for HTS filters since in planar technology there is always some leakage of signal directly from source to load. In the framework above this is represented as a coupling $k_{0,N+1}$.

It should be noted that a deeper explanation or even a proof of the statements and rules listed above is far beyond the scope of this review.

Now, other topologies besides the canonical one are considered. A further well-known topology is the inline-structure as represented in Fig. 3.7 (again for the 6-pole example filter). The resonators are arranged similar to a meander, and the close proximity of the resonators 1 to 4 and, respectively, 3 to 6 makes it easy to implement cross-couplings to get CQ- and CT-sections.

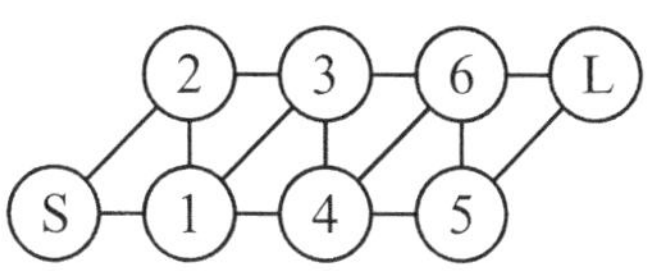

Fig. 3.7: Inline topology for the 6-pole example filter with response as depicted in Fig. 3.5.

One way to arrive at the coupling coefficients for this topology is to start with a matrix which is similar to the admittance nodal matrix Y from eq. 3.15 but which only contains coupling coefficients $k_{i,j}$ and resonant frequency deviations $B_{i,i}$. Note that here $0 < i,j < N+1$ holds, that means couplings to the source and load are not considered. This matrix is called coupling matrix and is denoted by $\mathbf{K}$:

$$\mathbf{K} = \begin{pmatrix} B_{1,1} & k_{1,2} & \cdot & k_{1,N} \\ k_{1,2} & B_{2,2} & \cdot & \cdot \\ \cdot & \cdot & \cdot & k_{N-1,N} \\ k_{1,N} & \cdot & k_{N-1,N} & B_{N,N} \end{pmatrix}. \tag{3.19}$$

If one considers the transform $\tilde{\mathbf{K}} = \mathbf{R} \cdot \mathbf{K} \cdot \mathbf{R}'$ to a new matrix, a *similarity transform* is achieved for $\mathbf{R}$ being a rotation matrix with determinant equal to one. In this case the eigenvalues of $\mathbf{K}$ and $\tilde{\mathbf{K}}$ are the same. As seen from eq. (3.19) $\mathbf{K}$ contains the coupling coefficients and the normalised resonator frequency deviations of the filter. The new matrix $\tilde{\mathbf{K}}$ describes a filter with the same poles and zeros (and thus with the same frequency response), but with new coupling coefficients (exception: $k_{0,1}$ and $k_{N,N+1}$ unchanged).

Therefore, if one set of coupling coefficients is known (and it was mentioned above that the canonical topology can always be derived in a straightforward manner), all other sets of coupling coefficients (as long as $k_{0,2} = k_{0,3} = ... = k_{0,N} = 0$ and similarly $k_{i,N+1} = 0$) can be found by applying similarity transformations.

However, in practice the difficulty lies in finding the right rotation angles such that the new coupling matrix has zeros at pre-determined places. Especially for the inline topology, closed formulas for the rotation angles have been published [35,37,38], but only for filters with a certain number of resonators. The larger this number, the more similarity transformations have to be applied successively, and the more complicated the formulas become. Therefore the usage of similarity transforms is restricted to a few cases in practice.

Table 3.2 presents four sets of coupling coefficients for the 6-pole filter example in inline topology (denoted Inline 1 – 4). The first three examples could have been derived via eq. $\tilde{\mathbf{K}} = \mathbf{R} \cdot \mathbf{K} \cdot \mathbf{R}'$, but actually they have been derived by using the optimisation procedure based on the inversion of the nodal admittance matrix as described above.

This topology represents an example where the set of coupling coefficients representing the same filter function (same pole-zero configuration) is infinite. Example „Inline 2" has been obtained by arbitrarily setting $k_{3,4}$ = 0.4, and „Inline 3" by even setting $k_{3,4}$ = 0. Furthermore, example „Inline 4" has been obtained by the same algorithm, but now also including a coupling from the source to the second resonator and from the second last to the load.

The designer of an advanced HTS filter must understand the various alternatives with respect to the topology and the strength of the coupling coefficients, all leading to the same filter function. At this step in the design procedure, the designer can account for the fact that in HTS technology certain coupling links as well as certain values for coupling coefficients can be better realised than others. At this point of our review paper, this statement appears of course to be quite general. However, in sub-section 4.2 where examples for realised HTS filters are presented, this statement will be further illustrated.

	Inline 1	Inline 2	Inline 3	Inline 4
k0,1	1.06622	1.06622	1.06622	1.00797
k1,2	0.66553	0.59320	0.51815	0.68506
k2,3	0.33198	0.26408	0.22664	0.45001
k3,4	0.57440	0.40000	0.00000	0.52094
k4,5	0.44685	0.37597	0.33464	0.78538
k5,6	0.80524	0.73720	0.65589	0.74264
k6,7	1.06622	1.06622	1.06622	1.01537
k1,3	0.53988	0.48876	0.28755	0.13234
k1,4	0.29110	0.47788	0.68409	0.26037
k3,6	0.24763	0.40419	0.59258	-0.24571
k4,6	-0.33078	-0.33512	-0.19441	-0.11526
k0,2	0.00000	0.00000	0.00000	-0.34760
k5,7	0.00000	0.00000	0.00000	0.32533
B1,1	0.01728	0.01728	0.01728	0.55378
B2,2	-0.77844	-0.84210	-0.88267	-0.52085
B3,3	-0.25271	-0.49598	-0.63113	-0.11602
B4,4	0.21428	0.41000	0.51377	-0.07042
B5,5	0.52900	0.64020	0.71215	0.39789
B6,6	0.01728	0.01728	0.01728	-0.49768

Table 3.2: Different sets of coupling coefficients for the 6-pole filter example (response in Fig. 3.5) in inline topology (Fig. 3.7).

Furthermore, there are very recent and highly interesting publications [39,40] on new filter topologies, an example is given in the following Fig. 3.8 (again, applied to the 6-pole example). Cameron has called this topology „cul-de-sac“ [39].

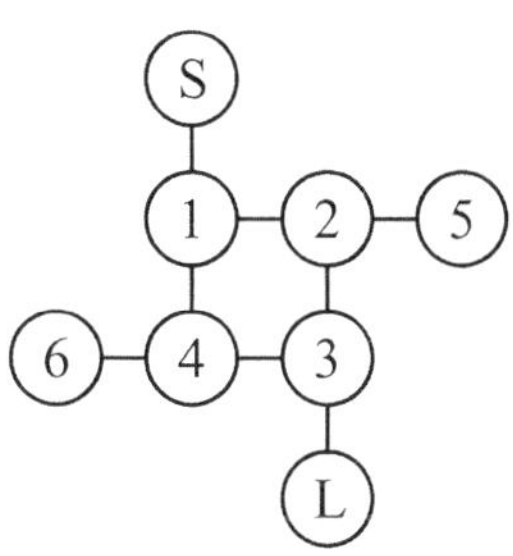

Fig. 3.8: 6-pole filter in cul-de-sac topology [39].

All the „old“ topologies have in common that there is one path from the source to the load which touches every resonator exactly once (this can be seen easily in Figs. 3.6 and 3.7). With the new topologies, this is no longer the case. They are especially suited for implementing asymmetric filter functions. Formerly, the asymmetry has led to a large number of cross couplings, and there were resonators with up to 4 couplings to other resonators (e.g. resonator 2 in Fig. 3.6) which is difficult to implement. In the new topology, the number of couplings is reduced. Furthermore, as seen in Table 3.3, all couplings have about the same strength, which is another advantage in the practical implementation.

k0,1	1.06622	B1,1	0.01728
k1,2	0.65442	B2,2	0.07318
k2,3	-0.65442	B3,3	0.01728
k3,4	0.62520	B4,4	-0.03205
k1,4	0.62520	B5,5	0.39111
k2,5	0.71233	B6,6	-0.72011
k4,6	0.53969		
k3,7	1.06622		

Table 3.3: Coupling coefficients for the example filter with response according to Fig. 3.5 in cul-de-sac topology (Fig. 3.8).

This sub-section has outlined a few aspects of the very vast field of filter design, due to limited space necessarily more by pure listing and less by consequent reasoning or even proving. As a conclusion, the relevance of the topics covered especially for HTS filters is again pointed out. Sub-section 4.2 will illustrate the importance by means of examples.

4. HTS FILTERS

This section continues along the line set by the previous section 3 where issues related to the approximation of filter specifications and the choice of the filter topology were highlighted. These two first steps (approximation and synthesis) are followed by the implementation in HTS technology. After a preliminary consideration of HTS resonators (building blocks of filters) in sub-section 4.1 a survey on filters in HTS technology is presented in sub-section 4.2. The purpose of this review is to demonstrate the ability of HTS technology to provide high-performance filters. This demonstration is based on examples which are each either representative for an important aspect of the technology or which have been driven to a limit with respect to the achieved performance.

4.1 HTS Resonators

Single resonators already reveal the major differences between HTS and conventional technology. The function of a resonator is to store electric and magnetic field energy. Thus, its structure (shape and material) serves to confine these fields to a finite volume. In case of a cavity resonator the fields are enclosed by metal walls. Dielectric resonators (DR) usually consist of a dielectric material with high permittivity and the confinement of the field is due to total reflection at its boundary. Hence, the fields are concentrated mainly inside the DR, with the remaining field distributed in the air around and decaying rapidly with distance away from the resonator. Many resonators are realized by means of transmission line sections. Folding of these transmission lines and employing a stepped impedance solution and/or filling with a high-permittivity helps to miniaturize the resonator size. In these structures magnetic and electric fields are "distributed" within the entire resonator volume (distributed resonator). If electric and magnetic field are stored in spatially separated substructures (capacitor and inductor) the resonator is called a lumped-element resonator [29].

Resonator structures possess a number of resonant frequencies. Typically, only one of these resonances (resonant frequency f_0) is used for the function of the resonator. Hence, the other resonances become unwanted "spurious resonances", except in case of dual- and multimode filter structures where more than one resonant mode of the individual resonator is utilised. The dissipative losses in a resonant mode are characterised by the *unloaded quality factor* Q_0 which is defined via the stored field energy W and the dissipated power P_{diss} as

$$Q_0 = 2\pi f_0 \frac{W}{P_{diss}} \ . \qquad (4.1)$$

The unloaded quality factor can easily be retrieved from a measurement of its insertion loss response [30].

For a resonator in HTS technology, the dissipated power is given as sum of three different loss contributions. Each contribution can be associated with a respective quality factor (Q_{diel}, Q_{cond} and Q_{hous}), leading to $1/Q_0 = 1/Q_{diel} + 1/Q_{cond} + 1/Q_{hous}$.

- There is *dielectric loss* contributing a Q_{diel} on the order of ($\alpha/\tan\delta$) where $\tan\delta$ denotes the loss tangent of the dielectric substrate and $\alpha \le 1$ the filling factor (close to one). Generally $\tan\delta$ is smaller than 10^{-5} for suitable substrate materials.
- In case of an open resonator (that means there is no housing enclosing the wafer) a loss contribution due to radiation would occur. However, in case of a housing (which is nearly always the case in practice) currents are induced in the inner housing walls which in turn produce ohmic losses (*housing losses*, Q_{hous}). For a given housing material the housing losses are strongly dependent on the size of the housing and the current pattern in the HTS circuit (here resonator). The reason for this can be understood by means of the following "housing-loss model" for microstrip structures: The volume between the upper surface of the substrate and the housing lid can be considered as waveguide shorted by the lid. Hence, the electromagnetic field in this volume can be represented by a superposition of waveguide modes (below and above cut-off) excited by the currents flowing in the planar HTS circuit. In order to minimise the losses in the enclosing normal conducting housing, size of the housing and the current density pattern in the HTS circuit should be chosen such that only higher-order evanescent waves are excited which rapidly decay with distance from the HTS circuit plane. A current density pattern with a large percentage of instantaneous current density flowing in the same direction (without compensation by currents in opposite direction) will excite the lower modes and therefore give rise to a high *housing-loss factor*. Opposed to that a current pattern where the instantaneous current density is characterised by strong local changes of the direction (e.g. meander-like pattern) within the circuit plane will be associated with a strong decay of the field and thus with a low housing-loss factor. This fact was experimentally studied in [41].

- There are losses due to current transport in the HTS structure, resulting in a finite Q_{cond} given by

$$Q_{cond} = \frac{\omega\mu_0 \int_V |H|^2 \, dV}{R_S \int_{HTS} |H|^2 \, dA} . \tag{4.2}$$

In the numerator of eq. (4.2) the integral is extended over the volume of the resonator, giving the stored energy W. In the denominator it is integrated over the surface of the superconducting material. R_s denotes the HTS surface resistance as introduced in sub-section 2.1 (eq. 2.2) . Thus the denominator gives the power dissipated in the HTS material. It is instructive to convert equation 4.2 into the following form [27],

$$Q_{cond} = \frac{\pi \cdot Z_0}{\lambda_0 R_S} l_C , \tag{4.3}$$

with Z_0 and λ_0 as the characteristic impedance of free space ($Z_0 = 377\ \Omega$) and the free space wavelength, respectively. The crucial parameter is l_C ("*effective size*", with the dimension of a length), which contains the above integrals and which is a figure of merit in the resonator design. Comparing two resonators which exhibit the same resonant frequency and which are using the same conductor material (same surface resistance R_S), their difference in quality factor is purely given by l_C. This effective size l_C is related to a characteristic geometric dimension of the resonator, e.g. for a conventional cavity resonator (fundamental mode) it is on the order of the wavelength λ_0. l_C decreases with decreasing resonator size, and for a planar resonator in microstrip technology it becomes as small as about the substrate height. Obviously, if resonant frequency and surface resistance R_S are kept constant, the achievable quality factor decreases with decreasing resonator size. However, by utilizing HTS material instead of conventional metals, the substantially decreased R_S keeps the quality factor high despite the miniaturisation(!). Moreover, the achievable Q-value for the highly miniaturised HTS resonator significantly exceeds the highest value obtained with a voluminous normal conducting cavity resonator.

In the frequency range around 3 GHz, a resonator in waveguide technology (cavity) reaches an unloaded quality factor of about $Q_0 = 15.000$. A resonator in microstrip technology (using normal conducting strip conductor and ground plane) provides a quality factor not higher than about $Q_0 = 150$. This can readily be understood with the reasoning from above, assuming a typical substrate thickness of 1 mm. Replacing the normal conductor with HTS material should- at a first glance- move the unloaded quality factor upwards by three orders of magnitude (see Fig. 2.1), that is to a value of $Q_0 = 150.000$. However, most planar HTS resonators do not reach such high values since Q_{diel} and especially Q_{hous} become limiting factors.

In the next part of this sub-section, several design alternatives for typical HTS resonators are presented and their respective advantages and drawbacks are highlighted. Most of these resonator structures are planar and the following Fig. 4.1 depicts some illustrative examples in microstrip technology which are further discussed below.

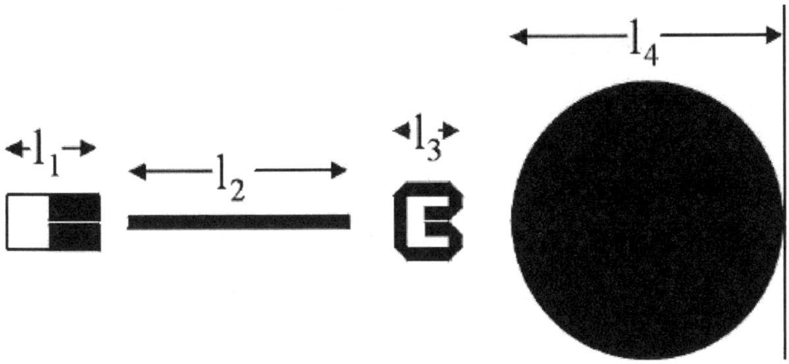

Fig. 4.1: Planar resonator configurations: From left to right (typical size given in brackets, where λ denotes the guided wavelength): lumped-element resonator ($l_1 < \lambda/6$), simple microstrip transmission line resonator ($l_2 = \lambda/2$), quasi-lumped element ($l_3 \approx\lambda/8$), and circular microstrip disk resonator ($l_4 >\lambda/2$).

The lumped-element resonator can be considered to consist of two transmission line sections, one with a very narrow and one with a wide strip width [29], representing inductance (stored magnetic field energy) and capacitance (electric field energy), respectively. As an advantage of this design the high degree of miniaturisation is noted, and also a large spurious-free frequency range. The latter statement implies that (unwanted) higher modes of the structure resonant at frequencies significantly higher than 2 times of the fundamental resonant frequency. Furthermore, there is extensive literature on filter design in terms of *LC*-circuits, and such a lumped-element resonator is therefore a natural candidate for a straightforward implementation of these design procedures. A disadvantage of this circuit is that achievable Q-values as well as power handling capability suffer from the narrow line width in conjunction with the fact that current density and also the fields are highly peaked at the strip edges. At 3 GHz and at 77 K, values of no more than Q_0 = 10.000 can be expected. The limitations in power handling capability can be expressed in terms of the maximum storable energy in a resonator W_{max}, or equivalently in terms of the maximum oscillating power P_{osc} (which is the term in the numerator of equation 4.1) [27], given by

$$P_{osc} = 2 \cdot \pi \cdot f_0 \cdot W_{\max} \approx (0.03\ldots 0.3) \cdot \frac{h \cdot w}{\sqrt{\varepsilon_{eff}}} \cdot (H_{\max})^2, \qquad (4.4)$$

with substrate height h and line width w in mm, maximum magnetic field $H_{\max}$ (before nonlinearities occur) in A/cm and the oscillating power in W. Recalling $\mu_0 H_{\max}$ = 6.4 mT at 77 K (sub-section 2.1) and using typical values h = 0.43 mm and ε_{eff} = 10, for a small strip width w = 0.1 mm the maximally storable oscillating power follows to be on the order of 1 - 10 W.

The next example for a resonator type is a simple open-ended microstrip transmission line section of length $\lambda/2$. Obviously this resonator exhibits an infinite number of higher resonant modes at (approximately) integer multiples of the fundamental frequency. Therefore, the spurious-free range is only up to $2 f_0$. The factor l_C (effective size) is remarkably higher (by about a factor 10) than for the previous example of lumped-element structures. Consequently, the unloaded quality factor is increased. It is, however, not as much increased as it could be expected from l_C, since for this example the housing-loss factor (current pattern with instantaneous current density in same direction) is relatively high. Thus, housing losses – described by Q_{hous} - become a limiting effect. At 3 GHz and at 77 K, values of $Q_0 = 40.000$ can be expected. In comparison to the lumped-element circuit the maximally storable oscillating power is increased (by the ratio of the line widths). Inserting a typical value of w = 1 mm, a maximally storable oscillating power on the order of 10 - 100 W can be obtained.

The two examples which have been discussed up to now represent the two extremes, namely „distributed" structures on one side and „lumped" circuits on the other side. Of course numerous mixed forms are possible. One of them is now dealt with as the next example [42], depicted as the 3rd resonator from the left in Fig. 4.1 (here referred to as hairpin resonator). It is obtained from the distributed transmission line resonator, discussed previously, by bending and folding. However, the two end sections are in close proximity, which corresponds to a capacitive loading, leading to a further miniaturisation of the resonator. Since the linewidth is in the same order than for the straight transmission line resonator, the power handling capability is also comparable. The advantage of this design is a significantly reduced housing loss factor which results in a superior unloaded quality factor. For a temperature of 77 K, the Q_0 -value is comparable to that of the straight transmission line resonator. For lower temperatures the simple transmission line resonator does not show significant improvement, since Q_{hous} is the limiting factor and the housing loss contribution does (because of normal conductivity) not significantly decrease with decreasing temperatures. In contrast, the hairpin resonator considered here, minimises the housing losses due to its symmetry in the current pattern: The current flow in the two arms of the resonator is in opposite directions, and therefore the induced current flow in the housing is fully cancelled on the axis of symmetry. Further cooling of this resonator allows for an unloaded quality factor which fully benefits from the unique loss behaviour of HTS, and at temperatures of 50 K values of $Q_0 > 100.000$ have been shown.

We now proceed to the circular microstrip disk resonator, the last example from Fig. 4.1. This resonator structure supports a multitude of modes, two of them are shown in the insert of Fig. 4.2 [27,43,44]. The TM_{010}-mode is of special interest since the current flows in radial direction only, and this in turn means that because of the absence of edge-parallel currents the problems due to peaked edge currents are completely avoided [44]. This special case (absence of peaked edge-parallel current) only occurs in TM_{0n0}-modes of circular disk and ring resonators and has a substantial impact on both the achievable unloaded quality factor and the maximum oscillating power. The reduction of HTS conductor losses due to the absence of peaked edge currents leads to a situation where for YBCO and temperatures below about 75 K the dielectric losses are dominating the HTS losses (except for sapphire). At 70 K and a frequency of 4 GHz the Q_0-values of about 200,000 and 70,000 were measured for disc resonators made from YBCO on 1 mm thick sapphire and Czochralski grown $LaAlO_3$, respectively [43]. By cooling down to 43 K even Q-values up to about 500,000 were observed. The impact on the power handling is even more impressive. In CW operation these resonators can handle a by (at least) a factor of 100 higher oscillating power than a resonator occupying about the same area, but possessing an edge-parallel peaked current density. This is illustrated in Fig. 4.2 where the dependence of the unloaded quality factor (normalised to its respective value at low fields) on the oscillating power is shown for both the fundamental TM_{110}-mode (with edge-parallel current) and the TM_{010}-mode (without edge-parallel current). Whereas no degradation of the unloaded quality factor (that is no onset of nonlinear behaviour) up to an oscillating power in excess of 10 kW is seen for the TM_{010} – mode, for the TM_{110}-mode nonlinear behaviour already starts at a level of less than 100 W.

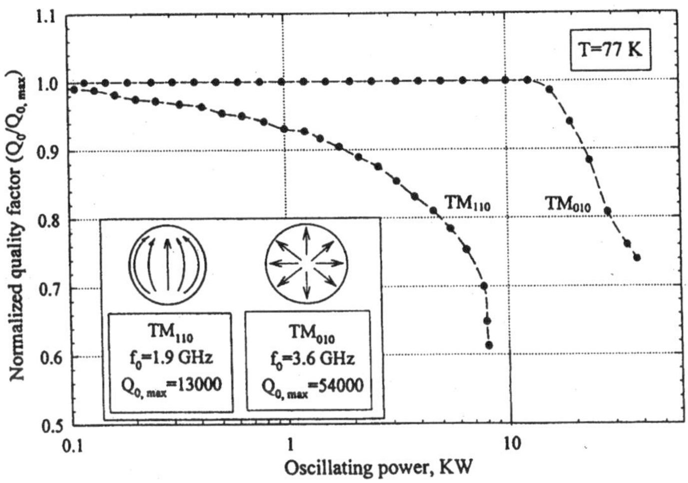

Fig. 4.2: Power-handling capability of two modes of the disk resonator [43,44,27].

The outlined advantages of the disk resonator are achieved at the cost of a significantly larger size (disk diameter about 1.22λ, with $\lambda = c/\sqrt{\varepsilon_r}\, f$) and the difficulty to achieve sufficiently strong mutual coupling between disk resonators placed on the same wafer. Because of this difficulty, disk resonator filters were realised by means of a 3-D structure with staked disk or ring resonators [43].

To conclude this sub-section, hybrid dielectric / HTS resonators are briefly considered. Dielectric resonators play an important role in the context of miniaturised filters: loading a conventional waveguide with a dielectric material miniaturizes the size (according to the dielectric constant and to the relative fraction of the electromagnetic energy stored in the dielectric) while maintaining the electrical properties (except for the behaviour at very high power levels) [45]. Dielectric resonator filters are therefore widely employed in various applications such as base stations for mobile or satellite communications. When the dielectric material is a single crystal such as sapphire, lanthanum aluminate or rutile, there is a huge improvement in loss behaviour upon cooling. However, cooling a conventional resonator loaded with a dielectric gives not yet access to improved quality factors, since the current flow in the metallic cavity surrounding the dielectric material does limit the loss behaviour. Full advantage of the superior behaviour of the dielectric is only gained when HTS is used to replace the conventional metal. Depending on the resonator shape and the used mode, it is sufficient to use HTS at places with high current flow and to leave conventional metal elsewhere. Comparing such a hybrid resonator to the examples above, the major difference is that the hybrid resonator is a three-dimensional structure and thus needs inherently more space. On the other hand, this is associated with a more homogeneous field distribution, keeping the factor l_C large. The highest Q_0 -value measured is 1.000.000 at 5.6 GHz, 50 K [46]. Hybrid resonators are therefore favourable candidates for low-noise oscillators to be used as frequency standards. A drawback is that the dielectric constant of single-crystalline material shows a significant variation with temperature, and in order to keep the resonant frequency of such a resonator constant, a precise control of the operating temperature is needed. For further information on the subject it is referred to the literature [47,48,49,50,51].

The resonators as dealt with in this sub-section already represent a simple version of a bandpass filter. At a two-port resonator, the frequency components which are "close" to the centre frequency f_0 are transmitted and others are rejected. However, this frequency response is not sufficient because of the broad transition band. Practical applications require more complex multi-resonant configurations, so that filters are generally composed of a number of resonators. This is the topic of the next sub-section.

4.2 HTS filters

This sub-section is devoted to implementation aspects in HTS filter technology. These aspects are highlighted by presenting a number of examples, which have all already been published elsewhere with the exception of the second example. Though the examples seem, at first sight, to be taken arbitrarily among the vast number of publications, altogether they are chosen such that the major aspects of HTS filters are covered. The reader can thus gain a comprehensive overview on the potential of HTS filter technology.

4.2.1 Coupling issues

A filter consists of a number of resonators mutually coupled in a suitable manner such that the prescribed frequency response is obtained. It was already shown that typical filter functions require the implementation of couplings with different signs and with magnitudes ranging from values close to zero up to maximum values on the order of one.

In principle, mutual coupling between planar resonators on a common wafer is accomplished by placing them in proximity such that their electric or magnetic field or both fields are "overlapping". It is a characteristic property of these planar structures that the decay of the field strength along directions in parallel to the wafer surface is rather slow. That means that even for a relatively large spacing (e.g. in the order of a wavelength) between two resonators significant coupling can occur. Because of this fact the realisation of filters which require some small coupling coefficients between resonators (as, e.g. , required for a narrow passband width) can become a very challenging task. Furthermore the performance of filters is often degraded by unwanted couplings between resonators which should (in theory) be uncoupled. A special case of unwanted coupling exists if in- and output are directly coupled, limiting the stopband attenuation of the filter.

To find a proper solution for all of these coupling issues is often regarded as the *major challenge in planar filter design and implementation*. Couplings can be foreseen and somehow accounted for in the design process only if a computer-based full-wave simulation and optimisation of the entire structure is employed, and this is obviously a very time- and memory-consuming process. Some general rules for keeping the undesired interaction small can be stated:

- Interactions originating from surface wave propagation are reduced if both the substrate height and the dielectric constant are small.
- Interactions based on propagating and evanescent housing modes are suppressed by reducing the cross-section of the housing.
- The spatial extension of the field of a resonator can in general be reduced if the instantaneous current flow in the resonator changes its direction as often as possible (e.g. meander structure). This requires strip conductors with very narrow widths.

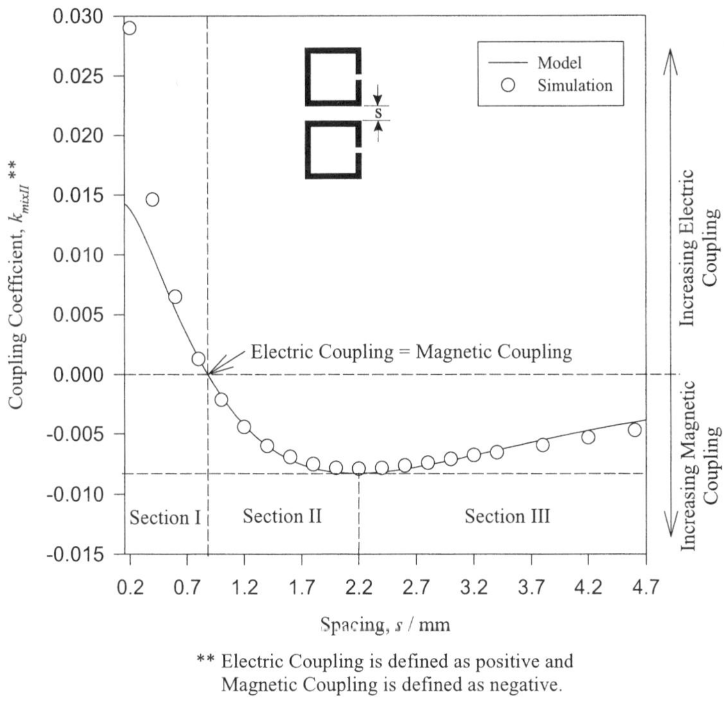

Fig. 4.3: Coupling coefficient between two resonators as a function of their distance, from [52] (© 2001 IEEE). The coupling is mediated by both electrical and magnetical fields and thus zero coupling is achieved for finite distance.

However, as discussed in the previous sub-section, these points have to be traded off. The benefit of a smaller linewidth and a thinner substrate is opposed to a high unloaded quality factor and a high power handling capability. A small dielectric constant of the substrate (if available with compatible substrate materials) is opposed to miniaturisation. A small height of the housing increases housing losses.

The two following examples were selected for a detailed discussion in order to demonstrate some clever ways for to cope with the long-range interaction, not sacrificing the advantage of distributed elements.

The graph in Fig. 4.3 is taken from [52] and shows the coupling coefficient between two resonators as a function of their spacing *s*. The resonators operate at 1.85 GHz and are arranged in such a manner that the coupling is caused by both electric and magnetic field: For small spacing (called "section I" in the graph) the electric field dominates and the coupling has a positive sign, while for a larger spacing ("sections II and III") the coupling is via the magnetic field and becomes negative. This arrangement takes advantage of the superposition of electric and magnetic coupling in such a way that very small coupling coefficients and even zero coupling is achieved without the need for large spacing between the resonators. (It should also be noted that the graph clearly demonstrates the above mentioned fact that the interaction between resonators in planar technology decays only slowly in space; the coupling converges hardly to zero on the right side of section III).

The next example is on a self-equalized 4-pole filter. These results were obtained by one of the authors (T. Kässer, [53]) and have not yet been published elsewhere.
Standard filter design as described in sub-section 3.3 has revealed the coupling coefficients to be:

$$k_{0,1} = k_{4,5} = 1.0509, \quad k_{1,2} = k_{3,4} = 0.9367, \quad k_{2,3} = 0.5217 \text{ and } k_{1,4} = 0.1903$$

From these coupling coefficients it is seen that the self-equalization is realised by the cross-coupling between resonators 1 and 4, having the same sign as the other (main) couplings. The left part of Fig. 4.4 depicts the geometry chosen for the realization.

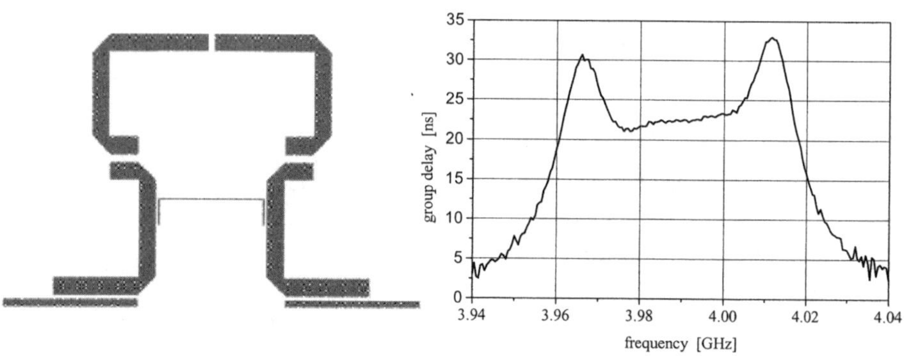

Fig. 4.4: 4-pole filter with group-delay equalization realised by positive cross-coupling.
Left: Circuit layout. Right: Measured group delay response [53].

The resonators are microstrip sections excited in their fundamental ($\lambda/2$) mode, and they are arranged in a clockwise direction with the in- and output lines in the lower left and right corners. The main couplings between these resonators are purely electric, due to the concentration of electric and absence of magnetic field at the open resonator ends. The cross-coupling between the first and last resonator is realized by means of a thin transmission line section (coupling line). The correct sign of the cross-coupling is adjusted by both the choice of the coupling between the resonators and coupling line and by the length of the coupling line (this line has half the length of the resonators and thus its first resonant frequency is twice as high as the centre frequency of the filter). On the right-hand side of Fig. 4.4 the measured group delay response of this filter is shown. Although the group delay is quite flat within the passband it possesses an unwanted "slope", which means an increase of the group delay with increasing frequencies. Results from simulation of this structure by means of a full-wave solver are in accordance with this measurement.

The explanation for this undesired slope is as follows: In the chosen geometry there is also an interaction between the first and third resonator and (due to symmetry) between the second and fourth resonator as well. At a first sight, the reason for this statement may seem unclear, but it will be justified in the following. Calculating the filter response on the level of a coupled-resonator model and assuming values for the coupling coefficients as given above (no coupling between 1 and 3 and 2 and 4) leads, of course, to a group delay response with no slope. However, the measured filter curve (that is with slope) can be reproduced by introducing the following additional couplings and changes in the resonance frequencies of the resonators:

$$k_{1,3} = k_{2,4} = 0.0302, \quad B_{1,1} = B_{4,4} = 0.0341, \quad B_{2,2} = B_{3,3} = -0.0151$$

The coefficients of the already existing couplings must additionally be slightly changed for achieving an exact reproduction of the measurement curve.

The trick to achieve a group delay with no slope is the following: Within the coupled-resonator model the filter is designed with a slope opposite to the measured slope. Consequently, in the real structure the unwanted slope is cancelled out. This idea is experimentally proved via the result shown in the right part of Fig. 4.5.

The group delay with the opposite slope can be achieved via introducing a further cross-coupling $k_{1,3} = -0.0604$, and the respective geometry is displayed on the left side of Fig. 4.5.

Again, this cross-coupling is achieved via a thin transmission line (coupling line). However, this time the cross- coupling has a different sign which is accounted for by the length of the coupling line now exceeding that of the resonators. The measurement curve in Fig. 4.5 (right) shows a group delay response without slope, and this result clearly justifies the above assumption. This example again strongly supports the statement on long-range interaction; though the respective resonators are separated by distances on the order of half the wavelength, the interaction is still strong enough to deteriorate the filter response.

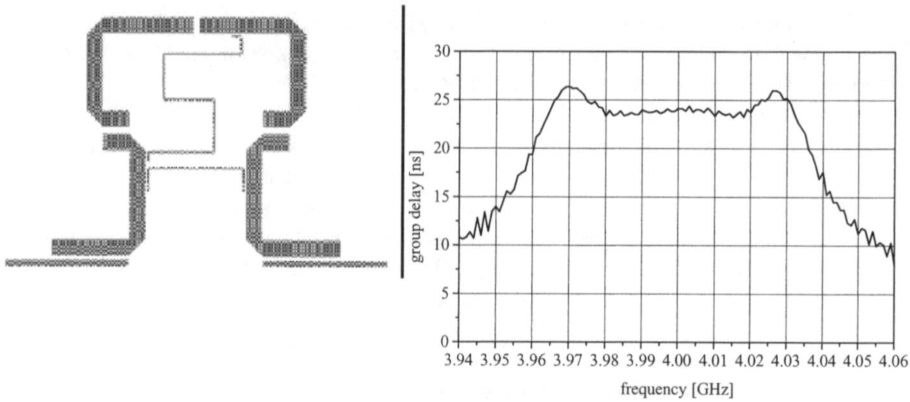

Fig. 4.5: 4-pole filter of fig. 4.7 improved by introducing pre-distortion via asymmetric cross-coupling; left: geometry, right: measurement of group delay [53].

There are further published results which are indicating principle approaches how to cope with unwanted coupling. In [54] a novel approach is introduced where coupling between closely spaced resonators is controlled by rotating the field polarisation from resonator to resonator and adjusting sign and magnitude of the coupling coefficients by a "disturbance" of the resonator symmetry.

4.2.2 Lumped element HTS filters

The next filter example is representative for the lumped-element technique. For a comprehensive overview on filters in lumped-element technology see [29]. The layout and the measurement curves for the selected example [55] are shown in Fig. 4.6 and 4.7.

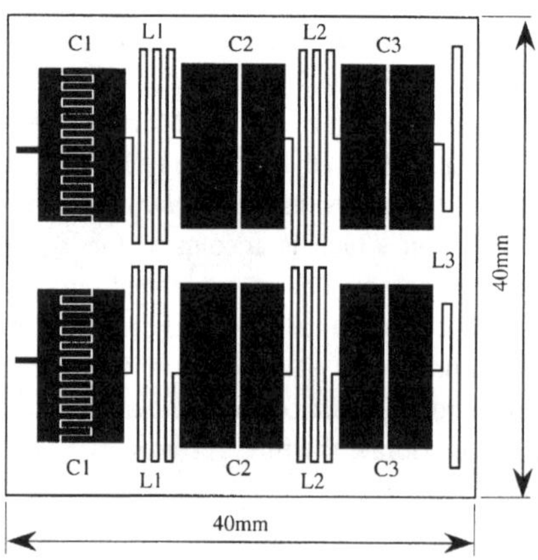

Fig. 4.6: Layout of a 5-pole lumped-element bandpass filter for a centre frequency of 246 MHz and a fractional bandwidth of 0.76 % [55].

A large number of different LC-circuits for the realisation of bandpass filters are known. However, due to restrictions in the realisable range of capacitance and inductance values, only a subset of these circuits can be implemented in planar HTS technology. The layout depicted in Fig. 4.6 (left part) is based on a circuit structure which turned out to be best suited for the realisation in planar HTS technology [55,29]. The structure with groundplane consists of a sequence of capacitive Π-sections (2 patches with gap between them) and series inductors. The coupling to the ports is provided by interdigital capacitors.

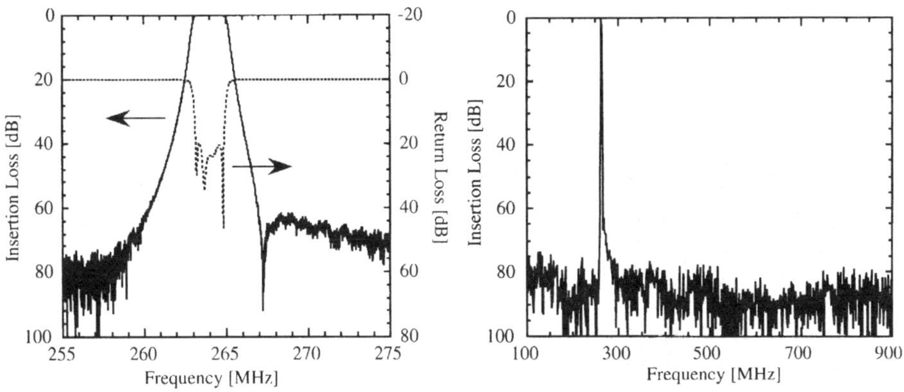

Fig. 4.7: Measured frequency response of the filter in Fig. 4.6. Left: Near-band behaviour; Right: Far-band behaviour.

The filter occupies an area of 40 x 40 mm^2 on a MgO substrate. At 70 K, the insertion loss was measured to be 0.1 dB and the return loss to be 20 dB. The layout clearly displays the use of very narrow and very wide strip conductors, corresponding to inductive and capacitive behaviour, respectively. One unique advantage of lumped-element technology, namely a very large spurious-free frequency range is impressively demonstrated; there is no disturbing mode up to frequencies of more than three times the centre frequency. The measured frequency response displays a transmission zero above the passband (267 MHz) which was most likely not intended in the design. This is a very typical effect arising from the undesired interaction between non-neighboured resonators; in this example, however, the transmission zero does not really render the filter unusable.

4.2.3 HTS filters with centre frequencies below 100 MHz

The lower limit for the operational frequency regime of planar HTS filters is obviously set by the fact that the required size of the components increases with decreasing centre frequency. This statement takes into account that e.g. 3-D coils and extremely narrow conductors (for large inductances) as well as patch capacitances with very thin isolating layers are not available in HTS technology. This problem was partially overcome by means of drop-in capacitors in the next filter example. This filter [56] from Superconductor Technologies Inc. (USA) exhibits the lowest centre frequency of all HTS filters known from literature. It is realized on a MgO substrate with a diameter of 2 inch. To allow a five-pole filter to fit on the wafer, each resonator was realized with a quarter-wavelength line folded as shown in Fig. 4.8 (left) and shorted on one side to ground. Furthermore, since with this approach a centre frequency smaller than 26 MHz could not be achieved, each resonator was additionally loaded with a commercial capacitance. The authors demonstrate that the losses are dominated by the losses in the commercial capacitor and not, as might be assumed, by the shorting to ground. Unloaded quality factors of 30.000 – 45.000 were deduced.

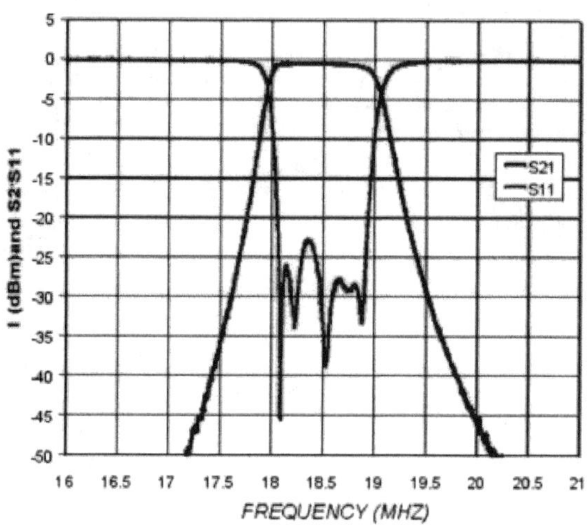

Fig. 4.8: Five-pole filter with centre frequency of 18.5 MHz (from [56], © 2000 IEEE); left: single resonator with size 30 mm x 45 mm; right: measured S-parameter data for a filter realized with Thallium-based HTS material, the insertion loss at centre frequency and at 77 K is 0.4 dB.

4.2.4 Highly selective HTS filters for application in terrestrial base stations of mobile communication systems

Many published filters are designed according to common specifications from mobile communications, and the following examples belong to this category. The first one is again from Superconductor Technologies Inc. [57]. As Fig. 4.9 clearly reveals, the filter is composed of 10 resonators which are realized as (quasi-) lumped-element resonators. Thin transmission lines provide cross-couplings and thus produce transmission zeros for improving the steepness of the skirts (see the measured results in Fig. 4.10). The filter was manufactured on a MgO substrate with a thickness of 20 mil and with thallium-based HTS films; the circuit size was 18 mm x 34 mm.

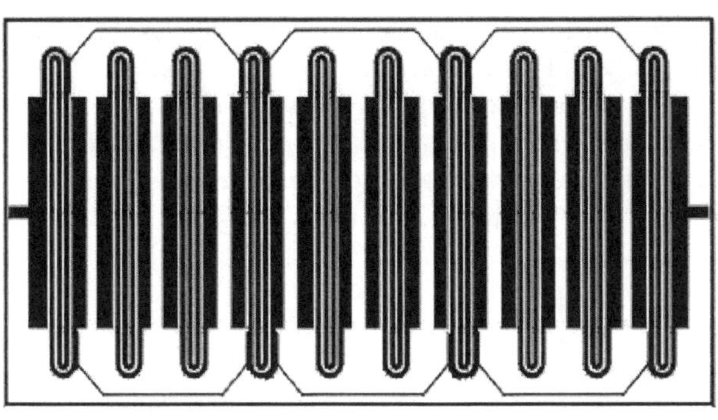

Fig. 4.9: Layout of a highly-selective ten-pole filter ([57], © 2000 IEEE).

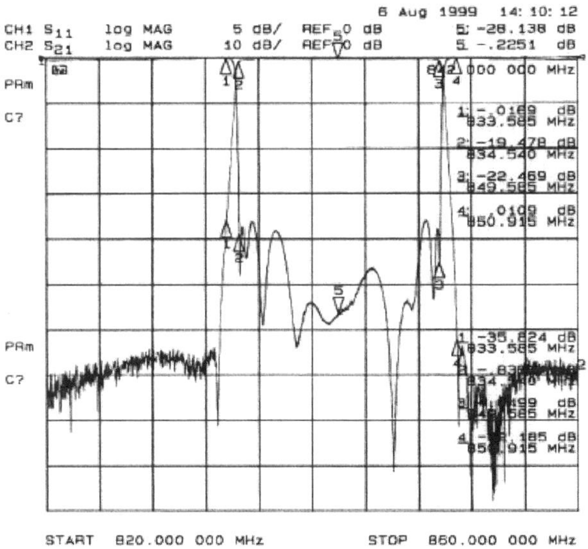

Fig. 4.10: Measured frequency dependence of insertion loss and return loss of the filter from Fig. 4.9 at a temperature of 77 K. The filter is centred at about 840 MHz and has a bandwidth of less than 20 MHz.

Other frequency bands to be used in 3rd generation mobile communication systems are slightly below 2 GHz. The next example [21] is a filter in Chebyshev design which possesses a high selectivity due to the large number of 32 resonators. It was developed by the Japanese Cryodevice Inc. (company no more active at present). The filter has a centre frequency of 1.93 GHz and a bandwidth of 20 MHz and is realized with YBCO films on a MgO substrate with 3 inch diameter. It is impressive to see that a HTS filter with such a very high number of resonators can still be well handled, as the measurement results in Fig. 4.11 clearly demonstrate. At a temperature of 70 K, the minimum insertion loss was found to be 0.32 dB; this value can, for a filter of such high-order, only be achieved if HTS technology is employed. Fig. 4.11 also displays the result of a two-tone measurement, and it is seen that the 3rd order intermodulation intercept point (IM3) is as high as 45 dBm. This last value is provided to prove that this filter provides a sufficiently large dynamic range, too.

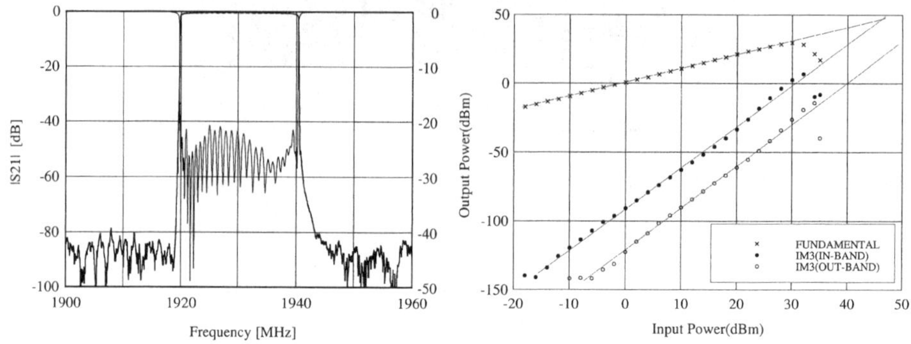

Fig. 4.11: Filter in Chebyshev-Design with 32 resonators ([21], © 2000 IEEE).
Left: Measured insertion and return loss. Right: Two-tone intermodulation measurement.

It was shown in section 3.2 that the selectivity of a filter can be increased by increasing the number of resonators (increased order), and/or also by introducing transmission zeros close to the passband. In the previous example, the number of resonators was chosen to be as high as 32. Though the measurement result is truly impressive, the number of resonators can hardly be surpassed for purely practical reasons.

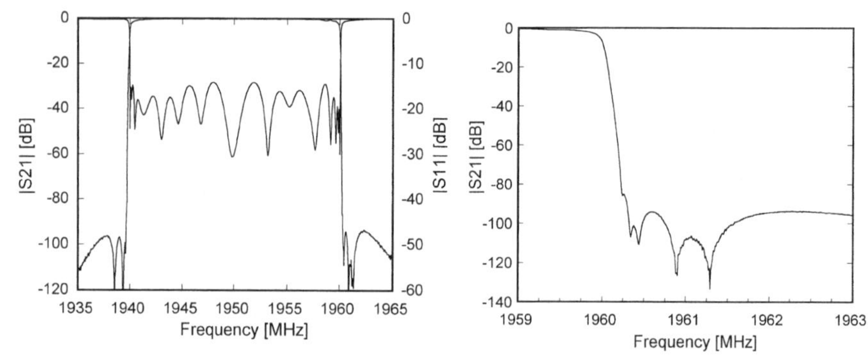

Fig. 4.12: Measurement results for a 22-pole filter with 10 transmission zeros ([58], © 2002 IEEE); Left: Insertion loss and return loss response. Right: Frequency response in the upper stopband. The 5 transmission zeros are clearly exposed.

The next example is a filter with the same bandwidth and same centre frequency, but now with 22 resonators which are cross-coupled such that 10 transmission zeros are achieved [58]. It was developed by Conductus Inc. (USA) which has merged with Superconductor Technologies Inc. in late 2002. The authors demonstrate that the steepness of the skirt of this filter surpasses that of a 50-pole Chebyshev (see Fig. 4.12). The filter is manufactured from the same materials as in case of the previous example, and at the same operating temperature of 70 K the steepness of the skirt is obviously superior. Furthermore, the filter now fits onto a smaller wafer of only 2 inch diameter because the number of resonators is reduced to 22.

The specifications for the already presented bandpass filters according to Figs. 4.11 and 4.12 correspond to 3rd generation W-CDMA (wide-band code-division multiple access) systems where the individual carriers (frequency bands) occupy a frequency bandwidth of about 5 MHz. Hence, the filters from Figs. 4.11 and 4.12 are designed to accommodate 4 carriers. If, by means of certain system considerations, the individual carriers are to be selected already in the RF regime, bandpass filters with a bandwidth of 5 MHz are needed. The next example is a filter with this very small bandwidth (0.25 % fractional bandwidth) which was developed by the German company Cryoelectra. As it was already discussed in sub-section 3.2, the required number of resonators (required order N) depends on the ratio $\Delta f / \delta f$ of the passband width to the transition width and for a given $\Delta f / \delta f$ elliptic and quasi-elliptic filters need a lower N than Chebyshev filters (see Fig. 3.2). Because of the latter fact, Cryoelectra has developed a new filter topology for the realisation of filters with elliptic response. These filters are composed of a cascade of dual- or triple-mode sections, with the multiple modes of a section providing parallel paths.

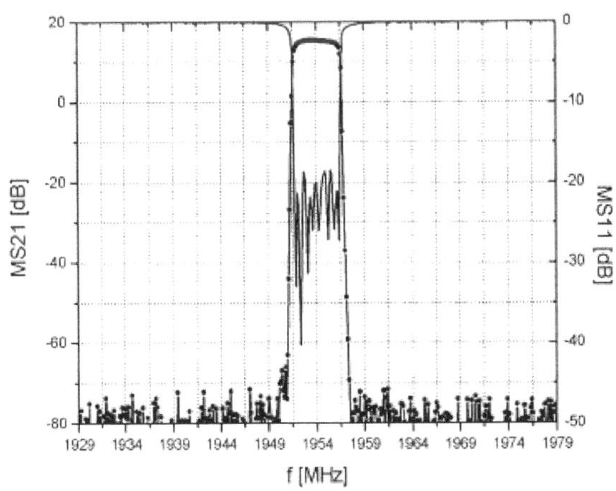

Fig. 4.13: Measurement results for a UMTS front-end composed of a 13-pole filter with 5 MHz in front of an LNA and followed by a 9-pole filter (5 MHz) bandwidth. Operating temperature at 70 K. Results from Cryoelectra (see also [59]).

Filters with 5 MHz bandwidth were built as a 9-pole and 13 pole filters (for a wider bandwidth, e.g. for $\Delta f = 20$ MHz filters with a comparable skirt steepness required 17 poles [59] according to Fig. 3.2). Despite this very small fractional bandwidth, very convincing results have been obtained. In conventional technology a filter with such a small fractional bandwidth would have a seriously degraded in-band behaviour. In Fig. 4.13 the measured frequency response for an entire UMTS front-end block composed of a 13-pole HTS filter (RF-BP1 in Fig. 5.2 a) in front of a low-noise amplifier (LNA) and a 9-pole HTS filter RF-BP2) is shown.

4.2.5 HTS filters for application in satellite communications

Now we turn to another major area of application for HTS bandpass filters, that is *satellite communications*.

The first example [60] is from the German company Tesat (formerly Bosch space division). Figure 4.14 shows the photograph of a 3-channel input multiplexer which was set up and fully space qualified. The channels are at 3.72, 4.08 and 4.12 GHz centre frequency and each consists of an 8-pole quasielliptic filter (large shaded area) with a relative bandwidth of about 1 % plus an external group delay equalizer (small shaded area), both in HTS technology. Furthermore the device comprises a number of circulators adapted to cryogenic temperatures, which serve for both splitting the broad frequency band (C-band) into the much smaller channels and also for coupling the circulators to the output ports of the filters. It is characteristic for these IMUX filters that not only the insertion loss response must obey severe specifications, but also the flatness of the group delay (see measurement curves in Fig. 4.15).

Fig. 4.14: Photograph of a space-qualified three channel input multiplexer with group delay equalization, company Tesat ([60], © 1999 IEEE).

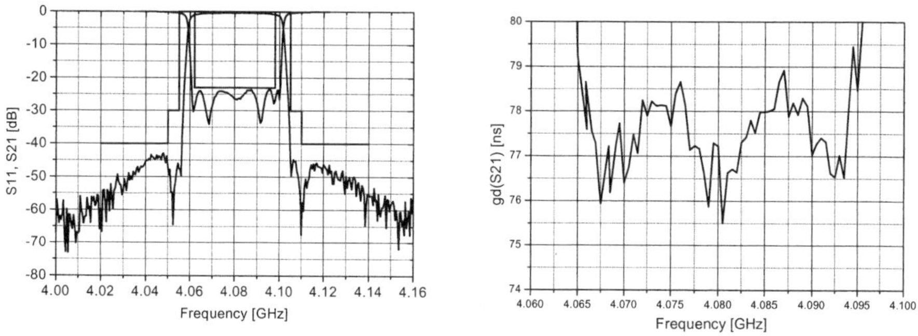

Fig. 4.15: Measurement curves from the middle channel of the IMUX from Fig. 4.14. Left: Insertion loss and return loss. Right: Group delay response on an enlarged scale. The group delay variation within the passband is only 3 ns.

Another superconductive input multiplexer has been set up by a consortium formed around the companies Com Dev, Lockheed-Martin and DuPont [61]. Their filters have electrical performances quite comparable to the previous example, but since the group-delay equalization is no longer formed with external equalizers but with a self-equalized approach the filters are further miniaturised.

Fig. 4.16: 60-channel superconductive input multiplexer for space applications ([61], © 2000 IEEE); the photograph does not show the cryocooler and the multi-layer insulation.

The importance of this work is that a very large number of filters (60) have been integrated, and therefore the benefit over conventional technology could be well verified. Comparing the cryogenic multiplexer to the conventional counterpart (which is dielectric technology in this application), savings in mass and volume of more than 50 % were demonstrated. Note that in this comparison, the cryogenic multiplexer comprises not only the 60 filters, but also the full cryogenic equipment including the cryocoolers and their electronics. Fig. 4.16 shows how the filters are arranged in 10 stacks with 6 filters each This is a favourable setup with respect to a minimised radiative heat input.

The previous two examples were on applications on the input side of a satellite transponder. In this case only very weak signals with low RF-power have to be handled by the filters. However, the transponders also comprise filters on the output side. The latter have to deal with RF-power levels often exceeding 100 W per channel. Although it was demonstrated that HTS resonators can handle such high power levels (see Fig. 4.5), for the following reasons it is rather difficult to implement complex HTS filters at the output side. To evaluate such an approach, one needs to consider the overall power balance (see also [62]). A 4-pole filter with a fractional bandwidth of 1 % at a centre frequency of 4 GHz and an unloaded quality factor of the single resonators of 15.000 is taken as a first example. When this filter handles a signal with an input power of 50 W, the power dissipation in the filter exceeds 1 W which is a significant load to the cooling system. For a typical cooler with an efficiency of 5 % at 77 K, already 20 W input power is needed, if only the dissipation of a single filter is accounted. However, when the unloaded quality factor is raised to 100.000, the same filter dissipates about 200 mW only, and the situation is significantly relaxed. The challenge is to transform the very high unloaded quality factors of single resonators to much more complex filter structures, taking into account that also the coupling and tuning structures have to be adopted to extremely low loss properties and a high-power handling capability as well.

All filter examples shown above are operated at frequencies below about 4.5 GHz. One reason for this is that the two major areas of application - mobile communications and the C-band in satellite communications - fall below this frequency boundary. The maximum frequency range at which the use of HTS technology is still advantageous over conventional technology depends, of course, on the type of application and the associated filter specifications, the major design goal, the number of filters needed etc. The range of 10 - 15 GHz is generally regarded as an upper bound. This is because at higher frequencies the advantage in loss behaviour of HTS in comparison to conventional metals becomes less pronounced (see Fig. 2.1). The advantage of miniaturization also diminishes since conventional filters are already sufficiently small at these frequencies. Furthermore, realisation and tuning of HTS filters becomes more difficult due to the fact that the sensitivity to geometrical deviations increases inversely proportional with the component size.

5. SYSTEM APPLICATIONS

5.1 General considerations

As it was figured out in the previous sections, due to the extremely low conductor losses in planar HTS thin-film structures at cryogenic temperatures, compact resonators in the lower GHz frequency range have become available which possess conventionally unachievable values for the unloaded Q on the order of (typically) $8 \cdot 10^4$ and even up to 10^6. If these high-Q resonators are combined with advanced filter topologies and design methods, compact filters with extremely high selectivity (steep skirts) as well as very low passband insertion loss can be realised. Whereas conventional filters are governed by the tradeoff in performance with respect to selectivity versus insertion loss, this tradeoff can practically be overcome by the utilisation of HTS filters.

However, RF filters are no end in itself. Their implementation is only justified if the superior properties on the component level can be transformed into a considerable benefit at system level. System benefit means either a higher system performance and/or functionality or reduction of system cost or a combination of both.

The superior features of HTS filters, namely a high RF selectivity combined with very low insertion loss, are of most interest for radio (wireless) systems, since wireless systems are especially susceptible to degradation from strong unwanted interference. Furthermore, an efficient usage of the limited spectrum requires placing different systems close to each other in the frequency domain. The other superior feature of HTS filters, namely the small size and mass can be transformed into an additional system benefit if a large number of filters are needed and can share a common cooling system.

Examples for wireless communication systems are terrestrial cellular and personal communication systems, wireless local area networks (WLAN), point-to-point radios (terrestrial and earth-satellite) and communication satellites. Examples for systems for other than communication purposes are radar, global positioning system (GPS), electronic support measures receivers (ESM), microwave radiometer for remote sensing and radio astronomical systems.

The function of all these systems can be decomposed in subfunctions performed by the corresponding subsystems. Examples for subsystems are RF front-ends which perform the subfunction of frequency down-conversion and pre-amplification and transparent repeaters which perform the subfunctions of channelisation, power amplification and recombination. Since the filters under consideration are always part of a subsystem it is of interest to first evaluate how and if superior properties on the component level transform into superior properties on the subsystem level, before the system benefits are discussed. An example may serve to illustrate the different views on these three different levels. The property "steep filter skirts" at the component level is related to the property "reduced intermodulation distortion originating from close-by interferers" on the subsystem level (here front-end = subsystem). This property in turn may e.g. in case of a cellular communication system be related to the system benefit "less dropped calls", if the system under consideration co-exists with another system occupying a close adjacent channel.

The benefit on system level depends on many conditions, like e.g. existing standards and market acceptance. Furthermore, there are typically many competing possibilities within a system to achieve a certain improvement. Consequently, a systematic and critical evaluation of system benefits must remain far beyond the scope of this review article. The main purpose of this section is therefore to provide a systematic discussion of the benefits on the subsystem level. This has the additional advantage that the conclusions apply to a larger variety of systems which have this type of subsystem in common. The section on real system applications will be restricted to a listing of features which are argued in the literature and at workshops to be beneficial. However, a critical evaluation of these discussed system benefits will not be given.

5.2 Subsystem considerations

5.2.1 Cryogenic front-end of superheterodyne single- and multicarrier receivers

In this section, we study the potential for improving the performance of receiver front-ends by replacing conventional RF filters with HTS filters and by operating the low-noise amplifiers (LNA) at cryogenic temperatures. This consideration is independent of the particular system application of the receiver. Therefore, results apply to all systems where selectivity, noise and dynamic range set the ultimate limits on the performance in terms of interferer resistance, operating range, power requirements etc., as well as on maximum data rate (communications) and resolution (radar).

The antenna receives electromagnetic radiation from different sources in a relatively broad frequency range. The contributions to the signal available at the antenna port can be classified as follows:

(i) "Useful signals" which contain the information to be recovered by the receiver system under consideration.

(ii) Manmade interference belonging to the same technical system (intra-system-interference) or to other technical systems (inter-system-interference).

(iii)Exterior noise, which is received by the antenna and can be described by the antenna noise temperature T_A.

In case of base station (BTS) receivers in a cellular communication system, useful signals are all (uplink) signals transmitted by those mobile stations (MS) which are associated with the corresponding cell and operator. Intra-system interferers are uplink signals from the same operator, but from different cells as well as downlink signals transmitted in the BTSs. Examples for inter-system-interferers are up- and downlink signals from other cellular operators as well as from other electromagnetic systems (TV, radar etc.). In case of a radar receiver for air surveillance the corresponding echoes from radar targets under consideration give the useful signals. Examples for interference are echoes from clutter (e.g. meteorological targets) or signals transmitted by non-cooperative transmitters.

The useful signals available at the antenna port are produced at the transmitter side by modulating (encoding) baseband signals onto high-frequency sine wave carriers. Each modulated carrier occupies a certain width Δf (carrier- or channel bandwidth) of the spectrum with channel centre frequencies f_{on}. The set of all receiver channels which belong to the considered system occupies a portion of the spectrum which can be characterised by the width ΔF (system bandwidth). As shown in Fig. 5.1 (part a), outside of this portion of the frequency spectrum with width ΔF out-of-band interference occurs. Additionally, external noise (not shown in Fig. 5.1) is superimposed. The function of the receiver is to select the appropriate frequency channels (carriers) and to recover the baseband information assigned to the receiver under consideration in the presence of interference and noise. Receivers can be categorised in single-carrier receivers (SCR) and multiple-carrier receivers (MCR). SCR receive only one frequency channel at one instant in time, but may switch between different channels. MCR must be capable to receive multiple channels simultaneously.

Here, the front-end of a classical superheterodyne receiver is discussed first. It allows studying the impact of superior filter properties onto the subsystem performance. Modifications of this receiver architecture will be considered later.

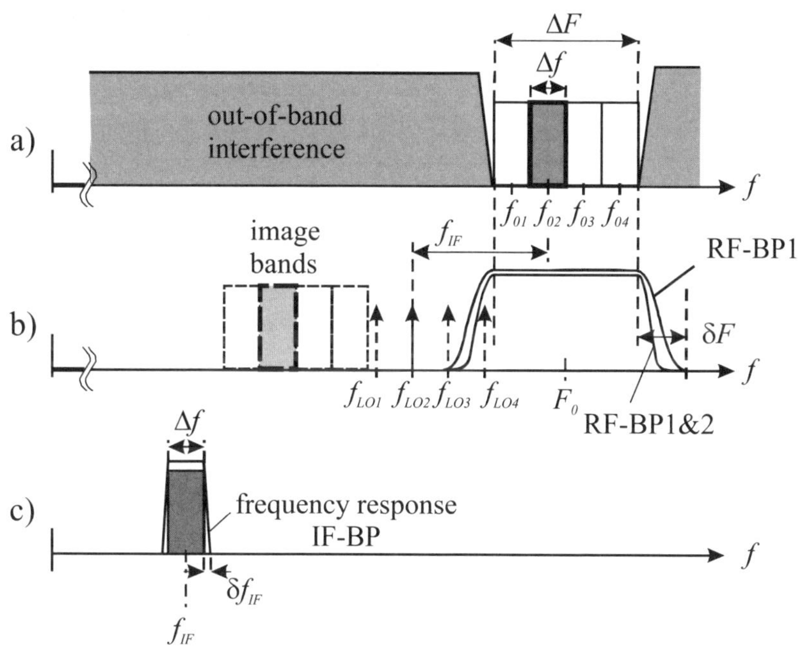

Fig. 5.1: Filtering functions of the front-end of a superheterodyne receiver: (a) Frequency spectrum at the antenna port with system band (width ΔF) comprising sub-channels of width Δf. Out-of-band interference. (b) LO frequencies $f_{LO,n}$ for downconversion of the different sub-bands and corresponding image bands. Frequency response of RF-BP1 and BP1&2. (c) Frequency response of IF-BP and frequency spectrum behind IF-BP.

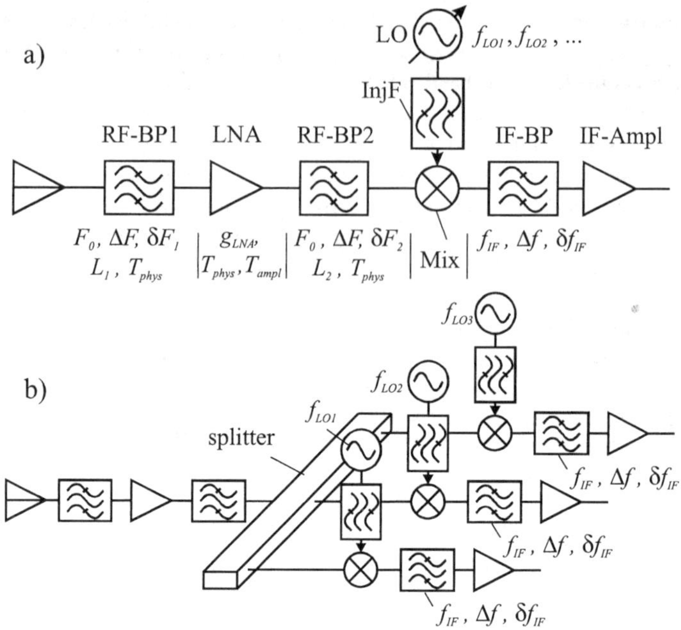

Fig. 5.2: Front-end architecture (a) for single-carrier receiver with switchable local oscillator (synthesiser) and fixed IF bandpass and (b) for multi-carrier receiver with parallel branches

There are 3 main functions provided by this functional block called "RF-front-end":

Downconversion: The received RF-signal is by means of analogue mixers (pumped by local oscillator LO) converted to a lower intermediate frequency regime (IF).

Analogue selection of frequency band(s): The width of the frequency spectrum received by the antenna is wider than the width occupied by the useful signal(s). Hence, one or multiple frequency channels have to be cut out.

RF amplification: At the very end of the analogue stages of a receiver the downconverted signal is digitised in an ADC. This can be done either at IF frequency, employing bandpass sampling or in the baseband with Nyquist sampling. This requires a power level of about 1 to 10 mW at the ADC input and therefore an overall amplification of typically 100 to 120 dB. To avoid amplifier instabilities this enormous gain has to be spread over the RF, IF and (for baseband sampling) the baseband. In order to reduce noise contribution by the first mixer, a part of this amplification has to be provided in the RF regime by means of a low-noise amplifier (LNA) placed in front of the first mixer.

There are two main reasons to perform the selection of narrow frequency bands (channel selection, function A) in the IF regime:

(a) Sharp filter skirts and hence narrowband bandpass filters can be more easily realized at IF.

(b) Frequency band selection for single-carrier receivers with variable centre frequency $f_{0,n}$ can be performed by means of a fixed IF filter (centre frequency f_{IF}, passband width Δf)

and a switchable local oscillator (synthesizer) with $f_{LO,n} = f_{0,n} \mp f_{IF}$. This is illustrated in Figs. 5.1 and 5.2 a).

(c) Simultaneous selection of multiple frequency bands of width Δf is accomplished by a power splitter and a bank of parallel branches (see Fig. 5.2 b) with different fixed LO frequencies and identical fixed IF filters.

Keeping these main functions of the analogue front-end in mind, the necessity for filters in the RF domain can be identified by assuming the existence of an IF bandpass (IF-BP) with bandwidth Δf and centre frequency f_{IF} behind the downconverter, but the absence of any RF filter in front of the (first) downconverter. In this case high frequency selectivity is provided by the IF-BP, but due to the mixing process the "image-problem" occurs.

Two frequency bands with centre frequencies $f_{LO,n} - f_{IF}$ and $f_{LO,n} + f_{IF}$ are simultaneously selected in the RF domain. RF signal components falling into these different frequency bands are converted into the same IF band where they are superimposed. If LO frequency and bandwidth are chosen such that the useful signal falls into one of the two RF bands it is converted into the IF regime, but unwanted RF signals falling into the other RF band (image band) are converted into the same IF band where they become unwanted interferers. This is illustrated in Fig. 5.1 a/b . Here the bands with $f_{0,n} = f_{LO,n} + f_{IF}$ are assumed to be the bands to be received, and in part b the corresponding image bands are depicted.

This problem can be overcome by inserting a bandpass filter in front of the first mixer (behind or in front of the LNA or a combination of both as shown in Fig. 5.2). Its passband width ΔF must cover all subbands and therefore the full system bandwidth. This function of RF bandpass filters is called "image rejection". Alternatively and/or additionally to image-reject filters a separation of used and image band can be accomplished by employing so-called image-reject mixers. However, the rejection with filters is potentially much higher than with this special type of mixers.

Whereas the rejection of signals received via the antenna and falling into the image bands can be performed by either RF-BP1 or RF-BP2 or by both, output noise of the LNA falling into the image bands can be rejected only by RF-BP2 behind the LNA (suppression of LNA image noise = function D).

From Fig. 5.1 b it is seen that the system bandwidth and the transition width δF of BP1 combined with BP2 sets a lower limit

$$f_{IF,\min} = 0.5\left(\Delta F + \delta F\right) \tag{5.1}$$

for the choice of the fixed intermediate frequency. Another function of RF bandpasses in front of the mixer can be to prevent the LO signal to pass back to the antenna where it is radiated. In order to ensure this function the lower limit

$$\tilde{f}_{IF,\min} = \Delta F + \delta F - 0.5\Delta f \tag{5.2}$$

applies, which is higher than the limit due to image rejection.

By taking into account that with HTS filters considerable smaller values of the transition width δF can be realised, it is seen from eqs.(5.1) and (5.2) that the freedom of choosing a lower first IF is gained. However, this is significant only in case of narrowband systems where with conventional filters the transition width becomes larger than the bandwidth ΔF.

Besides image rejection (functions B and D) and suppression of LO radiation leakage (function C) there are other (often most important) functions to be performed by RF filters in front of the first mixer. There is a necessity to prevent strong interfering signals from entering the LNAs and mixers, where they can cause intermodulation and saturation effects due to the nonlinear response of these active devices. The important intermodulation (IM) distortion problem is explained in Fig. 5.3. It is assumed that at the antenna port an out-of-band interferer with high spectral power density exists which is situated in frequency very close to the band edge of the used band with centre frequency F_0 and passband width ΔF.

As long as this interferer does not fall into an image band and the receiver would be perfectly linear, no problem would occur, since the interferer would be filtered out by the IF-filter behind the mixer. However, due to the nonlinear response of the LNA, IM products between different frequency components in the interferer spectrum are generated and cause the frequency width of this spectrum to grow into the used band. These in-band IM-products are not rejected by the IF filter and may be referred to as "IM noise". Fig. 5.3 depicts the impact of the steepness of the RF filter onto the magnitude and spectral width of the IM in-band noise. The lower left figure depicts schematically the spectral power density as a function of the frequency deviation from the centre frequency F_0 of the used band. The upper and lower figures at the centre of Fig. 5.3 show the frequency response of two bandpass filters with the same passband width ΔF and same stopband attenuation L_s, but with different skirt steepness characterised by different transition widths δF_1 and δF_2. As a result of this difference in steepness different results for the residual interferer spectral power density occur at point 2 (behind filter, but before LNA). Whereas this spectrum in front of the LNA (point 2 in Fig. 5.3) resides out of the band under consideration $|f - F_0| < \Delta F/2$, due to LNA nonlinearities the spectrum behind the LNA (point 3) grows into the considered band where it becomes IM noise. The power of this in-band IM noise depends on the 3rd power of the power of the residual interference signal and the spectral width of the IM noise depends on the spectral width of the residual interferer. This leads (see upper and lower figure on the right-hand side of Fig. 5.3) to a strong IM-noise reduction if a conventional filter with moderate steepness is replaced with a HTS filter with superior steepness. Note that in Fig. 5.3, part 3, the amplifier gain is ignored.

The lower IM in-band noise due to a strong out-of-band interferer close to the band edge allows to significantly reduce the guard bands between adjacent frequency channels, and thus a more efficient use of the rare frequency spectrum is achieved. At this point it must be recalled that that a steep skirt is always associated with a high group delay at the band edges (see sub-section 3.1). This in turn leads to a strong frequency variation of the group delay and hence to group-delay distortions of the signals. In a digital communication system, e.g., intersymbol interference is caused by group-delay distortion. Consequently, in systems which are sensitive to group-delay distortions the advantage of narrow guard bands can be employed only if a sufficient group delay equalisation is introduced. In systems where the signals are downconverted to IF and subsequently digitised, equalisation can alternatively be realised in the RF filter (self-equalised or separate RF equaliser) or by means of digital signal processing in the digital part.

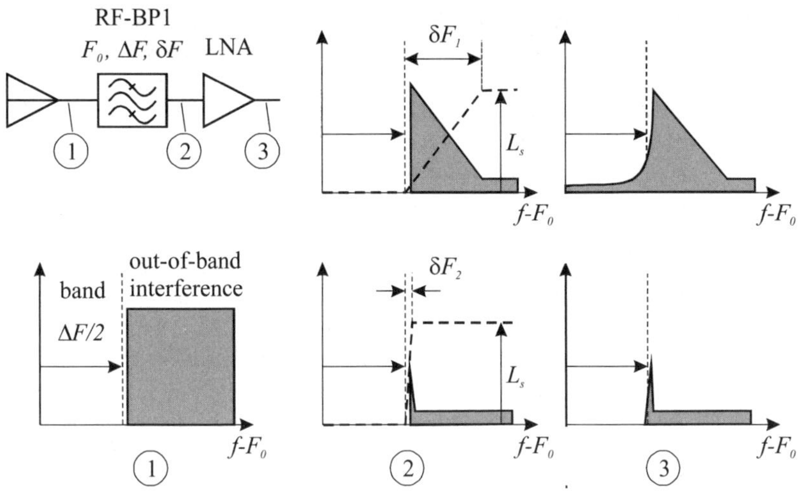

Fig. 5.3: Origin of in-band IM noise caused by strong out-of-band interferers and reduction of in-band IM noise due to an increased steepness of the filter skirt.

Besides the generation of IM products another effect of strong out-of-band interferers is that they reduce the effective gain for in-band signals (desensitisation) via saturation effects. In the extreme case saturation may lead to a complete "blocking". The threshold for blocking and desensitisation is considerably increased as a consequence of superior filter skirts.

By mixing of the second harmonics of the RF input signal f_0 with the second harmonics of the LO signal an unwanted IF signal is generated at the same frequency $\tilde{f}_{IF}$ as the wanted mixer response to an RF input signal with frequency $f_0 \pm \tilde{f}_{IF}/2$. Since the second harmonic of f_0 is not delivered by the antenna, but internally generated via the LNA nonlinearities, the suppression of this very special IF/2 effect requires a bandpass RP-BP2 to be placed in between LNA and mixer (function F).

Function	RF-BP1	RF-BP2	IF-BP
A: Frequency channel selection	-	-	X
B: Image rejection	X	X	-
C: Suppression of LO radiation leakage	X	X	-
D: Suppression of LNA image noise	-	X	-
E. Reduction of IM distortion and blocking via LNA	X	-	-
F: IF/2 suppression (internal second harmonics)	-	X	-

Table 5.1: Functions of RF filters RF-BP1 and RF-BP2 and IF filter IF-BP (see Fig. 5.2) in the front-end of a superheterodyne receiver

As a conclusion it can be summarised that 6 different filter functions A-F in the RF front-end were discussed. Table 5.1 lists these filter functions and indicates the corresponding filter. Whereas frequency channel selection is performed in the IF bandpass, the function E which is important to prevent IM inband noise and blocking by saturation can only be performed in the bandpass BP1 in between antenna and LNA. Image rejection for received signals can be realised by BP1 or BP2 or by a combination of both. Suppression of LNA image noise and of the IF/2 effect requires a RF bandpass BP2 behind the LNA.

The architecture of an RF front-end for a software-defined radio (SDR) can easily be derived from the classical architecture of Fig. 5.2. The significant difference in comparison to the classical structure is the fact, that frequency channel selection is now performed in the digital domain and can therefore be re-configured via software. From Fig. 5.2 one comes to the SDR architecture by using (1) a fixed LO instead of the switchable LO for SCR or the bank of LO for MCR and by replacing the IF filter with bandwidth $\Delta f < \Delta F$ by an IF filter with bandwidth ΔF. Now the superposition of signals from all sub-channels is passing through the IF-BP and is digitised. Channel separation is performed in the digital domain and the difference between single- and multi-carrier systems is reflected in the DSP algorithms only. All RF functions of the front-end remain the same, so that all subsystem benefits discussed above also apply to SDR systems.

Besides the superior resistance against strong out-of band interferers close to the band edge, cryogenic technology offers a significant *reduction of the receiver noise temperature* as well. The receiver noise temperature T_{rec} of a front end composed of (see also Fig. 5.2) bandpass BP1 (insertion loss L_1 in dB) , LNA (gain g_{LNA} and effective noise temperature T_{ampl} as function of the physical temperature T_{phys}), bandpass BP1 (L_2) and the converter section with mixer, IF-BP and IF-amplifier) with T_{conv} is given by

$$T_{rec} = \left(10^{0.1L_1} - 1\right) T_{phys} + 10^{0.1L_1} \left\{ T_{ampl}(T_{phys}) + \frac{1}{g_{LNA}} \left[\left(10^{0.1L_2} - 1\right) T_{phys} + 10^{0.1L_2} T_{conv} \right] \right\} . \tag{5.3}$$

Here for ease of presentation the same physical temperature T_{phys} is assumed for BP1, LNA and BP2.

If we compare a conventional front end at room temperature ($T_{phys} = T_0 = 290$ K) with conventional filters (insertion loss $\tilde{L}_1$ and $\tilde{L}_2$ dB, respectively) with a cryogenic front end, where the RF filters are replaced with HTS filters (insertion loss L_1 and L_2 dB, respectively) operated at a cryogenic temperature $T_{phys} = T_{cryo}$ (e.g. 70 K) and the LNA cooled down to the same cryogenic physical temperature, a reduction in the effective receiver noise temperature is achieved due to the combination of the following 3 effects (see eq.(5.3)):

(A) The effective noise temperature T_{ampl} of the LNA decreases with the lower operational temperature T_{phys} (see e.g. [63]).

(B) The thermal noise contribution of the RF filters decreases due to their reduced physical temperature T_{phys}.

(C) The reduced insertion loss L originating from the replacement of the conventional with HTS filter reduces both the thermal noise of the filters and the contribution of amplifier- and mixer noise to the overall receiver noise temperature T_{rec}.

Hence, the reduced receiver noise figure $F = (T_{rec}/T_0+1)$ of cryogenic front ends partially comes from the cryogenic ambient temperature (without use of HTS filters) and partially from the additional use of HTS filters. If, e.g. a conventional bandpass BP1 with $L_1 = 1$ dB in front of the LNA in cooled down from $T_{phys} = 290$ K to $T_{phys} = 70$ K (assuming no change in L_1) its contribution to the receiver noise temperature changes from 75 to 18 K. If this filter is replaced with a HTS filter at 70 K and L1= 0.1 dB one ends up with a noise contribution of 1.7 K. Furthermore, the reduction of insertion loss from 1 to 0.1 dB reduces the noise temperature contribution from all components behind BP1 (LNA, mixer etc.) by a factor of about 0.8. At this point a tradeoff in the design of cryogenic front ends should be mentioned. The cable which connects the (room temperature) input of the cryogenic encapsulation with the cooled HTS filters should have the lowest possible insertion loss to keep the noise figure low, but, on the other hand, is required to have a high thermal resistance to reduce the thermal load.

Further noise is contributed by the local oscillator. Due to oscillator phase noise the frequency spectrum of the LO output signal contains contributions which are separated from f_{LO} by $\pm f_{IF}$ spacing. These contributions are mixed with f_{LO} to produce noise at the IF frequency. This type of noise can be overcome by reducing the LO phase noise (e.g. with a high-Q HTS resonator) and additionally by inserting a bandpass filter (injection filter InjF) between the LO and the mixer (see Fig. 5.2).

5.2.2 RF-Selection and channelisation

The advantages of channel selection by means of IF filters in the downconversion stage, was discussed in detail in section 5.2.1, namely (i) the sharp skirts of IF filters (e.g. crystal or SAW filters) in comparison to the skirts of conventional RF filters and (ii) the ease of changing the RF centre frequency via changing the LO frequency.

However, there are situations where channel selection must be performed in the RF domain by means of RF filters.

The first situation (case A) is characterised by the fact that strong interferers within the system frequency band and with frequency changing with time are present. The disadvantage of the wide pre-select filter RF-BP1 with bandwidth ΔF (see Figs. 5.1 and 5.2) is that such an in-band interferer leads to a simultaneous desensitisation and blocking of all sub-bands within the system band due to nonlinearities in the LNA.

If in case of a single-carrier receiver (with variable centre frequency) the passband of the pre-select filter covers the instantaneous sub-channel band of bandwidth Δf only, the channel in use is not degraded if the in-band interferer falls into one of the other sub-bands (see Fig. 5.1 a). The required properties of such a filter (see Fig. 5.4a) are (I) very high selectivity and (II) tuneability and/or switchability between different sub-bands. HTS technology can help to meet requirement (I), but requirement (II) still needs quite a lot of progress in both basic technology aspect and implementation into filter structures. Competing approaches for switchable and tuneable HTS filters are [64] (a) ferroelectric [65] and/or ferrimagnetic [66] structures and furthermore mechanical solutions. The latter include MEMS [67] and piezoelectric actuators [68].

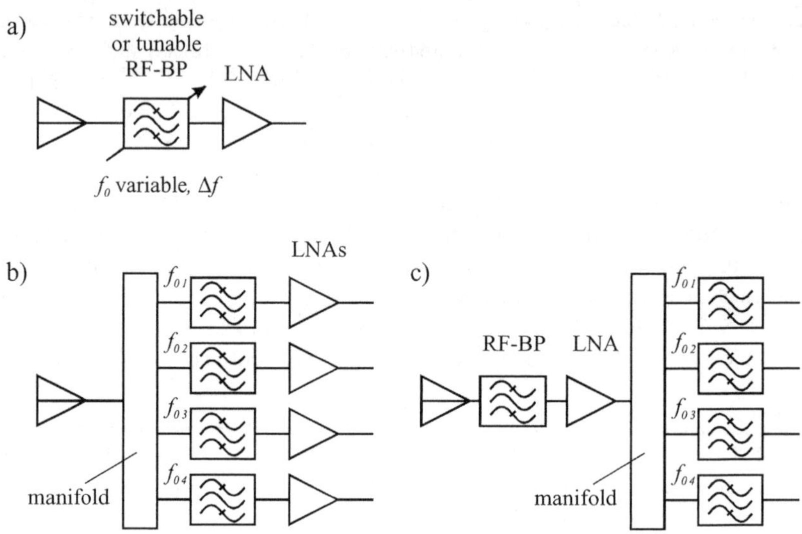

Fig. 5.4: Three different cases of RF channel selection: (a) Single-carrier receiver with switchable or tuneable pre-select filter with passband width equal to the width of a single sub-channel in front of the LNA (b) Multiple-carrier receiver with input multiplexer (IMUX) for channelisation of the system band of width ΔF into narrower subbands of width Δf. IMUX in front of LNAs. (c) Similar to (b), but with first LNA in front of IMUX, so that high resistance against inband interferers does not apply

In case of multi-carrier receivers a channeliser (input multiplexer, IMUX) in front of the first LNA is needed in order to achieve that an in-band interferer degrades only that sub-channel which corresponds to its instantaneous frequency. This channeliser is composed of a set of bandpass filters (channel filters). Because of mutual interaction these filters cannot just be connected in parallel. Decoupling of filters can in principle be performed by circulators or hybrid couplers, but low insertion loss requirements can only be met if a passive so-called manifold is used.

Another situation (case B) occurs in case of systems without downconversion to an IF band. Transparent repeater (e.g. in communication satellites) may serve as an example. In this case channelisation via parallel mixers (as in Fig. 5.2b) cannot be used. Hence, one is forced to use RF filter-banks (IMUX) for channelisation. Since in this case the motivation for the use of RF filter banks does not come from the interferer resistance point of view, pre-amplifiers as well as RF-RF frequency converters can be situated in front of the IMUX. Fig. 5.4 c depicts such a type of channeliser. If the frequency blocks corresponding to the sub-bands are well separated, the IMUX is referred to as non-contiguous. In a contiguous IMUX adjacent bands share a common crossover frequency point with (ideally) 3 dB insertion loss. Often, IMUXs are also called contiguous if the crossover frequency point lies in the transition region.

5.2.3 Transmitter combiner (output multiplexer)

If a number of narrower sub-bands have to be combined into a common port, e.g. a common antenna, the function of a transmitter combiner (output multiplexer, OMUX) is needed. If insertion loss associated with this function has to be minimised, combiner based on hybrid couplers and circulators have to be ruled out. The proper solution consists of a bank of filters connected by a manifold as shown in Fig. 5.5. In the design of the OMUX one should keep in mind that for M channels each with the same input power, the maximum instantaneous power scales with M^2.

The transmitter filter between OMUX and common antenna is required to further reject spurious frequencies outside of the frequency spectrum dedicated to the system.

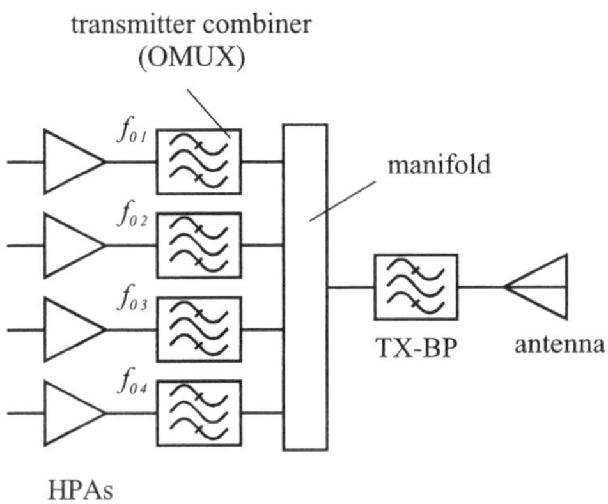

Fig. 5.5: Transmitter combiner (output multiplexer OMUX) behind a bank of high-power amplifiers (HPAs). Additionally a transmitter filter is shown.

5.3 System considerations

5.3.1 Terrestrial cellular communication systems

The subsystem performance of receiver front ends (see section 5.2.1) in base stations (base transceiver stations, BTS) for analogue (1G) and digital (2G, 2.5G and 3 G) cellular communication systems has a significant impact on system properties. At present cryogenic front ends [69,27] comprising HTS preselect filters and cooled LNAs are already commercially available. The number of HTS filters and cooled LNAs using a common closed cycle refrigerator equals the number of antenna ports. A BTS for 3 sectors with main and diversity antennas possesses 6 antenna ports and therefore 6 front ends in total.

The expected system benefit of these cryogenic front ends is derived from the superior subsystem performance as outlined in section 5.2.1, namely a (i) superior rejection of strong out-of-band interferers close to the band edge(s) of the used frequency band(s) and (ii) a reduced noise figure $F = T_{rec}/T_0 + 1$ of the receiver.

In a crowded frequency spectrum in urban areas, a superior out-of-band rejection will in general reduce the in-band intermodulation noise as well as the probability of blocking via saturation. This in turn decreases the bit-error rate and the probability of dropped calls. Out-of-band signals which are strong enough to generate in-band IM noise are e.g. from collocated transmitters (multiple antennas within the same small local area, "antenna farms") or from nearby mobile stations from other operators far away from their own BTS and thus operating at maximum transmit power.

However, a reliable quantitative evaluation of the system benefit of a superior out-of-band rejection requires a precise knowledge of the interferer scenario of the system under consideration. In this section we restrict ourselves to report some extreme interferer scenarios with very high demands on the steepness of the filter steepness.

A crucial interferer scenario occurs if in a relatively short distance from the BTS receiver under consideration mobile or stationary transmitters are using an adjacent frequency band. This problem is known to occur for the coexistence between FDD and TDD systems, if the FDD uplink band resides adjacent to the TDD band. Since the TDD systems are using the same frequency band for both transmission and reception by setting time intervals, they represent a strong out-of-band interference to FDD systems using different well-separated frequency bands for transmission and reception. A well-known extreme interference situation of this kind occurs in Japan where the 1.9 GHz frequency band of the personal handyphone system PHS which is a TDD system is very close to the frequency band allocated for the uplink of the 3rd generation IMT-2000 (e.g. DS-CDMA with FDD) system. Quite a large number of HTS filter demonstrators (see e.g. Figs. 4.11-4.13) were designed for this extreme PHS/ IMT-2000 situation. Similar problems may occur with the UMTS standard in Europe, where

Some BTS will have to accommodate the co-existence of wideband (W)-CDMA in paired frequency bands in the frequency-duplex mode (FDD), and TDMA/CDMA in unpaired bands in the TDD mode.

Two further examples which are highly demanding with respect to the selectivity of the RF pre-select filter are known to exist in the USA. First, Enhanced Specialized Mobile Radio (ESMR) systems are known to produce strong interference signals very close to the cellular bands . The second example is the "encapsulation" of a frequency block A within a frequency interval allocated for block-B operators. Due to their limited selectivity conventional pre-select filters used by block-B operators cannot reject the signals from the embedded block A, which causes a severe out-of-band interference problem. Highly selective HTS filters are available, where a narrowband band-reject filter for the A-block is combined with a bandpass such that a filter response with 2 passbands for the B-operator results.

Further potential system benefits for base stations (BTS) take advantage of the reduced noise figure $F = T_{rec}/T_0 + 1$ of the cryogenic front end. In order to evaluate the system impact of a reduced noise level, one has to take 4 different types of in-band signals into account, namely (i) the wanted signal, (ii) co-channel interference in FDMA/TDMA from remote cells using the same frequency band or multiple-access-interference (MAI) in CDMA systems with non-synchronised uplink, (iii) intermodulation-noise as discussed above and (iv) noise received by the antenna (described by T_A) and generated in the receiver (T_{rec}).

If one assumes a cable with an insertion loss of a dB to connect the antenna port with the receiver front end, the total spectral power density of the noise, formally transformed to the antenna port, is given by the product of Boltzman´s constant and the system noise temperature T_{syst} with

$$T_{syst} = T_A + \left(10^{a/10} - 1\right) T_0 + 10^{a/10} \left(F - 1\right) T_0 \quad . \tag{5.4}$$

The reduction in noise (contribution iv) resulting from replacing a conventional front end with F_{conv} with a cryogenic front end with F_{cryo} is described by the noise-reduction-factor Λ with

$$\Lambda = \frac{T_{syst}^{cryo}}{T_{syst}^{conv}} = \frac{1 + \left[10^{a/10} F_{cryo} - 1\right] T_0 / T_A}{1 + \left[10^{a/10} F_{conv} - 1\right] T_0 / T_A} > \frac{1 + \left[10^{a/10} - 1\right] T_0 / T_A}{1 + \left[10^{a/10} F_{conv} - 1\right] T_0 / T_A} \quad . \tag{5.5}$$

The lower limit for the noise-reduction-factor in eq. (5.5) is derived by assuming $F_{cryo} \to 1$ (0 dB). The antenna noise temperature which accounts for the received exterior noise is estimated to be given by

$$T_A = \eta T_0 + (1 - \eta) T_{sky} \tag{5.6}$$

with T_0 =290 K as room temperature and $T_{sky} \approx 10...20\ \text{K}$, and η denoting the effective absorption coefficient of the environment. If in a rough approximation η is assumed to be close to one, the antenna noise temperature approaches $T_A \approx T_0$, so that eq. (5.5) becomes

$$\Lambda = \frac{T_{syst}^{cryo}}{T_{syst}^{conv}} = \frac{F_{cryo}}{F_{conv}} > \frac{1}{F_{conv}} \quad . \tag{5.7}$$

The important result from eq.(5.7) is that independent of the insertion loss of the cable the noise-reduction-factor equals the ratio between the noise figures. Furthermore, this reduction factor cannot be smaller than $1/F_{conv}$. Thus, a significant reduction factor can be achieved only if the conventional system has a relatively high noise figure (e.g. $F_{conv} > 2$ corresponding to 3 dB).

In rural areas with low traffic the maximum cell size is noise-limited. This means that the maximum cell radius $R_{\max}$ is determined from the required signal-to-noise ratio at the BTS, the system noise temperature, the maximum transmitter power of the mobile station and the path-loss. Assuming a path-loss according to R^{-n}, the noise-reduction-factor transforms in an increased cell-size according to

$$R_{\max}^{cryo} = \Lambda^{-1/n} R_{\max}^{conv} \quad \text{with } n \approx 4 \tag{5.8}$$

With 3 dB improvement in noise figure ($\Lambda = 0.5$) the radius can be made about 19 % larger. This corresponds to a by 41 % increased cell area or a by a factor of about 2.3 smaller number of base stations. Instead of increasing the cell radius the coverage in strongly shadowed areas in buildings can be improved.

However, all these results for increased cell size and improved coverage apply to the uplink only. Hence, a real system benefit is achieved only if the system is uplink-limited and not downlink-limited. Furthermore, one must keep in mind that in areas with high traffic (typically urban areas) the maximum cell size is not determined by noise, but by interference and by the required capacity.

In case of W-CDMA systems with non-synchronised uplink and without multi-user detection the uplink signal of a considered user is superimposed by the signals of all other users in the cell and the adjacent cell. Via de-spreading these signals become noise-like "multiple-access-interference (MAI)". Considering the superposition of MAI and noise together with the limited transmitted power of the mobile stations, the maximum radius shrinks with the traffic within the cell. With a reduced receiver noise figure, for a given traffic the maximum radius becomes larger.

5.3.2 Receivers for radio astronomy in the lower GHz regime

The subset of radio astronomy receivers with an operational frequency range below about 10 GHz can significantly benefit from the unique properties of HTS filters. Because these receivers are designed to detect and analyse extremely weak signals, a strong demand for a very low receiver noise temperature T_{rec} has to be met.

It is important to recall that the sensitivity is determined by both receiver noise (T_{rec}) as well as background noise received by the antenna (T_A). Whereas for terrestrial communication systems (and for the earth-satellite uplink in space-born systems) the antenna noise temperature T_A is dominated by thermal noise contributions at room temperature (resulting in $T_A \approx 200 \ldots 300$ K), the antenna noise of radio astronomy antennas is dominated by the noise temperature of the sky, leading in the lower GHz range to $T_A \approx 10 \ldots 20$ K. The sensitivity improvement gained from a replacement of a room-temperature receiver by a cryogenic receiver is more significant for low than for high antenna noise temperatures. If e.g. for $T_A =$ 250 K the receiver noise temperature is reduced from 50 K to 10 K, the sensitivity is improved by 0.62 dB only. However for $T_A = 20$ K it is improved by 3.7 dB.

In order to achieve this low receiver noise temperature HEMT (high electron mobility) amplifiers as well as passive components in front of the amplifier (hybrid couplers, waveguide sections, etc.) are operated at cryogenic temperatures, typically at 15 K, and with these measures the receiver noise temperature can be reduced to values below 10 K. For cooled amplifiers, see e.g. [63].

Due to the high sensitivity and high gain of these radio astronomy front-ends, they become very susceptible to strong interference owing e.g. to radio stations, radar, GPS etc.. As it was explained in sub-section 5.2.1 this interference causes saturation and desensitisation of the receiver and this in turn requires filters which sufficiently suppress the interferer. However, the insertion loss of a filter in front of the first amplifier increases the receiver noise temperature (see sub-section 5.2.1) , such that (cooled) conventional filters are mostly placed behind the first amplifier where they prevent the following amplifier stages from saturation, but cannot prevent desensitisation of the first stage. HTS filters are very well suited to overcome this problem, since they provide both an extremely low insertion loss and steep filter skirts. The latter feature allows to operate the receiver for sensitive continuum measurements in a frequency range close to strong interferers. The fact that radio astronomy receivers are anyway employing a cryogenic front end and the small size of HTS filters are additional arguments which make HTS technology attractive for this kind of application.

5.3.3 Communication satellites

Communication satellites are an important field for the application of highly developed RF-equipment, with the number of satellites launched worldwide per year being between 20 and 30. (In 2002 this number dropped due to the general economic situation, but is expected to recover again.) The requirements in this field are especially demanding, e.g. with respect to a lifetime exceeding 12 years (remember that the time which has elapsed since the discovery of HTS materials hardly exceeds this limit!). Two experiments have demonstrated the robustness of HTS materials including the cryogenic setup in space [22,70] and one experiment has been set up and extensively qualified on ground, fully according to standard space procedures [23].

However, HTS systems are not in regular use in satellite communications due to the aforementioned requirements on reliability and due to the associated conservative attitude of the satellite operators. It is important to note in this context that the number of cryogenic systems on scientific satellites is increasing. There are about 15 cryocoolers being operated in space at present, and the growing amount of data on the reliability of such systems might smooth the way for future HTS RF-components in satellite communications. Though space provides a rather cold environment for satellites, on a satellite there are numerous devices (both RF-equipment and infrastructural) dissipating energy, and in general the temperature environment is between about –40°C and +80°C. Generally used frequency bands in satellite communications are the C-band (3.4 – 6 GHz), the Ku-band (10 – 14 GHz) and the Ka-band (18 – 32 GHz). Recalling the statements of section 4.2 on the frequency limits for HTS filters, the C-band is the first candidate for HTS components.

The simplest (but nevertheless often used) architecture is that of a transparent repeater. Its only task is to amplify a signal received from an earth station and to radiate it back to another (differently located) station. The amplification is performed in the RF-domain. Due to the tradeoff between efficiency and linearity, one has to cope with a significant non-linear response of the amplifiers. Hence, the broad frequency band must be split up in a number of narrower frequency channels each of which is amplified individually. This scheme looks as depicted in Fig. 5.4c, but with a larger number of channel filters. For the INTELSAT 8 satellite, as an example, 30 frequency channels within the C-band were in use. By using filters in HTS technology it is possible to perform the same RF-functionality, but with substantially reduced size and mass for the equipment even including the cryosystem (see the examples from Fig. 4.14 – 4.16). Saving a mass of 1 kg on a satellite launch transforms into savings of 50.000 US$. This application is often considered as the first candidate for introducing HTS filters on satellites.

Furthermore, by not only benefiting from the smaller mass and volume but also from the improved RF-properties of HTS, it can be envisaged to implement filters with sharper skirts and thus to make better use of the available frequency spectrum. Further possible components for HTS applications are the first RF-bandpass filters directly behind the antenna and in front of the LNA (see again Fig. 5.4c), but these are generally mounted very close to the receive antenna to keep the noise figure low. Close to the antenna there is hardly any space for implementing a cryocooler and the thermal radiation environment is also unfriendly towards such a cooler.

A third potential subsystem for application is located behind the RF-amplifiers where the frequency channels are recombined by means of an OMUX (Fig. 5.5). Here, power levels of up to 150 W per channel can occur (with a trend towards even higher values). A detailed tradeoff reveals two facts: At first, when replacing a conventional OMUX by a very low loss cryogenic OMUX, the miniaturisation is not as high as on the IMUX side since resonators suitable for handling high power levels are inherently larger (Fig. 4.1). At second, there is further benefit from the HTS OMUX: Due to the reduced insertion loss a substantial amount of amplifier power can be saved while still maintaining the same power level at the output [62]. However, as detailed in section 4.2 the challenge of transforming the achievable very high unloaded quality factors in HTS technology into more complex filter functions is still unresolved.

Besides the simple architecture of the transparent repeater, more elaborated regenerative transponders are in use which perform not only the function of signal amplification, but also some form of signal processing and restoration in the digital domain. A large variety of regenerative architectures is known, where various numbers of filters (in both RF- and IF-domain) are needed, and also various numbers of LNAs and local oscillators are in use. The potential advantages of HTS technology must be weighed with respect to the particular design goals for the individual architecture and is beyond the scope of this review.

REFERENCES

[1] H. Weinstock and M. Nisenoff, Eds., *Microwave Superconductivity*. Dordrecht: Kluwer Academic Publishers, 2001.

[2] M.J. Lancaster, *Passive Microwave Device Applications of High Temperature Superconductors*. Cambridge, MA: Cambridge Univ. Press, 1997.

[3] Z.Y. Shen, *High Temperature Superconducting Microwave Circuits*. Norwood, MA: Artech House, 1994.

[4] *IEEE Trans. Microwave Theory Tech. (Special Issue on Microwave and Communication Applications at Low Temperature)*, vol. 48, July 2000.

[5] R.R. Mansour, "Microwave Superconductivity," *IEEE Trans. Microwave Theory Tech.*, vol. 50, no. 3, pp. 750 - 759, March 2002.

[6] M. Tinkham, *Introduction to Superconductivity*. New York: Mc-Graw-Hill, 1996.

[7] M. Hein, *High-Temperature-Superconductor Thin Films at Microwave Frequencies*. Berlin: Springer-Verlag, 1999.

[8] www.theva.com

[9] N. Klein, H. Chaloupka, G. Müller, S. Orbach, H. Piel , "The effective microwave surface impedance of high-T_c thin films," *J. Appl. Phys.* 67, pp. 6940, 1990.

[10] T. Dahm, D.J. Scalapino, *Phys. Rev. B 60*, p.13125, 1999.

[11] R. Wördenweber, "Growth of high-Tc thin films," *Supercond. Sci. Technol.* 12, R86 - R102, 1999.

[12] R. Semerad, J. Knauf, K. Irgmaier, W. Prusseit, "RE-123 thin films for microwave applications," *Physica C* 378-38, pp. 1414-1418, 2002.

[13] H. Schneidewind *et al.*, "TlBaCaCuO-(2212) thin films on lanthanum aluminate and sapphire substrates for microwave filters," *Supercond. Sci. Technol.* 14, pp. 200 – 212, 2001.

[14] H. Schneidewind et al, "2212-Tl-Ba-Ca-Cu-O films on 3 inch sapphire wafers for high frequency filters," *IEEE Trans. Appl. Supercond.* 11, pp. 3106 – 3109, 2002.

[15] D.W. Face *et al.,* "Large area YBCO and TBCCO thin films for microwave and electronic applications," *Physica* C 357 – 360, pp. 1488 – 1494, 2001.

[16] Y.Y. Xie, J.Z. Wu, T. Aytug, A.A. Gapud, D.K. Christen, D.T. Verebelyi, and K. Song, "Fabrication and physical properties of large-area $HgBa_2CaCu_2O_6$ superconducting films," *Supercond. Sci. Technol.* 13, pp. 225-228, 2000.

[17] S. Chromik, A. Sin, V. Strbik, G. Plesch, P. Odier, and F. Weiss, "Hg-based cuprate superconducting thin films prepared on CeO_2 buffered R-plane sapphire substrate," *Supercond. Sci. Technol.* 14 , pp. 875-879, 2001.

[18] T. Stelzner, H. Schneidewind, G. Bruchlos, "Superconducting Hg-Ba-Ca-Cu-O thin films on lanthanum aluminate and sapphire substrates," *IEEE Trans. Appl. Supercond.* 2003, vol 13, 2003 (accepted for publication).

[19] www.ospreymetals.co.uk/low_expansion/.

[20] N.J. Parker *et al.*, "Tuning superconducting microwave filters by laser trimming," in *IEEE MTT-S Int. Microwave Symp. Dig.*, pp.1971 – 1974, 2002.

[21] G. Tsuzuki, "Superconducting filter for IMT-2000 band," *IEEE Trans. Microwave Theory Tech.*, vol. 48, pp. 2519 - 2525, 2000.

[22] M. Nisenoff *et al.*, "On-orbit status of the high-temperature superconductivity space experiment," *IEEE Trans. Appl. Superconduct.*, vol. 11, pp. 799-805, 2001.

[23] T. Kässer *et al.*, "Superconductors and cryotechnology for space communications - Adaptation of a new technology for applications," in *IEEE-MTT-S Int. Microwave Symp. Dig.*, pp. 657-660, June 2000.

[24] W. Prusseit, "Protective coatings for YBCO-thin film devices," *Inst. Phys. Conf. Ser. No 167*, pp. 69 – 72, 2000.

[25] J.D. Rhodes, *Theory of Electrical Filters*. London: John Wiley & Sons, 1976.

[26] W. Lyons *et al.*, "High-temperature superconductive wideband compressive receivers," *IEEE Trans. Microwave Theory Tech.,* vol. 44, pp. 1258-1278, 1996.

[27] H.J. Chaloupka, "Microwave applications of high temperature superconductors," in: H. Weinstock (ed.), *Applications of Superconductivity*, 295 - 383, Kluwer Academic Publishers, 2000.

[28] M. Klauda *et al.*, "Superconductors and cryogenics for future communication systems," *IEEE Trans. Microwave Theory Tech.,* vol. 48, pp. 1227 - 1239, July 2000.

[29] H. J. Chaloupka, S. Kolesov, "Design of lumped-element 2D RF Devices," in: H. Weinstock and M. Nisenoff (eds.), *Microwave Superconductivity*, pp. 205 - 238, Kluwer Academic Press, 2001.

[30] G. Matthaei, L. Young, E. M. T. Jones, *Microwave filters, impedance-matching networks, and coupling structures*. Norwood: Artech House, MA, 1980.

[31] H.C. Bell, "Coupled-resonator prototype synthesis," *IEEE MTT-S Int. Microwave Symp. Dig.*, Workshop Notes, June 2000.

[32] R.J. Cameron, "Fast generation of Chebyshev filter prototypes with asymmetrically-prescribed transmission zeros," *ESA Journal*, vol. 6, pp. 83 – 95, 1982.

[33] G. Macchiarella, "An effective technique for the synthesis of an equiripple low pass prototype filter with asymmetric frequency response and arbitrary transfer function zeros," *Proc. 25th Europ. Microwave Conf.*, vol. 2, pp. 722 – 724, Sep. 1995.

[34] G. Macchiarella, "A general approach to the design of bandpass cavity filters with multiple couplings," *IEEE Trans. Microwave Theory Tech.,* vol. 45, pp. 179 – 187, 1997.

[35] R.J. Cameron, "General coupling matrix synthesis methods for Chebyshev filtering functions," *IEEE Trans. Microwave Theory Tech.*, vol. 47, pp. 433 – 442, April 1999.

[36] R.J. Cameron, "General prototype network-synthesis methods for microwave filters," *ESA Journal*, vol. 6, pp.193 – 206, 1982.

[37] J.D. Rhodes, I.H. Zabalawi, "Synthesis of symmetrical dual mode in-line prototype networks," *Int. J. Circuit Theory and Applications*, vol. 8, pp. 145 – 160, 1980.

[38] R.J. Cameron, J.D. Rhodes, "Asymmetric realizations for dual-mode bandpass filters," *IEEE Trans. Microwave Theory Tech.*, vol. 29, No. 1, pp. 51 – 58, Jan 1981.

[39] R.J. Cameron, A. R. Harish, C. J. Radcliffe, "Synthesis of advanced microwave filters without diagonal cross-couplings," in *IEEE MTT-S Int. Microwave Symp. Dig.,* pp.1437 – 1440, June 2002.

[40] U. Rosenberg, S. Amari, "Novel design possibilities for dual-mode filters without intracavity couplings," *IEEE Microwave and Wireless Components Letters*, vol. 12, no. 8, pp. 296 – 298, August 2002.

[41] A. Baumfalk, M. Reppel, H. Chaloupka and S. Kolesov, "Investigations on the unloaded quality factor of planar resonators with respect to substrate materials and packaging," *Inst. Phys. Conf. Ser.*, no. 167, pp. 311-314, 2000.

[42] H. J. Chaloupka, "Application of high-temperature superconductivity to antenna arrays with analog signal processing capability," *Proc. 24th Europ. Microwave Conf., Cannes*, pp. 23 - 35, 1994.

[43] A. Baumfalk, H. Chaloupka, S. Kolesov, M. Klauda, C. Neumann, "HTS power filters for output multiplexers in satellite communications," *IEEE Trans. Appl. Supercond.*, vol. 9, no 2 and 3, pp. 2857 – 2861, 1999.

[44] H. Chaloupka, M. Jeck, B. Gurzinski and S. Kolesov, "Superconducting planar disk resonators and filters with high power handling capability," *Electronic Letters* , vol. 32, pp. 1735-1737, 1996.

[45] S. J. Fiedziusko *et al.*, "Dielectric Materials, devices and circuits," *IEEE Trans. Microwave Theory Tech.*, vol. 50, pp. 706 – 720, 2002.

[46] Z.-Y. Shen *et al.*, "High-Tc superconductor-sapphire resonator with extremely high Q-values up to 90 K," *IEEE Trans. Microwave Theory Tech.*, vol. 40, pp. 2424 – 2432, 1992.

[47] N. Tellmann *et al.*, "High-Q $LaAlO_3$ dielectric resonator shielded by YBCO-films," *IEEE Trans. Appl. Supercond.*, vol. 4, pp. 143 – 148, 1997.

[48] N. Klein *et al.*, "Properties and applications of HTS-shielded dielectric resonators: a state-of-the art report," *IEEE Trans. Microwave Theory Tech.*, vol. 44, pp.1369 – 1373, 1996.

[49] R.R. Mansour *et al.*, "A C-Band superconductive input multiplexer for communication satellites," *IEEE Trans. Microwave Theory Tech.*, vol. 42, pp. 2472 – 2479, 1994.

[50] S. Schornstein *et al.*, "High-temperature superconducting-shielded high power dielectric dual-mode filter for applications in satellite communications," in *IEEE MTT-S Int. Microwave Symp. Dig.*, p. 1319, 1998

[51] S. Vitusevich *et al.*, "An all-cryogenic low phase-noise hybrid K-band oscillator for satellite communications," in *IEEE MTT-S Int. Microwave Symp. Dig.*, pp. 1977-1980, 2001.

[52] K.S.K. Yeo *et al.*, "The design of microstrip six-pole quasi-elliptic filter with linear phase response using extracted-pole technique," *IEEE Trans. Microwave Theory and Tech.*, vol. 49, pp. 321 - 327, 2001.

[53] T. Kaesser, Internal report 2002 (unpublished).

[54] M. Reppel, H. Chaloupka, "Novel approach for narrowband superconducting filters," in *IEEE MTT-S Int. Microwave Symp. Dig.*, pp.1563-1566, 1999.

[55] K. Saito *et al.*, "246 MHz HTS lumped element bandpass filter," *IEICE Trans. Electron. E83-C*, SCE99-11, pp. 15 – 19, 2000.

[56] K.D. Mossman *et al.*, "A narrow-band HTS bandpass filter at 18.5 MHz," in *IEEE MTT-S Int. Microwave Symp. Dig.*, pp.653-656, 2000.

[57] K.F. Raihn *et al.*, "Highly selective HTS bandpass filter with multiple resonator cross-coupling," in *IEEE MTT-S Int. Microwave Symp. Dig.*, 2000

[58] G. Tsuzuki et al, "Ultra-selective 22-pole, 10-transmission zero superconducting bandpass filter surpasses 50-pole Chebyshev filter," *IEEE Trans. Microwave Theory Tech.*, vol. 50,pp. 2924 - 2929, Dec. 2002.

[59] S. Kolesov, B. *et al.*, "Cryogenic BTS receiver front end demonstrator," *Proc. of 30th Europ. Microwave Conf.*, Paris, vol. 3, pp. 230-232, Oct. 2000.

[60] M. Klauda, T. Kässer, C. Neumann, B. Mayer, C. Schrempp, A. Hoermann, A. Baumfalk, H. Chaloupka, S. Schornstein, N. Klein, "Superconductors and cryotechnology for future space communication technology - the BOSCH demonstrator experiment," in *IEEE MTT-S Int. Microwave Symp. Dig.*, pp. 1381-1384, June 1999.

[61] R.R. Mansour *et al.*, "A 60-channel superconductive input multiplexer integrated with pulse-tube cryocoolers," *IEEE Trans. Microwave Theory Tech.*, vol. 48, pp. 1171 - 1181, 2000.

[62] R.R. Mansour *et al.*, "Feasibility and commercial viability issues for high-power output multiplexers," *IEEE Trans. on Microwave Theory Tech.*, vol. 48, pp. 1199 – 1208, 2000.

[63] M.W. Pospieszalski, "Modeling of noise parameters of MESFET´s and MODFETS´s and their frequency and temperature dependence," *IEEE Trans. Microwave Theory Tech.*, vol. 37, pp. 1340 - 1350, Sept. 1989.

[64] D. Oates, "Tunable superconducting filters," *Workshop Notes of workshop "Cryogenic and Superconductor Microwave Subsystems", Applied Superconductivity Conference ASC*, Houston, 2002

[65] O.G. Vendik *et al.*, "High-Tc superconductivity: New applications of ferroelectrics at microwave frequencies," *Ferroelectrics*, vol. 144, pp.33-43, 1993.

[66] D.E. Oates, G.F. Dionne, "Magnetically tunable superconducting resonators and filters," *IEEE Trans. Appl. Supercond.*, pp. 4170-4175, June 1999.

[67] G.M. Rebeiz, J.B. Muldavin, "RF MEMS switches and switch circuits," *IEEE Microwave Magazine*, vol. 2, pp. 59-71, Dec. 2001.

[68] B.A. Aminov *et al.*, "High-Q tunable YBCO disk resonator filter for transmitter combiner in radio base stations," in *IEEE MTT-S Int. Microwave Sympos. Dig.*, pp. 363-366, 1998.

[69] R.B. Hammond *et al.*, "HTS wireless filters : Past, present and future performance," *Microwave J.*, vol. 41, no. 10, pp. 94-107, Oct. 1998.

[70] E. Polturak *et al.*, "Space-based high-temperature superconductivity experiment –design and performance," *IEEE Trans. Microwave Theory Tech.*, vol. 48, pp. 1289 - 129 , July 2000.

RESPONSE ANALYSIS AND MODELING OF HIGH TEMPERATURE SUPERCONDUCTOR EDGE TRANSITION BOLOMETERS

Mehdi FARDMANESH
Electrical and Electronics Engineering Dept., Bilkent University, 06800 Ankara, Turkey

1. INTRODUCTION

One of the promising devices made of high temperature superconducting (HTSC) materials are edge transition bolometers. Since the discovery of high-temperature superconductors, many works have been focused on the application of these materials in different types of bolometers for the near to far infrared wavelength regime [1]-[9]. They can be used to detect electromagnetic radiation over the whole spectrum from x-ray to the far-infrared [1], [9]-[13]. The superconductor bolometers typically consist of patterned thin or thick superconducting films deposited on crystalline substrates such as MgO, $SrTiO_3$, and $LaAlO_3$. Their operation is based upon their steep drop in the resistance, R, at their transition temperature, T_c. The detector is typically held at a temperature close to the middle of the superconducting transition, where the dR/dT is maximum. When the detector is illuminated its temperature rises by an amount proportional to the input radiation power resulting in a ΔR. The response obtained by the above mechanism is the so called the bolometric, or equilibrium response, as opposed to typically faster non-bolometric or intrinsic response caused by other mechanisms such as direct depairing. A typical response of an YBCO sample versus temperature at low frequencies is shown in Figure 1.1.

The responsivity versus modulation frequency and versus the temperature of edge-transition superconductive bolometers has been investigated and reported in different works [7]-[9], [14]-[18]. For example by use of a dc bias current the change in the resistance of the edge transition bolometer due to the absorbed radiation power can be measured. The detected signal obtained by this biasing can be bolometric, non-bolometric or a combination of both. Since the fabrication and testing of the first detectors, there have been many debates on the mechanism of the response in these devices. This is while the response at low frequencies of superconductor edge transition bolometer is commonly agreed to be mainly bolometric. But there have also been reports that the low frequency response of some samples is also enhanced strongly by an intrinsic response [19]-[21]. This has been reported to be dependent on bias

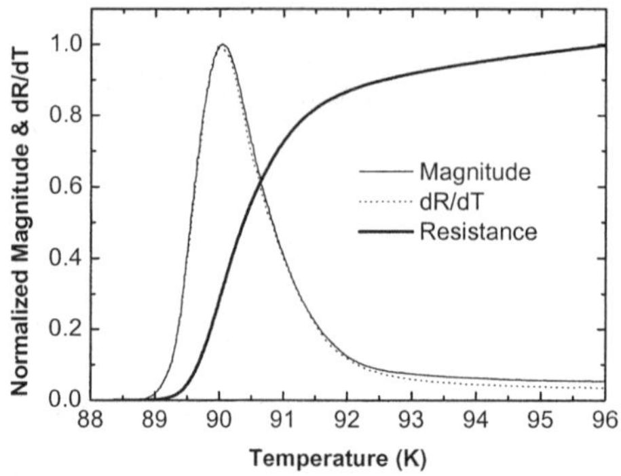

Figure 1.1. Normalized resistance and magnitude of the response at 20 Hz vs. temperature of 20μm×100μm area and 5μm wide meander line pattern on 1 mm thick $LaAlO_3$ substrate.

current, radiation, and temperature [19]-[21]. The interpretation and the analysis of the above is strongly dependent on the thermal properties of the substrate and its dimensions, the superconducting pattern, the modulation frequency regime with respect to the effective thermal diffusion length into the substrate, and electrical and the thermal operating conditions of the detectors [7], [18].

Major effects of the substrate-holder and substrate-film thermal boundary resistance on the response of this kind of bolometers are also reported [7], [9], [14]-[18]. For superconductive edge-transition bolometers with flat crystalline substrates directly in contact with the cold-finger, the effects of limited dc thermal conductance, G(0), and Joule heating on the response of the detectors have been misleading factor, resulting in unrealistic values for responsivities and detectivities [22]-[23]. This is particularly so for the low and midrange modulation frequency response of detectors with large area superconductive meander line patterns on crystalline substrates such as MgO, $LaAlO_3$, and $SrTiO_3$ [7], [24]. The above effects are strongly dependent on the dimensions of the pattern of the superconducting film with respect to that of the substrate, and the electrical connections [25]. The observed anomalies in studying these bolometers, also can also provide insight into the heat conduction process in these devices, and can be used to study both the superconducting films and heat flow across interfaces.

Thermal modeling is essential to analyze and predict the response of the bolometers to radiation signals. To perform a thermal design of a bolometer, it is also necessary to find the thermophysical properties of the superconducting film, substrate and the interfaces at the operating temperature. Most of the contemplated uses of superconducting bolometers involve mechanical chopping of the incident light. Hence frequencies of up to a few kHz are of interest for many applications. The magnitude and phase of the IR-response of YBCO superconductor detectors versus modulation frequencies from 0.5 Hz to 100 kHz are studied in this chapter, and found that the observed range of frequency dependence requires a model that includes the thermal boundary resistances, and proper consideration of different regimes of heat flow in the substrate. The edge transition bolometers can be classified as small area pattern (micro-bridges) and large area pattern samples with respect to the substrate dimensions and the substrate thermal parameters. The electrical connections can also be classified as voltage biased (current response) and current biased (voltage response) four (or two) probe configurations [25]. The samples considered in this chapter mostly carry large area superconductive patterns compared to the thickness of the substrates and the voltage-response of the bolometers is studied.

The thin film samples studied in this chapter are made of 120-550 nm thick superconducting $YBa_2Cu_3O_{7-x}$ films on crystalline $SrTiO_3$, $LaAlO_3$, and MgO substrates. The films are deposited using either off-axis dc planar magnetron sputtering or PLD techniques. Details on the deposition parameters and the properties of the films are contained in References [25]-[28]. The Y-Ba-Cu-O compound with 123 and 124 stochiometries are used for fabrication of the thick film devices, post annealed in argon and oxygen ambient resulting granular superconductive films with a 123 phase [29]. The thick film samples are screen printed patterns on polycrystalline MgO substrates using hybrid technology. Sample substrates have 0.25 mm to 1 mm thicknesses.

The investigated IR-response measurement versus temperature, and IR-response versus modulation frequency in this chapter are measured simultaneously avoiding any artifacts due to the shift in the T_c, either caused by the thermal cycling or the Joule heating in the device. These aspects will be discussed in details in section 6. The low noise measurements in four-probe configuration are performed using a current source, consisting of a battery, and a bias resistor. The investigated data here is taken using lock-in amplifiers allowing the measurements of both the magnitude and phase of the response.

After studying the response characteristics in the low and midrange frequencies of YBCO edge transition bolomters in section 2, a comprehensive model and its close form solution are discussed in section 3. Section 4 discusses the anomalies in the response and the deviation of the phase of the response versus temperature, from the expected values. Section 5 describes a physical thermal model for the heat propagation mechanism based on the electron- phonon interactions correlated to the absorbed radiation energy, explaining the reported anomalies in the phase of the response versus temperature discussed in the preceding section. Section 6 discusses the dc characteristics of the superconductor edge transition bolometers and the artifacts due to the Joule heating, ambient pressure, and the microstructure, which affect the responsivity of the detectors. The noise characteristics of the bolometers are studies and classified in section 7. Section 8 applies the results from the thermal modeling and study of the noise characteristics of the bolometers discussed in the preceding section to determine and control the responsivity and detectivity limits of the edge transition bolometers through the design parameters.

2. ANALYSIS OF LOW AND MIDRANGE MODULATION FREQUENCY RESPONSE AND THE INTERFACE EFFECTS

In the basic model for the bolometer [4], [7], the responsivity, r_V, defined as the ratio of the voltage signal to the input power, is:

$$r_V = \frac{\eta I}{G+j2\pi fC}\frac{dR}{dT}, \qquad (2.1)$$

where I is the dc bias current, dR/dT is the slope of the resistance versus temperature curve at the bias (or operating) temperature, and $\eta=W/P_t$, with W the incident power absorbed by the bolometer, and P_t the total incident radiation power at the surface of the detector. G is the total thermal conductance between the bolometer and its environment, and C is the heat capacity of the bolometer. A sample configuration in contact with the holder is shown in Figure 2.1.

As given in equation (2.1), r_V is a complex number with a magnitude of:

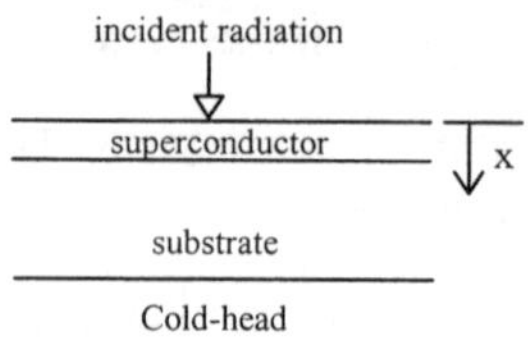

Figure 2.1. Typical experimental configuration of a superconductor bolometer on ceramic substrate, with contact to a holder (cold-head).

$$|r_v| = \frac{\eta I}{G\sqrt{1+(2\pi f\tau)^2}}\frac{dR}{dT} \qquad (2.2)$$

where $\tau = C/G$, and I is the constant dc bias current. Effects due to the variation of the bias current have been studied and can be found in Ref. [7]. If one assumes G and C are constant and independent of modulation frequency, then based on the above model, at frequencies where $2\pi f\tau \gg 1$, the voltage response of a bolometer exposed to incident radiation should scale as f^{-1}. The phase of the response in any frequency range is:

$$\theta = \tan^{-1}\left(\frac{-2\pi fC}{G}\right) = \tan^{-1}(-2\pi f\tau) \qquad (2.3)$$

For the studied samples in this chapter τ is typically on the order of 0.3 S, hence the $2\pi f\tau \gg 1$ regime holds at as low as a few Hz, where the model above works quite well [7]. However, at frequencies in the range of 100 Hz and above, the measured response of almost all of the investigated samples deviates significantly from the values predicted by the above model [7], [26]. Instead of a f^{-1} dependence, there is a $f^{-1/2}$ within the mid frequency range, as shown in Figure 2.2 for a sample with relatively large area pattern with respect to its substrate thickness. The thermal and geometrical parameters of this sample as well as other considered samples in this chapter are given in Table 2.1. To study this frequency behavior, an equivalent circuit model for the bolometric response to radiation signals for modulation frequencies ranging from dc to high values is considered. As shown in the circuit model in Figure 2.3, the response is determined by the total heat capacity of the film as a lumped element, and the film-substrate thermal boundary resistance as well as the thermal constants of the substrate and the interfaces.

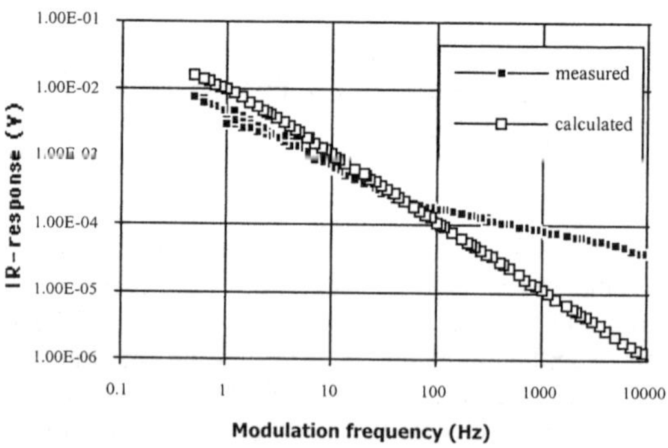

Figure 2.2. Measured and calculated IR-response versus modulation frequency of sample 064-04b, at 81.7 K with 0.68 mA dc bias current and 2.13 mW/cm^2 radiation intensity.

Table 2.1. The dimensions of the substrates are about 0.5 cm by 1 cm for all the samples. d_s is the thickness of the substrate, d_f is the thickness of the YBCO film, and A is the total area enclosed by the meander line pattern and is roughly twice the area of the meander line. The values of G are obtained at temperatures above T_c.

sample number	substrate material	d_f (nm)	d_s (cm)	A (cm^2)	area of the meander line	$T_{c\text{-onset}}$ (K)	R(300K) (k)	G (W/K)
064-02a	MgO	120-130	0.025	0.075	90μm×3.35cm	87	13.5	33×10^{-3}
064-02b	MgO	120-130	0.025	0.0168	50μm×1.9cm	86	11.2	15.5×10^{-3}
064-03a	MgO	170-180	0.05	0.075	90μm×3.35cm	82	7.7	11×10^{-3}
064-03b	MgO	170-180	0.05	0.0168	50μm×1.9cm	83	6.5	5.1×10^{-3}
064-01a	$SrTiO_3$	220-230	0.05	0.0168	50μm×1.9cm	82	10.0	3×10^{-3}
061-02a	$LaAlO_3$	540-560	0.05	0.075	85μm×3.35cm	80	4.4	8×10^{-3}
064-04b	$LaAlO_3$	190-200	0.05	0.075	90μm×3.35cm	84.5	17.5	7.8×10^{-3}

This section presents measurements and analysis of the magnitude and phase of the response of YBCO bolometers to radiation signals, with modulation frequencies up to 100 kHz. In this range of frequencies the only film property of importance is dR/dT. Behavior of the IR-response versus frequency to radiation signals in this section is considered to be independent of the radiation wavelength. The effect of Joule heating due to the dc bias current on the response of the samples is also negligible for the test currents used in this section.

A. Calculation of the response

To compare measured results to equation (2.2), one should first determine the dc thermal conductance of the YBCO film to the surrounding environment, G(0), which can be performed in several different ways [2], [7], [24]. The most convenient way is to make use of the slope of the resistance versus temperature curves of the films in the normal state near T_c [24]. At high bias currents, the resistance of the bolometer increases due to power dissipation heating the film. This creates a temperature gradient across the substrate and the thermal interfaces. The increase in the resistance of the film due to the Joule heating is [5], [7]:

$$\Delta R = I^2 R \frac{dR}{GdT} \qquad (2.4)$$

Using the measured values for R, I, and dR/dT, values of G are obtained and given in Table 2.1.

The heat capacity, C, and the time constant, τ, can be found by use of equation (2.3) as:

$$\tau = -(\tan\theta)/2\pi f = C/G, \qquad (2.5)$$

where θ is the measured value of the phase of the response. Using equation (2.5) and the

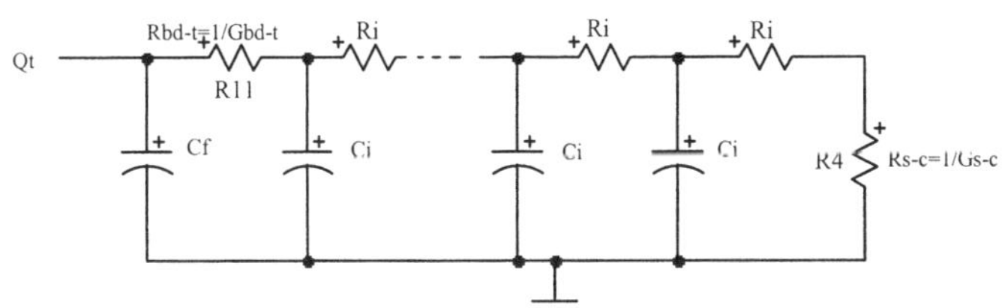

Figure 2.3. Circuit diagram of thermal model for bolometric response of a high temperature superconductor bolometer.

measured values of θ at low frequencies, τ is calculated for sample 064-04b and is found to be ~0.32 s at frequencies less than 2 Hz. Using the measured values of G=7.8 mW/K, τ=0.32 s, and dR/dT=3850 Ω/K (from R versus T curve), the response versus frequency of the sample is calculated for η=1 using equation (2.2), and is given in Figure 2.2. The calculation is done for 0.68 mA bias current, 2.13 mW/cm^2 incident radiation power with λ=0.85 nm and an effective absorbing area of 0.03 cm^2, corresponding to the values used in the measurement.

In the calculation of the response, the effect of Joule heating in the film, caused by the bias current, can also be considered. The change in the film resistance due to the input radiation power is augmented by the I^2R heating, and is a form of positive feedback. This can enhance the response resulting in an overall responsivity of:

$$r_{v\text{-}t} = \frac{r_v}{(1 - I\, r_v/\eta)}, \qquad (2.6)$$

where r_v is given as in equation (2.1). As observed in equation (2.6), there is a stability criterion in the voltage response, which has the same form as obtained for dc biasing at zero frequency [1], [7]. Since the bias current in the samples is on the order of 10^{-3} A, and the responsivity maximum is on the order of 10^0 V/W, even at low frequencies, the effect of the Joule heating is negligible as mentioned before. However, in very sensitive devices or at high currents, it could become a significant factor.

The difference between the measured and calculated response at low frequencies is mostly expected to be due to the assumption of η=1. Results similar to those reported by other groups [3], [10] have been obtained for the test samples. As reported earlier and shown in Figure 2.2, the model based on constant C and G, (equation (2.1)), can explain the response of the bolometers only for relatively low frequencies, where the response is limited by the thermal boundary resistance, $R_{s\text{-}c}$, between the substrate and cold head. A more accurate analysis model requires consideration of the thermal diffusion length for the radiation-induced temperature rise into the substrate, and both thermal boundary resistances. At very low frequencies the response does follow equation (2.1) [26], where C and G are constant, and G is interpreted to be mainly due to the substrate-cold head thermal boundary resistance, as verified below.

B. Calculations of the frequency dependent thermal parameters

To study the thermal response of the samples, one can model the substrates as a cylinder and use the heat diffusion equation. In this model the radiation absorbing area is at one end of the cylinder [7], [18]. This assumption will turn out to be valid since the film thickness of samples is much smaller than the substrate thick nesses and the patterns of the superconducting films are relatively large in comparison to the substrate thick nesses (see Table 2.1).

Solving the diffusion equation in the substrate material, $D\nabla^2 T = \frac{T}{t}$, as a function of the distance from the surface, x, and time, t, and knowing the temperature at the surface will result in [7], [18]:

$$T(x,t) = T_0 \exp\left[i\left(\left(\frac{i2\pi f}{D}\right)^{0.5} x - 2\pi f t\right)\right], \qquad (2.7)$$

Where T_0 is the temperature at x=0 and t=0, D= K_s/C_s has units of centimeter square per second in the CGS system, and is called the diffusivity of the material, and k_s and c_s are the thermal conductivity and the specific heat (per unit volume) of the substrate material, respectively.

A quantity L, called the "thermal diffusion length", can be used in equation (2.7), and is defined as:

$$L = \left(\frac{D}{\pi f}\right)^{1/2}. \tag{2.8}$$

It represents the characteristic penetration depth of the temperature variation into the substrate. Considering the thermal diffusion length as the effective length for heat flow into the substrate, the corresponding thermal conductance can be found from [2], [18]:

$$G = ak_S \frac{A}{L}, \tag{2.9}$$

Where, k_S is the thermal conductivity of the substrate material and "a" is a correction factor for the above approximation of the length of the heat flow. Substituting the value of L from equation (2.8) in equation (2.9), and using $D = \frac{k_S}{c_S}$:

$$G_S = a\,A\sqrt{c_S k_S \pi f} \tag{2.10}$$

Thus G_S will be a function of frequency and scales as $f^{1/2}$, increasing with the frequency. Similarly, the volume of the substrate in which the temperature variation occur changes with the frequency due to the thermal diffusion length. Hence, the total heat capacity, C_S, can be written as:

$$C_S = b\,c_S\,A\,L = b\,A\sqrt{c_S k_S/\pi f}, \tag{2.11}$$

where b is a constant to compensate for the drop of the amplitude of the temperature variation along L. Hence, C_S will also be a function of frequency and scale as $f^{-1/2}$. Under the above assumptions, τ will be:

$$\tau = \frac{C_S}{G_S} = \frac{b}{a\pi f}, \tag{2.12}$$

which will be frequency dependent, scaling as f^{-1}.

Considering frequencies low enough to allow the total heat conductance G_t G_S and the total heat capacity $C_t \approx C_S$, then use of equation (2.12) in equation (2.1), results in;

$$r_V = \frac{r'}{1+j2b/a}, \tag{2.13}$$

where $r' = \frac{\eta I}{G_t}\frac{dR}{dT}$. Equation (2.13) shows that the phase of the response is constant and frequency independent for the frequency range where the thermal diffusion length is smaller than the thickness of the substrate. But since G_S scales as $f^{0.5}$, r' will scale as $f^{-0.5}$, and the magnitude of the response will be frequency dependent. This is valid at frequencies where the thermal conductance of the YBCO film, G_f, and the film-substrate interface, $G_{bd}=1/R_{bd}$, are negligible compared to G_S.

The circuit diagram of the proposed thermal model of a superconducting bolometer is shown in Figure 2.3. In this model, the substrate-cold head thermal boundary resistance and the total YBCO film-substrate thermal boundary resistance are considered as $R_{S\text{-}c}$ and $R_{bd\text{-}t}$, respectively. The substrate is considered to be made of small increments (thin layers of substrate material), which can be shown, by a combination of R_i and C_i in the circuit diagram.

Now, for the overall frequency response analysis (from dc to high frequencies) assume that $C_f \ll C_s$ and $R_{bd\text{-}t} \ll R_s \ll R_{s\text{-}c}$ (or, $G_{s\text{-}c}=1/R_{s\text{-}c} \ll G_s \ll G_{bd\text{-}t}=1/R_{bd\text{-}t}$), which will be shown to be the case for the samples in table 2.1. Where $R_{bd\text{-}t}$ is the total thermal boundary resistance at the film-substrate interface. Based on these assumptions, at very low frequencies, the heat capacity of the film can be ignored compared to that of the substrate, and the total thermal conductance will be:

$$\frac{1}{G_t} = \frac{1}{G_{s\text{-}c}} + \frac{1}{G_s} + \frac{1}{G_{bd\text{-}t}} \tag{2.14}$$

Since $G_{s\text{-}c} \ll G_s(0) \ll G_{bd\text{-}t}$, we will have $G_t \approx G_{s\text{-}c}$ which is not frequency dependent. Then, the response can be obtained from [7]:

$$\Delta T_f = \frac{1}{G_{s\text{-}c}+j2\pi f C_s(0)} W, \tag{2.15}$$

Where ΔT_f is the temperature variation at the film and $\tau_0 = C_s(0)/G_{s\text{-}c}$. Then, ignoring the effect of the Joule heating, the responsivity of the bolometer is given by equation (2.1) with a frequency dependence of f^{-1}, and a phase angle of -90 degree, for frequencies at which $2\pi f \gg 1$, but not so high as to limit the thermal diffusion length.

When the frequency increases to the values where the thermal diffusion length in the substrate, L, becomes comparable to the thickness of the substrate, G and C will no longer be constant. Based on the above assumptions, $G \approx G_s$ and $C \approx C_s$, they are frequency dependent following equations (2.10) and (2.11) respectively, and the response is governed by:

$$r_v = \frac{\eta I}{G_s+j2\pi f C_s} \frac{dR}{dT}, \tag{2.16}$$

Where still $C_f \ll C_s$, $G_s \ll G_{bd\text{-}t}$. By applying the frequency dependence of the G_s and C_s in equation (2.16), the response will be in the form of equation (2.13), i.e. $r_v \sim f^{-1/2}$. As the frequency increases further, $G_s(f)$ increases further (or, $R_s(f)$ decreases) until it becomes comparable to $G_{bd\text{-}t}$ where C_f and $R_{bd\text{-}t}$ can no longer be ignored. At high enough frequencies, the G_s and C_s can be ignored and the response will be determined by G_{bd} (or $1/R_{bd}$) and C_f [7], [15], [17].

C. Frequency response analysis and the experimental results

Based on above analysis, the point at which the response start to scale as $f^{-1/2}$ occurs at frequencies where the thermal diffusion length, L, becomes comparable to the thickness of the substrate. Since from equation (2.8), L is proportional to the square root of the thermal diffusion constant, $D=k_s/c_s$, substrate materials with higher k_s, will have knee points in the

Table 2.2. Thermal characteristics of single crystal substrates

Substrate	k_s(W/K-cm)	c_s(J/K-cm^3)	D(cm^2/s)	$f_{L=0.05cm}$(Hz)
MgO	3[a]	0.53[a]	5.66	720
$LaAlO_3$	0.32[b]	0.59	0.55	70
$SrTiO_3$	0.052[c]	0.43[c]	0.12	15.4

[a]From Ref. [2] [b]From Ref.[17] [c]From Ref. [6]

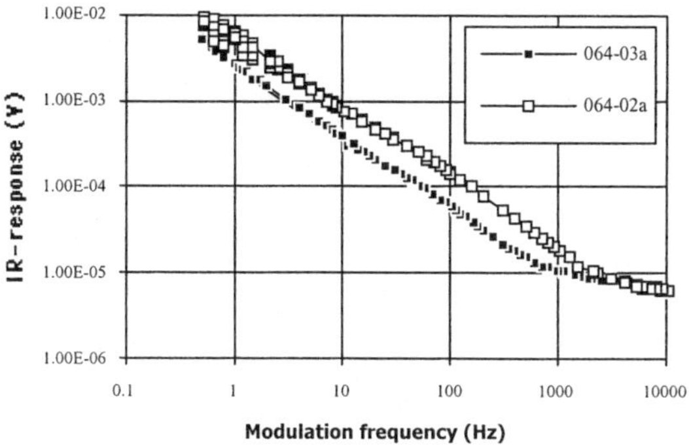

Figure 2.4. Measured IR-response versus frequency for 0.05 cm thick MgO substrate sample 064-03a and 0.025 cm thick MgO substrate sample 064-02a.

response at higher frequencies (assuming the substrate thick nesses are the same and c_S of the substrates are within the same range). The values of k_S, c_S and the calculated frequency of the knee point, f_L, of three different substrates used in this chapter are given in Table 2.2 The value of c_S for $LaAlO_3$ in Table 2.2 is derived using the phase of the low frequency response [7]-[26], and is consistent with the measured experimental values shown for MgO and $SrTiO_3$. The values of f_L are calculated by use of equation (2.8). For example, to find the frequency at which the thermal diffusion length is 0.05 cm for a $SrTiO_3$ substrate, one can use equation (2.8) in the form of:

$$f_L = \frac{D}{\pi L^2} \tag{2.17}$$

For the $SrTiO_3$ substrate from Table 2.2:

$$D = \frac{k_S}{c_S} = \frac{0.052}{0.43} = 0.121 \text{ cm}^2/\text{s} \tag{2.18}$$

Applying equation (2.18) and L=0.05 cm in equation (2.17), results in f_L= 15.4 Hz as given in Table 2.2. The measured responses versus frequency for 0.05 cm thick single crystal $LaAlO_3$, MgO, and $SrTiO_3$ substrates are shown in Figures 2.2, 2.4, and 2.5 respectively. From the figures, the knee points in the response versus frequency are about 15 Hz, 60 Hz, and 600 Hz, for 0.05 cm thick $SrTiO_3$, $LaAlO_3$, and MgO substrates, respectively, in good agreement with the calculated values in the table.

To verify the above analysis and to investigate the effect of the thickness of the substrate on the frequency response, we consider two samples with identical meander line patterns on 0.25 mm and 0.5 mm thick MgO substrates. The response of the samples 064-02a and 064-03a, are shown in Figure 2.4. As shown in the figure, the knee point is found to be dependent on the substrate thickness, and it is found not to be dependent on the pattern area. The knee point for the thinner substrate sample 64-02a occurs at higher frequencies, about 2.5 kHz to 3 kHz, whereas for the thicker substrate sample 64-03a, it occurs at about 700 Hz to 800 Hz. Setting the thickness of the substrate equal to L in equation (2.17), and using the values of c_S and k_S from Table 2.2 for MgO, knee points at frequencies of 720 Hz and 2.8 kHz are calculated for samples 64-03a (0.05 cm thick) and 64-02a (0.025 cm thick), respectively. The above calculated values of frequency agree with the break frequencies seen in Figure 2.4.

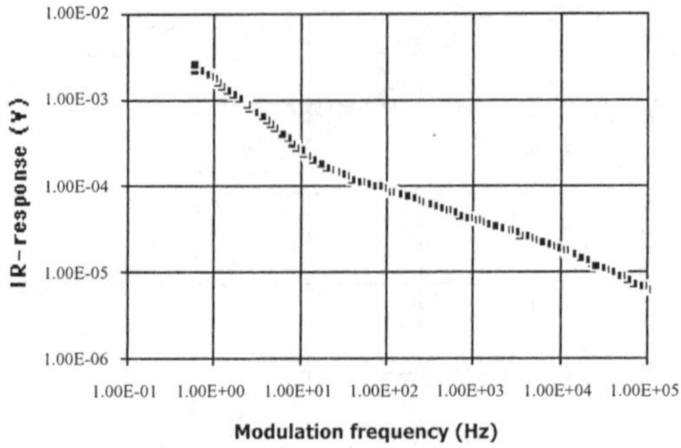

Figure 2.5. IR-response versus modulation frequency of $SrTiO_3$ substrate sample 064-01a measured at 0.25 mA bias currents.

From equation (2.13), for frequencies higher than f_L, the phase should be frequency independent. Figure 2.6 (next page) shows the measured phase of the response for sample 064-01a, which confirms the position of the break frequency discussed above. The flat region in Figure 2.6 extends only to about 10 kHz, but it should continue up to the frequencies where the response and its phase starts to be governed by the effect of R_{bd}, according to our model. Since the YBCO film is just a few 100 nm thick, the heat capacity of the film, C_f, is very low, and the phase of the response should approach 0° for the frequency range where the response is limited by the effect of R_{bd}, and continue until much higher frequencies. The frequency at which R_{bd}, the film-substrate thermal boundary resistance, becomes important varies with the thermal parameters of the substrate material. It is lower for substrate materials with higher k_s (assuming c_s of the substrates are within the same range) [7]. Using c_s and k_s from Table 2.2, and a typical value for R_{bd} as 1.1×10^{-3} K-cm^2/W [17], one can compare this with the frequency dependent thermal resistance of the substrate, as calculated from equation (2.10). The thermal resistance due to R_{bd} is equal to the thermal resistance of MgO, $LaAlO_3$, and $SrTiO_3$ substrates at modulation frequencies of 165.4 kHz, 1.413 MHz, and 11.67 MHz, respectively. These frequencies are calculated for the case that the corresponding thermal diffusion lengths are small compared to the width of the meander lines of the superconducting pattern, which is the case in the samples of Table 2.1. For smaller patterns, the spacing between the lines should also be considered. This will result in frequencies lower than above [7].

One can estimate the frequencies at which R_{bd} and c_s would change the phase, and this is 10 to 50 MHz range. Hence this cannot account for the drop in the phase seen in Figure 2.5. The drop in the phase at frequencies of about 10 kHz is also observed for samples with MgO and $LaAlO_3$ substrates. This departure from our model for the thermal response of the samples may be due to other effects, such as the system circuitry [7], or photo-induced conductivity changes in the YBCO film [34]. However, this analysis accurately accounts for the observed behavior below 10 kHz. To summarizes this section, the slope of the response changes at "knee" frequencies of about 15 Hz, 60 Hz, and 600 Hz for 0.05 cm thick $SrTiO_3$, $LaAlO_3$, and MgO substrates, respectively, from a f^1 to a $f^{0.5}$ dependence. From equation (2.1), one sees that a high G value, $G \gg 2\pi fc$, lowers the responsivity, and pins the phase of the response at

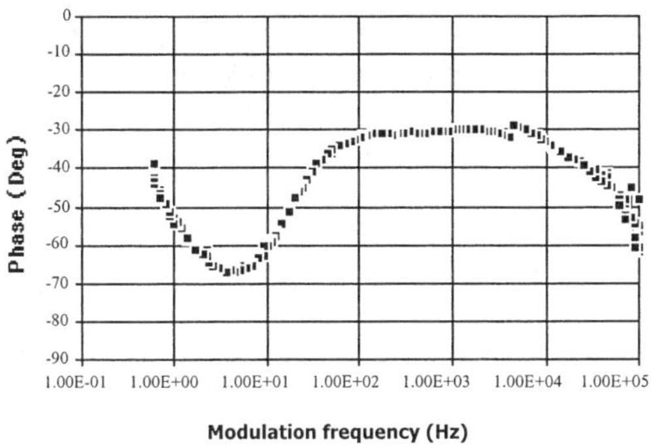

Figure 2.6. Phase of IR-response versus modulation frequency of sample 064-01a (0.05 cm thick $SrTiO_3$ substrate).

$\theta=0°$. A low G value, $G \ll 2\pi fc$, gives $\theta = -90°$, a f^{-1} dependence to the responsivity, and a magnitude inversely proportional to the heat capacitance of the bolometer. For this latter case a very thin substrate to reduce C is indicated. While G can be made very small by isolating the substrate (underneath the pattern) from the cold head, the minimum G(0) is needed to keep the sample in the superconducting state, and within a ΔT of its bias temperature. For the considered samples here with G's on the order of 10^{-2} W/K, Joule heating power on the order of $I^2R=10^{-2}$ W yields a shift temperature on the order of 1 K. Hence as a practical matter, a G much smaller than the values in table 2.1 would present a problem in maintaining the bias temperature for currents on the order of 1 mA.

3. ANALYTIC THERMAL MODELING FOR DC TO MIDRANGE MODULATION FREQUENCY RESPONSE

There has been different efforts on the response modeling of the edge-transition superconducting bolometers with and without the consideration of the effects of the thermal boundary resistance at the substrate interfaces and the self or Joule heating in the superconducting film [16], [35], [30]-[31]. Based on the above proposed effects of the introduced parameters, section 2 discussed successful analysis and prediction of the behavior of the response versus modulation frequency curves predicting the frequencies at which the slope of the curve changes (the knee frequency) [7], [16]. Though, one cannot predict and match the response in all the measured modulation frequency regimes using an analytic model. Using the analyses in section 2,one can predict and match the measured response of the samples at very low frequencies showing different considerable discrepancies at midrange and/or higher modulation frequencies [7], [16]. In many of the proposed models and the solutions, only the magnitude of the response have been considered and compared to the experimental values. This is while the phase of the response is a more sensitive characteristic of the response with respect to the model and the chosen or obtained thermal parameters of the devices [24]. In this section the results obtained from a recent closed form solution of a comprehensive one-dimensional thermal model is discussed. The results from the model are compared to the both measured values of the magnitude and phase of the response of the samples.

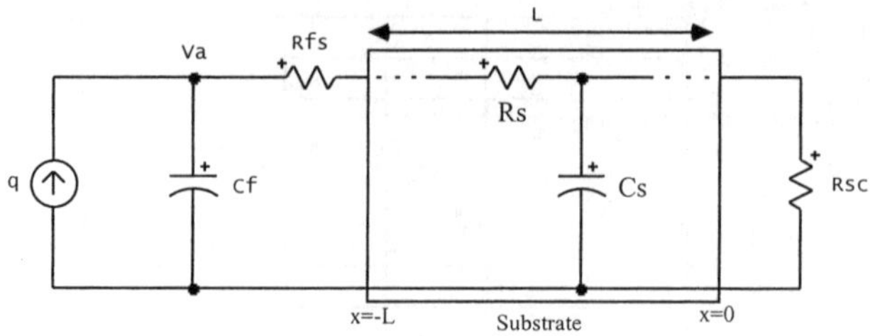

Figure 3.1. Equivalent thermophysical diagram of the bolometers in contact with the holder, Where q is input radiation power, C_f is lumped heat capacitance of the film, R_{fs} and R_{sc} are the thermal boundary resistance values at the substrate interfaces.

A. The Thermal Modeling

From the given dimensions in Table 2.1, the area of the superconducting pattern of the samples is large with respect to the substrate thickness. Hence a one-dimensional model is considered and applied for the studied samples while considering all the parameters for per unit area. Because of the partial lateral heat diffusion of the coming heat wave from the film into the substrate (the spreading resistance effect), a correction factor, α, is considered for applying to the measured R_{SC} values and is obtained and given for some of the samples in Table 3.1. The value of α increases with the increase of the thermal conductivity of the substrate material and being the highest for MgO. It increases with decrease of the superconducting pattern area with respect to the substrate thickness as given in the table to be higher for the small pattern film on $SrTiO_3$ compared to that of the $LaAlO_3$ substrate sample. As given in the tables, the thickness of the superconducting films of the samples is also very small compared to that of the substrate. This allows the consideration of a lumped film assumption on top surface of the substrate in the model.

Considering the superconducting pattern as a lumped film and substrate as a continues uniform media, the thermal equivalent circuit of the samples in contact with the holder can be represented as in Figure 3.1, where q is the absorbed radiation intensity at the surface of the sample. In this equivalent diagram, two thermal boundary resistances are considered at the substrate interfaces. The heat capacitance of the superconducting film is also considered to be a value representing the total heat capacity of the film per unit area. The thermal conductance of the superconducting film is calculated to be negligible compared to that of the substrate and the interfaces. For samples with thicker superconducting films as for the screen-printed materials [4], the limited heat conductivity of the superconducting material same as for the substrate, should also be considered. Based on the above considerations in the model as

Table 3.1. Dimensions and the obtained thermal parameters of the samples used in the model.

sample number	substrate material	D_S (cm)	C_S $(\frac{J}{k\text{-cm}^3})$	K_S $(\frac{W}{k\text{-cm}})$	R_{SC} $(\frac{K}{W\text{-cm}^2})$	α	d_f (nm)	A (cm^2)	R_{fs} $(\frac{mK}{W\text{-cm}^2})$	C_f $(\frac{\mu J}{k\text{-cm}^2})$
064-03a	MgO	0.05	0.53	3	6.82	2	170	0.075	4.4	642
064-01a	$SrTiO_3$	0.05	0.43	0.052	5.6	1.7	220	0.017	44	655
061-02a	$LaAlO_3$	0.05	0.59	0.16	8.82	1	550	0.075	27.5	131

shown in Figure 3.1, the derivation of the temperature variation and the corresponding thermal parameters for the samples under a constant bias current is presented in the following.

a) Temperature Variation in the Substrate

To approach the complete closed form solution to the model given in Figure 3.1, first one can consider the absorbed radiation power, q, is applied directly to the surface of the substrate. This is ignoring the effects of the surface boundary conditions of C_f and R_{fs}, which are representing the effect of the thermal parameters of the superconducting film at this stage. This particular assumption and the solution to it would be the case as for very low modulation frequencies. At low modulation frequencies, C_f is much smaller than the contributing heat capacity of the substrate, which can be determined from equation (2.17) [7]. Also the film-substrate thermal boundary resistance, R_{fs}, is much smaller than that of the substrate-holder and/or the low frequency thermal resistance of the substrate, $R_s=1/G_s$, determined from equation (2.10).

Based on the above assumptions and the steady state solution to general one dimension heat propagation equation, the temperature variation at the surface of the substrate, ΔT, can be obtained. Without the consideration of the effect of the superconducting film (i.e. C_f and R_{fs}) ΔT follows;

$$\Delta T= q\,\frac{\exp(\gamma L)+\Gamma\exp(-\gamma L)}{\exp(\gamma L)-\Gamma\exp(-\gamma L)}\sqrt{\frac{1}{j\omega c_s k_s}} \qquad (3.1)$$

where γ is the characteristic thermal impedance of the substrate material defined as;

$$\gamma=\frac{1+j}{\sqrt{2}}\sqrt{\frac{\omega c_s}{k_s}} \qquad (3.2)$$

And;

$$\Gamma=\frac{R_{sc}-\sqrt{\dfrac{1}{j\omega c_s k_s}}}{R_{sc}+\sqrt{\dfrac{1}{j\omega c_s k_s}}}, \qquad (3.3)$$

Where ω is the angular modulation frequency, and k_s and c_s are the heat conductivity and the heat capacity of the substrate materials, respectively. The value of R_{sc} can be obtained;

$$R_{sc}=\frac{1}{G(W/K)/A}, \qquad (3.4)$$

Where G is in units of W/K and is the total thermal conductance of the device at very low frequencies, G(0), and A is the film pattern area. This is based on the assumption of $G_{fs} \gg G_s \gg G_{sc}$ where $G_{fs}=1/R_{fs}$, G_s is the total heat conductance of the substrate, and $G_{sc}=1/R_{sc}$. This assumption is verified for the samples with dimensions close to that of table 2.1 and 3.1, which is also discussed in section 2 [7], [16], [22]-[24], [37].

b) Temperature Variation at the Film

To find the voltage response to the modulated radiation intensities, the temperature variation in the superconductive film of the samples can be obtained for all considered frequency ranges. This temperature variation corresponds to the temperature change across C_f in the equivalent circuit of Figure 3.1. Hence the only temperature gradient from top of the superconducting film into the substrate would be that across the film-substrate interface, R_{fs}.

Solving the one-dimensional thermal differential equation with the complete boundary conditions as shown in Figure 3.1 (i.e. with the consideration of C_f and R_{fs}), the temperature variation caused by the radiation absorption in the film follows;

$$\Delta T = q\left(\frac{\frac{\exp(\gamma L)+\Gamma\exp(-\gamma L)}{\exp(\gamma L)-\Gamma\exp(-\gamma L)}\sqrt{\frac{1}{j\omega c_s k_s}}+R_{fs}}{\frac{\exp(\gamma L)+\Gamma\exp(-\gamma L)}{\exp(\gamma L)-\Gamma\exp(-\gamma L)}\sqrt{\frac{j\omega}{c_s k_s}}\,C_f+1+j\omega C_f R_{fs}}\right). \tag{3.5}$$

Where, q is the absorbed radiation power in the film per unit area, and C_f and R_{fs} are the total heat capacity of the superconducting film and the thermal boundary resistance at the film-substrate interface per unit area, respectively.

The solution of equation (3.5) is up to frequencies where the thermal diffusion length into the superconducting film (for YBCO material) becomes smaller and/or comparable to the thickness of the film. This is while the penetration depth and/or the absorption coefficient length in the superconducting film is still considered to be much smaller than the film thickness. This frequency regime for the samples of II_Table 3.1 with a maximum film thickness of about 550 nm much higher than that considered here. As given in equation (3.5), at very low frequencies the denominator will approach one and the R_{fs} will become negligible compared to the first term of the numerator leading to the results of the one boundary condition problem as given in equation (3.1). This is because at very low modulation frequencies the temperature variation (heat wave) reaches the bottom of the substrate and hence the heat capacity of the film with respect to that of the whole substrate material under the pattern, and also the thermal boundary resistance at the film-substrate with respect to R_s and R_{sc} become negligible [7], [37].

c) The Voltage Response and Bias Current Dependence

Considering a uniform temperature variation in the film, caused by the absorbed radiation, ΔT, the voltage response in a dc current bias configuration can be obtained from;

$$\Delta V = I_b\,\Delta T\,\frac{dR}{dT}, \tag{3.6}$$

Where I_b is the dc bias current, dR/dT is the slope of the R versus T curve at the bias temperature as given in II_Table 3.1, and ΔT is the temperature variation in the superconducting film as given in equation (3.5). Equation (3.6) is valid within the temperature variation range where dR/dT can be considered to be constant. The maximum temperature variations of the films caused by the radiation is typically below about a few mK. The transition widths of the samples are typically of a few degrees of Kelvin and hence the variation of dR/dT and/or the non-linearity of equation (3.6) are negligible in the measured and calculated response. Though this would not be the case for devices exposed to high radiation intensities, which are normally voltage biased for further stability's [12]-[13]. The effect of the bias configuration on the response and the feedback effect of the Joule heating for each case of the voltage-biased and current-biased configurations are discussed elsewhere [25].

There is also a consecutive temperature variation due to ac Joule heating in the film caused by the resistance variation in the superconducting film due to the input radiation power. This effect on the response is in the form of a positive feedback in the used current-biased configuration. This ac Joule heating in a first order approximation can be considered either as a dependent radiation (or heating) source in parallel to the external radiation source, q, shown Figure 3.1 [40], or simply as an additional term in the overall responsivity of the sample discussed in section 2. The effect of ac Joule heating in the response of the samples studied in this work is negligible for the used biased currents.

B. Measured Frequency Response and the Results From the Model

The results from the model were compared to the experimental values for the samples given in II_Table 1. The frequency response behavior of the samples was found to be strongly dependent on the film and substrate dimensions as well as on the thermal parameters, while the observed dependencies are in good agreement with the thermal model. To calculate the response using equation (3.5), the dc thermal parameters of R_{fs}, R_{sc}, C_f, C_s, and K_s of the samples were determined. The measured values of the above parameters are given in II_Table 1. The values of R_{sc}, which are much higher than R_s and R_{fs}, are measured using the method of the resistive or the dc Joule heating in the film close to the transition temperature, T_c [5], [24], [39]. In this method a high bias current is passed through the film and the temperature rise due to the Joule heating is measured by measuring the resistance of the film. Then knowing the temperature rise and the Joule heating power the values of R_{sc} are obtained. Also the parameters such as the film-substrate thermal boundary resistance and the thermal parameters of the substrate or the heat capacity of the samples could be obtained and verified using the fit of the results from the model to the measured values. The above values in II_Table 1 are given per unit area as used in the one-dimensional solution of the model.

The results from the model were investigated for thin film samples on crystalline $SrTiO_3$, $LaAlO_3$, and MgO substrate materials. The measured and calculated magnitude and phase of the response of the meander line patterned YBCO films on the 0.05 thick $SrTiO_3$, $LaAlO_3$, and MgO substrate samples are shown in figures 3.2, 3.3, and 3.4 respectively. Calculated response using the basic R-C model in section 2 (equation (2.1)) is also plotted for samples 064-01a and 061-02Aa.

As discussed in section 2, the basic R-C model is as useful as the complete model in the very low frequency ranges where the thermal diffusion length into the substrate is much larger than the substrate thickness. This basic R-C model is appropriate for micromachined and/or freestanding film bolometers. The G and C values of such bolometers would be constant up to

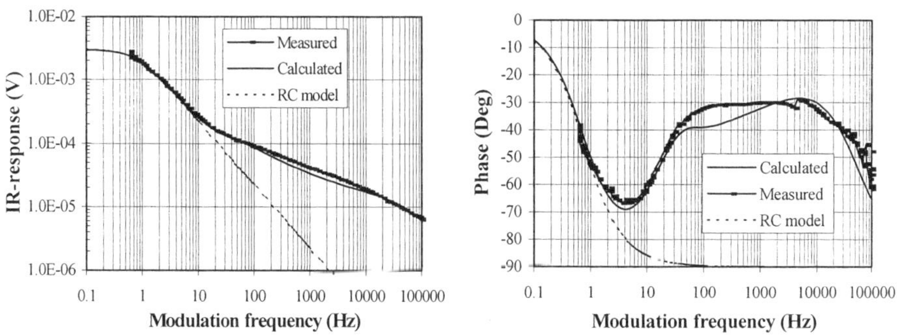

Figure 3.2. Measured and calculated magnitude (left) and phase (right) of response versus modulation frequency of the $SrTiO_3$ substrate sample 064-01a at 80 K and 250 mA dc bias current. The response is calculated from both the analytic model and that from the basic RC-model.

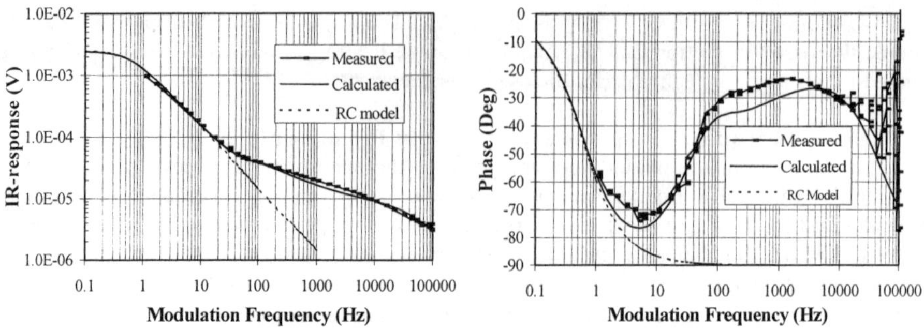

Figure 3.3. Measured and calculated magnitude (left) and phase (right) of response versus modulation frequency of the $LaAlO_3$ substrate sample 061.02a at 80 K and 500 μA bias current. The response is calculated from both the analytic model and that from the basic RC-model.

very high frequencies where the thermal diffusion length into the substrate or into the micromachined free standing base element of the bolometer becomes smaller or comparable to the thickness of the substrate or the base element [38]-[39]. In general the use of the basic R-C model is appropriate in the frequency regimes where one period of the modulation frequency is longer than the low frequency time response of the samples, τ_L. The τ_L is defined as $\tau_L = C(0)/G(0)$, where G(0) and C(0) are the total low or so called the dc thermal conductance and the heat capacity of the samples. In the following the results from the model to the measured values are compared and the effects of different thermal and dimensional parameters on the fit of the model to the experimental values are investigated. The comparison is done for both the magnitude and phase of the response in three regimes of low, mid, and high ranges of frequencies.

a) The Magnitude of Response at Low Frequencies and the Dependence on the Substrate Material and Dimensions

As given in Table 3.1, the substrate-holder thermal boundary resistance of the samples, R_{SC}, are given in per unit area and found using the measured dc thermal conductance of the samples using the shift in the temperature caused by Joule heating at elevated bias currents

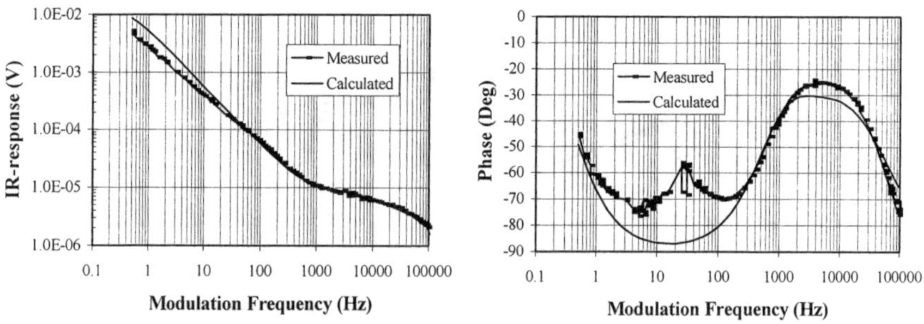

Figure 3.4. Measured and calculated magnitude (left) and phase (right) of response versus frequency of the MgO substrate sample number 064-03a at 79 K and 1 mA bias current. The response is calculated from both the analytic model and that from the basic RC-model.

[24]. As introduced before the factor α is defined for the correction of R_{SC} value and depends on the area of the pattern, A, the substrate thickness, d_s, and the thermal conductivity of the substrate materials, k_s. It increases by the increase of k_s and the increase of the ratio of d_s/A as expected. The requirement of the factor α for a better fit of the calculated response curves to the measured ones, is due to the spreading thermal resistance effect in the substrate material neglected by the one dimensionality of the model discussed here. Hence, for a relatively low thermal conductive substrate sample with a small d_s/A ratio, this factor is expected to approach one as observed in table 3.1. Using the known parameters for the films and the substrates and the obtained values for the α and the boundary resistance', the phase and magnitude of the response for the samples are calculated and plotted in figures 3.2 to 3.4.

As shown in the figures, at low modulation frequencies there is a discrepancy between the results from the model and the measured values. This discrepancy increases by the increase of the thermal conductivity of the substrate materials showing its maximum for the most conductive substrate material. This is due to the one dimensionality of the model ignoring the lateral heat diffusion in the substrate materials underneath the pattern. By comparing the magnitude of the response curves in Figures 3.2, 3.3 and 3.4 for $SrTiO_3$, $LaAlO_3$, and MgO respectively, the discrepancy is the most for the most thermal conductive MgO substrate sample. This is also consistent with the variation of α, which is the highest for the same sample. This discrepancy can be avoided by use of a three dimensional solution for the model, or use of an approximate cylindrical solution for a truncated cone structure with a linearly increasing base from the film to the bottom of the substrate. Such approximate solutions can also be applied using the simulation programs such as SPICE where the parameters can further be considered frequency dependent [7].

b) Phase of the Response and the High and low End Frequency Behavior

The phase of the response of the samples are much more sensitive to the values of the characteristic parameters of the samples. Particularly the drop of the phase at high-end frequency response has been the determining factor in finding the equivalent values of C_f and R_{fs} given in Table 3.1. While the low end calculated phase of the response is in a very good agreement with the experimental results for $SrTiO_3$ and $LaAlO_3$ substrate samples, there is a peak at the frequency range of 10 to 100Hz in the MgO substrate sample, which cannot be explained by this model. This behavior has also been observed for all other MgO substrate samples. This occurs at lower frequencies in thinner substrate samples e.g. at about 10 to 20 Hz frequency for 0.025 cm thick MgO substrate samples compared to that of a 0.5 cm thick MgO substrate sample 064-03a shown in Figure 3.4b. One speculated mechanism for this is possibly due to the reflection and interference effect of the acoustical phonons within the boundaries of the substrate underneath the superconducting pattern [24]. The possible effect of the superconducting material on the spectral phonon density in the substrate and its consequent effects is discussed in next sections.

c) Mid-Range Frequency Behavior of the Phase and Magnitude of the Response

There is a slight discrepancy in the midrange frequency response curves compared to that of the model. This is clearly seen in the plateau of the phase versus frequency curves for all the samples shown in Figures3.2 to 3.4. The discrepancy is also seen partly in the magnitude of the response curves in the same frequency regime. The measured response in this range is slightly higher in magnitude, and more in phase with the radiation signal than the expected values from the model.

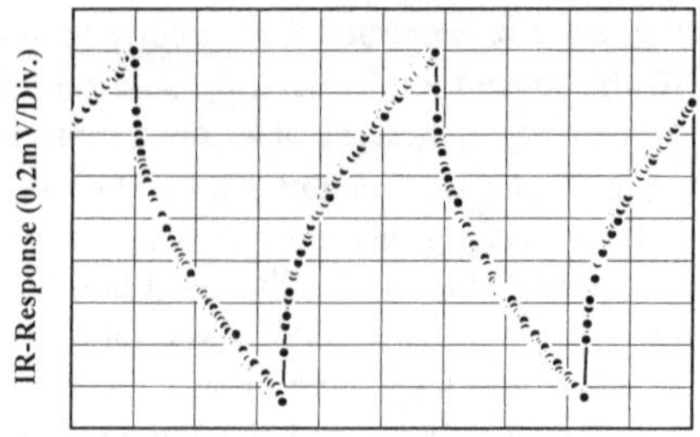

Figure 3.5. The steady state response versus time of the $SrTiO_3$ substrate sample 064-01a at 560 μA bias current, radiated by a 20 Hz square radiation signal with 2.13 mW/cm^2 maximum intensity.

The possible effect of the radiation absorption by the open areas of the substrate within the meander lines of the patterns has also been investigated as a possible source for the discrepancy. This consideration in the model Leeds to lower phase of the response increasing the observed discrepancy in the plateau region of the phase curves particularly for the $SrTiO_3$ substrate sample. The spectral absorption of the crystalline substrate materials from 600 nm to 50 μm are about 12%, 20%, and 8% for the 0.05 cm thick $SrTiO_3$, $LaAlO_3$, and MgO substrates at 850 nm radiation wavelength respectively. While the effect of the substrate absorption is negligible in the response of the samples measured at 850 nm radiation wavelength, it has a major effect in the measured spectral response of the samples at longer wavelengths [7], [9], [11], [37]. The effect of the substrate absorption in the spectral response of the samples is studied in next sections.

A small and relatively fast component in the time response of the samples also is reported which relatively increases compared to the total response as the modulation frequency increases. This cannot be explained by the thermal response as modeled in this work. The steady state time response of $SrTiO_3$ substrate sample 064-01a in the above mentioned frequency regime, is shown in Figure3.5. As observed from the figure there exist an abrupt jump and drop at the edge of the on and off switching times of the radiation source. The source is a fast near IR source, which is derived by a square wave voltage signal. The same type of the steady state time response is observed for all other samples in the same frequency regime. The effect of this fast component of the response is in the favor of the discrepancy (i.e. lowering the phase value and increasing the magnitude of the response from those calculated values by the model). This is possibly the non-thermal component of the response and being a possible source for the discrepancy that needs to be further investigated.

C. Response of thick film ETBs

Thick film IR detectors are typically fabricated by screen-printing technology using pastes manufactured from high-T_c superconducting powders on polycrystalline substrates such as MgO and Al_2O_3 [4]. The thickness of the superconducting films obtained by this method is in the range of a few 10s of micrometers, being about 30 μm for the studied samples here. The Y-Ba-Cu-O compound with 123 and 124 stoichiometries can be used for fabrication of thick film superconductive infrared detectors using the hybrid technology. The fabricated films are then post annealed in various ambient such as Oxygen at high temperatures resulting granular

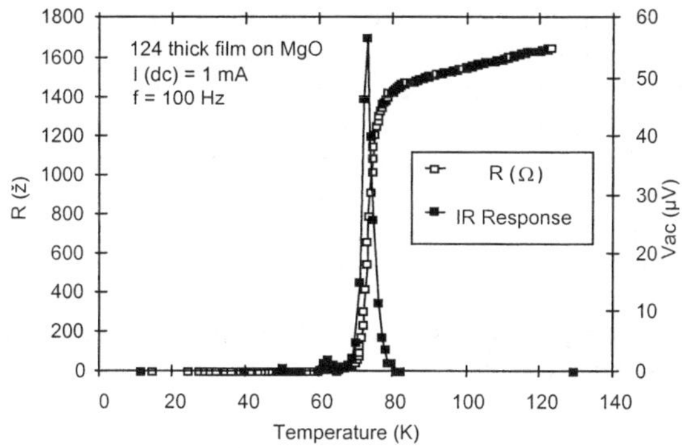

Figure 3.6. Resistance and magnitude of the response versus temperature of a 100 μm wide and 4mm×6mm total area thick film YBCO on MgO substrate sample.

superconductive films with a 123 phase [29]. The thick film samples here are screen printed on polycrystalline MgO substrates. The film geometry are a serpentine pattern, with a line of 250 μm width and 82.6 mm length, 0.89 mm pitch and 30-35 μm in thickness [4].

At sufficiently high annealing temperatures, about 850 °C in air, the detectors made of material with 124 phase of YBCO converts to the 123 phase plus a CuO species. The conversion of 124 material to 123 + CuO at high temperatures can give 123 material with higher critical currents [29], permitting a higher bias current for the bolometer. The CuO species, considered to be uniformly distributed in the bulk, can behave as pinning centers for fluxons and prevent flux flow in the material [24].

The IR-response versus temperature measurement on a 124-based sample (converted to 123 phase) is shown in Figure 3.6. The results of the IR-response vs. temperature measurement are superimposed on its R vs. T transition curve in the figure. It should be noted that there is a small peak at 60 K to 65 K in the response, which is below the region of its transition-temperature curve ($T<T_{c\text{-zero}}$). A similar feature is not reported in the IR-response of the sample at lower dc-bias current. This response, which shows a dependence on the dc-bias current through the sample, is attributed to a Kosterlitz-Thouless phase transition and the associated noise signal below $\sim T_c$ as discussed in section 7.

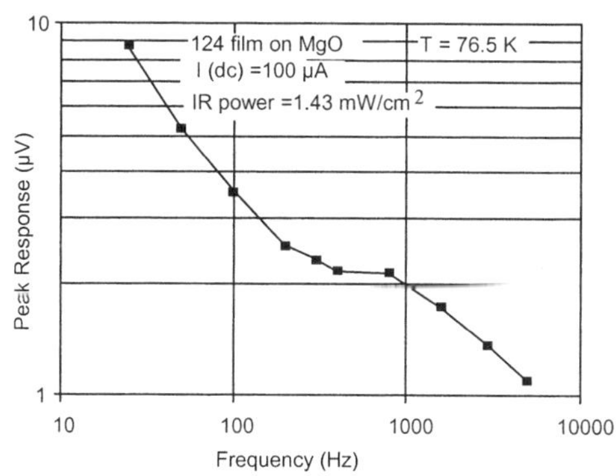

Figure 3.7. Magnitude of the response versus frequency of a 100 μm wide and 4mm×6mm total area thick film YBCO on polycrystalline MgO substrate sample.

An IR-response vs. frequency measurement on a thick film is shown in Figure 3.7. As shown in the figure, there are two breakpoints in the response. The first breakpoint, near 400 Hz, is attributed to the thermal time response, τ (C/K), of the substrate and matches the preceding analysis with respect to the effect of the substrate-holder thermal boundary resistance. The second breakpoint, near 800 Hz, is attributed to the thermal time response of the thick superconducting film with the consideration of the film-substrate boundary resistance. Considering this assumption, the second breakpoint is expected at much higher chopping frequency for thinner films as discussed earlier.

4. THE TEMPERATURE DEPENDENCE OF RESPONSIVITY AND DEVIATIONS FROM THE DR/DT CURVE

The phase and amplitude of the response to infrared signals of edge transition bolometers versus temperature is studied in this section. The measured amplitude and the phase of the response versus temperature of the devices deviate from that expected based on the discussed models in the preceding sections. In this regard a model is discussed here that can explain the observed discrepancy in the response versus temperature from the dR/dT curve in the studies samples in this chapter and some of other reported discrepancies from the bolometric response in the literature [22]-[23], [45]. Based on this model, the thermal constants, G, and C, change as the film goes into the superconducting state; G becomes strongly temperature dependent, and decreases with decreasing temperature. A probable cause for the change in G is a shift in the nature of the phonons emitted by the superconductor as it goes through T_C.

There is a dependence of the measured total dc thermal conductance, G(0), on the thickness of the substrate of the samples. Based on the preceding analysis, G(0) is always limited by the substrate/cold-head thermal boundary resistance at low modulation frequencies [7], [18]. This conclusion is confirmed by comparing the results to a model of the device's frequency response as discussed earlier. Also using the low frequency response of the bolometers, the specific heat and thermal conductance of the substrate material can be verified to confirm the a above approach [7], [18]. The study of this part has been focused on the characteristics of YBCO films on MgO substrates samples in response to infrared radiation with a wavelength of λ=0.85 nm emitted from a near IR-LED source.

A. Thermal conductance characterization

As discussed earlier, the thermal conductance of a bolometer is defined as the total heat conduction from the bolometer to its environment per degree of temperature difference. The amount of input power (heat), P_{in}, to a bolometer can be determined, and used to measure G from:

$$G=\frac{P_{in}}{\Delta T}, \tag{4.1}$$

where ΔT is the difference between the temperature of the bolometer and its environment. The thermal conductance of a superconductor bolometer can be measured in different ways, which are discussed here in more details. In each technique there are different kinds of possible error associated with the nature of the characterization [7], [24]. Heat loss by radiation from the superconducting bolometer (superconducting film and the substrate) in typical systems is normally negligible compared to the phonon heat conduction [7], [30].

One of the effects of the finite thermal conductance in a sample can be observed in the R versus T curves at different bias currents. The thermal conductance of the sample puts a limit on the maximum bias current that can be used in the measurement. The bias current at any temperature higher than $T_{c\text{-}zero}$, is limited by thermal runaway, characterized by the factor α defined as [1]:

$$\alpha = \frac{I^2 dR}{G(0)dT}\,; \qquad \text{with } \alpha<1 \text{ required for stability,} \tag{4.2}$$

where I is dc bias current and G(0) is the dc thermal conductance. The effect of instability in the temperature of the film, due to high bias currents, will appear as an unrealistically sharp transition in the R versus T curve [25], [42]. The instability and the shift in the $T_{c\text{-}onset}$ can be used to measure G(0).

At low bias currents the temperature of the superconducting film is very close to the temperature of the cold-head. As the bias current increases, a temperature gradient develops across the substrate and the interfaces due to Joule heating in the film. Since it is the temperature of the holder which is commonly controlled, this results in a shift of the bias point in the superconducting film towards $T_{c\text{-}onset}$ where dR/dT has lower values. The decrease of dR/dT lowers the detector response [25], which has been interpreted as the effect of the bias current on the intrinsic part of the response by some groups [22]. The drop of the response at high currents can be used as a different method to measure G. Knowing the dR/dT versus T curve, and the change in the response, the shift in the temperature of the superconducting film with respect to the holder can be found. Using the R and the bias current, the power dissipation in the film can be obtained. Then by use of the temperature shift and the dissipated Joule power in equation (4.1), the value of G(0) can be calculated [24].

The most convenient way to obtain G(0) is to make use of the slope of the R versus T of the films in the normal state near T_c as shown for sample 064-02b in Figure 4.1. At high currents, R increases due to the dc power dissipation heating the film. The relationship between the increase in the resistance of the film, R, and the bias current due to Joule heating is:

$$\Delta R = I^2 R \frac{dR}{GdT}\,. \tag{4.3}$$

Using the measured values for the dR/dT at the bias temperature of the holder, and the bias currents, values of G can be obtained for each data point. There is a slight error in the above

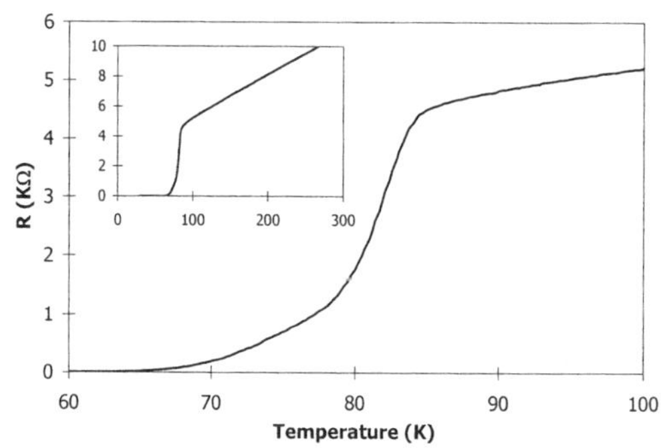

Figure 4.1. R versus temperature of sample 064-02b at 100 μA bias current and 2 K/min heating rate. Insert shows full temperature range.

method due to the change of the dR/dT at different temperatures.

Here one can use a method, which eliminates the above error. Considering Joule heating by the bias currents as the source for the shift of the film temperature, T_f, with respect to the temperature of the holder, G(0) can be found from:

$$G(0)=\frac{I^2R}{\Delta T} \tag{4.4}$$

where ΔT is the difference between T_f and the temperature of the cold-head. The resistance of the film, R, can be measured from V/I at each point. Using the R versus T curve for the region of interest ($90<T<120$K), and the measured value of R at different bias currents T_f can be obtained and used to determine ΔT. Then using ΔT and $P_{in}= I^2R$ in equation (4.4), values of G(0) can be obtained for each data point. This is the method used to find the averaged values given in III_Table 2.1. To measure G(0) relatively high bias currents should be used for ease in monitoring the temperature change in the film. However, high currents through the material can affect the electrical properties of the films shifting T_c to lower temperatures [25], [48].

B. Dependence of G on the YBCO film pattern and the substrate thickness

The values obtained for the dc or low frequency thermal conductance of the studied bolometers are about 2-3 orders of magnitudes lower than the values calculated from the properties of the substrate material, which can be obtained from:

$$G=k\frac{A}{L}\ . \tag{4.5}$$

Where k is the thermal conductivity of the substrate material and L and A are thickness of the substrate and area of the superconducting pattern. The substrate/cold-head thermal boundary resistance determines the measured G values in III_Table 2.1 [7], [25]. This can be found from the frequency response of the devices. At low frequencies, the thermal diffusion length into the substrate, L_f [2], [7], [18], [60], is larger than the substrate thickness, and G is constant with respect to f, and much lower than the values calculated for the substrate from equation (4.5). As the frequency increases, L_f becomes comparable or smaller than the substrate thickness changing G, which changes the slope of the response versus modulation frequency from a f^{1} to a $f^{0.5}$ dependence [2], [7], [16]. Then G is limited by the thermal conductance of the substrate material, and can be obtained from equation (4.5), where L is proportional to the thermal diffusion length, L_f [7], [18], [24].

The measured thermal boundary resistance at the substrate/cold-head interface can be explained by the thermal acoustic mismatch theory or Kapitza resistance [16], [30]. A thermal boundary resistance defined as the Kapitza resistance, is:

$$R_k=\frac{\Delta T}{Q'} \tag{4.6}$$

where Q' is the heat flux per unit area through the interface, and ΔT is the temperature difference across the boundary. This phenomenon is related to the reflection and refraction of the incident thermal phonons at the interface [30]. Based on the thermal acoustic mismatch model, R_k is temperature dependent with a dependence of T^{-3} [24], [30]. In the model, it is assumed that the heat is carried only by thermal phonons (only acoustic phonons are important at low temperatures), the two solids are perfect materials, and finally the interface is perfect or ideal [94], [95]. Based on the above assumptions the heat transfer through the interface is determined mainly by the acoustic impedance $\rho_{m1}S_1$ and $\rho_{m2}S_2$ of the two attached materials at the boundary, where ρ is the mass density and S is the velocity of sound in the material.

As shown in Figure 2.1, there are two interfaces to be considered in our experimental configuration. The film/substrate interface and the substrate/cold-head interface. The film/substrate interface and its corresponding thermal resistance, R_{bd}, have been investigated by different groups [15], [17], [30]. The effect of the R_{bd} on the response in the samples of table 2.1 is only at high frequencies, at which the thermal resistance of the substrate becomes comparable to the film/substrate thermal boundary resistance. This frequency is 53 KHz for 0.05 cm thick MgO substrate sample [4], [7], which is obtained from a typical value of $R_{bd}=1.1*10^{-3}$ K-cm^2/W [4], [15], [17]. Though G is limited by the substrate/cold-head thermal boundary resistance, $R_{s\text{-}c}$, at modulation frequencies less than 720 Hz, and 2.8 KHz, for 0.05 cm and 0.025 cm thick MgO substrate samples, respectively [4], [7].

However, a dependence of G(0) on the substrate thickness and the pattern area exists as given in Table 2.1. It should also be noted that the back of same substrates cannot be thoroughly cleaned due to the adhesion of silver epoxy applied to the back of the substrate for the in-situ deposition of the YBCO and this can cause a slight difference in the measured G compared to the G of a sample with a perfectly clean and smooth contact to the cold-head.

As given in Table 2.1, G(0) increases for thinner substrates, while it is still limited by the substrate/cold-head thermal boundary resistance. The dependence of the value of G(0) on substrate thickness can be attributed to scattering of phonons within the substrate which changes the angular distribution of the phonons at the substrate/cold-head interface. Phonons with a larger angle of incidence to the normal at the interface will be internally reflected at the boundary, lowering the value of G(0). This is also consistent with the observed dependence of the G(0) on the area of the superconducting pattern, it increases with an increase in the area of the pattern, suggesting that the conduction is mainly through the contact area below the pattern with direct on-line flow to the cold-head.

C. Effect of bias current on the response and dR/dT versus temperature

The shift of the temperature of the film with respect to the temperature of the cold-head at high bias currents, used to measure G(0)), can also be observed in the bolometric response versus T of the samples. The temperature of the peak in the response versus T curve shifts to

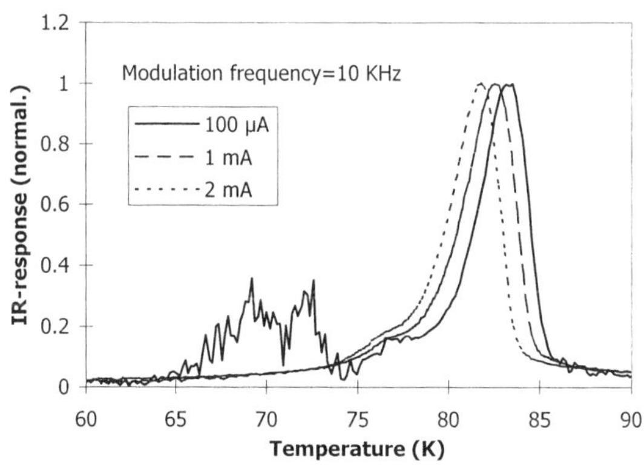

Figure 4.2. Normalized response of sample 064-02b, at 100 μA, 1 mA, and 2 mA bias currents with peak response of 386 nV, 3.28 μV, and 5.87 μV, respectively.

lower temperatures as higher bias currents are used. Though the response versus temperature will follow the measured R versus T curve (or the corresponding dR/dT curve), the magnitude of the response follows the real slope of the transition, the R versus temperature of the film (the slope at low bias currents) [7]. This effect for sample 64-02b is shown in Figure 4.2. The values of thermal conductance obtained by the shift in the peak of the response versus T of the samples, confirm the range of values given in Table 2.1. A slight change in G is expected, due to the difference of the temperature at the peak of the IR-response and the temperature at which the G values in Table 2.1 are measured. Also there is an error associated with the above method due to the continuous change of the temperature during the measurements [4].

As shown in Figure 4.2, there is a small peak at temperatures below the $T_{c\text{-}zero}$ for the response at 100 μA bias current, which is attributed to be the intrinsic noise of the sample. This excessive noise is in all samples that have weak links, or are granular and show a long tail in their R versus T curves. The peak noise in this region decreases with increasing bias current, while it widens on the lower temperature end. A detailed study of the different types of noise in the samples is given in later section.

D. Response versus temperature of the ETBs and deviations from the thermal model

Assuming the thermal parameters G and C to be independent of temperature as given in equation (2.1), the bolometric response is expected to follow the dR/dT versus T curve. Figure 4.3 shows the magnitude of the IR-response of sample 064-02b to a 20 Hz modulation frequency, and the corresponding dR/dT versus T curve. Both curves are normalized to their values at the $T_{c\text{-}onset}$. The dR/dT of the sample was obtained by use of the simultaneously measured R versus T. As observed in Figure 4.3, there is a slight discrepancy between the response and the dR/dT curve at the lower part of the transition region closer to $T_{c\text{-}zero}$. There have been several reports on such discrepancies [45]. The discrepancy in these samples is dependent on the modulation frequency of the signals, decreasing as the frequency increases and is smaller for samples with sharp transitions. This is verified for the frequency range of 20 Hz to 10 KHz [4].

The phase of the response of the samples also varies as a function of temperature. This variation indicates a change in the thermal constants of the bolometer. The measured phase of the response versus temperature of sample 064-02b at 20 Hz frequency, is given in Figure 4.4.

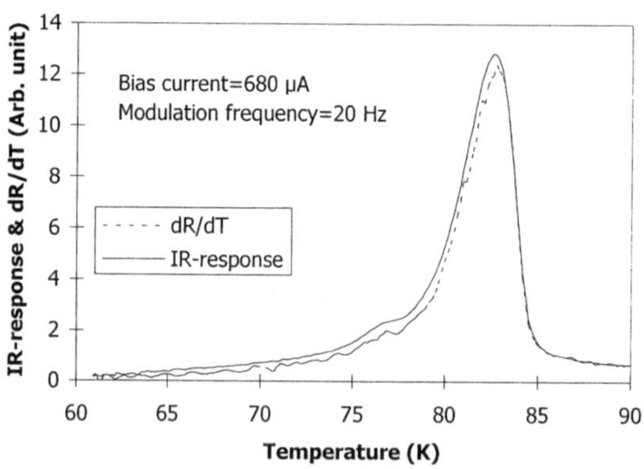

Figure 4.3. Normalized IR response and dR/dT vs. temperature of the sample 064-02b at 20 Hz and 680 μA bias current.

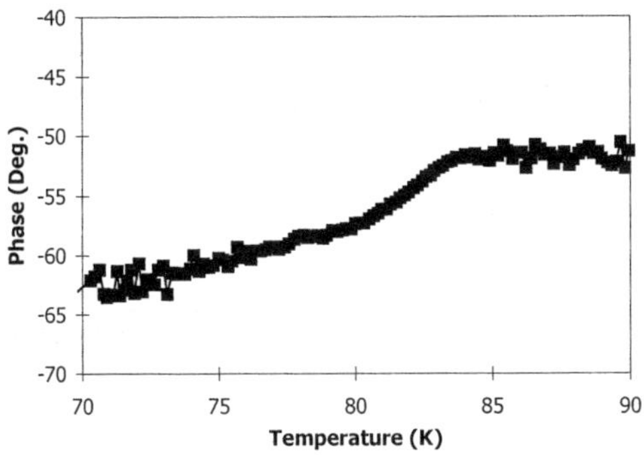

Figure 4.4. The phase of the IR-response vs. temperature of sample 064.02b at 20 Hz modulation frequency and 680 μA dc bias current.

The response at this frequency is mainly bolometric [7], in agreement with the response versus time of the sample [4]. The variation of the phase versus temperature in Figure 4.4 shows a dependence on the superconducting transition of the film. The magnitude of the phase of the response of the samples increases at the lower part of the transition. This variation of the phase versus temperature, which follows the superconducting transition, suggests a correlation between the thermal characteristics of the superconducting film and the response at low frequencies. It is already discussed that the response at low modulation frequencies is mainly governed by the thermal constants of the substrate material and substrate/cold-head interface i.e. τ=C/G with C determined from the heat capacity of the substrate and G the substrate/cold-head interface conductance. The same type of behavior of the phase of the response is observed for the other samples as well, and for samples with sharper transitions, the change is smaller [4]. One explanation for the change in the response is that the character of the phonons changes as the film goes into the superconducting state. In the normal state one expects the thermal phonons to be from the acoustic branch. In the superconducting state the IR-photons can break pairs, which upon recombination emit phonons that have energy of $2\Delta(T)$. These phonons have a different distribution than the thermal ones in the acoustic branches between longitudinal and transverse phonons, and may also include optical phonons. This is discussed in detail in the following section.

The increase of the amplitude of the response at the lower part of the superconducting transition at low frequencies, suggests a decrease in G, assuming that the heat capacity contributing to the response is a constant. From equation (2.1), a decrease in G should give an increase of the phase (lag), which is consistent with the measured increase of the magnitude of the phase given in Figure 4.4. The discrepancy in the magnitude of the response versus temperature of sample No. 064-02b is given in Figure 4.5. This shows that the deviation increases as the temperature drops. The variations in the ratio are a measure of the noise present. The measurements at lower temperatures are limited by the small signal, which decreases the signal to noise ratio.

E. Effect of variation of G on the response and its frequency dependence

Considering the discrepancy at 20 Hz in Figure 4.5, the response at 70 K is about 1.4 times higher than the expected values. This temperature is close to $T_{c\text{-zero}}$ and will be noted as

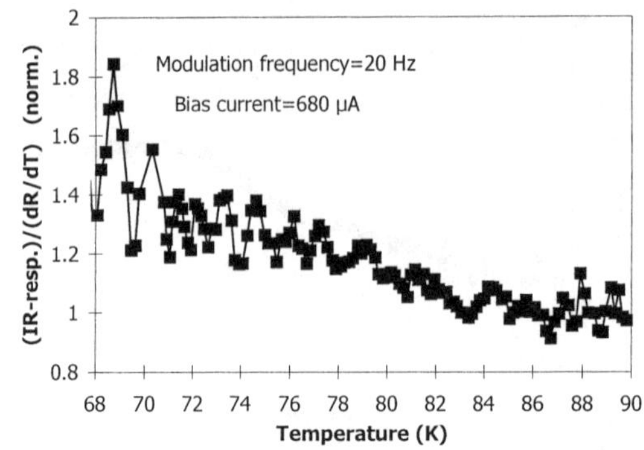

Figure 4.5. Deviation of IR-response vs. temperature from the dR/dT vs. temperature in Figure 4.4.

$T_{c\text{-low}}$. From Figure 4.4, the phase of the response, θ, at the onset temperature of ~85 K is about -51.5 ° and at the $T_{c\text{-low}}$ (~70 K) is about -62.5°. Then equation (2.1) results in:

$$-\tan\theta_i = \omega\,\tau_i\,, \qquad (4.7)$$

where θ_i is the temperature dependent phase of the response. Then:

$$\omega\,\tau_{c\text{-onset}} = 1.26 = \omega\, C_{c\text{-onset}}/G_{c\text{-onset}} \qquad (4.8)$$

Similarly for the phase at $T_{c\text{-low}}$ we have:

$$\omega\,\tau_{c\text{-low}} = 1.92 = \omega\, C_{c\text{-low}}/G_{c\text{-low}} \qquad (4.9)$$

Now, based on the above analysis, from the ratio of equation (4.8) and (4.9) one gets:

$$(\tau_{c\text{-onset}}/\tau_{c\text{-low}}) = 0.65 \qquad (4.10)$$

or:

$$\frac{G_{c\text{-onset}}\,C_{c\text{-low}}}{G_{c\text{-low}}\,C_{c\text{-onset}}} = 1.53. \qquad (4.11)$$

Assuming the heat capacity of the bolometer, which is mainly due to the substrate material at this frequency, drops from $T_{c\text{-onset}}$ to $T_{c\text{-zero}}$ according to its experimental results [61], inquires that the thermal resistance of the bolometer increases (G decreases) at lower temperatures, and consequently there would be an increase in the response which is consistent with the result in Figure 4.5. The variation in the responsivity, taking the changes in the C into account, can be obtained by use of equation (2.1) from:

$$\frac{r_v(\text{c-low})}{r_v(\text{c-onset})} = \frac{G_{c\text{-onset}}\sqrt{1+\omega^2\tau^2_{c\text{-onset}}}}{G_{c\text{-low}}\sqrt{1+\omega^2\tau^2_{c\text{-low}}}} \qquad (4.12)$$

By use of the results from Figure 4.3, and values in equation (4.8) and (4.9) in equation (4.12), a value of 1.89 can be obtained for $G_{c\text{-onset}}/G_{c\text{-low}}$. By applying this to equation (4.11), a value of 0.81 is obtained for $C_{c\text{-low}}/C_{c\text{-onset}}$ ratio. The change in the C can be attributed to the change of the effective heat capacity due to the change in the characteristics of the phonons and the their propagation and also the temperature dependence of the heat capacity of the substrate material as expected for a crystalline solid.

The thermal conductance at low frequencies is determined by the substrate/cold-head thermal boundary resistance, which can be strongly temperature dependent [24]. Based on above analysis, as the temperature decreases the substrate/cold-head thermal boundary resistance increases as expected for a solid/solid interface [24]. However, the phase of the response shows its strongest variation as the sample goes through the superconducting transition, indicating that these large changes are associated with differences between the normal and superconducting state, rather than the temperature dependence of the interface boundary resistance as modeled in the next section. At higher modulation frequencies, the thermal diffusion length into the substrate material becomes comparable to or smaller than the substrate thickness. Then the thermal conductance becomes independent of the substrate/cold-head interface, and is expected to be mainly governed by the thermal conductivity of the substrate material, k_S. The frequency at which the thermal diffusion length becomes equal to the substrate thickness of sample 064-02b is 2.8 kHz [4], [7]. For frequencies higher than this the G is governed by the k_S [4], [7]. The thermal conductivity of the substrate material can increase as the temperature decreases. This can compensate for the decrease in the C, reducing the discrepancy, and at higher frequencies it may even cause a decrease in the response at the $T_{c\text{-low}}$ compared to the response at the $T_{c\text{-onset}}$. Hence the response at higher frequencies can be closer to the dR/dT curve discussed in the following section.

5. EFFECTS OF SUPERCONDUCTIVITY TRANSITION AND THERMAL PARAMETERS ON THE PHASES OF THE RESPONSE

According to equation (2.1) the temperature–dependent phase of the response can be calculated by the formula,

$$\tan\theta = -\omega C / G \ . \qquad (5.1)$$

It is important to understand the factors that determine G and C of the bolometers as a function of the temperature, frequency, and other parameters of the bolometer configuration. The phase of the response to infrared signal of the edge-transition bolometers versus temperature at high and low modulation frequencies is studied in this section. A model is discussed that can explain the observed discrepancy in the response versus temperature discussed in previous section. According to the model, the thermal constants, G and C, change as a film goes into the superconducting state. This might also partly be accounted as for some observed non-bolometric component of the response in this type of bolometers.

As shown in Figure 5.1, there are two thermal boundary resistances at the substrate interfaces, one at the film interface and the other at the holder interface. There are also two major bulk areas, which constitute the overall heat capacity of the samples. One is due to the superconducting film, and one is due to the substrate material. In analyzing the different operation regimes of the edge-transition bolometer parameters, the frequency dependence of the thermal diffusion length also plays an important role as discussed earlier.

The phase of the response of the samples is more sensitive to the values of the characteristic parameters of the bolometers than the magnitude of the response. The phase of the response shows a strong variation as the sample goes through the normal-superconducting transition both at low and high modulation frequencies showing also modulation frequency dependence. At low frequencies, the phase of the response starts to decrease as the temperature decreases below about the onset temperature of the normal to superconducting transition (i.e. at $T<T_{c\text{-onset}}$). Opposite behavior exists at high modulation frequencies, i.e. the phase of the response starts to decrease at $T<T_{c\text{-onset}}$. This clearly indicates that, large changes in the phase and magnitude of response are mainly associated with differences in the conduction mechanisms in the normal and superconducting states, rather than the well known

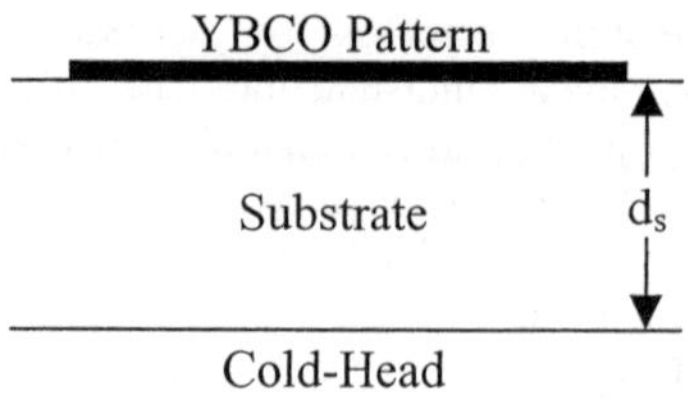

Figure 5.1. Lumped film sample configuration in contact with a cold-head. d_s is the substrate thickness.

temperature dependence of the interface boundary resistance. The mechanisms causing the above behaviors are discussed in the followings.

A. Low frequency regime

At low modulation frequencies the phase of the response versus temperature increases by decrease of the temperature starting at $T<T_{c\text{-onset}}$ as shown in the Figure 4.4. The respective amplitude of the response normalized to the measured dR/dT of the superconducting film at $T<T_{c\text{-onset}}$ for low modulation frequencies is shown in Figure 4.3. The amplitude of the response in Figure 4.3 increases above the expected value determined by the dR/dT of the film as the temperature is lowered toward the $T_{c\text{-zero}}$. Such behavior would be consistent with a decrease of the thermal conductance of the bolometer. The variation of the heat capacity at low temperatures, reveals typical T^3 dependence behavior [61]. At low modulation frequencies ($\omega = 20Hz$), the thermal conductance in the studied samples is determined by the substrate/cold-head interface conductance, or so called Kapitza boundary resistance, caused by the acoustic mismatch impedances of the interfaced materials [9], [18]. According to equation (2.17), by lowering the frequency the thermal diffusivity length into the substrate becomes longer than the substrate thickness and phonons emitted by the superconducting film into the substrate will reach substrate/cold-heat boundary.

The thermal boundary conductance at the substrate/cold-head interface can be explained by the theory of Kapitza conductance. It is well known [62], that Kapitza conductance is the derivative of the net heat flux transmitted across an interface with respect to the temperature difference between the two materials. Therefore, if one considers an interface between materials A and B, we can write G_K as follows:

$$G_K = \frac{1}{V}\frac{\partial}{\partial T}\sum_k \sum_j^A \hbar\omega_{kj} n(\omega_{kj}, T) v_{kjz} t_{kj} \tag{5.2}$$

$$= \frac{1}{V}\frac{\partial}{\partial T}\sum_k \sum_j^B \hbar\omega_{kj} n(\omega_{kj}, T) v_{kjz} t_{kj} \tag{5.3}$$

Where the labels A and B on the sums indicate that all quantities in each sum correspond to phonons incident on the interface from the materials A and B sides, respectively. In equations (5.2) and (5.3), ω_{kj} is the phonon frequency, v_{kjz} is the component of the phonon group velocity normal to the interface, $n(\omega_{kj},T)$ is the Bose-Einstein distribution function, t_{Kj} is the probability that a phonon of wave vector **k** and polarization j will be transmitted across the interface between materials A and B, and V is the volume. The Kapitza conductance can be calculated at low temperatures ($\hbar\omega_D / kT >> 1$) analytically, taking into account Debye phonon density of states, which can be obtained from;

$$D(\omega) = V^{-1}\sum_{kj}\delta(\omega - \omega_{kj}) = \omega^2 / 2\pi^2 c^3, \tag{5.4}$$

where c is the velocity of sound in the material. As a result from calculations, the thermal conductance has T^3 characteristics at low temperatures, and remains constant at high temperatures. The phonon system in each material in above calculations is considered to be at the equilibrium nature. However, in the calculation of the boundary conductance of the substrate/cold-head interface, one should take into account the phonon density of state spectrum of the non-equilibrium phonons, which propagates into the MgO substrate and reach boundary substrate/cold-head (diffusion regime of propagation of phonons). Hence should also be determined the differences in the photo-absorption process in normal and superconducting states should also be determinded. As shown by the analysis of thermalization and photo-absorption in $YBaCuO$ superconductors [63], interaction between hot quasi-particles and Cooper pairs condensate continues to divide the excess energy by forming three quasi-particles. The hot quasi-particles continue to break additional pairs by the electron-electron interaction process as they thermalize toward the energy gap by discrete energy steps. As a result there exists a characteristic feature- a jump at frequency $\omega = 2\Delta(T)$ due to activation of the recombination mechanism of phonon generation [64] (for $\omega \le 2\Delta(T)$ phonons are generated as a result of the energy relaxation of high energy quasi-particles). The phonon distribution function (product of $D(\omega)n(\omega,T)$) has fairly narrow maximum near $\omega = 2\Delta(T)$ [50], i.e. in the above calculation one should replace Debye phonon density of states by the Einstein phonon density function;

$$D(\omega) = \omega^2\delta(\omega - 2\Delta(T)). \tag{5.5}$$

Assuming that transmission probability and normal component of the group velocity to the interface to be independent of temperature, equation (5.2) can be rewritten as;

$$G_K = \frac{1}{V}\sum_j \int_0^{\omega_{Debye}} d\omega\hbar\omega_{kj} D(\omega)\frac{dn(\omega_{kj}, T)}{dT} v_{kjz} t_{kj}, \tag{5.6}$$

where for the $dn(\omega,T)/dT$ we have the following expressions:

$$\frac{dn(\omega,T)}{dT} = \begin{cases} \dfrac{k}{\hbar\omega}, \dfrac{\hbar\omega}{kT} << 1 \\ \dfrac{k}{\hbar\omega} x^2 \exp(-x), x = \dfrac{\hbar\omega}{kT} >> 1 \end{cases} \tag{5.7}$$

It is clear that at high phonon frequencies, $2\Delta(T)$ is about equilibrium temperature of the substrate T, and the contribution of the non-equilibrium phonons to the heat transfer has exponential behavior. As shown by the above calculations, at temperatures close to critical temperature $2\Delta(T) << T$, where $\Delta(T) = \Delta_0(1 - T/T_c)^{1/2}$ (standard BCS theory for example [65], low frequency phonons do not influence heat transfer process and boundary conductance at substrate/cold-head. Hence calculated normalized conductance based on the equation (5.5), (5.6), and (5.7) are in good agreement with experimental data for the thermal conductance versus temperature at low modulation frequency. The discrepancy between experimental data and calculations at temperatures close to T_c shown in_Figure 5.2, is associated with the tendency of the emitting phonon frequency to zero. Here we also consider superconducting

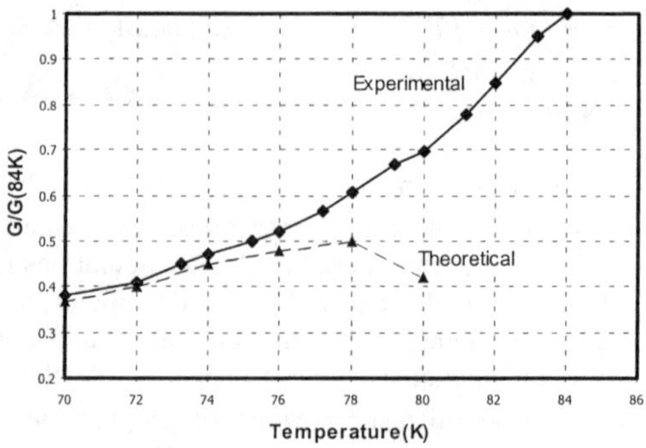

Figure 5.2. Calculated and measured normalized conductance of 200 nm thick YBCO film on MgO substrate versus temperature at low modulation frequencies.

materials without fluxons. It is clear that the temperatures close to the T_c can easily lead to Abrikosov vortexes in the films. In the equilibrium, fluxons are present in the film and move due to the thermal activations. The optical photons also create additional fluxons and the flux motion (creep and flow). This factor can lead to the no bolometric component in the optical response, not considered in the above calculations. The discrepancy between theory and experiment can also be related with flux dynamics near T_c. This argument is suggested by the experimental data from [9]. As mentioned before, for the samples with sharper transition, the change in the phase of the response and the discrepancies at low frequencies are smaller. One of the possible reasons for broadening of the transition into superconducting state is the penetration of the fluxons into the samples. Another possible improvement of the model is connected with using realistic behavior of the order parameter close to T_c and $d-wave$ characteristics of the pairing in $YBaCuO$ compounds [66]. On the other hand, influence of the order parameter relaxation near T_c on the bolometer characteristic should also be taken into account for further accurate calculations near critical temperature.

B. High frequency regime

The variation of the phase of the response versus temperature at high modulation frequency (e.g. $\omega = 10KHz$) is shown in Figure 5.3. At frequencies above the knee frequency, f_L, the phase of the response decreases as the temperature is lowered. To describe the heat transfer in this regime, one can imply the phonon radiation limit [62] (ballistic regime of propagation of phonons). In the high frequency regime all phonons from the superconducting film are assumed to transmit across the film-substrate boundary and absorbed in the substrate media. That is in equation (5.2) transmission probability t_{kj} can be considered one. Such approximation is valid due to low Kapitza resistance of the film-substrate boundary [18]. As shown by calculations [67], the thermal conductivity in high-T_c superconducting films has different behavior in contrast to conventional low temperature superconductors. In cuprate compounds thermal conductivity below T_c versus temperature has a peak near the T_c which is lowered by decreasing the film thickness [67]. The thermal conductance of the studied devices at high modulation frequencies is previously discussed to be mainly governed by thermal conductivity of the substrate material taking into account the processes in superconducting film. The thermal conductivity of the substrate can also increase as the temperature decreases

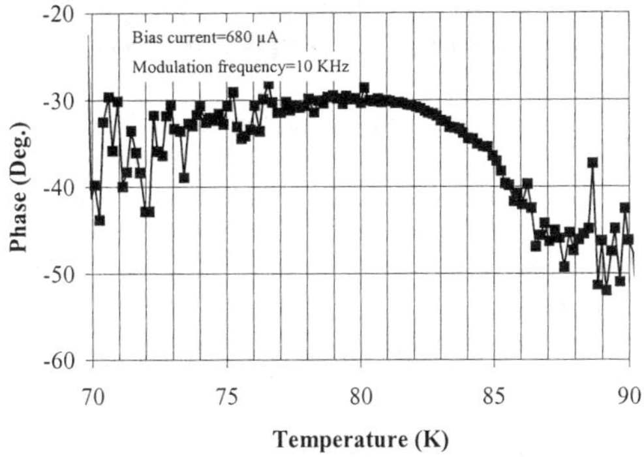

Figure 5.3. The phase of the response versus temperature at high modulation frequency of 10 KHz and 680 mA dc bias current.

below the $T_{c\text{-onset}}$. This is due to enhancement of the mean free path of phonons emitted from the superconducting film into the substrate. As a result the phase of the response in the temperature interval 75K<T<85K decreases and at lower temperatures T<75K increases. The corresponding amplitude of the response at high frequencies is shown in Figure 5.4. Response at temperatures T<75K can be closer to the dR/dT curve. The discrepancy in amplitude of response in the temperature interval 75K<T<85K can also be related to the variation of the thermal conductivity in the substrate material.

In brief, at low modulation frequencies, a shift of the position of the recombination phonons in the spectrum by changing the temperature leads to decreasing of the boundary conductance of the substrate/cold-head interface for $T<T_c$. This with variation of the heat capacitance versus temperature leads to the agreement between the theoretical and experimental results of temperature dependence of the response of the bolometers. However, such model is not adequate at temperatures close to the critical temperature at low modulation frequencies. One reason for such behavior is related with creating fluxons and their dynamics near T_c.

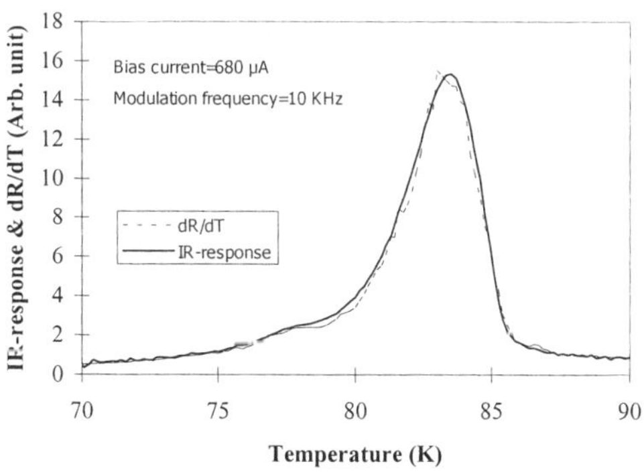

Figure 5.4. Amplitude of the response versus temperature at 10 KHz modulation frequency and 680 mA dc bias current.

6. DC CHARACTERISTICS OF THE BOLOMETERS AND ARTIFACTS RELATED TO JOULE HEATING AND THE FILM MICROSTRUCTURE

In most of the dc characterizations of the patterned superconducting films there can be considerable Joule heating associated with the resistance of the patterns at high bias currents. This power dissipation can cause a temperature gradient across the thermal conducting path. Hence there can be an error in the dc characterizations versus temperature caused by the above temperature gradient. This is because in most of the setups the temperature of the cold-finger (holder) or the underneath of the substrate is monitored [6]-[11], [19]-[24], [42]-[46] where the temperature of the film may be higher [11], [23]. At high bias currents this can cause a shift of $T_{C\text{-onset}}$ toward the lower temperatures in the measured R versus T curve as observed in many different works [21], [23]-[24], [42], [45], to be discussed later in this section. This can be either due to the temperature gradient across the film-substrate or the substrate-cold finger interfaces, depending on the film and the substrate dimensions. This effect has been a misleading parameter in the analysis of the dc characteristics and the T_C stability of the films [19], [20], [23], [42], [46].

Also effects of the mechanical stresses on the film under different atmospheres caused by the temperature gradients associated with Joule heating, has been a degrading factor for the quality evaluations of the samples [7], [47]. These effects are found to depend strongly on the dissipation of the Joule heating powers in the samples, hence making the dc thermal conductance, G(0), a determining factor for even dc characterizations.

One measure of the quality of the high-T_C superconducting films has been the presence of precipitation. The effect of the microstructure and the amount of precipitation in the film on the dc characteristics of the samples have been studied and reported by many different groups. Here we also discuss present a possible effect of the precipitation on the normal resistance of the samples due to their possible contributions in the electrical conduction at relatively high bias currents. The response of the samples in this study and their potential improvement are linked to the dc effects indicated.

A. Joule heating effects

The effect of Joule heating in the superconductive films, which is equal to I^2R, depends on the heat conduction path determining the temperature variations in the sample. Since the heat is generated in the film, the film-substrate contact area is one determining factor. Also the heat conduction through the substrate into the temperature reservoir is another factor to be considered. This also takes the thermal conductivity and heat capacity of the substrate materials into consideration. For the dc characterizations under stable conditions, the consideration of the heat capacitance of neither the film nor the substrate is necessary [7]. To study the effect of Joule heating in the samples, one can divide them into two main categories, with respect to the amount of the generated heat and mechanism of the heat conduction from the film to the temperature reservoir. Then the samples can be considered as small or large area patterns compared to the dimensions of the substrate. The large area patterns are normally made of a meander lines where the small area patterns can be considered for most of the micro bridges consisted of a strait narrow path of the superconducting film.

a) Small area patterns

For microbridges with dimensions small compared to the substrate dimensions, the generated temperature gradient due to the Joule heating is mostly expected to be at the film-substrate interface or across a thin layer of the substrate at the vicinity of the superconducting path [23], [30]-[50]. The reasons for this are because only a small area of the substrate underneath the pattern conducts the heat, where the substrate would behave like a heat bath for the generated heat in the pattern, because of its large dimensions compared to that of the pattern. In samples with above configurations, the heat conductivity of the substrates is the determining factor in temperature gradient. For R versus T measurements of such microbridges, the temperature rise in the film can cause a thermal runaway over heating the whole film if a constant current source is used [1], [5], [24], [44]. If the measurement is monitored by a voltage controlled source then the overall heating will be limited by a local thermal runaway creating a hot spot in the bridge [24], [51]-[52]. Based on thermal stability, the hot spot would be at a point of the bridge where it can have the maximum thermal conductance, G. Hence, despite of some expectations [24] it won't happen in the middle of the bridge. Taking the thermal conduction through the film into the consideration as well, the hot spot would be located at either end of the microbridge where it has the maximum thermal conductance as reported [24] The samples studied here are not among the above category.

b) Large area patterns

For the samples with dimensions of patterned superconducting film comparable to that of the substrates, the temperature gradient is mainly across the substrate-cold finger interface [7], [16], [24]. Hence the whole pattern of the above samples is at the same temperature under any stable circumstances [7], [24]. The thermal boundary resistance at this interface (substrate-cold finger) is the determining factor in the overall thermal conductance of our samples [7], [24]. In such samples, there won't be a hot spot. The heat conduction is blocked mainly by the substrate-cold finger interface, giving an almost uniform heat distribution over the superconducting pattern. This is while the total thermal resistance of the samples is a few orders of magnitudes higher than the values just across the substrates [16], [24]. So, the whole superconducting film can be considered at the same temperature to a very good approximation.

At high bias currents there can be a stable heating by the complete path of the pattern creating a higher temperature in the film where the monitored temperature of the cold-finger may show lower values. This temperature rise in the film is proportional to the resistance of the film given as [7], [24];

$$\Delta T=\frac{RI^2}{G(0)}\cdot \qquad (6.1)$$

In R versus T characterizations using a constant current source, this can cause a false sharper transition at high currents. At values close to $T_{c\text{-onset}}$, as resistance of the film increases, the Joule heating increases and creates higher ΔT with respect to the temperature of the cold-finger. Figure 6.1 shows this effect for one of the samples.

This false sharp transition by shifting the measured $T_{c\text{-onset}}$ to lower temperatures is interpreted to be an intrinsic shift by some groups [19], [23], [42], [46]. This shift increases the dR/dT value of the film obtained from the measured R versus T curves which is measured to be different from the real R versus T of the film at high bias currents. One way to investigate this effect is by use of the responsivity of these samples to radiation intensities (i.e. measuring ΔR due to known values of absorbed radiation) [7]. The responsivity of such patterns used as bolometers, calculated based on this unreal high dR/dT, will lead to misleading values much higher than the possible values for such configurations [22]-[23]. In granular samples with wide normal to superconducting transition, the widening of the

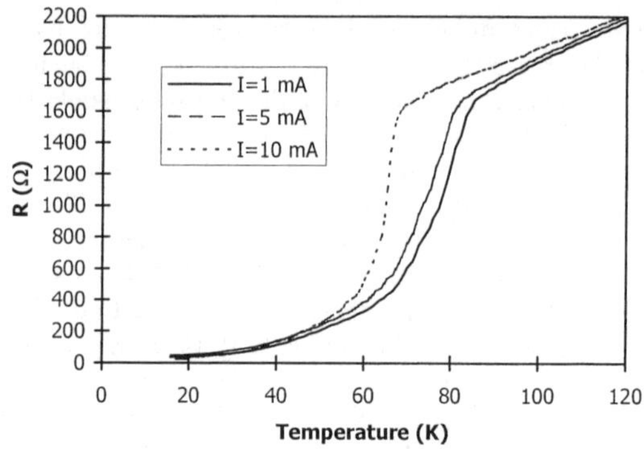

Figure 6.1. R versus temperature of $LaAlO_3$ substrate sample 057-04b at 1 mA, 5 mA, and 10 mA dc bias currents, and 2 K/min heating rate.

transition due to high bias currents dominates the shift of the $T_{C\text{-onset}}$ to the lower temperatures, lowering the maximum magnitude of the dR/dT curve showing the expected changes of the dR/dT at higher bias currents.

B. Effect of ambient atmosphere and the Joule heating

Some samples show T_Cs sensitive to bias currents shifting to lower temperatures, after being exposed to high currents. Such a shift would dependent on the maximum applied current determining the maximum temperature gradient in the device due to the Joule heating. This extreme change of T_C has been mostly observed in samples with low and relatively sharp transition temperatures [24], [68]. The transition region at the T_C of $YBa_2Cu_3O_{7-x}$ films on $LaAlO_3$, and $SrTiO_3$ substrates are typically sharper than the films on MgO substrates where the T_C of the later one is reported to be more stable [7]. When these type of samples with extreme low T_C due to high bias currents are exposed to room temperature at atmosphere pressures, the T_C of them reaches higher values close to the initial values measured in first cycles under low bias currents. This is interpreted partly to be caused by the strain under vacuum where the strain is due to the temperature gradient in the film and the substrate. This temperature gradient is caused by Joule heating at relatively high bias currents in the normal state and in the transition region close to the $T_{c\text{-onset}}$.

The strains under vacuum can cause release of oxygen (outgasing) in imperfect $YBa_2Cu_3O_{7-x}$ crystalline structures, decreasing the T_C by decrease in the over all oxygen content. This effect can be observed when the sample is biased at a constant temperature of the cold-finger, under variable applied bias currents. This configuration is used for the measurements of the dc thermal conductance of the samples [5], [7], [24]. The result of such a measurement for sample 064-01a is shown in Figure 6.2, showing both changes due to the temperature rise in the film with respect to the cold-finger and the changes of T_C due changes of the oxygen stoichiametry (shift of T_C) of the sample. As observed in the figure the later one that happens at bias currents over about 5 mA in this sample is not recovered when the temperature rise due to Joule heating is not present anymore (i.e. $I \approx 0$). Considering the measured 3 mW/K for the G(0) of this sample the temperature rise caused by the Joule heating at a bias current of 5 mA is about 37 K. Some samples with initially unstable T_C are reported

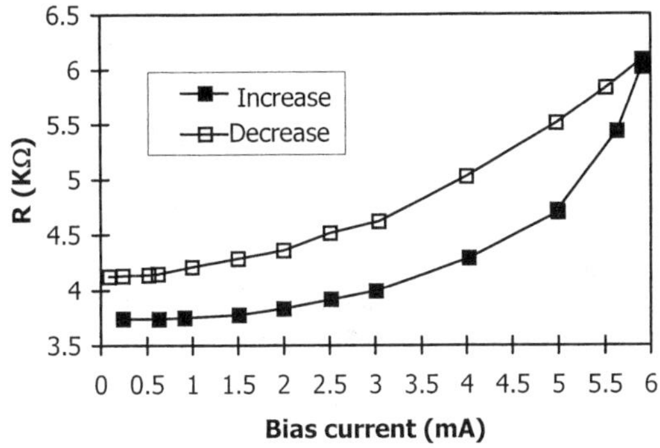

Figure 6.2. R vs. I_{bias} of sample 064-01a at a bias temperature of 97 K, measured in two directions of increasing (filled squares), and decreasing (empty squares) I_{bias}.

to reveal a stable and higher T_C after being exposed to high mechanical stresses under ambient atmosphere [25]. This is interpreted to be due to the release of undesired strains existing in the film from the growth of the superconducting crystalline film on the substrate.

C. Effect of microstructure and bias current

There are also repeatable and stable changes in the magnitude of the normal resistivity of some samples in the R versus T measurements, which happens by changing the bias current as shown in Figure 6.3. This change of the magnitude of resistance versus bias current is apart from that due to the shift of T_C and the changes discussed before. Two major effects are the cause for this type of dependence of the normal resistance, R, on the bias current that is observed even for the samples with stable T_C. Some of these changes are due to the instrumentation in the R versus T characterization systems using constant current bias. This is partly caused by non-ideality of the current sources specifically in the systems where the current sources are made of a resistor and a battery. This configuration is essential for ultra low noise measurements of the signal, which is discussed in next section.

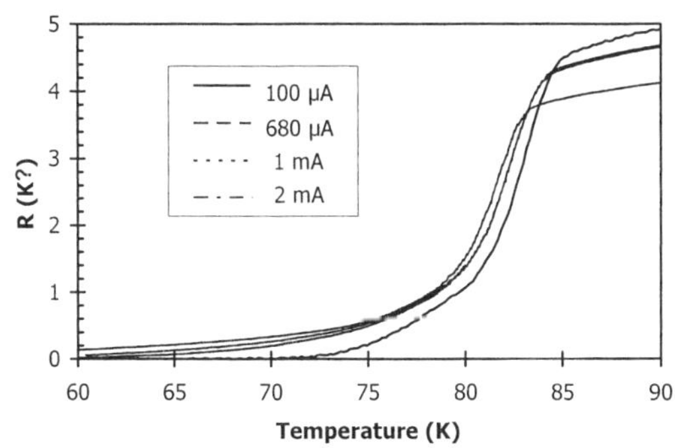

Figure 6.3. R versus temperature of sample 064-02b at 100 μA, 680 μA, 1 mA, and 2 mA dc bias currents with 2 K/min heating rate.

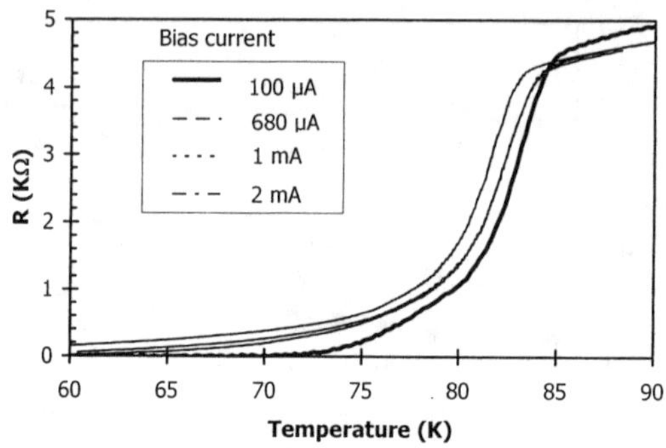

Figure 6.4. R vs. temperature curves of sample 064-02b (in Figure 5) at 100 μA to 2 mA I_{bias}, corrected for errors caused by the impedances of the source and the device.

Knowing the output impedance of the current source and the relative changes of R versus T at the transition, this error can be corrected. The R versus T of sample 064-02b in Figure 6.3 corrected for this effect is shown in Figure 6.4. As shown in Figure 6.4 there still exist a dependence of R on bias current showing a nonlinear decrease of the normal resistance when the current is increased. This decrease of the R at higher bias currents (at the same T) is reverse of the expected change due to shift of T_C caused by Joule heating.

As shown in Figure 6.4, there is a change in the normal state R much deviating from the predicted values due to the shift in the R versus T curve. This lowering of the normal state resistance at higher bias currents is observed in highly granular samples. There can be at least two possible mechanisms for this phenomenon. One can to be due to high voltage drops across the precipitation that may have comparable resistance with the shunt superconducting paths in the normal state. Then the nonlinear dependency is due to possible minimum threshold voltage values required at the boundaries for passing the current through the potential barriers at the boundary (some kind of metal-semiconductor tunneling mechanism and or metal-semiconductor junction). Another can be due to a possible tunneling mechanism at the week links mainly created during the patterning process. Due to high precipitation in the film there can be some undesired etching around the large particles of the precipitation. The further increase of current ($I>100$ μA) in Figure 6.4 which is not showing further considerable changes in the normal state resistance, is consistent with the above analysis. After the above considerations there still is a shift in the $T_{C\text{-onset}}$ to lower temperature at higher bias currents, which, as discussed before, is due to the Joule, heating in the film and the limited G.

As observed in Figure 6.4, there is also an excessive decrease in $T_{c\text{-zero}}$ at high bias currents, which increases the width of the transition region. This excessive widening of the transition which occurs in highly granular samples can dominate the sharpening of the transition caused by the shift of the $T_{C\text{-onset}}$ to lower values due to the Joule heating, increasing the over all width of the measured transition. The dR/dT of the R versus T curves in Figure 6.4 are shown in Figure 6.5. As is observed in the figure, there is a shift of $T_{C\text{-onset}}$ of about 2 degrees at 2 mA bias current toward the lower temperature. The sharpening of the transition region due to the shift of $T_{c\text{-onset}}$ to lower values in the measured R versus T curves does not have observable effect on the real magnitude of the dR/dT at any point, which is investigated using the measured bolometric response of the samples. Though the measured

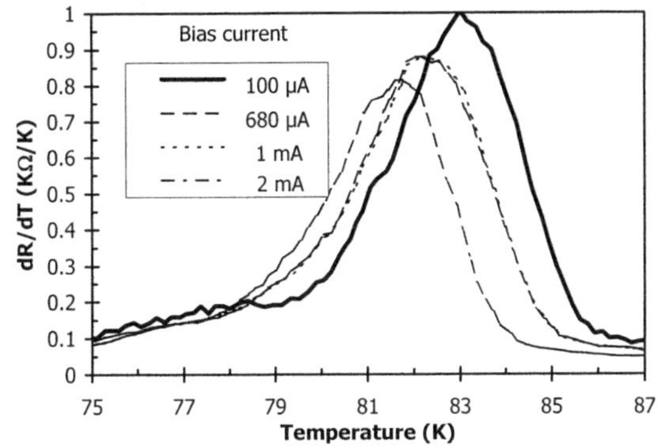

Figure 6.5. dR/dT versus temperature of sample 064-02b at 100 mA to 2mA dc bias currents, derived using the R versus T curves with a 2 K/min heating rate.

shift at high bias currents due to the granularity of the film lowering the $T_{c\text{-zero}}$, is real and intrinsic lowering the real peak magnitude of the dR/dT of the samples that was also investigated using the bolometric response of the samples, discussed in following section.

D. Determining dc characteristics using current normalized response

For the measurement of the bolometric response versus temperature of the cold-finger at different bias currents (i.e. the temperature of the film can be different due to Joule heating), the lowering of the $T_{c\text{-zero}}$ at higher bias currents should cause a decrease in the current normalized response of the bolometers [24]. Where the shift due to the Joule heating only shifts the measured temperature of the onset of the transition and the peak temperature of the response (and also the dR/dT of the film) with respect to the temperature of the cold-finger (i.e. the monitored temperature). This temperature shift due to the Joule heating is toward lower temperatures and is higher at higher bias currents and higher temperatures. This shift can be found and corrected for, using equation (6.1). The current normalized response of sample No. 064-02b with the above characteristics is shown in Figure 6.6. This shows the real

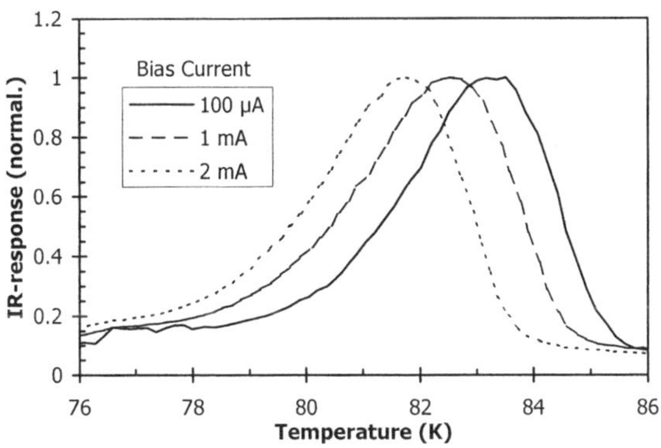

Figure 6.6. Normalized response of sample 064-02b, at 100 μA, 1 mA, and 2 mA, and 10 KHz modulation frequency and 2 K/min heating rate.

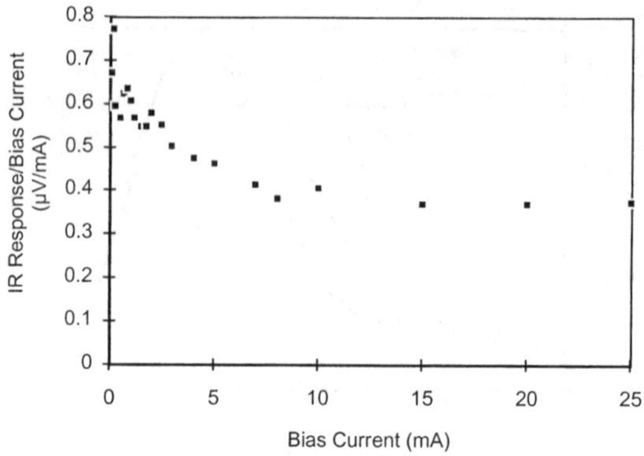

Figure 6.7. Current-normalized responsivity of a highly granular-film bolometer vs. dc bias currents at 50 Hz modulation frequency = 50 Hz.

dR/dT curve at different bias currents of Figure 6.4, since for the bolometric response we have:

$$r_v = \eta I \frac{dR}{dT} \tag{6.2}$$

As shown in Figure 6.6, the peak of the response is shifted to lower temperatures due to Joule heating shifting the $T_{C\text{-onset}}$ to lower values. The magnitude of peak of the response measured as 386 nV, 3.28 μV, and 5.87 μV for 100 μA, 1 mA, and 2 mA bias currents respectively, shows a slight lowering due to the shift of $T_{c\text{-zero}}$ to lower temperatures following the changes in the maximum dR/dT curves of Figure 6.5. These changes in the magnitude of the response due to real changes of dR/dT and the peak temperature caused by the temperature rise of the film with respect to the substrate or the cold finger is interpreted as an effect of intrinsic response of the high-T_C superconducting bolometers by some groups [22]-[23], [46]. Based upon the above analysis, such an interpretation is not warranted. The misinterpretation of the shift of $T_{C\text{-onset}}$ and the dc characteristic of the samples can also mislead the responsivity calculations by two orders of magnitude higher, as obtained by some groups [23]. It can also cause the equilibrium response of such a devices to be mistaken by no equilibrium responses [22].

The dc characteristics of the thick films are more sensitive to both the environments and the bias currents. This is mainly to high granularity of this type of films since they are fabricated by post-annealed screen-printed pastes made of powders on polycrystalline substrates. [29], [4] As shown in Figure 6.7, the current-normalized responsivity of such detectors drops with increase of the bias current up to current densities even near 800 A/cm^2. This effect is attributed to the granular structure of the films. The superconducting material at the grain boundaries can be partially driven to the normal state, resulting in a decrease in the sensitivity of the film [25].

7. NOISE CHARACTERISTICS AND SPECTRAL DETECTIVITY OF YBCO SUPERCONDUCTOR BOLOMETERS

The ultimate sensitivity of a superconductor infrared bolometer is determined by its noise characteristics, which is dependent on the operating temperature, bias current, and modulation frequency [7]. A parameter such as responsivity is important in that one wishes to detect the smallest possible signals, limited by the noise from the detector called intrinsic noise. One criterion for the responsivity is to be large enough to increase the signal and the intrinsic noise from the detector, above that from the system, such as the noise from preamplifier, circuit elements, etc. Then the parameter noise equivalent power (NEP) together with the responsivity determine the directivity, D* as given in equation (7.1). The NEP is defined as the input signal power that would give the noise observed in the output. Ultimate performance in a detection system is achieved when the noise generated by the background photon flux, incident on the IR-detector is larger than any thermally generated noise within the detector [1], [9], [11].

The detectivity of bolometers is basically a function of the bias current, the absorbing area (and its absorption coefficient), and the bias temperature through the slope of the resistance versus temperature curve, dR/dT. The detectivity versus temperature can be written as a function of the responsivity and the noise as:

$$D^* = \frac{\sqrt{A.\Delta f}\,.\, r_V}{V_n} \qquad (7.1)$$

where A is the radiation absorbing area of the detector, r_v is the responsivity, V_n is the voltage noise, and Δf is the frequency range used to obtain the noise measurement.

A. Noise in high T_C superconducting materials

There has been considerable work on the noise characteristics of the superconducting materials, most of which has been focused on the low frequency or 1/f flicker noise. The study of noise in the high T_c materials may give important information about the physics of the transition [69]-[70]. This includes the onset of the several stages of the transition, the intergrain critical currents, the effect of a magnetic field, the role of inhomogeneities, and other kinds of disorder, and the dynamics of the vortex motion in the material [71]. Noise measurements due to conductivity fluctuations (or resistance fluctuations) reported by different groups [42], [49] have given the following main results; (a) no noise is found in the superconducting state well below T_C, (b) in the normal state near T_c, the noise is large compared to metal-insulator composites, and (c) the noise was reported to increase in the transition region [71].

The effect of a defect can be modeled as a two-level fluctuator, (TLF), moving between two energy states separated by an energy barrier [72]. Then the system undergoes transitions between the two states generating a random telegraph signal (RTS). Among the various mechanisms that can be responsible for such a signal in a current biased high T_C film, are charge trapping and flux hopping mechanisms, which can have different origins. The noise voltage generated by two-level fluctuators is frequently observed at temperatures right below T_c, and seems to be one of the universal features of these materials [73]-[74].

An anomalous increase of low frequency, 1/f, noise for epitaxial, and granular high temperature superconductors has been reported for temperatures close to the superconducting transition temperature [75]-[79]. There is general agreement on the observation of excess noise in granular, or a and c axes mixed orientation YBCO superconducting films, compared to high quality epitaxial c-axis oriented films [1], [45], [76], [77]. Proposed interpretations of the above phenomenon have been based on two ideas, a percolating network made of superconductor and normal elements [80]-[82], and the random motion of magnetic flux tubes [81]-[85].

Above T_c, the Cooper pairs have a finite lifetime and the conductivity is the sum of the conductivity of the normal electrons and the conductivity of Cooper pairs. Hence the conductivity fluctuations in this region can be interpreted to be due to the fluctuations in mobility and/or density of the carriers. Following the above discussion, in the lower part of the transition, another type of conductivity fluctuation is reported to be dominant in comparison to the fluctuations in number or type of carriers, and is interpreted as fluctuations of the volume fraction of the superconducting phase [86].

One of the major sources of noise in these material is the motion of flux tubes or current vortices, which can be generated by an external magnetic field, by the self field produced by current in the material, or spontaneously as pairs of opposite oriented flux tubes. The noise study in the low field, or isolated vortex regime, in high T_c materials has shown the importance of vortex motion in low frequency noise, which is proposed to be due to thermally activated two-level or diffusive vortex motions [87]-[88].

Other mechanisms such as phase slip, a strong Lorentz-force due to an applied current, and localized weak pinning centers, have also been reported to be strongly related to the above noise characteristics [89]-[90]. It can be argued that the corresponding voltage noise at the transition temperature could be generated by the rearrangement of vortices collectively pinned by the film microstructure, which has been shown to be affected, and partly controlled by the external field [91].

In this part, results from the measurements of the noise with different frequency, temperature, and bias current dependences are discussed. Noise measurements versus temperature and bias currents are studied, and two major possible governing mechanisms for the noise behavior of the samples are discussed. For a $SrTiO_3$ substrate sample both the spectral response and noise is given and from these the spectral detectivity of the sample is obtained. From the above results and the frequency response of the sample one can project a maximum detectivity for an optimally designed bolometer [7]. In addition, the noise seen in all the regions can serve as a sensitive characterization tool for superconducting thin films [11].

B. Noise characteristics of the detectors

Simultaneous measurement of the IR-response and the resistance versus temperature, show that the bolometric response closely follows the dR/dT curve of the samples. The results for sample 064-02b are given in Figure 7.1. The response in this range of bias currents,

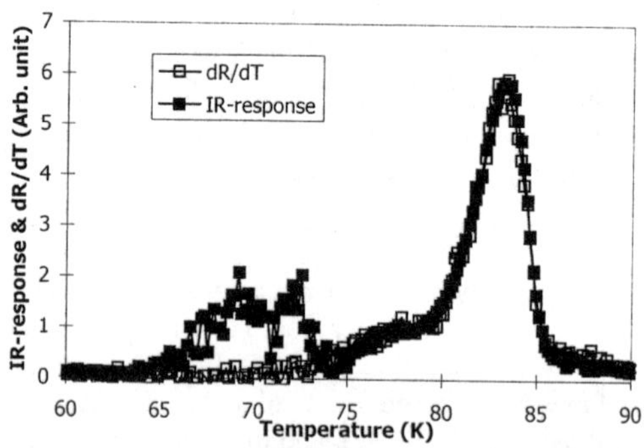

Figure 7.1. IR-response and dR/dT vs. temperature of sample 064-02b, at 100 μA bias currents, and 10KHz modulation frequency.

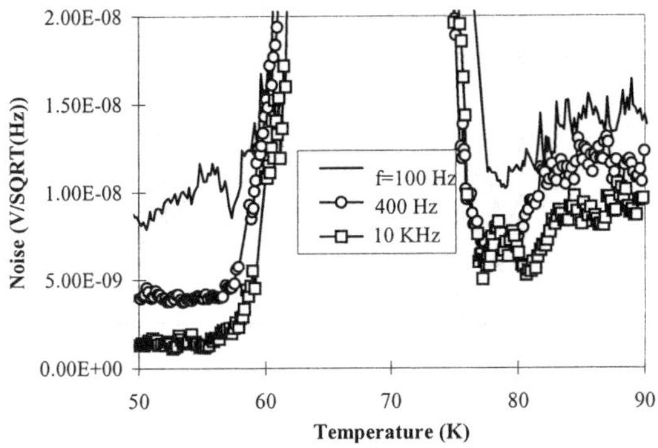

Figure 7.2. Noise voltage vs. temperature of sample 064-02b at 100 μA bias current.

frequency, and radiation intensity is expected to follow the dR/dT curve [24]. As observed in Figure 7.1, a noise voltage signal, which is comparable to the bolometric response signal, occurs at temperatures close to and below $T_{c\text{-zero}}$, and for low bias currents is larger than the bolometric response. This signal exists even in the absence of the optical radiation, and its magnitude is also significantly affected by the intensity of the radiation. Measurements of the noise voltage versus temperature of the samples, in the four probe configuration, reveals that this voltage signal is an excess noise voltage in the material that occurs at and below $T_{c\text{-zero}}$.

The results for above noise at different frequencies and temperatures, is given in Figure 7.2. From the noise characteristics versus temperature shown in the figures, the noise in the material can be divided into two major kinds. One is the excess noise at and below $T_{C\text{-zero}}$, and the other is the noise that follows the resistance versus temperature transition at and above $T_{C\text{-onset}}$. As shown in Figure 7.2, the excess noise at the lower part of the transition is not significantly affected by the frequency and has the characteristics of a white noise. The magnitude of this noise is a few orders of magnitudes higher than the calculated Johnson noise, and phonon noise [7]. The Johnson noise is calculated for 1 Hz bandwidth from

$$V_{John} = \sqrt{4K_B TR} \qquad (7.2)$$

Where, K_B is the Boltsman constant (1.38×10^{-23}), T is the temperature, and R is the resistance of the sample at the temperature indicated. The resistance and thermal conductance of the samples are given in VI_Table 2.1. The phonon noise was calculated from:

$$V_{phonon} = r_v \times \sqrt{4GK_B T^2} \qquad (7.3)$$

where, G is the thermal conductance of the samples.

The dependence of this excess noise on the bias currents at 10 kHz, is given in Figure 7.3. As shown in the Figure 7.3, the width of the temperature region of the noise increases with an increase in the bias current and goes to lower temperatures at higher bias currents. This is can be due to the weak links and the corresponding SNS junctions. The broadening of the noise with an increase in the current can be interpreted as due to the effect of the bias current on the weak links in the film. In a granular film, or a film with weak links, additional regions at the grain boundaries or the weak links, can be driven into the normal state at lower temperatures by the increase of the current through the film. Based on the properties of the superconducting material, and the relations between J_C, T_C, and H_C, the broadening of the noise versus temperature, or in other words the lowering of the onset temperature of the noise, should also

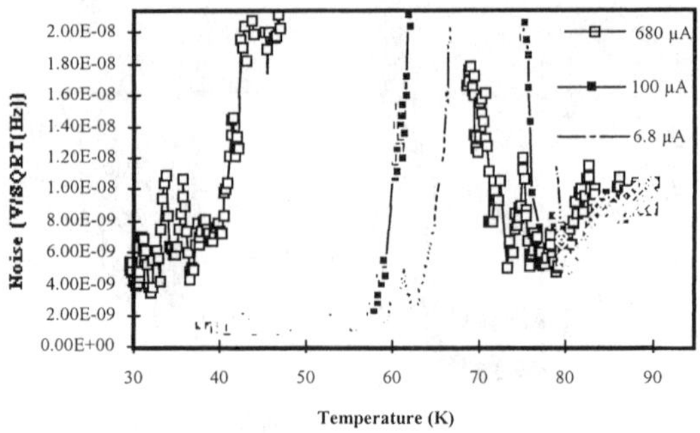

Figure 7.3. Noise voltage vs. temperature of sample 064-02b at 6.8 μA to 680 μA and 10 KHz modulation frequency.

occur by applying a magnetic field. The temperature onset of the noise can be considered as the temperature at which the first normal links appear in the path of the current as the temperature is increased from low temperatures. The effect of the magnetic field on this type of noise has been studied by other groups for high T_C superconducting films at low frequencies [91].

The onset of the excess noise for sample No. 064-02b at different bias currents is obtained from the noise versus temperature of the sample and given in Figure 7.4 The values of $T_{C\text{-zero}}$ of the sample obtained from the R versus T curves, are also plotted for different bias currents. Based on this assumption the onset temperature of the noise in the above sample versus bias current should have a behavior similar to the critical current density, J_C, versus temperature, and Figure 7.4 shows such a correlation. Based on the above, this noise should exist even at zero bias currents, and not to be changed considerably by currents lower than some value, where the normal regions are mainly determined by the temperature. This is observed in the noise measurements versus temperature at the very low bias current close to 6.8 μA, which is the lowest bias current at which the reported measurement is done.
One explanation for the existence of the excess noise at very low bias currents or zero bias

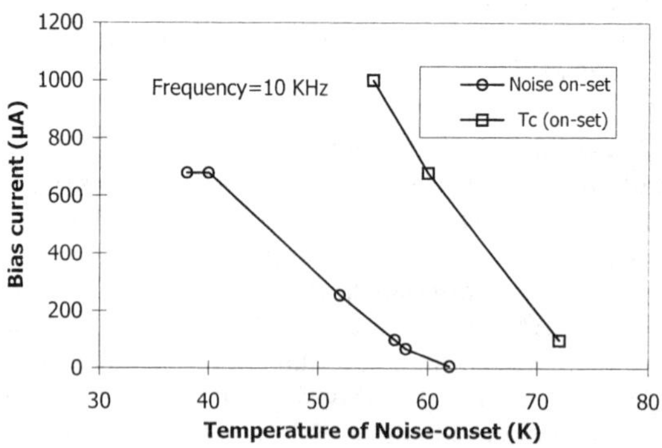

Figure 7.4. The $T_{c\text{-onset}}$ and on-set temperature of the excess noise vs. bias current of sample 64-02b.

current is as follows. Any two neighboring superconducting grains, with a normal region in-between or a gap, can be considered as a Josephson junction (SNS junction). Then there can be a supercurrent associated with the junction as a result of the tunneling mechanism between the two superconducting grains. A granular film or material can be considered as a many junction media. Near T_C these junctions can be considered in series and parallel combinations, which may not be shunted by any complete superconducting path. In such a configuration, there can be voltage fluctuations due to the fluctuating super current, even with zero bias current through the material. The onset of this noise can also be shifted toward lower temperatures by applying a bias current through the material, since only the normal regions between the grains that have high Josephson currents will carry super currents. There have also been reports on the effect of the patterning process, and the dimensions of the patterns of a design on the noise performance of the materials, which is in agreement with the above analysis. [1], [45], [91].

The effect of the bias current on the noise in the normal region above $T_{c\text{-onset}}$ of the material has also been extensively studied. This noise is reported to be higher than the Johnson noise, and is current dependent at low frequencies as reported by different groups [1], [11], [45], [91]. The dependence of the noise on the bias current suggests that the noise can be due to the resistance fluctuations in the material. This noise is linearly proportional to the bias current. This suggests that the noise in the normal state is caused by resistance or conductivity fluctuation in the film and also suggests that the resistance (or conductivity) fluctuation is not dependent on the bias current. As discussed previously, above the transition temperature, the Cooper pairs have a finite lifetime, and the conductivity is the sum of the conductivity of the normal electrons and the conductivity of Cooper pairs, which can fluctuate randomly. Hence, the conductivity fluctuations in this region can be interpreted to be due to the fluctuations in the density of the different type of the carriers [92].

There has been another noise characteristic, observed at temperatures right below $T_{c\text{-zero}}$, in sample No. 064-03b, which is less granular than sample 064-02b [24]. This kind of noise peaks right below the onset temperature of the excess noise. Figure 7.5 shows the noise versus temperature of this sample. As observed in the figure the sample has excess noise the same as sample 064-02b, which is associated with the long tail of the R versus T curve due to the granularity of the sample. As shown in Figure 7.5 the noise-peak below $T_{C\text{-zero}}$, can increase with an increase in the bias current, and approaches zero for zero bias current. For very large

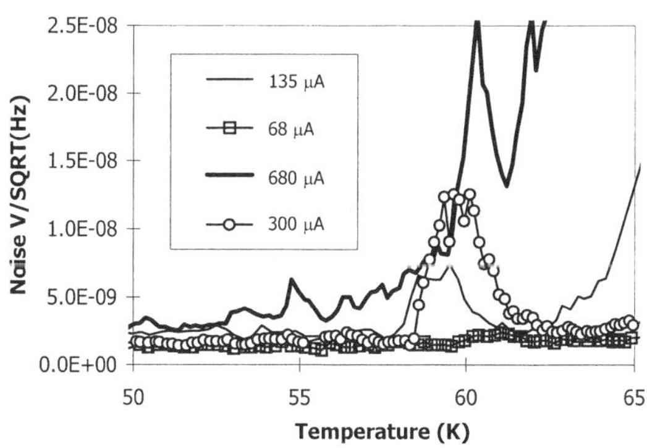

Figure 7.5. Noise voltage vs. temperature of sample 064-03b at 10 KHz.

bias currents the excess noise extends over the whole region creating other small peaks, at even lower temperatures. In the temperature region below $T_{c\text{-}zero}$, and low bias currents, a continuous superconducting path is expected to exist between the current and voltage contacts of the superconductor film in the four-probe configuration. Consideration of this, and the strong dependence of the noise on the bias current, suggests vortex motion as the source for noise in this region, which can be due to thermally activated two-level or diffusive vortex motion [87]-[88] that may also be enhanced by a strong Lorentz-force due to an applied current [89]-[90].

The noise versus temperature, for sample 064-01a, is shown in Figure 7.6. This sample has a relatively sharp transition and is less granular compared to the MgO substrate samples. Noise in this sample at high frequencies (10 kHz and 14 kHz) has shown a very small peak at $T_{c\text{-}zero}$, and follows the R versus T curve in the upper part of the transition toward the normal state. The noise behavior in the normal state of this sample is the same as observed in MgO substrate samples. The magnitude of the peak in the noise of sample 064-01a at $T_{c\text{-}zero}$ is very small, a few nanovolts, and its maximum for the maximum used bias current of 680 μA, is less than 10 nanovolts-$Hz^{-0.5}$. The noise in this sample is strongly dependent on the frequency, and increases rapidly for the frequencies below about 400 Hz, as shown in Figure 7.6. This type of noise is not observed in MgO substrate samples. As can be observed in the Figure 7.6 the peak of the noise increases as the frequency decreases, and it is also found to be current dependent. The frequency and current dependence of the above noise, suggests that conductance fluctuations are the source of the noise in the transition region of this sample. This type of fluctuation in the lower part of the transition is interpreted to be dominated by the conductivity fluctuations, due to the variations in the volume fraction of the superconducting phase, in the current path through the material [86].

So four major types of noise can be identified according to where they occur in temperature relative to the $T_{c\text{-}onset}$ and the zero resistance temperature, $T_{c\text{-}zero}$, and their dependence on the frequency and bias current. They are also associated to with the granularity of the superconducting films, which is partly related to the substrate material used. From the above discussion a specific cause for each type of noise is associated and the results can be classified as follows. (i) In the normal state with temperature $T > T_{c\text{-}onset}$, noise with a magnitude that is consistent with thermal (Johnson) noise exits, and depends linearly on bias

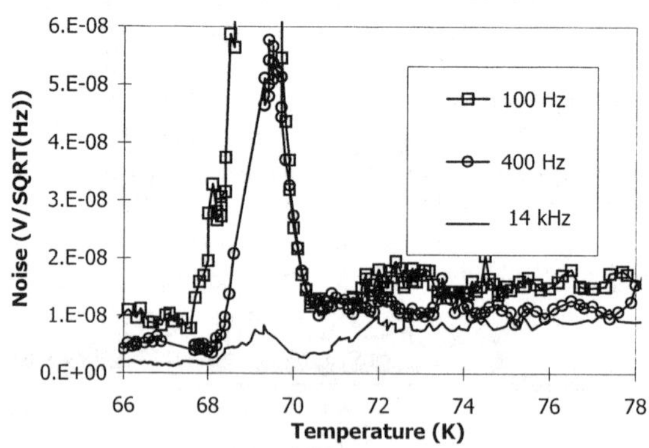

Figure 7.6. Noise voltage vs. temperature of sample 064-01a at 100 μA bias current.

current above a threshold value at low frequencies. The suggested noise source is conductivity fluctuations due to the Cooper pairs. (ii) A second type of noise exists above $T_{c\text{-}zero}$ in granular films. With increasing the bias current, its magnitude increases, and it shifts to lower temperatures; however the noise magnitude becomes constant as the current goes to zero while is weakly dependent on the frequency for above 400 Hz. This is suggested to be caused by voltage fluctuations in the superconductor-normal-superconductor junctions at grain boundary. (iii) One type of noise occurs below $T_{c\text{-}zero}$ with peaks at various temperatures. With increasing the bias current the peaked noise spreads to lower temperatures, but the noise goes to zero as the current goes to zero. This can be caused by the magnetic flux tube motions. (iv) A noise signal occurs between $T_{c\text{-}zero}$ and $T_{c\text{-}onset}$ and is present in al bolometers. Its suggested cause is fluctuations in the volume fraction of the superconducting phase along the current path. This type of noise is the limiting factor in the detectivity of the edge transition bolometers as it occurs in the range of interest for the bias temperature of the devices.

8. CONTROL OF RESPONSIVITY AND DETECTIVITY THROUGH SUBSTRATE PROPERTIES

The thermal resistances at the substrate-cold finger and the superconductor-substrate interfaces need to be used in the design considerations to improve the detectivity, D*. For an optimal design, both electrical and thermal properties of the detectors should be considered. The optimal design parameters are also strongly dependent on the application and the operating circumstances, mainly the modulation frequency. In this section we discuss the analysis for the response and D* of samples with large area patterns on crystalline MgO, $LaAlO_3$, and $SrTiO_3$, under dc bias current in the four probe configuration.

A. Effects of substrate properties and biasing on the responsivity and detectivity limits

The responsivity of a bolometer is defined as the ratio of the voltage response (in current biased detectors) to the radiation power received by the detector [7]-[24]. The equilibrium responsivity of thin film superconductive edge transition bolometers is mainly a function of the substrate properties, since the mass of the substrate material is usually much larger than the other masses in this type of the detectors. The thermal and structural characteristics of the substrate also affect the electrical properties of the superconducting film, which is considered later in this section as well [95]-[98]. The electrical biasing is also a major determining factor in the response, and it needs to be considered before further discussion of the effects of substrate properties on the response.

a) Responsivity and effects of biasing configurations

Self or Joule heating is a major factor in the bolometric response, since the bolometer operates within the resistive transition region of the superconductor near the maximum in dR/dT. The ac component is caused by the change in the operating point resistance due to the input radiation power, and produces positive feedback for current biased (dc) detectors in the four-probe measurement configuration. Taking the above effect into account in the equation for the responsivity of bolometers [7], the overall frequency dependent responsivity, given in equation (2.6) can also be obtained directly from [9];

$$r_V = \frac{hI_b}{G_t + j2\pi f C_t - I_b^2(dR/dT)} \frac{dR}{dT}, \qquad (8.1)$$

Where I_b is the dc bias current, η, is the fraction of the incident power absorbed by the bolometer (the absorption coefficient), dR/dT is the slope of the resistance versus temperature

(R vs. T) curve at the bias point G_t and C_t are the frequency dependent total thermal conductance and heat capacity of the bolometer, respectively, which are mainly governed by the substrate properties in most of the cases [7]. As shown in equation (8.1), there is a stability criterion in the voltage response, which has the same form as obtained for dc biasing at zero frequency [7].

For dc voltage biased detectors (constant voltage across the samples), the response is in the form of current variations, and the Joule heating in the film produces negative feedback [12]. This decreases the responsivity, changing the sign of the Joule heating term ($I_b^2(dR/dT)$) in equation (8.1). This negative feedback can improve the dynamic range of voltage-biased detectors by stabilizing the bias temperature at high radiation intensities. This is when the sample is biased such that the Joule heating is comparable to the total absorbed radiation power.

*b) Effects of substrate dependent film properties on the detectivity, D**

The detectivity, D*, of the bolometers is a function of responsivity, the absorbing area (and its absorption coefficient), and the intrinsic noise in the device. The D* of a edge transition superconductive bolometer versus temperature is a function of the voltage responsivity and the noise as defined in equation (7.1). Using equation (8.1), D* might be increased by an increase in the responsivity by using higher dc bias currents. This depends on the dependence of the ultimate responsivity and the noise voltage on the bias current, as discussed in the following.

When ac Joule heating is negligible, as is the case for the currents in our samples, the responsivity increases linearly with I_b as given in (8.1). Hence D* can be increased by increasing I_b if the signal to noise ratio also increases. This requires the noise increases with I_b at a rate less than that of the responsivity. Noise in $YBa_2Cu_3O_{7-x}$ samples has a dependence on the bias current and the modulation frequency, and also is strongly temperature dependent as discussed in the preceding section. The $T_{c\text{-onset}}$ is considered here to be the point above which the resistance starts to show linear behavior versus temperature, and $T_{c\text{-zero}}$ the point below which the resistance drops below 1% of its value at $T_{c\text{-onset}}$. While both T_c values are dependent on the bias current, $T_{c\text{-zero}}$ of the studied is decreased versus bias current depending on the granularity of the films [7]. The measured voltage noise versus temperature at different I_b values in the region of interest for one of the samples (064-02b) made of 120-130 nm thick patterned granular $YBa_2Cu_3O_{7-x}$ film on crystalline MgO substrate is shown in Figure 8.1.

The noise from about the middle of the transition to $T_{c\text{-onset}}$ is reported to be dependent on the bias current, I_b, scaling as [9];

$$V_n = C_1 + C_2 I_b \qquad (8.2)$$

Where C_1 and C_2 are constants and differ from one sample to another, with the $C_1 > C_2 I_b$ relation just below $T_{c\text{-onset}}$ [11]. The noise in the $LaAlO_3$ samples is reported to be typically higher in this region. The high level of noise in films on $LaAlO_3$ substrates can be due to the high granularity (twined grains) typical of this type of the substrate material. Considering the current dependence of the noise, substrate samples having sharper transitions (higher dR/dT), which lead to lower I_b values for the maximum responsivity, are more favorable in this respect. The highest dR/dT is also found to be at the lower end of the above noise temperature region [11]. Hence, since the responsivity is proportional to I_b, the D* of samples increases by increasing I_b if the sample is biased in the above temperature region. This increase is limited

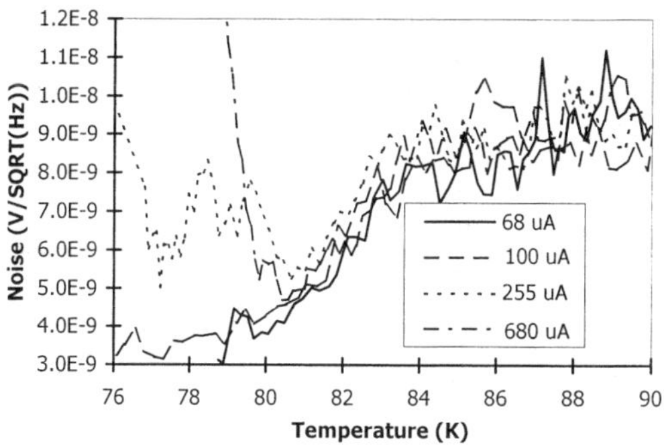

Figure 8.1. Noise vs. Temperature of sample MgO substrate 064-02b at 10 KHz for 68, 100, 255, and 680 μA bias currents.

by the thermal runaway due to dc Joule heating in the device, and the limits to the responsivity which are discussed later in this section.

c) Effects of substrate properties and bias current on responsivity limits

Due to the limited thermal conductance of the detectors, which is mainly determined by the substrate properties, dc Joule or self heating is a major limiting factor to the use of high bias currents to obtain higher responsivity values, and hence higher values of D*. Self-heating can cause an excessive temperature rise, ΔT_{dc}, in the film with respect to the temperature of the sample holder where the temperature sensor is normally placed. The resulting ΔT_{dc} may be mainly across the substrate material, or across the substrate-cold finger thermal boundary resistance. This is determined by the relative dimensions of the superconducting film pattern compared to the thickness of the substrate. The relative dimensions of the film and the substrate determine the dominating factor limiting the dc thermal conductance, G(0), as discussed in details in sections 4 and 5.

For patterns with dimensions much smaller than the substrate thickness, d_S, G(0) is limited by the lateral heat flow through the superconductive film and heat diffusion into the substrate material [5]-[6], [13], [14], [54]. This is highest for the MgO substrate samples since they have the highest thermal conductivity. In samples with pattern dimensions comparable to, or larger than their substrate thickness, G(0) is limited by the thermal resistance at the substrate boundaries, and is not strongly dependent on the substrate material [24],[16]. In the studied samples, ΔT_{dc} is mainly at the substrate-cold finger interface, since the dimensions of the meander line patterns are comparable to that of the substrate [7]-[24].

Regardless of the dominating factor determining the total thermal conductance, G_t, the dc bias current at any temperature higher than $T_{c\text{-zero}}$ is limited by the thermal runaway characterized by the factor α defined in equation (4.2). As given in equation (8.3), for substrates yielding films with sharper transitions, the ultimate allowed I_b would be lower. The effect of an instability in the temperature of the film due to high bias currents, will appear as an unrealistically sharp transition in the R versus T curve [23], [7], [42]. The increase of the response due to the sharper transition, dR/dT, dominates the effect of this decrease in the maximum allowed I_b in (4.2), giving an overall increase in the responsivity for films with sharper transitions. The maximum bias currents, $I_{b\text{-max}}$, of the characterized samples obtained

Table 8.1. Thermal and electrical parameters, and dimensions of the superconducting film and substrates of the samples. D*s are obtained for 100 Hz modulation frequency and 20 μm radiation wavelength. r_{max} is the maximum zero modulation frequency response, and $I_{b\text{-}max}$ is the maximum possible bias current limited by the thermal runaway obtained for an stability factor of α=0.4.

sample number	substrate material	C_s ($\frac{J}{k\text{-}cm^3}$)	K_s ($\frac{W}{k\text{-}cm}$)	d_s (cm)	G(0) ($\frac{mW}{K}$)	dR/dT ($\frac{\Omega}{K}$)	d_f nm)	A (cm^2)	$I_{b\text{-}max}$ (mA)	r_{max} ($\frac{V}{W}$)	D* ($\frac{cm\text{-}\sqrt{Hz}}{W}$)
064-01	$SrTiO_3$	0.43	0.052	0.05	3	1800	220	0.017	0.66	400 η	3.6*10^7
064-02b	MgO	0.53	3	0.025	15.5	950	120	0.075	2.08	127 η	0.3*10^6
064-03a	MgO	0.53	3	0.05	11	1000	170	0.075	1.66	151 η	-
064-04	$LaAlO_3$	0.59	0.16	0.05	7.8	2200	190	0.075	1.0	274 η	1.8*10^6

for α=0.4 are given in Table 8.1. Considering the maximum I_b determined by equation (8.3), the magnitude of maximum responsivity follows;

$$|r_v|_{max} = \frac{h}{G_t}\sqrt{\frac{\alpha G(0)}{1+(2\pi f t)^2}\frac{dR}{dT}}, \tag{8.4}$$

Where $\tau = C_t/G_t$ [7]. Hence from equation (8.4), devices on substrates with lower thermal conductivities and lower heat capacities, and allowing higher dR/dT in the transition region will yield the highest responsivities. This favors $SrTiO_3$ substrate devices compared to MgO and $LaAlO_3$ ones, for the same superconducting film and substrate dimensions. This is confirmed by the measurements in the modulation frequency range of the responsivity governed by equation (8.4). The r_{max} obtained for the studied samples, and their superconducting pattern dimensions, are given in Table 8.1.

As given in equation (8.4), $|r_f|_{max}$ can be increased by increasing G(0), scaling as $G(0)^{0.5}$. This is when the value of G_t at the operating frequency is different from that of G(0) which occurs at midrange or higher frequencies [7]. At very low frequencies, lower than a critical value, f_l, G_t might becomes equal to or comparable to G(0). The value of f_l depends on the substrate properties, and is the frequency at which the thermal diffusion length into the substrate approaches the substrate thickness. Then $|r_f|_{max}$ is decreased by increasing G(0), scaling as $G(0)^{-0.5}$. Variation of the determining factor for G_t affects the slope of the response-magnitude versus modulation frequency. For $f<f_l$ the response scales as f^{-1}, and for $f>f_l$ scales as $f^{-.5}$ [2], [7]. This is a function of substrate properties and is discussed in the following.

d) Substrate material dependent substrate-thickness effects on the responsivity limits

Considering the thermal resistance at the film-substrate interface to be negligible compared to that of the substrate material and the substrate cold-finger interface in the low and midrange frequency [7]-[24], the overall thermal conductance, G_t, is then determined by the thermal diffusion length of the absorbed energy (by the superconducting film) into the substrate, L_f, as defined by equation (2.8). Considering L_f as the effective length for the ac heat flow into the substrate, in the modulation frequency range where L_f is less than the substrate thickness (i.e. $f>f_l$), G_t will be determined by the substrate material, and increases with the frequency, scaling as $f^{0.5}$ [7]. Hence, for a detector being operated above f_l, the substrate thickness doesn't affect the responsivity, and $|r_f|_{max}$ is governed by equation (8.4), with the substrate thermal parameters the determining factors. This is valid for frequencies at

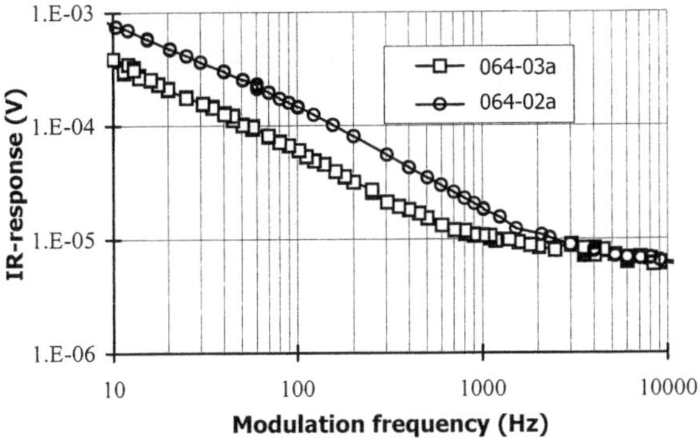

Figure 8.2. IR-response vs. modulation frequency for 0.05-cm thick MgO substrate sample 064-03a and 0.025-cm thick MgO substrate sample 064-02a at 1 mA bias current.

which the thermal conductance of the superconductive film and the film-substrate interface are negligible [7].

When the frequency decreases below the values where L_f becomes comparable to or larger than the substrate thickness (i.e. $f<f_l$), G_t approaches G(0) and the substrate-cold finger boundary resistance, $R_{s\text{-}c}=1/G_{s\text{-}c}$, will be the determining factor in the total thermal conductance. In the studied samples $R_{s\text{-}c}$ values are found to be much larger than the thermal resistance of the substrates, giving $G(0)\cong G_{s\text{-}c}$. For $(2\pi f\tau)^2>>1$, and still valid in the above frequency range, and $G_t\cong G(0)$ in equation (8.4), the maximum responsivity follows;

$$|r_v|_{max}=\frac{h}{2\pi fC_t}\sqrt{\alpha G(0)\frac{dR}{dT}}. \tag{8.5}$$

Hence, under the above conditions and within this frequency regime ($f<f_l$), thinning the substrate will increase r_v by decreasing C_t, as long as the bias current does not exceed the critical current value of the superconducting film, I_c, assuming G(0) constant. The thermal boundary resistance at the substrate-cold-finger interface, 1/G(0), also decreases by thinning the substrate regardless of the substrate materials. [24]. For large area patterns the decrease in G(0) is approximately proportional to the substrate thickness, d_s. Hence, by considering C_t to be also proportional to the substrate material's thickness, thinning the substrate is expected to increase $|r_v|_{max}$ by about $(d_{si}/d_{sf})^{3/2}$ where d_{si} and d_{sf} are the initial and final substrate thick nesses, respectively.

The effect of the modulation frequency on the response in different frequency regimes for two samples with identical meander line patterns on 0.025 cm and 0.05 cm thick crystalline MgO substrates are shown in Figure8.2. At high frequencies where L_f is less than the substrate thickness for both samples, the responsivities are equal and the response is determined by the characteristics of the superconducting pattern and the substrate material. For lower frequencies where L_f is larger than the thickness of the substrates, the response of the sample with thinner substrate (064-02a) is higher. This is mainly due to differences in C_t value, being approximately proportional to the thickness of the substrates since the bias current is same. The maximum responsivity of the 0.025 cm thick substrate sample (064-02a) is about 2.8 times than the 0.05 cm thick substrate sample (064-03a) due to also the differences in G(0) values as well as the C_t values.

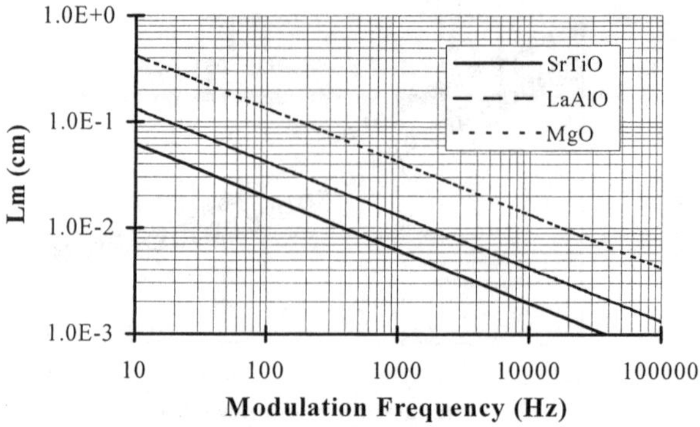

Figure 8.3. Calculated L_m, versus modulation frequency for bolometers on crystalline $SrTiO_3$, $LaAlO_3$, and MgO substrate materials.

If a detector is used in a bridge (micro-bolometer) configuration [99], G_t will be primarily determined by the lateral heat flow through the substrate. Since d_S is still much larger than the film thickness, the resulting heat flow through the film is negligible. In this configuration, thinning the substrate will also increase r_v, mainly by lowering C_t, [9], [13], [99]. Thinning the substrate in this case will cause a decrease in the G(0) of the detector by reducing the area of the lateral thermal conducting path through the substrate material. This gives an increase in $| r_v |_{max}$ lower than the expected value, and scaling as $(d_{si}/ d_{sf})^{0.5}$. The minimum required thickness of the substrate, L_m, versus modulation frequency beyond which the responsivity increases by further thinning is given for the three substrate materials in Figure 8.3.

B. Spectral responsivity and substrate material dependence

The spectral responsivity of sample 064-01a measured up to 20 μm, at two bias currents, is shown in Figure 8.4. The sample is made of a 220-230 nm YBCO superconducting film on a 0.05 cm thick, crystalline, $SrTiO_3$ substrate. The superconductive pattern is a 50 μm wide

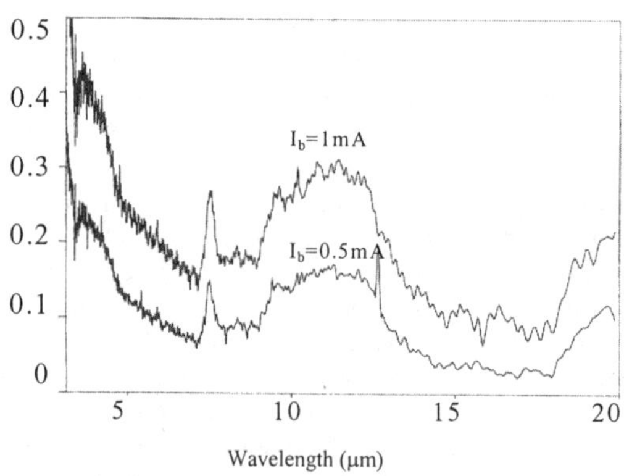

Figure 8.4. Spectral response of $SrTiO_3$ sample 064-01a at 0.5 mA and 1 mA bias current.

and 1.9 cm long meander line with a total area of 0.0168 cm^2 [11]. The magnitude of the response in Figure 8.4 is proportional to I_b, while the spectral response follows the spectral absorption of the substrate material as shown in Figure 8.5. This is due to the absorption of radiation in the open areas of the substrate between the meander lines, with the flow of heat to the superconductor. The measured spectral absorption of the $SrTiO_3$ substrate is slightly different from that obtained using the classical dispersion theory [41], [100]-[101] particularly showing a strong peak at about 7.5 μm radiation wavelength, as observed in both Figure 8.4 and Figure 8.5.

As shown in Figure 8.4, the response of the $SrTiO_3$ substrate sample also follows the trend for the absorption in thin film and single crystal $YBa_2Cu_3O_{7-x}$ material, increasing at shorter wavelengths [10]-[33]. Using the ratio of the change in the response due to the substrate absorption with respect to that of the film, the fraction of the absorbed power by the substrate can be obtained at any wavelength. As derived for sample 064-01a ($SrTiO_3$ substrate), about 50% of the response at 10 μm and over about 75% of the response at 20 μm is due to the absorption of the incident radiation onto the open areas of the substrate. The effect of the substrate absorption on the response dominates at some longer wavelengths (i.e. 20 μm) where the absorption by the superconducting film is even further reduced [10]-[33]. The absolute value of the responsivity for the above sample is reported at 7.39 μm with a 500 K blackbody source. A response of 1.582 μV is measured for a radiation intensity of 5.64×10^{-4} W/cm^2 at 400 Hz modulation frequency, and 300 μA dc bias current. A responsivity of 0.5 V/W and a D* of 0.72×10^6 cm-$Hz^{0.5}$/W are obtained under above conditions.

C. Engineering changes for higher D*

As an example of engineering changes required to increase D* using the analysis presented in this work, an analysis for the sample on $SrTiO_3$ substrate is discussed. From the spectral response in Figure 8.4 and by use of a maximum bias current of 0.66 mA from Table 8.1, a D* of about 3.6×10^7 cm-$Hz^{0.5}$/W is obtained for the $SrTiO_3$ substrate sample (064-01a), at 20 μm wavelength, and 100 Hz modulation frequency. The spectral detectivity, at a constant bias current and temperature, follows the spectral response of the detector for a constant radiation intensity. The D* of some samples at 20 μm radiation wavelength is given

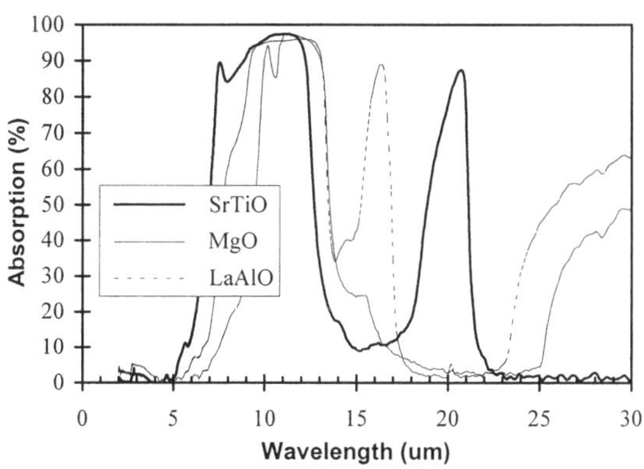

Figure 8.5. Measured spectral absorption of 0.05-cm thick crystalline $SrTiO_3$, MgO, and $LaAlO_3$ substrates.

in Table 8.1.

To increase D*, considering the 400 Hz and 100 Hz modulation frequency points in Figure 8.3 as an example, the substrate of sample 064-01a should be thinned to below 0.01 cm and 0.02 cm respectively. The substrate can be thinned using wet etching after the patterning process for the superconductive film. Thinning the substrate will also increase G(0), while it is still limited by R_{s-c}. Based on above analysis, by thinning the substrate down to 0.005 cm, with an increase of the total area of superconductive pattern up to 0.09 cm^2, a D* on the order of 1×10^9 cm-$Hz^{0.5}$/W at 20 μm for 200 nm thick $YBa_2Cu_3O_{7-x}$ superconducting films on crystalline $SrTiO_3$ substrates will be obtained. Also as observed from equation (8.2) and Table 8.1, the detectivity of samples can be enhanced by a sharper transition (favored by the substrate lattice structure), since it allows lower $I_{b\text{-}max}$ values, and hence lower current dependent noise in the region.

9. SUMMARY AND CONCLUSIONS

To explain the behaviors of edge transition bolometers, it is necessary to include in the modeling, the effects of thermal boundary resistances, temperature dependence of the thermal parameters of the devices, the frequency dependence of the thermal diffusion length into the materials, noise, and artifacts due to the bias configurations and ambient atmosphere. Such models are studied that explain the modulation frequency dependence in the low and midrange frequencies within the range of interest for contemplated applications of bolometers.

For the samples with large area superconducting film patterns compared to the substrate thickness, the thermal boundary resistance at the substrate-holder of the samples is considerably higher than that of the other sources as film-substrate boundary resistance, the substrate material resistance, and the bulk film thermal resistance. This is the case only at very low modulation frequencies. In such a configuration the response of the samples follow a basic two-lumped element R-C model. Then, G is mainly determined by the thermal boundary resistance at the substrate-holder interface, and C is mainly due to the heat capacity of the substrate materials. But this is up to frequencies where the thermal diffusion length, L, into the substrate material is more than the substrate thickness, L_s. Several methods for measuring the thermal conductance G, has been introduced.

The increase of the modulation frequency beyond the values where $L<L_S$, causes a change in the slope of the response magnitude versus modulation frequency compared to that of the basic R-C model. This happens at a "knee" frequency, f_L, which is dependent on the substrate material and its thickness. This leads to different regime of frequency response behavior. For frequencies less than f_L, the response scales as f^{1} following the response for a basic R-C model. For frequencies higher than f_L, the response scales as $f^{0.5}$due to the variation of the effective penetration depth of the temperature variation into the substrate. This is up to the frequencies where the thermal resistance of the superconducting film-substrate boundary, R_{fs}, is negligible compared to the frequency dependent effective thermal conductance of the substrate materials, R_s. In the regime of the $f^{0.5}$ dependence for the responsivity, thinning the substrate underneath the pattern to increase the responsivity is dependent on the thermal diffusion length.

Thinning the substrate of the samples to increase the maximum responsivity, $|r_v|_{max}$, depends on the variations of the total thermal conductance, G_t, and the total heat capacitance of the device, C_t. The critical required thinning of the substrates below which the responsivity increases is dependent on the substrate materials and the modulation frequency, f, scaling as f

$^{0.5}$. In low enough frequency regime, where the thermal diffusion length into the substrate is higher than the substrate thickness, $L_f > d_s$, thinning substrate of the detector increases $| r_v |_{max}$ by a factor of $(d_{si}/ d_{sf})^{3/2}$, where d_{si} and d_{sf} are the initial and final substrate thicknesses, respectively. In the higher frequency regime where $L_f < d_s$, thinning the substrate of the detector increases $| r_v |_{max}$ by about $(d_{si}/ d_{sf})^{0.5}$.

The film-substrate thermal boundary resistance, R_{bd}, is not effective on the IR-response in the thin film samples with in the studied modulation frequencies here. Considering a film thickness of 200 nm, the R_{bd} would be effective above about 100 kHz, 1 MHz, and 10 MHz, in MgO, $LaAlO_3$, and $SrTiO_3$, respectively which is mainly determined by the thermal parameters of the substrate material. The R_{bd} is effective at much lower frequency range of KHz for the screen-printed thick film samples, which have values of thickness typically in the range of 10s of micrometers. To improve the responsivity of the bolometer in the various regimes of frequency, for a fixed transition region width, thinner films with narrow line width and a long meander line are favored (to increase dR/dT). For high absorption in the film, a minimum thickness of about 150 nm is also needed [I_7]. The response of the thick film Y-Ba-Cu-O bolometers mostly exhibits a bolometric response at their transition temperature. This is while small response, relative to the bolometric response at the transition of the films, is reported at temperatures right below the $T_{c(zero)}$ under certain conditions. This response is attributed to the noise signal due to the Kosterlitz-Thouless effect, which is strongly dependent on the bias current. The thick film samples are typically made of post-annealed screen-printed films and hence are highly granular.

A one-dimensional model based on the above facts, which is for bolometers with superconducting film-patterns larger or comparable to the substrate thickness, is described. The model while being a useful tool for design optimizations for such devices could also be used to monitor the thermal parameters of the detectors and the used materials. A very good fit between the both measured magnitude and phase of the response to that obtained from the model is obtained for the samples from very low modulation frequencies of 0.5 Hz up to about 100 kHz, the studied range in this chapter. The phase of response versus modulation frequency of the bolometers is a much more sensitive characteristic to the thermal parameters compared to the magnitude of the response. Using the model and from the low and midrange frequency of the response, the heat capacity and thermal conductivity of the used substrate materials and the substrate-holder thermal boundary resistance can also be obtained.

Simultaneous measurements of the IR-response and dR/dT versus temperature reveal a discrepancy between the magnitude of the response and the measured dR/dT of the bolometers, which is frequency dependent. This is verified to be apart from the artifacts caused by the Joule heating due to the bias current and the microstructure of the films. Experimental data on the phase of the response variation suggest a change of the thermal conductivity of the substrate material as a source for the reported discrepancy. This is through the radiation limit in phonon propagation. Due to enhancement of the mean free path of phonons emitted from the superconducting film, the limit thermal conductance at $T < T_{c\text{-}onset}$ is increased. The influence of the superconducting state on the thermal parameters of the bolometers is also conducted. Different modulation frequency regimes is analyzed and good agreement between the theoretical calculations with experimental data is obtained in particular for low modulation frequencies at temperatures closer to $T_{c\text{-}zero}$.

The noise characteristics of one sample can't be fully predicted, and may differ strongly from one sample to another. Four different noise characteristics are identified at the transition temperature of the superconductor. Since a superconducting bolometer is operated at the maximum dR/dT, the noise characteristics in the transition region are of main interest in the design of devices. Two types of the four identified noise signals are identified to be in the transition region of interest. One is excess noise over a region at $T_{c\text{-zero}}$, which has the characteristics of a white noise. It peaks at $T_{c\text{-zero}}$ and widens with an increase in the bias current, extending to lower temperatures. This noise is attributed to the weak links and the corresponding Josephson junctions (SNS) at the grain boundaries. The onset temperature of this type of noise versus bias current has a behavior similar to the critical current density, J_C, versus temperature. This is observed in samples, which are highly granular and have a wide transition.

The second type of noise peaks at the middle of the transition and is strongly dependent on the bias current and frequency. Hence, this is attributed to conductivity fluctuations in the region of the superconducting transition. This type of fluctuations is interpreted to be dominated by the conductivity fluctuations, due to the variations in the volume fraction of the superconducting phase, in the current path through the material.

The substrate absorption is also a determining factor in the spectral responsivity, and hence in the detectivity of the detectors. This is in particular when the open areas of the substrate are comparable or larger than the absorbing surface area of the superconducting film. For an YBCO edge-transition bolometer operating at wavelengths from the near infrared to the far infrared, the spectral absorption of the substrate material is a major factor since the absorption of the superconducting films is strongly reduced compared to its values at shorter radiation wavelengths. This property of the substrates can be used for selective wavelength sensitivity for an edge-transition superconductor infrared detector.

In conclusion, the studied models and the analysis in this chapter can be used to monitor the thermal parameters of the superconducting film, substrate, and the thermal boundary resistance values of an edge-transition superconductive bolometer. Also the mechanism of heat flow caused by the absorbed radiation in the samples can be investigated by use of the results from the analysis in this kind of devices. The response versus modulation frequency of the bolometers is the most informative data in this respect. The response of the bolometers with high quality large superconducting pattern areas compared to the substrate thickness is concluded to be mainly bolometric at low and midrange modulation frequencies. Consideration of a three dimensional solution to the model would be required for further accurate response analysis of samples with small superconducting pattern areas compared to the substrate thickness.

REFERENCES

[1] P. L. Richards, J. Clark, R. Leoni, Ph. Lerch, S. Verghese, M. R. Beasley, T. H. Geballe, R. H. Hammond, P. Rosenthal, and S. R. Spielman, "Feasibility of the high T_C superconducting bolometer", *Appl. Phys. Lett.*, vol. 54, no. 3, pp. 283-285, 1989.

[2] Qing Hu and P. L. Richards, "Design analysis of a high T_C superconducting microbolometer", *Appl. Phys. Lett.*, vol. 55, no. 23, pp. 2444-2446, 1989.

[3] J. Schutzmann, W. Ose, J. Keller, K. F. Renk, B. Roas, L. Schultz, and G. Saemann-Ischenko, "Far-infrared reflectivity and dynamical conductivity of an epitaxial $YBa_2Cu_3O_{7-\delta}$ thin film", *Europhys. Lett.*, vol. 8, pp. 679-684, 1989.

[4] M. Fardmanesh, M. Ihsan, A. Rothwarf, K. Scoles, and K. Pourrezaei, " Thick and thin film Y-Ba-Cu-O infrared detectors", *American Institute of Physics Proceeding*, no. 251, pp. 681-691, 1991.

[5] M. Nahum, Qing Hu, P. L. Richards, S. A. Sachtjen, N. Newman and B. F. Cole, " Fabrication and Measurement of High T_c Superconducting Bolometers", *IEEE Trans. on Mag.*, vol. 27, pp. 3081-3084, 1991.

[6] U. P. Oppenheim, M. Katz, G. Koren, E. Polturak and M. R. Fishman, "High temperature superconducting bolometer", *Physica C*, vol. 178, pp. 26-28, 1991.

[7] M. Fardmanesh, A. Rothwarf, and K. J. Scoles, "Low and Midrange Modulation Frequency Response for YBCO Infrared Detectors: Interface Effects on the Amplitude and Phase", IEEE Trans. on Appl. Supercond., Vol. 5, no. 1, pp. 7-13, March 1995.

[8] S. Zeuner, W. Prettl, and H. Lengfellner, " Fast thermoelectric response of normal state $YBa_2Cu_3O_{7-d}$ films", Appl. Phys. Lett., Vol. 66, no. 14, pp. 1833-1835, 1995.

[9] M. Fardmanesh and A. Rothwarf, "Control of superconductive edge transition $YBa_2Cu_3O_{7-x}$ bolometers responsivity and detectivity through substrate properties," *Applied Optics,* Vol. 38, No. 19, 1999.

[10] Z. M. Zhang, B. I. Choi, T. A. Le, M. I. Flik, M. P. Siegal and J. M. Phillips, "Infrared refractive index of thin $YBa_2Cu_3O_7$ superconducting films", *J. Heat Transfer*, vol. 114, pp. 644-652, 1992.

[11] M. Fardmanesh, A. Rothwarf, and K. J. Scoles, "Noise characteristics and detectivity of $YBa_2Cu_3O_7$ superconducting bolometers: Bias current, frequency, and temperature dependence," J. Appl. Phys., Vol. 79, no. 4, pp. 2006-2011, 1996.

[12] K. D. Irwin, G. C. Hilton, D. A. Wollman, and John M. Martinis, "X-ray detection using a superconducting transition-edge sensor microcalorimeter with electrothermal feedback, " *Appl. Phys. Lett.* Vol 69, no. 13, pp. 1945-1947, 1996.

[13] G. Darcy Poulin, Jean Lachapelle, Steven H. Moffat, Frank A. Hegmann, and John S. Preston, "Current-Voltage characteritics of dc voltage biased high temperature superconducting microbridges," Appl. Phys. Lett., Vol. 66, no. 19, pp. 2576-2578, 1995.

[14] A. V. Sergeev, A. D. Semenov, P. Kouminov, V. Trifonov, I. G. Goghize, B. S. Karasik, G. N. Gol'tsman, and E. M. Gershenzon, "Transparency of a $YBa_2Cu_3O_7$-film/substrate interface for thermal phonons measured by means of voltage response to radiation," Phys. Rev. B, Vol. 49, no. 13, pp. 9091-9096, 1994.

[15] C. D. Marshal, I. M. Fishman, R. C. Dorfman, C. B. Eom, and M. D. Fayer, "Thermal diffusion, interfacial thermal barrier, and ultrasonic propagation in $YBa_2Cu_3O_{7-x}$ thin films: Surface-selective transient-grating experiments", *Physical Review B*, Vol. 45, no. 17, pp. 10 009-10 021, 1992.

[16] P. E. Phelan, "Thermal response of thin-film high-T_c superconductors to modulated irradiation," J. of Therm. Phys. Heat Transfer, Vol. 9, no. 3, pp 397-402, 1995.

[17] M. Nahum, S. Verghese, and P. L. Richards, "Thermal boundary resistance for $YBa_2Cu_3O_{7-x}$ films", *Appl. Phys. Lett.*, Vol. 59, no. 16, pp. 2034-2036, 1991.

[18] M. Fardmanesh, "Analytic Thermal Modeling for dc to Midrange Modulation Frequency Response of Thin Film High-Tc Superconductive Edge-Transition Bolometers," *Applied Optics; Optical Technology & Biomedical Optics*, Vol. 40, No. 7, pp. 1080-1088, 2001.

[19] S. Bhattacharya, X. X. Xi, M. Rajeswari, C. Kwon, S. N. Mao, Qi Li, and T. Venkatesan, "Optical response of an ultrathin film and a large-angle grain-boundary in superconducting $YBa_2Cu_3O_{7-x}$," J. Appl. Phys., Vol. 62, no. 26, pp. 3510-3512, 1995.

[20] Kin Li, and Joseph E. Johnson, "Influence of substrate and biasing current on response of YBCO microbolometers," IEEE Trans. on Appl. Supercond., Vol. 3, no. 1, pp. 2111-2114, 1993.

[21] F. A. Hegmann, and J. S. Preston, "Identification of nonbolometric photoresponse in $YBa_2Cu_3O_{7-x}$ thin films based on magnetic field dependencd," Appl. Phys. Lett., Vol. 62, no. 10, pp. 1158-1160, 1993.

[22] J. H. Hao, F. Q. Zhou, H. D. Sun, X. J. Yi, and Z. G. Li, "Responsivity calculation and measurement of YBaCuO optical detector," *IEEE Trans. on Appl. Supercond.*, Vol. 3, no. 1, pp. 2167-2169, 1993.

[23] Hsiung Chou, ,H. Z. Chen, M. T. Hong, Y. C. Chen, and T. C. Chow, "Bolometric detection in a precipitation free $YBa_2Cu_3O_{7-x}$ film at 77 K," *Appl. Phys. Lett.*, Vol. 68, no. 19, pp. 2741-2743, 1996.

[24] M. Fardmanesh, A. Rothwarf, and K. J. Scoles, "$YBa_2Cu_3O_{7-x}$ infrared bolometers: temperature dependent responsivity and deviations from the dR/dT curve.", *J. Appl. Phys.*, Vol. 77, no. 9, pp. 4568-4575, 1995.

[25] M. Fardmanesh, K. J. Scoles, and A. Rothwarf, "DC charateristics of patterned $YBa_2Cu_3O_{7-x}$ superconducting thin film bolometers: Artifacts related to Joule heating, ambient pressure, and microstructure," *IEEE Trans. on Appl. Supercond.,* Vol. 8, no. 2, pp. 69-78, 1998.

[26] M. Fardmanesh, M. Ihsan, K. Scoles, and A. Rothwarf, "Thin film YBCO infrared detector design and characterization", *American Institute of Physics Proceeding*, no. 273, pp. 142-154, 1992.

[27] M. Ihsan, M. Fardmanesh, A. Rothwarf, K. Pourrezaei, and E. Lenzing, "Design and performance of microstrip resonators using YBCO thin films made by in-situ off-axis sputtering", *American Institute of Physics Proceeding*, no. 273, pp. 122-131, 1992.

[28] J. Schubert, M. Siegert, M. Fardmanesh, W. Zander, M. Prompers, Ch. Buvhal, Judit Lisoni, C. H. Lei, "Superconducting and Electro-optical Thin Films Prepared By Pulsed Laser Deposition Technique," *Appl. Surface Science*, Vol. 168, pp. 208-214, 2000.

[29] M. Fardmanesh, K. Scoles, A. Rothwarf, E. Sanchez, E. Brosha, P. Davies, "ANNEALING STUDY OF YBa2Cu3Ox AND YBa2Cu4Ox SCREEN PRINTED THICK FILMS", Amer. Inst. Phys. Proc., No. 251, pp. 283-291, 1991.

[30] Flik, M. I., Phelan, P.E., and Tien C. L., "Thermal model for the bolometric response of high T_c superconducting films to optical pulses", *Cryogenics*, vol. 30, pp. 1118-1128, 1990.

[31] Neff, H., "Modeling and optimization of high-T_c superconducting bolometers", *J. Appl. Phys.*, vol. 69, no. 12, pp. 8375-8379, 1991.

[32] D. Robbes, P. Langlois, C. Dolabjian, D. Bloyet, J. F. Hamet, and H. Murray, "YBCO microbolometer operating below T_c: A modelization based on critical-temperature dependence", *IEEE Trans. on Applied Superconductivity*, vol. 3, no. 1, pp. 2120-2123, 1993.

[33] D. Van der Marel, H.-U. Habermeier, D. Heitmann, W. Konig, and A. Wittlin, " Infrared study of the superconducting phase transition in $YBa_2Cu_3O_{7-x}$", *Physica C*, vol. 176, pp. 1-18, 1991.

[34] S. L. Bud'ko, H. H. Feng, M. F. Davis, J. C. Wolfe, and P. H. Hor, "Persistent photoconductivity in insulating and superconducting YBaCuO thin films: temperature and spectral dependence", *Physical Review B*, vol. 48, pp. 16707-16713, 1993.

[35] K. Fushinobu, P. E. Phelan, K. Hijikata, T. Nagasaki, and M. I. Flik, "Thermal analysis of the performance of a high-T_C superconducting bolometer", J. Heat Transfer, Vol. 116, pp. 275-279, 1994.

[36] A. Frenkel, "Mechanism of nonequilibrium optical response of high-temperature superconductors", Physical Review B, Vol. 48, no.13, pp. 9717-9725, 1993.

[37] M. Fardmanesh, A. Rothwarf, and Kevin J. Scoles, "The responsivity and detectivity limits for patterned $YBa_2Cu_3O_{7-X}$ superconductive IR-detectors," *6th Int. Superconductive Electronics Conference proc.,* Vol. 3, pp. 399-401, 1997.

[38] Agha Jahanzeb, Christine M. Travers, Zeynep Celik-Butler, and Donald P. Butler, "A semiconductor YBaCuO microbolometer for room temperature IR imaging," IEEE Trans. on Electron Devices, Vol. 44, no. 10, pp. 1795-1801, 1997.

[39] Christine M. Travers, Agha Jahanzeb, Donald P. Butler, and Zeynep Celik-Butler, "Fabrication of semiconducting YBaCuO surface-micromachined bolometer arrays," J. Microrelectro-mechanical Systems, Vol. 6, no. 3, pp. 271-276, 1997.

[40] Z. M. Zhang and A. Frenkel, "Thermal and nonequilibrium responses of superconductors for radiation detectors," J. Superconductivity, Vol. 7, no. 6, pp. 871-883, 1994.

[41] K. kamaras, K. L. Barth, F. Keilmann, R. Henn, M. Reedyk, C. Thomsen, M. Cardona, J. Kircher, P. L. Richards, and J. L. Stehle, "The low-temperature infrared optical functions of SrTiO determined by reflectance spectroscopy and spectroscopic ellipsometry," *J. Appl. Phys.*, Vol. 78, no. 2, pp. 1235-1240, 1995.

[42] D. Janik, D. May, H. Wolf, and R. Schneider, "High-T_c transition-edge bolometer for detecting guided millimeter waves," IEEE Trans. on Appl. Supercond., Vol. 3, no. 1, pp. 2148-2151, 1993.

[43] M. I. Flik, and C. L. Tien, J. Appl. Phys., Vol. 67, no. 1, pp. 362-370, 1990.

[44] P. L. Richards, S. Verghese, T. H. Geballe, and S. R. Spielman, "The High-T_c superconducting bolometer," IEEE Trans. on Mag., Vol. 25, 2, pp 1335-1338, 1989.

[45] E. Zeldov, N. M. Amer, G. Koren, and A. Gupta, "Nonbolometric optical response of epitaxial films," Phys. Rev. B, Vol. 39, no. 13, pp. 9712-9714, May 1989.

[46] Q. P. Li, "Theory of effect of uniaxial-stress on Tc in ,YBa2Cu3O7" Physica C, Vol. 209, no. 4, pp. 513-515, 1993.

[47] Q. Xiong, Y. Y. Xue, Y. Y. Sun, P. H. Hor, C. W. Chu, M. F. Davis, J. C. Wolfe, S. C. Deshmukh and D. J. Economou, "Effect of pressure on the critical current-density of YBa2Cu3O7 thin-films," Physica C, Vol. 205, no. 3, pp. 307-310, 1993.

[48] B. H. Moeckly, D. K. Lathrop, and R. A. Buhrman, "Electromigration study of oxygen disorder and grain-boundary effects in $YBa_2Cu_3O_7$ thin films," Physical Review B, Vol. 47, no. 1, pp. 400-417, 1993.

[49] A. V. Sergeev, and M. Y. Reitzer, "Photoresponse mechanisms of thin superconducting films and superconducting detectors," Int. J. of Modern Phys. B, Vol. 10, no.6, pp. 635-667, 1996.

[50] A. V. Gurevich and R. G. Mints, "Self-heating in normal metals and superconductors," Rev. Mod. Phys., Vol. 59, no. 4, pp. 941-999, 1987.

[51] A. Sh. Fix, I. L. Maximov, K. V. Morozov, and V. V. Opsipov, IEEE Trans. on Appl. Supercond., Vol. 3, p. 1608, 1993.

[52] W. M. Huber, B. Arendt, P. G. Huggard, and W. Prettl, "Square-law Josephson detection of far infrared radiation with current-biased granular high-Tc thin films," Supercond. Sci. Tech., Vol. 8, no. 10, pp. 769-773, 1995.

[53] Sung Joon Bae, Seungbum Hong, Eunah Kim, Woosuck Shin, Chang Hyun Cho, Young Kun Kim, Yong Soo Bea, Tae-Hyun Sung, and Kwangsoo No, "Effect of texturing and microbridge length on the responsivity of YBa2Cu3Ox thin film," Jpn. J. Appl. Phys., Vol. 35, no. 3, pp. 1716-1719, 1996.

[54] Q. Li, D. B. Fenner, W. D. Hamblen, and D. G. Hamblen, Appl. Phys. Lett., 62, p. 2428, 1993.

[55] J. C. Braunas and B. Lakew, Appl. Phys. Lett., 64, 777 (1994).

[56] B. J. Berkwitz, A. S. Hirahara, K. Char and E. N. Grossman, Appl. Phys. Lett., 69, p. 2125, 1996.

[57] N. Bluzer, J. Appl. Phys., 78, p. 7340, 1995.

[58] H. Kraus, Supercond. Sci. and Technology, 9, p. 827, 1996.

[59] S. F. Karmanenko, A. A. Semenov, Ia Khrebtov, Vn Leonov, TH Johansen, Yu M. Galperin, A. V. Bobyl, A. I. Dedoboretz, M. E. Gaevski, A.V. Lunev, R. A. Suris, Supercond. Sci. and Technology, 13, p. 273, 2000.

[60] T.-L. Hwang, S. E. Schwarz, and D. B. Rutledge, Appl. Phys. Lett., 34, p. 773, 1979.

[61] T. H. K. Barron, W. T. Berg, and J. A. Morrison, Royal Society Proc., A 250, p70, 1959.

[62] E. T. Swartz and R. O. Pohl, Rev. Mod. Phys., 61, p. 605, 1989.

[63] N. Bluzer, Phys. Rev. B, 44, p. 10222, 1991.

[64] J.-J. Chang and Scalapino, Phys. Rev. B, Vol. 15, no. 5, 1977.

[65] M. Tinkham, Introduction to superconductivity (Kreger, Malabar, 1985).

[66] D. Pines , Physica B, 199, p. 300, 1994.

[67] R.A.Richardson, S.D.Peacor, C.Uher, F.Nori J. Appl. Phys., 72, p. 4788, 1992.

[68] M. Fardmanesh, I. N. Askerzade, “Temperature dependence of phase of the response of YBCO Edge Transition Bolometers: effects of superconductivity transition and thermal Parameters,” Superconductor Sci. Technol., Vol. 16, No. 1, pp. 28-32, 2003.

[69] M. Tinkham, and C. J. Lobb, Solid State Phys. 42, p. 91, 1988.

[70] B. W. Ricketts, R. Driver and H. K. Welsh, Solid State Commun. 67, p. 133, 1988.

[71] L. B. Kiss, P. Svedlindh, L. Lundgren, J. Hudner, H. Ohlsen, L. Stolt, and Z. Gingl, Physica B 172, pp. 441-444, 1991.

[72] A. Frenkel, M. A. Saifi, T. Venkatesan, P. England, X. D. Wu, and A. Inam, J. Appl. Phys. 67, p. 3054, 1990.

[73] P. Dutta, and P. M. Horn, Rev. Mod. Phys. 53, p. 497, 1981; M. B. Weissman, Rev. Mod. Phys. 60, p. 537, 1988.

[74] G. Jung, S. Vitale, J. Konopka, and M. Bonaldi, J. Appl. Phys. 70, p. 5440, 1991.

[75] K. H. Han, M. K. Joo, H. J. Shin, Sung-Ik Lee, and Sung-Ho Suck Salk, IEEE Transaction on Applied Superconductivity, 3, 1, pp. 2947-2949, 1993.

[76] L. B. Kiss, T. Larsson, P. Svedlindh, H. Ohlsen, M. Ottosson, J. Hudner, and L. Stolt, Physica C, 207, pp. 318-332, 1993.

[77] R. D. Black, L. G. Turner, A. Mogro-Campero, T. C. McGee, and A. L. Robinson, Appl. Phys. Lett. 55, 21, pp. 2233-2235, 1989.

[78] J. Hall, H. Hickman, and T. M. Chen, Solid State Communications, 76, 7, pp. 921-922, 1990.

[79] R. W. Zindler, and R. R. Parsons, J. Vac. Sci. Technol. A 10, 5, pp. 3322-3324, 1992.

[80] L. B. Kiss, P. Svedlindh, L. Lundgren, J. Hudner, H. Ohlsen, L. Stolt, Solid State Commun. 75, p. 747, 1990.

[81] G. Forgacs, L. S. Schulman, L. B. Kiss, P. Svedlindh, and L. Lundgren, Physica C 177, p. 67, 1991.

[82] M. Celasco, M. Lo Bue, A. Masoero, P. Mazzetti, and A. Stepanescu, IEEE trans. Appl. Supercond., AS-31, p. 111, 1993.

[83] A. Maeda, Y. Nakayama, S. Takebayashi, K. Uchinokura, Phys. C, 160, p. 443, 1989.

[84] A. Maeda, H. Watanabe, I. Tsukada, and K. Uchinokura, in *Noise in Physical Systems and 1/f Fluctuations*. Edited by T. Musha, S. Sato, and M. Yamamoto, p. 115, Ohmsha, Tokyo, 1991.

[85] A. E. Khalil, Physica C, 196, p. 357, 1992.

[86] Yi Song, A. Misra, P. P. Crooker, and J. R. Gains, Phys. Rev. B 45, p. 7574, 1992.

[87] M. J. Ferrari et al., Phys. Rev. Lett. 64, p. 72, 1990.

[88] C. T. Rogers, K. E. Myers, J. N. Eckstein, I. Bozovic, Phys. Rev. Lett. 69, p. 160, 1992.

[89] H. J. Jensen, A. Brass, and A. J. Berlinsky, Phys. Rev. Lett. 60, p. 56, 1988.

[90] H. J. Jensen, y. Brechet, and A. Brass, J. Low Temp. Phys. 74, p. 293, 1989.

[91] E. R. Nowak, N. E. Israeloff, and A. M. Goldman, Submitted to physical Review Letter, 1993.

[92] M. Tinkham, Introduction to Superconductivity. McGraw Hill, New York, 1975.

[93] Z. M. Zhang, B. I. Choi, T. A. Le, M. I. Flik, M. P. Siegal and J. M. Phillips, 1991 ASME Winter Annual Meeting, Atlanta, GA, December 1-6, 1991.

[94] K. E. Goodson and M. I. Flik, "Electron and Phonon Thermal Conduction in Epitaxial High-Tc Superconducting Films," National Heat Transfer Conference, Minneapolis, Minnesota, July 1991.

[95] M. I. Flik, Z. M. Zhang, K. E. Goodson, M. P. Siegal, and J. M. Phillips, Phys. Rev. B, 46, pp. 5606-5614, 1992.

[96] R. T. Collins, Z. Schlesinger, F. Holtzberg, and C. Feild, Phys. Rev. Lett., 63, pp. 422-425, 1989.

[97] S. L. Cooper, G. A. Thomas, J. Orenstien, D. H. Rapkine, M. Capizzi, T. Timusk, A. J. Millis, L. F. Schneemeyer, and J. V. Waszczak, Phys. Rev. B, 40, pp. 11358-11361, 1989.

[98] Julia M. Phillips, "Substrate selection for high-temperature superconducting thin films," *J. Appl. Phys.*, Vol. 79, no. 4, pp.1829-1848, 1996.

[99] T. G. Stratton, B. E. Cole, P. W. Kruse, R. A. Wood, K. Beauchamp, T. F. Wang, B. Johnson, and A. M. Goldman, "High-temperature superconducting bolometer," *Appl. Phys. lett.*, Vol. 57, no. 1, pp. 99-100, 1990.

[100] A. S. Barker, Jr., and M. Tinkham, "Far-infrared ferroelectric vibration mode in $SrTiO_3$" *Physical Review.*, Vol. 125, no. 5, pp. 1527-1530, 1962.

[101] J. C. Galzerani R. S. Katiyar, "The infrared reflectivity in SrTiO and the antidistortive transition," *Solid State Communications.*, Vol. 41, no. 7, pp. 515-519, 1989.

HIGH TEMPERATURE SUPERCONDUCTING CABLES

Ole TONNESEN* and Jacob OSTERGAARD**

*Oersted-DTU, Section Electric Power Engineering Building 325,
Technical University of Denmark, DK-2800 Kgs. Lyngby, DENMARK

**DEFU, Rosenorns Alle 9, DK-1970 Frederiksberg C, DENMARK

1.INTRODUCTION

The research and development of superconducting cables started in the late 1960s when the low temperature superconducting alloys became available [1]. The motivation for this development in the 1970s and 1980s was the increasing need for transmission systems with high rated power. Later on the increasing environmental concern about the overhead lines especially in urban areas made it desirable to find solutions where cables could transfer large amounts of power.

Conventional cables have limited power ratings due to the conductor and dielectric losses. The range of transmission capacities in naturally cooled 110 kV cables are up to about 200 MVA, at 400 kV it can reach up to about 800 MVA. For higher power ratings the use of water cooling can increase the levels up to 500 MVA at 110 kV and 1500 MVA at 400 kV, cf. [2].

System capacity targets in the 1970s were set at very high levels in anticipation of continued growth of transmitted power. This required new cable designs. A cable with significantly smaller ohmic and dielectric losses was needed.

There were principally two concerted projects carried out on the development of low-temperature metallic superconducting cables, which lead to practical demonstration projects. One was done at the Brookhaven National Laboratory [3], and the other at Arnstein in Austria [4].

The main disadvantages emerged as the very high costs of the low temperature superconductors (LTS) conductor, the thermal insulation system and the cooling system. As a result of this the costs of the LTS cable system could only be justified for cable systems with extremely high transmission capacities (> 4000 – 5000 MVA) which is considered way in advance of reality.

Therefore the low temperature superconducting cable technology development was abandoned with the discovery of the high temperature superconductors with critical temperatures above 77 K.

After 1986 the application of high temperature superconductors (HTS) in power transmission cables returned as one of the main research and development areas around the world.
The development of the HTS cable was done along with the development of the HTS conductor. Small samples of Bi-2223/Ag HTS wire were first developed in 1989 followed by a 10 m long Bi-2223/Ag wire in 1990. In 1993 100-meter length of Bi-2223/Ag wire was manufactured and in 1994 a 1 kilometre Bi-2223/Ag wire that carried over 20 kA/cm^2 was manufactured.

During the 1990s the quality of the Bi2223/Ag HTS wire has been under a contineous development and today for example engineering current densities of about 16 kA/cm^2 at 77 K and self field for 200 meter long commercially available multifilament tapes have been reported. This correspond to a critical current of about 135 A per tape. This value is projected to rise to above 150 A by year 2004 [5].

Also the costs of the Bi-2223/Ag conductor are changing with time. Way back the cost of a long length Bi2223/Ag tape was about 1000 USD/kAm, today (October 2002) it is available at 200 USD/kAm, and the potential lower limit is about 10-20 USD/kAm [5]. For comparison, the cost for copper is 9.56 USD/kAm and for aluminium 1.60 USD/kAm based on current densities used in conventional transmission cables.

In the mid 1990s the so called coated conductor emerged. This material is a tape shaped wire that show promise as a next generation wire technology. These tapes are made by depositing thin films of HTS material onto flexible substrates. Different processes are under development.

Engineering critical current densities of 42 kA/cm^2 at 77 K for YBCO tapes manufactured by the IBAD-PLD process have been reported for short lengths [6]. It is expected that this second generation of superconducting coated conductors in the future will technically and economically be a competitor to the Bi2223/Ag materials. The expected costs depend on the manufacturing process. For the RABitS substrate and MOD HTS YBCO processes the price performance is expected to be < 10 USD/kAm.

The continuous development of the conductor materials means that the prototype superconducting cables that have been developed up till now by no means can be considered to be optimal solutions. The still changing conductor parameters will have an influence on the design and economy of the cables to be developed in the future. Therefore the evaluation of the prototype superconducting cables can only be considered as intermediate results, improvements do still have a potential.

During the 1990s a number of prototype HTS cables were developed and demonstrated in laboratory scale along with other superconducting devices such as transformers and fault current limiters. However, the (optimistic) projected market for these products has not yet showed up which means that here in the beginning of the new millennium it seems, that there is a sort of intermission in the development projects among the leading companies. Generally grants for the further development work are limited. The technology push is still prevalent, but the commercial market pull is missing.

So far HTS ac-cables have been developed, but HTS dc-cables are still missing. The HTS cables are considered as equipment for the transmission system due to their high power rating. Therefore they have been accepted by i.e. CIGRE (International Conference on Large High Voltage Electric Systems) as a relevant study subject despite many of the demonstration cables are using voltages in the distribution range.

There are still works to be done especially concerning the cable accessories in the form of cable end terminations and splicings. Also improvements of the thermal insulation system of the cables have large potential possibilities. Also the cooling systems do need to be developed with regard to costs and reliability.

2.HTS CABLE DESIGNS

Up till now two main different designs of high temperature superconducting cables have been developed and demonstrated. They are named the room temperature dielectric (RTD) design [34] - [38], and the cryogenic dielectric (CD) design [9] - [33], corresponding to the location of the main phase insulation relative to the cryostat i.e. the insulation can be placed inside the cryostat at 77 K (CD) or outside the cryostat at "room" (ambient) temperature (RTD). The CD design can consist of 3 single core cables or alternatively a 3-core cable where all three phases are assembled in one cable cryostat whereas the RTD design consists of 3 separate phases.

Recently a third design has emerged, the so called tri-axial cable system [7] and [8]. Here three concentric HTS phases are separated by layers of cold dielectric tapes i.e. the construction belong to the CD design group.

The main construction parts in the CD and RTD designs are common. They consist of a cooling duct, an HTS cable conductor, electrical main insulation, thermal insulation (cryostat), screen and an outer jacket. The cable conductor is build up of a number of layers of Bi2223/Ag tapes. Liquid nitrogen circulates through the cable in one or more cooling channels and cools the part of the cable, which is located inside the thermal insulation.

2.1 THE CRYOGENIC DIELECTRIC (CD) DESIGN

The CD design can be made as a 3 phase cable where all three phases are assembled in one cable with one common cryostat or alternatively as three separate phases with induvidual cryostats.

In Fig. 1 the principle of a single phase construction is outlined. The cable is constructed from the inside with an inner pipe, which often is called the former. In this case the former has a porous wall allowing the liquid nitrogen to impregnate the surrounding materials i.e. the superconductor layers and the cryogenic dielectric.
Liquid nitrogen is pumped under pressure through the former. At the end of the cable (not shown on the sketch) the liquid nitrogen is returned under the inner surface of the cryostat in the opposite direction. The flow of the liquid nitrogen in two opposite directions inside one cryostat may cause problems with the resulting temperature profile along the cable. Due to the thermal contact between the two flows a hot spot may result in long cables. One solution to this problem is to avoid using a hollow former with liquid nitrogen flow, whereby the coolant only flow in one direction [29].

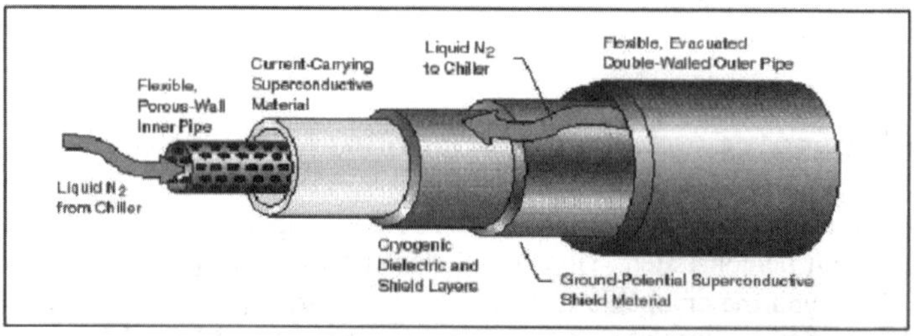

Fig. 1. Single phase of an HTS cable with cryogenic dielectric design. From Southwire 2001 by permission.

In the CD design a superconducting layer for transport current, the phase conductor is placed concentric with a superconducting screen. The phase conductor is separated from the screen by the main electrical insulation.

The phase conductor as well as the screen are build up of a number of superconducting layers which are helically wound along the former axis onto the former c.f. Fig. 2. The magnetic field generated by the transport current in the phase will be shielded by an induced current in the shield. In the shield a return current is induced which is equal and opposite to the current in the phase conductor. Therefore induced ac losses (hysteretic losses) in neighbour phases as well as eddy currents and inductive coupling with other metal parts are avoided. By connecting the three shield layers at the end of the cables the net current from the shields to ground is zero when balanced three phase currents are applied.

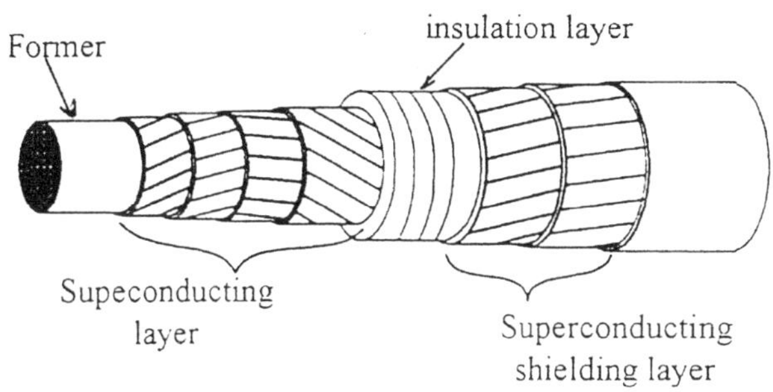

Fig. 2. Arrangement of the superconducting cable conductor, main insulation and the corresponding superconducting screen in a CD cable

The losses in the CD cable are mainly due to the hysteretic losses in the cable conductor and in the screen plus the dielectric losses in the main insulation and the losses due to the heat leak in. This means that the CD cable design will enable the cable to provide design currents up to an estimated range of 10 kA_{rms} , which correspond to powers in the GVA range. Thus with this technology underground cables can transfer power quantities that only is possible with overhead lines.

The electric main insulation at 77 K requires comprehensive development work to ensure its long time withstand strength from a mechanical, chemical and electrical point of view.
Normally a cable should be able to be in operation for more than 30 years without failure.
Up till now all CD designs employ a composite laminar or taped dielectric system as for example Polypropylene Laminated Paper (PPLP), which is impregnated with liquid nitrogen. This insulation system has low dielectric losses. The withstand strength of the insulation system has to be extrapolated in time to ensure a low probability of dielectric failure in a 30 year time span.

In summary, the main features of the CD design are:

The superconducting screen eliminates the magnetic stray field around the cable.

The cable can transfer very high currents up to an estimated range of 10 kA_{rms} .

Due to the superconducting screen the cable can be constructed with three adjacent phases in one common cryostat.

The drawbacks are:

The cable has high costs due to the superconducting screen.

The construction is more complicated than for the RTD design.

The thermal losses at 77 K are higher than for the RTD design.

2.2 THE ROOM TEMPERATURE DIELECTRIC (RTD) DESIGN

The RTD design is based on 3 separate phases i.e. the system consists of 3 single phase RTD cables which should be placed with a given distance (~ 0,5 m) between the phases in order to reduce the losses and increase the power transfer capacity, cf. Chap.4, Fig. 11.

In Fig. 3 the principle of a single phase construction is outlined. The cable is constructed from the inside with a former. The former may be tight to the coolant in which case the room between the superconductor and the inner cryostat wall is either evacuated or filled with an auxiliary gas. The losses in the cable are led away by radial conduction through the HTS conductor layers and the former wall.

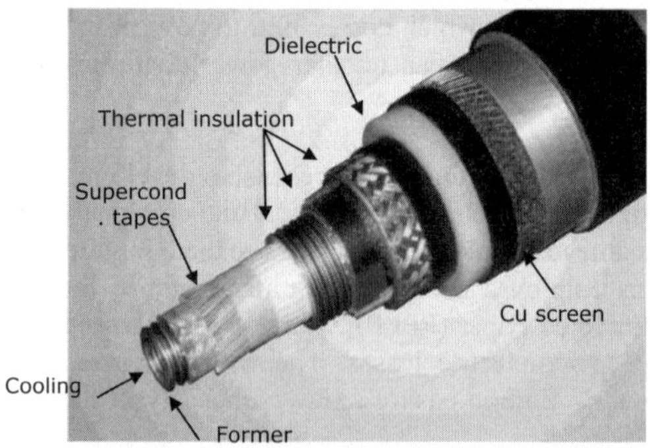

Fig.3 Single phase of an HTS cable with room temperature design.
From NKT Research by permission.

In other designs (also including the CD designs) a perforated or spiral former are used to allow embedding of the superconductors with the liquid nitrogen to increase the heat transfer [39]. However, when the BSCCO/silver tapes are exposed to liquid nitrogen for long periods of time the liquid may penetrate through small cracks in the silver covering and into the interior of the tapes. When this occurs a large volume expansion within the HTS tape will develop if the conductor is heated to room temperature. The result of this phenomenon is a considerable degradation of the critical current of the Bi2223/Ag tape, which will have an influence on the current carrying capability of the whole cable. One solution to this problem is to laminate the HTS tapes by adding thin layers of stainless steel reinforcement to both sides of the HTS tapes and seal the two edges [40].

For the cable design in Fig. 3 liquid nitrogen is pumped under pressure through the former. At the end of the cable (not shown on the figure) the liquid nitrogen is returned through the former of another phase. One solution to a cooling asymmetry in a cable system with 3 phases is to pump the liquid nitrogen forward in two parallel phases and return in the third phase as discussed later under cooling systems. The phase conductors are built up of a number of superconducting layers that are helically wound along the former axis onto the former, which may be a flexible tube c.f. Fig. 4.

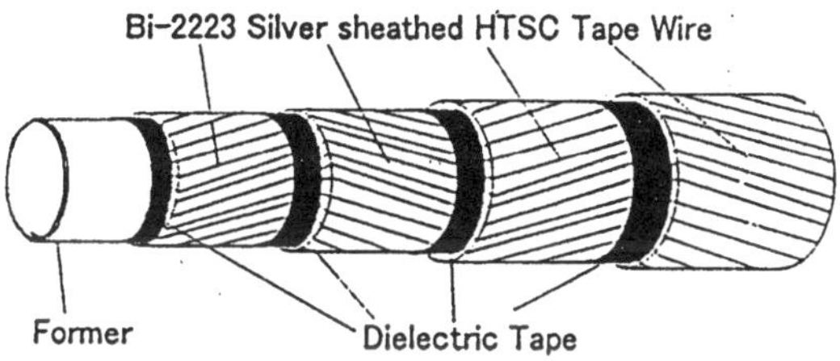

Fig. 4. Arrangement of the superconducting cable conductor in a Room Temperature Dielectric (RTD) cable

The winding angles are adopted in order to fulfil mechanical and electrical requirements. The number of layers needed is determined by the power rating of the cable. Typically the number of layers is chosen so the critical current of all the superconducting layers is above the nominal ac-current by use of a safety margin factor.

The losses in the RTD HTS cable are higher than in the CD HTS cable due to the lack of the superconducting return conductor. This results in higher electrical losses in the phase conductor and in metallic components surrounding the cable. Therefore, the RTD HTS cable is limited to carry 2-3 times the current of a conventional cable. The absence of a superconducting shield in the RTD HTS will result in lower material costs.

The thermal insulation of the cable consists of an outer and inner cryostat wall made of corrugated stainless steel. Super insulation, in the form of thin plastic films with metallic insulated aluminium, are placed between the two walls of the cryostat together with spacers. The vacuum in the cryostat together with the reflecting films in all minimise heat radiation, convection and conduction from the surroundings. As an example the thermal losses in such a cable are of the order 1.5 W/m [35].

The main electrical insulation is placed outside the cryostat i.e. it is kept at the ambient temperature independently of the load of the cable. (In conventional cables the maximum load of the cable is determined by the maximum allowable temperature rise in the dielectric insulation system). Therefore the insulation system can be fabricated by use of conventional techniques such as triple extruded polyethylene [35], oil impregnated PPLP [41] and EPR [42]. The insulation system does not need an extensive research and development work as for the CD HTS cable design.

Just as in conventional cables a metallic screen and a plastic sheath are placed as the outer layers.

In summary, the main features of the HTS RTD design are:

The insulation system is based on standard technology

The thermal losses are low

The costs of a RTD cable are lower than for a CD cable

Whereas the drawbacks are

There is a magnetic stray field around a RTD cable

The power rating (current) is limited

A minimum distance between the phases is needed

2.3 The triaxial cold dielectric cable design

Recently the socalled triaxial cable design has been under development for HTS cables [7], [8].
Here all three phases are contained in one cable making the design very compact as shown in Fig. 5. The cable is constructed from the inside with an inner tube (the former) with a porous wall allowing the nitrogen to impregnate the surrounding i.e. the superconducting phases and the cryogenic main insulation layers. Liquid nitrogen is pumped under pressure through the former. At the end of the cable (not shown on the sketch) the liquid nitrogen is returned under the inner surface of the cryostat in the opposite direction.

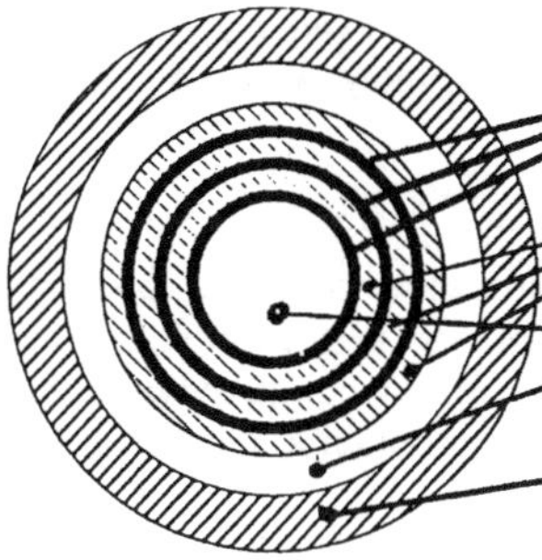

Fig. 5. Principle design of a CD HTS triaxial cable.

The phase conductors are constructed with a number of superconducting layers and the three phases are placed concentric relative to one another separated by the main electrical insulations as shown in Fig. 6.

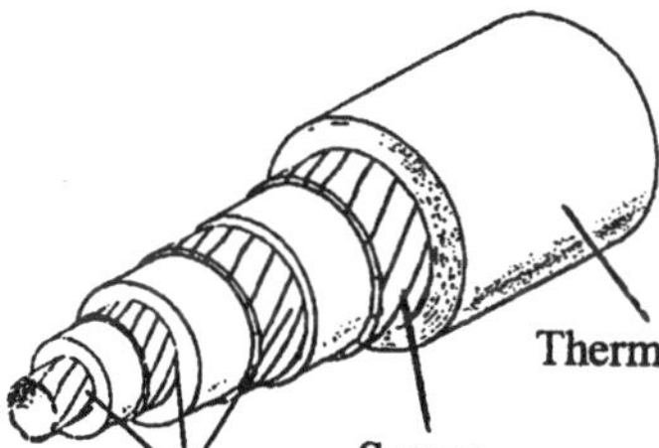

Fig.6. Arrangement of the triaxial HTS cable conductors. Also shown is a superconducting shielding layer.

For a balanced current system the superconducting shielding layer can be replaced with a metal screen.

The cable system thermal losses will be lower than in other HTS cables due to the compact geometry.

3.Cable accessories

At the ends of the cable circuit, terminations are required to make connections to the other parts of the power systems. For longer distances joints are required.

Because accessories constitute an important part of the cable system, they must exhibit performance at least equal to that of the cable. Thus, they are always included in the pre-qualification and type tests on cable systems.

For a superconducting cable the termination will generally consist of 3 main parts: A termination for the electrical insulation, a transition zone where the cable superconductor is connected to a metallic lead conductor from 77K and to the ambient temperature, and a connection for the circulating liquid nitrogen. Auxiliary cooling systems that are separated from the cooling system of the cable, may be used to obtain a more efficient cooling of the termination.

In Fig. 7 an example of a cable termination for a CD HTS cable is shown. The HTS cable conductor tapes are soldered onto the copper lead. In this transition region the conductors at high voltage are insulated with cone shaped paper insulation.

The liquid nitrogen in the cable termination is separated from the cooling and impregnation liquid nitrogen in the cable i.e. two cooling circuit systems are used as shown for the whole cable test system in Fig. 8.

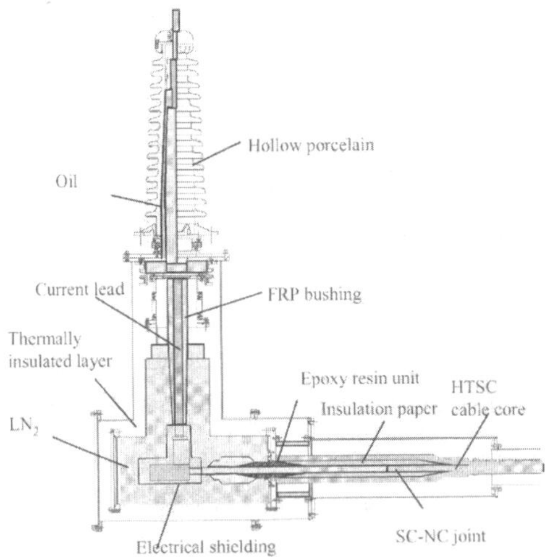

Fig. 7. Principle of a cable termination for a CD HTS cable with 66 kV and 1 kA. [29] by permission.

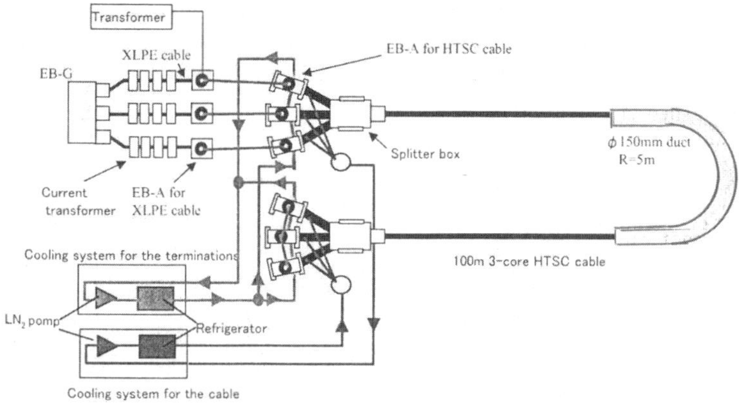

Fig. 8. Schematic view of a 100 m HTS cable test system. The cable and the 6 terminations have two distinct cooling systems.[29], by permission.

The accessories for the RTD cable design are more like a conventional design. An example of the layout of a RTD termination is shown in Fig. 9.

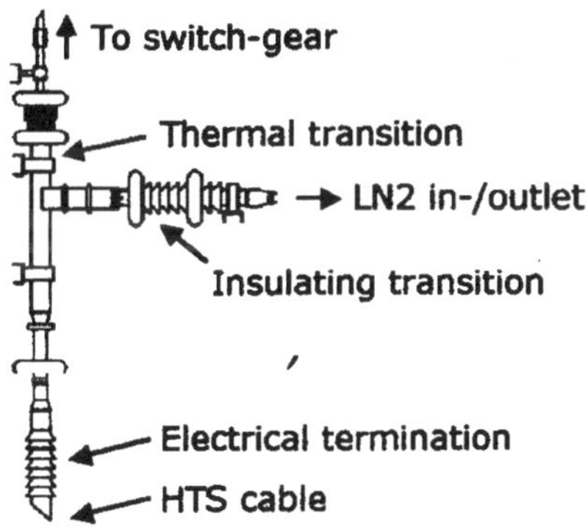

Fig. 9. Principle of a cable termination design for a RTD HTS cable. [35], [44], [45].

The electrical termination is designed as a conventional electrical termination. An outlet for the liquid nitrogen with an insulating transition has been designed and developed for this purpose. The thermal transition from 77K to ambient temperature is optimized in such a way that minimum heat will flow into the cryostat at nominal current load. Note that no auxiliary cooling system has been used specific for this design.
The termination including the cryostat is on high voltage potential whereas the liquid nitrogen pump and cooling system are at ground potential. A connection, which is thermally and electrically insulated and allows liquid nitrogen transfer is therefore needed. This can be

solved by the use of an insulating foam (i.e. Expancel) as a thermal insulation. A withstand strength of about 6 kV/mm makes it possible to use the foam filled with small cavities (diameter ~ 50μm) in the electric field without inception of partial discharges [45].
An example of this design is shown in Fig. 10.

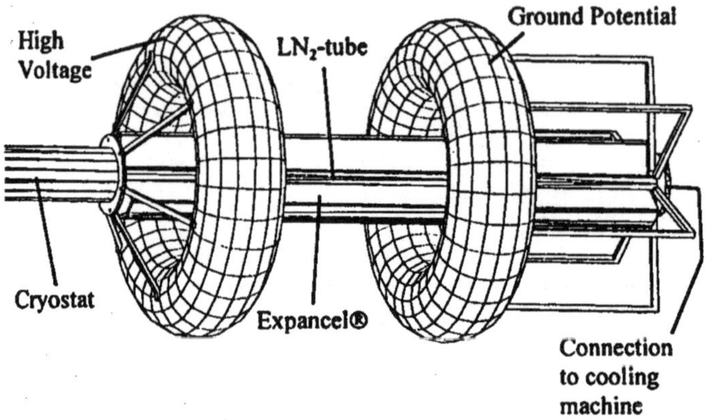

Fig. 10. Thermal and electrical insulation of the connection for the outlet of liquid nitrogen at high potential to a cooling machine at ground level. [44].

The current lead with its thermal conduction through the metallic current lead from ambient temperature to liquid nitrogen temperature as well as resistive losses in the current lead in the terminations is a major contributor to the thermal losses in the cable.

An optimal current lead design results in the following minimal power load to the liquid nitrogen [46], [47].

$$W_{t,\min} = I \cdot \left(2\rho \int_{77K}^{300K} \kappa(T)dT \right)^{0.5}$$

where ρ is the electrical conductivity of the current lead material and κ(T) is the thermal conductivity of the current lead material. If the current lead is made of brass (which is one of the best alternatives) the minimal loss per kA in the termination is 43 W/kA. Minor losses will occur in joints due to connections etc.

For a 30 kV RTD HTS cable this loss is about 100 W per termination at 77K and 2 kA compared to a loss in the cable of about 1-2 W/m per phase.

With a penalty factor of 10 (10 W at ambient temperature must be applied to remove 1 W at 77K), about 1 kW cooling power at ambient temperature must be applied to cool one termination.

4.ENERGY LOSS COMPONENTS IN CABLES

For conventional copper and aluminium power transmission ac cables the energy losses are well understood and described in i.e. international standards [48]. The main loss elements are: conductor losses, induced losses in the sheath, induced losses in metal pipes (for pipe-type cables) and dielectric losses.

The conductor losses are due to the resistive (ohmic) losses in the conductor material. The losses can be calculated by the equation:

$$W_c = R_c \cdot I^2$$

where R_c is the ac resistance and I is the rms current.

The induced losses in a metallic sheath in conventional cables arise due to an alternating magnetic field in and around the sheath. The losses can be reduced by transposition of the sheath at certain length intervals along the cable. The induced losses are proportional to I^2 [48].

The induced losses in a pipe are similar to the induced losses in the sheath.

The dielectric losses are losses in the electric insulation material system of the cable. They arise from the electric field in the insulation and they are independent on the load current in the cable. The dielectric losses can be calculated as

$$W_d = \omega\, C U_0^2 \tan\delta \qquad \text{where} \qquad C = \frac{2\pi\varepsilon_r\varepsilon_0}{\ln\left(D_0 / D_i\right)}$$

where $\omega = 2\pi f$, C is the capacitance per unit length of the cable, U_0 is the rms phase voltage, and $\tan\delta$ is the loss factor for the insulating material. The capacitance is determined by the permittivity ε and the outer and inner diameter D_o and D_i of the insulation.

The maximum current (losses) in a conventional cable is determined by the maximum allowable temperature rise in the insulation. For a typical direct laid 3-phase single-core cable (typically used in Europe) the full load losses are in the order of 30-40 W/m and phase.

For HTS cables the losses are fundamentally different from the losses in conventional cables. Even though the losses are very small they have to be removed by the internal cooling liquid nitrogen i.e. the losses at around 77K have to be removed by a cooling system at ambient temperature. The efficiency of this cooling process is typically 10 % as mentioned above.

The losses in an HTS cable system can according to the origin of the losses be divided into the following components:

HTS cable conductor ac loss

Thermal leak from ambient through the cryostat into the cold region of the cable

Induced losses in the shield of an HTS cable with RTD design

Induced losses in the metal parts of the thermal insulation of an HTS cable with RTD design

Dielectric losses in the main insulation system

Hydraulic losses in the liquid nitrogen

Pumping losses

Losses in joints and cable end terminations

Losses in the former

HTS cable conductor ac losses have been studied extensively in the past and the research and development on this subject is still under progress (2002). In Vol. 32 in the present series of Studies of High Temperature Superconductors, M. Däumling et al. have presented a comprehensive treatment of "AC loss in superconducting power cables" [49]. (Here 42 further references about the subject are listed).

High precision and reliable measurement of ac losses is a nontrivial task [50], [51], [52], [31]. and no widely accepted physically based model taking into account the complicated geometrical properties exists. The occurrence of 50 % difference or even more between measurements and model calculations is not unusual.

In spite of this, several models are often used to estimate the losses. This includes for example the mono block model [55], the UCD model [56], and the network model [57].

The ac losses of HTS cable conductors of lengths 1, 3 and 10 m have been measured electrically and calorimetrically [52], [53], [54] . Until now, only a few have attempted to measure the ac losses in a three phase configuration [58], [59].

For a 3 phase RTD cable system the phases are normally separated. However, additional ac-losses will occur due to the overlap of the magnetic fields from the neighbouring phases. In [60], a 10 m long test bed for three phase calorimetric loss measurements was constructed. The test bed was versatile and therefore it was possible to vary the inter phase distance from about 9 to 50 cm.

In Fig. 11 the ac losses are shown as a function of the load current and for different phase distances.

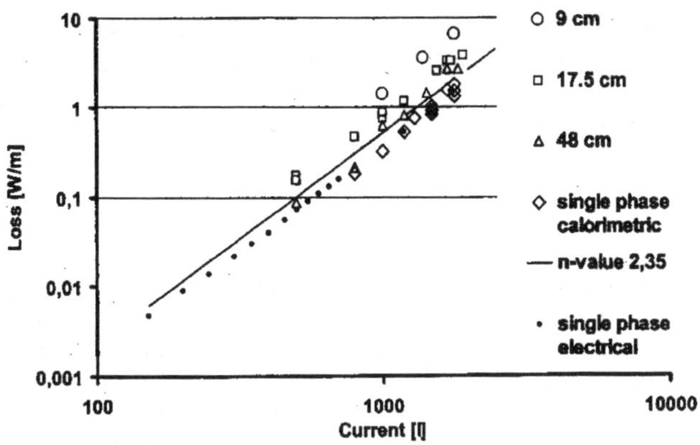

Fig. 11. AC loss in W/m of a 10 m long HTS cable conductor versus the load current and the inter phase distance. The line corresponding to the n value 2.35, which is shown for comparison, [60].

The results show that the HTS cable loss increases when the distance to the neighbouring phases are decreased. When a single phase is compared to a 3 phase configuration with a phase distance of 50 cm the ac loss of the cable HTS conductor is increased with 50 %. Further reduction of the phase distance to about 10 cm will increase the ac loss with a factor 4 compared to the single-phase loss.

Thermal leak from ambient through the cryostat into the cold region of the cable

In an HTS cable there will exist a thermal leak through the thermal insulation due to non ideal insulation and the substantial temperature difference ΔT between the surroundings at ~ 300K and the liquid nitrogen at ~ 77K.

If super-insulating foils in the cryostat vacuum suppress the heat radiation sufficiently well then the thermal losses can be calculated by

$$W_{th} = \frac{2\pi\,\lambda\,\Delta T}{\ln(D_o / D_i)}$$

where λ is the equivalent thermal conductivity of the thermal insulation and D_o and D_i are the outer and inner diameter of the thermal insulation.

The insulation has to be made as a vacuum insulation in order to sufficiently reduce the thermal leak. Depending on the outer and inner diameter of the thermal insulation and the level of vacuum pressure obtained, thermal conductivities in the range 1-2x10^{-4} $Wm^{-1}K^{-1}$ can be obtained. This results in losses of the order of 1.5 W/m and phase. The losses are independent on the current rating of the cable.

Induced losses in the shield occur in RTD HTS cables as it does in conventional cables. Therefore the models used for conventional cables can be used for calculating the losses [64].

For CD cables the induced currents will result in ac losses in the HTS shield as in the HTS cable conductor and therefore the same types of models used for HTS cable conductors can also be used ere.

Dielectric losses in the main insulation system occur in HTS cables as in conventional cables. These losses are independent upon the load current and can therefore be considered as no load losses. For a RTD cable these losses are dissipated to the ambient, whereas the dielectric losses in a CD cable has to be removed by the liquid nitrogen. Therefore, for a CD cable it is important to use a low loss dielectric insulation system. An example is the polypropylene/paper (PPLP) insulation impregnated with liquid nitrogen.

Hydraulic losses in the liquid nitrogen occur due to flow friction from the forced cooling with the liquid nitrogen. The friction losses can be calculated by

$$W_{hyd} = \frac{{}^{dm}\!/_{dt} \cdot \Delta p}{\gamma_{LN}}$$

where dm/dt is the mass flow rate, Δp is the pressure drop of the flow and γ_{LN} is the mass density of liquid nitrogen (809 kg/m^3) . For longer length cables a pressure drop of 2-5 bar/km can be expected. The friction losses have to be removed by the liquid nitrogen cooling system.

The pumping losses are due to the forced cooling by the pressurized liquid nitrogen. An external pumping system will consume energy corresponding to the hydraulic losses and include the losses in the circulating system itself. These losses are denoted pumping losses.

Losses in joints and terminations arise from the 2x3 cable end terminations at the cable ends and from cable joints at each cable section, c.f. Chapter 3.

Losses in the former

Depending on whether the former is made of a good or poor electrical conductor, eddy currents in the former itself may also contribute significantly to the ac losses of the cable. Each superconducting layer conducting an ac current contributes to an axial field (if helically wound) in the centre of the former. The resulting magnetic field obtained from a multitude of layers is said to be unbalanced if this is nonzero.

The unbalanced field may occur due to an uneven current distribution between two or more layers. Such an axial field would induce eddy currents in an electrically conducting former.

It has been shown experimentally [61] that former materials with low resistivity may under certain circumstances significantly contribute to the total ac losses in a superconducting cable compared with the hysteretic losses which are about 0.2 - 1 W/m for a 2-5 kA_{rms} HTS cables [51],[62],[63].

The total losses in HTS cables can be calculated for the contributions listed above. In [65] a calculation model is described which has been applied on a real case study. The principle of the calculation is illustrated in the flow diagram shown in Fig. 12.

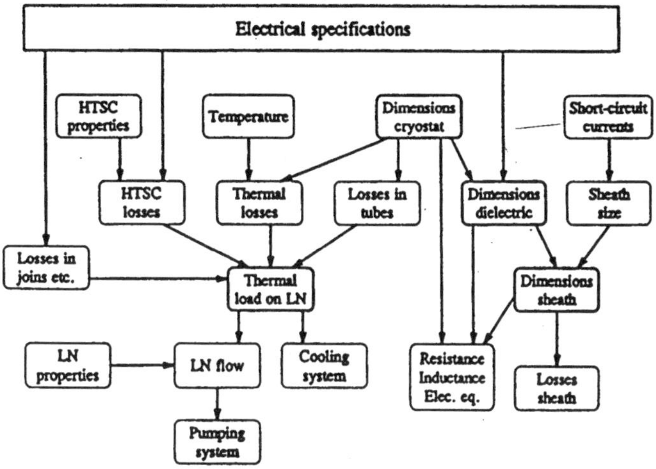

Fig. 12. Flow chart of the calculation procedure of an HTS cable [65].

On the basis of electric specifications and properties of the materials, the physical, thermal and electrical properties of the cable are calculated.
As an example the results of a calculation of a 4 km 450 MVA 132 kV 3 phase single core HTS cable system with RTD design is shown in Fig. 13.

At full load (100 % nominal current) three major loss components dominate: The thermal leak, ac conductor loss and induced loss in the shield whereas at 50 % load only the thermal leak is dominating.

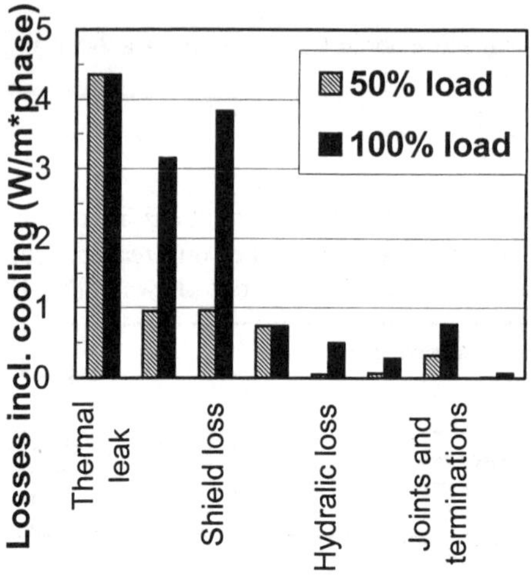

Fig. 13. Distribution of losses in a 450 MVA 132 kV HTS cable system with room temperature dielectric design at 50 and 100 %load. The efficiency of the cooling machine is included in the loss components [47].

In Fig. 14 the total losses for the same cable system is shown as a function of the load current, and the result is compared with the losses in a comparable conventional cable system with 450 MVA rated power.

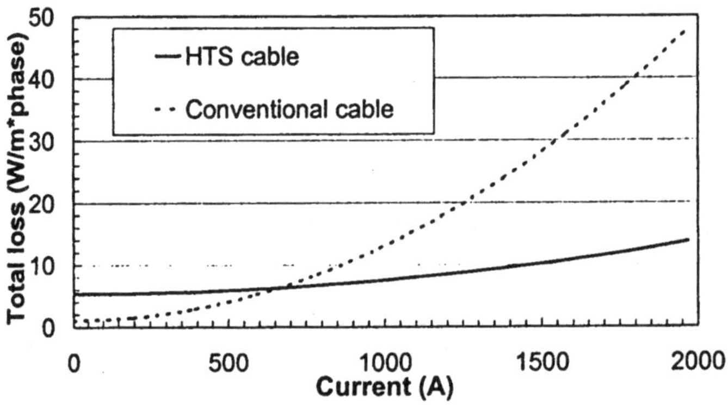

Fig.14. Losses in an HTS cable system and a conventional cable system Both systems have a rated power of 450 MVA at 132 kV. The conventional system consists of two parallel cable systems [41].

The conventional system consists of two parallel systems (6 parallel single core cables) to provide the same power transmission. The losses in the HTS cable at rated current 2 kA_{rms} are ~ 25 % of the losses in the conventional cable.

At no load the losses in the HTS cable are larger than the losses in the conventional cable. The breakeven point is at about 650 A load current which corresponds to ~33 % load of the cable system.

In summary, the HTS cables are dominated by no load losses due to a non ideal thermal insulation. The load-dependent losses due to ac conductor losses and induced losses are relatively small. Thus from a loss point of view HTS cables are most feasible at connections with high load currents during long periods.

5.THE COOLING SYSTEM

Conventional, naturally cooled cable systems can at best barely achieve 50 % of the transmission capacity of overhead lines of the same voltage level. If therefore the electrical energy delivered by overhead lines is to be taken further by conventional cables and with the same transmission performance, the only remaining option is the installation of double or multiple parallel systems and/or the use of artificial cooling. Therefore, artificially forced cooling is by no means a new phenomenon in the transmission technology.

A distinction is made between three fundamentally different cooling methods according to the region of application in the cable and the means of heat of removal i.e. **indirect surface cooling** where the cable surface is cooled indirectly through the installation of pipes carrying cooling water that are laid parallel to the cable system and in the same trench, **direct surface cooling** where the cable surface is in direct contact with the coolant enclosed in a pipe and **direct conductor cooling** provided by forced oil or water circulation in a hollow conductor.

From a physical point of view, direct conductor cooling is the most attractive solution for increasing capacity in cable systems since it removes the heat losses directly at its most important source. The internal and external thermal resistances of the cable and the surrounding soil therefore become almost completely irrelevant.

Conductor cooling can represent a viable alternative for special applications that involve extreme power transmission capacities over a short cable section. If oil is used as the coolant, then the capacity of a 1 km long three phase system with conductor cooling can be increased between 5-7 times that of a system with natural heat removal; if water is used as the cooling medium the transmission capacity can even be increased by a factor of 10.

However, this type of forced cooling also involves the highest level of technical and financial resources and this factor is ultimately responsible for the fact that up to now no commercial transmission line with direct conductor cooling has ever been implemented anywhere in the world [2].

The customers prefer maintenance free cables such as XPLE (cross linked polyethylene) cables, which will have an expected lifetime of 30 years or more. A prerequisite for the acceptance of a superconducting system is that it has better technical performance and lower costs. The reliability and maintainability of such a cable system must fulfil the customer requirements.

The attraction of the HTS superconductors lies especially in their high critical temperatures, which allow operation at 77 K, which is the boiling point of liquid nitrogen compared to 4 K for liquid helium used for the low temperature superconductors.

This development has led to simpler and cheaper cooling systems with higher efficiency and simpler thermal insulation systems. For example with an efficiency $\eta = 28$ % of the cooling system the penalty factor for the HTS system at 77 K is 10 compared to 241 for a low temperature system at 4 K, and for $\eta = 10$ % the corresponding penalty factors are 28 respectively 688. Furthermore, liquid nitrogen has much lower costs (~ 10 cents per litre) than liquid helium (~ 5 $ per litre) and it is available in unlimited quantities.

For superconducting cables the need for cooling capacity is of the order 1 W/m per phase. For long cables with cooling stations at 4-10 km intervals the cooling capacity for a 3-phase system is about 10 – 30 kW.

Today a number of commercial cooling systems are available. Generally the cooling systems either have a relatively low efficiency and/or a large need of maintenance. The use of today technology will require a backup system to ensure high reliability. A cooling system for a superconducting cable plant has to be specially designed for the specific application. Furthermore, superconducting systems could represent a step backwards for network operators (as compared to conventional extruded cables) as the new technology will, at the very least, require some maintenance of the nitrogen cooling system. In the event of a line failure in the HTS cable, unavailability would be longer due to the time required to repair and re-cooling of the cable system.

Thus, further research and development of cooling systems for superconducting cables are needed to ensure high reliability, low need for maintenance, high efficiency and low costs.

An example of a cooling system for a 30 m HTS cable that has been taken into operation in the network [35] is shown in Fig. 15. Liquid nitrogen at 65 K – 77 K is pumped forward through two of the single core cables and returns through the third single core cable.

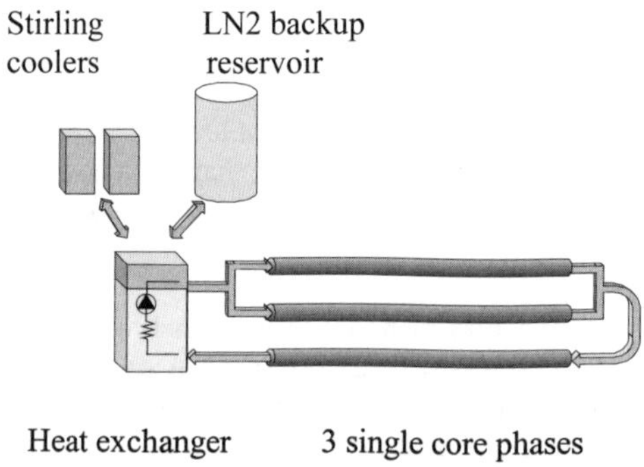

Fig. 15 Principle of a prototype cooling system [35].

The main cooling system consists of two Stirling cycle coolers. Each cooling machine can remove 1.1 kW at 77 K. A backup cooling system with liquid nitrogen boil off and a liquid nitrogen reservoir has also been installed. The passive backup cooling system automatically takes over if the main cooling system fails. Thus, continuous operation of the cable system is ensured in this way.

In order to avoid cavitations in the liquid nitrogen and to facilitate start up, the circulation pump is placed immediately after the heat exchanger in the heat exchanger tank. The storage tank with pressurised nitrogen gas is connected to the system just ahead of the heat exchanger. Thus, any nitrogen gas that enters to compensate for volume changes has time to condensate into the liquid phase in the heat exchanger.

6. RESEARCH AND DEVELOPMENT HTS CABLE PROJECTS

Around the world there are identified about 24 ongoing and finished HTS cable development projects from 11 countries representing a total development cost of estimated 143 millions USD.
Some of the major projects are listed in table 1.

Project	Rated power	Cable length	Design	Time period	Comments
Nexans, SUPER-POLI project [18],[19],[66]	40 MVA (1 GVA)	2 m (200m)	20 kV CD 3 phases	1998-2001	Development of a cable design with inherent current limiting properties due to the transition out of super-conductivity.
NKT cables [34],[35]	104 MVA	30 m	30 kV RTD 3 phases	1997-2003	Taken into commission May 2001 in a substation, Copenhagen Energy. Planned to be in commission until February 2003. This is the first HTS Cable installed in an operating Utility Network supplying 50000 customers.
Pirelli/Detroit Project [36],[37],[38],[39]	100 MVA	120 m	24 kV RTD 3 phases	1999-2003	Installed at a 24 kV substation at Detroit Edison. The project has been delayed due to leakage problems in the cryostats.
Pirelli/EPRI [39]	400 MVA	60 m	115 kV RTD 1 phase	1995-1999	Tested in laboratory at EPRI Inclusive a cooling system, a joint and an outdoor termination.
Southwire [25]	27 MVA	30 m	12,4 kV CD 3 phases	1998-2001	The worlds first HTS cable system which supply an industrial customer the Southwire Cable factory in Carrolton, Feb.2000
Sumitomo/TEPCO [28],[29]	114 MVA	100 m	66 kV, CD 3 phases	1999-2003	Compact cable under test by CRIEPI Long-term tests began June, 2001. Electrical properties such as critical current, ac loss and lightning withstand strengths have been evaluated.
IGC Super-Power	72 MVA	400 m	34.5 kV, CD 3 phases	2001-2004	To be installed as a conventional cable transmission line.

Pirelli/Long Island (DOE press release 24 Sept. 2001)	77 MVA	750 m		2001-	To be installed at Long Island Power Authority. The project shall result in accurate service data, which can be used for further grid analysis.
Southwire/NKT/ AEP		330 m	13.2 kV RTD triax 3 phases	2002-2005	To be installed at a substation at American Electric Power. The design of the cable may be a triaxial configuration with 3 concentric phases in one cryostat.

Table 1. Major HTS cable development projects. All the cables are cooled with LN2.
From Table 1 it is seen that the developed prototypes have cable lengths of about 30-120 m with rated currents of 1-2 kA and rated voltages 12-66 kV. The next generation cable projects initiated in the years 2001-2002 will have cable lengths of about 300-700 m.

Fig. 16. The SouthWire HTS cable installation in Carrolton GA.
27 MVA, 12.4 kV, 3 single phases with CD design. With permission.

7.POTENTIAL APPLICATIONS AND FUTURE MARKET FOR HTS CABLES

Superconducting cables have advantages such as high transmission capacity, compact design, low losses, low reactive power generation and no thermal and magnetic (EMC) interaction with the surroundings. HTS cables have potential applications such as:

- Network configurations with reduced voltage level
- HTS cable series connection with overhead lines
- Connection to generators including inherent fault current limiting

- Retrofitting with higher transmitting power in existing ducts
- High power DC connections with low voltage and high current
- EMC neutral HTS cable installations
- DC backbones
- Environmental thermal neutral installations

Network configurations with reduced voltage level

One of the potential applications is to reduce the electric power system costs by re-design of new or expanded substations. By using HTS cables with high amperage ratings it will be possible to eliminate one or two transformation steps in the transmission and distribution network.

This solution could be relevant for high voltage transmission into towns or in the network itself, where a voltage level and the corresponding voltage transformation can be eliminated, as for example by transmission of 200 MVA at 20-60 kV level instead of 120-150 kV level.

It is anticipated that this application will lead to a reduction of the investment costs for buildings, transformers, switchgears, measurement transformers etc.

As for conventional cables the use of high current and low voltage in the transmission system has constraints for long length cable systems due to the critical length and the phase shift of the cable [67],[68],[69].

HTS cable series connection with overhead lines

The insertion of one HTS cable in series with an overhead line is possible due to the matching power capabilities of the two systems. The use of conventional cables will require more parallel systems in order to match the power capabilities. This potential solution has become a highly topical subject in recent years due to environmental considerations. It is increasingly difficult to obtain permission to use overhead lines in both urban and rural scenes. Therefore new lines will mainly be installed as cables and existing lines are progressively exchanged by cable systems.

Recently EDF has studied the prospects for application of HTS cables in electric power networks [74]. It is concluded that superconducting cables could be competitive with conventional underground solutions for interconnection links, although no particular space constraints are expected in this case. For power feeding of urban areas, local competition for subsoil use appears as a prerequisite for the interest of superconducting transmission lines.

Connection to generators including inherent fault current limiting

HTS cables can be used to connect power plant generators with the grid. At the same time there is a possibility to build in current limiting properties in the cable [18]. Using this technique the short circuit current will be limited at the same time as the grid impedance is low. This is advantageous with respect to voltage regulation and stability.

Retrofitting with higher transmitting power in existing ducts is possible with the HTS technology. The outer diameter of the HTS cables can be made equal to or smaller than for existing cables whereby the HTS cables with higher power rating can be directly installed (retrofitted) into existing ducts [31]. Thereby the installation of high power transmission cable systems in urban areas can be realised without extensive installation work such as excavation.

High power DC connections with low voltage and high current

Projects with large offshore wind turbine farms (up to 500 MW) are in progress. Today 130 MW farms are implemented. Connecting the parks to the power grid with conventional ac transmission technique is difficult due to non-controllable power flow and voltage stability problems.

The high current capability of HTS DC cables and the low cost power electronics for low voltage application are the key elements in a new transmission concept for grid connection of wind turbine farms in remote locations. Furthermore, a common cooling of the HTS power cable and the power electronics can reduce the total transmission losses in such a system.

In Fig. 17 an example of the transmission concept is shown. A 150 MW offshore wind turbine farm is connected to the grid through a ± 7.5 kV, 10 kA HTS DC cable. Each wind turbine or group of wind turbines is connected to the HTS cable with a conventional sea cable matched to the current rating of the particular wind turbine(s).

The voltage source commutated (VSC) converter on the shore can be cooled with liquid nitrogen, which is also shared with the cooling system for the HTS cable. This solution eliminates the need for a power transformer at sea due to the low transmission voltage.

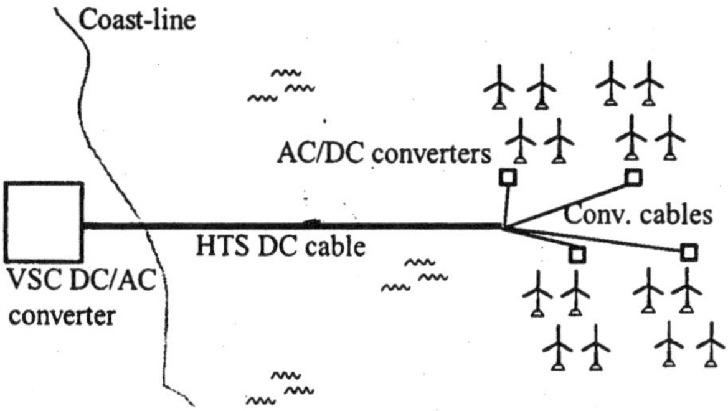

Fig. 17. Outline of transmission concept based on a high current low voltage HTS dc cable connection, simple ac/dc converters in the wind turbines and liquid nitrogen cooled VSC based converter.

Compared to a conventional dc solution, the overall transmission losses are reduced by a factor of two or more. At the same time it is possible to utilise the benefits from the VSC converter to give an extended control of the active and reactive power fed into the grid. Furthermore, the wind turbines and the offshore converters can be integrated, resulting in wind turbines with variable speed. This leads to less mechanical stress on the construction (extended lifetime) and increased power production.

It is expected that the system concept can eliminate problems regarding voltage stability, provide a more flexible and controllable power flow and finally provide a cost effective transmission concept [70].

EMC neutral HTS CD cable installations can be used in urban areas where the electromagnetic compatibility can be a problem due to electromagnetic emissions at the power frequency from a high voltage power transmission lines. The 50 or 60 Hz field may interfere with computer screens and other sensitive apparatus.

Also, the effects of magnetic (and electric) fields on biological systems and human beings are a subject of concern and investigation [71] together with environmental considerations. Therefore, overhead power lines in urban areas are today increasingly exchanged by cable installations. However, dependent on the configuration of the cable system, electromagnetic fields in neighbourhood of the cables will still be present. The use of cable systems with HTS CD cables will completely eliminate such problems and result in a high flexibility with regard to the choice of the cable routing.

DC backbones in the form of dc connections with high power dc connections.
For example, it is anticipated that it is possible to transmit about 2000 MW at ± 50 kV.

Environmental thermal neutral installations are a result of the thermal insulation in the HTS RTD and CD cables. The outer surfaces of the cables are at ambient temperature, and therefore the cable systems can be located anywhere without regard to heat dissipation from the cable in contrast to conventional transmission cables. The environmental heating effect from the cable system is completely eliminated.

8. FUTURE MARKET FOR HTS CABLES

During the 1990s a large number of HTS development projects concerning power application of superconductivity were started. These projects aimed at the development of transformers, motors, generators, cables, fault current limiters and superconducting magnetic energy storage (SMES) with HTS (LTS) conductors and flywheels.

It was anticipated at that time, that the market for these products would begin in the new millennium, however it turned out that this was not the case.

Apart from the SMES (LTS) apparatus, none of the HTS apparatus have found commercial application in the electric power grid today. The technology push has been substantial but the market pull is missing although a number of potential applications have been identified.

Today some of the large development projects have been stopped as they are waiting for new, better and low cost HTS conductors. For some applications the breakeven costs are estimated to about 50 USD/kAm. Today the costs for Bi2223/Ag tapes are about 200USD/kAm whereas the projected minimum costs in the future will be $\sim$ 20 USD/kAm, and correspondingly for coated conductors $\sim$ 10 USD/kAm.

The use of superconducting devices may be realised separately for each product or introduced as a superconducting system consisting of for example an HTS transformer, an HTS fault current limiter and an HTS cable.

The HTS cables developed hitherto have demonstrated the applicability of the technology. Further development work is necessary to develop joints, short circuit protection [72], thermal losses and cooling systems. Cable lengths of several kilometres have to be constructed and tested with regard to reliability, thermal contraction, thermal insulation etc.

The acceptance of the HTS technology by the electric power industry is a prerequisite for the market development. The obvious weak link in an HTS cable system is the cooling system. As it contains rotating parts in the pumps and cooling machines a maintenance programme is absolutely necessary and it must be carried out about once a year using the cooling technology of today.

In case of an error in the system redundant cooling power must be available. Furthermore the operation of a superconducting cable will require special trained personal regarding the operation of the cooling system. The development of steering, control and monitoring auxiliary equiptment may eliminate the need for operators. For cable damages during operation an emergency strategy for the repair must be developed.

It is difficult to estimate the development of the future market for HTS cables. Although the HTS technology is considered as promising quantitative estimates of how well and how quickly real HTS devices will be introduced on their markets are lacking.

In a recent study [73] a method for estimating future markets in the U.S. for high temperature superconducting power devices has been presented. In this study a spreadsheet analysis is used to project the sales market for HTS power devices.

In Fig. 18 the market penetrations for HTS devices are shown corresponding to a specific choice of input parameters.

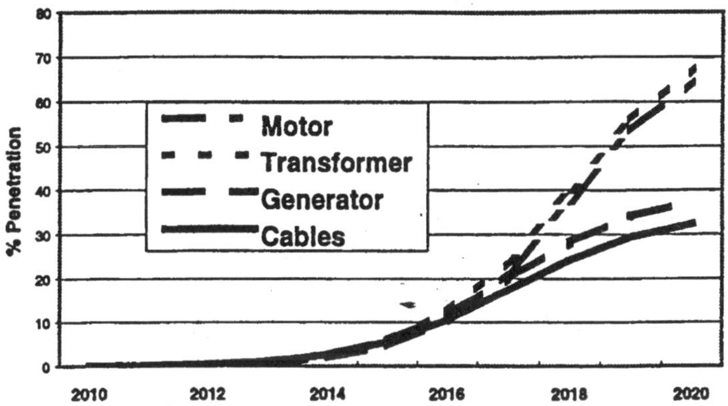

Fig. 18. Market penetration curves for HTS motors, transformers, generators and cables for a given numeric input. [73], with permission.

It is seen that the market only will start to grow after year 2010 and that the market penetration saturates after 2020 with an asymptotic market capture for cables at about 35 %.

The time span from today till the start of the market after year 2010 could lead to an intermediate pause in the development work due to lack of subsidy and grants. This situation may even influence the manufacturers of the superconducting materials as the sales of superconductors for electric power applications in large quantities will be absent in the next decade.

One result of the above study is that obviously research and development is still needed to

- Improve the cryogenic efficiency
- Improve the current carrying capacity of HTS conductor
- Reduce the HTS conductor cost and cryogenic cost

Further extension of the market would require new breakthroughs in the elaboration process of the HTS conductor as well as a large cost decrease of cryogenic components. Another possibility in the long range would be to identify new high power applications of superconductivity as a consequence of a progressive evolution of network architecture. Here power network deregulation could play a role [74].

References

[1] J. Gerhold, "Power transmission", Handbook of applied superconductivity IOP Publishing Ltd 1998 pp1627-1689

[2] E. Peschke, R. V Olshausen, "Cable systems for high and extra-high voltage", Publicis MCD Verlag1989, pp 178-180.

[3] E.B.Forsyth, R.A.Thomas, "Performance summary of the Brookhaven superconducting power transmission systems", Cryogenics, Vol.26, 1986, pp. 599-614.

[4] P.A. Klaudy, F.Gerhold, "Practical conclusions from field trials of a superconducting cable", IEEE Trans. on Magnetics, Vol. MAG-19, 1983, pp.656-661.

[5] G.J. Yurek Presentation at the International Superconductor Industry Summit, ISIS-10, Santa Fe, New Mexico, 14-16 March 2002.

[6] S.R. Foltyn et al., "Toward a 1000 A coated conductor: Increasing the I_c of thick BCO films", Los Alamos National Laboratory, New Mexico, US, presentation at Superconductivity programme for electric systems, Annual Peer Review, 17-19 July 2000.

[7] P.W. Fisher et al., "Design and analysis of a tri-axial cable system", Applied Superconductivity Conference ASC 2002, August 4-9, 2002.

[8] J.W. Lue et al., "AC tests of a 5 m triaxial HTS cable", Applied Superconductivity Conference ASC 2002, August 4-9, 2002.

[9] T.P. Beales et al. "A dc transmission cable prototype using high temperature superconductors", Supercond. Sci. Technol. 1996, pp 43-47.

[10] Y.B. Lin et al., "Development of HTS transmission power cable", IEEE Transaction on Applied Superconductivity, Vol. 11, No.1, March 2001, pp 2371-2374.

[11] Y.B. Linet al., "Development of a 1000 A class bismuth based HTS model cable", Advances in Cryogenic Engineering, Vol. 45.

[12] K. Goto et al., "Development of HTS cable with Bi2223 transposed segment conductors", ISS 2000.

[13] S. Nagaya, "High T_c superconducting cable with Bi-2223 transposed segment conductors", Fujikura Technical Review, 2002.

[14] K.Miyoshi et al., "Design and production of High-T_c superconducting power transmission cable", IEEE Transaction on Applied Superconductivity, Vol. 11,No. 1, March 2001, pp 2263-2366.

[15] O. Miura, "66 kV- 2kA peak load test of high T_c superconducting model cable", Cryogenics, Vol. 36, No. 8, 1996, pp 589-598.

[16] Mukoyama et al., Advances in Superconductivity, IX, Vol. 2, 1997, p1337f.

[17] K.C. Seong et al., "Study on the application effects of HTS power cable in Seoul", IEEETransaction on Applied Superconductivity, Vol. 11, No. 1, pp. 2367-2370.

[18] T. Verhaege et al., "European superconducting power link project", JiCable, Vol.1, 1999, pp.115-120.

[19] High T_c Update, Vol.13, No. 10, May 15, 1999.

[20] M- Nassi, "HTS prototype for power transmission cables:Recent results and future programs", Inst. Phys. Ser. No. 167, 2000, pp23-28.

[21] P.L. Ladie, "Latest development of high temperature superconductor (HTSC) cable systems", JiCable, Vol. 1, 1999, pp. 97-102.

[22] High Tc Update, March 2000.

[23] Transmission & Distribution World, March 2000.

[24] Ladie et al., "Pirelli-EDF development on superconducting cables", JiCable, Vol.1, 1999, pp. 103-108.

[25] J.P. Stovall et al., "Installation and operating of the Southwire 30-meter high temperature superconducting power cable", IEEE Transaction on Applied Superconductivity, Vol. 11, No. 1, March 2001, pp. 2467-2472.

[26] "2001 Annual Peer Review of the US Department of Energy Superconductivity for Electric System Program", 2001 Peer Review Report, U.S.Department of Energy.

[27] Superconductivity News Update, Department of Energy, 17 April 2002.

[28] S. Honja et al., "High T_c superconducting power cable development", Physica C, 2001, pp. 1234-1240.

[29] T. Masuda et al., "Development of a 100 m 3-core 114 MVA HTSC cable system", Physica C 372-376 (2002), pp. 1580-1584.

[30] T. Shibata et al., "Development of high temperature superconducting power cable prototype system", IEEE Trans. on Power Delivery, Vol. 14, No. 1, January 1999.

[31] T. Masuda et al., "Experimental results of 30 m, 3-core HTSC cable", Physica C 372-376 (2002), pp 1555-1559.

[32] S. Honjo et al., "High T_c superconducting power cable development", Physica C, 2001, pp 357-360.

[33] T. Shibata et al., Advances in superconductivity VIII, Vol. 2, 1996, p. 1303.

[34] D. Willen et al., "First operation experiencea from a 30 kV 104 MVA HTS power cable installed in a utility substation", Physica C 372-376 (2002), pp 1571-1579.

[35] J. Ostergard and Ole Tonnesen, "Design, installation and operation of world's first high temperature superconducting power cable in a utility network", CIGRE 2002 Session, Paris France.

[36] "HTS cables ready for full-scale demonstrations", Superconducting Week, Vol. 11, No. 3, Fall 1998.

[37] N.J. Kelly et al., "Field demonstration of a 24 kV warm dielectric HTS cable", IEEE Transaction on Applied Superconductivity, Vol. 11, No. 1, March 2001.

[38] M. Nassi, "HTS prototype for power transmission cables: Resent results and future programs", Inst. Phys. Ser. No. 167, pp. 22-28.

[39] N. Kelley et al., "Application of HTS wire and cables to power transmission: State of the art and opportunities", IEEE PES WM January 2001.

[40] L. Maur et al., "Long length manufacturing of high performance BSCCO-2223 tape for the Detroit Edison Power Cable Project", IEEE Transaction on Applied Superconductivity, Vol. 11, No. 1, March 2001

[41] M. Nassi et al., " Qualification results of a 50 m – 115 kV warm dielectric cable system", IEEE Transaction on Applied Superconductivity, Vol. 11, No. 1, March 2001, pp. 2355-2358.

[42] Norman Steve et al., "High temperature superconducting cable field demonstration at Detroit Edison", Physica C 354 (2001) pp. 49-54.

[43] M. Hazeyama et al., " V-t characteristics for partial discharge inception of high temperature superconducting power cable", Physica C 372-376(2002), pp.1551-1554

[44] C. Rasmussen, A. Kuhle, O. Tonnesen, "Design of a termination for a high temperature superconducting power cable", IEEE Transaction on Applied Superconductivity, Vol. 9, No.2, June 1999, pp 1273-1276.

[45] C. Rasmussen, T. Holboll, Electrically and thermally insulated joint for liquid nitrogen, Inst. Phys. Conf. Ser. No 167, 2000 IOP Publishing Ltd, pp 1139-1142.

[46] Y.L. Buyanov et al., "A review of current leads for cryogenic devices", Cryogenics VOL. 15, No. 4, 1975, pp 193-200.

[47] Jacob Ostergaard, Jan Okholm, Karin Lomholt, Ole Tonnesen, "Energy losses of superconducting power transmission cables in the grid", IEEE Transactions on Applied Superconductivity, Vol. 11, No. 1, March 2001, pp 2375-2378.

[48] "Calculation of the continuous current rating of cables (100% load factor)", IEC Standard Publication 287, 1982.

[49] M.Daumling et al., "AC loss in superconducting power cables", Studies of High Temperature Superconductors, A.V. Narlikar (ed.) 2000.

[50] A.Kuhle et al., "Measuring ac-loss in high temperature superconducting cable-conductors using four probe methods", IEEE Transaction on Applied Superconductivity, Vol: 9, No.2, June 1999, pp 1169-1172.

[51] S.K.Olsen et al., "Test of a 10-meter long, $2kA_{rms}$ high temperature superconducting cable conductor model based on Bi-2223/Ag-alloy tapes", Inst. Phys. Conf. Ser., No. 167, 1999, pp 1151-1154.

[52] S.K.Olsen et al., "Alternating current losses of a 10 metre long low loss superconducting cable conductor determined from phase sensitive measurements", Supercond. Sci. Technolol. 12, (1999) pp 360-365.

[53] S.K. Olsen et al., " Loss and inductance investigations In a 4 layer superconducting prototype cable conductor", IEEE Transaction on Applied Superconductivity, Vol. 9, No. 2, June 1999, pp.833-836.

[54] C. Træholt et al., "Calorimetric measurements of losses in HTS cables", IEEE Transaction on Applied Superconductivity, Vol.11, No. 1, March 2001, pp. 1777-1780.

[55] G. Vellego, P. Metra, Supercond. Sci. Technol., 8, 1995, pp. 476

[56] S.Mukoyama et al., "50-m long HTS conductor for power cable", IEEE Transaction on Applied Superconductivity, Vol. 7, N0. 2, 1997, pp 1069-1072.

[57] M. Daumling, " A model for the current distribution and ac losses in superconducting power cables, Cryogenics, Vol. 39, 1999, pp 759-766.

[58] J.O. Willis et al., " Multiphase ac loss mechanisms in HTS prototype multistrand conductors", IEEE Transaction on Applied Superconductivity, Vol.11, No.1, 2001, pp. 2188-2191

[59] J.O. Willis et al., " Single and multi-phase ac losses in HTS prototype power transmission conductors", IEEE Transaction on Applied Superconductivity, Vol.1, No.2, 1999, pp. 778-781.

[60] C. Træholt, E. Veje, O. Tønnesen, "Electromagnetic losses in a three phase high temperature superconducting cable determined by calorimetric measurements, Physica C 372-376, (2002), pp 1564-1566.

[61] C.Træholt et al., "The electrical aspects of the choice of former in a high T_c superconducting power cable.

[62] M. Leghissa et al. " Development of HTS tapes and multistrand conductors for power transmission cables", Inst. Phys. Conf. Ser. No. 158, 1997, pp. 1191-1194.

[63] K. Sato et al., "HTS large scale application using BSCCO conductor", IEEE transaction on Applied Superconductivity Vol.7, No. 2, 1997, pp. 345-350.

[64] R. Bartnikas, K.D. Srivastava, "Power cables and communication cables", IEEE Press Series on Power Engineering/Mc Graw Hill, 2000, pp.466

[65] J.Østergaard, "Superconducting power cables in Denmark – a case study", IEEE Transaction on Applied Superconductivity, Vol. 7, No. 2, June 1997, pp.719-723.

[66] T.Verhage et al.," European project on a self-limiting superconducting power link", Inst. Phys. Conf. Ser. No. 167, 2000, pp. 943-948.

[67] O.Tønnesen et al., "Power applications for superconducting cables", Inst. Phys. Conf. Ser. No 167, 2000, pp.1103-1108.

[68] "Superconducting cables impact on network structure and control", CIGRE Task Force 38.01.11, Bulletin No. 199, February 2002.

[69] "Superconducting cables impact on network structure and control", Elektra 58, No. 200, February 2002.

[70] J. Østergaard et al., "A new concept for superconducting DC transmission from a wind farm", Physica C 372-376 (2002) pp. 1560-1563.

[71] F.S. Barnes, "Typical electric and magnetic field exposure at power line frequencies and their coupling to biological systems", Biological Effects of Environmental Electromagnetic fields, Editor M. Blank, Washington DC, ACS Books 1995.

[72] K.H. Jensen at al., "Short circuit experiments on a high T_c superconducting cable conductor", Physica C 372-376 (2002), pp. 1585-1587.

[73] T.P. Sheahen et al.," Method for estimating future markets for high temperature superconducting power devices", IEEE Transactions on applied superconductivity, Vol. 12, No. 2, June 2002, pp. 1784-1789.

[74] P. Manuel et al., "Prospects for application of high temperature superconductors to electric power networks", Physica C 372-376 (2002) pp. 1591-1597.

SUBJECT INDEX

Printing: Mercedes-Druck, Berlin
Binding: Stein+Lehmann, Berlin